ZINN & THE ART OF MOUNTAIN BIKE MAINTENANCE

ZINN & THE ART OF MOUNTAIN BIKE MAINTENANCE

The World's Best-Selling Guide to Mountain Bike Repair

6TH EDITION

LENNARD ZINN

Illustrated by Todd Telander and Mike Reisel

BOULDER, COLORADO

Zinn & the Art of Mountain Bike Maintenance, 6th Edition

velopress®

3002 Sterling Circle, Suite 100

Boulder, CO 80301–2338 USA

VeloPress is the leading publisher of books on endurance sports. Focused on cycling, triathlon, running, swimming, and nutrition/diet, VeloPress books help athletes achieve their goals of going faster and farther. Preview books and contact us at velopress.com.

Distributed in the United States and Canada by Ingram Publisher Services

Library of Congress Cataloging-in-Publication Data

Name: Zinn, Lennard, author.

Title: Zinn & the art of mountain bike maintenance : the world's best-selling guide to mountain bike repair / Lennard Zinn.

Other titles: Zinn and the art of mountain bike maintenance | Zinn and the art of mountain bike maintenance

Description: 6th edition. | Boulder, Colorado : VeloPress, [2018] | Earliest edition bears title: Zinn and the art of mountain bike maintenance.

Identifiers: LCCN 2017058778 | ISBN 9781937715472 (pbk. : alk. paper)

Subjects: LCSH: Mountain bikes—Maintenance and repair—Handbooks, manuals, etc.

Classification: LCC TL430 .Z56 2018 | DDC 629.28/772—dc23

LC record available at https://lccn.loc.gov/2017058778

This paper meets the requirements of ANSI/NISO Z39.48-1992 (Permanence of Paper).

Cover and interior design by Erin Farrell / Factor E Creative

Cover photographs by Brad Kaminski

Front cover bike built by Lennard Zinn

Custom paint job on front cover bike by Spectrum Paint & Powder Works

1983 Ritchey "Faux Lugs" Competition courtesy of The Pro's Closet, Boulder, CO, www.theproscloset.com

Illustrations by Mike Reisel and Todd Telander

18 19 20 / 10 9 8 7 6 5 4 3 2

To Sonny, my wife, without whose support I could not have written this book

A TIP OF THE HELMET TO . . .

My heartfelt thanks go out to Todd Telander and Mike Reisel, for producing illustrations that make the procedures more intelligible and beautiful. This book would be a lot poorer and less useful without their clear and detailed drawings.

My everlasting appreciation goes out to my editors and in-house support system, Ted Costantino, Dave Trendler, Renee Jardine, and, before them, Charles Pelkey and Mark Saunders, for separating the wheat from the chaff, adding more wheat when necessary, and making sure people know about this book; to Terry Rosen, for bugging me to write this book for so many years; to Mike Sitrin, formerly of VeloPress, for doing the same and promising to publish the first edition when I did; to Felix Magowan and John Wilcockson, for sharing their vision, extending financial support, and giving encouragement; to John Muir and Robert Pirsig, for writing their own great books that inspired this effort. Special thanks to everyone at VeloPress for the countless efforts to improve it over the years.

For technical assistance with the details, thanks to Wayne Stetina, Steve Hed, Chris Rebula, Ryan Lalic, Nick Wigston, Bill Mead, Nick Legan, JP Burow, Ken Beach, Portia Masterson, Charlie Hancock, Sander Rigney, Doug Bradbury, Paul Turner, Mike McAndrews, Steve Rempel, and Chris DiStefano, and to Scott, John, and Rusty at Louisville Cyclery (Louisville, Colorado) as well as to folks at Shimano, SRAM, RockShox, Pedro's, Effetto Mariposa, NoTubes, Manitou, Cane Creek, Chris King, Fox, FSA, Park Tool, Hayes, Magura, DT Swiss, Thomson, ENVE, Spectrum, Deda, Salsa, Mavic, Selle San Marco, SQ Lab, Intense, Specialized, Cannondale, and Ritchey.

I also want to thank my entire family for support and inspiration: Sonny, my wife, for almost four decades of unconditional support; Emily and Sarah, my daughters, for showing me that books can be written, completed, and published at a prolific rate; Dad and Mom, for encouraging me my whole life; Rex and Steve, for offering suggestions; Kai, Ron, and Dad, for being authors themselves and an inspiration to me; and Marlies, for taking the kids when I needed it when writing the first edition. Thanks, Sarah, for proving Groucho Marx right.

CONTENTS

I.1 Believe it or not, you will be able to put all of this back together!

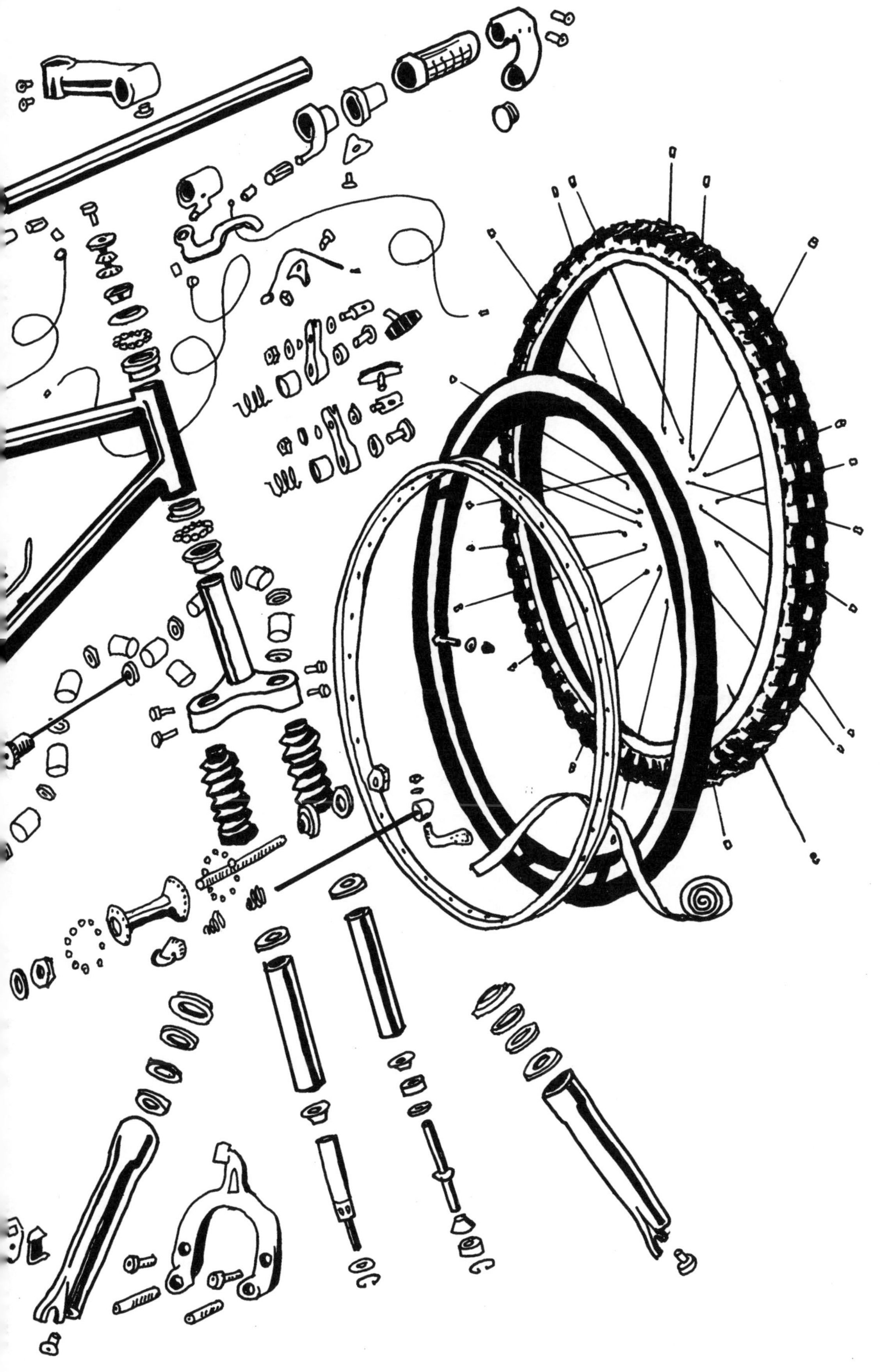

INTRODUCTION

Peace of mind isn't at all superficial, really. It's the whole thing. That which produces it is good maintenance; that which disturbs it is poor maintenance. What we call workability of the machine is just an objectification of this peace of mind. The ultimate test's always your own serenity. If you don't have this when you start and maintain it while you're working, you're likely to build your personal problems right into the machine itself.

—ROBERT M. PIRSIG,
ZEN AND THE ART OF MOTORCYCLE MAINTENANCE

This book is intended for people who have an interest in maintaining their own mountain bikes. I have written it for mountain bike owners who do not think they're capable of working on their own bikes, as well as for those who do and who want the how-to details at their fingertips.

In *Zen and the Art of Motorcycle Maintenance*, the late Robert Pirsig explores the dichotomy between the purely classical and purely romantic views of the world, a dichotomy that also applies to mountain biking. Riding a mountain bike is generally a romantic experience of emotion, inspiration, and intuition, even when solving the complex physics of how to negotiate a technical section of trail without putting your foot down. Mountain bike mechanics, however, is a purely classical structure of underlying form dominated by reason and physical laws. The two practices—mountain bike riding and mountain bike maintenance—fit eloquently together. Each is designed to function in a particular way, and to have one without the other would be missing out on half the fun.

The romantic can appreciate how success at bike mechanics requires that the procedures be done with love, without which the care you imagined putting into your mountain bike would be lost. And even the pure romantic can follow the simple step-by-step procedures and "exploded" diagrams in this book (of which Fig. I.1 is an extreme example and is the only one not intended to be simple and clear!) and discover a passion for spreading new grease on old parts.

Zinn & the Art of Mountain Bike Maintenance is organized in such a way that you can pick maintenance tasks appropriate for your level of confidence and interest. The repairs in these pages require no special skills to perform; anyone can do them. It takes only a willingness to learn.

Mountain bikes are admirably resilient machines. You can keep one running a long time just by changing the tires and occasionally lubricating the chain. Chapter 2 is about the most minimal maintenance your bike requires. Even if that is the only part of this book you end up using, you'll have gotten your money's worth by avoiding some unpleasant experiences out on the trail.

This book is intended for home enthusiasts, not professional mechanics. For that reason, I have not included the long and precise lists of parts specifications that a shop mechanic might need. Nonetheless, when combined with a specification manual, this book can be a useful, easy-to-follow reference for bike shop mechanics, too.

WHY DO IT YOURSELF?

There are a number of reasons why you would want to maintain your own mountain bike. Obviously, if done right, it is a lot cheaper to do yourself than to pay someone else to do it. This is certainly an important factor for those riders who live to ride and have no visible means of support. Self-maintenance is a necessity for that crew.

As your income goes up and the time available to maintain your bike goes down, this becomes less and less true. If you're a well-paid professional with limited free time, it probably does not make as much economic sense to maintain your own bike. Yet you may find that you enjoy working on your bike for reasons other than just saving money. Unless you have a mechanic whom you trust and to whom you take your bike regularly, you are not likely to find anyone else who cares as much about your bicycle's smooth operation and cleanliness as you do. You may also need your bike fixed faster than a local shop can do during its busy season. And you need to be able to fix mechanical breakdowns that occur on the trail.

It is a given: Breakdowns will happen, even if you have the world's best mechanic working on your bike. For this reason, it takes away from my enjoyment of a ride if I have something on my bike that I do not understand well enough to know whether it is likely to last the ride or how to fix it if it does not.

There is an aspect of bicycle mechanics that can be extremely enjoyable in and of itself, almost independent of riding the bike. Bicycles are the epitome of elegant simplicity. Bicycle parts, particularly high-end components, are meant to work well and last a long time. The best ones are designed and engineered by people who care deeply about them and how they work. With the proper attention, these parts can shine both in appearance and in performance for years to come. There is real satisfaction in dismantling a filthy part that is not functioning well, cleaning it up, lubricating it with fresh grease, and reassembling it so that it works like new again. Knowing that I made those parts work so smoothly—and that I can do it again when they get dirty or worn—is rewarding. I am eager to ride hard to see how they hold up rather than being reluctant to ride for fear of breaking something.

Also, if you share my stubborn unwillingness to throw something out and buy a replacement simply because it has quit working—be it a leaky Waterpik; a torn tent; a duffle bag with a broken zipper; or an old car, dishwasher, clock, or chainsaw that is no longer running well—then this book is for you. It is satisfying to keep an old piece of equipment running long past its prime, and it's a great learning experience!

There is also something very liberating about going on a long ride and knowing that you can fix just about anything that might go wrong with your bike out on the trail. Armed with this knowledge (which begins with learning to identify the parts of a mountain bike, shown in Fig. I.2) and the tools to put it into action, you will have more confidence to explore new areas and to go farther than you might have otherwise.

To illustrate, an experience from way back in 1995 comes to mind, when I took a day to ride the entire 110-mile White Rim Trail loop in Utah's Canyonlands National Park. It is quite dry and desolate out there, and I was completely alone with the sky, the sun, and the rocks for long stretches. I had a good mileage base in my legs, so I knew I was physically capable of doing the ride during the limited daylight hours of late October. I had checked, replaced, or adjusted practically every part of my bike in the weeks before the ride. I had also ridden the bike on long rides close to Moab in the preceding days and knew that it was in good running order. Finally, I added to my saddlebag tool kit a few tools that I do not ordinarily carry.

I knew that there was very little chance of anything going wrong with my bike, and with the tools I had, I could fix almost anything short of a broken frame on the trail. Armed with this knowledge and experience, I really enjoyed the ride! I stopped and gawked at almost every breathtaking vista, vertical box canyon, colorful balanced rock, or windblown arch. I took scenic detours. I knew that I had a good cushion of safety, so I could totally immerse myself in the pleasure of the ride. I had no nagging fear of something going wrong to dilute the experience.

Confidence in your mechanical ability allows you to be more courageous about what you will try on trails. And armed with this confidence, you'll be more willing to share your love of the sport with less experienced riders. Bringing new people along on rides is a lot more fun if you know that you can fix their bikes and they won't be stranded with a junker that won't roll.

HOW TO USE THIS BOOK

Skim through the entire book. Skip the detailed steps, but look at the exploded diagrams and get the general flavor of the book and what's inside. When it is time to perform a particular task, you'll know where to find it, and you'll have a basic idea of how to approach it.

Along with illustrators Todd Telander and Mike Reisel, I have done my best to make these pages as understandable as possible. Exploded diagrams are purposefully used instead of photographs to show more clearly how each part goes together. The first time you go through a procedure, you may find it easier to have a friend read the instructions out loud as you perform the steps.

Obviously, some maintenance tasks are more complicated than others. I am convinced that anyone with an opposable thumb can perform virtually any repair on a bike. Still, it pays to spend some time getting familiar with the really simple tasks, such as fixing a flat, before throwing yourself into complex jobs, such as servicing a suspension fork.

Tasks and tools required are divided into three levels indicating their complexity or required proficiency. Level 1 tasks need level 1 tools and require of you only an eagerness to learn. Level 2 and level 3 tasks also have corresponding tool sets and are progressively more difficult. All repairs mentioned in this book are classified as level 1 unless otherwise indicated. Tools are shown in Chapter 1. The section at the end of Chapter 2, "Performing Mechanical Work: A General Guide" (2-19), is a must-read; it states general policies and approaches that apply to all mechanical work.

Each chapter starts with a list of required tools in the margin. If a section involves a higher level of work, there will be an icon designating the level and tools necessary to perform the tasks in that section. Tasks and illustrations are numbered for easy reference. For instance, "3-6" means "see Section 3-6 in Chapter 3." Illustrations are referred to as "Figures," for instance, "Fig. 3.3."

At the end of some chapters there is a troubleshooting section. This is the place to go to identify the source of a certain noise or particular malfunction in the bike. There is also a comprehensive troubleshooting guide in Appendix A.

There is a wealth of other valuable information in the appendices. Get used to using them; many tasks will be simplified.

Appendix B has complete gear charts for the three most common mountain-bike wheel sizes, and it also includes instructions on calculating your gear with non-standard-size wheels. Appendix C is an extensive section on selecting a properly sized bike and positioning it to fit you. Appendix D lists the tightening specifications of almost every bolt on the bike in the Torque Table. As bike parts become ever lighter and made out of ever more exotic materials, tightening them to the recommended torque spec becomes ever more important. The glossary is a comprehensive dictionary of mountain bike technical terms. There is a separate index of the key words as well as of the illustrations in the book if you want to quickly check and see what something looks like.

THE MOUNTAIN BIKE

This (Fig. I.2) is the creature to which this book is devoted (in this case, a "hardtail" with cantilever brakes). All of a mountain bike's major parts are illustrated and labeled here. Take a minute to familiarize yourself with these parts now, and refer back to this diagram whenever necessary.

The mountain bike comes in a variety of forms, from models with rigid frames and forks (Fig. I.3), to hardtails (front suspension only—Fig. I.2), to models with front- and rear-suspension systems (Fig. I.4). They can come with rim brakes (Figs. I.2, I.3, and I.5) or disc brakes (Fig. I.4).

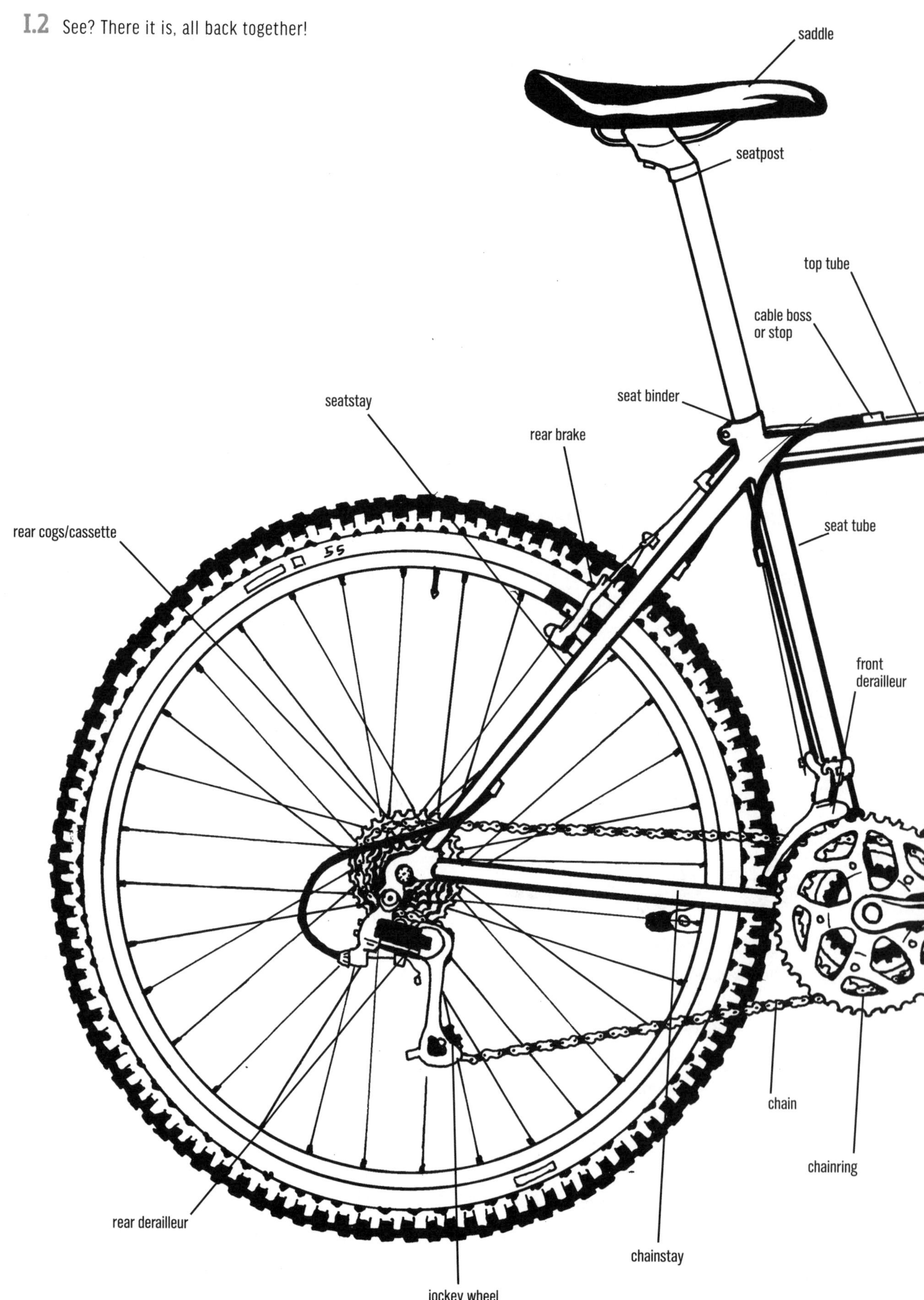

I.2 See? There it is, all back together!

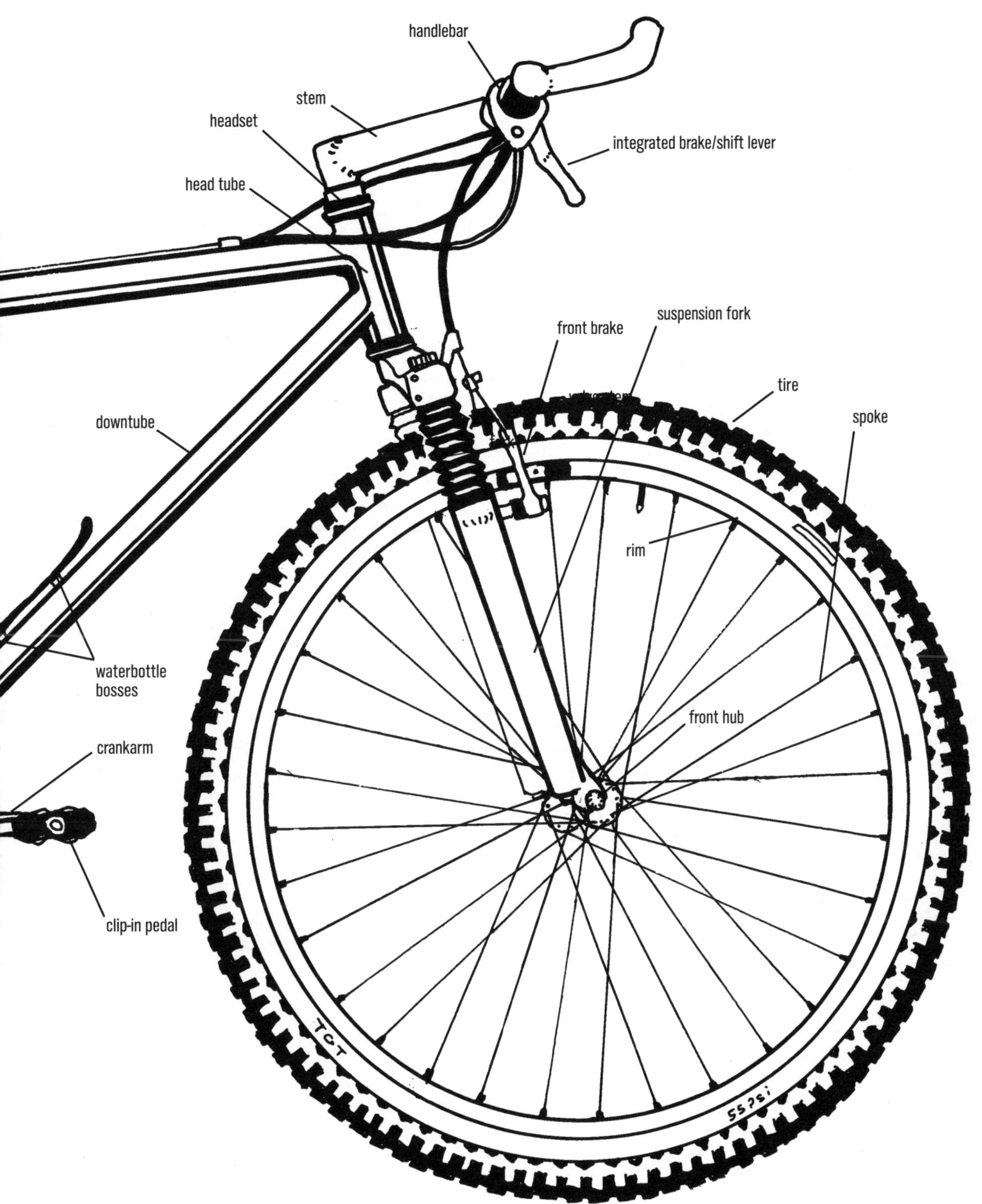
handlebar
stem
headset
integrated brake/shift lever
head tube
front brake
suspension fork
tire
spoke
downtube
rim
waterbottle
bosses
front hub
crankarm
clip-in pedal

A mountain bike generally comes with knobby tires in a 26-inch, 27.5-inch, or 29-inch diameter, and fat bikes have their own fatter and taller tires. Smaller 24-inch wheels and tires are found on small mountain bikes. Tire widths and shapes vary and include everything from studded snow tires to smooth street tires. This book also covers "hybrid" bikes (Fig. I.5), which are a cross between road bikes and mountain bikes.

No matter how a mountain bike is configured, even those who see themselves as having no mechanical skills will be able to tackle problems as they arise if they study the steps necessary to properly maintain and repair their bike. With a little bit of practice and a willingness to learn, your bike will transform itself from a mysterious contraption seemingly too complicated to tamper with to a simple, very understandable machine that can be a genuine delight to work on. Just allow yourself the opportunity and the dignity to follow along, rather than deciding in advance that you will never be able to do this. All you have to do is follow the instructions and trust yourself.

So, set aside your self-image as someone who is "not mechanically oriented" (and any other factors that may stand in the way of your making your mountain bike ride like a dream), and let's start playing with your bike!

I.3 Fully rigid

I.4 Fully suspended

I.5 Hybrid

Behold, we lay a tool here and on the morrow it is gone.

—THE BOOK OF MORMON

1

TOOLS

You can't do much work on a bike without tools. Still, it's not always clear exactly which tools to buy. This chapter identifies the tools you should consider owning on the basis of your level of mechanical experience and interest.

As I mentioned in the Introduction, the maintenance and repair procedures in this book are classified by their degree of difficulty. Nearly all the repairs in this book are classified as level 1, unless otherwise indicated. The tools for levels 1, 2, and 3 are pictured and described in the following pages. Lists of the tools needed in each chapter are shown in the margin at the beginning of each chapter.

For the uninitiated, there is no need to rush out and buy a large number of bike-specific tools. With only a few exceptions, the Level 1 Tool Kit (Fig. 1.1A) consists of standard metric tools. This kit is similar to the collection of tools I recommend later in this chapter to carry with you on rides (Figs. 1.5 and 1.6), though in a sturdier and more durable form. The Level 2 Tool Kit (Fig. 1.2) contains several bike-specific tools, allowing you to do more complex work on the bike. The tools in the Level 3 Tool Kit (Fig. 1.3) are extensive (and expensive), and they ensure that your riding buddies will show up not only to ask your sage advice but also to borrow your tools.

After that, if you really want to go all out and be set up like a pro (and have a line of mechanics waiting to borrow your tools), you can splurge on the set shown in Figure 1.4. If you loan tools, you might consider marking your collection and keeping a file of who has what to help recover those items that might otherwise take a long time finding their way back to your workshop. You would be surprised how easy it is to forget who has one of your seldom-used tools when you need your snapring pliers or a metric tap.

1-1

LEVEL 1 TOOL KIT

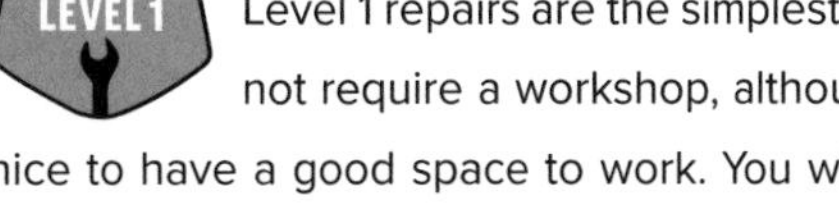

Level 1 repairs are the simplest and do not require a workshop, although it is nice to have a good space to work. You will need the following tools (Fig. 1.1A):

- **Tire pump** with a gauge and a valve chuck to match your bike's tubes (either Presta or Schrader valves—see Fig. 1.1B; many pumps will fit both). A spare rubber insert for the chuck is a good idea; these wear out.

1.1A Level 1 Tool Kit

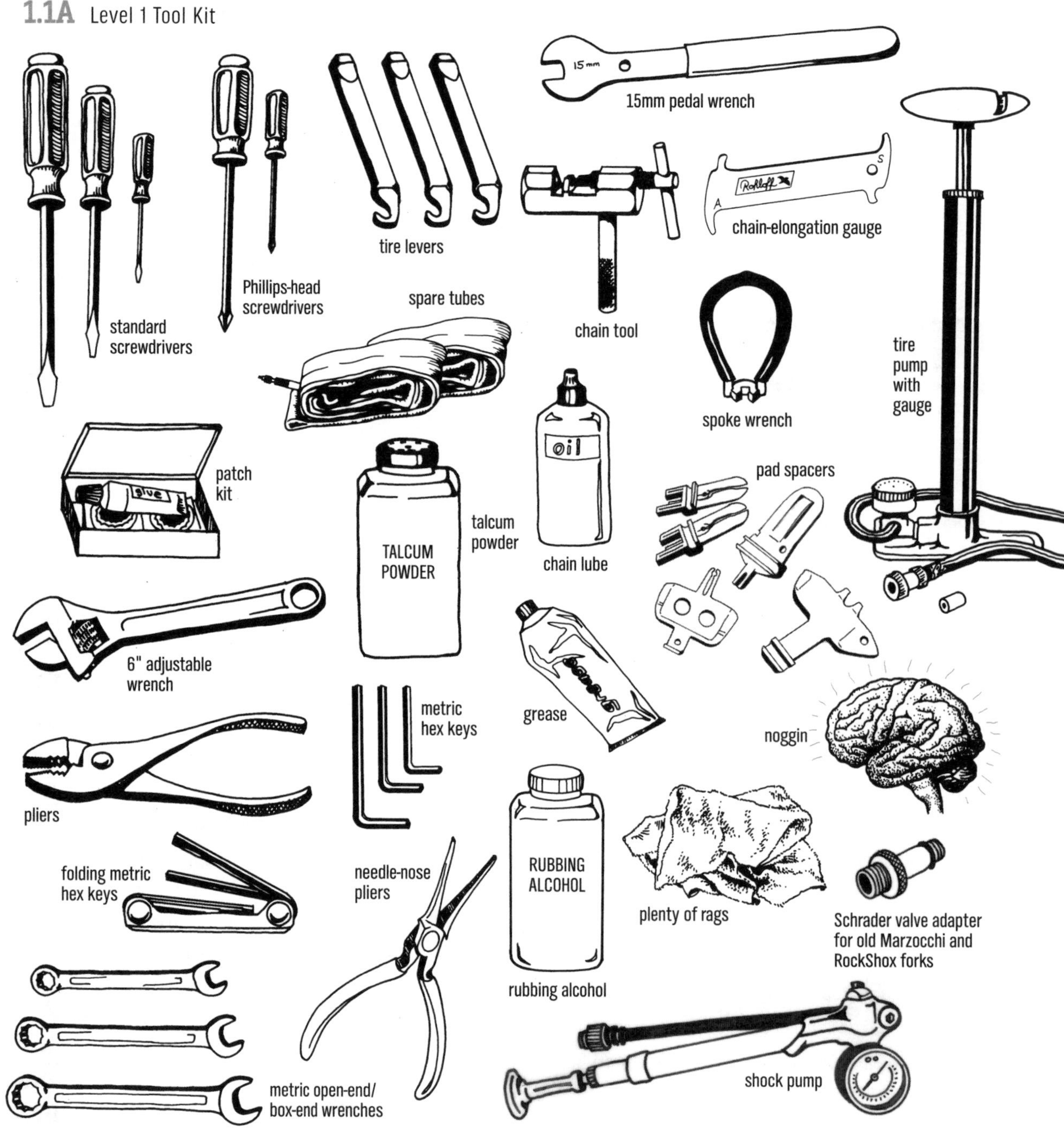

- **Standard screwdrivers:** small, medium, and large (one of each).
- **Phillips-head screwdrivers:** one small and one medium.
- Set of three plastic **tire levers** (Fig. 7.5).
- At least two **spare tubes** of the same size and valve type as those on your bike.
- Container of **talcum powder** for coating the inside of tires. Do not inhale this stuff.
- **Patch kit.** Choose one that comes with sandpaper, not a metal scratcher, and patches with soft orange rubber backing to the black rubber (Fig. 7.10). At least every year and a half, check that the glue has not dried up, regardless of whether the tube has been opened or not. On rides, you may as well take a little packet of glueless patches; they don't work as well as standard patches, but if the glue in your patch kit has dried up, you'll be glad you have them.

- One 6-inch **adjustable wrench** (a.k.a. Crescent wrench).
- **Pliers:** regular and needle-nose.
- Set of **metric hex keys** (a.k.a. Allen wrenches or hex wrenches) that includes 2.5mm, 3mm, 4mm, 5mm, 6mm, 8mm, and 10mm sizes. Folding sets are available and work nicely to keep your wrenches organized. But the folding variety aren't strong enough or long enough in the big sizes (6mm and up); big bolts require more leverage so you'll want the full-size model. I also recommend buying extras of the 4mm, 5mm, 6mm, and 8mm sizes.
- Set of **metric open-end/box-end wrenches** that includes 7mm, 8mm, 9mm, 10mm, 13mm, 14mm, 15mm, and 17mm sizes.
- **15mm pedal wrench** (Fig. 13.3)**.** This is thinner and longer than a standard 15mm wrench and thicker and stronger than a cone wrench. Your bike's pedals may accept only a 6mm or 8mm hex key (Fig. 13.4), so you may not need this tool.
- **Chain tool** for disconnecting and reconnecting chains (Figs. 4.10 and 4.11). Older chain tools may be too wide for the narrow chains on newer bikes; read the Pro Tip in Chapter 4 before buying one.
- **Chain-elongation gauge** to monitor the condition of the chain (Figs. 4.5 and 4.6).
- **Spoke wrench** that matches the size of the nipples on your bike's wheels.
- **Pad spacers** for disc brakes to prevent pushing the pads out too far when the wheel is out. Sometimes these have an integrated bleed block and hose-clamping groove, which are required for cutting and bleeding hydraulic brake hoses.
- Tube or jar of **grease.** I recommend using bicycle grease; however, if you already have some automotive grease, you can use it on everything except suspension forks and shocks or in twist shifters.
- Drip bottle of **chain lubricant** (Fig. 4.1). Please choose a nonaerosol; it is easier to control, uses less packaging, and wastes less in overspray.
- **Rubbing alcohol** for cleaning disc-brake pads, rotors, shocks, and internal parts and for removing and installing handlebar grips.
- A lot of **rags!**

1.1B Valve types

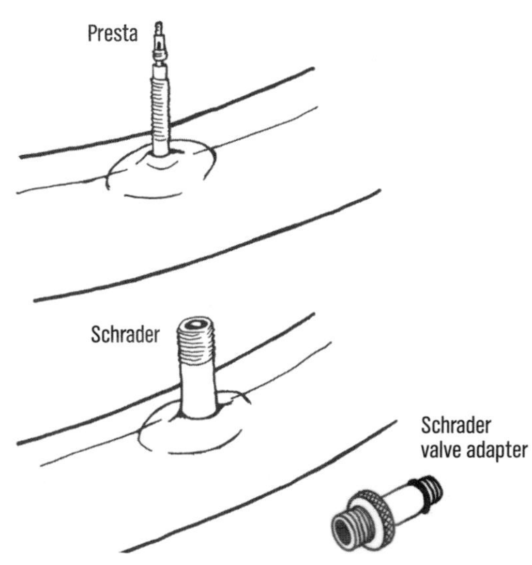

Other useful items:

- If you have an air-sprung suspension fork or rear shock, you need a **shock pump**. Get one with a no-leak head (Fig. 16.11), and get the adapter you need if your bike's fork requires either a ball needle (Fig. 16.12) or a **special adapter** to insert inside a sunken Schrader valve.

1-2

LEVEL 2 TOOL KIT

LEVEL 2

Level 2 repairs are a bit more complex, and I recommend that in addition to assembling these tools you create a well-organized workspace with a shop bench. Keeping your workspace organized is probably the best way to make maintenance and repair easy and quick. You will need the entire Level 1 Tool Kit (Fig. 1.1A) plus the following tools (Fig. 1.2):

- **Portable bike stand.** Be sure that the stand is sturdy enough to remain stable when you're really cranking on the wrenches. If for some reason you can't clamp your bike's seatpost, you will need a bike stand that holds the bike by the bottom bracket and the front or rear end with one wheel out; see the one in Figure 1.4.
- **Shop apron** (to keep your nice duds nice).
- **Tire pressure gauge.** It is more accurate than a pump gauge and a must for getting pressure exact for technical riding.

- **Hacksaw** with a fine-toothed blade.
- **Box-cutter knife** (Fig. 11.32) or **razor blades.**
- **Files:** one round and one flat, not too coarse.
- **Cable cutter** for cutting coaxial shift cable housing without crushing it as well as for cutting brake and shifter cables without fraying.
- Set of **metric socket wrenches** that includes 7mm, 8mm, 9mm, 10mm, 13mm, 14mm, and 15mm sizes.
- **Torx keys,** which look like hex keys with star-shaped tips. Torx T10, T25, and T30 are common sizes on modern bikes.
- **Crank puller** for removing crankarms (Fig. 11.6) if you have an old, three-piece crankset. Its push rod is sized for either square-taper spindles (Fig. 11.20) or ISIS or Octalink spindles (Figs. 11.21 and 11.22); get the right one for your crankset.
- **Chainring-nut tool** for holding the nut while you tighten or loosen a chainring bolt (Fig. 11.9).
- **Chainring-cassette removal tools,** if you happen to have old Shimano Octalink–style HollowTech I cranks (Fig. 11.12).
- **Bottom-bracket tools.** For external-bearing cranks (Fig. 11.2), you'll need an oversize **splined wrench** to remove the cups (Fig. 11.14); this will also remove some Center Lock rotor lockrings (Fig. 9.8). For some Shimano cranks, you'll also need a **little splined tool** to tighten the left crank's adjustment cap. To fit smaller external-cup sizes without having to buy a wrench or socket for each size, use a **splined step-down insert** to plug into a standard external-bearing splined tool. For sealed-cartridge bottom brackets (Figs. 11.20 and 11.21), you'll need a **splined bottom-bracket socket** (Fig. 11.28); if your bike has an ISIS or Octalink splined-spindle bottom bracket, you'll need one with a bore large enough to swallow the fatter spindle (Fig. 11.21), and if your bike has a square-taper cartridge bottom bracket (Fig. 11.20), the bigger-bore tool (Fig. 11.21) will work on both types. And for old-style cup-and-cone and adjustable cartridge-bearing bottom brackets (Figs. 11.23 and 11.24), you'll need a **lockring spanner** and a **pin spanner** to fit the bottom bracket (Fig. 11.31).
- **Snapring pliers** (Fig. 11.27) for BB30 cranks (Fig. 11.15) and other unthreaded bottom brackets with snapring grooves and for use in removing snaprings from suspension forks, pedals, and other parts.
- **Cone wrenches** for loose-bearing hubs (Fig. 8.9). The standard sizes are 13mm, 14mm, 15mm, and 16mm, but check which size you need before buying.
- Medium **ball-peen hammer.**
- Two **headset wrenches.** Be sure to check the size of the headset nuts (Fig. 12.30) before buying these. This purchase is unnecessary if your bike has a threadless headset and you don't plan to work on old bikes. Some suspension forks have crown nuts that fit headset wrenches.
- Medium **bench vise,** bolted securely to the bench; especially useful for working on rear shocks (Figs. 17.9, 17.11, and 17.14).
- **Cassette lockring tool** for removing cogs from the rear hub (Figs. 8.27 and 8.28) as well as some Center Lock rotor lockrings (Fig. 9.7).
- **Chain whip** for holding cogs while loosening the cassette lockring (Fig. 8.28) or a **Pedro's Vise Whip,** which holds the cog more firmly (Fig. 8.27) and won't fall off like a chain whip does when the freewheel rotates, saving your knuckles when the lockring breaks free.
- **Channellock pliers.**
- **Splined pedal-spindle removal tool** (Fig. 13.9).
- **Tweezers.**
- Fine (180 grit) **drywall sanding screen** for sanding disc-brake pads.
- **Valve core removers** for both Schrader and Presta valves. These are used for tire service and shock service.
- **Tire sealant** for setting up tubeless tires (Fig. 7.17) or installing into inner tubes for puncture protection.
- Fine-tipped **grease gun.**
- **Assembly paste.** Especially for seatposts (Fig. 14.7).
- Tube of **silicone-based grease** if you have Grip Shift shifters.
- **Nonlithium suspension grease** for front and rear shocks and pivots.
- **Threadlock fluid** for keeping bolts tight that have a tendency to unscrew.
- **Penetrating oil** or **ammonia** for freeing stuck parts.
- **Sound system** laden with good tunes.

1.2 Level 2 Tool Kit

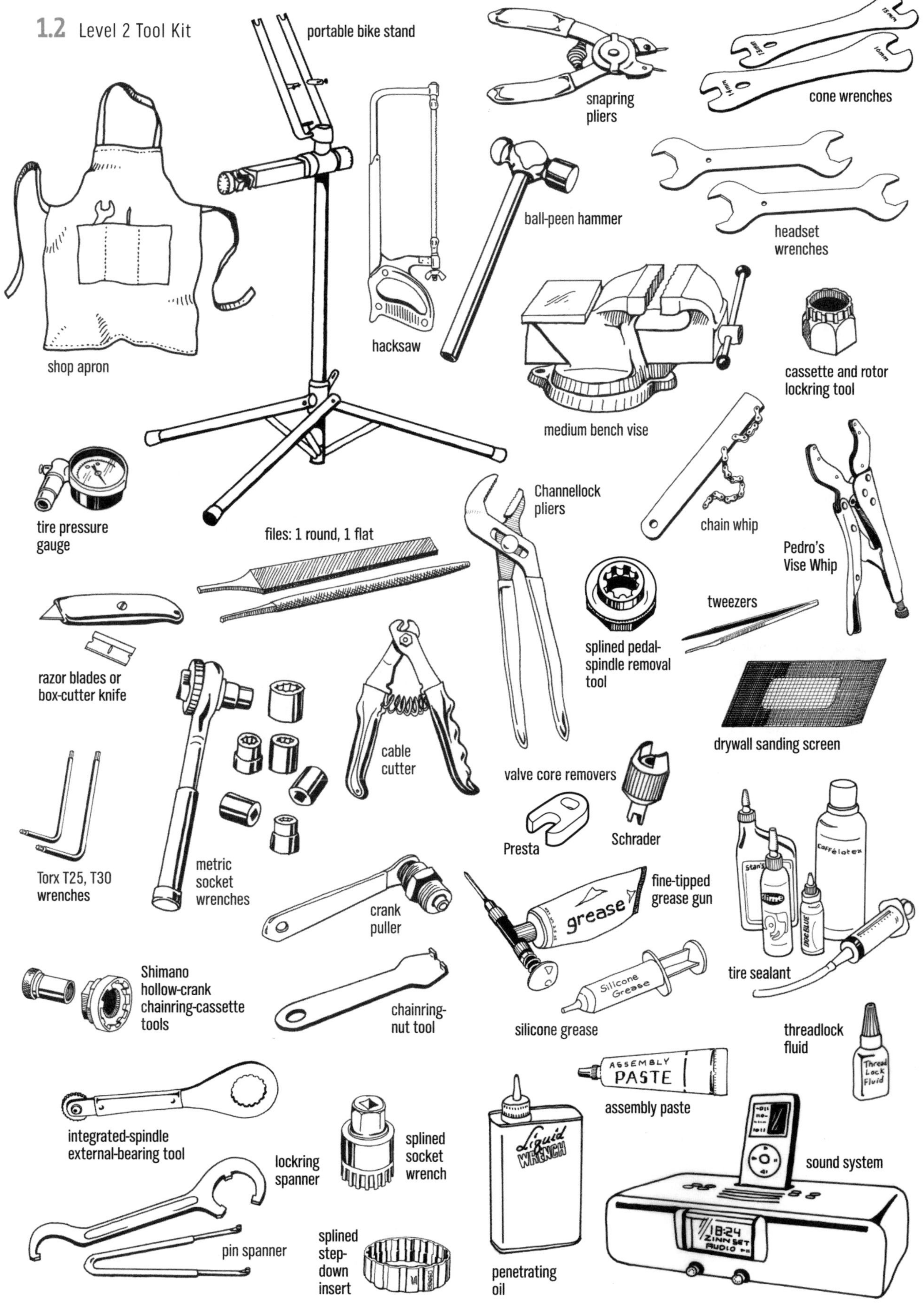

- You'll also want such stuff as tape, safety glasses, and rubber dish gloves or a box of cheap latex gloves. Buckets, large brushes, sponges, cog picks, degreaser, and dish soap or bike cleaner also will serve you well for cleaning a dirty machine rapidly.

1-3

LEVEL 3 TOOL KIT

LEVEL 3

If you are an accomplished level 3 mechanic, you can build a complete bike from a bare frame. That is assuming, of course, that the following tools (Fig. 1.3) are neatly organized in your shop:

- **Parts-washing tank.** Please use an environmentally safe degreaser. Dispose of used solvent responsibly; check with your local environmental safety office.
- Large **bench-mounted vise** to free stuck parts and press in others (Figs. 17.9, 17.11, and 17.14).
- **Master link pliers** (Fig. 4.23).
- **Shop chain tool** that works on 6-speed to 11-speed or 12-speed chains.
- **Headset press.** A simple, inexpensive press can press in any size headset (Fig. 12.44) and can install press-in bottom brackets as well (Fig. 11.25).
- **Fork-crown race punch** (a.k.a. slide hammer) for installing the fork-crown race of the headset (Fig. 12.42); its size depends on the size of the steering tube at its base.
- **Headset cup remover rocket** (Figs. 12.37 and 12.38). This tool also removes PF30 bottom brackets.
- **PF24 (BB86/92) bottom-bracket-remover rocket.** This is a smaller version of the above tool for fitting through smaller bearings.
- **Star-nut installation tool** for threadless headsets (Fig. 12.15).
- Large **ball-peen hammer.**
- **Soft hammer.** Choose a rubber, plastic, or wooden mallet to prevent damage to parts.
- **Large sockets.** To work on a suspension fork air spring or damper, depending on fork model and stanchion diameter, a 22mm, 24mm, 26mm, or 28mm six-point socket ground flat on the end is a must (Fig. 16.8).
- **Torque wrenches** for checking proper bolt tightness. Following manufacturer-specified torque settings prevents parts from stripping, breaking, creaking, or falling off while you are riding. Ideally, you want a small torque wrench for small bolts and a big, long torque wrench for large bolts.
- **Torx** and **metric hex square-drive bits.**
- Instead of the Torx and metric socket-wrench bits just mentioned, you can invest in a full selection of ¼-inch **hex drive bits** in **Torx, metric hex,** Phillips, flat-blade, and a tiny **Giustaforza torque wrench** (in place of a smaller square-drive torque wrench) or a **Prestacycle mini ratchet handle,** all of which can come in handy for small bolts.
- **Park IR-1 or IR-1.2 internal wire routing kit.** This is how you get shift and brake cables, shift and brake cable housings, and hydraulic hose through the frame easily (Figs. 6.16, 6.17, and 6.20).
- Set of **metric taps** that includes 5mm × 0.8mm, 6mm × 1mm, and 10mm × 1mm sizes for fixing mangled frame threads.
- **Truing stand** for truing (Figs. 8.2 and 8.3) and building wheels (Chapter 15).
- **Through-axle adapters for truing stand.** To hold a through-axle wheel securely while truing the wheel and the rotor.
- **Dishing tool** for checking that a wheel is properly centered (Figs. 15.23 and 15.24).
- **Spoke wrenches** of all sizes.
- **Splined** or other **specialty spoke wrench** for wheels with splined or oversized nipples, bladed spokes, and/or internal nipples (see other specialty spoke wrenches in Fig. 1.4).
- **Rotor truing gauge** that attaches to truing stand for checking rotor alignment (not shown here; see Fig. 9.16).
- **Rotor-alignment forks** for straightening out-of-true disc-brake rotors (Fig. 9.16) and getting rid of that annoying brake rub noise.
- **Bleed kit** (not shown here; see Figs. 9.17 to 9.26) for bleeding hydraulic brakes and hydraulic-release dropper posts—the type varies with brand. All hydraulic brakes require some sort of tube and may require only a squeeze bottle of fluid rather than a syringe.

1.3 Level 3 Tool Kit

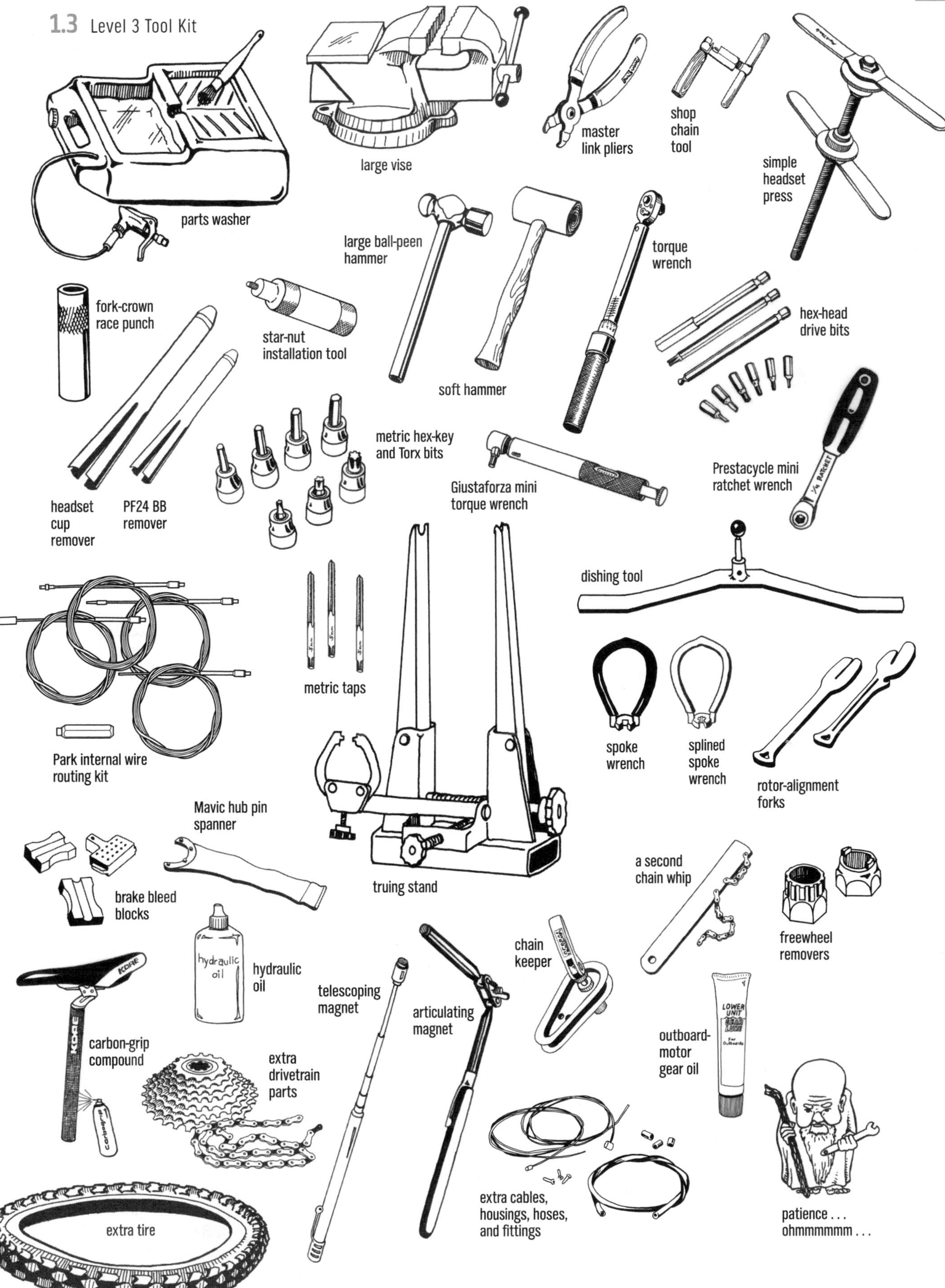

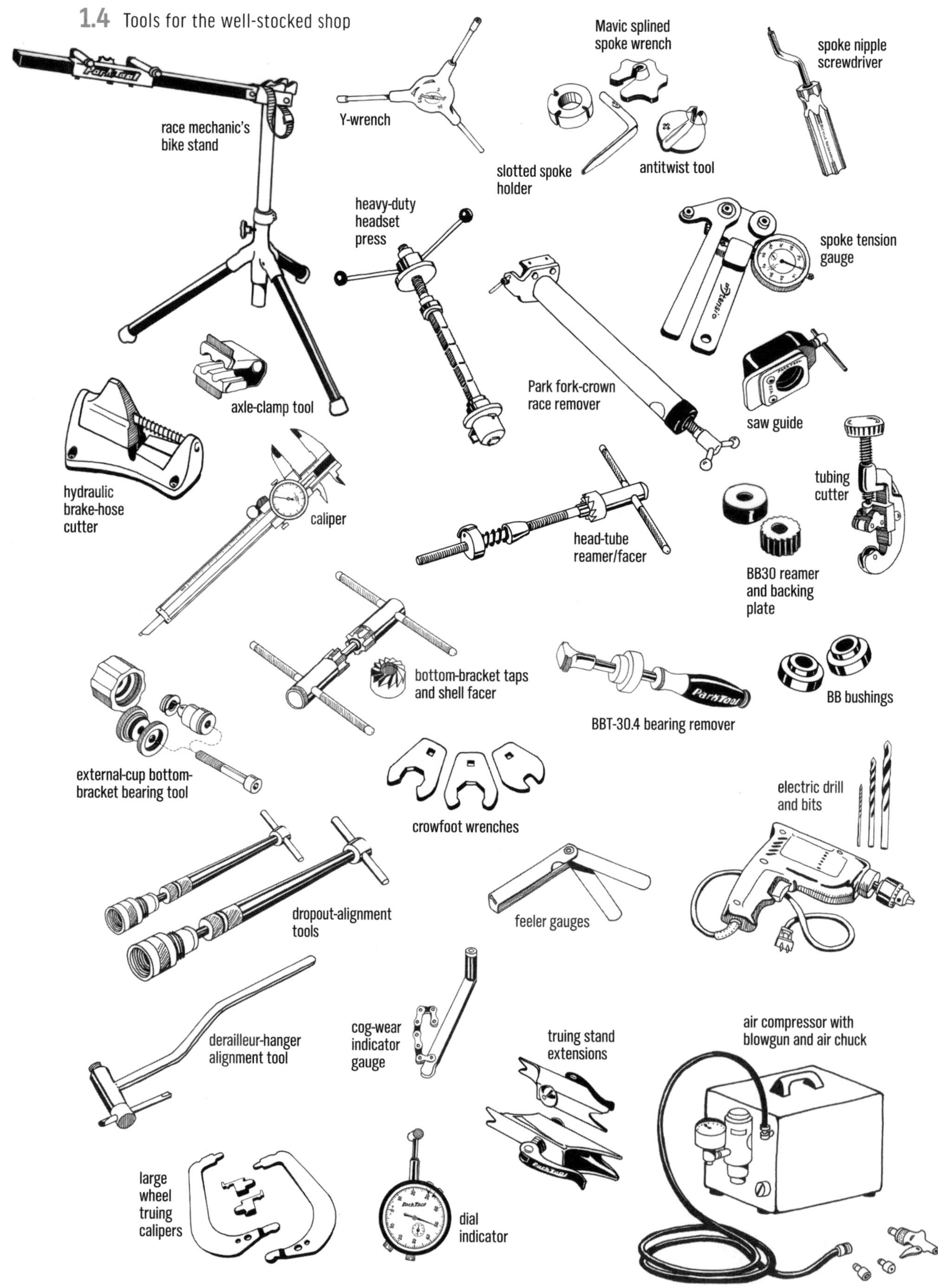
1.4 Tools for the well-stocked shop
race mechanic's bike stand
Y-wrench
Mavic splined spoke wrench
spoke nipple screwdriver
slotted spoke holder
antitwist tool
heavy-duty headset press
spoke tension gauge
Park fork-crown race remover
saw guide
axle-clamp tool
hydraulic brake-hose cutter
caliper
tubing cutter
head-tube reamer/facer
BB30 reamer and backing plate
bottom-bracket taps and shell facer
BBT-30.4 bearing remover
BB bushings
external-cup bottom-bracket bearing tool
crowfoot wrenches
electric drill and bits
dropout-alignment tools
feeler gauges
derailleur-hanger alignment tool
cog-wear indicator gauge
truing stand extensions
air compressor with blowgun and air chuck
large wheel truing calipers
dial indicator

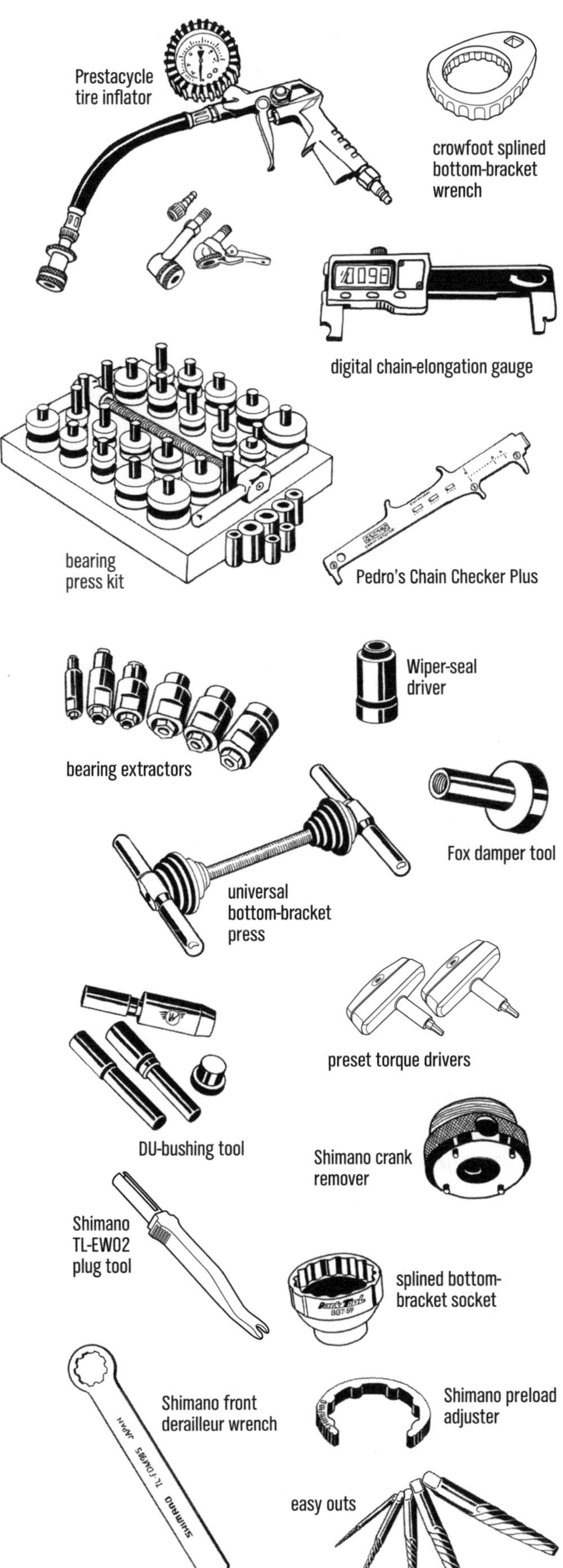

- **Brake bleed block** to keep pistons pushed back while bleeding disc brakes; you can substitute **plastic grooved blocks,** which are equally thick and which you also need for clamping hydraulic hose when inserting barbed fitting.
- **Pin spanner** for adjusting Mavic hubs.
- **Telescoping** or **articulating magnet** for picking up dropped parts or small tools.
- **Chain keeper** (attaches to dropout to run chain over for cleaning drivetrain with wheel off).
- An extra **Vise Whip** or **chain whip** for disassembling old-style 6- and 7-speed cogsets or freewheels.
- **Freewheel removers** for Shimano, Sachs, and Sun-Tour freewheels.
- One healthy dose of **patience** and an equal **willingness to work** and rework jobs until they have been properly finished.

Other items you might like to have on hand:

- **Spare parts** to save you from last-minute runs to the bike shop, such as several sizes of ball bearings, zip-ties, spare cables, cable housing, and a lifetime supply of those little cable-end caps. Keep on hand spare tires, tubes, chains, master links, and cogsets. If you expect to be working on suspension forks, rear shocks, and hydraulic brakes, be sure to have spare hoses, seals, and fittings.
- **Carbon-grip compound** for clamping carbon seatposts (Fig. 14.7) and handlebars.
- **Various fluids.** Special hydraulic brake fluids, hydraulic suspension oils and greases, threadlock fluid, titanium antiseize compound, outboard-motor gear oil, or specialty freehub lubricants are required for some jobs.

1-4

NOW, IF YOU REALLY WANT A WELL-STOCKED SHOP

The following tools (Fig. 1.4) are not even part of the Level 3 Tool Kit and are not often needed for home bike repairs. That said, they sure do come in handy when you need them.

- Euro-style race team **mechanic's bike stand** that supports the bottom bracket and clamps either the fork ends or the rear dropouts without the wheel on. This is required for bikes with integrated seat masts or seatposts that for some reason cannot be clamped in a bike stand (only clamp a dropper post on its large diameter, below its collar).

- Long **Y-wrench** (a.k.a. three-way spoke wrench) with square-drive, 5mm, and 5.5mm sockets for tightening spoke nipples internal to a deep rim, or a specialty wrench for working a specific type of internal nipple.
- **Antitwist tool** for preventing twisting bladed (aero) spokes during truing.
- **Splined spoke wrench** for adjusting spokes on Mavic tubeless wheels.
- **Slotted spoke holder** for preventing flat spokes from rotating during wheel truing.
- **Bent-shaft spoke nipple screwdriver** for achieving faster spoke lacing when building wheels (Chapter 15).
- **Spoke tension gauge** for checking for proper spoke tension and thus ensuring long wheel life.
- **Axle-clamp tool** for clamping the end of a hub axle in a vise.
- **Hydraulic brake-hose cutter** for getting an optimal square cut and for reducing the likelihood of fluid leaks.
- Heavy-duty shop-grade **headset press.** Faster and easier to use than the inexpensive one in Figure 1.3, this tool is for pressing in headsets (Fig. 12.44) and bottom brackets into threadless bottom-bracket shells (Fig. 11.25).
- **Park universal fork-crown race remover.** This tool allows you to remove headset fork-crown races from any fork without pounding at them with a hammer and screwdriver and marring the fork crown.
- **Saw guide** for cutting threadless steering tubes off straight.
- **Tubing cutter** for cutting handlebars off straight without a hacksaw. Do not use it for cutting carbon handlebars, though; you'll need a carbon saw for that.
- **Caliper** with vernier, dial, or digital measurement for measuring parts in order to optimize function.
- **Head-tube reamer and facer** for keeping both ends of the head tube perfectly parallel and of the proper inside diameter for the headset cups.
- **BB30 reamer and backing plate** for perfecting the fit of bearings in BB30 shell. Use with head-tube reamer and facer handle.
- **English-threaded bottom-bracket tap set** for cutting threads in both ends of the bottom-bracket shell while keeping the threads in proper alignment.
- **Bottom-bracket shell facer.** Like a bottom-bracket tap, this tool cuts the faces of the BB shell so they are parallel to each other.
- **Park BBT-30.4 bearing remover** for BB30. Various **bushings** also allow pressing in BB30, BB86, BB90, BB92, and BB95 bearings with a **bearing press, headset press,** or **vise** (Fig. 11.25).
- **Splined bottom bracket sockets** make tightening cups with a torque wrench far easier. A full set would fit cups of 39mm, 41mm, 44mm, and 46mm diameter.
- **Crowfoot wrenches.** Turn big **headset nuts** and **pedal spindles** and **splined bottom-bracket cups** to precise torque settings. Have the crowfoot at 90 degrees to the torque wrench handle to achieve the torque setting displayed on the wrench's gauge (i.e., if the crowfoot is extended straight ahead or back, it multiplies or reduces the torque setting shown on the wrench handle, but at 90 degrees, it does not significantly change the torque setting).
- **Electric drill** with drill bit set.
- **Dropout-alignment tools** (a.k.a. tip adjusters). You need two: one for each dropout or fork end (Figs. 16.48 to 16.50, 17.29).
- **Derailleur-hanger alignment tool** for straightening the hanger after you shift it into the spokes or crash on it (Fig. 17.4).
- **Cog-wear indicator gauge** for determining if cogs are worn out.
- **17mm, 20mm, 21mm, 22mm, 23mm, 24mm, 28mm cone wrenches** (not shown) for loose-bearing through-axle hubs.
- **Feeler gauges** for precisely adjusting postmount disc brakes.
- **Air compressor** with quick-connect fitting on the hose for a blowgun and a Prestaflator. It is useful for lots of things, including overhauling disc brakes and seating tubeless tires.
- **Prestacycle Prestaflator tire inflator with gauge,** a quick-connect to clip onto the hose, and **air chucks** for Presta and Schrader valves. If you are using an

air compressor, this tool delivers fast and accurate inflation.

- **Large wheel truing-stand calipers (Park #238K)** to upgrade Park TS-2 or 2.2 to handle 29-inch wheels with big tires.
- **Truing-stand extensions with integrated through-axle adapters (Park TS-2EXT.2)** to upgrade Park TS-2 or 2.2 for truing super-tall wheels (36-inch, 32-inch) or fat-bike or 29er wheels with inflated tire. Also accepts through-axle wheels without installing a big Phillips screwdriver or a through-axle.
- **Dial indicator** for attaching to rotor-truing gauge to more accurately check rotor alignment (Fig. 9.15).
- **Digital chain-elongation gauge** to precisely monitor chain length over time.
- **Pedro's Chain Checker Plus** isolates pin wear to ensure that what you are measuring is actually chain elongation and not roller wear; Shimano TL-CN41 is similar.
- **Bearing-press kit** for installing cartridge bearings in hubs, rear-suspension pivots, and bottom brackets. Includes bushings for all bearing sizes found on hubs and suspension; get bushings for all bottom brackets, or get a **universal bottom-bracket press** with stepped bushings to fit either the ID or OD (inside diameter or outside diameter) of every type of bottom-bracket cartridge bearing; Wheels Mfg. shown.
- **Bearing-extractor kit** for removing bearings from blind holes; Wheels Mfg. shown.
- **Wiper-seal driver** for installing wiper seals atop suspension-fork lower legs.
- Fox **damper-removal tool** for freeing damper shaft from hole in bottom of lower fork leg.
- **DU-bushing tool** for removing and installing bushings in rear-shock eyelets; Wheels Mfg. shown.
- **Fixed-torque drivers** are preset for 4 N-m, 5 N-m, or 6 N-m. They ratchet freely when their torque setting

1.5 Tools to take on all rides

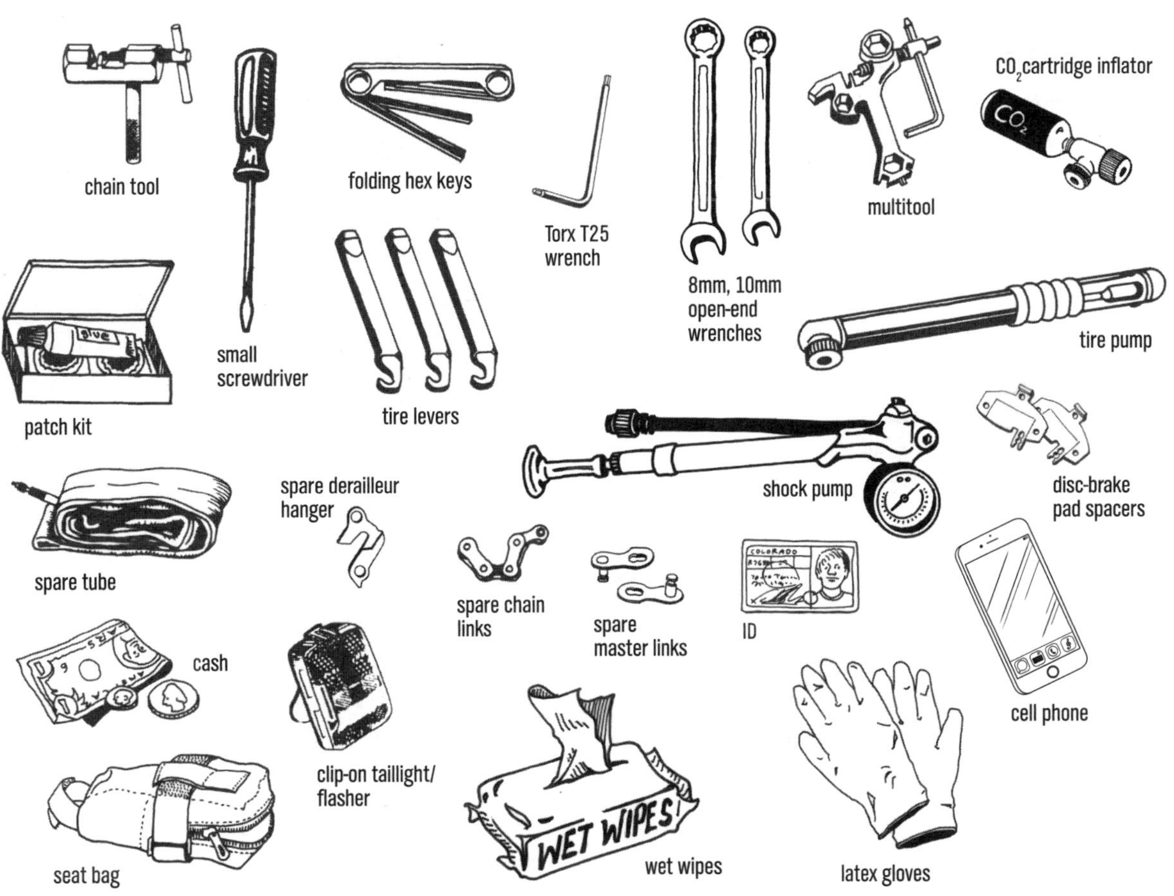

is reached, so there's no way to over-torque. The magnetic socket tip holds standard ¼-inch hex bits. Speeds accurate stem/handlebar installation.

- **Shimano TL-EW02 plug tool** for plugging in and unplugging Di2 electronic-shift wires.
- **Shimano TL-FDM905 Di2 front-derailleur-cage wrench** to replace the front-derailleur cage, which is available as a separate part.
- **Shimano TL-FC35 crank-removal tool** and **TL-FC17 bearing-preload adjustment tool** for XTR FC-M970 crank.
- **Easy outs** (spiral-flute screw extractors) for removing broken screws and rear shock spacers.

1-5

SETTING UP YOUR HOME SHOP

I recommend keeping this area clean and very well organized. Lame as this sounds, remember that a clean shop is a happy shop! Make it comfortable to work in and easy to find the tools you need. Hanging tools on a pegboard or slatboard or placing them in bins or trays is an effective way to maintain an organized work area. Being able to find the tools you need will immensely increase the enjoyment of working on a bike. It is harder to do a job with love if you're frustrated about not being able to find the cable cutter. Placing small parts in one of those benchtop organizers with several rows of little drawers is another good way to keep chaos from taking over.

1-6

TOOLS TO CARRY WITH YOU WHILE RIDING

a. For most riding

Keep all of the following stuff (see Fig. 1.5) in a bag under your seat or in a hydration pack or fanny pack. The operative words here are "light" and "serviceable." Many of these tools are combined in multitools. Make sure you try all tools at home before depending on them on the trail.

- **Chain tool** that works (try it before an emergency occurs).
- Small **screwdriver** for derailleurs and other parts.
- Compact set of **hex keys** (also called hex wrenches or Allen keys) that includes 2.5mm, 3mm, 4mm, 5mm, 6mm, and 8mm sizes.
- **Torx T25 wrench** for 6-bolt disc-brake rotors and some SRAM handlebar controls and chainring spiders.
- 8mm and 10mm **open-end wrenches** if riding an old bike with hex-head bolts.
- A good **multitool** to replace some or all of the preceding items but with less weight and bulk.
- **Tire pump** and/or **CO_2 cartridge inflator** with a spare cartridge. Larger pumps are faster than itty-bitty mini-pumps. Make sure the pump or cartridge is set up for your bike's type of tire valves.
- **Patch kit.** You'll need something after you've used your spare tube. Check at least every year and a half to make sure the patch kit glue has not dried up. Carry glueless patches or foam insulating tape as well.
- At least two plastic **tire levers,** preferably three.
- **Shock pump.** If the fork requires a pump adapter, make sure that you carry one.
- Two plastic **pad spacers** if your bike has hydraulic disc brakes (in case you get rescued and have to throw your bike in a vehicle with the wheels off; spacers prevent the pistons from coming out too far if a brake lever is inadvertently squeezed).
- **Spare tube.** Make sure the valve matches the ones on your bike and pump, and check that the Presta valve collar nuts on the wheels are loose enough to unscrew by hand out on the trail. Keep the tube in a plastic bag to prevent deterioration and to protect it from the sharp tools in your bag.
- **Spare derailleur hanger** that fits your frame in case you crash on the drive side.
- **Spare chain links** and two **spare master links** that match the chain width you're using (i.e., 8-, 9-, 10-, 11-, or 12-speed). If you're using a Shimano chain, you can instead carry at least two subpin rivets; master links are preferable for on-trail repairs, though.
- **Identification.** A driver's license or a Road ID.
- **Cell phone.** Like I need to tell anyone these days to carry one along!
- **Cash,** for obvious reasons, and as a temporary patch for sidewall cuts in tires.

- **Taillight** that you can clip on or, better yet, leave mounted on your bike in case you stay out after dark.
- **Wet wipes** or **latex gloves** to keep your hands clean.
- **Food and water.**

b. For long or multiday trips

The items in Figure 1.6 are, of course, in addition to proper amounts of food, water, and extra clothes, as well as in addition to the tools shown in Figure 1.5.

- **Spoke wrench** sized to your bike's spoke nipples.
- **Spare spokes.** Innovations in Cycling sells a really cool folding spoke made from Kevlar. It's worth getting one for emergency repairs on a long ride.
- Another **spare tube.**
- **Sealant-filled quick aerosol inflator** to rehabilitate a tube or tubeless tire with a slow leak.
- Small plastic bottle of **chain lube.**
- Small tube of **grease.**
- Compact 15mm **pedal wrench** unless the pedals don't have wrench flats. One with a headset wrench on the other end is handy for a threaded headset.
- **Pliers,** useful for innumerable purposes.
- **Wire** and/or a small **bungee cord,** which can be very handy for all kinds of things.
- **Duct tape.** It's like The Force. It has a light side and a dark side, and it holds the universe (and sometimes your bike or your shoes) together.
- **More money,** or its **plastic equivalent,** which can get you out of lots of scrapes.
- **Matches,** because you never know when you might be stranded overnight.
- A lightweight, aluminized, folding **emergency blanket.**
- **Rain gear.**
- **Satellite communicator.** The Garmin InReach shown sends and receives texts and emails, sends SOS signals, receives weather forecasts, and allows others to track you.
- **Headlight.** This can be a lightweight unit that clips onto the handlebar or a headlamp with a strap that fits over your helmet. An extra battery for it is a good idea too.
- Small **flashlight.** This can be a little LED type—just something to help find things in the dark if your headlight dies.

1.6 Tools for extended backcountry riding

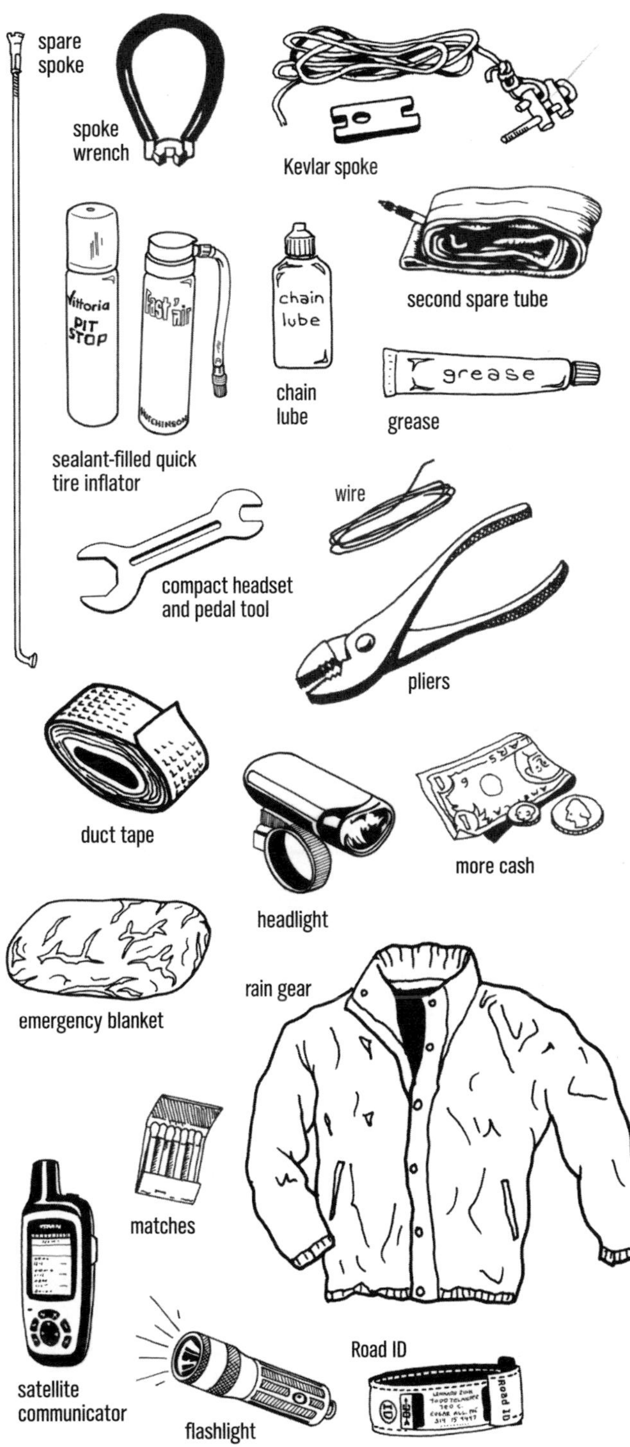

NOTE: *Read Chapter 3, which covers emergency repairs, before embarking on a lengthy trip. And if you are planning a bike-centered vacation, be sure to take along a Level 1 Tool Kit in the car, some headset wrenches, and various incidentals such as duct tape and sandpaper.*

Everything should be made as simple as possible, but not simpler.

—ALBERT EINSTEIN

2

BASIC STUFF

PRERIDE INSPECTION, WHEEL REMOVAL, AND GENERAL CLEANING

TOOLS

Chain lubricant
Rags

OPTIONAL

Solvent (citrus-based)
Chain-cleaning tool
Old water bottle
Bucket(s)
Large sponge
Large and small brushes
Dish soap
Chain keeper

Making sure your bike is safe is essential, so developing the habit of checking your bike before heading out on a ride is a good idea. Performing a preride inspection regularly could help you avoid delays due to parts failure. I won't even mention the injury risks you face by riding a poorly maintained bike.

LEVEL 1

Unless you always have a mechanic with you, you also need to know how to take the wheels off and put them back on, or you won't be able to transport your bike or effectively deal with minor annoyances such as flat tires or jammed chains. And if you do absolutely nothing else to your bike, keeping the chain and a few other parts clean will enhance enjoyment of each ride. This chapter's three cleaning and maintenance procedures are easy to perform and fundamental to keeping your bike running smoothly.

All home mechanics in particular should read "Performing Mechanical Work" in this chapter.

The work in this chapter requires no special tools beyond level 1. If you have never maintained your own bike, you may find the procedures a bit challenging at first. But with repetition, your confidence will grow and you'll soon find yourself tackling more advanced repairs. So have at it, and enjoy your bike's improved performance!

2-1

DOING THE PRERIDE INSPECTION

1. **Check to be sure that the quick-release (QR) levers or axle nuts are tight.** They secure the front and rear hub axles to the dropouts.
2. **Check the brake pads for excessive or uneven wear.** On disc brakes, this requires looking down into the slot in the caliper to see the pads—the pad material should be at least the thickness of a dime. On rim brakes, make sure that the molded-in grooves in the pads are not worn off.
3. **Check that the brakes are securely mounted.** On disc brakes, grab the caliper and try to twist it. On rim brakes, grab and twist the brake pads and brake arms. Make sure all bolts are tight.
4. **Squeeze the brake levers.** With rim brakes, this should bring the pads flat against the rims (or slightly toed-in) without hitting the tires. In the case of disc brakes, this should bring the pads against the rotor. Make certain that you

cannot squeeze the levers all the way to the handlebar (see 10-2 and 10-4 for brake-cable tension adjustment or 10-22 for hydraulic disc-brake bleeding, after first making sure that the disc-brake pads are in place and in good condition).

5. **Spin the wheels.** Check for wobbles while sighting along the rims, not the tires. (If a tire wobbles excessively on a straight rim, it may not be fully seated in the rim; check it all the way around on both sides.) Make sure that the rims do not rub on the brake pads.
6. **Check the tire pressure.** On most mountain bike tires the proper pressure is between 30 and 60 pounds per square inch (psi), although tubeless tires are able to run well below 30 psi. On fat-bike tires, pressure is 5–8 psi, with pressures as low as 2 psi for riding in snow. Look to see that there are no foreign objects sticking in the tire. If there are, you may have to pull the tube out and repair or replace it. If you are getting flat tires frequently, take a look at the section on tire sealants (i.e., the goop inside the tube that fills small holes), 7-9.
7. **Check the tires for excessive wear, cracking, or gashes.**
8. **Be certain that the handlebar and stem are tight.** Check that the stem is lined up with the front tire.
9. **Check that the gears shift smoothly.** The chain should not skip or shift by itself. Make sure that each click of the shifter moves the chain over one sprocket, starting with the first click. Make sure that the chain does not overshift the smallest or biggest rear cog or the smallest or biggest front chainring.
10. **Check the chain for rust, dirt, stiff links, or noticeable signs of wear.** The chain should be clean and lubricated. (Be cautious about overdoing it with the lube, though. Gooey chains pick up lots of dirt, particularly in dry climates.) The chain should be replaced on a mountain bike about every 500 to 1,000 miles of off-road riding or every 2,000 miles of paved riding; see 4-6.
11. **Apply the front brake, and push the bike forward and backward.** The headset (Fig. 12.1) should be tight and should not make clunking noises or allow the fork any fore-and-aft play.

If all these things check out, go ride your bike! If not, check the table of contents or Appendix A, go to the appropriate chapter, and fix the problems before you ride.

2-2
REMOVING THE FRONT WHEEL

You can't transport your mountain bike easily if you can't remove the front wheel, because removing the front wheel is required for mounting your bike on most roof racks and for shoehorning the bike inside your car. As outlined in the following sections, wheel removal involves releasing the brake and opening one of the following: the hub quick-release skewer, the through-axle mechanism, the bolt-on skewer, or the axle nuts on the low-end models.

If your bike has a single-leg fork (i.e., a Cannondale Lefty), wheel removal is different; see section 2-7.

2-3
RELEASING THE BRAKE

a. Disc brakes

Most disc brakes (Fig. 9.1) allow the disc to fall away without any extra step to release the pads, as the caliper is bolted to the fork and the disc slips in and out of it easily.

IMPORTANT: *Do not squeeze the lever of a hydraulic disc brake when there is neither a disc nor a travel spacer between the pads; otherwise, the pistons can pop out too far, and you won't be able to get the rotor in between the pads without pushing the pistons back with a tool. (Travel spacers are flat plastic pieces that usually come with the brakes, but not necessarily with a new bike; they act like the disc to keep the brake pistons from traveling too far. You can cut your own spacer from a piece of corrugated cardboard and slip it between the pads when the wheel is out.)*

b. Rim brakes

Most rim brakes have a mechanism to release the brake arms so that they spring away from the rim (Figs. 2.1–2.2), allowing the tire to pass between the pads. If yours does not, you will have to deflate the tire. V-brakes (a.k.a. "sidepull cantilevers"—Fig. 2.1) are released by pulling the end of the curved cable-guide tube (a.k.a. the "noodle") out of the horizontal link atop one of the brake arms while either holding the link or squeezing the pads against the rim with the other hand (Fig. 2.1). Most cantilever brakes (Fig. 2.2) and U-brakes (Fig. 10.39) are released by pulling the enlarged head of the straddle cable out of a notch in the top of the brake arm while holding the pads against the rim with the other hand (Fig. 2.2).

c. Rare rim-brake types

Roller-cam brakes (Fig. 10.40) are released by pulling the cam down and out from between the two rollers while holding the pads against the rim. Many linkage brakes (Fig. 10.38) are released in the same way as V-brakes or cantilever brakes. Releasing hydraulic rim brakes (Fig. 10.33) usually requires detaching the U-shaped brake booster connecting the piston cylinders together, if installed, followed by unscrewing or quick-releasing one wheel cylinder.

2-4

DETACHING A WHEEL WITH A QUICK-RELEASE SKEWER

This is easy, and you don't need a tool.

1. **Pull outward on the quick-release lever to open it** (Fig. 2.3).
2. **Unscrew the nut on the opposite end of the skewer's shaft.** Loosen until both the nut and the head of the skewer clear the fork's wheel-retention tabs. Most mountain bike forks have a wheel-retention system consisting of nubs or bent tabs on the fork ends (also known as dropouts) or an axle washer with a bent tooth hooked into a hole in the dropout. These systems prevent the wheel from falling out if the quick-release loosens.
3. **Pull the wheel off.**

NOTE: *Some bikes have non-quick-release bolt-on skewers (Fig. 2.4). The wheel is removed by unscrewing the skewer with a 5mm hex key until the head and the nut clear the wheel-retention tabs on the dropouts.*

2.1 Releasing the noodle from the link on a V-brake

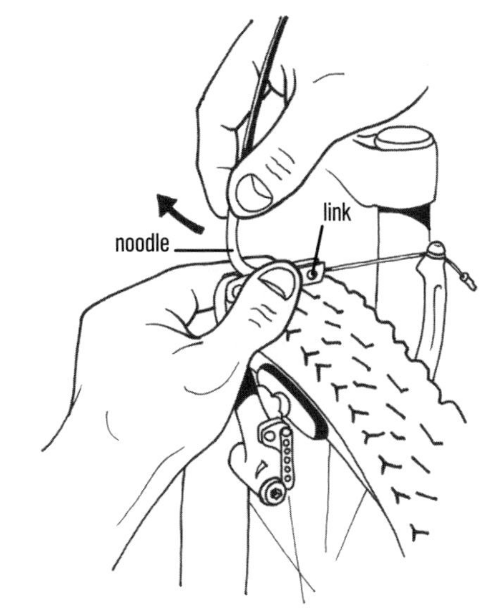

2.2 Releasing a cantilever brake

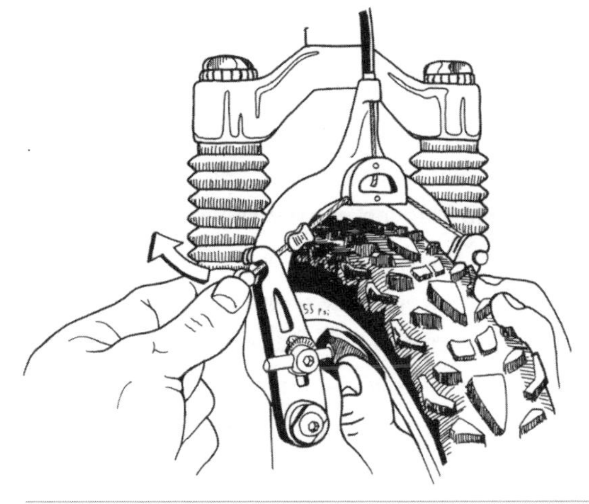

2.3 Opening a quick-release skewer

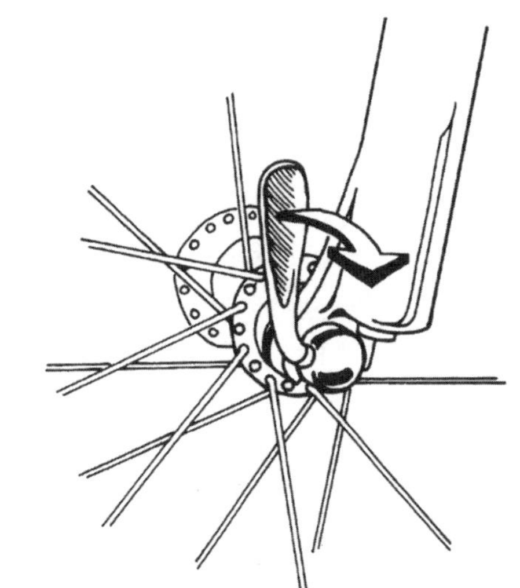

2.4 Bolt-on skewer

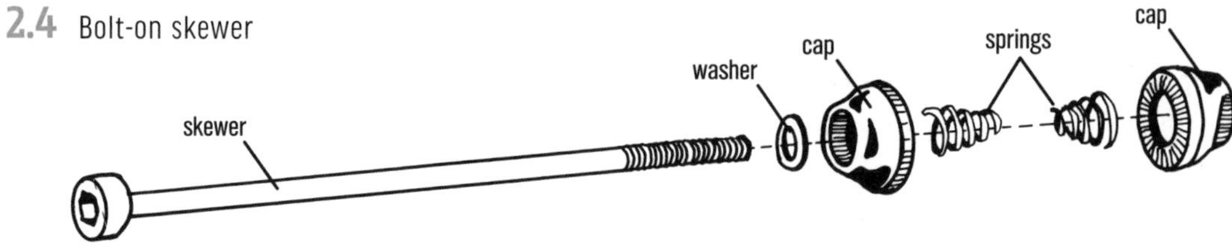

2-5

DETACHING A WHEEL WITH A THROUGH-AXLE

Through-axles are extra-long hub axles that fit directly through the hub-cartridge bearings and screw or clamp directly into the dropouts; the axle must be fully removed before you can take off the wheel. Front through-axles are generally 15mm or 20mm in diameter. They stiffen the fork against lateral and twisting flex, and they offer a higher degree of safety against the wheel falling out of the fork than a quick-release skewer does. In particular, the closed ends of the fork dropouts prevent the braking force applied to the hub rotor from pushing the axle out of the dropouts, even if the securing system is loose.

A fork with a through-axle will offer improved tracking and steering as well as smoother up-and-down action than one with a quick-release skewer. A through-axle is a necessity on an "upside-down" fork—a fork whose lower legs are its inner legs, motorcycle style. Because the wheel moves up and down with the inner legs, which slide up and down in the fixed upper outer legs, it is not possible to have a brace between the lower legs (a brace adds considerable lateral and torsional stiffness to a standard telescoping bicycle suspension fork; the upside-down fork compensates for this by having large-diameter outer legs that resist flex). The through-axle is all that ties the lower legs of an upside-down fork together.

Through-axle systems vary—there are bolt-on and quick-release versions—but they share some common traits. The axle comes as a part of the fork, not as a part of the wheel. All 15mm through-axle hubs have the same (15mm) inside diameter (ID) of their bearings and the same position of the disc-brake rotor relative to the caliper mounting tabs. Ditto for 20mm or 24mm hubs.

Traditional through-axles (Fig. 2.5) generally resemble a long bolt that clamps into the fork's dropouts. The head is too large to pass through the hub bearings and the dropouts; it snugs up against the drive-side fork dropout. The ends are usually round, but they may also be hex-shaped. A clamp-in through-axle usually has some sort of bolt system on the opposite end from the head to draw the ends toward each other, but some of these axles are threaded and simply screw into the opposite dropout. Pinch bolts (or a pair of quick-release levers) tighten the dropouts around the ends of the axle once it is fully installed.

Quick-release through-axles (Fig. 2.6) have a lever on the end to release them. Except on a Manitou fork, you can use the lever as a handle to screw the axle out of or into the opposite dropout. When you flip the lever closed, it either expands the axle inside its through-hole in the dropout to secure it, or squeezes the dropouts against the hub-axle ends. A quick-release through-axle can be as quick to use as a quick-release skewer and a standard hub.

2.5 Tightening and loosening a traditional 20mm through-axle

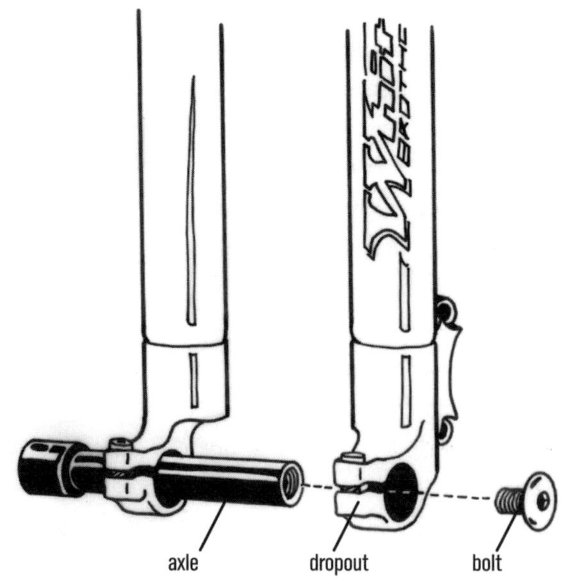

2.6 Tightening and loosening a RockShox Maxle

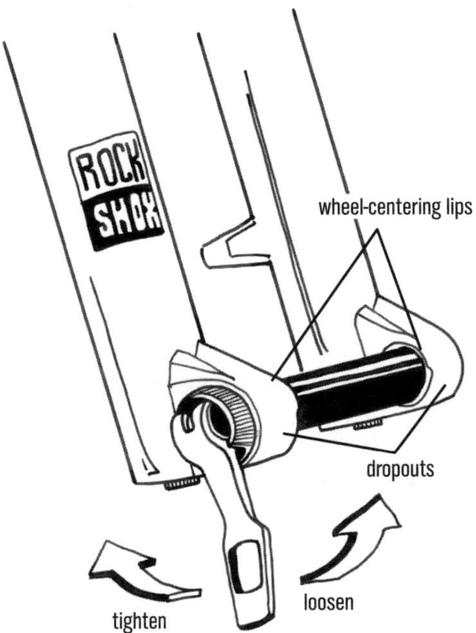

2.7 Loosening an axle nut

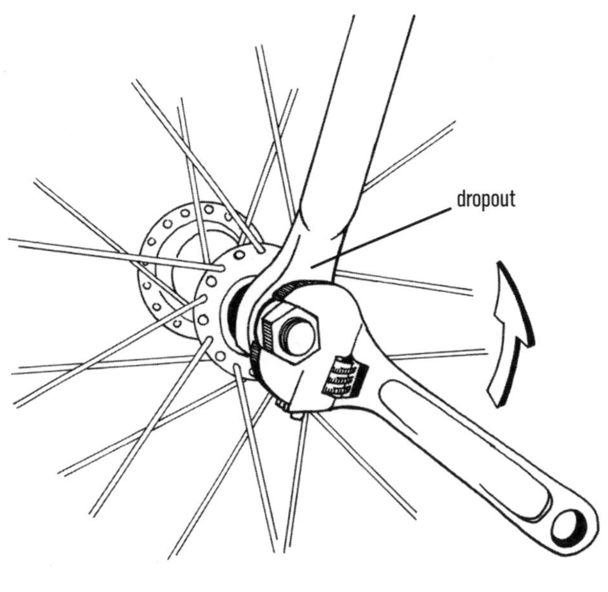

A through-axle does complicate mounting the fork on a roof rack's fork mounts. You may need to purchase adapters to accommodate and secure the fork.

To remove a quick-release through-axle on all forks except Manitou, flip open the lever fully. If there is a cutout in the ring under the lever as in Figure 2.6, engage the lever into that cutout (in the fully open position) to ensure that the lever unscrews the axle, rather than spinning freely. Rotate the lever counterclockwise to unscrew the axle, and pull the axle out.

To remove a Manitou QR15 HexLock Thru Axle, flip open the lever fully (it will go from pointed straight up to pointed straight down) and rotate it counterclockwise 90 degrees (a quarter turn) so that it is pointed straight forward; an internal spring will cause the lever to pop out a few millimeters. Grasp the lever and pull the axle straight out.

To remove a traditional draw-bolt through-axle (Fig. 2.5), loosen whatever clamp bolts or levers are securing the bolt head, and unscrew the draw bolt. Loosen the pinch bolts to free the axle head, and pull the axle out to that side.

To remove a threaded, clamp-in through-axle, loosen all of the clamp bolts on both dropouts. Unscrew the entire axle and pull it out.

2-6

DETACHING A WHEEL WITH AXLE NUTS

1. **Unscrew the nuts on the axle ends** (usually with a 15mm wrench—Fig. 2.7). The nuts unscrew counterclockwise ("lefty loosey, righty tighty").
2. **Loosen the nuts enough to clear the retention tabs on the dropouts.** Most mountain bikes have some type of wheel-retention system consisting of nubs or bent tabs on the dropouts or an axle washer with a bent tooth hooked into a hole in the dropout. These systems prevent the wheel from falling out if the axle nuts loosen.
3. **Pull the wheel out.**

2-7

DETACHING A WHEEL ON A CANNONDALE LEFTY FORK

1. **Remove the disc brake caliper.** It will likely require a 5mm hex key or a Torx T25 wrench (Fig. 2.8A). The mounting tabs are slotted, so you only need to loosen the bolts, not remove them.
2. **Unscrew the axle bolt.** The axle bolt will require a 5mm or 6mm hex key (Fig. 2.8B). This procedure

2.8A Cannondale Lefty wheel removal: loosen, but do not remove, the brake caliper mounting bolts and slide the caliper off the mounting tabs

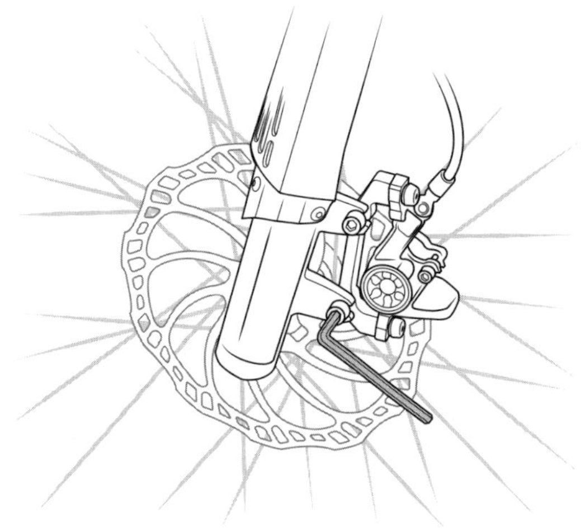

2.8B Cannondale Lefty wheel removal: unscrew the axle bolt

2.8C Cannondale Lefty wheel removal: slide the wheel off the fork axle

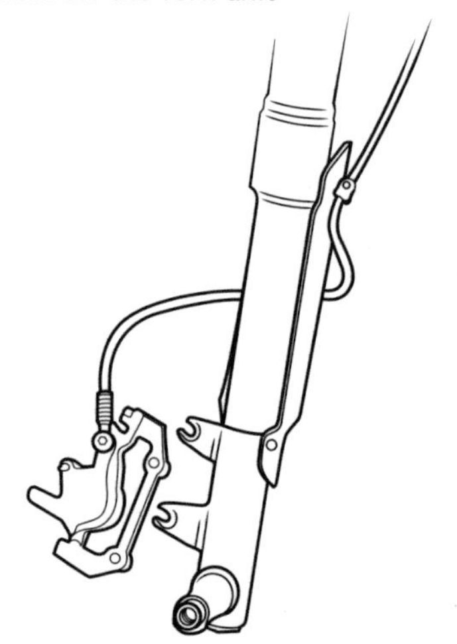

pulls the wheel right off without any further encouragement from you (Fig. 2.8C).

NOTE: *On reinstallation, grease the bearing seats on the axle (the thing sticking out from the fork). Slide the hub back on, line it up, and tighten the bolt. Mount the brake again, ensuring that you keep the same spacers, if any, between it and the mounting tabs on the fork.*

2-8

INSTALLING THE FRONT WHEEL

With a disc brake, lower the dropouts down onto the hub ends so that the slot in the brake caliper attached to the fork slides down over the rotor (the big disc attached to the wheel). Make sure that the rotor does not dislodge either pad. Allow the bike's weight to sit on the hub, ensuring that the axle ends are fully seated in the dropouts.

With rim brakes, leave the brake open and lower the fork onto the wheel so that the bike's weight pushes the dropouts down onto the hub axle. This action will seat the axle fully into the fork and center the rim in the fork.

Continue with the appropriate hub-securing step. Ensure that the axle ends or hub ends are fully seated in the dropouts. Some older through-axle forks don't have lips on the medial sides of the dropouts (Fig. 2.5) to center the wheel; in that case, you must manually hold everything yourself while you line up the hub, axle, and dropouts (2-9).

2-9

TIGHTENING THE QUICK-RELEASE SKEWER

The quick-release skewer is not a glorified wing nut and should not be treated as such.

1. **Hold the quick-release lever in the open position.**
2. **Finger-tighten the opposite-end nut.** Snug it up against the face of the dropout.
3. **Push the lever over** (Fig. 2.9) **to the closed position.** It should now be at a 90-degree angle to the axle. If done right, you should have needed a good amount of hand pressure to close the quick-

2.9 Tightening the quick-release

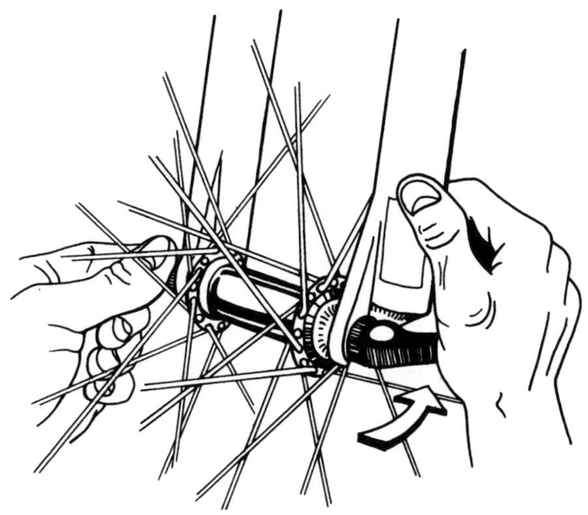

release lever properly; the lever should have left its imprint on your palm for a few seconds.

4. **If the lever does not close tightly enough, open the lever again, tighten the end nut one-quarter turn, and close the lever again.** Repeat until tight.
5. **If the lever cannot be pushed down perpendicular to the axle and the nut is therefore too tight, open the quick-release lever, unscrew the end nut one-quarter turn or so, and try closing the lever again.** Repeat this procedure until the quick-release lever is fully closed and snug. When you are done, the lever should be pointing straight up or toward the back of the bike so that it cannot hook on obstacles and be accidentally opened.
6. **Double-check that the axle is tight.** Try to knock the wheel out by banging on top of the tire with your hand.

2-10

TIGHTENING A THROUGH-AXLE

To install a through-axle, slide the wheel into the fork (sans axle), aligning the rotor between the pads of the disc-brake caliper. If the dropouts have little inboard lips that sit on the hub ends (Fig. 2.6), everything will be lined up to insert the axle; if not (Fig. 2.5), you'll have to carefully hold the bike up as you line up the hub and dropouts to install the axle. Push the axle through from the side with the bigger hole in the dropout.

On a QR15 axle other than a Manitou, tighten the axle into the opposite fork leg using the quick-release lever as a handle; if there is a cutout in the ring under the lever (Fig. 2.6), flip it open fully and twist until the lever engages the cutout so that the axle rotates along with the lever. Periodically flip the lever over and check for tightness; keep screwing the axle in until the lever takes a firm push to flip it into the tightened position. When properly adjusted, the lever should not only leave a momentary imprint in the palm of your hand to indicate proper tension, but it should also be pointed up, parallel to the fork leg. If proper tension and the lever being pointed up do not occur simultaneously, then you will need to adjust the thread. To do so, remove the fixing screw and triangular washer securing the notched thread-adjustment ring (Fig. 2.10A). Unscrew the axle a few turns and then push inward on the lever end to pop the thread-adjustment ring out of its notches (Fig. 2.10B). Turn the ring the appropriate direction to

2.10A QR15 axle adjustment: loosen the fixing screw and triangular washer

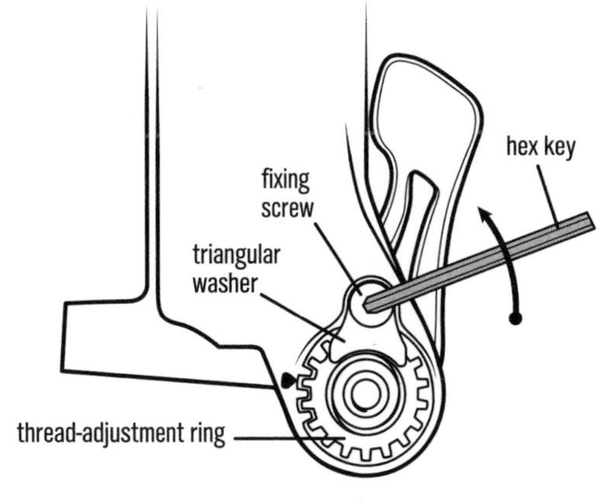

2.10B QR15 axle adjustment: pop out the thread adjustment ring

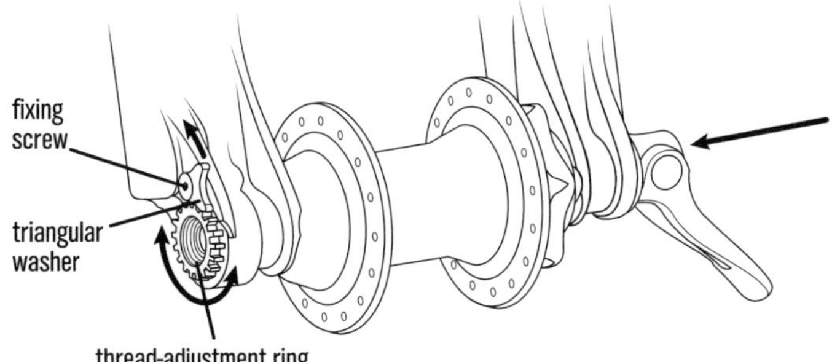

loosen or tighten the adjustment of the QR lever as needed so that it is at optimal tension when it is in the straight-up position. Replace the triangular washer and fixing screw. Check that the lever and axle are tight and that the hub is secure in the dropouts.

A Manitou QR15 HexLock through-axle (Fig. 2.11) has a hex shape on either end of the axle that mates with the same shape inside of the dropouts, and a key on the end opposite the lever to engage the opposite dropout. Orient the axle so that the laser-etched instructions along its length face up, toward the handlebar. With the lever flipped open (down), slide the axle straight in from the same side you pulled it out. While pushing inward to compress the spring and engage the key in the opposite dropout, rotate the lever 90 degrees (from pointed straight ahead to pointed straight down). Flip the lever over to the closed position to secure the wheel in the dropouts. If it requires insufficient pressure to push the lever over to its closed position (it didn't give you the satisfaction of providing enough resistance to temporarily indent and squeeze the blood out of the palm of your hand), flip the lever open (down) again, unscrew (counterclockwise) the adjustment ring under the lever a bit, and flip the lever back up to see if it is now tight enough. Similarly, tighten (clockwise) the adjustment ring with the lever open if it is too hard to flip the lever up to the closed position.

For a screw-in through-axle, with or without clamps on the dropouts, tighten the axle with a hex key or Torx key of the appropriate size. If the fork has pinch bolts on the dropouts, tighten the bolts on the threaded end first. Push down on the handlebar a few times to compress the fork and settle the head end, and then tighten the bolts that pinch the dropout on the head end.

On most traditional (usually 20mm) through-axles, tighten the draw bolt from outside the dropout into the thin end of the axle (Fig. 2.5). You may need to snug up a clamping bolt on the head end a bit so that the axle does not spin when you tighten the draw bolt. Once the axle is seated, tighten all of the clamping bolts on both dropouts.

Automobile roof racks and truck-bed carrying racks present a challenge for a through-axle fork. If the rack clamps the dropouts, you must first install a through-axle adapter such as the Hurricane Components Fork Up, which is a tube welded to two slotted plates. The slotted plates clamp onto the fork mount like standard QR fork dropouts, and you push the axle through the 15mm- or 20mm-ID tube once you set the fork over it.

2-11

TIGHTENING BOLT-ON SKEWERS

Hold the end nut with one hand and tighten the skewer with a 5mm hex key. If you have a torque specification for the skewer and a torque wrench, use them. If you have a torque wrench and no spec, Control Tech, a company that used to make lots of these skewers,

2.11 Manitou QR15 HexLock through-axle

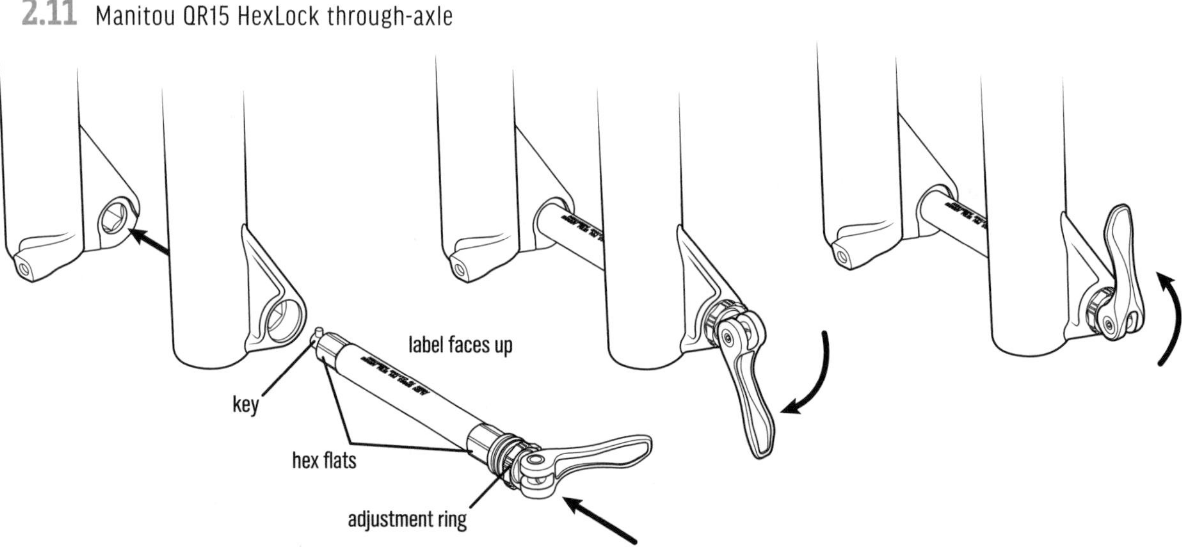

recommends 65 inch-pounds (in-lbs) of tightening torque for steel bolt-on skewers and 85 in-lbs for titanium versions; I recommend following those guidelines. If you do not have a torque wrench, you can approximate the correct torque by using a short hex key and tightening as tightly as you can with your fingers. You want the wheel secure, but you also don't want to snap the skewer by overtightening it as a result of leaning on a long hex key with all your might.

2-12

TIGHTENING AXLE NUTS (MASS-MERCHANT BIKES)

Snug up the nuts clockwise (opposite direction of Fig. 2.7) with a wrench (usually 15mm). Alternate sides, snugging up the nuts bit by bit until they are quite tight.

2-13

CLOSING THE BRAKES

With a disc brake, there is no need to tighten the brakes after putting the wheel in, because you didn't loosen them to get the wheel out. You are ready to ride.

The steps required to close rim brakes are the reverse of what you did to release them.

1. **Hook up the brake cable.** With a V-brake (sidepull cantilever), hold the link in one hand, pull the noodle back, push the cable coming out of the noodle into the slot in the end of the link, and then pop the end of the noodle back into the slotted hole (Fig. 2.1 in reverse).

   With a cantilever or U-brake, hold the brake pads against the rim with one hand, and hook the enlarged end of the straddle cable back into the end of the brake arm with your other hand (Fig. 2.2 in reverse).

   Do the reverse of 2-3 to reconnect the more rare types of brakes, or find the brake type at the beginning of Chapter 10 and read up on it.
2. **Check the brakes.** Squeeze the lever and make sure the cable doesn't slip. Lift the front end of the bike and spin the front wheel, gently applying the brakes several times. Check that the pads are not dragging, and recenter the wheel (or adjust the brakes as described in Chapter 10 for your type of brake).

   If the spring balance in the brakes is off, one pad may rub the rim when the rim is centered in the fork; in that case, skip to 10-10 (V-brake) or 10-16 (cantilever) for adjustment procedure.

   A pad may also rub because the fork or wheel is misaligned and there is nothing wrong with the brakes; they are centered. To eliminate the rub, either you'll need to off-center the brakes by adjusting spring tension in the arms as in 10-10d (V-brake) or 10-18 (cantilever), or you'll need to hold the rim centered between the brake pads when securing the hub. These should be only temporary fixes; if it's the wheel that's off, get the wheel trued or, better yet, true it yourself (8-2). You will know whether the wheel is off if it sits off-center one way when you tighten it in and then sits off-center to the other side when you flip it around and tighten it in again. If the wheel sits off-center both ways, the fork could be off. You can check fork alignment; that procedure is in Chapter 16.

   If everything is reconnected and centered properly and the brakes don't rub, you're done.

2-14

REMOVING THE REAR WHEEL

Removing the rear wheel is done in the same way as removing the front (2-2 to 2-5), with the added complication of the chain and cogs.

1. **Before releasing the hub axle, shift the chain onto the smallest rear cog.** Lift the rear wheel off the ground and shift while turning the cranks.
2. **Release the rear brake (only required with rim brake).** Follow the same procedure as with the front wheel.
3. **On a clutch rear derailleur, free or lock the jockey-wheel cage.** With your hand, pull the bottom of the cage of a SRAM X-Horizon rear derailleur down and forward to put slack in the chain. Push the cage-lock button (Fig. 2.12) to lock the cage in this loose-chain position. With a Shimano rear derailleur with

2.12 Engage the lock on the cage of a SRAM rear derailleur that has a clutch

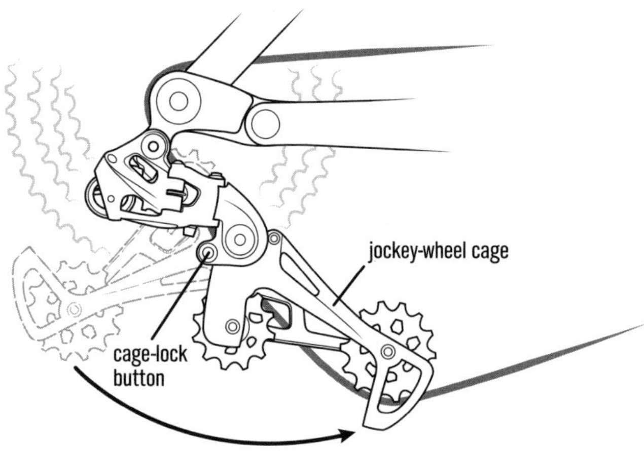

2.13 Removing and installing the rear wheel

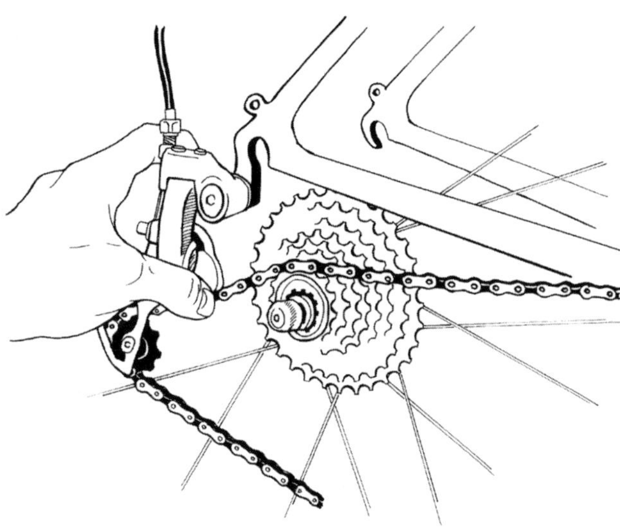

a lever switch above the lower knuckle, switch the lever to the off position (forward) in order to free the lower knuckle before removing the rear wheel.

4. **Release the hub.** Depending on axle type, free or remove the axle from the rear dropouts following the above steps for removing the axle from the front dropouts.
5. **Push the wheel out of the rear dropouts.** You'll first need to move the chain out of the way. This is usually a matter of grabbing the rear derailleur and pulling it back so that the jockey wheels (pulley wheels) move out of the way, while pushing forward on the quick-release or axle nuts with your thumbs and letting the wheel fall as you hold the bike up (Fig. 2.13). If the bottom half of the chain catches the wheel as it falls, jiggle the wheel while lifting it to free the cogs from the chain.

2-15

INSTALLING THE REAR WHEEL

1. **Shift the rear derailleur to high gear.** It will be in its outermost position, under the smallest cog (Fig. 5.4).
2. **Check the cage lock or lever switch.** On a clutch rear derailleur, if the cage isn't locked (SRAM) or freed (Shimano), lock the SRAM cage or flip the Shimano lever switch off now. Follow step 3 above (in section 2-14).
3. **Slip the wheel in between the seatstays and chainstays (the "swingarm" on a full-suspension bike).** Maneuver the upper section of chain onto the smallest cog (Fig. 2.13). If the bike has a rear through-axle, you'll need to have pulled that out first.
4. **Set the bike down on the rear axle.** As you let the bike drop down, guide the disc-brake rotor up between the pads in the caliper or guide the tire up between the rim-brake pads. Pull the rear derailleur back with your right hand, and pull the hub ends back into the dropouts with your index fingers. Your thumbs push forward on the rear dropouts, which should now slide over the hub ends. (With a quick-release skewer, if the axle does not slip into the dropouts, you may need to unscrew the quick-release skewer nut farther or spread the dropouts apart or squeeze them toward each other to get them to fall between the quick-release ends and the axle ends.)
5. **Check that the axle or hub ends are fully seated in the dropouts.** This should center the wheel.
6. **Secure the axle.** Tighten the quick-release skewer, through-axle, bolt-on skewer, or axle nuts in the same way as explained for the front wheel.
7. **Release the rear-derailleur cage lock, if applicable.** Simply pull forward (counterclockwise) on the bottom of the jockey-wheel cage, and the button

should release. Let go, and the cage should snap back and tension the chain.

8. **Reconnect the rear brake (rim brake only).** Follow the same procedure as you did on the front wheel. You're done. Go ride your bike.

2-16

CLEANING THE BICYCLE

Most cleaning can be done with soap, water, sponges, and brushes. Soap and water are easier on you and on the earth than stronger solvents, which are generally needed only for the drivetrain, if at all.

Avoid using the high-pressure sprayers you find at pay car washes to clean your bike. The soaps can be corrosive, and the high pressure forces water into bearings, pivots, and frame tubes, causing extensive damage over time. If you do use a pressure washer (these are often set up for participants to use at races and bike festivals), never point it at the bike from the side, as it can blow the bearing seals inward; instead, point the washer in the plane of the bike from the top, bottom, front, and back.

If the bike is really dirty, you can start by washing it with a hose while the wheels are on or off. A car-washing brush that you screw onto the end of the hose works well for this. If the weather is cold, wear appropriate clothing.

Scrubbing the bike is easier if you use a bike stand. In the absence of a bike stand, you can hang the bike from a garage ceiling with rope, stand it upside down on the saddle and handlebar, or stand it vertically, balanced on the front of the fork and the handlebar with the front wheel removed.

1. **Remove the wheels.**
2. **Secure the chain (optional).** If the bike has a chain hanger (a little nub attached to the inner side of the right seatstay, a few centimeters above the dropout), hook the chain over it. If not, pull the chain back over a dowel stick (Fig. 2.14), an old rear hub secured into the dropouts, or a chain keeper (Fig. 1.3).
3. **Fill a bucket with hot water and liquid dish soap.**
4. **Scrub the entire bike and wheels.** Use a stiff nylon-bristle brush for tough dirt and hard-to-reach places and a big sponge for gentler cleaning on accessible areas. Leave the chain, cogs, chainrings, and derailleurs for last.
5. **Rinse the bike.** Hose it off or wipe it with a wet rag.
6. **Check for remaining grime.** ProGold Bike Wash, Pedro's Green Fizz, or equivalent cleaners can remove stuff stuck to the frame and get it sparkling.

Avoid getting water in the bearings of the bottom bracket, headset, pedals, and hubs. Also avoid getting water into the lip seals of suspension forks, as well as any pivots or shock seals on rear-suspension systems. In addition, most metal frames and forks have vent holes

2.14 Wiping the chain

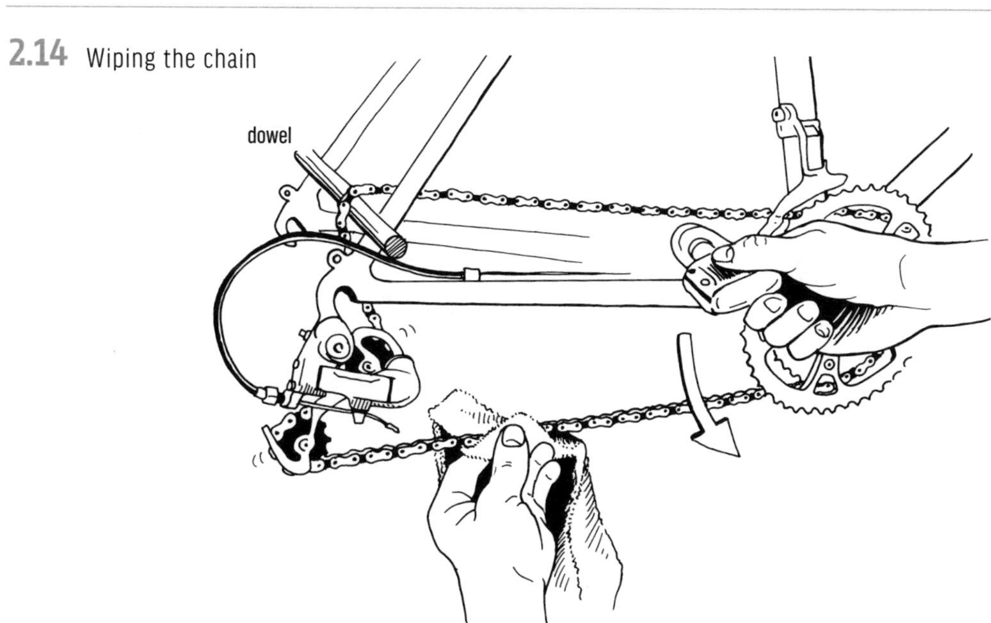

in the tubes to allow hot gases to escape during welding. The holes are often open to the outside on the seatstays, fork legs, chainstays, and seatstay and chainstay bridges. Avoid getting water in these holes, especially if you use a high-pressure washer. Taping over the vent holes before you wash the bike is a good idea, and you can leave that tape in place indefinitely to protect the frame on rainy or muddy rides.

2-17

CLEANING THE DRIVETRAIN

The drivetrain consists of an oil-covered chain running over the gears and derailleurs. Because the drivetrain is totally exposed to the elements, it picks up lots of dirt.

The drivetrain is also what transfers your energy into the bike's forward motion, which means that it should be kept clean and lubed so that it can move freely. Frequent cleaning and lubrication will keep the drivetrain rolling well and will extend its life as well.

The drivetrain can often be cleaned sufficiently by using a rag and wiping down the chain, derailleur jockey wheels, and chainrings. You might want to wear rubber gloves for this. If you keep the gloves, rag, and lube near where you store your bike, you will tend to clean it before or after almost every ride and be able to do so without dirtying your hands. It's a good habit to develop, because the drivetrain will last much longer and perform better with regular attention.

Lubricating the chain regularly is also important; see 4-1.

1. **Wipe the chain.** Turn the cranks while holding a rag in your hand and grabbing the chain (Fig. 2.14).
2. **Wipe the jockey wheels.** Holding a rag in one hand, squeeze the teeth of the jockey wheels between your index finger and thumb as you turn the cranks with the other hand (Fig. 2.15).
3. **Floss the cogs.** Slip a rag (or, better yet, a piece of Gear Floss absorbent string, available at bike shops) between the cogs of the freewheel, and work it back and forth to clean each cog (Fig. 2.16).
4. **Wipe down the derailleurs and the front chainrings with the rag.**

The chain will last much longer if you perform this sort of quick cleaning regularly, followed by dripping chain lube on the chain (see 4-1) and then doing another light wipe-down. You'll also be able to skip those heavy-duty solvent cleanings that are necessary when a chain gets really grungy. You can get it just as clean as with a solvent if you wipe the chain down thoroughly after lubricating it and clean between all of the outer link plates with cotton swabs (they won't fit between the inner link plates, but the roller will spin with them, and they will get clean that way).

You can also remove mud packed into derailleurs and cogs with soapy water and a scrub brush. The soap will not dissolve the dirty lubricant all over the drivetrain, but the brush will smear the lubricant all over the bike

2.15 Cleaning jockey wheels

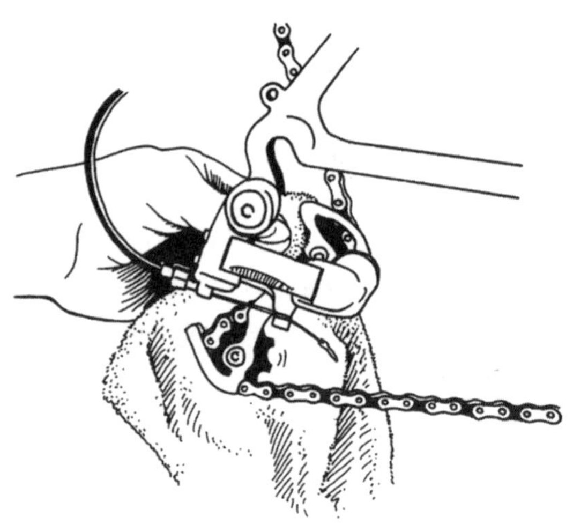

2.16 Cleaning the cogset

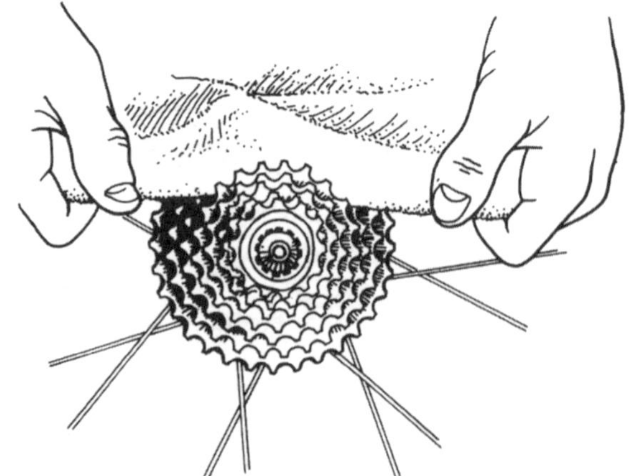

if you're not careful. Use a different brush from the one you use for cleaning the frame, because the bristles of this brush will become black and oily. Follow the brushing with a wipe-down of the frame, using a cloth.

2-18

CLEANING THE CHAIN WITH SOLVENT

LEVEL 1

If you frequently wipe off the chain and lubricate it (Fig. 4.1)—before or after every ride is ideal—you can minimize the need for solvent cleaning, with its associated disposal and toxicity problems. If you determine that using a solvent is necessary, work in a well-ventilated area, use as little solvent as possible, and pick an environmentally friendly product. Using one of the many citrus solvents on the market will minimize the danger if you breathe the fumes or get the stuff onto and into your skin, and will reduce a major disposal problem. If you are using a lot of solvents, organic ones such as diesel fuel can be recycled and therefore may be preferable to citrus solvents, as long as you protect yourself from the fumes with a respirator and dispose of the solvent properly if you are not recycling it.

Because all solvents suck the oils out of your skin, I recommend using rubber gloves, even with "green" solvents. You might also want to rub some skin lotion into your hands before starting, which will help keep your skin from absorbing solvent and will also make your hands easier to clean afterward.

A self-contained chain cleaner with internal brushes and a solvent bath is a quick and convenient way to clean a chain (Fig. 2.17), and this type of cleaning can be done without risk of later chain breakage caused by opening and closing the chain. A nylon brush or an old toothbrush dipped in a solvent is good for cleaning cogs, pulleys, and chainrings, and it can be used for a quick cleaning of the chain as well.

One way to thoroughly clean the chain is to remove it and put it into a solvent bath, but I do not recommend this unless you have a chain with a master link. Opening a standard chain by pushing rivets in and out with a chain tool is hard on the chain and can lead to breakage while riding. Because 9-, 10-, 11-, and 12-speed chains must be progressively narrower to deal with narrower cog spacing resulting from fitting ever more cogs into the space between the drive-side dropout and the hub flange, only a very small length of each rivet protrudes from the chain plates. Master links for chains narrower than 9-speed often are not supposed to be openable. And with a pre-Quick Link Shimano chain (no master link), even if you use Shimano's special link pin to reassemble the chain, you are still weakening the chain and can bring on breakage under shifting load.

2.17 Using a solvent-bath chain cleaner

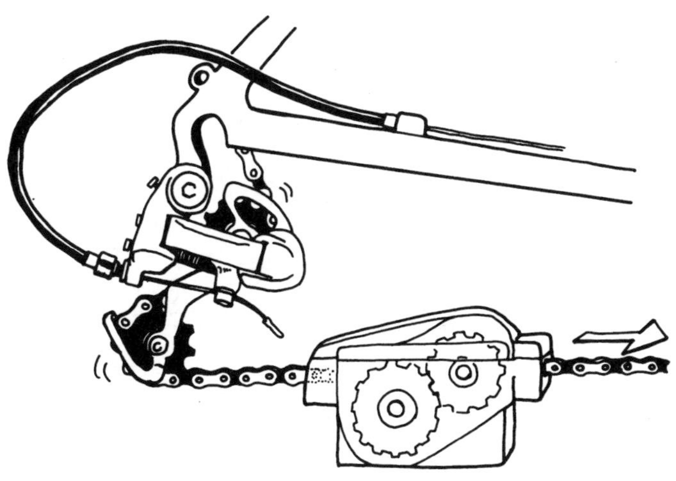

If it is absolutely necessary to remove the chain to clean it, here is the procedure:

1. **Remove the chain.** Follow the directions in 4-7 or, ideally, those in 4-12 for using a master link.
2. **Put the chain in an old water bottle that is about one-fourth full of solvent.**
3. **Shake the bottle vigorously to clean the chain.** Do this close to the ground in case the water bottle leaks.
4. **Hang the chain to dry.**
5. **Install the chain on the bike.** Follow the directions in 4-8 to 4-10, or, better yet, get a master link and use it to put the chain together (4-12).
6. **Lube the chain.** Drip chain lubricant into each of the chain's links and rollers as you turn the cranks to move the chain past the drip bottle. Drip lube on the moving chain by gently squeezing the bottle

2.18 Dripping chain lube along the top edges of the chain to get it between the plates at each rivet

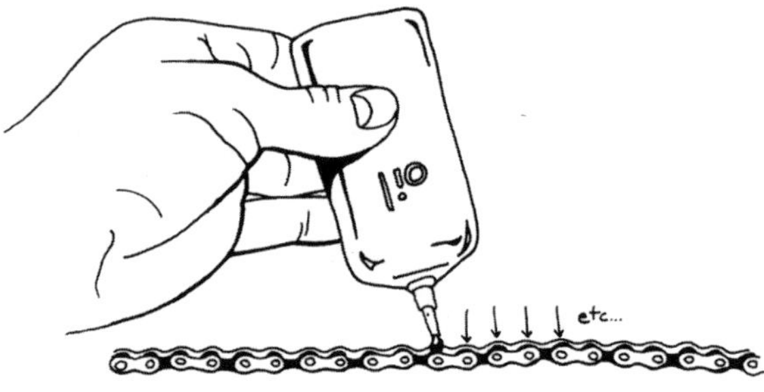

with the tip on each top edge of the chain (Fig. 2.18) for a couple of turns of the crank on each edge. See 4-1 for more information.

7. **Lightly wipe down the chain with a rag.**
8. **Deal with the solvent.** You can reuse much of the solvent by allowing it to settle in a clear container over a period of days or weeks. Decant and save the clear stuff, and dispose of the sludge.

To avoid repeating this procedure frequently, after every ride or two from now on, use a rag to wipe the jockey wheels (Fig. 2.15), chainrings, front derailleur, and chain (turn the crank to pull it through a rag you clasp around it, as in Figure 2.14; there's no need to remove the wheel), and then lubricate the chain (Fig. 2.18). Keep dish gloves and a rag near where you store your bike so that you can clean the drivetrain quickly on your return from a ride without having to scrub dirty oil off your hands afterward.

A clean bike invites you to jump on it and will feel faster. Corrosion problems are minimized, and you can see other problems as they arise. A clean bike is a happy bike.

2-19

PERFORMING MECHANICAL WORK: A GENERAL GUIDE

a. Threaded parts

All threads must be prepped before tightening. Depending on the bolt in question, prep with lubricant, threadlock compound, or an antiseize compound. Clean off excess thread-prepping compound to minimize dirt attraction.

1. **Lubricated threads:** Most threads should be lubricated with grease or oil. If a bolt is already installed, you can back it out and drip a little chain lube on it and then tighten it back down. Lube items such as crank bolts, pedal axles, cleat bolts on shoes, derailleur- and brake-cable anchor bolts, and any control-lever mounting bolts.
2. **Locked threads:** Some threads need to be locked to prevent the bolts they are on from vibrating loose. These are bolts that need to stay in place but are not supposed to be tightened down fully for some reason or other, usually to avoid seizing a moving part, throwing a part out of adjustment, or stripping threads in a soft material. Examples of bolts of this type are derailleur limit screws, jockey-wheel center bolts, brake-mounting bolts (for rim brakes), and spokes. Use Loctite, Finish Line Threadlock, or an equivalent; use Wheelsmith SpokePrep or an equivalent on spokes. If the wheel has extremely high spoke tension on both sides, the spokes will stay in adjustment when you use grease on the threads instead of SpokePrep or an equivalent. Some spoke nipples made by DT Swiss come with threadlock compound already inside them.
3. **Antiseize threads:** Some threads have a tendency to bind up and gall, making full tightening as well as extraction problematic. They need an antiseize compound on them to prevent galling. Examples of this kind of thread are a bolt threaded into a titanium part—including any parts mounted to titanium frames, such as bottom-bracket cups—and any titanium bolt. Use Finish Line Ti-Prep or an equivalent antiseize formulation.

IMPORTANT: *Never tighten a titanium bolt into a titanium part. Even with an antiseize compound, a titanium bolt will almost certainly gall and rip the titanium part when you try to remove the bolt.*

Wrenches (Fig. 2.19) must be fully engaged before tightening or loosening.

1. **Hex keys** (also called "Allen wrenches" or "hex wrenches") and **Torx wrenches** must be fully inserted into the bolt head, or the wrench and/or the bolt hole will round off. A good example is a shoe-cleat bolt; wash and scrub dirt and rocks out

2.19 Types of wrenches

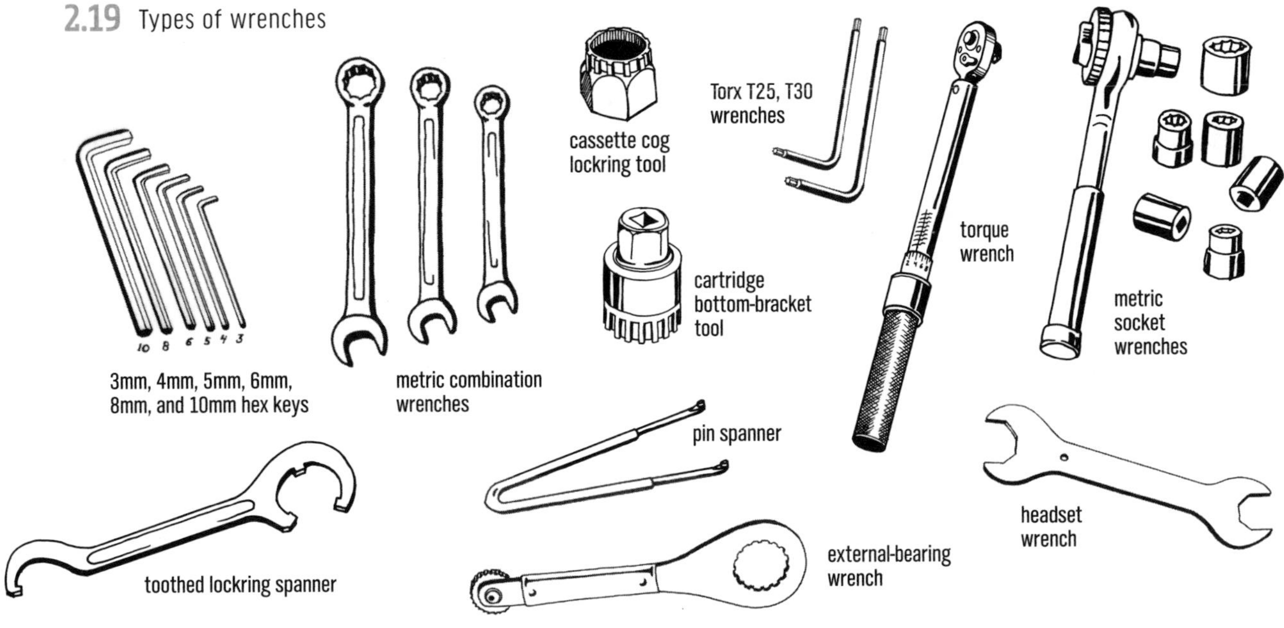

of shoe-cleat bolts before tapping the hex key in fully with a hammer so that it engages maximally.

2. **Open-end, box-end, and socket wrenches** must be properly seated around a hex bolt, or the bolt head will round off. A good example of a fitting that's easy to ruin if the wrench is not fully seated is an aluminum headset nut.
3. **Splined wrenches** (Torx, cog lockring, and bottom-bracket cup tools) must be fully engaged. If they are not, the splines will be damaged or the tool will snap. If you strip the splines in a cassette lockring, you will not be able to get it off.
4. **Toothed-lockring spanners** need to stay lined up on the lockring. If they slide off, they will not only tear up the lockring; they will also damage the frame paint. Such a lockring can be found on a bottom-bracket adjustable cup.
5. **Pin spanners** need to be fully seated in the pinholes in the part being turned in order to prevent the holes in the part from slipping out and being damaged. You can find pinholes in some bottom-bracket adjustable cups, crank-bolt collars, and cartridge-bearing hubs. Make certain that the pinholes are clean and that the spanner pins are fully engaged before exerting any force on the wrench.

b. Tightening torque

A full list of specific tightening torques can be found in Appendix D. To best understand them, it helps to know a little about metric bolt sizes, particularly as they are used on bikes.

The designation "M" in front of the bolt size number means millimeters and refers to the bolt shaft, not to the hex key that turns it: An M5 bolt is 5mm in diameter, an M6 is 6mm, and so on, but the designation may not have any relationship to the wrench size. For instance, an M5 bolt usually takes a 4mm hex key (or in the case of a hex-head style, an 8mm box-end or socket wrench). However, M5 bolts on bicycles often accept different wrench sizes than are normally found on M5 bolts. Bolts that attach bottle cages to the frame are M5, and although some accept the normal 4mm hex key, many have a rounded cap head and take a 3mm hex key. The bolts that clamp a front derailleur around the seat tube, or that anchor the cable on a front or rear derailleur, are also M5, but they instead take a bigger than standard hex key size, namely, 5mm. The big single bolts found on old stems and some seatposts take lots of different bolt sizes (M6, M7, and even M8), but usually only one wrench size (6mm hex key).

Generally, tightness can be classified into four levels:

1. **Snug** (10–30 in-lbs or 1–3 N-m [Newton-meters in SI units]): Small setscrews (such as Grip Shift mounting screw), bearing-preload bolts (as on threadless-headset top caps), and screws going into plastic parts need to be merely snug.
2. **Firmly tightened** (30–80 in-lbs or 3–9 N-m): Small bolts—often M5 size—such as shoe-cleat bolts, cable anchor bolts on brakes and derailleurs, small stem bolts, brake-lever-clamp bolts, and some disc-brake-caliper mounting bolts need to be firmly tightened.
3. **Tight** (80–240 in-lbs or 9–27 N-m): Wheel axles, old-style single-bolt stem bolts (M6, M7, or M8), most M6 disc-brake-caliper mounting bolts, seatpost binder bolts, and seatpost saddle-clamp bolts need to be tight.
4. **Really tight** (280–600 in-lbs or 31–68 N-m): Crankarm bolts, pedal axles, cassette lockring bolts, and bottom-bracket cups are large parts that need to be really tight.

c. Cleanliness

1. Do not expect parts to work simply because you've squirted or slathered lubricant on them (meanwhile patting yourself on the back for maintaining your bike). The lube will pick up lots of dirt and get very gunky.
2. Do not expect parts to work if you just wash them but don't lubricate them. They will get dry and squeaky.

d. Test-riding

Always test-ride the bike after adjusting it in the bike stand. Parts behave differently under load.

2-20

PERIODIC MAINTENANCE SCHEDULE

If you follow this guide, your bike will last longer, and you will have less need for the emergency repairs in Chapter 3.

You may find it strange that this schedule is arranged in order of number of rides, rather than on mileage. While both are somewhat arbitrary, mileage makes sense for a bike that is used in dry, clean conditions (a road bike), but it does not make as much sense for a mountain bike ridden on dirt trails. A short ride in the rain on a sandy or muddy trail has a lot more impact on a bike's need for maintenance than does a 100-mile ride on dry pavement. What really makes sense on a mountain bike is the number of hours it is ridden in dirty conditions. But who wants to keep track of hours of riding, and separating out which hours were spent on paved roads and which hours on dirt? So number of rides is what I settled on.

The interval periods are by no means written in stone; they depend on the bike, the trail conditions, the duration of the rides, and how you ride. In case it is not obvious, a bike used in wet conditions will require more frequent maintenance than one used only in dry, clean conditions; think about weighting rides on trails more heavily than those on pavement, and the rougher, muddier, and sandier the ride, the more heavily it should be weighted. Also realize that a bike that's in bad shape will need more frequent attention to provide hassle-free riding than one that is in good shape.

Use this as a guide to support you; don't add to the stress that your bike riding is intended to relieve by worrying that you're already a few rides past your 10-ride maintenance interval, and early tomorrow morning you're heading out on a long ride with some buddies. If you've kept up with it in the past, it probably can wait another ride or two.

After dripping oil on a moving part, wipe the area to remove excess oil, and then wipe it again after you ride and dust has stuck to it. Lubing is always better than not lubing, even if you leave excess around.

Each maintenance task on this schedule is followed by the section in this book where you can find the instructions for doing it.

You may have other tasks that you want to add to this list; that's why I've added extra lines.

BEFORE EVERY RIDE

1. **Pull the brake levers.** Make sure each brake is working, hits the rotor or rim properly before the lever comes to the handlebar, and the brake quick-release (if applicable) is not open.

2. **Check that quick-release hub skewers, through-axles, or axle nuts are tight.**
3. **Inspect the tires.** Look for cuts, bulges, and worn tread.
4. **Check tire pressure.** Ideally use a tire gauge, but at least squeeze each tire to ensure that it has adequate pressure.
5. **Check shock pressure.** If revising your shock settings based on your last ride, pump to the new settings.
6. **Check for slop in dropper post activation.** Make sure it releases early in the lever throw and returns snappily.
7. **Look over the entire bike.** Scan for anything out of the ordinary, like paint cracks or bulges, frayed cables, rust.
8. ______________________________
9. ______________________________

AFTER EVERY RIDE

1. **Wipe the suspension shafts and pivots, dropper-post shaft, chain, chainrings, derailleurs, and cogs with rags, and lubricate the chain** (4-1). Wash bike if needed.
2. **Wipe off the bike.** Look for damage to the frame or fork. Check the fork stanchions and rear shock shaft for scratches.
3. **Adjust derailleurs** (Chapters 5, 6) **and/or brakes** (Chapters 9, 10), if they were not working ideally.
4. **Look for the source of noises.** Investigate any rattles, rubbing noises, or creaks you may have noticed during the ride.
5. **Take notes about performance issues you noticed during the ride.** Specifically, if you want more or less shock pressure or tire pressure or want to speed up or slow down compression damping or rebound damping, write that down while it's fresh in your mind.
6. **If you've ridden in wet conditions, remove the seatpost and turn the bike upside down to drain water.** Grease and reinstall the seatpost (Chapter 14) the following day.
7. ______________________________
8. ______________________________

EVERY 5–10 RIDES

1. **Check chain wear with a chain-elongation gauge** (4-6). Replace chain if wear exceeds acceptable elongation.
2. **Inspect brake pads for wear. Replace if needed** (Chapters 9, 10). Check for proper pad alignment with rotor or rim.
3. **Clean drivetrain and entire bike** (Chapter 2).
4. **Lube pedal bindings.** Drip light oil on pedal springs. Drip wax-based (dry) lube on cleat-contact areas (Chapter 13).
5. **Lubricate mechanical disc brakes.** Drip light oil on lever pivots and the arm pivot on mechanical disc-brake calipers. Be careful not to get any oil on the rotor or pads.
6. **Replace tire if tread wear is excessive or you see other tire damage.** Inspect rim strip after removing the tire (Chapter 7).
7. **Push rims back and forth to check for play in hub bearings.** Correct if loose (Chapter 8).
8. **Check suspension for wear.** Check for pivot wear by pushing and twisting rear suspension arms back and forth. Check for slop in fork bushings. Check for sticking in fork and rear shock by slowly compressing suspension and seeing whether it moves down smoothly or in steps.
9. **Lubricate suspension pivots.** Drip oil on bushing pivots. Peel off the cartridge-bearing covers and grease the pivot bearings whenever they get noisy or loose. Using a grease gun, lubricate pivots that have grease fittings (Chapter 17).
10. **Check for slop in dropper post.** Twist saddle back and forth. Adjust as needed (Chapter 14).
11. **Check crank bolt(s) with torque wrench.** Tighten as needed (Chapter 11).
12. **Check crankset for bearing play and freedom of rotation.** Push crankarms laterally and spin them with the chain off. Adjust or replace bottom bracket as needed (Chapter 11).
13. ______________________________
14. ______________________________

EVERY 20–40 RIDES

1. **Check that frame pump works.** Or check that CO_2 cartridges and inflator are in good condition.
2. **Check condition of spare inner tube.** Also check for presence of appropriate tools in seat bag (1-6).
3. **Drip chain lube on front and rear derailleur pivots.**
4. **Overhaul derailleur jockey wheel bushings and seals.** If the derailleur has cartridge-bearing jockey wheels, check for smooth action and regrease if needed after removing bearing covers (Chapter 5).
5. **Peel off bearing covers on integrated-spindle bottom brackets and add grease** (Chapter 11).
6. **Clean and lubricate hydraulic brake pistons.** Using the correct brake fluid for the brake, clean the outside of the hydraulic disc pistons in the caliper one at a time after pushing each piston farther out by pulling the lever while holding the other piston in (Chapter 9).
7. **Check brake movement.** Check disc-brake operation for smooth return of brake pads. Check rim brakes for free rotation around frame and fork pivots and for a snappy return.
8. **Check rotor trueness and condition or rim brake-track wear.** Replace rim if wear indicator dictates it (Chapters 9 and 10).
9. **Check wheel trueness.** Correct as needed (Chapter 15).
10. **Check rims for cracks, particularly at the spoke holes.** Replace rim if cracks exist (Chapter 15).
11. **Service front and rear shocks.** Perform rear shock air-sleeve service (Chapter 17). Drain and replace fork fluid and replace seals.
12. **Overhaul pedal bearings.** Pull out the axles and regrease (Chapter 13).
13. **Lubricate shift and brake cables** (5-19, 10-5).
14. **Remove and regrease seatpost.**
15. **Check shoe cleats for wear.** Replace if needed (Chapter 13).
16. ______________________________
17. ______________________________

EVERY 80–150 RIDES

1. **Send fork and rear shock to rebuild facility for damper service.**
2. **Overhaul bearings.** These bearings can be in hubs (Chapter 8), pedals (Chapter 13), bottom brackets (Chapter 11), and headsets (Chapter 12). Clean and lube loose ball bearings and replace or grease cartridge bearings in these parts and in suspension pivots if they are worn, tight, or grinding or they exhibit play.
3. **Pull cantilever and V-brakes off their pivot bosses.** Regrease them and reassemble them.
4. **Bleed hydraulic brakes.** More often if brake performance drops or lever pumps up (Chapter 9).
5. **Replace shift and brake cables and housings** (5-6 to 5-18).
6. ______________________________
7. ______________________________

EVERY 3–6 YEARS

1. **Replace handlebar.**
2. **Replace stem.**
3. **Replace fork and rear shock.**
4. **Replace seatpost.**
5. **Replace saddle.**

Always carry a flagon of whiskey in case of a snakebite, and furthermore, always carry a small snake.

—W. C. FIELDS

3

EMERGENCY REPAIRS

HOW TO GET HOME WHEN SOMETHING BIG BREAKS OR YOU GET LOST OR HURT

TOOLS

Carry-along tools (Fig. 1.5)
Extended-trip carry-along tools (Fig. 1.6)

I've included this chapter so that you do not face disaster if you have a mechanical problem on the trail. If you ride your bike out in the boonies, sooner or later you will encounter a mechanical problem that has the potential to turn into an emergency. The best way to avoid such an emergency is to plan ahead and be prepared before it happens. Proper planning involves steps as simple as bringing along a few tools, spare tubes, and a little knowledge.

If something breaks while you are on the trail, the procedures in this chapter will help you to deal with it, whether or not you have all of the necessary tools. You always have the option of walking, but this chapter is designed to get you home pedaling.

Finally, you may find yourself with a perfectly functioning bicycle and still be in dire straits because you're lost, you've bonked (i.e., your body has run out of fuel), or you've become injured on the trail. Carefully read the final part of this chapter for pointers on how to avoid getting lost or injured and what to do if the worst does happen.

If this chapter does nothing other than alert you to some of the dangers facing you out in the backcountry, then perhaps you'll prepare for them and this chapter will have accomplished its purpose.

3-1 RECOMMENDED TOOLS TO TAKE ON RIDES

The take-along tool kit for your hydration pack or seat bag is described in section 1-6 (Fig. 1.5). If you're going to be a long way from civilization, take along the extra tools recommended for longer trips (Fig. 1.6).

NOTE: *As you will see in this chapter, a chain tool is one of the handiest items you can take along. As the ad says, "Don't leave home without it." A master link is very good to have with you as well.*

3-2 PREVENTING FLAT TIRES

Your first line of defense to avoid flats is to always have good tires on your bike. Check them regularly for wear, cracking, and tread cuts. Coat tires you don't ride often with 303 Aerospace Protectant or Armor All to prevent ozone cracking.

Using tire sealant inside your tubes can greatly reduce the potential for flat tires, as can installing tubeless tires. Using them together—sealant with tubeless tires—almost makes punctures a thing of the past.

Modern sealants are generally thin, liquid latex solutions that coagulate at a puncture; they have largely displaced more traditional viscous liquid sealants with chopped fibers in solution that plug holes. Sealant can be injected into an existing tube or a mounted tubeless tire through the valve, after removing the valve core. Installing sealants is covered in 7-9, and tubeless tires are discussed in 7-2 and 7-7. You also can purchase inner tubes with a sealant already inside.

If there is sealant in the tube or tubeless tire and the tire still gets low owing to a small hole through the tread (this is most likely to happen when you stop riding shortly after getting a puncture), put more air in and spin the wheel or ride for a couple of miles to get the sealant to flow out to the hole. A large hole will not be filled, although amazingly big holes can be plugged enough to get you home if you locate where the sealant is squirting out through the tire. Rotate the wheel so that the puncture is at the bottom and wait. The sealant may pool enough there to plug the hole. Add more air and continue riding. Note that if the hole is on the rim side (from a snakebite puncture or a protruding spoke end, for example; see 3-3), the sealant will not flow to it.

I do not recommend the old technology of stiff, sharp-edged plastic tire liners placed between the tire and tube. They are so stiff that they decrease traction and cornering ability, and they can slip sideways where they not only stop protecting the tube but can even cut or abrade it. There are, however, lighter and less stiff tire liners made of tightly woven Kevlar and other fibers (SpinSkins is one brand). They are fairly expensive, though.

3-3

FIXING FLAT TIRES

a. With a spare or a patch kit

LEVEL 1

Simple flat tires are easy to deal with. The first flat you get on a ride, whether in a standard inner tube and tire or in a tubeless tire, is most easily fixed either by injecting sealant and compressed gas together through the valve from a sealant-filled aerosol injector (7-9b, Fig. 7.19), or by installing your spare tube (7-6 describes how to do it). To install a tube in a tubeless tire, simply remove the tire, remove the valve stem from the rim, install a new inner tube, and reinstall the tire. If you can't manage to remove the tire bead from the rim with your fingers alone and don't have any tire levers, you can use the wheel quick-release skewers as levers.

Check your spare tube before leaving home to make sure that it holds air and that you can loosen the valve nut by hand out on the trail (assuming it has a Presta valve; see Fig. 1.1B or Figs. 7.2 and 7.3). If your bike has tubeless tires, take the additional step of making sure that the knurled retainer nut that holds the valve stem to the rim can also be loosened by hand out on the trail, in case you have to install a spare tube.

Before installing the new tube, make sure you remove all thorns or glass from the tire and feel around the inside of the tire for any other sharp objects and remove them as well. You will almost always find something sticking through the tire that has caused the flat, although it may be hard to find. If you don't find it and remove it, it will cause a new flat when you reinflate the tire.

Some flats happen without foreign objects sticking through the tire. Sometimes inner tubes just fail, particularly near the valve on the rim side and particularly if you have rim brakes and do a steep descent on a hot day. Another type of failure is the pinch flat, which is caused by the tire hitting a hard, sharp edge with insufficient air pressure in the tire to prevent the inner tube from being pinched between the tire and the rim. It will be apparent if you find two adjacent holes on the top and bottom of the tube (a.k.a. a snakebite).

Check the rim to see that your flat wasn't caused by a protruding spoke or nipple, a metal shard from the rim, or the edge of a spoke hole protruding through a worn rim strip. The rim strip is the piece of plastic or rubber that covers the spoke holes in the well of the rim (Fig. 6.13), and it prevents punctures to the underside of the tube. Many rim strips are totally inadequate, being either too narrow or prone to cracking or tearing. Also, metal hunks left from the drilling of rims during manufacture can work their way out into the tube. Endeavor to eliminate these problems before leaving on a backcountry ride by shaking out any metal fragments and using good rim strips or a couple of layers of fiberglass-reinforced packing tape (with superstrong lengthwise fibers) as rim strips. If the hole in the tube is on the rim side, tire sealant will not fill the hole because the liquid will be thrown to the outside, toward the tread, when the wheel turns.

After you run out of spare tubes, use your patch kit to patch additional punctures (find the details in 7-4 and 7-5).

In the absence of patches, you can patch punctures with insulating tape (the type used around doors). It's not a bad idea to stick some insulating tape to a frame tube for this kind of occurrence.

b. Without spare tubes or patches

If you're out of spare tubes and patches, you can tie a knot in the inner tube, pump it back up, and ride it home. You'd be amazed how well this works, and it's quick. Simply fold the tube at the puncture and tie an overhand knot with the folded end (Fig. 3.1). To maximize the length of inflatable inner tube, minimize the length of the folded end sticking out of the knot.

Knotting the tube works fine—and you can ride the bike as if nothing had happened—as long as there is only a single puncture in the tube, a snakebite (pinch flat), or multiple punctures all within a few inches of each other that you can seal off with a single knot. Obviously, if you have widely spaced punctures, multiple knots will seal off sections of the tube from air, leaving them flat.

3.1 Tying a knot in a punctured inner tube to seal off the hole

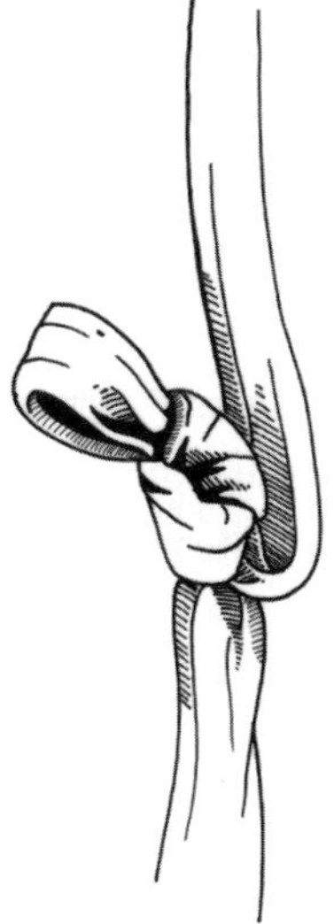

If tying off the tube won't work because the hole is at the valve, or because the tube has more than one hole in it, continue to the next section.

c. Without any way to inflate the tube or a section of a knotted tube

If you are without a pump or air inflation cartridges, or if there are multiple knots in the tube or a hole at the valve and no patches, or if the valve is broken, you are going to have to ride home without air in the tube or in a section of it. But riding a flat for a long way will destroy the tire and will probably damage the rim, too. You can minimize that damage by filling the space in the tire with grass, leaves, or similar materials. Pack the stuff in tightly, and then remount the tire on the rim. This fix should make the ride a little less dangerous by minimizing the flat tire's tendency to roll out from under the bike in a turn.

d. When a sidewall is torn

Rocks and glass can cut tire sidewalls. The likelihood of sidewall problems is reduced if you do not venture into the backcountry on old tires with abraded, rotten, or otherwise weakened sidewall cords. If the tire's sidewall is torn or cut, the tube will stick out, and on a tubeless tire, no amount of sealant will close the leak. Patching the tube or installing a new one isn't going to solve the problem, either. Without reinforcement, the tube will soon blow out through the sidewall gash.

Begin by looking for something to reinforce the sidewall (Fig. 3.2). Dollar bills work surprisingly well as tire boots (i.e., temporary internal casing reinforcements). The paper is pretty tough and should hold for the rest of the ride if you are careful. (I told you that cash would get you out of bad situations. Just don't try putting a credit card in there; tire cuts don't take American Express or Visa!) Business cards are a bit small but are better than nothing. You might even try an energy bar wrapper. A small piece of an old tire sidewall or lawn-chair webbing cut into an oval, or a strip of duct tape, might be a good addition to your patch kit for this purpose. You get the idea.

1. **Lay out some cash.** Place the bills or other reinforcement inside the tire over the gash (Fig. 3.2), or

3.2 Temporary fix for a torn tire casing

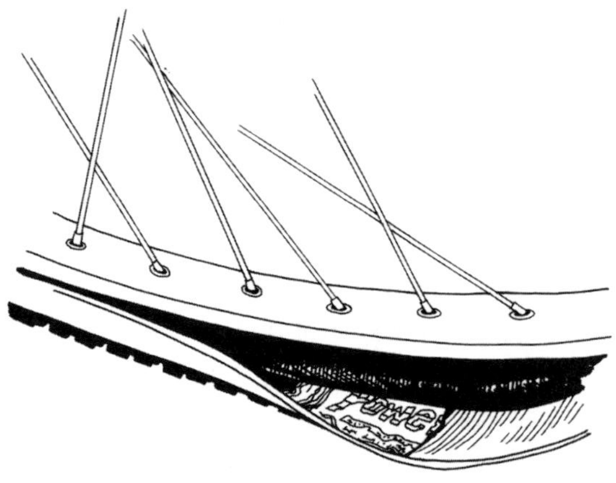

wrap it around the tube at that spot. Place several layers between the tire and tube to support the tube and prevent it from bulging out through the hole in the sidewall. If you only have one bill, fold it in half.

2. **Put a little air in the tube.** This holds the makeshift reinforcement in place.
3. **Mount the tire on the rim.** The procedure is in 7-6 (starting with step 3) and in Figures 7.13 to 7.15. Make sure that the tire is seated and the boot is still in place.
4. **Inflate the tube to about 40 psi.** Much lower than 40 psi will allow the boot to move around and may also lead to a pinch flat if you're riding on rocky terrain. In any case, this type of fix is not a perfect solution, so you will need to check the boot periodically to make certain that the tube is not bulging out again.

3-4

FREEING A JAMMED CHAIN AND A TWISTED LINK

If the chain gets jammed between the chainring and the chainstay, it may be hard to get it out if the clearance is tight. You may find that you tug and tug on the chain, and it won't come out. Well, chainrings flex, and if you apply some mechanical advantage, the chain will come free quite easily. Just insert a screwdriver or similar thin lever between the chainring and the chainstay, and pry the space open while pulling the chain out (Fig. 3.3).

3.3 Freeing a jammed chain

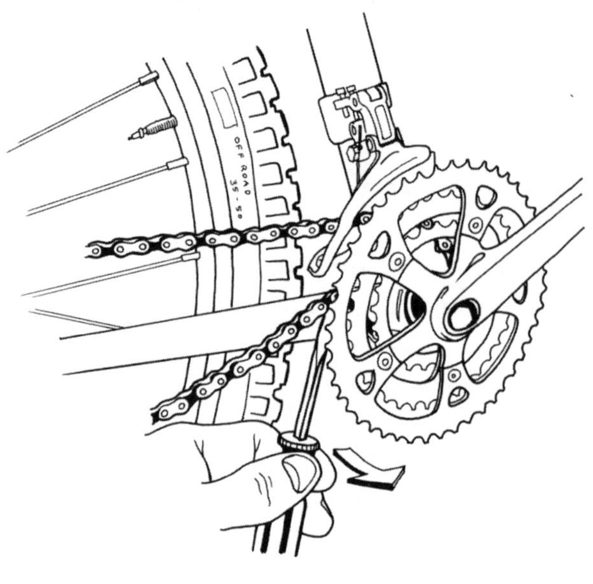

3.4 Fixing a broken chain

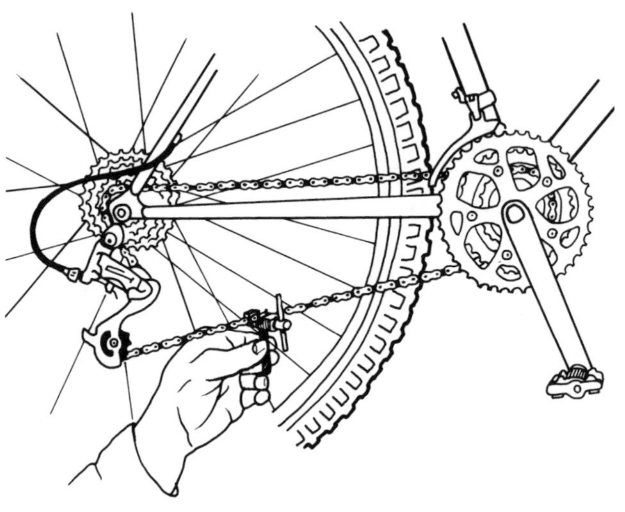

You will probably be amazed at how easy this is, especially in light of how much hard tugging would not free the chain.

If you still cannot free the chain, disassemble it. Open the chain with a chain tool (4-7 or 4-11), pull it out, and put it back together as in Figure 3.4 (4-9 to 4-11). You can push out a pin on a 9-, 10-, 11-, or 12-speed chain, push it back in, and get home. But do not ride it any longer than you absolutely have to, because that link will be very weak; the plate will be prone to pry off the end of the rivet upon shifting.

If you have nine or more cogs on the rear wheel, you really need to install a master link (4-12) to be able to disassemble and reassemble the thin chain on your bike and expect to use it without worry for a prolonged period. If you already have a master link (Fig. 3.5) in the chain, it will make unjamming it easy; otherwise, carry one that is the proper width for your chain. SRAM, KMC, FSA, and Wippermann (Figs. 4.22, 4.23, and 4.24) make master links for most chain widths.

Once you get rolling again, you may find that if you continue to pedal a split second too long after the chain starts to jam, you will have a twisted chain link (Fig. 3.6). A twisted link will continually pop off the cogs and chainrings and won't stay in gear. Once you find the twisted link, it will be obvious why it was popping out of gear—the chain will be running along nicely with the sides of the links vertical, and all of a sudden you will see some links leaning off to the side.

Untwisting such a link is easy if you have two pairs of pliers, two adjustable wrenches, or one of each. With the tools, just grab the links on either side of the twisted one at the rivet pins, and twist them to straighten the link.

Without pliers or wrenches, you can still untwist the link without having to bail on your ride. Shift to the smallest rear cog and flip the chain off the innermost chainring so it drops around the bottom-bracket shell and has no tension on it. Fold the chain at the twisted link so that link alone is at the top, horizontal. Grasp the vertical sections of chain running up to it on either side, and pull one hand toward you and push the other hand away from you to untwist the link (Fig. 3.7). Repeat until the twist is gone.

3.5 SRAM, Shimano, KMC, or FSA master link

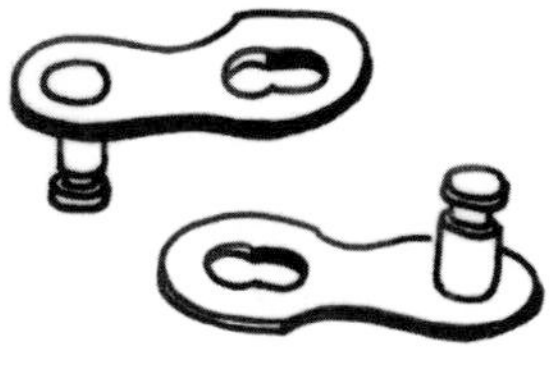

3.6 Twisted chain link

3.7 Untwisting a twisted chain link without tools

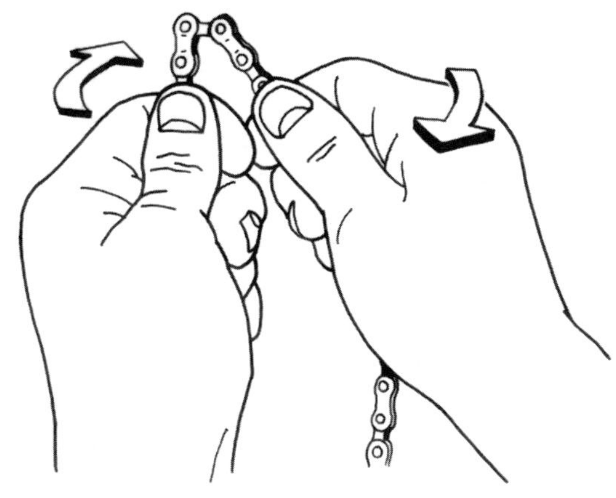

3-5

FIXING A BROKEN CHAIN

LEVEL 1

Chains can break when you are mountain bike riding, usually while shifting the front derailleur under load. The side force of the derailleur cage plate pushing laterally against the chain, coupled with the high tension, can pop a chain plate off the end of a rivet. As the chain rips apart, it can cause collateral damage to other parts. The open chain plate can snag the front-derailleur cage, bending it or tearing it off, or it can jam into the rear dropout. When a chain breaks, the end link is certainly shot, and some others in the area may be as well.

1. **Remove the damaged links with the chain tool.** (You or your riding partner did remember to bring a chain tool, right?) Again, the procedures for removing the damaged links and reinstalling the chain are covered in 4-7 to 4-11. If only the outer chain plates of a single link are damaged, you can install a master link (Fig. 3.5), producing a permanent fix that doesn't shorten the chain. Remove the damaged outer plates by pushing out the remaining rivet attaching them to the chain, and install a master link (4-11), noting proper orientation if you're using a Wippermann master link (Fig. 4.24).

2. **Replace the damaged links if you can.** Did you bring along extra chain links? If so, replace the same number you removed. If not, you'll need to use the chain in its shortened state. It will still work, but be careful to avoid shifting into the big chainring–biggest cog combination. Otherwise, you can rip up the rear derailleur if the chain is not long enough to encompass this span with a little slack.
3. **Join the ends and connect the chain** (Fig. 3.4). The procedure is in 4-10 (4-12 with a master link). Some lightweight chain tools and multitools are more difficult to use than a shop chain tool. Some flex so much that it is hard to keep the push rod lined up with the rivet. Others pinch the plates so tightly that the chain link binds up. It's a good idea to discover these things before you need to perform repairs on the trail. Try out the tool at home or at your local bike shop. This way you'll know what you're getting into before you reach the trailhead.

3-6

STRAIGHTENING A BENT WHEEL

LEVEL 2

If the rim is banging against the brake pads or, worse yet, against the frame or fork, pedaling becomes very difficult. This can result from a loose or broken spoke or a badly bent, or even broken, rim. The following sections detail four ways to deal with this problem.

a. Tightening loose spokes

If the wheel has a loose spoke or two, the rim will wobble all over the place, and if the tire is hitting the frame, or the rim is hitting the brake pads so hard that you can't ride home on it, you'll want to straighten the wheel before continuing.

1. **Find the loose spoke (or spokes) by feeling all of them.** The really loose ones, which would cause a wobble of large magnitude, will be obvious. If you find a broken spoke, skip to the next section (b). If you discover no loose or broken spokes, skip ahead to d.
2. **Get out the spoke wrench that you carry for such an eventuality.** (If you don't have one, skip to c.)

3.8 Tightening and loosening spokes

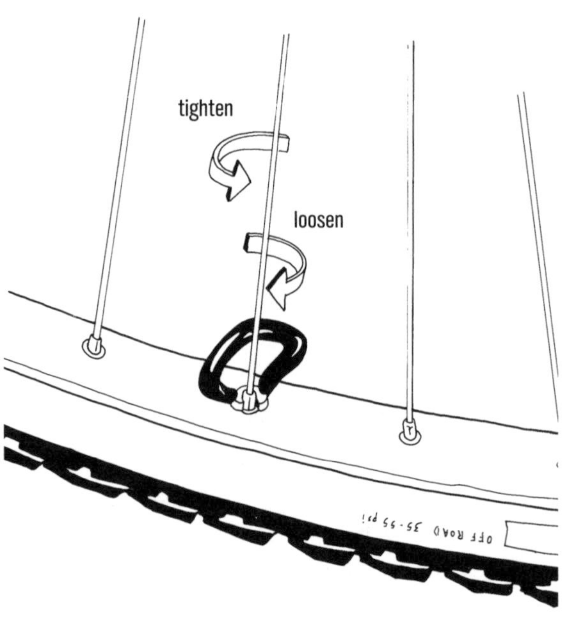

3. **Mark the loose spokes.** Tie blades of grass, sandwich bag twist-ties, tape, or the like around them so that you can still find the ones that originally caused the problem as the wheel becomes more true.
4. **Tighten the loose spokes** (Fig. 3.8). True the wheel, following the procedures in 8-2.

b. Fixing and replacing broken spokes

If you break a spoke, the wheel will wobble wildly and the tire will hit the chainstay, preventing you from riding farther unless you want to wear out the tire and wear into the chainstay.

1. **Locate the broken spoke.**
2. **Remove the remainder of the spoke.** Remove both the piece going through the hub and the piece threaded into the nipple. If the broken spoke is on the drive side of the rear wheel, you may not be able to remove it from the hub, because it will be behind the cogs. If so, wrap it around a neighboring spoke (Fig. 3.9), thus preventing it from slapping around, and skip down to step 6.
3. **Get out your spoke wrench.** (If you have no spoke wrench, skip to c.)
4. **Replace the spoke if you can.** If you brought a spare spoke of the right length or a Kevlar replacement spoke (mentioned in 1-6b), you're in business;

3.9 Wrapping a broken spoke

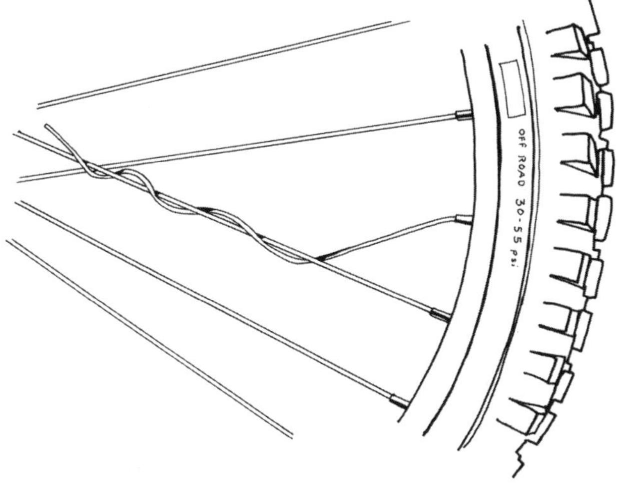

if not, skip to step 6. Put the new spoke through the hub hole, weave it through the other spokes the same way the old one was, and thread it into the spoke nipple that is still sticking out of the rim. If the new spoke looks just like the other spokes, mark it with a pen or tie blades of grass, sandwich bag twist-ties, tape, or the like around it to keep track of it as the wheel becomes more true. If you are using a Kevlar replacement spoke, thread it through the hub hole, attach the ends to its included spoke stub, adjust the ends to length, tie them off, and tighten the spoke nipple.

5. **Tighten the nipple on the new spoke.** Use the spoke wrench (Fig. 3.8), checking the rim clearance with the brake pad as you go. Stop when the rim is centered between the pads at that point, and finish your ride. Skip to step 7.
6. **If you don't have a spare spoke but you do have a spoke wrench, bring the wheel into rideable trueness by loosening the spokes on either side of the broken one.** These two spokes come from the opposite side of the hub and will let the rim move toward the side with the broken spoke as they are loosened. A spoke nipple loosens counterclockwise when viewed from its top (i.e., from the tire side—see Fig. 3.8). Ride home conservatively, as this wheel will rapidly get worse as the loose spokes loosen up even more.
7. **Once at home, replace the spoke.** Follow the procedure in 8-3, or take the spoke to a bike shop for repair. If you break a spoke more than once on a wheel, you should replace all of the spokes on the wheel, and you may need to replace the rim as well (Chapter 15 will lead you through rebuilding the wheel).

c. Opening the brake to get home

If the rim is banging the brake pads, but the tire is not hitting the chainstays or fork legs, just open the brake so that you can get home. If you have disc brakes, you won't need to do anything with a wobbly wheel that's not hitting the frame other than ride back carefully.

1. **Loosen the brake cable.** Decrease its tension by screwing in (clockwise) the barrel adjuster on the brake lever (Fig. 3.10). Remember that braking on that wheel is greatly reduced or nonexistent, so ride slowly and carefully.
2. **If the rim is still banging the brake pads, loosen the brake cable at the clamp on the brake (usually using a 5mm hex) and then clamp the cable back down.** You can also un-hook the cable as you would when removing the wheel (2-3, Figs. 2.1 and 2.2), but the brake arm on some cantilever brakes can flip around into the spokes as you ride, so I don't recommend it. You now have no brake on this wheel; ride carefully and walk the bike through difficult sections.
3. **If the bent wheel still won't turn, remove both brake arms from the cantilever posts, put them in your pocket, and pedal home slowly.** You will usually need a 5mm hex key for this. Do not attempt to ride a bike with brakes still attached to the frame or fork but disconnected from the cable. The brake arms will flap around as you ride and may get caught in the spokes, which could crack the seatstay or fork, not to mention your head, in a heartbeat.

If you want to straighten the wheel without using a spoke wrench, follow the procedures for dealing with a bent rim in d. Recognize that if you bend the rim by smacking it on the ground to correct for a loose or broken spoke, you will permanently deform the rim. Try

3.10 Brake-lever adjusting barrel

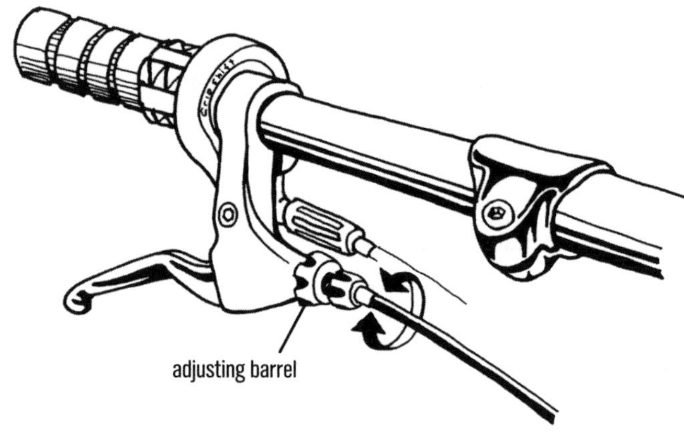

3.11 Fixing a bent rim

to get home without resorting to this maneuver so that you don't have to replace the rim.

d. Fixing a bent rim

If the rim is only mildly out of true and you brought a spoke wrench, you can fix the rim. The procedure for truing a wheel is explained in 8-2.

If the wheel is really whacked out, spoke truing won't do much. To get the rim to clear the brakes so that you can pedal home, follow the steps in c above.

If the wheel is bent to the point that it won't turn even when the brake is removed, beat it straight as long as the rim is not broken.

1. **Find the area that is bent outward the most.** Mark it on the outward side.
2. **Hold the wheel by its sides.** Have the bent part at the top with the mark facing away from you. Make sure to leave the tire on and inflated.
3. **Smack the bent section of the rim against flat ground** (Fig. 3.11).
4. **Put the wheel back in the frame or fork.** See if anything has changed.
5. **Repeat the process until the wheel is rideable.** You may be surprised how straight you can get a wheel this way. Of course, you can also make it a lot worse if you hit it too hard or at the wrong spot.

3-7

STRAIGHTENING A BENT BRAKE ROTOR

LEVEL 1

You can snag a disc-brake rotor on a rock, root, branch, or another bike's rotor (in the vehicle on the drive to the trailhead) and bend it. Generally, you can straighten a bent rotor easily enough with your hands (leave it attached to the wheel) and get it to pass through the brake caliper well enough to get back to your car or home. It may rub and make noise, but you can ride safely to where you need to go.

If the rotor is bent so badly that you cannot straighten it enough to get it to pass through the caliper, you can remove either the caliper or the rotor for your (slow and careful) ride back to the trailhead. For the caliper, you'll most likely need a 5mm hex key. For the rotor, you'll need either a Torx T25 wrench or a cassette remover and a big wrench, but you always carry one with you in your hydration pack or seat bag (Fig. 1.5), right?

If you remove either the rotor or the caliper and you have hydraulic brakes, I recommend that you wedge something between the brake pads, such as a piece of corrugated cardboard (or, ideally, a disc-brake pad spacer; it's wise to carry one). That way, if you accidentally pull that brake lever, you won't push the pistons out too far, at which point you would be stuck with the tough job of pressing them back in before you could fit a rotor between them again (see section 9-12).

If you do remove the caliper, make sure that it cannot get caught in your wheel as you ride. On a cable-actuated disc brake, you can disconnect the cable, stuff the caliper in your pocket or pack, and pull out or tie up the cable. On a hydraulic brake, tie or tape the caliper to your frame.

Once home, flip to 9-16 on truing rotors to come up with a more lasting fix.

3-8

REPAIRING A DAMAGED FRONT DERAILLEUR

If the front derailleur is mildly bent, straighten it with your hands or leave it until you get home. If it has simply rotated around the seat tube (the chain, your foot, or a pants leg can catch it and turn it), reposition the front derailleur so that the cage is just above and parallel to the chainrings, and then tighten the derailleur in place, usually with a 5mm hex key. If the derailleur is broken or so bent that you can't ride, you will need to remove it or route the chain around it as described in the following discussions.

a. With only a screwdriver

1. **Get the chain out of the derailleur cage.** To do this, open the derailleur cage by removing the screw at its tail (Fig. 3.12). If for some reason the derailleur cage can't be opened this way, open the chain with a chain tool or by hand at a master link (4-7 or 4-12).
2. **Bypass the derailleur.** Put the chain on a chainring that does not interfere with it (either shift the derailleur to the inside and put the chain on the big chainring, or vice versa).

b. With hex keys and a screwdriver (or a chain tool or master link)

1. **Detach the derailleur from the seat tube.** This usually takes a 5mm hex key.
2. **Remove the screw at the tail of the derailleur cage** (Fig. 3.12).

3.12 Opening the front-derailleur cage

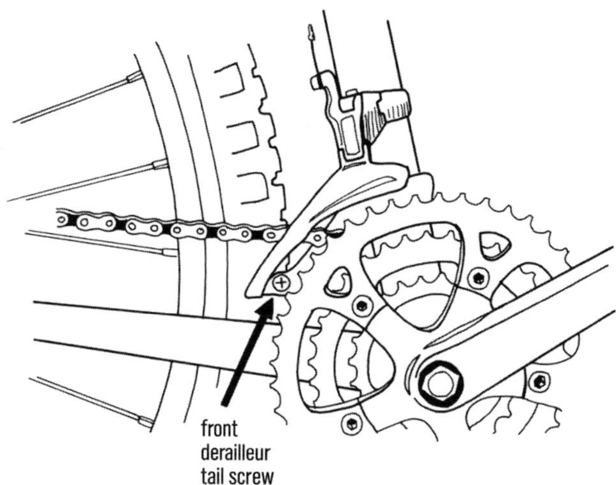

3. **Separate the derailleur from the chain.** Pry open the cage to do it. You could also disassemble the chain, pull it out of the derailleur, and reconnect it (4-7 to 4-11), but you will shorten the chain's life in the process unless you do it at a master link.
4. **Manually put the chain on.** Choose whichever chainring is best for the ride home. If in doubt, put the chain on the middle one.
5. **Tie the cable up.** You don't want it to catch in a wheel.
6. **Stuff the derailleur in your pocket.**

   Now you can ride home.

3-9

REPAIRING A DAMAGED REAR DERAILLEUR

LEVEL 1

If the rear derailleur gets bent just a bit, you can probably straighten it by hand enough to get home. If it's damaged more than this and you need only to descend back to your home or car, it's your lucky day. Remove the chain with a chain tool (4-7) or by hand if the chain has a master link (4-12), and coast back down.

If a jockey wheel fell out, you may be able to fix it as described in 3-10. Or if the return spring breaks and the chain hangs slack, you can try the fix in 3-11.

The most common rear-derailleur damage occurs when the derailleur gets caught in the spokes. If that happens, you almost invariably bend the derailleur as well as the rear-derailleur hanger—the part that hangs down from the dropout to which the derailleur attaches (Fig. 5.2). For this reason, always carry a spare derailleur hanger with you (Fig. 1.5) when riding a mountain bike. (Old steel and titanium frames don't have replaceable hangers, but almost all aluminum and carbon and any current frames do; make sure you get the correct one that fits your bike.) If you can bend the derailleur and the hanger back enough to ride home, great. If not, or if you break the hanger by bending it back, you'll need to replace the hanger; remove the rear wheel and derailleur to do it.

Once the new hanger is on, check to see that the derailleur jockey wheels line up vertically under the cogs. More often than not, the jockey-wheel cage will be bent in toward the spokes, making it shift poorly and making it likely to go into the spokes again. Often in these cases, the derailleur jockey-wheel cage has gotten bent inward, and as the wheel's spokes have dragged the cage hard around with them, the cage or upper knuckle has rotated beyond its stop. To get the derailleur working again, bend the cage straight and then wind the cage or knuckle back into position past its stop. If it's the stop on the cage and it is removable with a screwdriver, this is easy. If not, flex or bend the tab stop on the upper knuckle plate or cage until it allows the cage or entire derailleur to rotate back; a pair of pliers, screwdriver, or small hammer may do the trick. Often, you break the (already bent and weakened) tab off doing it, but the derailleur will still work. Once the derailleur has been twisted back into position with the chain routed properly, you may have to stick something like a screwdriver or spanner hook into the derailleur jockey-wheel cage to give you the leverage you need to straighten it in all three dimensions.

If the rear derailleur has been bent beyond fixing or is broken, then you will not be able to continue pedaling with the chain routed through it. To pedal home, route the chain around the derailleur, effectively turning your bike into a single-speed for the duration of your ride (Fig. 3.13). Watch out if you don't have a hardtail, though, because the only rear-suspension bikes that can be set up as a single-speed bike are unified-rear-triangle bikes, where the dropouts are attached directly to the bottom-bracket shell via rigid chainstays, or concentric

3.13 Bypassing a damaged rear derailleur

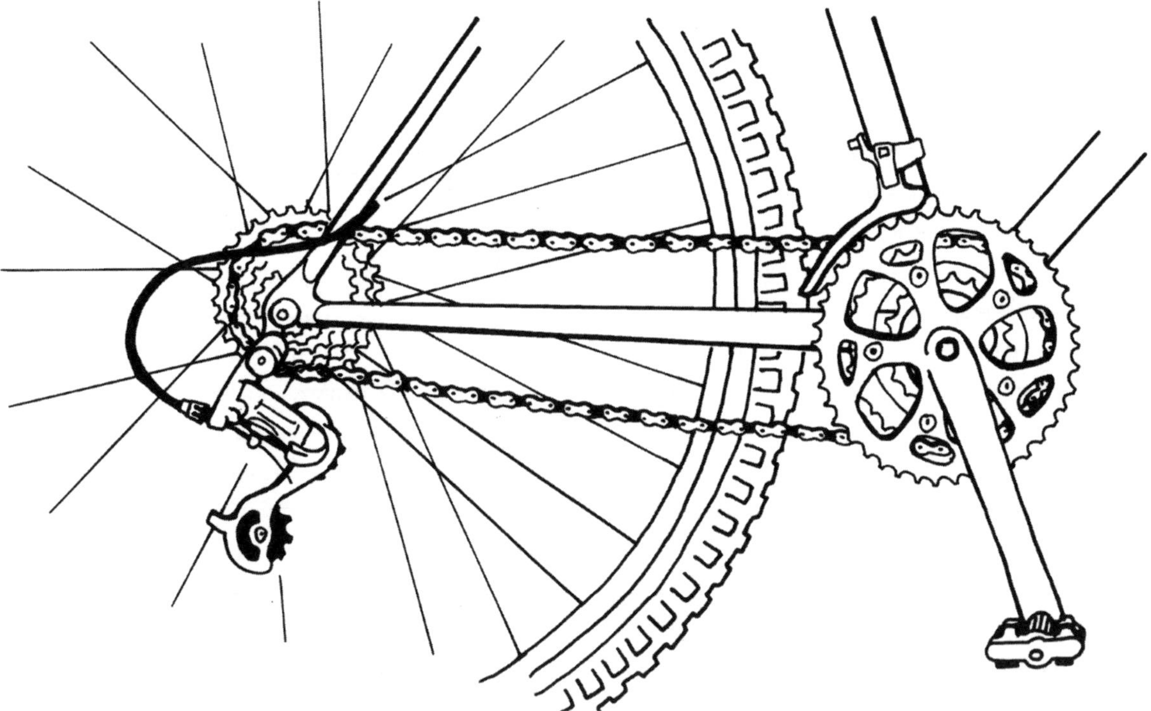

pivot bikes whose main pivot is centered on the bottom bracket. Other rear-suspension systems will alternately yank on and slacken the chain as they move. Taking the derailleur out of the equation means that there is nothing to take in or let out the slack. If you have a lockout on the rear shock, you can use that and continue with the following instructions. Otherwise, you must try another way to repair the derailleur, or you will be walking home.

1. **Open the chain and pull it out of the derailleur.** Use a chain tool (4-7), or do it by hand if the chain has a master link (4-12).
2. **Shift the front derailleur to a selected chainring.** Pick the gear combination in which you think you can make it home most effectively. Be aware that the chain line must be tight and aligned straight (i.e., the chain must parallel the frame), or the chain will fall off the cog and the chainring and frustrate any attempts to pedal.
3. **Choose a rear cog and wrap the chain around it.** Bypass the rear derailleur entirely.
4. **Remove any overlapping chain.** Make the chain as short as you can, where you can still get the ends together. Then, if the bike has horizontal dropouts, push the wheel forward in the dropouts to get a bit more slack.
5. **Connect the chain.** Use a chain tool as described in 4-10. Pull the wheel back in the dropouts as far as you can to tension the chain.

   Now ride home.

3-10

COMPENSATING FOR A LOST REAR-DERAILLEUR JOCKEY WHEEL

If you can find the jockey wheel and the bolt, just reassemble the parts onto the derailleur (see 5-33).

If you find the jockey wheel but not the bolt, you can reattach the wheel with one of the bolts holding a water-bottle cage to the frame (provided you did not try to save weight by using short bottle-boss bolts!). The thread may be the same, although if the bolt is too long, you will need to be careful that you don't shift the derailleur inward far enough to catch it on the spokes as you ride. Otherwise, try wire or a zip-tie.

If you cannot find the jockey wheel, you can still rig up the derailleur to work or at least to pedal without shifting. If you lost the upper jockey wheel and the derailleur is a type with the same bolts top and bottom, then put the lower wheel on top first. Now, if you found the bolt for the lost wheel, just tighten the bolt back in where it was, making sure the chain is routed over it in the normal fashion as if the jockey wheel were still in place on the derailleur. If the bolt is also missing, you can still rig up a fix. Collect three threaded collars from the Presta valves on both of your wheels and from your spare tube. String them up between the cage plates with a twist-tie, wire, or zip-tie.

3-11

WORKING WITH A BROKEN REAR-DERAILLEUR RETURN KNUCKLE SPRING

If the spring in the rear derailleur's lower knuckle breaks or gets dislodged, it will not twist the jockey-wheel cage and pull tension on the chain. If you have a bungee cord, hook it to the derailleur's jockey-wheel cage and loop it around the end of the quick-release skewer. Reverse the skewer so that the lever is on the drive side, pointed back. Hook the other end of the bungee to wherever you can, such as a water-bottle cage or the seat tube, to maintain good tension.

3-12

RIDING WITH A BROKEN FRONT-DERAILLEUR CABLE

The chain will be on the inner chainring, and you will still be able to use all of the rear cogs. Depending on which chainring you want for your return ride, pick one of these three options:

1. **Leave it on the inner ring and ride home.**
2. **Tighten the inner derailleur limit screw until, with luck, the derailleur sits over the middle chainring** (Fig. 3.14). Leave the chain on the middle ring and ride home.
3. **Bypass the front derailleur by removing the chain from the derailleur and putting it on the big chainring.** You can do this either by opening the

3.14 Tightening the inner front-derailleur limit screw

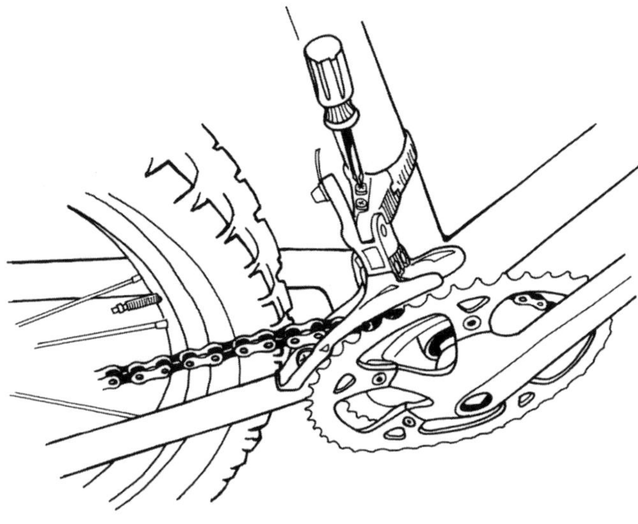

3.15 Broken rear-derailleur cable—option 2

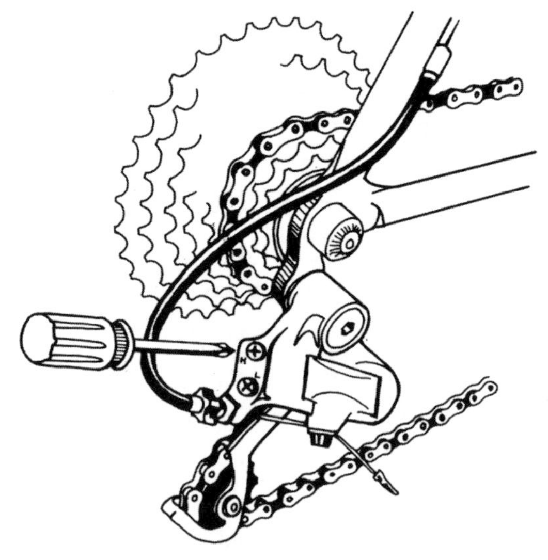

derailleur cage with a screwdriver (Fig. 3.12) or by disconnecting and reconnecting the chain with a chain tool or at a master link (4-7 and 4-9 to 4-11 or 4-12).

3-13

RIDING WITH A BROKEN REAR-DERAILLEUR CABLE

The chain will be on the smallest or largest rear cog, depending on the type of derailleur, and you will still be able to use all three front chainrings. Most derailleurs use spring tension to move the chain to the small cog, so that is where the chain most likely will be. Conversely, Shimano Low Normal (current) or Rapid Rise (late 1990s) rear derailleurs use spring tension to move the chain to the largest cog, so a broken cable will leave it stuck there. You have three options:

1. **Leave the chain on the cog it's on and ride home.**
2. **With a standard rear derailleur, move the chain to a larger cog, push inward on the derailleur with your hand, and tighten the high-gear limit screw on the rear derailleur (usually the upper of the two screws) until it lines up with a larger cog** (Fig. 3.15). Move the chain to that cog and ride home. With a Shimano Low Normal or Rapid Rise rear derailleur, move the chain to a smaller cog, push outward on the derailleur with your hand, and tighten the low-gear limit screw on the rear derailleur (usually the lower of the two screws) until it lines up with a smaller cog. Move the chain to that cog and ride home. You may have to fine-tune the adjustment of the derailleur limit screw to get it to run quietly without skipping.

3.16 Broken rear-derailleur cable—option 3

3. **If you do not have a screwdriver, you can push inward on a standard rear derailleur while turning the crank with the rear wheel off the ground to shift to a larger cog.** Jam a stick in between the derailleur-cage plates to prevent the derailleur from moving back down to the small cog (Fig. 3.16).

Don't try this with a Shimano Low Normal or Rapid Rise rear derailleur, because the stick will be too close to the spokes for comfort.

3-14

GETTING HOME WITH A BROKEN BRAKE CABLE OR A BLOWN HYDRAULIC BRAKE HOSE

Walk home or ride slowly and carefully home if the trail is not dangerous.

3-15

RIDING WITH A FLAT SUSPENSION FORK

If you have a blown or leaking air-spring fork and can't pump it, there's not much you can do. You will just have to ride back with it that way. If the fork has a lockout lever or compression-damping adjustment, tighten it down to minimize bottoming out. Go slowly and keep your weight back.

3-16

RIGGING BROKEN SEAT RAILS OR A SEATPOST CLAMP

If you can't tape or tie the saddle back on, try wrapping your gloves or some clothing over the top of the seatpost to pad it. Sticking an inverted water bottle over the top of the post also might make it rideable. Otherwise, remove the seatpost and ride home standing up.

3-17

RIDING WITH A BROKEN SEATPOST

If the seatpost shaft breaks, you can splint it internally with a stick, tape it up, and ride very carefully. A better solution is to ride home standing up.

3-18

RIDING WITH A BROKEN HANDLEBAR

It's probably best to walk home. You could splint the broken handlebar by jamming a stick inside and wrapping it with duct tape. If the break is right next to the stem clamp, you could also loosen the clamp, move the handlebar over so that the break is inside the clamp, and retighten it. In either case, you must ride very carefully. The stick could easily break or the clamp could let go of the broken handlebar, leaving you with no way to control the bike. A sudden collision of your face with the ground would follow. From what I've heard, that can be a painful experience. In fact, now that I think of it, you might want to just walk your bike home.

3-19

FIXING A BROKEN LINKAGE BOLT ON REAR SUSPENSION

Try sticking a hex key into the hole where the bolt was and tape it in place.

3-20

DEALING WITH A SEIZED FREEHUB OR FREEWHEEL

If the rear cogs will not freewheel, you cannot coast. If you do stop pedaling, the forward-turning cogs will pull the slack chain around and rip up the rear derailleur. Try squirting some chain lube into the front and back of the freewheel mechanism. If you have no chain lube and the temperature is above freezing, try squirting water in to restore the freewheeling action. If the temperature is below freezing, you can sometimes free a frozen freehub by peeing on it. Hey, don't laugh. It's warm!

No matter what liquid you use for lubrication, you may need to hit the freehub with a stick to get it to turn.

3-21

AVOIDING GETTING LOST OR HURT AND DEALING WITH IT IF YOU DO

Mountain biking in the backcountry can be dangerous. You need to prepare properly and take personal responsibility for your own and others' safety when riding in deserted country. Two deaths near Moab, Utah, in the summer of 1995 highlight the risks facing anyone who rides off into the backcountry.

The two who died in Moab were riding the popular Porcupine Rim Trail. They got lost on the descent off Porcupine Rim, missed the turn into Jackass Canyon, and then headed into Negro Bill Canyon—which divides the Porcupine Rim Trail from Moab's most famous ride, the Slickrock Trail. It may come as a surprise that people could die and go undiscovered for 17 days so close to a main highway into town (which was right below them) and to two heavily traveled trails. But since they hadn't told anyone of their plans, no one in town noticed when they did not return.

Their parents, not hearing from them for a few days, called the sheriff, and a search was mounted.

Once the riders got lost, they abandoned their bikes and tried to walk down to the road, instead of riding back the way they had come. That road and the Colorado River are very close as the crow flies and are visible at a number of points but, owing to the numerous cliffs, are quite difficult to reach. The two climbed, fell, or slid down to a ledge from which they apparently were unable to climb either up or down, and there they slowly perished from exposure.

They died on this ledge in such a way that they were very difficult to spot from the air. They had placed no items to indicate their positions to airborne spotters. Had searchers found their bikes, they could have concentrated the search on a small area. Regrettably, thieves took their bikes and helmets and did not alert the authorities.

Eventually, a helicopter searcher saw the bodies on the ledge, and a Forest Service ranger rappelled 160 feet down to them. He was able to then walk out unaided, indicating that perhaps the riders were so injured, exhausted, delirious, or hypothermic that they had been unable to take the same route out.

Cliffs, steep hills, and an array of other natural features can also pose a risk. Some years ago, another Moab rider barely managed to jump off his bike before it went hurtling over the edge of a cliff and dropped some 400 feet. Anyone who has ridden much in the canyon country of the Southwest can tell you that there are countless other trails near cliff edges that present a similar threat.

Even in seemingly safe areas, the risks can be high. Pro rider Paul Willerton came close to meeting his end on a relatively standard, cliff-free, but isolated trail near Winter Park, Colorado. Unable to walk after crashing and breaking his leg, Willerton had to drag himself many miles using only his arms.

We all tend to think that nothing like this will happen to us. But it can happen, and far too easily. That shouldn't discourage you from riding in the backcountry, but it should encourage you to think and utilize the following 12 basic backcountry survival skills. They could make the difference between life and death.

1. **Always take plenty of water.** You can survive a long time without food but not without water.
2. **Tell someone where you are going and when you expect to return.** If you know of someone who is missing, call the police or sheriff.
3. **If you find personal effects on the ground, assume that someone could be lost or in trouble.** Report the find and mark the location.
4. **If you get lost, backtrack.** Even if going back is longer, it is better than getting stranded.
5. **Don't go down something you can't get back up or go up something you can't get back down.**
6. **Bring matches, extra clothing and food, a flashlight, and perhaps an aluminized emergency blanket in case you have to spend the night out or need to signal searchers.**
7. **If the area is new to you, go with someone who is familiar with it or take a map and compass—and know how to use them.**
8. **Wear a helmet.** It's hard to ride home with a cracked skull.
9. **Bring basic first aid stuff and bike tools, and know how to use them well enough to keep yourself and your bike going.** Test your portable tools at home before packing them into your spares kit.
10. **Walk your bike when it's appropriate.** Falling off a cliff is a poor alternative to taking a few extra seconds or displaying less bravado. Try riding on difficult sections of trail to improve your bike handling, but if the risk of falling off a cliff is great or a mistake could leave you injured a long way from help, find another place to practice those moves.
11. **Don't ride beyond your limits if you are a long way out.** Take a break. Get out of the hot sun.

Avoid dehydration and bonking by drinking and eating enough.

12. **Teach your friends all these things.**

These rules are in addition to the following International Mountain Bike Association Rules of the Trail, which we would all do well to adhere to:

1. **Plan ahead.**
2. **Always yield trail.**
3. **Never scare animals.**
4. **Ride on open trails only.**
5. **Control your bicycle.**
6. **Leave no trace.**

Keep in mind that your decisions affect yourself and could affect your riding partners, your families, and countless others. Recognize that endangering yourself can also endanger the person trying to rescue you. Search-and-rescue parties are usually made up of helpful people who will gladly try to save you, but no one appreciates being put in harm's way unnecessarily.

In summary, make appropriate decisions when cycling the backcountry. Learn survival skills and prepare well. Understand that even though you have a $4,000 bike and are riding on popular trails, you are not immune to danger. When ignorance makes us oblivious to danger, it sadly becomes the danger itself.

A chain is only as strong as its weakest link.

—ANONYMOUS

A sausage is only as good as its last link.

—BLUTO

4

CHAINS

TOOLS

- Chain lubricant
- 12-inch ruler
- Chain tool ("chain breaker" or "chain cutter")
- Lots of rags
- Rubber gloves

OPTIONAL

- Chain-elongation indicator
- Master link pliers
- Solvent (citrus-based)
- Self-contained chain cleaner
- Old water bottle
- Caliper
- Pliers
- Solvent tank
- Rohloff cog-wear indicator

A bike chain is a simple series of links connected by rivets. Rollers surround each rivet between the link plates and engage the teeth of the cogs and chainrings. It is an extremely efficient method of transmitting mechanical energy from the pedals to the rear wheel. In terms of weight, cost, and efficiency, the bicycle chain has no equal, and—believe me—people have tried without success to improve on it for years.

To keep your bike running smoothly, you have to take care of the chain. It needs to be kept clean and well lubricated in order to transmit your energy efficiently and shift smoothly. Chains should be replaced frequently to prolong the working life of other, more expensive, drivetrain components because a chain gets longer as its internal parts wear, thus contacting the gear teeth differently than intended.

Today's derailleur chains require more care to avoid breakage because the rivets are shorter and the outer plates thinner than when bikes had only five rear cogs. This makes proper connection especially important.

To isolate specific chain issues, check out "Troubleshooting Chain Problems" at the end of this chapter (4-14 through 4-16).

CHAIN SERVICE AND ASSEMBLY

4-1

LUBRICATING THE CHAIN

When lubricating the chain, use a lubricant intended for bicycle chains. If you want to get fancy about it, you can assess the type of conditions in which you ride and choose a lubricant intended for those conditions. Some lubricants are dry and pick up less dirt in dry conditions. Some are sticky and therefore less likely to wash off in wet conditions. Still others claim to be metal conditioners that actually penetrate and alter the surface of the metal. Some lubes run faster than others in chain-friction tests, but that may or may not say anything about longevity. Be aware that some wax-based lubricants don't protect well; the chain tends to look clean and doesn't pick up dirt because the wax continually flakes off, but chain life with such lubricants can be short. Find a lube that works for you and the conditions you ride in.

Chain lubes generally come in spray cans and squeeze bottles. Sprays should be avoided for regular maintenance chores because they tend to spew too much oil over everything, including in the air where you can inhale it. The chain needs oil only between contacting parts. On the outside, a thin film will keep corrosion at bay; more than that will only attract dirt and gunk without improving the function of the chain.

1. **Drip a small amount of lubricant across each roller** (Fig. 4.1). Periodically move the chain so that you can easily access the links you are working on. To speed the process, turn the crank slowly while dripping lubricant onto the chain as it goes by. Yes, this method will cause you to apply excess lubricant, which will pick up more dirt. But overlubricating is far preferable to not lubricating, and if you wipe and lube the chain after each ride or two, it won't build up excessive grime.
2. **Wipe off the chain lightly with a rag.** In wet conditions, expect to use more lubricant (after every ride or even during a long, rainy ride).

4.1 Lubing the chain

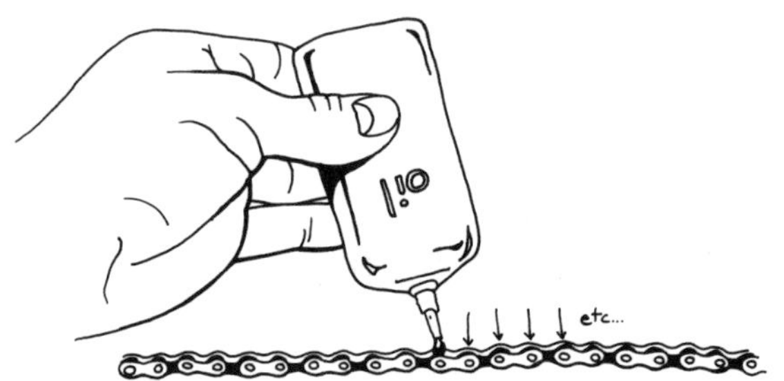

4-2

CLEANING THE CHAIN BY FREQUENT WIPING AND LUBRICATION

You can clean the chain in a number of ways. The simplest method is to wipe it down frequently, lubricate it, and then wipe off the excess lube. If you do this before every ride, you will never need to clean the chain with a solvent. The lubricant softens the old sludge buildup, which is driven out of the chain when you ride. Of course, the lubricant also picks up new gunk, but if you wipe it off before it's driven deep into the chain and if you relubricate the chain frequently, it will stay clean and supple. Chain cleaning can be performed as follows with the bike standing on the ground or in a bike stand:

4.2 Wiping the chain with a rag

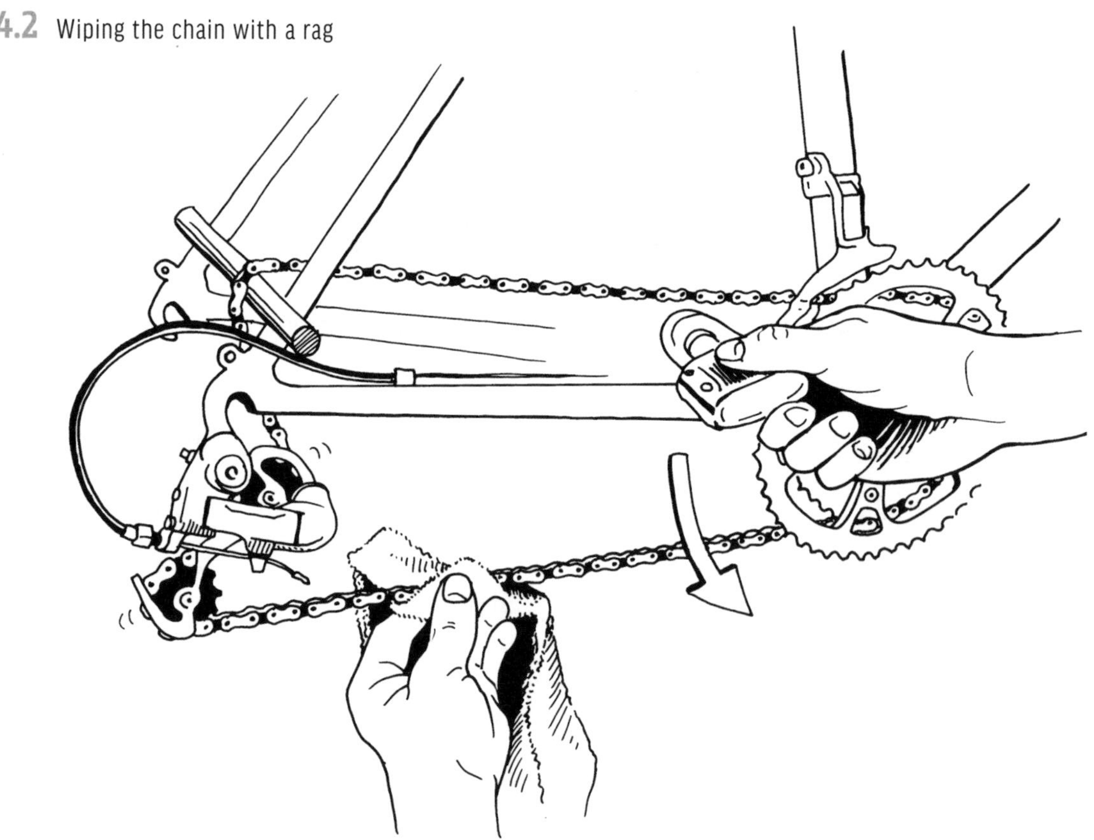

1. **Grab the chain with a rag.** Grasp the lower length of the chain (between the bottom of the chainring and the rear-derailleur lower jockey wheel).
2. **Turn the crank backward a number of revolutions.** Pull the chain through the rag (Fig. 4.2). Periodically rotate the rag to present a cleaner section of it to the chain.
3. **Lubricate the chain as in 4-1.**

To encourage regular care, leave a pair of rubber gloves, a rag, and some chain lube next to your bike. Then, whenever you return from a ride, put on the gloves, wipe and lube the chain, and put your bike away. It takes maybe a minute, your hands stay clean, and your bike is ready for the next ride. If you can find time to take a shower after you ride, you can find time for this. Wipe the chainrings, cogs, front derailleur, and jockey wheels (Fig. 4.3) while you're at it, and the entire drivetrain will always work ideally.

4-3
USING CHAIN-CLEANING UNITS

Several companies make chain-cleaning units that scrub the chain with a solvent while it is on the bike. These chain cleaners are generally made of clear plastic and have two or three rotating brushes that scrub the chain as it moves through the solvent bath (Fig. 4.4). The units offer the advantage of letting you clean the chain without removing it from the bike. Unless the chain has a master link that allows it to be separated easily, regularly removing the chain with a tool that drives a pin out of a link shortens the chain's life significantly; moreover, repeatedly opening a 9-, 10-, 11-, or 12-speed chain can result in the chain breaking under high load, thereby driving your foot, and perhaps your entire body, into the ground. Installing a master link in a chain that does not have one is a good idea; see section 4-12.

Most chain-cleaning units come with a nontoxic, citrus-based solvent. For your safety, and other environmental reasons, I strongly recommend that you purchase nontoxic citrus solvents for the chain-cleaning unit, even if it already comes with a petroleum-based solvent. If you recycle used diesel fuel, then go ahead and use it as solvent. In any case, wear gloves and glasses when using any solvent, and make sure you have good ventilation if you're doing this indoors.

Citrus chain solvents often contain some lubricants, so they won't dry the chain out. The combination of lubricant and solvent is why diesel fuel has had such a following as a chain cleaner. A really strong solvent without lubricant (acetone, for example) will displace the oil from inside the rollers. The solvent will later evaporate, leaving a dry, squeaking chain that is hard to rehabilitate. The same can happen with a citrus-based solvent that does not include a lubricant if no lubricant is separately applied, especially if the chain is not allowed to dry completely. The procedure for using a chain-cleaning unit is straightforward:

4.3 Wiping the jockey wheels with a rag

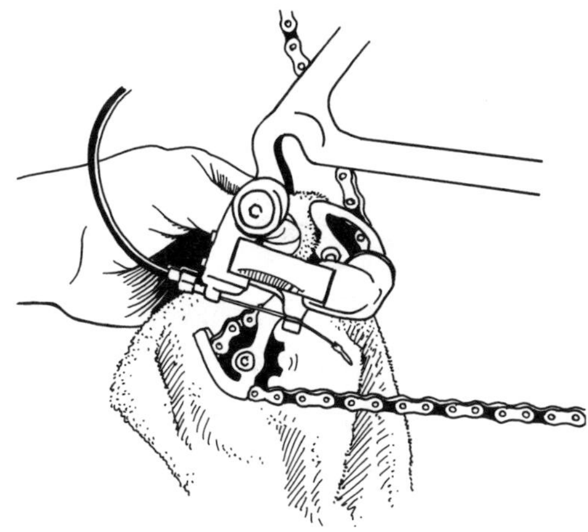

4.4 Using a solvent-bath chain cleaner

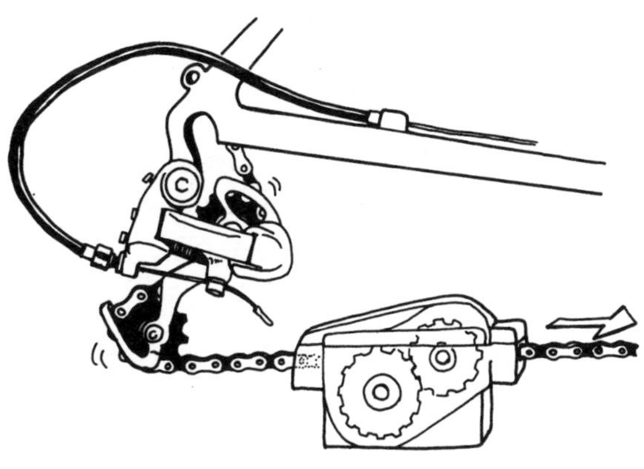

1. **Remove the top and pour in the solvent up to the fill line.**
2. **Place the chain-cleaning unit up against the bottom of the chain.** Reinstall the top so that the chain runs through it (Fig. 4.4).
3. **Turn the bike's crank backward.** Run the chain through the unit's brushes until it is clean.
4. **Remove the unit from the chain.**
5. **Lubricate the chain as in 4-1.**
6. **Let the solvent settle, decant the clear portion, and discard the sludge.**

4-4

REMOVING AND CLEANING THE CHAIN

You can also clean the chain by removing it from the bicycle and cleaning it in a solvent. I do not recommend this procedure unless the chain has a master link, because repeatedly disassembling the chain by pushing rivets in and out weakens it.

Mountain bike chains are prone to breakage because of the conditions in which they are used, but chain breakage is even more of an issue because of the narrow width of 9-, 10-, 11-, and 12-speed chains. A chain that breaks during riding generally does so when you shift the front derailleur while pedaling hard. This technique can pry a link plate open so that the head of a rivet pops out of the plate, tearing the chain apart. Chain disassembly and reassembly expand the size of the rivet hole where you put the chain together, allowing the rivet to pop out more easily. Shimano supplies special subpins for reassembly of its chains. These are meant to prevent this problem, but the chain is still not as strong there as if you had left the original pin in place.

A master link can avoid the problem of repeatedly opening and reassembling the chain. Master links are standard on SRAM, Wippermann, Taya, KMC, FSA, and, now, Shimano chains. An aftermarket master link, like Lickton's SuperLink or one from any of the brands listed above, can also be installed into any chain so long as you make sure that the master link is the right width (that is, use a 9-speed link for a 9-speed chain, and so on).

If you do disassemble the chain (see 4-7 or 4-12 for instructions), you can clean it well, even without a solvent tank. Just drop the chain into an old jar or water bottle half filled with solvent. Using an old water bottle or jar allows you to clean the chain without touching or breathing the solvent—something to be avoided even when you are using citrus solvents.

The procedure for cleaning the chain without using a chain-cleaning unit could not be simpler:

1. **Remove the chain from the bike** (4-7 or 4-12).
2. **Drop it into a water bottle or jar.**
3. **Pour in enough solvent to cover the chain.**
4. **Shake the bottle vigorously.** Keep it low to the ground in case the top pops off or the jar breaks.
5. **Hang the chain to air-dry.**
6. **Reassemble it on the bike** (see 4-8 to 4-12).
7. **Lubricate it as in 4-1.**

Whatever you do, don't leave the chain to soak for extended periods in citrus-based solvents, because these are water based and will cause the chain to oxidize (rust), making it move with more friction and be more prone to breakage. (Some people believe in having two chains they rotate on and off the bike, leaving one soaking in solvent while the other one is on the bike. Although this would work with diesel fuel as the solvent, it won't work with water-based solvents. In any case, you gain nothing by soaking the chain for extended periods, so just don't do it.)

After removing the chain, allow the solvent in the bottle or jar to settle for a few days so that you can decant the clear stuff and use it again. I'll say this throughout the book: Use a citrus-based solvent. It is not only safer for the environment; it is also gentler on your skin and less harmful to breathe. Wear rubber gloves when working with any solvent, and use a respirator meant for volatile organic compounds if you are not using a citrus-based solvent. There is no sense in fixing your bike to go faster if you end up becoming a slow, sickly bike rider.

4-5

REPLACING THE CHAIN

As the rollers, pins, and plates wear out, the chain lengthens. That, in turn, hastens the wear and tear on other drivetrain parts. An elongated chain concentrates the load on each individual gear tooth, rather than distributing it over all of the teeth that the chain is wrapped around, and as a result the gear teeth become hook-shaped and the tooth valleys become wider. If such wear has already occurred, a new chain will not solve the problem. A new chain will not mesh properly with deformed teeth, and it is likely to skip whenever you pedal hard. At that point, you'll need to replace the rear cogs, the front chainrings, and possibly the chain yet again, depending on how long you tolerate the skipping. So before all of that extra wear and tear hits your pocketbook, get in the habit of checking the chain on a regular basis (4-6) and replacing it as needed.

How long it takes for the chain to wear out will vary, depending on chain type, maintenance, riding conditions, and strength and weight of the rider. In general, mountain bike chains live short, hard lives; figure on replacing the chain every 500 to 1,000 miles, especially for bikes ridden in dirty conditions by a large rider. Lighter riders staying mostly on paved roads can often extend replacement time to more than 2,000 miles. Instead of counting miles, though, the best way to monitor the chain's condition is with a length gauge, as covered next.

4-6

CHECKING FOR CHAIN ELONGATION

a. Chain-elongation gauges

LEVEL 1 The simplest accurate method for checking chain elongation is to use a chain-elongation gauge. Make sure you check a number of spots on the chain; you'll find variation.

The Rohloff Caliber 2 gauge (Fig. 4.5) is simple, quick, and reliable. It's a go/no-go gauge. Brace the hook end against a chain roller, and if the opposing curved tooth falls completely into the chain so that the length of the tool's body contacts it, the chain is shot. If the chain is still in good shape, the curved tooth will not go all of the way in. The tooth marked "S" is for checking a chain running strictly on steel rear cogs, and the tooth marked "A" is for checking aluminum and titanium cogs, but I use just the A side. I find that if the A edge comes down to the chain and I replace it right then, I get almost infinite life out of my chainrings and cogs, even titanium ones. That's worth it to me.

4.5 Checking chain wear with the Rohloff gauge. If the curved tooth with the S (steel cogs) falls completely into the chain, replace the chain (A is for aluminum cogs).

4.6 Using a ProGold chain gauge

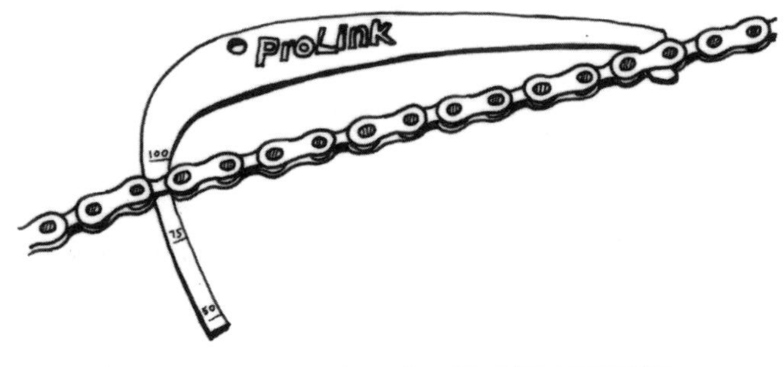

The Park CC-3.2 and the Bike Hand chain-wear indicators are also double-sided gauges. They also indicate less elongation on one side than the other, and you use them in the same way as the Rohloff Caliber 2.

The ProGold chain gauge (Fig. 4.6) is also quick and accurate. Brace the hooked end against a chain roller and let the long tooth drop into the chain. If it drops in close to the 90 percent mark, that is equivalent to the A side of the Rohloff dropping down flush with the chain.

Shimano's chain elongation gauge, Pedro's Chain Checker, and Topeak's Chain Hook & Wear Indicator also check for 0.75 percent chain elongation, which is between ¹⁄₁₆ and ⅛ of an inch of chain growth over 12 inches. They are different from the gauges mentioned above in that they do not push opposing rollers apart. Instead, you don't drop their go/no-go tooth into the chain until you have already braced two other teeth between rollers. This ensures that you're actually measuring the spacing between trailing sides of rollers. On the other hand, they are less quick and require touching the chain with your fingers. The Pedro's and Topeak tools have the additional feature of a chain hook on the opposite side—two hooked prongs that pull the chain ends and release the tension at the connection point so you can more easily push in a connector pin or assemble a master link.

Feedback Sports makes a digital chain gauge shown in Figure 1.4 and also offers it under the KMC brand. With it you can monitor chain wear precisely over time.

There are other chain-elongation gauges as well from Park, Wippermann, and other brands. Some gauges are definitely less convenient to use than the Rohloff and ProGold quick go/no-go indicators.

b. Ruler method

Another way to measure chain wear is with an accurate ruler. Chains are measured on an inch standard and should measure ½ inch between adjacent rivets (and nominally have 3⁄32-inch-wide rollers on derailleur chains and often ⅛-inch-wide rollers on single-cog bicycles). There should be exactly 12 links in 1 foot, where each complete link consists of an inner and an outer pair of plates (Fig. 4.7).

1. **Set one end of the ruler on a rivet edge.**
2. **Measure to the rivet edge at the other end of the ruler, 12 links away.** The distance between these rivets should be 12 inches exactly. If it is 12⅛ inches or greater, replace the chain; if it is 12 1⁄16 inches or greater, replacing it is a good idea (and a necessity if you have any titanium or alloy cogs or an 11-tooth small cog). Some chain manufacturers recommend replacement if elongation is 1 percent, or ½ inch in 50 complete link pairs (50 inches), which is a little less than ⅛ inch over 12 link pairs (1 foot). If the chain is off the bike, you can hang it next to a new chain for comparison; if the used one is more than a third of a link longer for the same number of links, I recommend replacing it.

If you always replace the chain as soon as it becomes elongated beyond the spec I've indicated on these chain-elongation gauges, you will replace at least three chains before needing to change the cogs.

4.7 One complete chain link

4-7

REMOVING THE CHAIN

LEVEL 1

The following procedure applies to all standard derailleur chains except those with a master link. Master-link-equipped chains include all SRAM, Wippermann, KMC, FSA, Taya, and 2017 and later Shimano chains. All of these chains snap open by hand at the master link (see 4-12). If need be, they can also be opened at any other link with the use of a chain tool as described here.

1. **Place any link over the back teeth on a chain tool** (Fig. 4.8).
2. **Tighten the chain-tool handle clockwise.** Push the link rivet out unless you don't have a master link or a Shimano chain and a new subpin for it. In that case, leave 1mm or so of rivet protruding inward from the chain plate to hook the chain back together when reassembling.
3. **Separate the chain.** Flex it away from the pushed-out pin if you left the stub in. If you pushed the pin all the way out, the two ends will just pull apart, but you won't be able to reconnect them without a subpin or a master link.

4.8 Removing a chain rivet

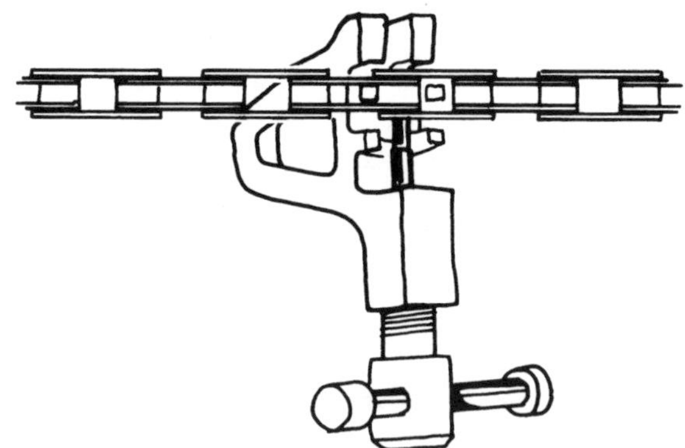

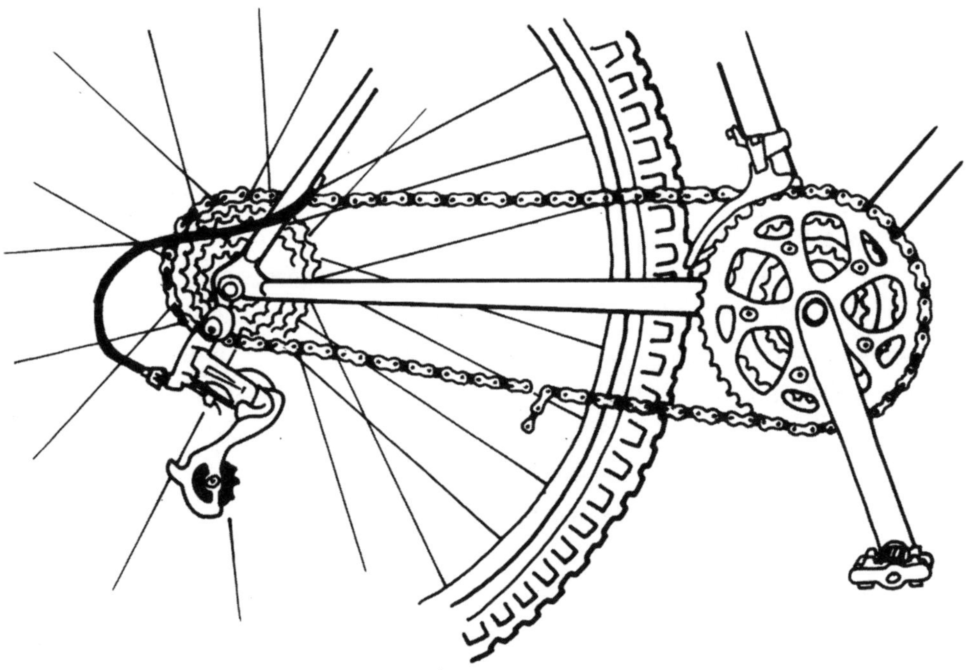

4.9 Determining chain length

4-8

DETERMINING CHAIN LENGTH

a. Chain length with a derailleur

If you are putting on a new chain, determine how many links you'll need in one of two ways: (a) Under the assumption that your old chain was the correct length, compare it with the new one, and use the same number of links. (b) If you have a standard long-cage mountain bike rear derailleur on your bike, wrap the chain around the big chainring and the biggest cog without going through either derailleur. Bring the two ends together until the ends overlap; one full link (Fig. 4.7—a complete link pair) should be the amount of overlap (Fig. 4.9).

Use one and a half more links of overlap for a 1×11 system with a SRAM X-Horizon, roller clutch straight parallelogram rear derailleur or any system with a huge rear cog (40-tooth or larger). In other words, you'll overlap at least two full links, but both links will end in an inner link, because you'll be connecting them with a master link, thus making the actual overlap at least 2.5 links.

Remove the remaining links and save them in your spare-tire bag so that you have spares in case of chain breakage on the trail.

b. Chain length with a single-speed or internal-gear rear hub

Bikes made for single-speed or Rohloff and other internal-gear hubs have a system to vary the distance from the crank to the rear hub. Bikes not made for these hubs can be fitted with adapters to tension the chain.

The simplest made-for-single-speed system consists of rear dropouts with long, horizontal slots in which you can slide the wheel back and forth and clamp it in place where you choose. Some of these systems have adjuster screws with locknuts on them; you push the wheel forward or backward by turning the adjuster screw(s) and locking it into place with the locknut(s).

Sliding rear dropouts (Fig. 14.3) are a more elegant system for varying drivetrain length on single-speed/internal-gear frames. The axle ends sit in near-vertical dropout slots, but to tension the chain, you loosen the bolts securing the dropouts to the ends of the chainstays. Slide them back and forth while the wheel is already clamped in, and lock it in place by tightening the bolts securing the sliding dropouts to the frame.

Another method is an eccentric bottom bracket; the bottom-bracket shell is oversized and the bottom-bracket bearing and axle assembly mounts into a tunnel bored off-center in an aluminum cylinder (I'll call this the eccentric cylinder). The eccentric cylinder clamps into the oversized bottom-bracket shell, either by means of a slot in the shell closed by pinch bolts, setscrews on the shell driven into the eccentric cylinder, or a sliding wedge piece in the eccentric cylinder. You tension the chain by loosening the pinch bolts, setscrews, or wedge bolts and rotating the eccentric cylinder (usually with a pin tool in a pair of holes in the face of the eccentric cylinder) to move the crank toward or away from the rear hub.

With any of these systems, first set the distance from the wheel to the crank at its minimum by sliding the wheel or the sliding dropouts all of the way forward or by rotating the eccentric bottom bracket so that the bottom-bracket spindle is as close to the back of the bottom-bracket shell as possible. Wrap the chain around the cog and chainring, and make it the minimum length that still allows you to connect the ends. Once the chain is connected, tighten the chain by sliding the wheel back or by moving the bottom bracket forward, depending on the system you have. You don't want the chain to be taut like a drumhead, as this would create excess pedaling resistance as well as premature wear of the hub and bottom-bracket bearings. You also don't want the chain so loose that it can fall off. A good test is to push the chain over gently with your thumb just behind the chainring while turning the crank. If you can derail the chain this way, tighten it more.

To adapt a standard frame to use a single-speed or internal-gear hub, you can install a chain tensioner to pull a slack chain taut. Often called a "singleator," after the Surly model of that name, it is usually a spring-loaded arm with a single jockey wheel or roller that bolts into the rear-derailleur hanger on the dropout. The jockey wheel or roller pushes up or down on the chain coming to the bottom of the rear cog to keep it tight (two springs are often supplied so that you can choose if you want the roller to push up or down on the chain). The jockey-wheel shaft is often free to move laterally in order to center the roller or jockey wheel on the chain.

Another type is a U-shaped clamp with a roller bolted through the ends of the U. The unit clamps around the chainstay and can be slid back and forth to vary the pressure of the roller on the chain; one of these is pictured at the center of the chainstay in Figure 5.45.

For single-speed adapters with a single roller, make the chain just long enough to give the minimum amount of slack when wrapped around the chainring and cog without the roller in place. The chain tensioner will do the rest.

A twin-jockey-wheel type (the Paul Melvin, Rohloff Twin Pulley, or Shimano Alfine), which cannot be used with a fixed gear or a coaster brake, also bolts to the rear-derailleur hanger and has a pair of jockey wheels in a spring-loaded cage that force a Z-bend in the chain to keep it tight. It is set up like a fixed-position rear derailleur and consequently requires more chain length than a single-roller tensioner. You want enough chain length that the jockey wheels are not stretched out in a straight line but also not so much that they cannot take up all the slack. There is a lot of latitude here.

4-9

ROUTING THE CHAIN PROPERLY

1. **Shift into small-small.** Shift the derailleurs so that the chain will rest on the smallest cog in the rear and on the smallest chainring up front.
2. **Guide the chain up through the rear derailleur.** Start with the rear-derailleur pulley that is farthest from the derailleur body (this will be the bottom pulley once the chain is taut). Go around the two jockey pulleys. Make sure the chain passes inside of the prongs on the rear-derailleur cage.
3. **Guide the chain over the smallest rear cog and through the front-derailleur cage.**
4. **Wrap the chain around the smallest front chainring and bring the chain ends together** (Fig. 4.10). When connecting the chain, it is easiest if you completely remove the tension from it by pushing it off the inner chainring to the inside.

4-10

CONNECTING A 5-, 6-, 7-, OR 8-SPEED CHAIN (WITHOUT A MASTER LINK OR A SPECIAL CONNECTING PIN)

A "standard" chain—that is, a non-Shimano chain without a master link—is a disappearing breed, as newer mountain bikes have more rear cogs and correspondingly narrower chains.

NOTE: *If you have a Shimano chain or a chain with a master link, go to 4-11 or 4-12 as appropriate. Except as a short-term emergency fix to get home, don't connect it by using the original rivet (as described in this section); otherwise, you could be injured if the chain breaks.*

ANOTHER NOTE: *This section applies only to wider chains, such as 5-, 6-, 7-, and 8-speed chains. Never use the same pin (except in an emergency out on the trail) on a 9-, 10-, 11-, or 12-speed chain or on any master-link-equipped chain or Shimano chain.*

Connecting a chain without a master link or a Shimano subpin is much easier if the link rivet that was partially removed when the chain was taken apart is sticking out toward you. Positioning the link rivet this way allows you to use the chain tool in a much more comfortable manner (driving the rivet toward the bike as in Fig. 4.10, instead of from the wheel side).

1. **Push the ends of the chain together.** Snap the end link over the little stub of pin you left sticking out to the inside between the opposite end plates. You will need to flex those end plates apart as you push the same link in to get the pin to snap into the hole.
2. **Push the rivet through with the chain tool** (Fig. 4.11). Make the same amount protrude on either side.

4.10 Chain assembly

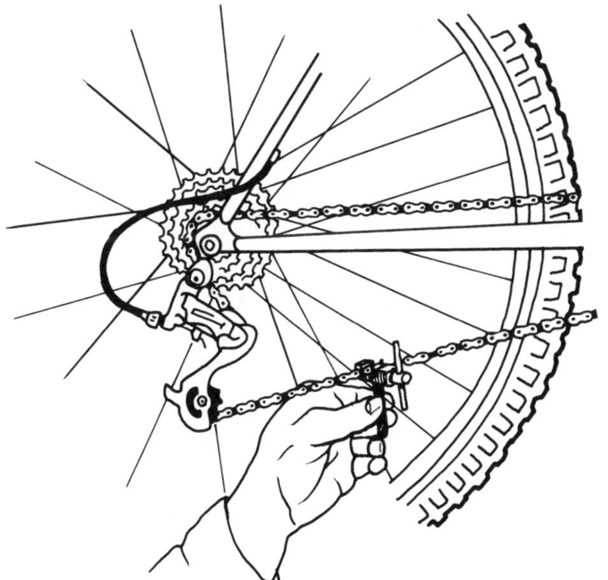

3. **Fold the link.** The link will likely be stiff because the outer plates are being pushed closer together than they were meant to be. If this link does not fold as easily as the surrounding links (Fig. 4.12), continue on with step 4. If it folds freely, you're done.
4. **Put the link over the set of teeth on the tool closest to the screw handle** (Fig. 4.13). If your chain tool has only one set of teeth, free the stiff link by flexing it laterally with your fingers (Fig. 4.14) instead.
5. **Push the pin a fraction of a turn to spread the plates apart.**

4.11 Replacing a chain rivet

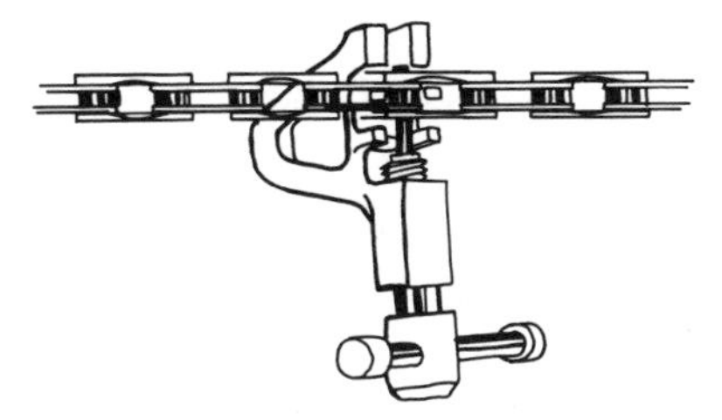

4.12 Stiff link

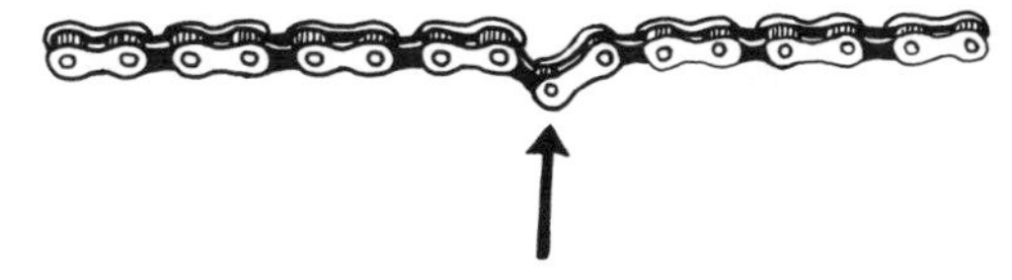

4.13 Loosening a stiff link

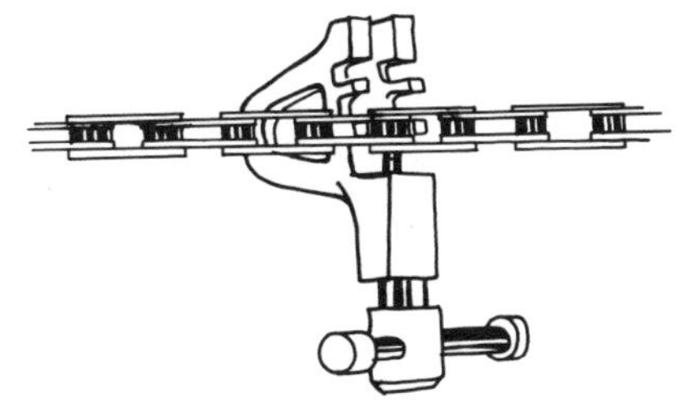

4.14 Loosening stiff link(s) by hand

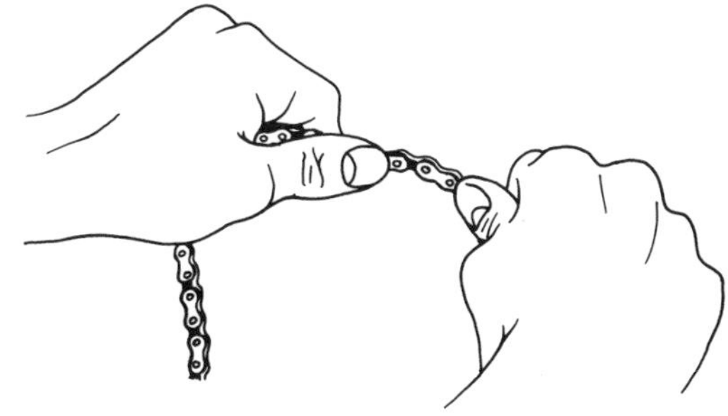

4-11

CONNECTING A SHIMANO CHAIN

Shimano chains prior to the 2017 introduction of the Quick Link have a special connecting pin ("subpin" in Shimano-speak) to ensure a strong chain connection. Insert the subpin in the same direction that you pushed out the old rivet.

Post-2010 10-speed Shimano chains (Shimano introduced 10-speed mountain bike drivetrains and chains for the 2011 model year) are asymmetrical due to the different shifting requirements on each side of the chain for climbing up the cogs versus climbing up the chainrings. Make sure the Shimano logo is to the outside; that is how you know it is oriented properly.

1. **Get out your Shimano subpin.** The subpin is twice as long as a standard rivet and has a breakage groove at the middle of its length. It looks like a black (for 8-speed) or silver (for 9-speed) rivet with a second segment ending in a pointed tip; the main identifying characteristic of the 10-speed pin is a set of two lines around the circumference of the leading edge of the pin. The 11-speed pin is gray but otherwise nondescript, other than the fact that the part that stays in the chain is quite short—only 5.5mm. One or two subpins come with a new Shimano chain. If you are reinstalling an old Shimano chain, get a new subpin at a bike shop. If you don't have a subpin and are going to connect it anyway, follow the procedure in 4-10, but get a new chain soon, because the chain is far more likely to break than if it had been assembled with the proper subpin.
2. **Remove any extra links.** Push the appropriate rivet completely out. Remove extra lengths at the end of the chain ending in an inner link.
3. **Line up the chain ends.** Orient the chain so that the open end of the outer link leads the chain over the top of the chainring, and overlap its holes with those of the inner link at the opposite end.
4. **Push the subpin in with your fingers, pointed end first.** It will go in about halfway.
5. **With the chain tool, push the subpin into place** (Fig. 4.15). Push until there is only as much left protruding at the tail end as the other rivets in the chain.
6. **Break off the leading half of the subpin.** Use the hole in the end of a Shimano chain tool or a pair of pliers (Fig. 4.16).
7. **Check the link's freedom of movement.** If the chain kinks at the link you just assembled (Fig. 4.12), push the link rivet in a little deeper (Fig. 4.11) if it isn't sticking out as much on the backside. Or, if you pushed the link rivet in too far, push it back a hair from the other side with the chain on the teeth closer to the screw (Fig. 4.13 if you're using a chain tool with two sets of teeth). Note that Shimano chain tools (Figs. 4.17 and 4.18) do not have two sets of teeth because it is usually hard to push a Shimano subpin in too far; it will usually be easy, hard, easy, hard, and then very hard to push in as you turn the chain-tool handle, and you will tend to stop before pushing it too far. Otherwise, carefully flex the chain back and forth with your thumbs at the stiff rivet (Fig. 4.14).

4.15 Driving in a Shimano subpin with a chain tool

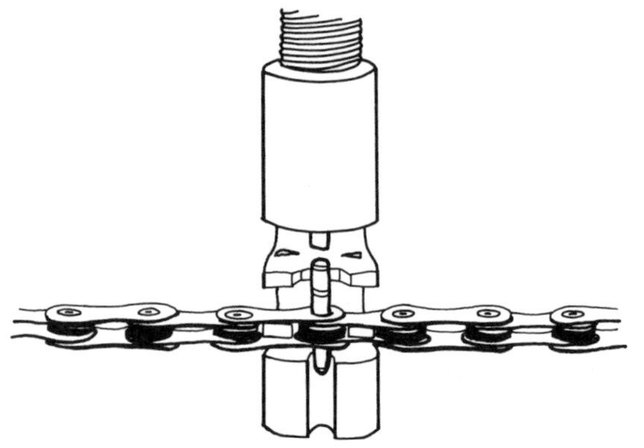

4.16 Snapping the end off a Shimano subpin

PRO TIP — CHAIN TOOLS

IF YOU RIDE A LOT, you will change the chain frequently. It then becomes worthwhile to have a good chain tool (i.e., chain-breaker or chain-cutter tool). If you currently just have a cheap little chain tool, like the one in Fig. 4.11, you will be glad you made the investment to upgrade. A good chain tool is easier to use; lines up the chain better so the pin goes in straighter; lasts longer; properly fits the latest, narrowest chains; and it will have a replaceable drive pin.

Most high-quality chain tools are backward-compatible with older, wider chains. Shimano, for instance, comes out with a new chain tool every time it changes the number of cogs on the rear cassette. Its 11-speed shop tool, the TL-CN34 (Fig. 4.17), works not only on the chain it is contemporary with (11-speed) but also on 6-, 7-, 8-, 9-, and 10-speed chains as well. It also works on 12-speed SRAM chains. Similarly, Shimano's simple consumer TL-CN23 tool (Fig. 4.18) is meant for Shimano 10-speed chains but also for 6-speed chains and every chain in between.

The TL-CN34 resembles its predecessors, the TL-CN32, TL-CN31, and TL-CN30 shop tools (10-, 9-, and 8-speed tools, respectively, all backward-compatible with the wider chains that came before), with spare drive pins stored in a compartment in the handle and four locating teeth in a row to hold the chain. These four teeth, extending to either side of the tool body, secure the chain better than the two teeth of most chain tools. Pedro's Pro 1.0 chain tool (Fig. 4.19) for 1-speed to 10-speed chains also has these extra two locating teeth extending to either side. Pedro's Tutto (meaning "everything") and 11-speed Pro chain tools work on all single-speed through 11-speed chains and incorporate an anvil to flare the connector pin on Campagnolo 11-speed road chains (so you can fix your friends' road bikes). Park's CT-4.3 and CT-6.3 and Topeak's All Speeds chain tools also share these features.

The Rohloff Revolver chain tool (Fig. 4.20) has been around for a couple of decades and continues to work on modern chains as well as 20-year-old ones; it has a thumbscrew that tightens down against the chain and secures it and a revolving plate with different patterns on it to peen the end of the rivet. Like the Shimano TL-CN34, the Rohloff works on SRAM 12-speed chains.

Park's CT-3 (Fig. 4.21) is an older, standard shop chain tool, with both a front set of teeth and a back set of teeth, for prying a link apart to free a stiff link, as in Figure 4.13.

While you can get away with using an older chain tool on ever-narrower chains for a little while, eventually you will bend and break the chain-retention teeth in the tool. As chains become narrower, tool manufacturers move the supporting teeth on newer chain tools closer to the tool's back plate to fully support the rear outer link plate while driving the pin in. When you use an

Continues >

4.17 Shimano TL-CN34 6-11-speed chain tool, which also works on SRAM 12-speed; the TL-CN33 and TL-CN32 (6s-10s), TL-CN31 (6s-9s), and TL-CN30 (6s-8s) tools look identical

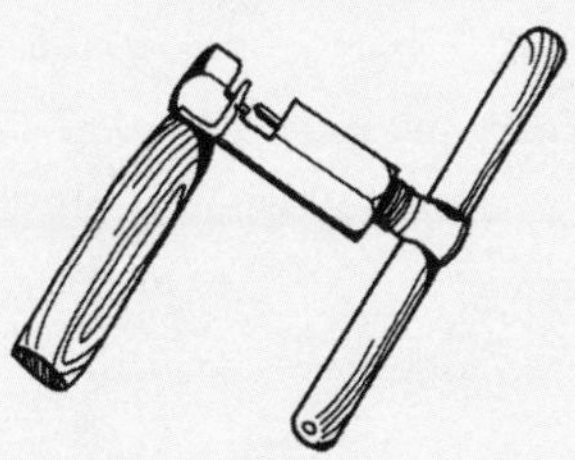

4.18 Shimano TL-CN23 6-10-speed chain tool; the TL-CN22 (6s-9s) and TL-CN21 (6s-8s) tools look identical

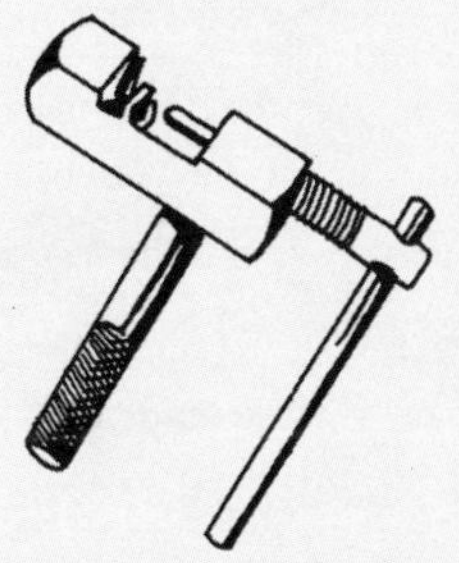

4.19 Pedro's Pro 1.0 1-10-speed chain tool

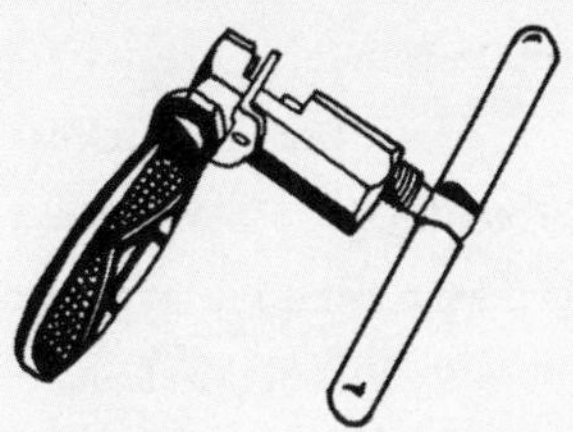

PRO TIP — CHAIN TOOLS, CONT.

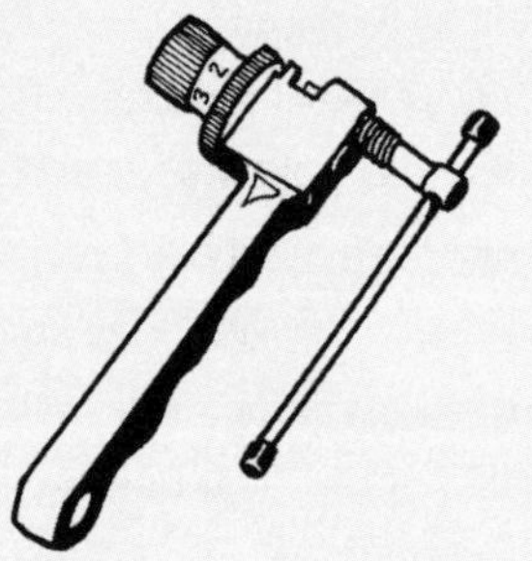

4.20 Rohloff Revolver 1-12-speed chain tool

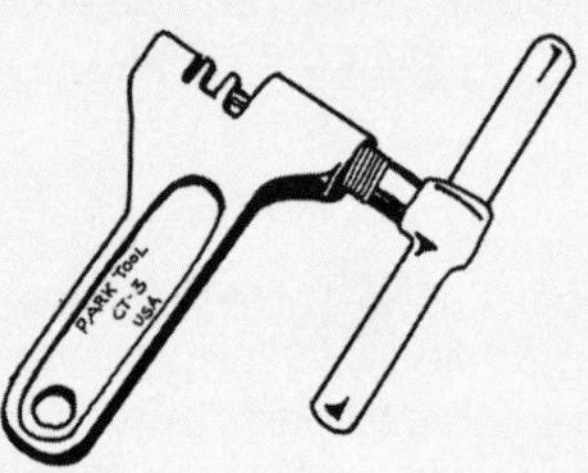

4.21 Park's CT-3 6-9-speed chain tool

older-generation chain tool (for wider chains) on a newer (narrower) chain, you laterally load the chain-locating teeth. A two-generation-older chain tool will not seat the chain connector pin, so don't use it. Ideally, it is best to get the latest tool to do the best job with the latest chains, and it will also be compatible with older (wider) chains. The Park CT-4.3 has no chain-retention teeth, instead utilizing a pocket molded into the tool body for the chain link to nest into, so it should still work without sustaining damage if chains continue to get narrower than 12-speed.

I think that if you are careful, you need only one good tool that is at most one generation back (i.e., it is meant for at least 11-speed chains), and you can use it on any chain up through 12 speeds. By "careful," I mean that you must make sure that the connecting pin and the holes are all lined up perfectly (which the Rohloff, Shimano, Park CT-4.3, and Pedro's Pro tools definitely help guarantee). I also mean that you must make sure that you stop at the right point and do not go too far or not far enough. For this, you need a feel for the loose-tight-loose-tight pressure changes as you push a Shimano connecting pin into place, as well as an eye for when the pin is protruding (or recessed) the same amount on both faces of the chain.

NOTE: *If you have an 11-speed chain and an older chain tool, you may find that the prongs in the tool to hold the chain are too far from the backing plate of the tool and will get bent. Shimano's TL-CN34 (Fig. 4.17) works on all chains, including 12-speed SRAM. Many other tool brands and models also work; see the Pro Tip on chain tools.*

4-12

CONNECTING AND DISCONNECTING A MASTER LINK

a. SRAM (Sachs) PowerLink and PowerLock, Shimano Quick Link, FSA Drive Link, Lickton's SuperLink, and KMC Missing Link

These links all work the same way. The master link is made up of two symmetrical link halves, each of which has a single pin sticking out of it. There is a round keyhole in the center of each plate that tapers into a slot on the end opposite the pin.

SRAM (which purchased Sachs) licensed Lickton's SuperLink design (Fig. 4.22), and the Missing Link works the same way. The SRAM 10-, 11-, and 12-speed PowerLock links and the discontinued KMC Missing Link II are not supposed to be openable (SRAM PowerLock master links with an "M" or "N" stamped on them were recalled in late 2009) but can be opened with master-link pliers.

Connecting

1. **Put the pin of each half of the link through the hole in each end of the chain.** One pin will go down and one up (Fig. 4.22).

4.22 SRAM PowerLock and PowerLink (also Sachs PowerLink), Shimano Quick Link, KMC Missing Link, FSA Drive Link, and Lickton's SuperLink master links

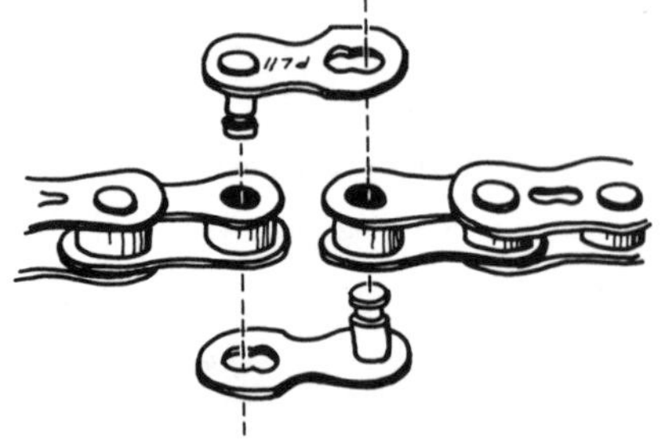

2. **Pull the links close together, and insert each pin into the keyhole in the opposite plate.**
3. **Tug on the chain.** This slides the groove at the top of each pin to the end of the slot in each plate.

NOTE: *Orient a SRAM 12-speed PowerLock link like the Wippermann ConneX link below, with its convex edges facing inward, toward the sprocket teeth.*

Disconnecting

1. **Squeeze the master-link plates toward each other.** This will free the plate from the grooves in the pins.
2. **Push the chain ends toward each other.** This brings the pins to the center hole in each plate. If you have a pair of master-link pliers, which makes this job far easier, use them to grab the two rollers through which each pin of the master link is inserted (Fig. 4.23). Squeeze the pliers at the same time you squeeze the link plates toward each other with your fingers. The link will come right apart. Master-link pliers are one of the slickest tools in existence; with them you can easily open SRAM 10-, 11-, and 12-speed PowerLock master links, which are not supposed to be openable.

NOTE: *Without master-link pliers, it is often hard to open an old, dirty master link. The problem may seem to be that you don't have enough hands. Try squeezing the link plates toward each other with a clothespin or a pair of Vise-Grip pliers set on very low pressure, or use a pair of Channellock pliers diagonally across from the trailing edge of one master link plate to the leading edge of the opposite one to disengage the link plates from the pin grooves while pushing the ends toward each other. In an emergency, you may have to just open the chain somewhere else, reassembling it using a second master link or with the link pin, as in 4-10. BikeTube makes a pair of tire levers that double as master-link pliers to ensure that you can always get a link open when out on the trail.*

3. **Pull the two halves of the master link apart.**

4.23 Using Park master-link pliers

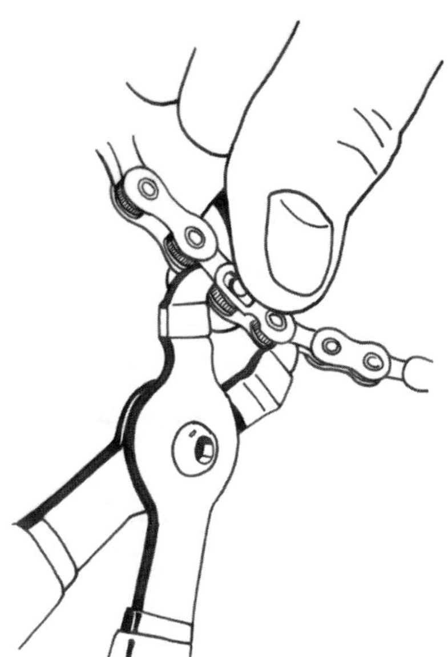

b. Wippermann ConneX link

The Wippermann link works much the same way as the SRAM PowerLink, but unlike other master links, the edges of the link plates are not symmetrical. This means that there is a definite orientation for the link, so make sure that you don't install it upside down.

Orient the ConneX master link so that its taller convex edge is away from the chainring or cog (Fig. 4.24). The link plate is bowl-shaped, and if you have it upside down so that the convex bottom of the bowl faces the cog or chainring, then when it is on a 10-, 11-, 12-, or maybe even 13-tooth cog, the convex edge will ride up on the spacer between the cogs, lifting the rollers out of the tooth valleys and causing the chain to skip under load. Another way to think about this orientation is to notice that the pair of connected holes on each plate (into which you push the pin) forms a heart shape. When the chain is on the top of the cog or chainring, make sure that the heart is right side up as in Figure 4.24.

4.24 Wippermann ConneX link (note that high bump faces away from the chainring)

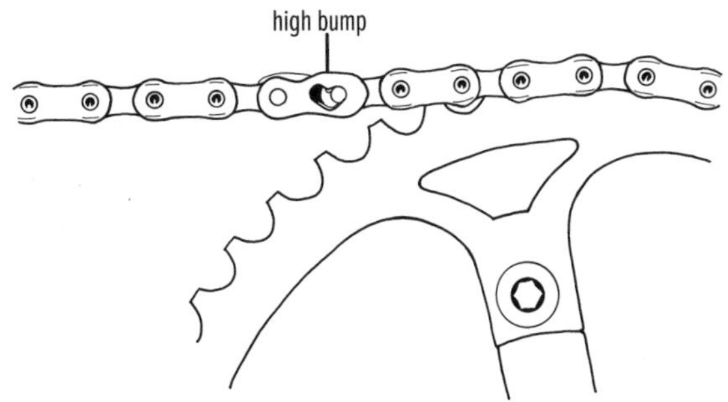

4.25 Taya "Sigma Connector" master link

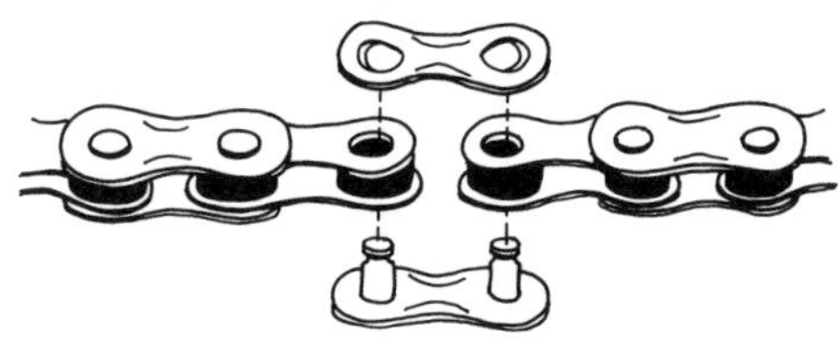

Remove and install the ConneX link the same way as the SRAM PowerLink in 4-12a, while making sure that the convex link edge is facing outward from the chain loop (Fig. 4.24) as described in the previous paragraph; you want the long concave edge to run over the spacers on the smallest cogs without lifting the chain.

c. Taya master link

Connecting

1. **Connect the chain ends with the plate that has two rivets sticking out of it** (Fig. 4.25).
2. **Snap the outer master-link plate over the rivets and into their grooves.** To facilitate hooking each keyhole-shaped hole over its corresponding rivet, flex the plate with the protruding rivets so that the ends of the rivets are closer together.

Disconnecting

1. **Flex the master link so that the pins come closer together.**
2. **Pull the plate with the oval holes off the rivets.**

4-13

INSTALLING A BELT DRIVE

The Gates Carbon Drive toothed belt (Fig. 4.26) works only on single-speed bikes and bikes with internal-gear rear hubs. The belt is a continuous loop that cannot be opened, so installing it requires a frame whose right dropout, chainstay, or seatstay will come apart. The bike must also have a means to tension the belt by adjusting the distance between the rear hub and the crank, such as the systems described in 4-8b. Like any single-speed system, the Gates belt will work only on hardtail frames or full-suspension frames in which the distance from the bottom bracket to the rear hub does not change as it absorbs bumps. The only suspension systems like this are concentric-pivot systems (where the rear swingarm pivots around the bottom bracket), unified-rear-triangle systems (where the bottom bracket and dropouts are rigidly attached to each other), and softtails (where the suspension movement works by means of flex in the chainstays).

The belts come in only certain lengths denoted by the number of teeth on the belt. You cannot adjust the belt length; you have to get the right length belt for your frame and front and rear belt sprockets. Gates has online charts for determining length at carbondrivesystems .com. Keep in mind that the number of teeth on the sprockets does not equate to standard chainring and cog sizes, because there are more belt sprocket teeth per inch than chainring teeth. As with a chain drive, the gear ratio is proportional to the ratio of front sprocket size to rear sprocket size.

The belt comes coiled in a thin box. Taking it out of the box requires some finesse because if you inadvertently put a sharp bend in the belt while trying to uncoil it, you will break the carbon fibers that give the belt its tensile strength. Pull the belt open slowly, and then give it a little flip. If this is not enough to uncoil the belt without putting a sharp bend in it, look at the carbondrivesystems.com support page for links to video and photographic sequences of the uncoiling method.

4.26 Gates Carbon Drive belt system

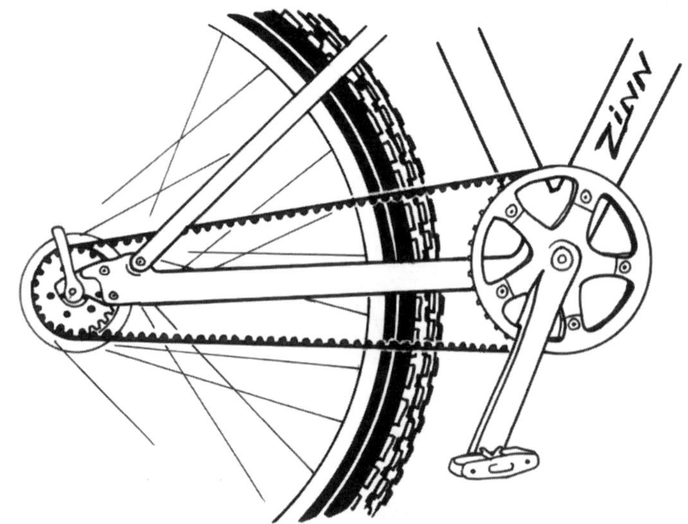

As in 4-8b, first set the distance from the wheel to the crank at its minimum by sliding the wheel or the sliding dropouts all the way forward or by rotating the eccentric bottom bracket so that the bottom-bracket spindle is as close to the back of the oversized bottom-bracket shell as possible. Slip the belt around the front and rear sprockets from the side. Tighten the belt by sliding the wheel back or by moving the bottom bracket forward, depending on the frame you have.

The best way to determine proper belt tension is to use a Carbon Drive tension gauge. Without the gauge, push down on the belt halfway between the front and rear sprockets. The belt should deflect approximately ½ inch with 5–10 pounds of force.

4.27 Chain suck

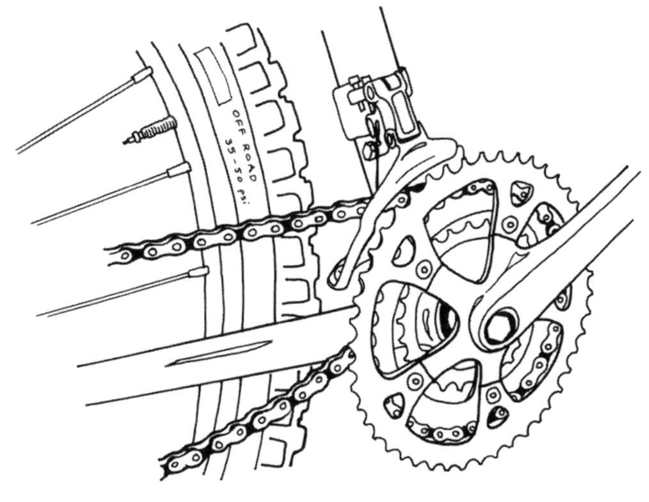

TROUBLESHOOTING CHAIN PROBLEMS

4-14

CHAIN SUCK

Chain suck occurs when the chain does not release from the bottom of the chainring and pulls up rather than running straight to the lower rear-derailleur jockey wheel. The chain will come around and get sucked up by the inner or middle chainring until it hits the chainstay (Fig. 4.27). Sometimes the chain becomes wedged between the chainstay and the chainring (if it does, free it with a screwdriver as in Fig. 3.3).

A number of things can cause chain suck. To eliminate it, try the simplest methods first and move down this list if it persists:

1. **Clean and lube the chain, and clean the chainrings.** See whether the chain now runs properly. A rusty chain will take longer to slide off the chainring than will a clean, well-lubed chain.
2. **Check for tight (or stiff) links** (Fig. 4.12). Slowly turn the crank backward, and watch the chain move through the derailleur jockey wheels. A stiff link will cause the lower jockey wheel to jump forward. Loosen stiff links by flexing them side to side with your thumbs and fingers (Fig. 4.14) or by using the back teeth on a chain tool that has them (Fig. 4.21). Set the stiff link over the back teeth closest to the screw handle (Fig. 4.13), and push the pin a fraction of a turn to spread the link.
3. **Check that there are no bent or torn teeth on the chainring.** Replace the chainring, or try straightening any broken or torn teeth you find by using pliers or by filing away rough, bent-over edges.
4. **Replace the inner chainring with a thin one that's chrome-plated or made of stainless steel.** The thin, slick teeth will release the chain more easily.
5. **Increase the rear derailleur's pivot-spring (p-spring) tension** (see 5-2g). This will increase the tension on the lower run of the chain as it comes off the bottom of the chainring.
6. **Get an anti-chain suck device that attaches under the chainstays.** Ask your bike shop about what is available.

4-15

SQUEAKING CHAIN

Squeaking is caused by dry or rusted surfaces inside the chain rubbing on each other.

1. **Wipe down the chain** (Fig. 4.2), **lubricate it** (Fig. 4.1) **with a wet lube, and go for a test ride.** Ride for a half hour or more, and then wipe and lubricate the chain again and ride another half hour or more. Do not use a wax-based lubricant; using one previously may have brought on the dry chain chirp in the first place.

2. **If the squeak is not gone, replace the chain.** If the initial remedy does not work, the chain is probably too dry and rusted deep inside. Chains often don't heal from this condition. Life is too short, and bike riding is too joyful to put up with the sound of a squeaking chain.

4-16

SKIPPING CHAIN

There can be a number of causes for a chain to skip and jump as you pedal.

a. Stiff links

1. **Turn the crank backward slowly to see if a chain link is stiff** (Fig. 4.12). A stiff link will be visible because it will be unable to bend properly while going through the rear-derailleur jockey wheels. The link will deflect the jockey wheels when passing through.
2. **Loosen stiff links by flexing them from side to side between the index finger and thumb of both hands** (Fig. 4.14) **or by using the second set of teeth on a chain tool that has back teeth** (Fig. 4.21). Set the stiff link over the teeth closest to the screw handle, and push the pin a fraction of a turn to spread the link (Fig. 4.13).
3. **Wipe down and lubricate the chain.**

b. Rusted chain

A rusted chain will often squeak as well as skip. If you watch it move through the rear derailleur, many links will appear tight; they will not bend easily and will cause the jockey wheels to jump back and forth.

1. **Lubricate the chain with wet lube** (Fig. 4.1). You want a lube that penetrates, not a dry lube. Ride a few miles, and see if the chain action improves.
2. **If the chain doesn't improve, replace it.**

c. Worn-out chain

If the chain is worn out, it will be elongated and will skip because it does not mesh well with the cogs. A new chain will fix the problem unless the worn chain was used long enough to ruin some cogs.

1. **Check for chain elongation as described in 4-6.**
2. **If the chain is elongated beyond the specifications discussed in 4-6, replace it.**
3. **If replacing the chain does not help or makes matters worse, see the next section.**

d. Worn cogs

If you just replaced the chain and it is now skipping (despite the rear derailleur being in adjustment; see 5-2), at least one of the cogs is probably worn out. If this is the case, the chain will probably skip on the cogs you use most frequently and not on others. However, if the chain skips only on the smallest cog or two and you have a Wippermann chain or a 12-speed SRAM chain, check that you have not installed the master link upside down (see 4-12b).

1. **Check each cog visually for wear.** If its teeth are hook-shaped, the cog is shot and should be replaced. Rohloff makes a simple HG-IG-Check tool (pictured in Fig. 1.4) that checks for cog wear by putting tension on a length of chain wrapped around the cog. If the last chain roller on the tool hooks on the tooth and resists your attempt to flip it in and out of the tooth pocket while the tool handle is under pressure, or, worse, if the entire measurement chain except the first roller slides easily away from the cog teeth while the handle is under pressure, the cog is worn out. Note that this tool has limited applicability to mountain bikes because it only works on cogs smaller than 21 teeth.
2. **Replace the worn cogs (or the entire cogset or freewheel).** See "Cog Change" in 8-11 and 8-13.
3. **Replace the chain.** An old chain will wear out new cogs rapidly.

e. Misadjusted rear derailleur

If the rear derailleur is misadjusted or bent, it can cause the chain to skip by lining up the chain between gears.

1. **Shift back and forth and pedal backward.** The rear derailleur should shift equally well in both directions. The chain should pedal backward without catching.
2. **Adjust the rear derailleur.** Follow the procedure described in 5-2.

f. Sticky shift cable

If the shift cable does not move freely enough to let the derailleur's spring return the chain to be lined up under the cog, it will jump off under load. Frayed, rough, rusted, or worn cables or housings will cause the problem, as will overly thick cables or kinked or sharply bent housings. Replacing the shift cables and housings (5-2) should eliminate the problem.

g. Loose rear-derailleur jockey wheel(s)

A loose jockey wheel on the rear derailleur can cause the chain to skip by letting it move too far laterally.

1. **Check that the bolts holding the jockey wheel to the cage are tight.** Use the appropriate hex key (usually 3mm).
2. **Tighten the jockey-wheel bolts if necessary.** Hold the hex key close to its bend so that you don't have enough leverage to overtighten the bolts. If the jockey-wheel bolts loosen regularly, put Loctite locking compound on their threads.

h. Bent rear derailleur or rear-derailleur hanger

If the derailleur or derailleur hanger is bent, adjustments won't work. You will probably know when it happened, too. The bending occurred either when you shifted your derailleur into your spokes, when you crashed onto the derailleur, or when you pedaled a stick or a tumbleweed through the derailleur.

Unless you have a derailleur-hanger alignment tool and know how to use it (Fig. 17.4), take the bike to a shop and have it checked, and have the dropout-hanger alignment corrected. Most modern bikes have a replaceable (bolt-on) right rear dropout and derailleur hanger, which you can purchase from the manufacturer or from wheelsmfg.com and bolt on yourself.

If a straight derailleur hanger does not correct the misalignment, the rear derailleur is bent. This is generally cause for replacement of the entire derailleur (see 5-1). With some derailleurs, you can just replace the jockey-wheel cage, which is usually what is bent. If you know what you are doing and are careful, you can sometimes straighten a bent derailleur cage back with your hands. This seldom works well, but it's worth a try if your only other alternative is to replace the entire rear derailleur. Just make sure you don't bend the derailleur hanger in the process.

i. Worn derailleur pivots

If the derailleur pivots are worn, the derailleur will be loose and will move around under the cogs, causing the chain to skip. The solution is to replace the derailleur.

j. Bent rear-derailleur mounting bolt

If the mounting bolt is bent, the derailleur will not line up straight. To fix the derailleur, get a new bolt and install it following the instructions in 5-1. Be sure to observe how the spring-loaded assembly goes together during disassembly to ease reassembly.

k. Missing or worn chain rollers

A chain that passes the elongation tests mentioned in 4-6 can still skip because here and there one of the cylindrical rollers has broken and fallen off its rivet or is so worn that it is spool-shaped. If you don't happen to check that particular link with the chain-elongation gauge, you may miss broken rollers. The width of the gauge is the same as between the inner plates, so the gauge won't catch worn-out, spool-shaped rollers either, because it will ride up on the edges of the rollers and not fall down into the center of the narrower waist of the worn roller. You might never know the chain is shot without inspecting every link.

l. Inverted master link

If you have a Wippermann chain or a SRAM 12-speed chain and have the master link upside down (described in 4-12b), the taller link edge will ride up on the spacers between the smallest cogs, lift the rollers off the cog, and cause the chain to skip. Remove, invert, and reinstall the master link as described in 4-12b.

Most Americans want to be somewhere else, but when they get there, they want to go home.

—HENRY FORD

5

CABLE-SHIFT TRANSMISSIONS

FRONT AND REAR DERAILLEURS, CABLES, AND SHIFTERS

TOOLS

2mm, 3mm, 4mm, 5mm, and 6mm hex keys
Torx T25, T30 keys
Flat-head and Phillips screwdrivers, small and medium
Pliers
Indexed-housing cutter
Cable cutter
Grease
Chain lubricant
Rubbing alcohol
Torque wrench (esp. digital or beam-type)

OPTIONAL

Park Tool IR-1 or IR-1.2 Internal Cable Routing Kit

There is nothing like having derailleurs that work smoothly, predictably, and quietly under all conditions. Knowing that you can shift whenever you need to inspires confidence when riding on difficult singletrack sections of trail. It is also a lot more pleasant to ride through beautiful terrain without the grinding and clunking noises of an out-of-whack derailleur.

Improperly adjusted rear derailleurs are a common problem, which is surprising because derailleur adjustments are easy provided the equipment is clean and in good working order. With a few simple tweaks of the limit screws and the cable tension, you're on your way. Once you see how easy it is to keep derailleurs in tune, you will be able to keep yours in adjustment all the time.

THE REAR DERAILLEUR

The rear derailleur is one of the more complex parts on a bike (Fig. 5.1). It moves the chain from one rear cog to another, and it also takes up chain slack when the bike bounces or the front derailleur is shifted.

The rear derailleur bolts to a hanger on the rear dropout (Fig. 5.2). Two jockey wheels (pulley wheels) hold the chain tight and help guide the chain as the derailleur shifts. Depending on the model, a rear derailleur has one or two springs in the lower, or both upper and lower, knuckles that pull the jockey wheels tightly against the chain, creating a desirable amount of chain tension.

Except on Shimano's reverse-action Low Normal or (older) Rapid Rise derailleurs, increasing the tension on the rear-derailleur cable moves the derailleur inward toward the larger cogs. When the cable tension is released, a return spring between the derailleur's two parallelogram plates pulls the chain back toward the smallest cogs. By connecting the return spring to the other pair of corners of the rear derailleur's parallelogram linkage, Low Normal and Rapid Rise derailleurs work in exactly the opposite fashion. In fact, Low Normal refers to exactly this phenomenon—that is, the derailleur's "normal" position, when the cable tension is removed, is in the low-gear position (large cog) rather than in the high-gear position found in traditional rear derailleurs. (Just to clarify some Shimano terminology, Top Normal refers to a standard rear derailleur that moves to the smallest cog [top gear] when there is no cable tension. As I said,

5.1 Rear derailleur, exploded view

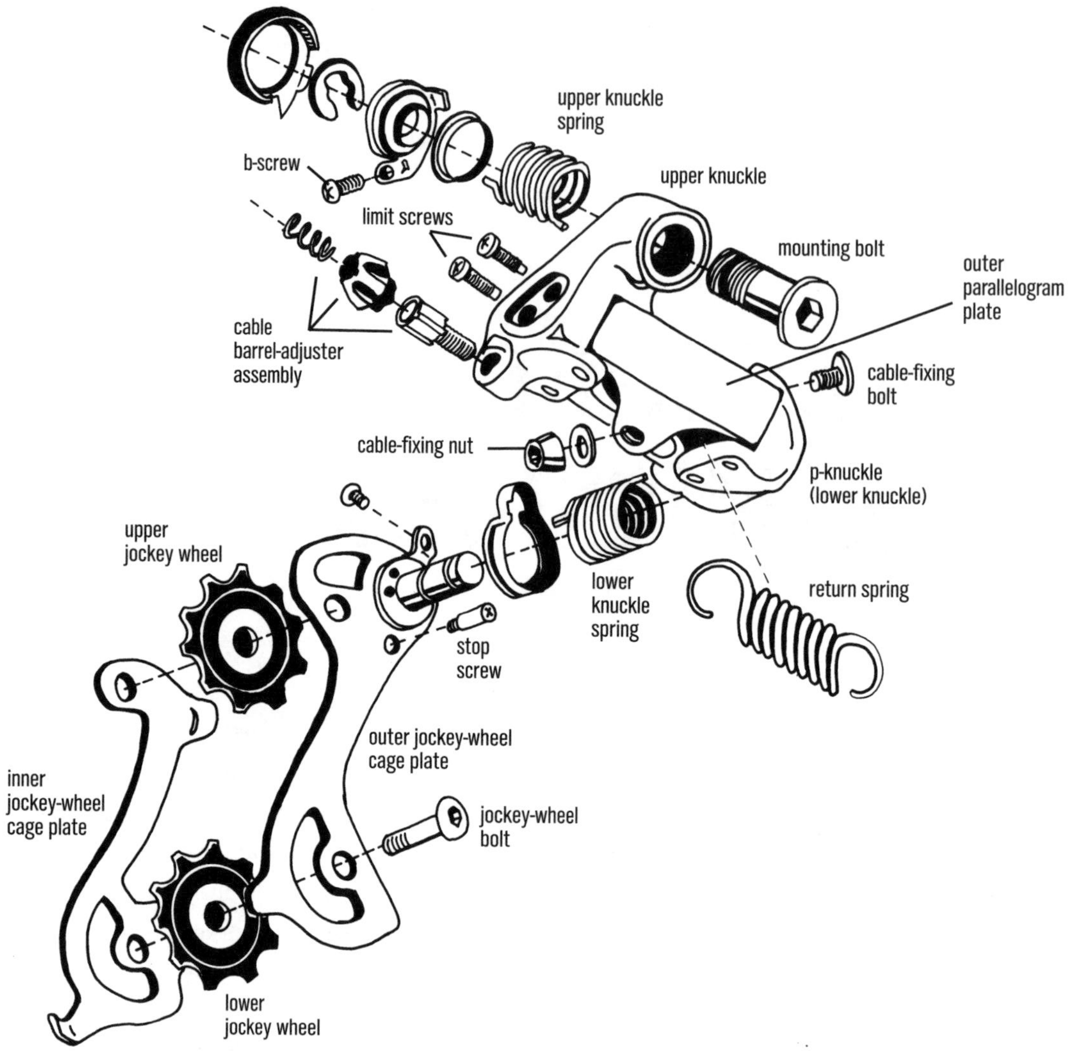

upper knuckle spring
upper knuckle
b-screw
limit screws
mounting bolt
outer parallelogram plate
cable barrel-adjuster assembly
cable-fixing bolt
cable-fixing nut
p-knuckle (lower knuckle)
upper jockey wheel
lower knuckle spring
return spring
stop screw
outer jockey-wheel cage plate
inner jockey-wheel cage plate
jockey-wheel bolt
lower jockey wheel

Low Normal refers to a rear derailleur that moves to the largest cog [low gear] when there is no cable tension. And Shadow refers to a Shimano low-profile rear derailleur with a thin body and a very short mounting bolt to keep the derailleur close to the bike and away from rocks and branches. A thin, sacrificial b-link bolts to the dropout and houses the short mounting bolt. The b-link [Fig. 5.4] sets the derailleur back in proper position, but it also is designed to snap off and save the derailleur in the event of a hard impact.)

The chain length, the balance between the springs in the upper and lower knuckle pivots, and the b-screw (Fig. 5.2) adjustment determine how closely the derailleur tracks the cogs during its lateral movement and how well it keeps the chain from bouncing off the front

5.2 Right rear dropout details

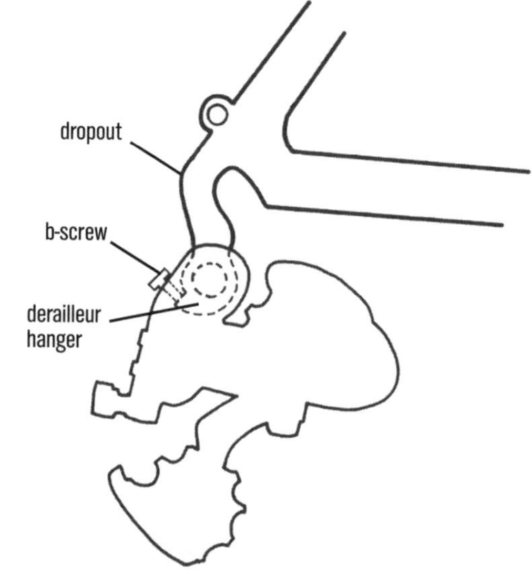

5.3 Rear-derailleur adjustments

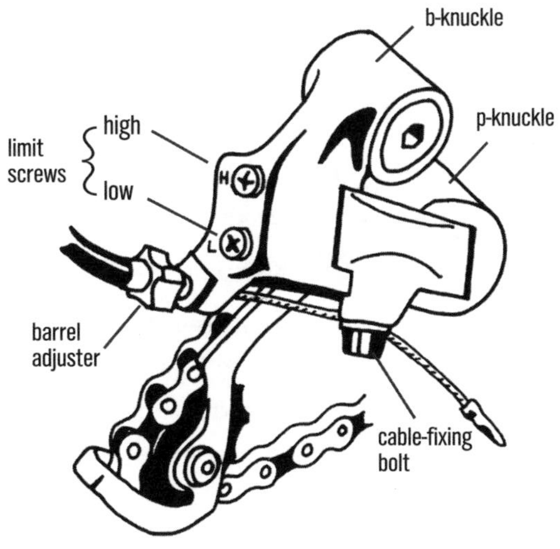

5.4 Shimano Shadow rear derailleur with sacrificial b-link offsetting it to the rear

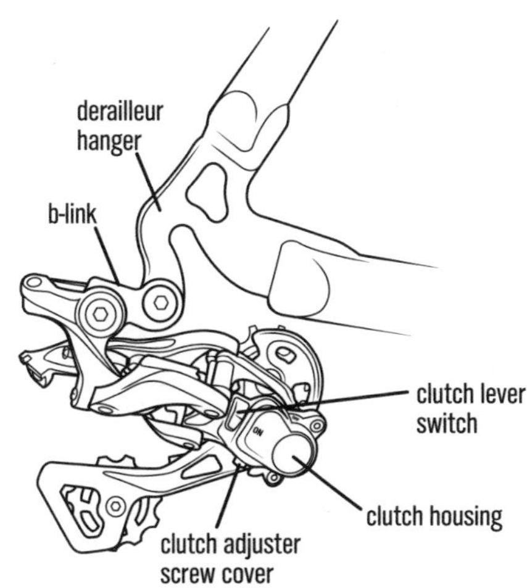

chainrings when the bike hits bumps. The two limit screws on the rear derailleur (Fig. 5.3) prevent the derailleur from moving the chain too far to the inside (into the spokes) or to the outside (into the dropout). In addition to limit screws, most rear derailleurs have a cable-tensioning barrel adjuster located at the back of the derailleur, where the cable enters it (Fig. 5.3). This barrel adjuster can be used to fine-tune the shifting adjustment to land the chain precisely on each cog with each click of the shifter. Rear derailleurs also often have a tensioning screw (the b-screw) at the back of the derailleur (Fig. 5.2) that rotates it around its mounting point to control the space between the bottom of the cogs and the upper jockey wheel.

5-1

INSTALLING THE REAR DERAILLEUR

a. For all derailleurs except axle-mounted Shimano Saint (see 5-1b for those)

LEVEL 1

1. **Grease the derailleur's mounting bolt threads.** If you have a derailleur with a clutch lever switch (Figs. 5.4, 5.5) at the lower knuckle, make sure it is in the off position.
2. **Line up the derailleur for mounting.** Rotate the derailleur back (clockwise) so that the b-screw (shown in Fig. 5.1) on the upper knuckle ends up behind the flat on the back of the derailleur hanger on the right rear dropout (Fig. 5.2). Some inexpensive derailleurs do not have a b-screw; instead, they just have a nonadjustable tab extending inward where the b-screw would be. Make sure this tab is behind the flat on the derailleur hanger; this also applies to Shimano low-profile Shadow rear derailleurs (Fig. 5.4) whose b-link (Fig. 5.5 inset) has such a tab on it.
3. **Tighten the mounting bolt until the derailleur is snug against the hanger.** Torque is 8–10 N-m.
4. **Route the chain through the jockey wheels and connect it.** Make sure that it is the correct length (see 4-8 to 4-11).
5. **Install the cables and housings** (see 5-6 to 5-18).
6. **Pull the cable tight with a pair of pliers, and tighten the cable-fixing bolt.** This will require either a 5mm hex key, a Torx T25, or an 8mm box wrench.
7. **Follow the adjustment procedure described in the next section.**

b. Axle-mounted Shimano Saint derailleurs

Original Shimano Saint rear derailleurs (short-cage and super stiff—for rough, gravity-riding applications) do not mount to the dropout derailleur hanger. Rather, they mount to the axle of the rear hub. Current Saint derailleurs share a Shadow design (Fig. 5.4) and a removable b-link (Fig. 5.5 inset) with many Shimano derailleurs, though with a circlip securing the bracket axle, and will bolt directly to a direct mount (Fig. 5.5) once the b-link is removed.

5.5 Installing Shimano rear derailleur on direct mount

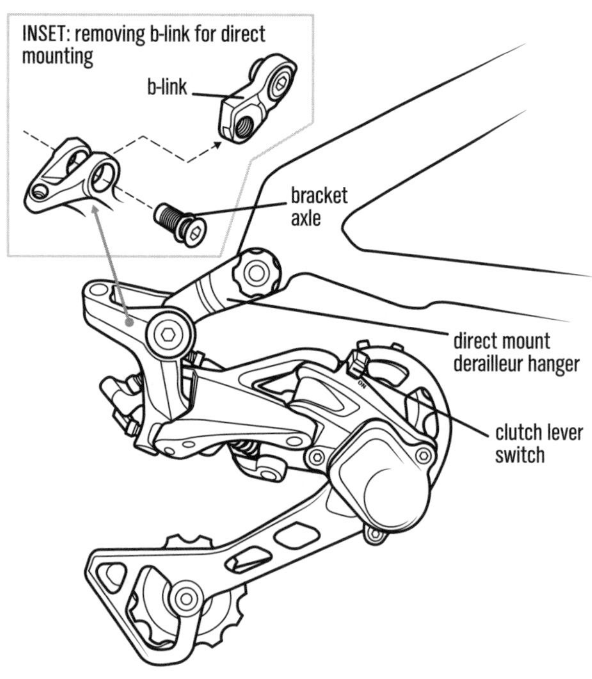

1. **Install the rear wheel into the dropouts and push the axle through it from the nondrive side.** Make sure the axle is seated fully into the dropouts.
2. **Grease the threaded, drive-side end of the axle.**
3. **Install the derailleur onto the axle.** There is a hub-axle fixing nut built into the derailleur's upper knuckle. Make sure that the derailleur's end stopper is in the dropout slot adjacent to the hub axle and that the b-screw is behind the dropout-hanger tab (same as shown in Fig. 5.2).
4. **Route the chain through the jockey wheels and connect it.** Make sure that it is the correct length (see 4-8 to 4-11).
5. **Install the cables and housings** (see 5-6 to 5-18).
6. **Pull the cable tight with a pair of pliers, and tighten the cable-fixing bolt.** This will require either a 5mm hex key, a Torx T25, or an 8mm box wrench.
7. **Follow the adjustment procedure described in the next section.**
8. **With a 2mm hex key, turn the bump stopper screw on the front of the upper knuckle (it is on the opposite side of the knuckle from the b-screw) to stop the derailleur so that the lower pivot cannot hit the chainstay.** Pull up on the lower jockey wheel to swing the derailleur up as high as it will go, and turn the bump stopper screw to stop it when the lower derailleur-pivot body is 5–10mm below the chainstay.

5-2

ADJUSTING THE REAR DERAILLEUR AND RIGHT-HAND SHIFTER

Perform all of the following derailleur adjustments with the bike held in a stand or hung from the ceiling. That way, you can turn the crank and shift gears while you put the derailleur through its paces. After adjusting the derailleur off the ground, test the shifting while riding. Derailleurs often perform differently under load than in a bike stand.

Before starting, lubricate or replace the chain (see Chapter 4) so that the whole drivetrain runs smoothly. To remove and install the rear wheel, follow the instructions in 2-13 and 2-14, paying particular attention if the derailleur has a chain-tension clutch.

a. Limit-screw adjustments

The first and most important rear-derailleur adjustment is setting the limit screws. Properly set, these screws (Fig. 5.3) should make certain that you will not ruin the frame, rear wheel, or derailleur by shifting into the spokes or by jamming the chain between the dropout and the smallest cog. It is never pleasant to see your expensive equipment turned into shredded metal. All it takes to turn these limit screws is a small screwdriver. Remember, it's "lefty loosey, righty tighty" for turning these screws.

b. High-gear limit-screw adjustment

This screw limits the outward movement of the rear derailleur. You will tighten or loosen this screw until the derailleur shifts the chain to the smallest cog quickly but does not overshift.

How do you determine which limit screw works on the high gear? Often, it will be labeled H, and it is usually the upper of the two screws (Fig. 5.3). If you are not certain, just try both screws. Whichever screw, when tightened, moves the derailleur inward when the chain is on the smallest cog is the one you are looking for.

On most derailleurs, you can also see which screw to adjust by looking in between the derailleur's parallelogram side plates. You will see one tab on the back end of each plate. Each tab is designed to hit a limit screw at one end of the movement. Shift into the highest gear and notice which screw is touching one of the tabs; that is the high-gear limit screw. The procedure for the limit-screw adjustment is as follows:

1. **Shift the chain to the large front chainring.**
2. **Shift to the smallest rear cog** (Fig. 5.6). Do this gently in case it overshifts into the dropout.
3. **If there is hesitation dropping to the small cog, adjust the cable tension.** Don't touch the limit screws yet. Notice that the upper jockey wheel in Figure 5.6 is not lined up directly under the smallest cog; either the cable tension, the high-gear limit screw, or both, can stop it from moving outboard enough to be directly under the cog.
   (a) With a traditional rear derailleur, loosen the cable a little to see if it is stopping the derailleur from moving out far enough. Do this by turning the barrel adjuster on the derailleur or shift lever clockwise (when viewed from the end of the barrel adjuster, as if it were a screw viewed from the top) or by loosening the cable-fixing bolt on the derailleur, creating some slack in the cable, and retightening the bolt.
   (b) With a Low Normal or Rapid Rise rear derailleur, you are pulling cable (rather than releasing cable), and because you can keep pulling more, if the chain won't drop to the small cog, the limit screw is stopping it, so go to step 4.
4. **If the chain still won't drop without hesitation to the smallest cog, loosen the high-gear limit screw.** Loosen it one-quarter turn at a time, continuously repeating the shift, until the chain reliably drops quickly and easily.
5. **If the derailleur goes past the smallest cog, tighten the high-gear limit screw.** Tighten it one-quarter turn and redo the shift. Repeat until the derailleur shifts the chain quickly and easily into the highest gear without throwing the chain into the dropout.

5.6 High gear

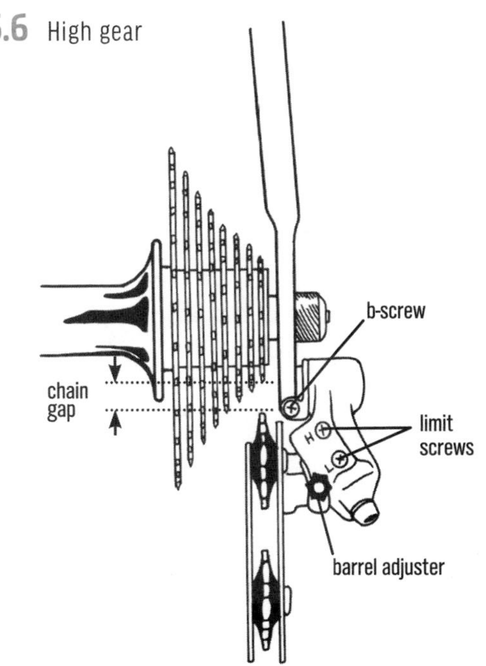

5.7 Low gear

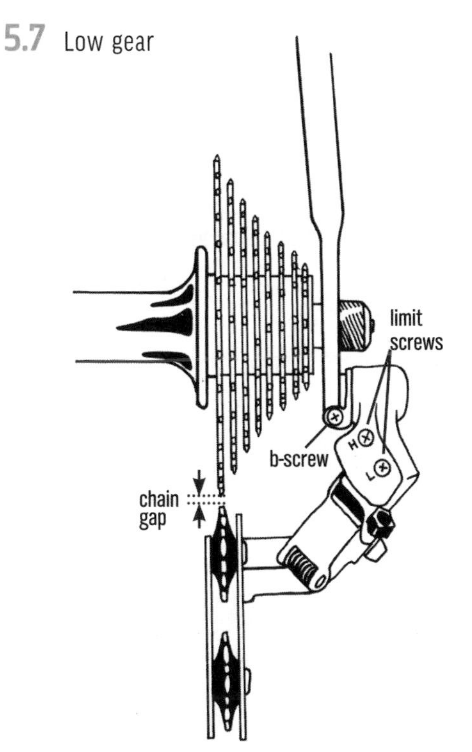

c. Low-gear limit-screw adjustment

This screw stops the inward movement of the rear derailleur, preventing it from going into the spokes. This screw is usually labeled L, and it is usually the bottom screw (Fig. 5.3). You can check which one it is by shifting to the largest cog, maintaining pressure on the shifter, and turning the screw to see if it changes the position of the derailleur.

1. **Shift the chain to the inner chainring on the front.**
2. **Shift the rear derailleur to the largest cog** (Fig. 5.7). Do it gently in case the limit screw does not stop the derailleur from going into the spokes.
3. **Tighten the low-gear limit screw if the derailleur touches the spokes or shoves the chain over the largest cog.**
4. **Loosen the low-gear limit screw if the derailleur cannot push the chain onto the largest cog.** Loosen it one-quarter turn until the chain shifts easily up to the cog but does not brush the spokes. On a Low Normal or Rapid Rise rear derailleur, you must first check that the cable tension is not so high that it prevents the chain from getting to the large cog. Loosen the cable by turning the barrel adjuster on the derailleur or shift lever clockwise (when viewed from the end of the barrel adjuster, as if it were a screw viewed from the top), or by loosening the cable-fixing bolt, creating some slack in the cable, and retightening the bolt.

d. Cable-tension adjustment: Indexed rear shifters

With an indexed shifting system (one that "clicks" into each gear), the cable tension determines whether the derailleur moves to the proper gear with each click.

1. **With the chain on the large chainring in the front, shift the rear derailleur to the smallest rear cog.** Keep clicking the shifter until you are sure it will not let out any more cable (or it will not pull any more cable if you have a Low Normal or Rapid Rise rear derailleur).
2. **Shift one click in the other direction.** This should move the chain smoothly to the second cog.
3. **Adjust the cable tension if the chain climbs slowly or not at all to the second cog.** On a traditional rear derailleur, increase the tension in the cable by turning either the derailleur-cable barrel adjuster (Fig. 5.3) or the shifter barrel adjuster (Fig. 5.8) counterclockwise (when viewed from the end of the barrel adjuster, as if it were a screw viewed from the top). Turn the opposite way for a Low Normal or Rapid Rise derailleur. Note that recent SRAM rear derailleurs have no barrel adjuster; make all cable-tension adjustments at the shifter.

If you run out of barrel-adjustment range and the cable is still not tight enough, retighten the shifter and derailleur barrel adjusters clockwise, to one turn from where they stop, loosen the cable-fixing bolt, and pull some of the slack out of the cable. Tighten the cable-fixing bolt and repeat the adjustment. If the derailleur has no barrel adjuster, use the barrel adjuster on the shifter.

5.8 Shifter barrel adjuster for cable tension

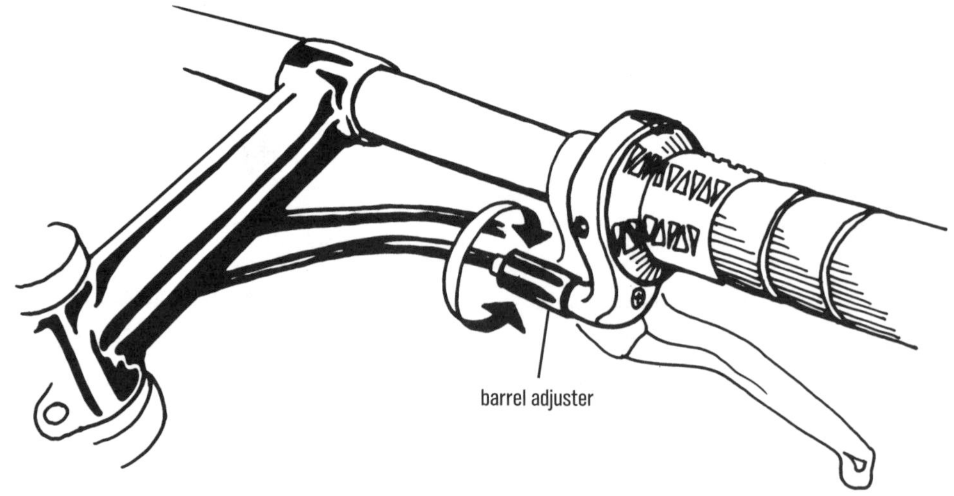

4. **Adjust the cable tension if the chain overshifts the second cog.** On standard derailleurs, decrease the cable tension by turning one of the barrel adjusters clockwise. With a Low Normal or Rapid Rise derailleur, increase the tension by turning one of the barrel adjusters counterclockwise. Again, always determine clockwise and counterclockwise from the position of the end of the barrel adjuster where the cable housing enters.
5. **With the chain on the middle chainring in the front (or on either chainring of a double crank), shift the rear derailleur to one of the middle rear cogs.**
6. **Shift the rear derailleur back and forth a few cogs.** Check for precise and quick movement of the chain from cog to cog. Fine-tune the shifting by making small adjustments to the cable-tensioning barrel adjuster on the shifter or rear derailleur.
7. **With the chain on the large chainring in the front, shift the rear derailleur to the largest rear cog.** Shift up and down one click in the rear, again checking for symmetry and precision of chain movement in either direction between the two largest cogs. Fine-tune the barrel adjuster until you get it just right.
8. **Go back through the gears.** On a bike with a triple crank and the chain in the middle chainring in front, the rear derailleur should shift smoothly back and forth across all the cogs. With the chain on the big chainring of either a triple or a double crank, the rear derailleur should shift easily on all but perhaps the largest one or two cogs in the rear. With the chain on the inner chainring, the rear derailleur should shift easily on all but perhaps the two smallest cogs.
9. **Skip to step f.**

NOTE: *If the shifter barrel adjuster does not hold its adjustment, derailleur performance will steadily worsen as you ride. This problem has occurred in early XTR shifters, which have no springs or notches to hold the adjuster in place, as do most barrel adjusters. If you have this problem, there are a couple of things you can do besides getting a new shifter. One is to put Finish Line Anti Seize Assembly Lube or a similar anti-seize compound on the threads to create a bit more friction; I have found this to be a temporary fix only. Applying Loctite or scoring the threads crosswise may also help, although I have not tried either method. One foolproof solution, though a bit of a hassle, is to keep the shifter barrel adjuster turned all the way in and make all cable-tension adjustments with the barrel adjuster on the rear derailleur.*

ANOTHER NOTE: *If the cable tension is okay in the mid-sized cogs, too high in the large cogs, and too low in the small cogs (or the opposite with a Low Normal or Rapid Rise rear derailleur), then the rear derailleur is moving more than one cog spacing with each click of the shifter. The problem could be that you are using an 8-speed shifter with a 9-speed cogset or another mismatched combination.*

If the shifter and cogs are both for the same number of gears, the problem could be that some of the spacers between the cogs are too thin, making the entire stack of cogs narrower than it should be. This is a rare, albeit not unprecedented, situation when you are using components from a single manufacturer that have all been designed to work together, but it is not uncommon when you are mixing components from different manufacturers.

You can remedy this spacing problem without buying a new cogset, however. Remove the cassette lockring (see 8-11), and pull apart those cogs that are separable from each other. Trace around one of the spacers on a piece of aluminum you have cut from a beer or pop can. Follow the tracing with a box-cutter knife or similar implement to cut out a spacer shim. Fine-tune its shape until the shim will slip over the freehub body. Reassemble the cogset onto the freehub, placing the shim between a spacer and a cog where it seems that the shifting starts to get thrown off. If this improves shifting but does not completely fix the problem, try adding another beer-can shim. A beer-can shim is approximately 0.1mm thick, and I have seen 0.3mm variation between the total stack height of a SRAM and a Shimano 11–34 9-speed cogset, for example. Experiment with various positions for the shim(s) within the cogset until you have optimized shifting performance.

5.9 Chain-gap adjustment of an early SRAM ESP derailleur

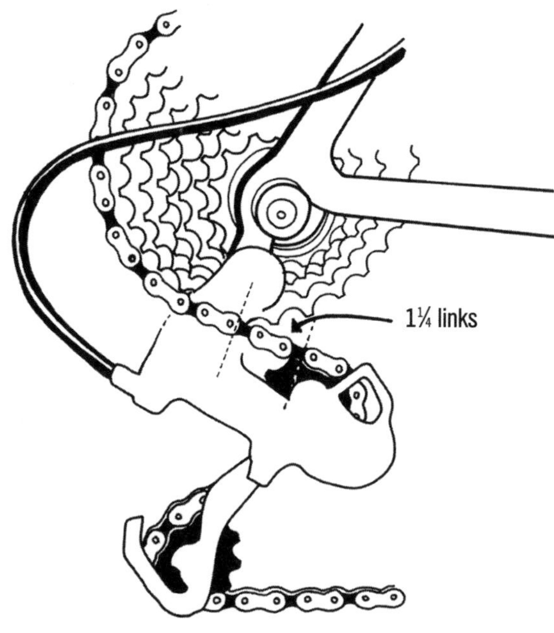

e. Cable-tension adjustment: Nonindexed rear shifters

If your bike does not have indexed shifting, adjustment is complete after you remove the slack in the cable. With proper cable tension, when the chain is on the smallest cog, the derailleur should move as soon as the shift lever does. If there is free play in the lever, tighten the cable by turning the cable barrel adjuster on the derailleur or shifter counterclockwise. If the rear derailleur and shifter don't have barrel adjusters, loosen the cable-fixing bolt, pull some slack out of the cable with pliers, and retighten the clamp bolt.

In most cases, you can stop after adjusting the limit screws and cable tension, but there is more you can do if you are a stickler for optimum performance. If you are still having shifting trouble, proceed to at least step f, and maybe to step g and its following note as well. And if you have a problem with the chain bouncing off, attend to step g.

f. Chain gap: The b-screw adjustment

You can get a bit more shifting precision by adjusting the small screw (b-screw—see Fig. 5.2) that changes the derailleur's position against the derailleur-hanger tab on the right rear dropout. Viewing from behind with the chain on the inner chainring and largest cog (Fig. 5.7), adjust the screw so that the upper jockey wheel is close to the cog but not pinching the chain against the cog. For most derailleurs, the chain gap between the top of the jockey wheel and the bottom of the cog that the chain is on should be about 6mm. Repeat with the chain on the smallest cog (Fig. 5.6). You'll know that you've moved the jockey wheel in too closely when it starts making noise or bumps up and down against each successive tooth.

SRAM suggests setting the b-screw on its early ESP derailleurs with the chain on the middle chainring and largest cog. Viewing from the drive side, turn the screw so that the length of chain across the chain gap (from where the chain leaves the bottom of the cog to its first contact at the top of the upper jockey wheel) is 1 to 1¼ links (Fig. 5.9), where one link is a complete male-female link pair (Fig. 4.7). Generally, SRAM derailleurs for bikes with two or more front chainrings from 2000 on specify a 6mm vertical distance from the top of the upper jockey wheel to the bottom of the large cog (the chain gap depicted in Fig. 5.7), when in the small chainring–large cog combination.

Shimano rear derailleurs generally specify a chain gap of 5–6mm if the maximum rear cog size is 42 teeth. If the largest cog is 46 teeth, the chain gap should be set at 8–9mm.

NOTE: *On Shimano Shadow low-profile rear derailleurs (Fig. 5.4), the b-screw pushes against the b-link (Fig. 5.5 inset) attached to the derailleur's upper pivot, which bolts to the dropout derailleur hanger, rather than against the derailleur hanger itself.*

SRAM X-Horizon rear derailleurs for single-chainring drivetrains (1×7, 1×11, or 1×12) are specified with a chain gap of 15mm. The upper knuckle doesn't pivot, so the lower knuckle on these derailleurs holds a fixed distance from the chainstay, rather than being able to swing forward and back about the upper knuckle. The offset of the upper jockey wheel from the lower knuckle's axis is larger than normal and is intended to maintain a constant chain gap relative to every cog (Fig. 5.10), so you can adjust the chain gap on any cog.

5.10 Fixed position of lower knuckle and consistent chain gap throughout gear range on SRAM X-Horizon derailleur

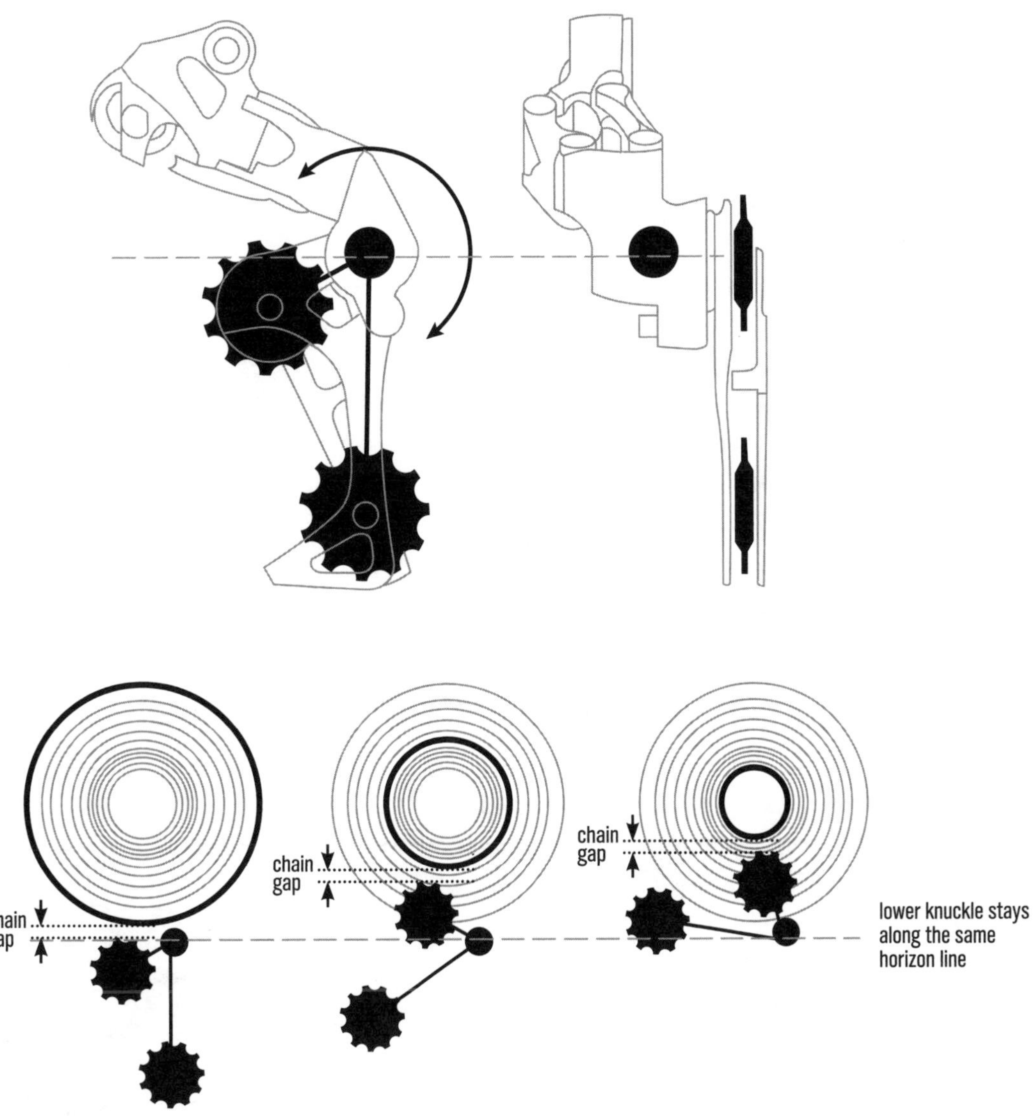

g. Lower-knuckle pivot-spring tension adjustment

Modern mountain-bike derailleurs generally have a clutch at the lower pivot to minimize bouncing of the chain on rough trails, rather than depending on the tension in the lower-knuckle spring. It makes the jockey-wheel cage hard to rotate, which is why you need to turn off the clutch (Fig. 5.4) or lock it out (Fig. 2.11) when removing and installing the rear wheel (2-13, 2-14). To adjust a Shimano clutch, see "h" below; do this rather than adjusting the spring tension.

Tension in the lower pivot spring controls chain bounce in older derailleurs; the lower pivot spring twists the derailleur forward and puts pressure on the chain through the jockey wheels. By increasing the lower-knuckle p-spring tension, you can bring the upper jockey wheel closer to the cogs and increase tension in the lower run of the chain. This procedure will reduce the chain bouncing over rough terrain. It does put more drag (i.e., friction) on the chain, but it can be well worth doing for gravity-driven riding.

This is a complex adjustment requiring disassembly of the derailleur pivot, so be sure it is justified. Also, if you were thinking about replacing the steel pivot bolt with a lightweight aluminum one, now is the time to do it.

On post-1997 Shimano XT, LX, and STX rear derailleurs, and a year or two later with other models, there is a setscrew on the side of the lower pivot (Fig. 5.11) that makes it possible to disassemble the pivot without disconnecting the derailleur from the cable and the chain. Nevertheless, I recommend removing the derailleur first. You would have to unscrew the mounting bolt anyway, and the derailleur will get so twisted around that it will be hard to tell which way is up with the cable and chain connected. You could end up turning the jockey cage in the wrong direction and deforming the spring so that it would not fit back in the knuckle.

The setscrew on the lower pivot engages a groove in the pivot shaft to keep it from pulling apart. Remove the screw with a 2mm hex key (Fig. 5.11), and pull the jockey cage away from the spring. Put the spring in the next spring hole to increase its tension (Fig. 5.12), push the pivot assembly back together, and replace the setscrew. Shimano derailleurs come with the p-spring in the low-tension hole.

Increasing p-spring tension on older Shimano derailleurs is more complicated because there is no setscrew on the lower-knuckle housing. After removing the derailleur from the bike, remove the tall stopscrew that prevents the jockey cage from twisting all the way around (Fig. 5.13). Remove the upper jockey wheel, and unscrew the pivot bolt from the back with a 5mm hex key (Fig. 5.14). Pull the derailleur cage off the end of the spring, and move the end of the spring into the other spring hole. Wind the jockey-wheel cage back around,

5.11 Removing the p-knuckle setscrew with a 2mm hex key

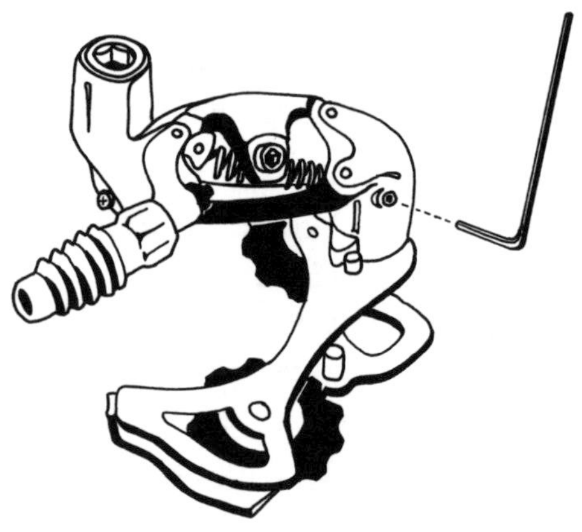

5.12 Increasing the p-spring tension by putting the spring end in a different hole

5.13 Removing the stopscrew that prevents the jockey cage from twisting all the way around

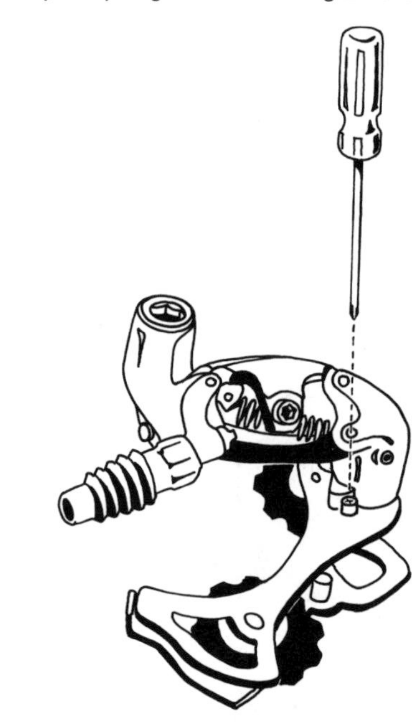

5.14 Unscrewing the pivot bolt with a 5mm hex key to pull the derailleur cage off the p-spring

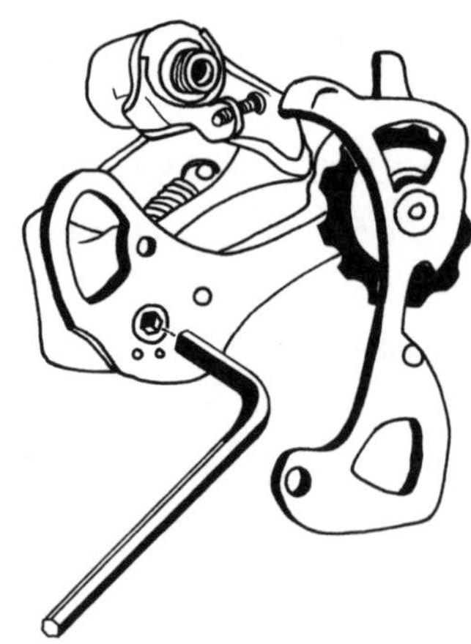

screw it all back together with the pivot bolt, and replace the stopscrew.

NOTE: *If you cannot get the rear derailleur to shift well, or if it makes noise in even mild cross-gears no matter what you do, or if it throws the chain off despite your best efforts, take a close look at the bike from the rear and make sure that neither the derailleur nor the derailleur hanger on the frame dropout is bent. If the parts are undamaged, refer to the chainline discussion in the troubleshooting section at the end of this chapter to eliminate the shifting problem.*

h. Adjust Shimano chain stabilizer (clutch) mechanism

A tighter clutch increases chain retention—you want this in mud as well as in bumpy gravity riding. On the other hand, a tighter clutch makes for a heavier downshift (to larger cogs) and harder multi-shifting. Turn the clutch off with the clutch lever switch (Figs. 5.4, 5.15) for light, quick shifting feel when riding on smooth surfaces.

1. **Turn the clutch lever switch to "on."** It is atop the clutch housing (Figs. 5.4, 5.5, 5.15).
2. **Open the cover over the adjustment screw.** It's on the bottom of the clutch housing (Fig. 5.15). Older models have a plastic cap (Fig. 5.15A); turn it counterclockwise one-eighth turn to line up the notches in the cover with the tabs on the housing, and pull it straight off. Newer models have a rubber cover you just flip open (Fig. 5.15B); its retaining leash keeps it from getting lost. If the derailleur is missing its cover, get a new one! Without the cover, the clutch will fill with water and dirt and will rapidly cease to function.
3. **Turn the adjuster screw.** Reach in where the cover was with a 2mm hex key to turn the screw; clockwise tightens the clutch. Apply very low torque on this screw; just turn it with the short end of the hex key with fingertips or with a tiny torque wrench, not exceeding 0.25 N-m of torque. The tighter the clutch, the less the chain can bounce when hitting bumps, and the harder the push it takes to shift.
4. **Check the adjustment with a torque wrench.** This is most easily done with a digital or beam-type torque wrench, but you can also use a clicker-type torque wrench by trial and error. Older models require a 5mm hex key; newer models take a Torx T30 key. The hex or Torx hole is on the inner cage plate, adjacent the upper jockey-wheel bolt (Fig. 5.58). With the Torx T30 or 5mm hex bit in the hole, turn the wrench handle clockwise (Fig. 5.16). With a digital or beam-type torque wrench, read the setting when the jockey-wheel cage starts to rotate. With a clicker-type torque wrench, set it to 4 N-m, and turn it clockwise. If the jockey-wheel cage rotates, back off the torque setting and try again. Repeat until it just doesn't turn the cage; the proper clutch torque will be the last setting on your wrench before this final try. If the cage instead

5.15A Opening plastic twist-type clutch-adjustment-screw cover

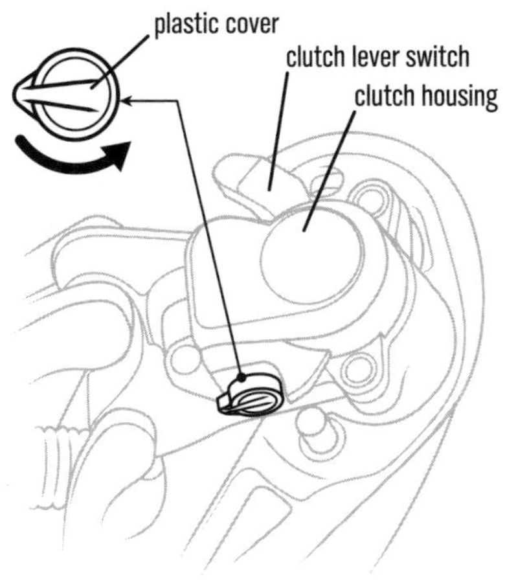

5.15B Opening leashed, rubber clutch-adjustment-screw cover

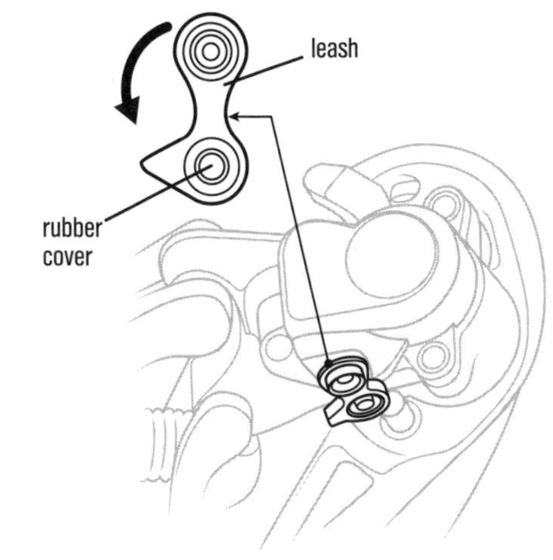

5.16 Checking clutch torque

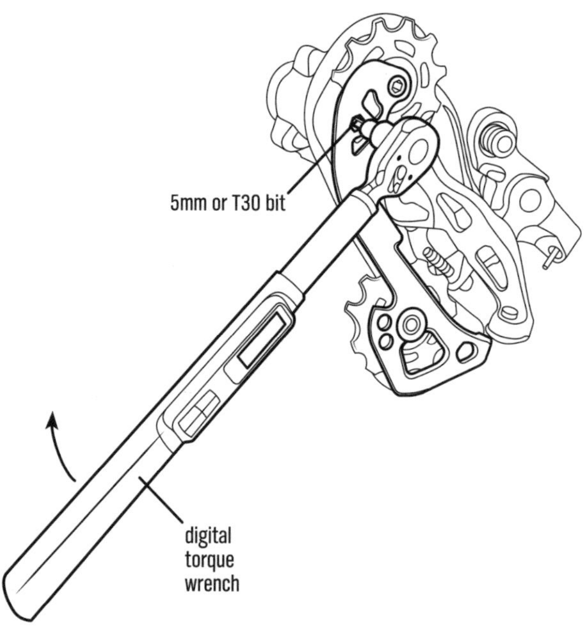

doesn't rotate at your first torque setting, increase the torque and do this trial and error in an increasing direction.

5. **Set the clutch torque.** It should be in the range of 3.5–5.4 N-m; gently turn the clutch adjuster screw with the 2mm hex key clockwise to increase clutch torque, and vice versa to decrease it. If you can't get to the clutch-torque setting you want without tightening the tiny adjustment screw tighter than 0.25 N-m, there must be grease on the clutch shaft (a no-no) or the cam is in the wrong position; overhaul the clutch (5-36) to fix it.
6. **Replace the adjustment-screw cover.** If there is a rubber leashed cover (Fig. 5.15A), flip it closed. With a plastic cover (Fig. 5.15B), line up the notches on the cover and the tabs on the housing, and turn the cover clockwise until its central rib is parallel to the face of the clutch housing.

THE FRONT DERAILLEUR

The front derailleur moves the chain between the chainrings. The working parts consist of a cage to enclose the chain, a linkage, and an arm attached to the shifter cable. The front derailleur is attached to the frame, usually by a clamp surrounding the seat tube (Fig. 5.17). Some frames do not have room for a front-derailleur band clamp, either because there are rear suspension parts in the way, the seat tube is nonstandard in shape or angle, or the chainstays attach where the front derailleur clamp would need to be. Front derailleur models for bikes like this are called direct-mount (Figs. 5.18, 5.19, 5.21) and bolt, often via an adapter, directly into threaded mounts on the side of the seat tube or the swingarm (rear-suspension member), or they attach to the face of the bottom-bracket shell (Fig. 5.20). Shimano bottom-bracket-mounted front derailleurs are called E-type, and some mount not only to the face of the bottom bracket but also to a braze-on boss (or band clamp adapter) on the seat tube (Fig. 5.20).

Standard seat-tube band clamps also vary. They need to be the correct diameter for the seat tube (or supplied with shims to make them fit smaller seat tubes),

5.17 Down-swing front derailleur

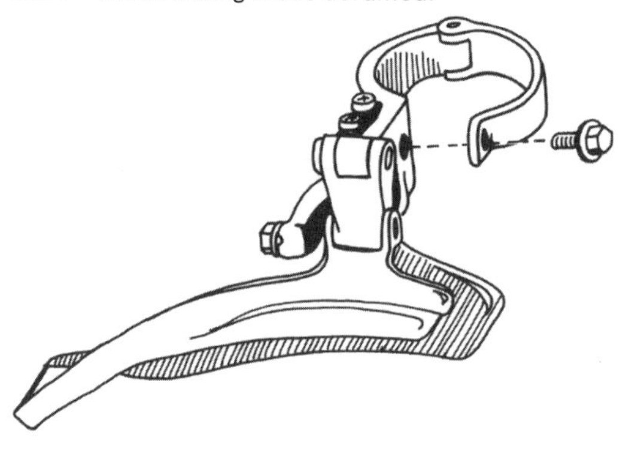

5.18 SRAM direct-mount top-pull top-swing front derailleur

5.19 Shimano direct-mount down-swing front derailleur

5.20 Shimano bottom-bracket-mount (E-type) top-swing front derailleur and band-clamp adapter (bottom left)

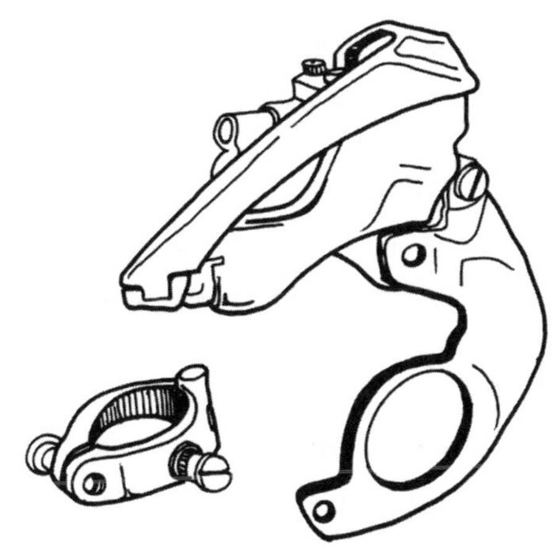

5.21 Shimano direct-mount side-swing front derailleur

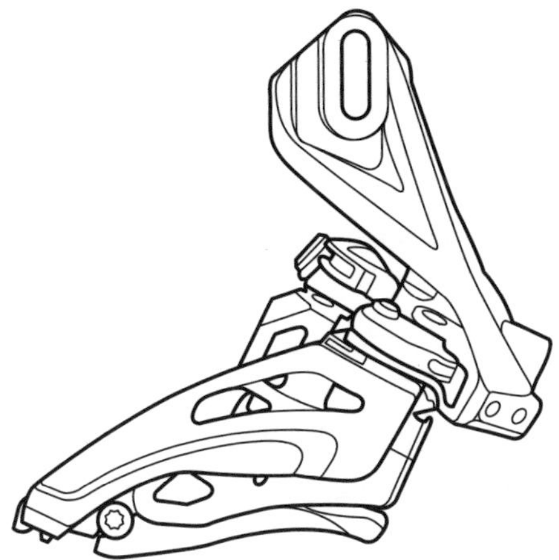

and they can also have a high clamp or a low clamp relative to the derailleur mechanism.

Front derailleurs vary in pivot geometry as well. A top-swing front derailleur has a band clamp or bolt-on plate that is lower than the height of the top of the front derailleur's cage, and the derailleur's activation linkage and pivots are behind the cage. Figure 5.20 shows a bolt-on E-type top-swing front derailleur, whereas Figures 5.44A and B depict a top-swing clamp-type derailleur from the back. A traditional or down-swing front derailleur has a band clamp or bolt-on mounting plate well above the cage, and the large outer linkage plate and cage pivot are above the cage (Figs. 5.17, 5.19, and 5.23, 5.24). A side-swing front derailleur (Fig. 5.21) has the pivots behind the cage and the cable-mounting lever arm directly above them, providing greater tire clearance by moving the cable and actuation arm farther forward, away from the tire knobs. Which style to choose depends largely on the bike's frame design, as often a pivot or shock mount on the frame, or short chainstays, will preclude the use of some styles.

The direction that the cable pulls also distinguishes front derailleurs from each other. A bottom-pull front derailleur requires the cable to come from below, generally by wrapping underneath the bottom-bracket shell. A top-pull front derailleur is activated by a cable pulling on it from above, generally after the cable runs along the top tube (in this case, the cable housing generally terminates at a cable stop on the back of the seat tube). Many modern front derailleurs have a dual cable-pull option; the cable can come from above or below—either straight to the cable-fixing bolt from above or over the top of a rocker-arm assembly, wrapping down again to attach at the bolt (see Figs. 5.44A and B).

5-3

INSTALLING THE FRONT DERAILLEUR: BAND TYPE

LEVEL 1

1. Clamp the front derailleur around the seat tube. Make sure that you have either the right-size band clamp for the seat tube or the correct shims in place under it to make it fit the seat tube.

Some front derailleurs have a size-specific band-clamp diameter matching that of the seat tube, but modern Shimano and older SRAM front derailleurs work on more than one seat-tube diameter. They have a band clamp for a 35mm (1⅜-inch) diameter seat tube, but they come with shims for either 31.8mm (1¼-inch) or 28.6mm (1⅛-inch) seat tubes. In the case of Shimano front derailleurs, C-shaped aluminum shims are held in place by C-shaped plastic brackets that clip within the circle of the front-derailleur band clamp.

2. **Adjust the height and rotation as described in 5-5a.**
3. **Tighten the clamp bolt** (Fig. 5.17).

5-4

INSTALLING THE FRONT DERAILLEUR: DIRECT MOUNT AND BOTTOM-BRACKET MOUNT

a. Direct mount

LEVEL 2

Full-suspension bikes often need special front-derailleur mounting, because a clamp-on front derailleur might interfere with the rear swingarm (the moving chainstay on a rear-suspension bike) as it moves through its travel or because the seat tube is not round. One solution is to bolt the front derailleur directly to either the seat tube or the swingarm (Figs. 5.18, 5.19, 5.21); a mounting bracket with a couple of bolt holes or a tongue-and-groove arrangement and a single bolt will be welded onto or machined into the seat tube or the front of the swingarm. In the case of a swingarm mount, the front derailleur will move forward and back over the chainring as the suspension moves.

Direct-mount front derailleurs that bolt to the frame are simple to mount (just bolt them on!), but they are generally chainring-size-specific as well as frame-specific. For instance, SRAM XX direct-mount front derailleurs are marked 39 or 42, depending on whether they are for the 26–39 or 28–42 chainring set, but they also come in S1, S2, and S3 configurations depending on the mounting system on the frame, and there is a top-pull and a bottom-pull model in each configuration as well. The derailleur for a Specialized Epic with a 28–42 crankset, for example, will be stamped S1 42 between the mounting holes on the derailleur back plate, and it will be bottom-pull.

b. Bottom-bracket mount

Shimano E-type (i.e., bottom-bracket face-mounting type) front derailleurs (Fig. 5.20) are direct-mount front derailleurs that mount to a plate that fits between the bottom-bracket cup and the bottom-bracket shell. They are more complicated to mount than simply bolting the derailleur to threaded holes in the frame. To install one, do the following:

1. **Remove the bottom bracket** (see 11-12 and 11-14).
2. **Slip the (E-type) derailleur bracket over the right-hand bottom-bracket cup.** Start the cup into the bottom-bracket shell a few threads.
3. **Attach the bracket to the seat tube.** With less expensive models, place the bracket's C-shaped stabilizer around the seat tube. This fixes the rotational adjustment. With a high-end E-type front derailleur (Fig. 5.20), loosely screw the mounting bolt into the frame's special braze-on boss designed for it. If the frame does not have the braze-on boss, a separate seat-tube band clamp with a threaded hole in the side is used (see Fig. 5.20). The circular band will need to be bent to fit if the frame has an oval-shaped seat tube.
4. **Tighten the right-hand bottom-bracket cup against the bottom-bracket face.**
5. **If applicable, tighten the mounting bolt into the braze-on boss (or band-clamp hole).**
6. **Complete the bottom-bracket installation** (see 11-7 to 11-11).

NOTE: *There are no (or limited) height and rotational adjustments on these derailleurs, and they must be used with the chainring size for which they were intended. They can be turned only slightly to line up better with the chain.*

Some E-type front derailleurs have two mounting-bolt holes to allow for two possible outer chainring sizes. The derailleur's rotational adjustment can be fine-tuned without the braze-on boss, because the band clamp can be twisted around the seat tube a few degrees.

5-5

ADJUSTING THE FRONT DERAILLEUR AND LEFT-HAND SHIFTER

a. Position adjustments

With a seat-tube-clamp front derailleur, the position is adjusted with a 5mm hex key, a Torx T25, or an 8mm box wrench on the band-clamp bolt. Direct-mount and Shimano E-type front derailleurs have little or no vertical or rotational (twist about the seat tube) adjustments.

1. **Position the height of the derailleur.** Set the outer cage about 1–2mm (1⁄16–1⁄8 inch) above the highest point of the outer chainring (Fig. 5.22). On full-suspension frames, deflate the rear shock and swing the swingarm up and down to make sure the derailleur cage will not hit the chainring. Out of the box, new Shimano front derailleurs have a piece of clear tape on the cage illustrating the proper gap, with teeth drawn onto it to line up with the chainring teeth. They also have a plastic block called a Pro-set alignment block installed that forces the derailleur out to the high-gear position. Leave that block in place for this step, and remove it for the second part of step 2 (and from then on).
2. **Position the rotation of the front derailleur about the seat tube.** The outer plate of the derailleur cage should be parallel to the chainrings (or to the chain in the lowest and highest gears) when viewed from above. Check this by shifting to the big chainring and smallest cog and sighting from the top (Fig. 5.23). Most derailleurs need the outer face of the cage to be exactly parallel to the chainring; check this by measuring the space between the cage and the inner side of the crankarm as it passes by. The cage of some derailleurs flares wider at the tail, and the cage should be parallel to the chain in the lowest and highest gears. The outer tail of the derailleur cage on these models needs to be out a bit from parallel to the plane of the frame in order to parallel the chain. Similarly, when on the inner chainring (Fig. 5.24) and largest cog, the inner-cage plate should parallel the chain, making the tail a bit in from parallel with the plane of the frame.

5.22 Proper clearance

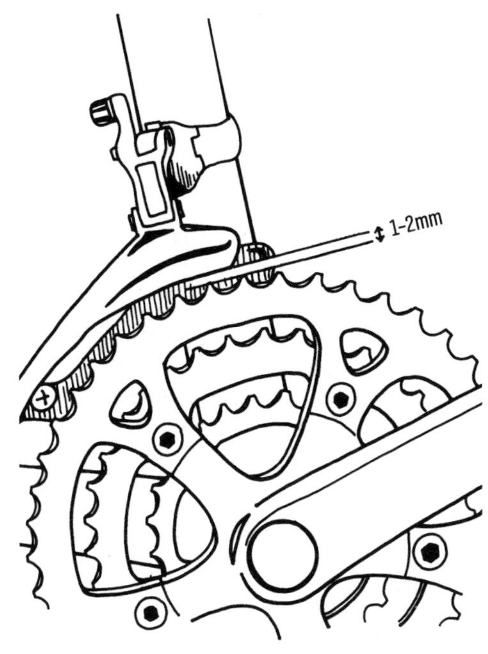

5.23–5.24 Proper cage alignment

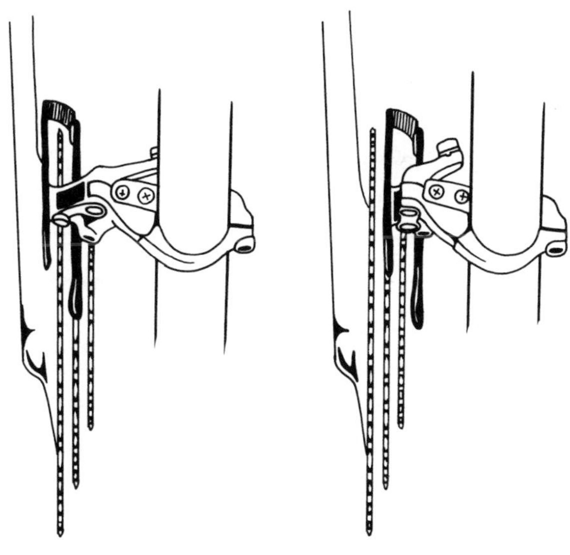

b. Limit-screw adjustments

The front derailleur has two limit screws that stop the derailleur from throwing the chain to the inside or outside of the chainrings. These are usually labeled L for low gear (small chainring) and H for high gear (large chainring) (Fig. 5.25). On most derailleurs, the low-gear screw is closer to the frame.

5.25 Limit screws

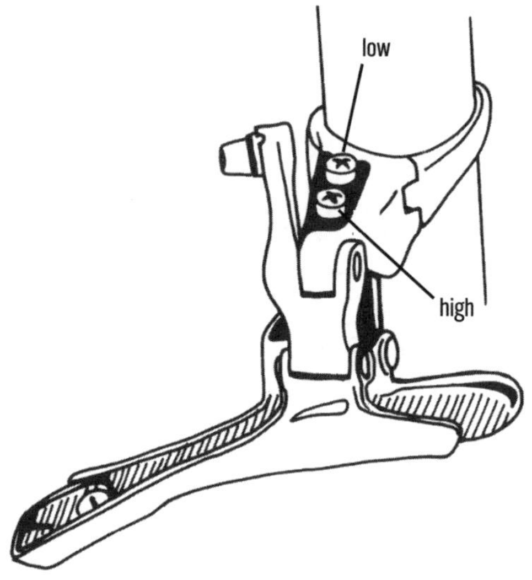

If in doubt, you can determine which limit screw controls which function by the same trial-and-error method outlined for the rear derailleur. Shift the chain to the inner ring, and then tighten one of the limit screws. If tightening that screw moves the front derailleur outward, then it is the low-gear limit screw. If turning that screw does not move the front derailleur, then the other screw is the low-gear limit screw.

c. Low-gear limit-screw adjustment

1. **Shift back and forth between the middle and inner chainrings.**
2. **Tighten the low-gear limit screw if the chain drops off the little ring to the inside.** Tighten it (clockwise) one-quarter turn and try shifting again.
3. **If the chain does not drop easily onto the inner chainring, loosen the low-gear limit screw.** Loosen it (counterclockwise) one-quarter turn and repeat the shift. Make sure that overly high cable tension is not preventing the derailleur from reaching the inner limit screw; if it is, reduce the cable tension by turning the shifter barrel adjuster clockwise.

d. High-gear limit-screw adjustment

1. **Shift the chain back and forth between the middle and outer chainrings.**
2. **Tighten the high-gear limit screw if the chain jumps over the big chainring.** Tighten it (clockwise) one-quarter turn and repeat the shift.
3. **If the chain is sluggish going up to the big chainring or does not go up at all, loosen the high-gear limit screw.** Loosen it (counterclockwise) one-quarter turn and try the shift again.

e. Cable-tension adjustment

1. **With the chain on the inner chainring, remove any excess cable slack.** Turn the barrel adjuster on the shifter (as shown in Fig. 5.8, except on the left shifter) counterclockwise (determine rotation direction by looking at the adjuster from the end from which the cable housing emerges, as if you were looking at the top of a bolt). Or tighten the cable without the barrel adjuster: Loosen the cable-fixing bolt, pull the cable tight with pliers, and tighten the bolt.
2. **Check that the cable is loose enough.** It should allow the chain to shift smoothly and repeatedly from the middle to the inner chainring.
3. **Check that the cable is tight enough.** The derailleur should start to move as soon as you move the shifter.

NOTE: *This tension adjustment will work for indexed as well as friction shifters. With indexed front shifting, you may want to fine-tune the barrel adjuster to avoid noise from the chain dragging on the derailleur in some cross-gears or to get more precise shifting.*

ANOTHER NOTE: *Some front derailleurs have a cam screw at the end of the spring to adjust spring tension. For quicker shifting to the smaller rings, increase the spring tension by turning the screw clockwise.*

NOTE ON SHIFTING TROUBLE: *If you cannot get the front derailleur to shift well, or if it rubs in cross-gears no matter what you do, or if it throws the chain off despite your best efforts, refer to the chainline discussion in the troubleshooting section at the end of this chapter.*

5-6

REPLACING AND LUBRICATING SHIFT CABLES AND HOUSINGS

LEVEL 2

For derailleurs to function properly, they must be connected to clean, smooth-running cables (also called inner wires). Because of all the muck that you encounter while riding a mountain bike, you need to regularly replace those cables. As with replacing a chain, replacing cables is a maintenance operation, not a repair operation. Do not wait until cables break to replace them. Replace any cables that have broken strands, kinks, or fraying between the shifter and the derailleur. Also replace housings (also called outer wires) if they are bent, mashed, or just plain gritty, or if the color clashes with your bike (this is really important).

5-7

BUYING CABLES

1. **Buy new cables and housing with at least as much length as the ones you are replacing.**
2. **Make sure that the cables and housing are for indexed systems.** These cables will stretch minimally, and the housings will not compress in length. Protected by an external vinyl sheath, indexed housing is not made of steel coil as brake housings are; it is made of parallel (coaxial) steel strands of thin wire. If you look at the end, you will see numerous wire ends sticking out, surrounding a central low-friction plastic tube (make sure the housing you buy has this plastic liner) (Fig. 5.26).
3. **Buy cable crimp caps and tubular cable-housing end ferrules** (Fig. 5.26). Caps prevent cable fraying and ferrules prevent housing from kinking at the cable stops, shifters, and rear derailleur. You might also consider getting ferrules that have a long nose that sticks out through the slot in a cable stop on the frame. A tubular rubber dust shield comes with them and slips onto the long nose from the opposite side of the cable stop to prevent entry of dirt and water into the housing.

NOTE: *If your frame has internal cables that run bare inside the frame, and especially if you don't have the Park IR-1 or IR-1.2 magnetic internal-routing tool (Fig. 5.38), buy thin plastic tubing that fits the cables along with the new cables and housings to reduce installation time (5-15). Before you pull out the old cable, slide a long piece of tubing onto the old cable and through the frame as a guide for the new cable. After the new cable is in, you can pull the tube back out. Make sure that if the new internal shift cables cross each other inside the down tube (they will do so if the housings cross in front of the head tube) that they cross a maximum of once. Multiple crosses (i.e., the cables are wound around each other) not only will result in undue friction, but shifting one lever will move both derailleurs.*

5.26 Cable-housing types and end caps

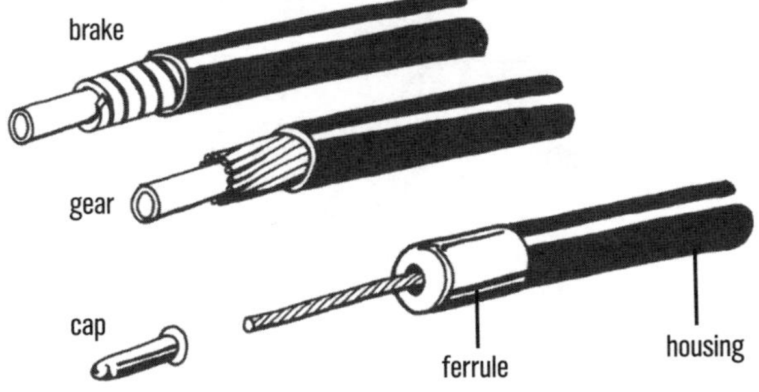

5-8

CUTTING THE HOUSING TO LENGTH

1. **Use a special cable-housing cutter.** Park, Shimano, Pedro's, and Wrench Force (see 1-2) make these cutters. Standard side cutters for cutting wire will not cut index-shift housing, and they won't do a good job on cables, either.
2. **Cut the housing to length.** Cut to the same lengths as your old ones if you liked their function; otherwise, cut housing so that the housing sections curve smoothly from cable stop to cable stop and so that turning the handlebar does not pull or kink them. Allow for enough length at the rear derailleur so that the derailleur can freely swing back (Fig. 5.27) and forth (Fig. 5.28 for chainstay cable; Fig. 5.41 for seatstay cable). Current SRAM and old Sachs D.I.R.T. rear derailleurs use a short housing section without a cable loop (the bare cable runs over a pulley), and they are particularly sensitive to housing length. The housing should curve gently into its receptacle without being so short that it limits derailleur movement or so long that it has a sharp bend; the latter is particularly important to watch for with a seatstay cable as the derailleur swings backward. Make sure your cut is square.

5.27–5.28 Correct housing length

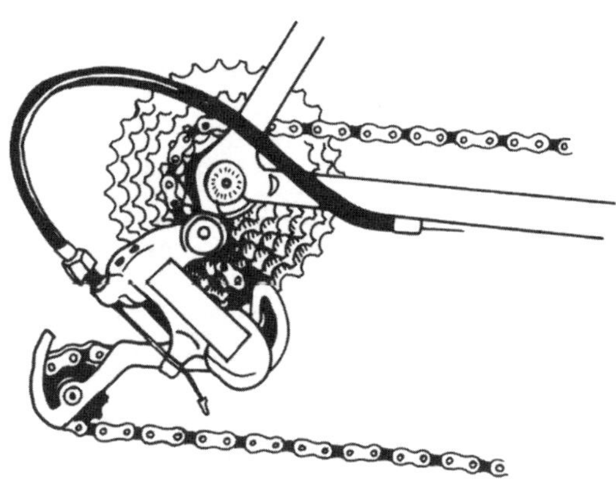

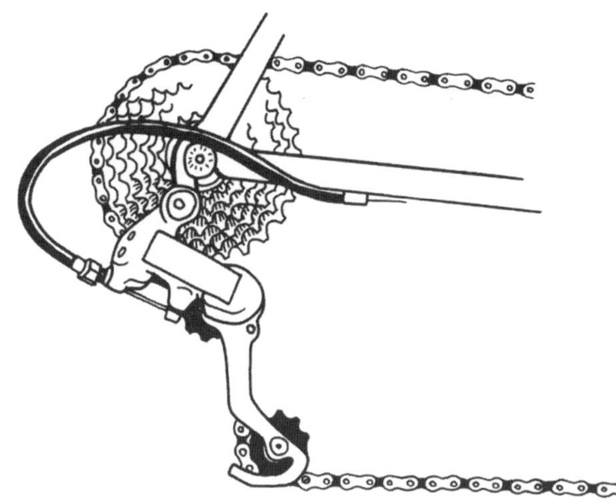

3. **With a sharp tool such as an awl, open each plastic sleeve end that has been smashed shut by the cutter.**
4. **Place a ferrule** (Fig. 5.26) **over each housing end.**

5-9

REPLACING CABLE IN SRAM TRIGGER LEVERS, OLDER SHIMANO RAPIDFIRE LEVERS, OR THUMB SHIFTERS

1. **Disconnect the old cable at the derailleur, and clip off the end cap.**
2. **Shift to the lever setting that lets the most cable out.** On most systems, this will be the highest-gear position for the rear shift lever (small cog) and the lowest for the front (small chainring). On Low Normal or Rapid Rise systems, however, this will be the low-gear position front and rear.
3. **Push out the old cable and recycle it.** You may need to uncover the cable head first. On some SRAM trigger levers, open the rubber cable-change flap. On many recent SRAM trigger levers, you must remove the cover over the cable-release lever (upper lever); this may require a Torx wrench. Once the cover is off, push on the cable as you pry the cable head outward with a pick or knife tip from its pocket under the flat spring coil (Fig. 5.29). On many Shimano Rapidfire levers, unscrew the large plastic plug that covers the cable-access hole if one is present.
4. **Slide in the new cable.** Push the cable into the recessed hole in which the cable head seats and out through the barrel adjuster (Figs. 5.29 to 5.31). Replace the lever cover if you removed it.
5. **Route the cable to the derailleur.** With external cables, guide the cable through each housing segment (making sure each housing segment has a ferrule on the end; see Fig. 5.26) and cable guide and cable stop. With internal cables, see section 5-15. The frame's slotted cable stops allow you to slip the cable and housing in and out from the side. If present, replace the plastic plug or rubber flap over the cable-access hole in the shifter.

5.29 Prying the cable head from its pocket on a SRAM XX trigger lever

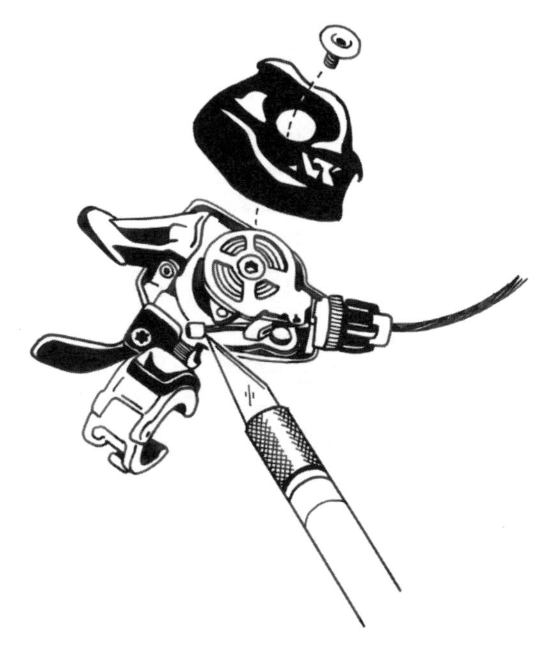

5.30 Rapidfire shifter

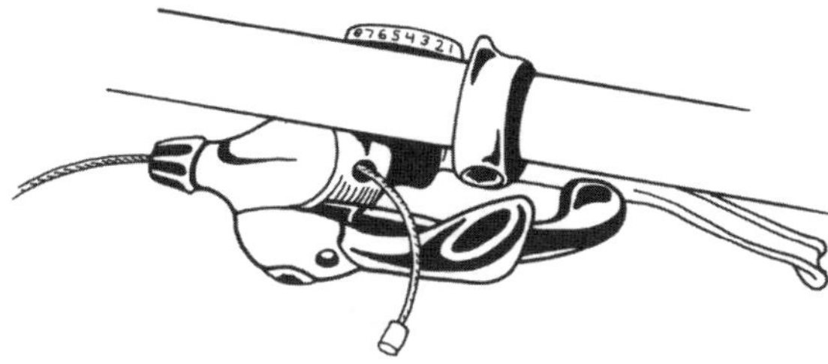

5.31 Thumb shifter

5-10

REPLACING SHIFT CABLE IN SHIMANO DUAL CONTROL INTEGRATED SHIFTER/ HYDRAULIC DISC-BRAKE LEVERS

These models are 2003 (and later) XTR, 2004 (and later) XT, and 2005 (and later) LX, and any one may be used with the Saint group. Note that cable change is different (and more complicated) for the front shifter than for the rear one.

When moved laterally downward with the palm side of the fingertips, the brake lever on these units (Fig. 5.32) pulls shift cable. When the brake lever is moved laterally upward with the back (fingernail) side of the fingertips, it releases shift cable. (Of course, when the brake lever is pulled straight back toward the handlebar, it pushes on the hydraulic fluid to apply the disc brakes.)

a. Either lever

1. **Disconnect the old cable at the derailleur, and clip off the end cap.**
2. **Shift to release the maximum amount of cable.** Flip the underside of the brake lever or the auxiliary thumb-release lever repeatedly until the gear-indicator needle reaches the L position inside the indicator window.

5.32 Replacing the cable in a left Shimano XT Dual Control shifter/hydraulic brake lever

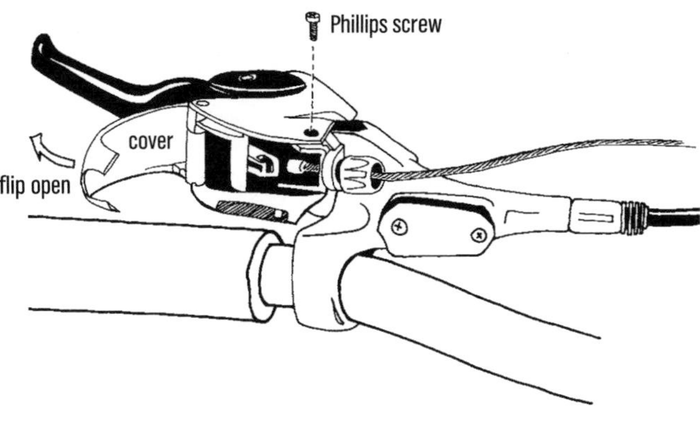

b. Rear (right) lever

1. **Remove the large plastic Phillips screw plug.** It's on the opposite side of the lever body from the barrel adjuster.
2. **Push the cable out.** The end should pop right out of the hole the screw plug was in. You may need to coax it a bit with a paper clip or thin knife to guide it out of the hole while you push the cable.
3. **Slide the new cable straight in.** Go through the unplugged hole in the cover and straight out through the barrel adjuster.
4. **Route the cable to the derailleur.** With external cables, guide the cable through each housing segment (making sure each housing segment has a ferrule on the end; see Fig. 5.26) and cable guide and cable stop. With internal cables, see section 5-15.
5. **Replace the plastic plug in the shifter's cable-access hole.**

c. Front (left) lever

1. **Open the cover concealing the cable hook.** This is quite ingeniously designed and consequently simple to perform. On pre-2008 models, remove the Phillips screw on the cover adjacent to the barrel adjuster (Fig. 5.32); it is on the opposite side of the shifter housing from the gear-indicator window. Do not remove the other two, smaller Phillips screws also on the same plastic cover piece. Flip open the shifter-housing cover; it has the Shimano logo on it and hinges from its end opposite the barrel adjuster. More recent dual-control levers with the

drum-shaped body have a simple flip-open cable-access cover held down by a plastic screw plug.

2. **Push the old cable free.** While jiggling it to free it from the cable hook, pry up on the cable where it exits the hook to flip its head out of the hook (Fig. 5.32). Pull the old cable out.
3. **Engage the head of the new cable.** Snap the head of the new cable into the cable hook.
4. **Shift the lever to its maximum cable pull position.** With the palm side of your fingers, repeatedly shift the brake lever laterally through all of its clicks.
5. **Push the cable out through the barrel adjuster.**
6. **Close the cover and tighten the screw (gently!).**
7. **Route the cable to the derailleur.** With external cables, guide the cable through each housing segment (making sure each housing segment has a ferrule on the end; see Fig. 5.26) and cable guide and cable stop. With internal cables, see section 5-15. The frame's slotted cable stops allow you to slip the cable and housing in and out from the side.

5-11

REPLACING 1996–2000 SHIMANO XTR RAPIDFIRE CABLE

The 1996–2000 Shimano XTR shifters have a plastic cover over the wire-end hook (Fig. 5.33) and also have a slotted barrel adjuster and shifter body.

1. **Shift the smaller (upper and forward) finger-operated lever until it lets all the cable out.**
2. **Turn the shifter barrel adjuster so that its cable slot is lined up with the slot in the shifter body** (Fig. 5.33). The slot is on the opposite side from the gear indicator. Barrel adjusters for 1999–2000 XTR are in a plastic housing that pulls out of the lever body; with this type, pull the entire barrel adjuster straight out.
3. **Unscrew the Phillips-head screw on the plastic cover.** It will not come completely out (to avoid losing it), because it is retained in the cover by a plastic ring. Open the cover.
4. **Pull out the old cable.** Pull it down out of the slot, and then pull its head out of the hook.

5.33 Replacing the cable in a 1996-2000 Shimano XTR Rapidfire SL shifter (cover removed)

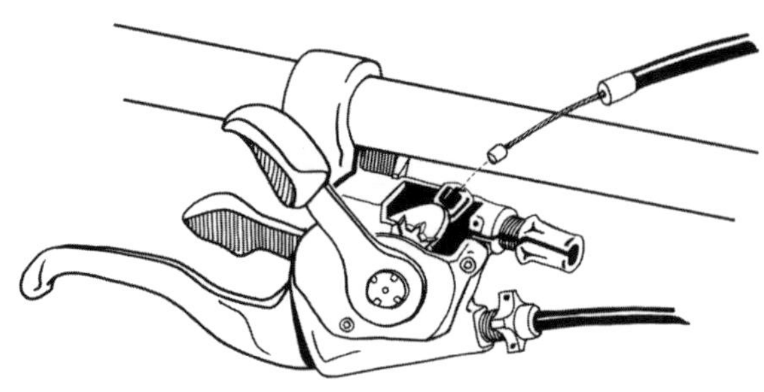

5. **Install the new cable.** Slip the cable head into the cable hook (Fig. 5.33), and pull the cable into the slot. Turn the barrel adjuster so that the slots no longer line up. Or with 1999–2000 XTR, push the barrel adjuster back in.
6. **Close the cover and tighten the screw (gently!).**
7. **Route the cable to the derailleur.** With external cables, guide the cable through each housing segment (making sure each housing segment has a ferrule on the end; see Fig. 5.26) and cable guide and cable stop. With internal cables, see section 5-15. The frame's slotted cable stops will allow you to slip the cable and housing in and out from the side.

NOTE: *Replacing the thin cables connected to the XTR Rapidfire Remote bar-end-mounted shifters (which were manufactured for only a couple of years) requires buying the thin double-headed cables and housings from Shimano. The small heads simply slip through the holes in both sets of shift levers (on the handlebar and on the bar end) from the backside. Install the remote shifter's little plastic caps onto the metal cable heads to keep them from pulling back through. That's it.*

GRIP SHIFT

NOTE: *Many Grip Shifts snap together with their accompanying SRAM grip, and the grip itself is the retainer holding the shifter together. If you use a different grip next to the Grip Shift, you must install the retaining washer (Fig. 5.49) packaged with the Grip Shift into the end of it. If you don't, there is nothing to keep the shifter grip from sliding off the shifter body, and you could inadvertently pull it apart when it is not butted up*

5.34 Cable change in post-2012 Grip Shift

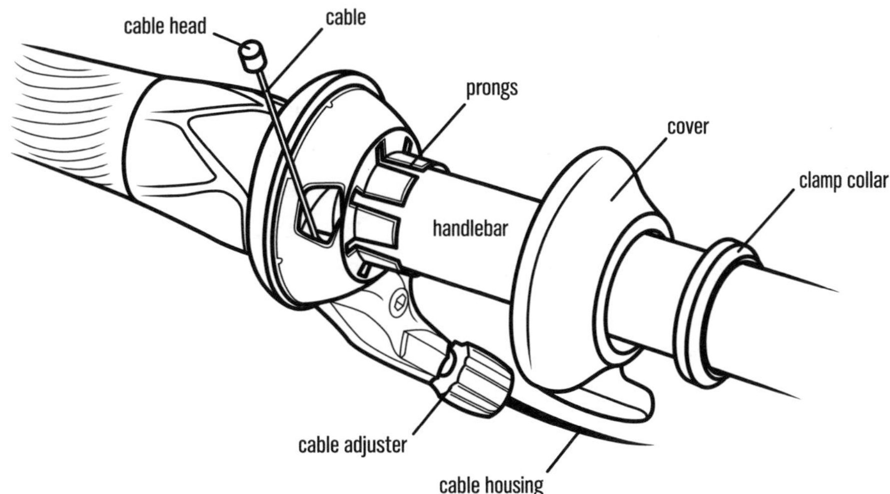

against a grip. For instance, this situation could arise if you were to move the shifter away from the brake lever in 5-12, below, by sliding the grip off the bar and then moving the shifter outward. If the shifter comes apart, the worm spring (Fig. 5.50) will become dislodged, and you'll have to reconnect it to the two parts before you can put them back together.

5-12

REPLACING CABLE IN POST-2012 X-ACTUATION SRAM GRIP SHIFT

These shifters have three rows of ball bearings inside the shifter body.

1. **Move the brake lever away from the shifter.** Loosen the brake-lever clamp bolt and slide the lever inboard on the handlebar at least 10mm away from the Grip Shift.
2. **Remove collar.** With a 2.5mm or 3mm hex key, loosen the shifter's clamp collar and slide it off the prongs on the inboard end of the shifter (Fig. 5.34) and toward the brake lever.
3. **Remove cover.** Slide the cover, which surrounds the inboard end of the shifter and the underside of the cable spout, inboard on the handlebar (away from the shifter).
4. **Twist shifter to let out maximum cable.** Shift a rear shifter to high gear (smallest cog) or a front shifter to the inner-chainring position. This reveals the cable head through the hole in the shifter body.
5. **Push out the old cable.** Recycle or otherwise discard the old cable.
6. **Push in new cable.** If you pull off the barrel adjuster, the cable slides through more easily, and then you can feed it through the adjuster and into the end of the cable housing with ease; put the barrel adjuster back on once the cable is in place. Otherwise, it helps to put a little bend in the end of the new cable. Unlike original Grip Shift (Fig. 5.31), there is no need to pull off the housing cover to get at the cable track (you would have to use a very tiny Torx key to remove the housing-cover screw shown in Fig. 5.30).
7. **Replace cover and clamp collar.** Tighten the little pinch bolt on the collar (not too tight!).
8. **Replace brake lever.** Slide the brake lever back to its original position on the handlebar and tighten its clamp bolt.
9. **Route the cable to the derailleur.** With external cables, guide the cable through each housing segment (making sure each housing segment has a ferrule on the end; see Fig. 5.26) and cable guide and cable stop. With internal cables, see section 5-15. The slotted cable stops on the frame allow you to slip the cable and housing in and out from the side.

5-13

REPLACING CABLE IN 1998–2012 SRAM GRIP SHIFT AND SACHS TWIST SHIFTERS

Changing cables is easy on these Grip Shift and Half Pipe shifters. The high-end twist-shifter models got this feature in 1998, and by 2000 all SRAM shifters had the quick-cable-change system.

1. **Disconnect the cable at the derailleur, and cut off the end cap.**
2. **Twist the shifter to let out the maximum amount of cable.** Shift to the highest number on the rear shifter and to 1 on the front.
3. **Get at the end of the cable.** This differs by model. Some have no cover over the cable head; it is exposed as soon as you shift the lever to let the cable out fully. Some have a rectangular rubber hatch that you pull open (Fig. 5.35). While pushing back on the cable, you then pry up the plastic tab that obscures half of the cable head. Use a small screwdriver. The cable should pop out from under the hook on the tab. Other high-end SRAM shifters have a plastic hatch that you slide off to the side. Once the hatch is off, you will see the cable head on the front shifter; on the rear shifter, you'll see a small setscrew that you remove with a 2.5mm hex key to reveal the cable head. On inexpensive shifters, peel back the corner of the rubber grip cover and you will see the cable head.
4. **Push out the old cable.**

5.35 SRAM Half Pipe twist shifter

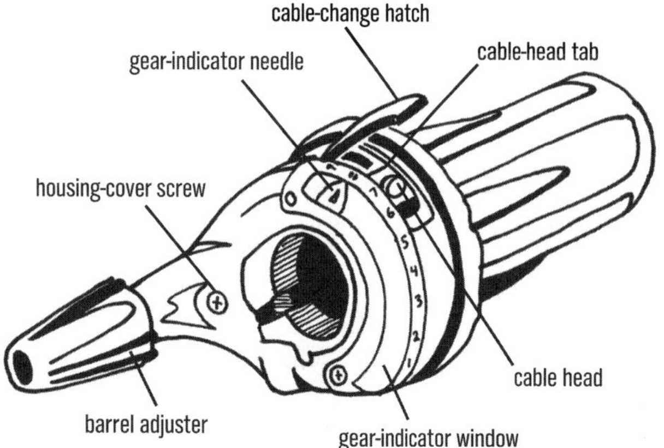

5. **Slide in the new cable and pull it snug.** Replace the setscrew, if present, and carefully screw it in until it contacts the cable head. If the screw feels as though it has stripped the threads, it hasn't—it just has gone in beyond the threads and is turning freely. No worries.
6. **Push the cover that conceals the cable head back into place.** Again, yours might not have a cover.
7. **Route the cable to the derailleur.** With external cables, guide the cable through each housing segment (making sure each housing segment has a ferrule on the end; see Fig. 5.26) and cable guide and cable stop. With internal cables, see section 5-15.

5-14

REPLACING CABLE (AND OVERHAULING), 1992–1998 GRIP SHIFT

Prior to 1998, all Grip Shifts had to be disassembled to replace the cable; 1998–2000 inexpensive shifters still require disassembly for cable changing.

1. **Disconnect the derailleur cable.**
2. **Move the brake lever inboard to make room for pulling the grip apart.** If a bend in the handlebar or other obstruction prevents this, you must roll or slide the handlebar grip away from the Grip Shift to allow room for the shifter to slide apart.
3. **Remove the triangular plastic cover holding the two main sections together** (Figs. 5.36 and 5.37). Use a Phillips screwdriver.
4. **Pull the outer shifter section away from the main body to separate it from the inner housing.** Watch for the spring (Figs. 5.36 and 5.37) to ensure that it does not fall out. The spring can be nudged back into place if it does.
5. **Pull the old cable out and recycle it.**
6. **Clean and dry the two parts.** A rag and a cotton swab are usually sufficient. Finish Line offers a cleaning and grease kit specifically designed for Grip Shift shifters. A really gummed-up shifter may require a solvent and compressed air to clean and dry it.
7. **Lubricate the inner housing tube and spring cavity, all cable grooves, and the indexing notches**

5.36 1992–1998 Grip Shift right shifter

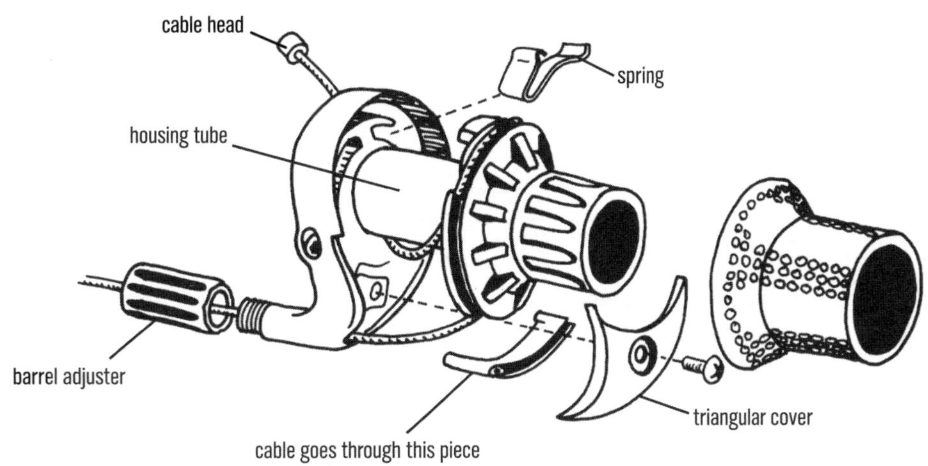

5.37 1992–1998 Grip Shift left shifter

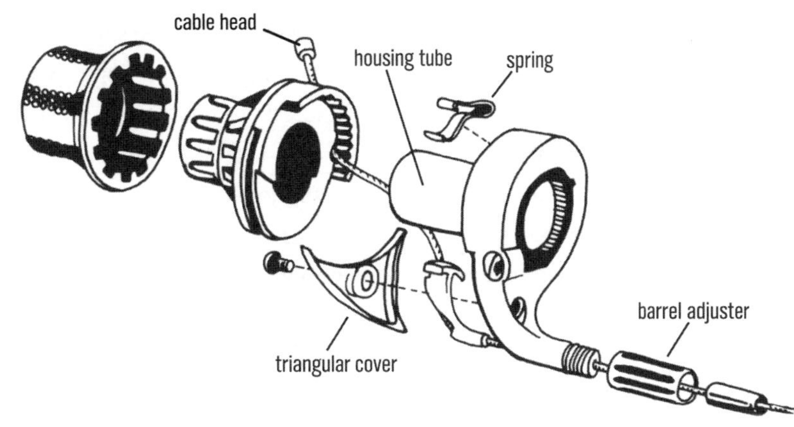

in the twister. Use a nonlithium grease such as SRAM Jonnisnot or Finish Line Teflon fluorocarbon grease.

8. **Thread the cable through the hole, seating the cable end in its little pocket.**
9. **Route the cable through the shifter.** For the rear shifter, loop the cable once around the housing tube, and exit it through the barrel adjuster (Fig. 5.36). For the front shifter, the cable routes directly into its guide (Fig. 5.37).
10. **Make sure the spring is in its cavity in the housing.** Hold the spring in with a small amount of grease if need be.
11. **Slide the outer (twister) body over the inner tube.** Be sure that the shifter is in the position that lets the most cable out (on models with numbers, line up the highest number with the indicator mark on rear shifters; use lowest number on front shifters).
12. **Lift the cable loop into the groove in the twister** (Fig. 5.36). Push straight inward on the twister as you pull tension on the cable exiting the shifter. The twister should slide in until flush under the housing edge; you may have to jiggle it back and forth slightly while pushing in to get it properly seated.
13. **Replace the cover and screw.**
14. **Check that the shifter clicks properly.**
15. **Slide the grip back into place.**
16. **Route the cable to the derailleur.** With external cables, guide the cable through each housing segment (making sure each housing segment has a ferrule on the end; see Fig. 5.26) and cable guide and cable stop. With internal cables, see section

5-15. The frame's slotted cable stops on the frame allow you to slip the cable and housing in and out from the side—use them that way to save yourself some effort.

5-15

ROUTING INTERNAL CABLES

Often, internal-cable frames have a removable plastic cover over guides under the down tube just ahead of the bottom-bracket shell that allows access to the shift cables. They also usually have at least one cable-entry point with either a rubber grommet or a removable plastic step-down stop or insert for the cable housing. This allows you to pull out the grommet or stop and fish the cable through a large hole in the frame, rather than a tiny one sized to the cable alone.

As I mentioned in the Note in 5-7, if you are replacing a cable, you can slide a thin plastic tube over the cable and into the frame until it sticks out the other end; tape it down on both ends. You may be able to find this plastic tubing at bike shops, or you can strip the outside cover and wires off the shift-cable housing, leaving only the liner. Alternatively, electronics stores sell heat-shrink tubing in various sizes and colors or colored plastic tubing for insulation or wire color-coding; it's often dubbed "spaghetti." Then, after the plastic tubing is in and taped down, you can pull the old cables out while leaving the tubing in place, and slide the new cables in through the tubing. Easy schmeezy; the new cable is then routed through the frame, and you can pull the plastic tube out and leave the new cable in.

On the other hand, if you yanked the old cable out before doing this, or if you are dealing with a new frame without cables already running through it, I highly recommend using Park Tool's IR-1 (Fig. 5.38) or IR-1.2 (Fig. 6.16) Internal Cable Routing Kit. This kit works for routing cables, cable housing, electronic wires, and hydraulic hose inside of a frame. See 6-6 for routing electronic-shift wires with it.

Without the Park IR-1 tool (the IR-1.2 has an additional cable specifically for Shimano Di2 electronic wires), you will be stuck with trying to fish cables through by themselves (put a bend in the end so that you can twist the cable when the end gets to the hole until it pops out) or with a stiff wire bent into a hook at the tip shoved through and taped to the cable. Another method is to crimp a piece of string to the end of the cable with a cable-end crimp cap (Fig. 5.45), drop the string into the hole in one end of the frame, and suck its end out the other hole with the narrow edging tip of a vacuum cleaner (tip the frame in the bike stand to get the assistance of gravity). Or, since the cables you will be installing are generally magnetic, there is a chance that you can drag the tip of the cable through the frame with a large magnet on the outside of the frame. It's tough to do this without getting the cable hung up inside, however. Allow extra time to avoid frustration with any of these methods.

The Park IR-1 magnet kit comes with a large magnet and three coated shift cables (the IR-1.2 has a fourth cable) with a cylindrical magnet attached to the head end. I'll call the "fishing cable" the one that has nothing on the other end. The kit includes two (or three, with the IR-1.2) different lead cables. One is for guiding cables or electronic wires and has a flexible rubber hollow receiver on the end that will stretch over a Di2 electronic connector, or over the head of a shift or brake cable. Park's other lead cable has a conical threaded steel barbed end that can be screwed into the end of

5.38 Park IR-1 internal-routing cable and magnet kit

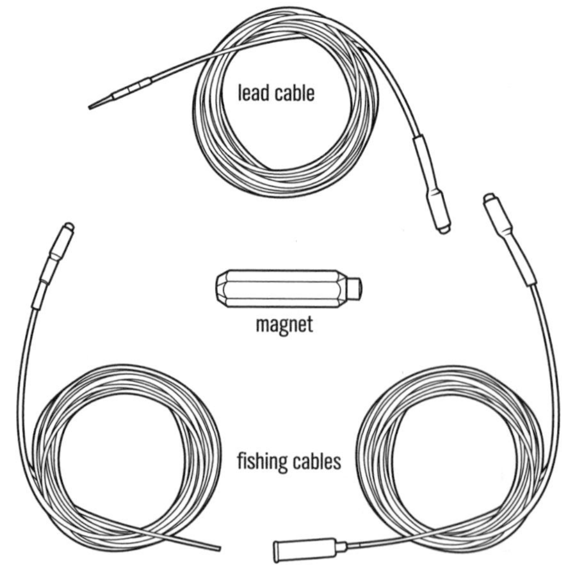

5.39 Cable-entry hole, insert, and lead cable

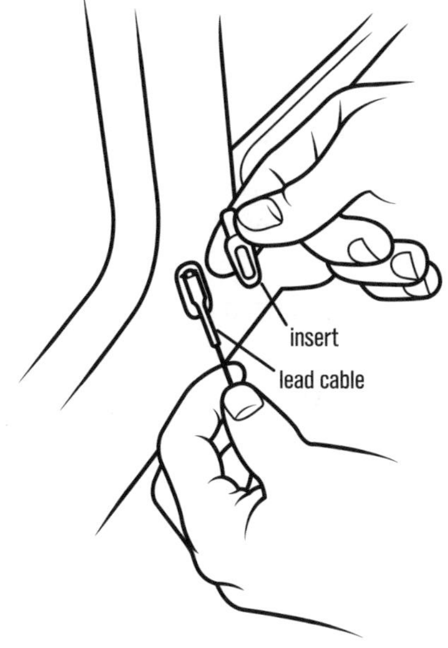

5.40 Down-tube cable holes above bottom bracket with cover removed and fishing cable entering opening

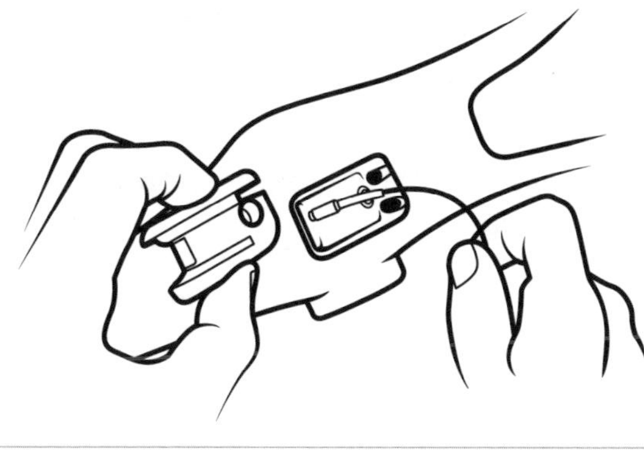

either cable housing or hydraulic hose, allowing the user to easily pull either of those through a frame. The lead-cable magnets repel each other and are attracted to the large magnet and the fishing-cable magnet.

1. **Push the magnet end of one of the two Park IR-1 lead cables down the cable-entry hole at the front of the frame.** You must first pull the rubber grommet, plastic cable stop, or insert out of the frame to have access to the large hole it was filling at the front of the frame.
2. **Push the magnet end of the Park fishing cable up from the large opening at or near the bottom-bracket shell.** You must first remove the cover over the large port in either the down tube or the bottom-bracket shell. The two magnets will stick together inside the down tube.
3. **Pull the lead cable out from the hole at the front.** Pull carefully and jiggle it as the magnet end arrives at the hole so that you pull the magnet end of the fishing cable out with it.
4. **Stick the tip of the new cable to the magnet at the end of the fishing cable sticking up out of the frame hole.** The "tip" is the derailleur end of the cable—not the head end. Before connecting it to the magnet, the new cable should already be routed through the shifter, housing, ferrule, and frame hole plug/cable stop you removed.
5. **Push the new cable into the frame hole.** Allow it to push the Park fishing cable out of the bottom bracket port. Gently pull on the fishing cable until the tip of the new cable emerges.
6. **Route the new cable through the bottom-bracket cable guide.** Make sure the cable passes under the loop(s) in the guide, if present. If it is a front-derailleur cable, you can now thread it up and out so it emerges near the base of the seat tube and skip to step 11. If it is a rear-derailleur cable, continue with step 7.
7. **Push the magnet end of one of the Park lead cables down the cable-entry hole at the dropout.** You must first pull the plastic cable stop or insert out of the chainstay to access the large hole it was filling.
8. **If the magnet end of the cable doesn't emerge at the exit port, guide it out.** You may use the large magnet to guide it out. Or you can push the magnet end of the fishing cable into the chainstay until the two magnets stick together, and then carefully pull both magnets out with the fishing cable. On a carbon hardtail, the cable may go right back in through the same large port. On full suspension frames, you may have to run the cable through the cable housing around the outside of the bottom bracket; it will likely terminate at inserts, stops, or grommets covering entry holes into the down tube and the chainstay.
9. **Stick the tip of the new cable to the magnet at the end of the lead cable.** It will be sticking out

of a hole in either the chainstay or the bottom-bracket shell.

10. **Push the new cable through the chainstay.** Allow it to push the Park lead cable out of the hole at the dropout. Gently pull on the fishing cable until the tip of the new cable emerges.
11. **Replace the cover over the bottom bracket guides and the cable stop(s) or insert(s) at the front of the frame (and at the rear).** You did it!

NOTE: *Make sure that if the new internal shift cables cross each other inside the down tube (they will do so if the housings cross in front of the head tube) that they cross a maximum of once. Multiple crosses (i.e., the cables are wound around each other) not only will result in undue friction, but shifting one lever will move both derailleurs.*

5-16

ATTACHING CABLE TO THE REAR DERAILLEUR

1. **Put the chain on the cog that the derailleur return spring pulls it to.** With anything but Low Normal or Rapid Rise derailleurs, put the chain on the smallest cog. With Low Normal or Rapid Rise derailleurs, start with the chain on the largest cog.
2. **Route the cable.** Run the cable from the shifter barrel adjuster through each of the housing segments or through the frame as in 5-15 until you reach the cable-fixing bolt on the derailleur.
3. **Click the rear shifter to the setting that releases the most cable.**
4. **Tighten the barrel adjuster(s) in all the way, and then back it (them) out one turn.** The shifter has a barrel adjuster, and the rear derailleur may (Fig. 5.41) or may not (Fig. 5.42) have one as well.
5. **Pull the cable taut.** Find the groove in the washer under the cable-fixing bolt to determine the proper cable path; it could be on either side of the cable-fixing bolt (Fig. 5.41 or 5.42).
6. **Tighten the cable-fixing bolt.** This usually requires a 4mm or 5mm hex key, a Torx T25 key, or an 8mm box wrench.

5-17

ATTACHING CABLE TO THE FRONT DERAILLEUR

1. **Shift the chain to the inner chainring so that the derailleur moves farthest to the inside.** This ensures that the maximum amount of cable is available to the derailleur.

5.41 Attaching a rear-derailleur cable

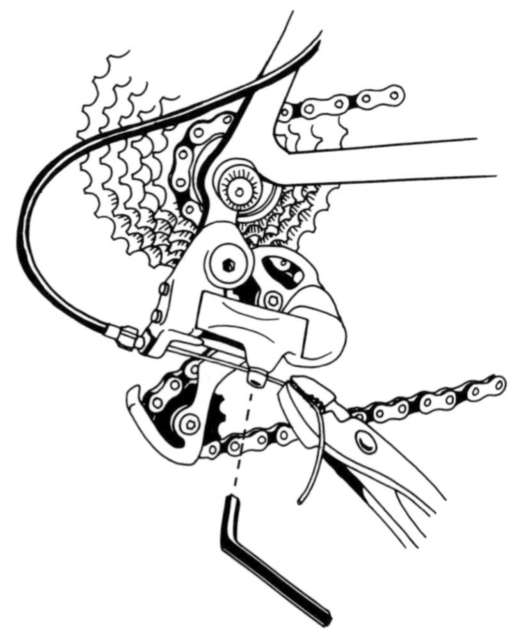

5.42 Cable routing to a SRAM X-Horizon 1X rear derailleur; note the absence of a barrel adjuster

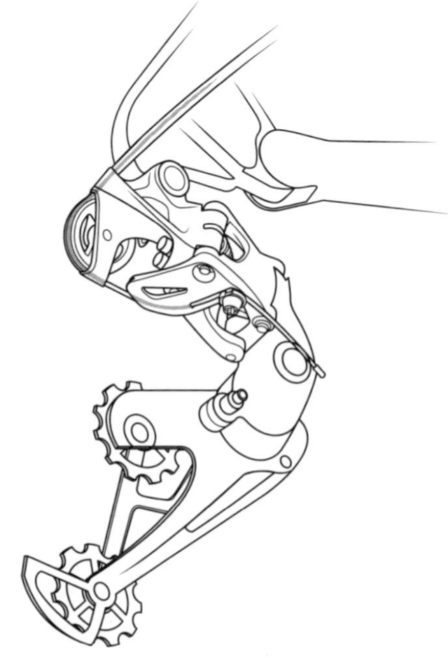

2. **Tighten the shifter barrel adjuster in all the way, and then back it out one turn.**
3. **Hook up the cable.** Place the cable into its groove under the cable-fixing bolt on the derailleur arm while pulling the cable taut with pliers (Fig. 5.43), and tighten the bolt with a 4mm or 5mm hex key, a Torx T25 key, or an 8mm box wrench, as required. Make sure you do not hook up a top-pull-style front derailleur from the bottom, or vice versa.

Many front derailleurs now work both top-pull and bottom-pull style. They are actually top-pull front derailleurs, because you hook up directly to the cable-fixing bolt when the cable comes from the top (Fig. 5.44A). But if the frame routes the cable to the front derailleur from the bottom, it is no problem; you simply run the cable up, over, and around a pivoting rocker arm to the cable-fixing bolt located on its outboard end (Fig. 5.44B). Note that some derailleur models require housing to run the full length of the cable to the front derailleur, terminating at a housing stop forged into the derailleur body.

5-18

ADDING FINAL CABLE TOUCHES

A high-quality cable assembly includes the cable-housing end ferrules throughout, ideally ones with rubber seals to keep contamination out of the housings, and crimped cable caps. Clip cables about 1–2cm past the cable-clamp bolts before crimping the cable caps in place (Fig. 5.45).

5.43 Pull the cable taut before tightening the clamp bolt with a hex key

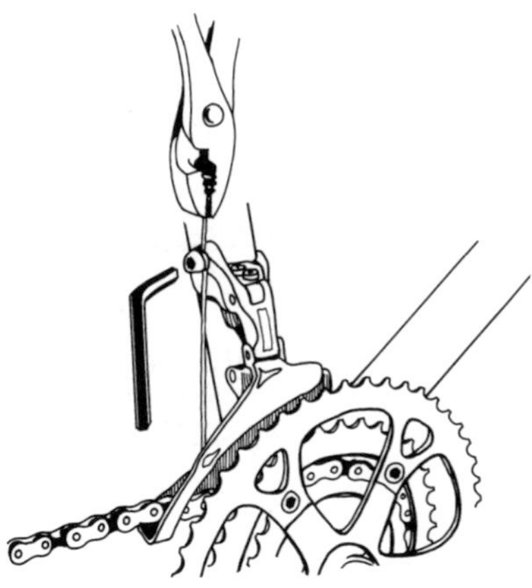

5.44A Over-the-top-tube (top-pull) cable routing to a Shimano low-clamp top-swing front derailleur

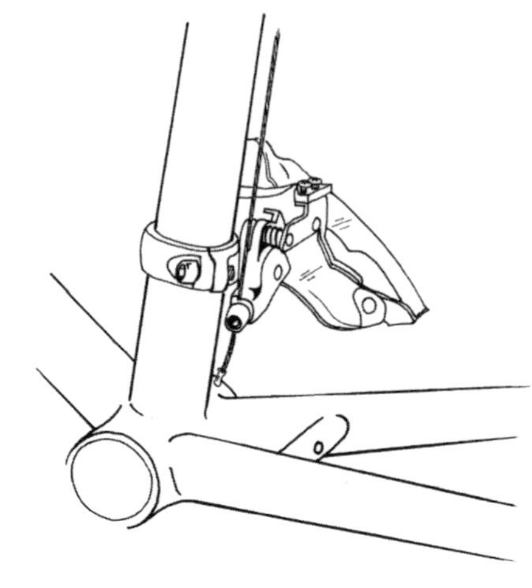

5.44B Under-the-bottom-bracket (bottom-pull) cable routing to a Shimano top-swing front derailleur

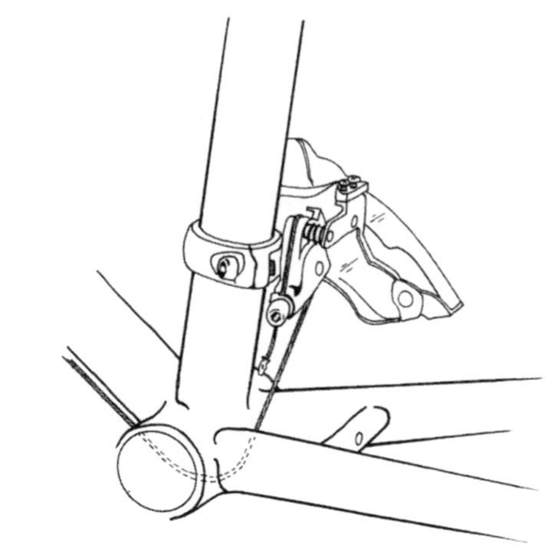

5.45 Crimping a cable-end cap

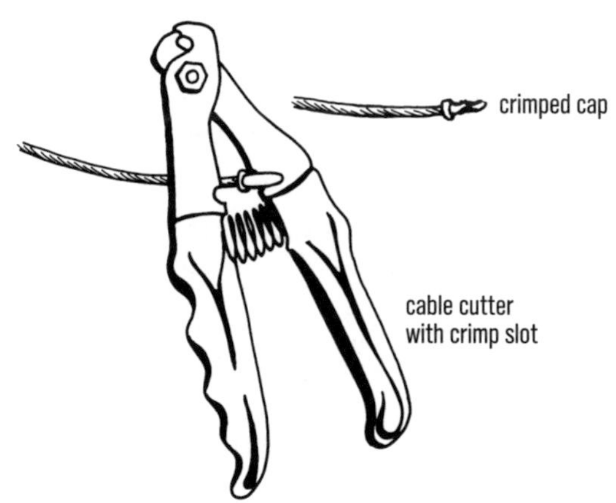

5-19

LUBRICATING CABLE

New cables and housings with plastic liners do not need to be lubricated. Used cables and housings can be lubricated with chain lubricant. Standard grease sometimes slows cable movement, so various manufacturers recommend (and some supply) their own molybdenum disulfide grease for cables.

1. **Pull the housing segments out of their slotted cable stops.** There is no reason to disconnect the cable at the derailleur. If the bike has internal cables, you will have to do this less frequently, but since you'll have to pull the entire cable out to lubricate it, you may as well replace it (5-15).
2. **Use lubricant to coat the areas of the cable that will be inside the cable housing segments.**

NOTE: *If the bike has external cable routing but does not have slotted cable stops, you may as well replace the cables and housings because an old cable will have a frayed end and will be hard to put back through the housing after lubrication. That's another reason to keep extra cables, housing ferrules, and cable ends in your workshop.*

5-20

REDUCING CABLE FRICTION

In addition to replacing old cables and housings with good-quality cables and lined housings, you can take other steps to improve shifting efficiency.

1. **Route the cable so that it makes smooth bends and so that turning the handlebar does not increase the tension on the shift cables.** (This is the most important friction-reducing step.)
2. **Choose cables that offer low friction.** Die-drawn cables, which have been mechanically pulled through a small hole in a piece of hard steel called a die, move with lower friction than standard cables. Die-drawing flattens all of the outer strands and smoothes the cable surface. Thinner cables and lined housings with a large inside diameter also reduce friction.
3. **Increase the size of the derailleur-return spring to quicken shifting when the cable is released (as opposed to pulled).** You can buy a stiffer spring designed specifically for some Shimano derailleurs at your bike shop.

SHIFTERS

Twist shifters (Grip Shift by SRAM is the predominant one), Rapidfire levers, thumb shifters (Figs. 5.46 to 5.48), SRAM trigger levers (Fig. 5.29), and Shimano's post-2003 Dual Control brake and shift levers (Fig. 5.32) all move the derailleurs, but each goes about it in a very different way.

5.46 Grip Shift

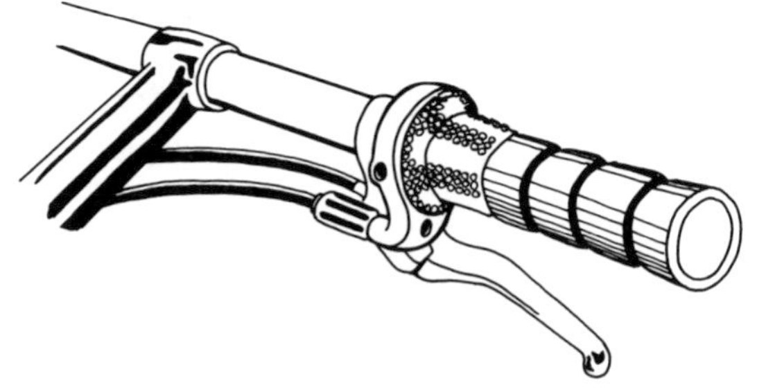

5.47 Rapidfire shifter

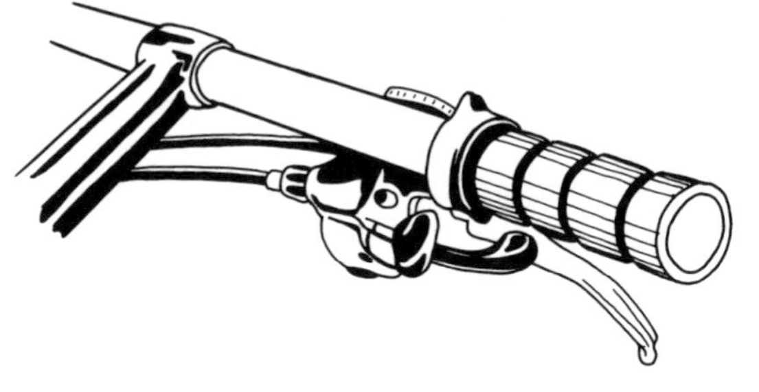

5.48 Thumb shifter

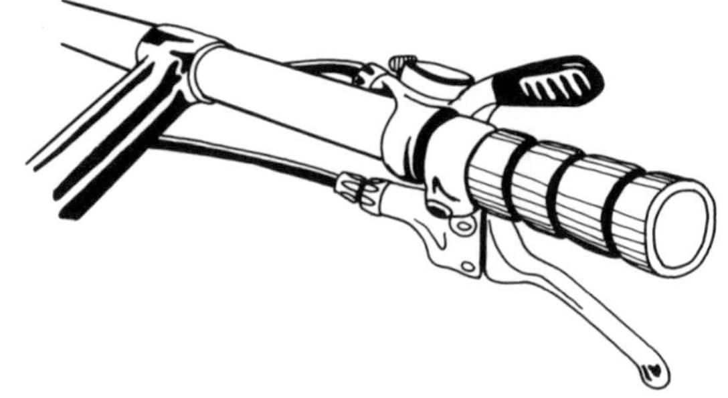

5-21

REPLACING AND INSTALLING BAR ENDS AND GRIPS

LEVEL 1

Replacing shifters requires removing at least grips and often bar ends and brake levers as well. Shifters are generally labeled right and left, but if you're in doubt, you can tell which is which because the right one has a lot more clicks. This is not true of friction shifters.

1. **If bar ends are installed, remove them.** This usually requires a 5mm hex key.
2. **Remove the grips.** If you see clamp bolts at either end of the grips, just loosen the bolt(s) and slide the grip off. Remove simple rubber grips by rolling them back or lifting them away from the handlebar at either end, squirting water or rubbing alcohol underneath, and twisting back and forth while sliding them off; alternatively, use the compressed-air method mentioned in the next step. If you are not planning to reuse the grips, you can cut them off.
3. **Install the new grips.** If the grips have bolts, simply slide them on and tighten the clamp bolts. When installing simple rubber grips, squirt rubbing alcohol inside to make them slide on easily; they will slip for a number of days until the alcohol evaporates. You can also lubricate them with spray adhesive, and they will slide right on and not move once the glue sets.

   If the grips have closed ends, you can punch a hole in the ends and put the grips on by inflating them with an air compressor. Because of the length of the handlebar, it will take two people to span the reach and install the grips. Cover the opposite end of the handlebar (the palm of your helper's hand works well), and push the grip with the little hole in it onto the handlebar (at your end) an inch or so. Press the compressed-air-gun tip against that hole, and the grip will balloon up and slide right on. Then push the fully closed grip an inch or so onto the opposite end of the handlebar, and again press the air-gun tip against the first grip's hole. This will inflate both grips so that they can easily be pushed on. Removal can be done the same way, in reverse.

5-22

REPLACING A SHIMANO SHIFTER AND INTEGRATED BRAKE LEVER

If you are replacing the entire brake lever and shifter unit, proceed as follows (Fig. 5.46).

1. **Remove the old brake lever and shifter.**
2. **Slide the new brake lever and shifter onto the bar.** Make sure that you put the right shifter on the right side.
3. **Slide the grip back into position.**
4. **Mount the bar end if you have one.**
5. **Rotate the brake lever to the position you like.**
6. **Tighten the brake-lever fixing bolt.**

5-23

REPLACING THE SHIFTER UNIT ON SHIMANO INTEGRATED BRAKE AND SHIFT LEVERS

These instructions apply to Shimano Rapidfire shifters (Figs. 5.30, 5.33, and 5.47).

1. **Click the shifter to release the maximum amount of cable.**
2. **Unbolt the old shift lever from the brake-lever body it's attached to.**
3. **Position the new shifter exactly as the old one was positioned.**
4. **Replace the bolt and tighten it.**
5. **Install the cable.** Instructions in 5-9, 5-10, or 5-11.

NOTE: *For Rapidfire Remote installation, mount the remote lever onto the tip of the bar end with a 2mm hex key. Hook up the cables as described in 5-11.*

5-24

REPLACING SRAM GRIP SHIFT, HALF PIPE, AND OTHER TWIST SHIFTERS

See Figure 5.34 for SRAM Grip Shift and Figure 5.49 for Half Pipe Grip Shift.

1. **Remove the old shifter.** Replace the brake lever if you had to remove it to get the old shifter off.
2. **Loosen and slide the brake lever inward.** This allows room for the shifter.

5.49 Squeeze the two tabs together to release the Half Pipe retaining washer

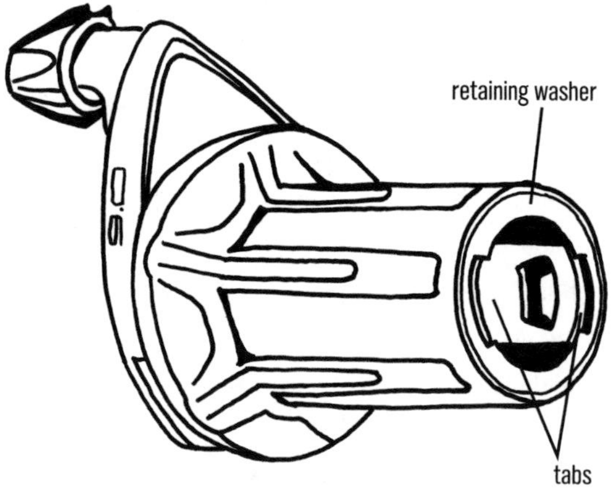

3. **Slide the appropriate (right or left) new shifter on with the cable-exit barrel pointing inward.**
4. **Slide the plastic washer on over the handlebar that separates the grip from the Grip Shift.** This step is not necessary with Half Pipes or with 2001 (and later) Shorties, as the plastic washer is integrated within the shifter.
5. **Replace the grip (and bar end).**
6. **Butt the Grip Shift up against the plastic washer or the grip.** Rotate the shifter until the cable-exit barrel is oriented so that it will not interfere with the brake lever.
7. **Tighten the mounting bolt to the handlebar.** You will need either a 2.5mm or 3mm hex key.
8. **Slide and rotate the brake lever to the position you like, and tighten it down.**
9. **Install the cable.** If you need to install the cable into the new shifter, see 5-12, 5-13, or 5-14.

5-25

REPLACING TOP-MOUNTED THUMB SHIFTERS OR TRIGGER LEVERS NOT INTEGRATED WITH THE BRAKE LEVER

Thumb shifters are shown in Figure 5.48.

1. **Remove the bar end, grip, brake lever, and old shifter.**
2. **Slide on the replacement shifter.**
3. **Slide on the brake lever, grip, and bar end.**
4. **Tighten the bar end and brake lever in the positions you want.**
5. **Tighten the shifter in the position that is comfortable for you.** Make sure the position allows easy access to the cable barrel adjuster and free cable travel. Some SRAM trigger levers allow attachment to various MatchMaker clamps (Fig. 5.29); more options for positioning the lever to your preference are available with this design.
6. **Install the cable.** Instructions in 5-9.

SHIFTER MAINTENANCE

5-26

SRAM GRIP SHIFT

A Grip Shift unit (Figs. 5.36, 5.37, 5.50, and 5.51) requires periodic cleaning and lubrication, as described in this section. Clean only with soap and water, and lubricate only with nonlithium grease, such as Grip Shift Jonnisnot or Finish Line Teflon fluorocarbon grease.

In 2012, SRAM reintroduced Grip Shift for its high-end XX and X0 groups, and this time, the shifters rotate on three sets of ball bearings, rather than just sliding a plastic shell around a plastic core as on prior versions. While the older plastic-on-plastic systems require frequent maintenance to keep them working well with hard use in dirty environments, the ball-bearing system is much more robust and requires little maintenance.

If shifting has become poor on Half Pipes or other Grip Shifts with the quick-cable-change feature, try replacing the cable and housing first before disassembling the shifter for cleaning.

a. Overhaul original short Grip Shifts

The exploded diagrams in Figures 5.36 and 5.37 and the text in 5-14 detail how to take apart, clean, and grease an original (1992–98) short-length Grip Shift. With these older models, so long as you have the shifter disassembled, you might as well replace the cable (5-14), because shifter disassembly is required for that task as well.

b. Overhaul long (Half Pipe) Grip Shifts and post-2001 Shortie Grip Shifts (does not include post-2012 XX1, X01, XX, and X0 Grip Shifts)

The 2001 (and later) Shortie follow the same disassembly procedure as Half Pipes (Fig. 5.49).

1. **Twist the shifter away from you to let out all the cable.** Turn it to "1" on the front shifter and "9" on the rear.
2. **Remove the cable** (5-12) **and the fixed grip** (5-21). Note that, unlike the older Grip Shift models, you do not remove the triangular housing cover to get the twist grip off.
3. **Remove the shifter from the handlebar by using a 3mm hex key on the setscrew.**
4. **Squeeze the tabs on the end of the shifter housing tube together, and remove the plastic retaining washer from the end** (Fig. 5.49). You may have to pry inward with a screwdriver on each tab.
5. **Separate the grip and remove the long worm spring.** Slowly slide the grip outward as you twist the shifter gently away from you in the cable-release direction. If you pull too fast, the long worm spring inside (Fig. 5.50) may pop out.
6. **Remove the two Phillips screws, and pop off the clear plastic window covering the gear-indicator needle** (shown in Fig. 5.35).
7. **Pull straight out on the gear-indicator needle** (shown in Fig. 5.35), **and remove it.**
8. **Pull the small leaf spring** (Fig. 5.50) **up and out using needle-nose pliers or tweezers.** The spring looks like a flat, bent piece of steel; note its position before removing it.
9. **Remove the housing-cover screw** (shown in Fig. 5.35). This takes a Phillips screwdriver or a 2.5mm hex key.
10. **Slide the outer parts (the housing cover and cable spool) away from the housing** (Fig. 5.50). Be careful not to break the tab on the housing cover. To get the housing cover to come up, you may need to push up on it with the end of a paper clip through the hole for the housing-cover screw.
11. **Clean all of the parts with soap, water, and clean rags.** Do not use solvents.

5.50 Half Pipe springs

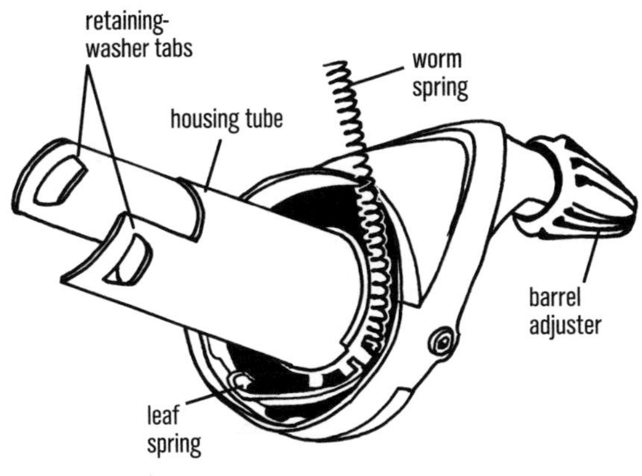

5.51 Half Pipe disassembled

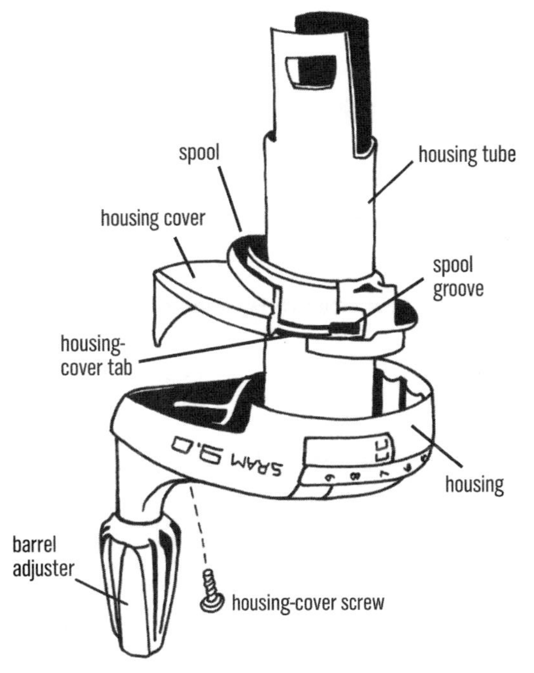

12. **Lubricate the housing tube, the cable track, and all detents for the springs.** Use a nonlithium grease such as SRAM Jonnisnot or Finish Line Teflon fluorocarbon grease.
13. **Slide the housing cover and spool back down onto the housing tube.** First slip the cover's tab into the spool's groove (Fig. 5.51). Replace the housing-cover screw.
14. **Set the leaf spring in place into the spool** (Fig. 5.50). Carefully press down on one end of the

spring as you work to seat the other end in place in the same position it was in before.

15. **Install one end of the worm spring onto its tab** (Fig. 5.50). Slide the outer grip onto the tube, angle the other end of the spring up toward the grip, and pop the other end of the spring onto the tab inside the grip.
16. **Slide the grip into place.** Rotating the grip slightly away from you to compress the worm spring, slide the grip inward until it engages in the housing.
17. **Replace the plastic retaining washer over the tabs on the housing tube** (Fig. 5.49).
18. **Rotate the shifter away from you to the position that would let out the most cable.**
19. **Peer into the curved slot that the gear-indicator needle slides in.** If you followed step 18, you should see the receptacle for the needle at one end of the slot. Slip the gear-indicator needle (Fig. 5.35) back into its receptacle, replace the plastic window, and install the two screws that hold it on.
20. **Install the shifter onto the handlebar using a 3mm hex key.** Check that it works.
21. **Install a new cable (5-12) and connect it to the derailleur.**

    Your bike is now ready to ride.

c. Grip replacement on post-2012 X-Actuation XX1, X01, XX, and X0 Grip Shifts

This ball-bearing system requires little maintenance. The shifter grip can be replaced if the rubber wears out. Follow the instructions in 5-12, except pull the grip and the Grip Shift off the handlebar, rather than working with the parts still on the bar.

1. **Loosen the pinch bolts holding on the shifter and grip.** Loosen the bolt on the shifter collar (Fig. 5.34) and on the end of the grip. Slide the Grip Shift and grip off of the bar.
2. **Pull the long grip away from the Grip Shift.**
3. **With a pick, remove the collar from the outboard end of the Grip Shift.**
4. **While rotating the shifter forward, pull the short grip away from the shifter body.** Rotate in the cable-release direction. Be careful of the long worm spring connecting the two parts (it looks like the one in Fig. 5.50 and serves the same purpose).
5. **Find the spring tab on the shifter body and the new grip where the worm spring fits.** Grease the spring groove with non-lithium grease (SRAM Jonnisnot).
6. **Insert the worm spring onto the spring tab on the shifter body.** Holding the spring in place, install the worm spring against the grip's spring tab.
7. **Push the grip straight toward the shifter body until they are flush.** Do not rotate the grip during this step or the spring will fly.
8. **Push the collar into the end of the grip.** This will secure the new grip onto the shifter body.
9. **Reassemble the cover, collar, and grip.** Snap the long grip onto the Grip Shift. As in 5-12, reinstall the cover and the clamp onto the inboard end of the shifter.

5-27

SHIMANO RAPIDFIRE SL, RAPIDFIRE PLUS, AND DUAL CONTROL LEVERS

Shimano Rapidfire SL (Fig. 5.33), Shimano Rapidfire Plus (Fig. 5.47), and Shimano Dual Control shifters (Fig. 5.32) are not designed to be disassembled by the consumer. Squirting a little chain lube inside every now and then is a good idea, though. If a Rapidfire lever stops working, you'll need to get a new shifter unit. On an integrated Rapidfire unit, the brake lever does not need to be replaced; just bolt the new Rapidfire shifter to it (5-22). If you manage to break a Shimano Dual Control shifter, you will be replacing the entire (expensive!) hydraulic brake master-cylinder and shifter-lever unit.

Sometimes the gear-indicator unit on a Shimano shifter stops working, and it can even jam the lever and stop it from reaching all the gears. This problem was most common in the 1993 and 1994 models. The indicator can be removed from the shifter with a small screwdriver. The indicator's little link arm needs to be stuck back into the hole it came from. Once the indicator jams, you can expect it to happen again; eventually, you will want to replace the lever or dispense with the indicator.

Once you get used to shifting Shimano Dual Control levers solely with the front and back of your fingers, moving the brake-lever blade laterally in either direction, you can remove the optional auxiliary release lever that allows you to shift with your thumb to let cable out. There is a little screw holding it on that you simply unscrew from the nut and the circlip. If you still have the rubber cap, you can push it over the sharp metal tab that is now exposed.

5-28

ORIGINAL SHIMANO RAPIDFIRE

Shimano's first attempt at a two-lever mountain shifter did not work very well. If you are having trouble with yours, I recommend throwing it out and getting a new system. You know you have original Rapidfire if the thumb operates both the down- and upshift levers. Newer Rapidfire SL and Rapidfire Plus levers (Figs. 5.33 and 5.47) have a thumb-operated cable-pull lever and a finger-operated cable-release lever.

5-29

THUMB SHIFTERS

Indexed (click) thumb shifters (Fig. 5.48) are not to be disassembled beyond removing them from their clamp. Periodic (semiannual or so) lubrication with chain lube is recommended and is best accomplished from the backside once the shifter assembly is removed from its clamp (Fig. 5.52).

Frictional (non-clicking) thumb shifters can be disassembled, cleaned, greased, and reassembled. Put the parts back the way you found them. You can avoid the hassle of disassembly by squirting in some chain lube instead.

5-30

CHAIN-RETENTION DEVICES FOR DOWNHILL, FREERIDE, SLALOM, AND JUMP BIKES

Modern chain guides mount on ISCG (International Standard Chain Guide), which is also known as ICMS (International Chainguide Mounting Standard), integrated onto the exterior of the bottom-bracket shell (Fig. 5.53). These are three mounting tabs with M6×1 threaded holes; they radiate out from the face of the bottom-bracket shell at angles and distances from center prescribed by the international standard. The fixed-position mounts eliminate the twisting and spacing problems that can occur without them.

5.52 Indexed thumb shifter, exploded view

5.53 ISCG mounts on the frame's bottom-bracket shell

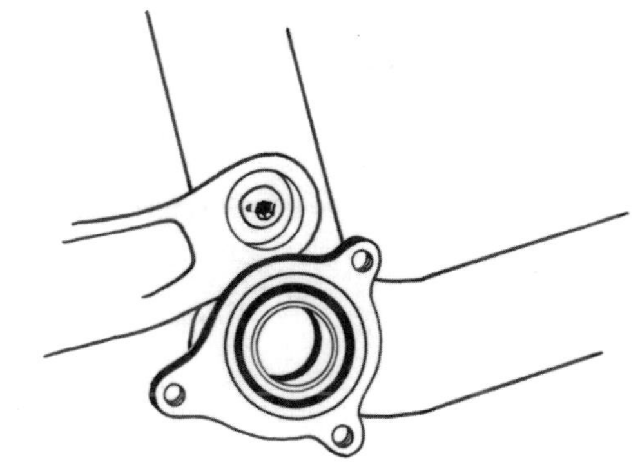

The frame must also have ISCG mounts if you wish to use a Truvativ HammerSchmidt crankset, which is a crank with an internal two-speed transmission.

Most chain guide systems (Fig. 5.54) consist of a mounting plate that bolts to the ISCG tabs with adjustable-position enclosed rollers that are positioned at about 11 o'clock and 7 o'clock on the top and bottom of a bash ring. A bash ring looks like a chainring without teeth, and it mounts to the crank spider where the outer chainring would ordinarily be mounted.

Tension arms, rollers, giant front-derailleur cages (of which an early version is illustrated in Fig. 5.55), and other chain-retention systems have become a part of downhill racing, because of the extraordinary demands the sport places on the drive system. Downhill-specific products are outside the scope of this book, but once assembled on the bike, most of these drivetrain items will be completely obvious in their function and maintenance.

5.54 MRP chain guide/bash ring system

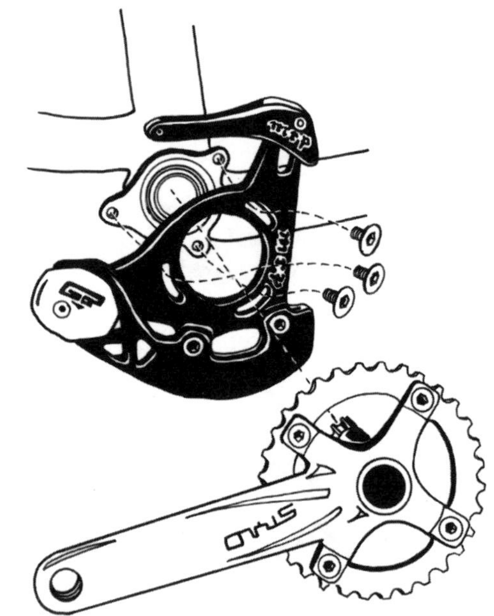

5-31

BELT DRIVE

The Gates Carbon Drive belt-drive system (Fig. 4.26) is becoming more popular on single-speed bikes and bikes with internal-gear rear hubs. Keys to making this system work are described in 4-13.

5-32

ROHLOFF AND OTHER INTERNAL-GEAR HUBS

Internal-gear systems offer the nice advantage of being able to shift when you're not pedaling and not having derailleurs get clogged with mud or caught on things. Bikes using internal-gear rear hubs require a frame with a system like those described in 4-8b to tension the chain either by adjusting the distance between the rear hub and the crank or by using a chain tensioner with a roller pushing on the chain. And as I say in 4-8b, the only full-suspension bikes you can use them on without a chain tensioner are those in which the distance from the bottom bracket to the rear hub does not change as it goes over bumps. The only full-suspension bikes that are thus qualified are concentric-pivot systems (where the rear swingarm pivots around the bottom bracket), unified-rear-triangle systems (where the bottom bracket

5.55 This sucka ain't goin' nowhere

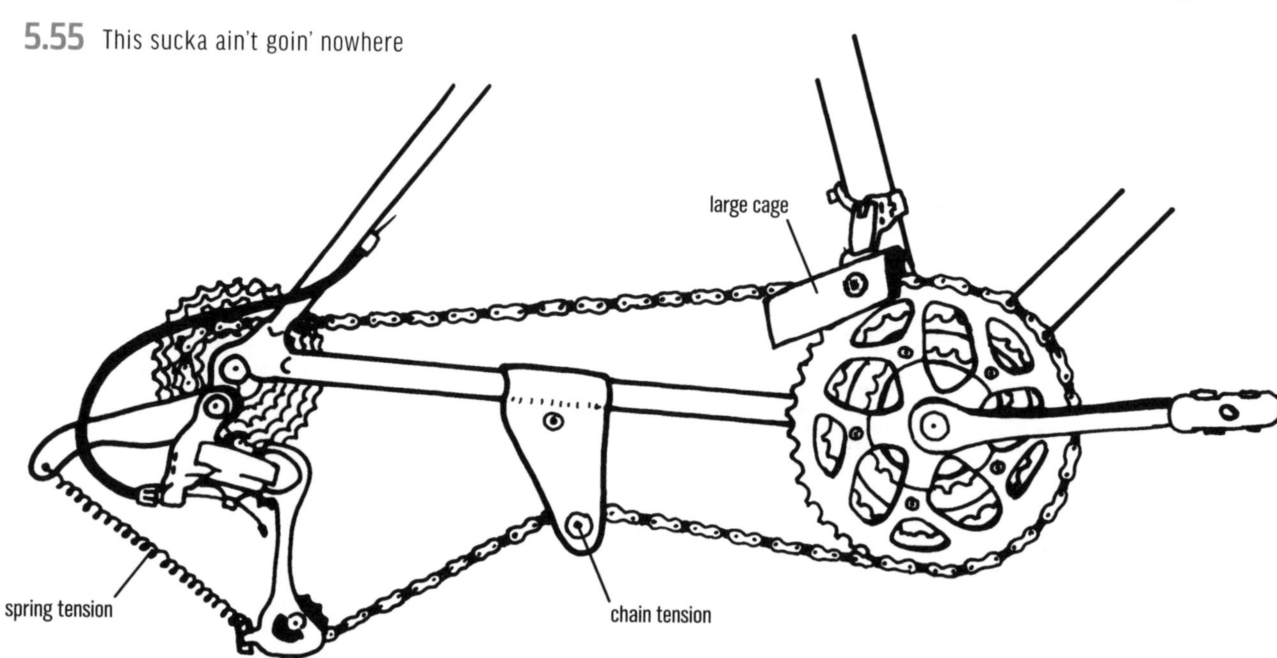

and dropouts are rigidly attached to each other), and softtails (where the suspension movement works by means of flex in the chainstays).

The hub shell of an internal-gear hub rotates at a different speed from its cog in all gears except the one-to-one gear—the gear in which the cog and hub shell are directly linked. In a gear that multiplies the rotation of the hub from that of the cog (a high gear), a torque is generated trying to twist the axle forward. In a gear that reduces the rotation of the hub from that of the cog (a low gear), a torque is generated trying to twist the axle backward. Internal-gear hubs thus require a method to anchor the hub axle to the frame more securely than by simply clamping the axle ends, as on a normal hub. The wider the gear range, the higher this torque is twisting the axle, and the more secure must be the clamping method to resist it. The most basic system to counteract the torque on the axle is a long, flat torque arm made of steel plate that bolts to a clamp wrapped around the left chainstay, similar to what you would see on a bike with a coaster brake or your grandfather's old British three-speed.

The Rohloff SpeedHub is the internal-gear hub of choice for mountain biking because of its wide range as well as its robust construction. This hub, available in myriad variations to anchor its internals, and the plate on the left end of the hub can be interchanged with different torque arms to fit different bikes. The slickest anchoring system for the Rohloff SpeedHub is a round plate with a little rectangular block sticking up from it that is an inch out radially from the axle stub in the center. That little rectangular stub on the plate is the torque arm, and all you need is a left-side dropout with a slot long enough to secure it. Another system for fitting a Rohloff hub onto a mountain bike with disc-brake mounts is to use a Rohloff SpeedBone, which is a torque arm that bolts to the disc-brake mounts on the left dropout.

The Rohloff SpeedHub has a dual-cable twist shifter that moves it through all of its 14 speeds. The gear ratios are spaced evenly, approximately 13.5 percent apart, and their range spans approximately the same gear range as an older 3×9-speed mountain bike system.

Even though the Rohloff shifter seems to click into each gear, the shift indexing is inside the hub, not in the shifter. There is no wear on the indexing system, nor is there a need to adjust it.

The most common SpeedHub variation used on mountain bikes has a detachable external cable box. The two cables enter the cable box on the left side of the SpeedHub through barrel adjusters. A slew of countersunk holes around the exterior of the hub end plate allow you to rotate the attachment position of the cable box so that the cables enter their barrel adjusters smoothly, whether you are routing to it over the top of the top tube and seatstays or underneath the down tube and chainstays. Using the barrel adjusters, adjust the cable tension so that there is 2mm of rotational play in the shifter between gear clicks.

To most easily remove the wheel and to ensure that the cable box, shifter, and hub internals agree on what gear they are in when you put the wheel back in, always remove and install the wheel with the shifter in gear position 14. Because you can shift the bike while it is stopped, this is not a problem. Unscrew the knurled thumbscrew on the face of the cable box, and pull off the cable box. Now you can remove the wheel as you would a normal rear wheel. You can interchange cogs by holding the cog with a Pedro's Vise Whip or a chain whip while unscrewing the lockring with a wrench on a Rohloff lockring remover (Fig. 5.56).

To reinstall the wheel, drop it in with the chain on the cog as you would with a normal single-speed rear wheel.

5.56 Removing the cog from a Rohloff SpeedHub

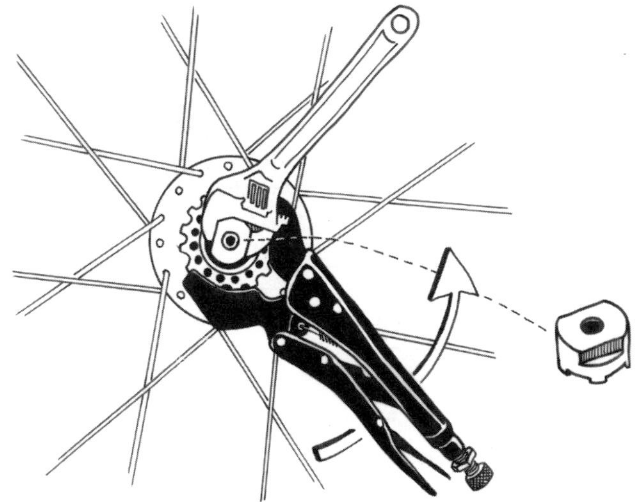

If you forgot to shift to position 14 (all 14 gear positions are numbered on the twist shifter) before removing the wheel, you can get into that gear now. To do so, turn the shifter to position 14 and turn the hexagonal peg on the hub plate that the cable box engages with (it looks like a nut sticking out of the center of a round brass flange) to the last gear position by turning it counterclockwise through all of its clicks with an 8mm box wrench. (If, when you start riding, all 14 gears are not usable, you mounted the cable box somewhere in the middle of the gear range, and you need to pull off the cable box, twist the shifter to position 14, and turn the hexagonal peg as I just described before reinstalling it.)

Set the cable box in place over the hexagonal peg so that the two locating pegs insert into the two holes in the back of the cable box. Twiddle the shifter back and forth around position 14 until the cable box drops into place over the hexagonal peg. Tighten the knurled thumbscrew to secure the cable box. You're now ready to ride.

The hub needs to be filled with 25mL of Rohloff oil, which must be changed every 3,000 miles. You also should put a drop of oil on the hexagonal peg and the hole into which it engages in the cable box every few times you remove the wheel. Otherwise, maintenance is generally minimal aside from normal cable maintenance.

DERAILLEUR MAINTENANCE

5-33

MAINTAINING JOCKEY WHEELS

LEVEL 1

The jockey wheels (a.k.a. "guide pulleys"—Fig. 5.57) on a derailleur will wear out over time. On a mountain bike that sees heavy use, standard jockey wheels should be overhauled every 200–500 miles. That should take care of the gunk that the chain and the trail regularly deliver to them. The mounting bolts on jockey wheels also need to be checked regularly. If a loose jockey-wheel bolt falls off while you are riding, you'll need to follow the procedure for a broken rear derailleur on the trail, described in 3-9.

Some expensive guide pulleys have cartridge bearings (Figs. 5.57, bottom, and 5.58), whereas standard ones (Fig. 5.57, top and center) have a center bushing

5.57 Jockey wheels, exploded view

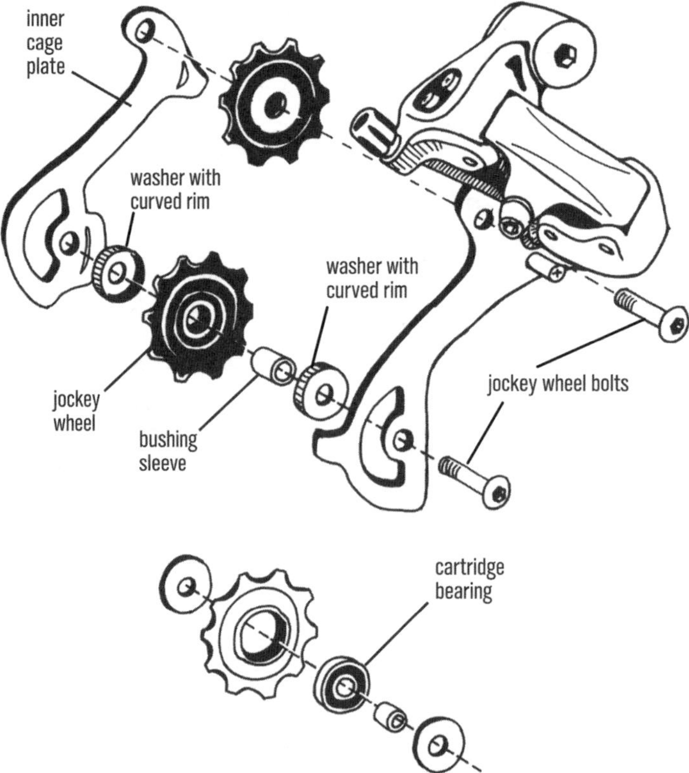

sleeve made of either steel or ceramic (SRAM upper jockey wheels have an oversized steel center bushing). A washer with a curved rim facing inward (visible in Fig. 5.57, center) is usually installed on both sides of a standard jockey wheel. Some guide pulleys also have rubber seals around the edges of these washers.

5-34

OVERHAULING STANDARD JOCKEY WHEELS

1. **Remove the jockey wheels.** Remove the bolts (2.5mm or 3mm hex key) that hold them to the derailleur (Figs. 5.57, 5.58), and catch the parts as they fall.
2. **Wipe all parts clean with a rag.** A solvent is usually not necessary but can be used.
3. **If the teeth on the jockey wheels are broken or worn off, replace the wheels.**
4. **Smear grease over each bolt and sleeve and inside each jockey wheel.**
5. **Reassemble the jockey wheels on the derailleur.** Be sure to properly orient the inner cage plate (the larger part of the cage plate should be at the bottom jockey wheel). Some Shimano jockey wheels have a rotation-direction arrow, so check orientation when installing.

5-35

OVERHAULING CARTRIDGE-BEARING JOCKEY WHEELS

If the cartridge bearings (bottom, Fig. 5.57; both in 5.58) in high-end jockey wheels do not turn freely, they can usually be overhauled.

1. **Remove the jockey wheels.** Remove the bolts (2.5mm or 3mm hex key) that hold them to the derailleur (Figs. 5.57, 5.58), and catch the parts when they fall.
2. **Pry the plastic cover off one or, preferably, both sides of the bearing** (Fig. 8.20). Use a single-edge razor blade or box-cutter knife.
3. **With a toothbrush and solvent, clean the bearing.** Use a citrus-based solvent, and wear gloves and glasses to protect skin and eyes.
4. **Blow the solvent out with compressed air or your tire pump, and allow the parts to dry.**
5. **Squeeze new grease into the bearing, and replace the covers.**
6. **Reassemble the jockey wheels on the derailleur.** Be sure to orient the inner cage plate properly (the larger part of the cage plate should be at the bottom jockey wheel).

5.58 Shimano rear-derailleur clutch and lower pivot assembly, exploded

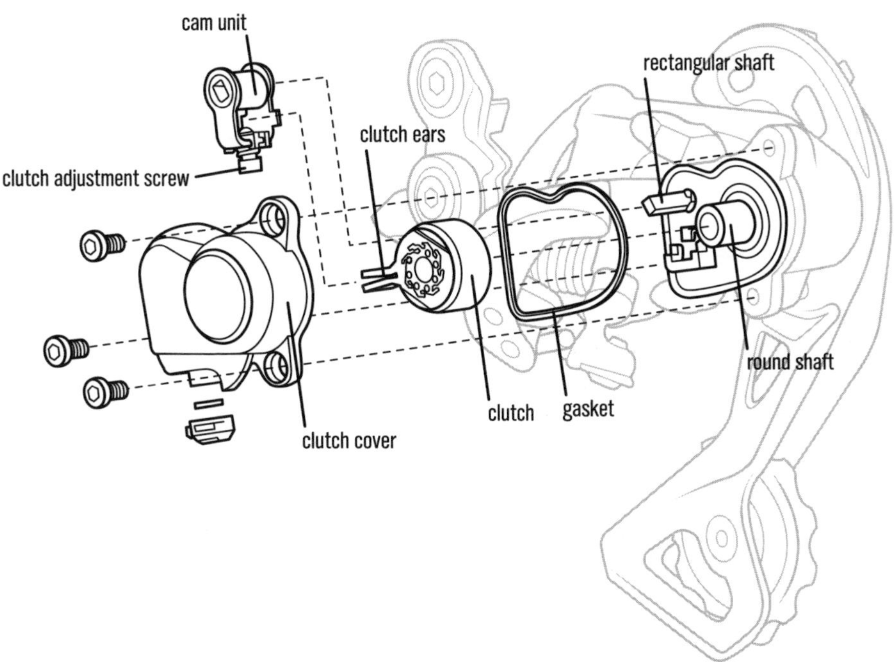

5-36

OVERHAULING A SHIMANO CHAIN STABILIZER (CLUTCH)

The derailleur does not need to be removed; this task is perhaps performed most easily while the derailleur is on the bike. Avoid dropping the three little screws that hold the clutch cover on.

If you ride your full-suspension bike a lot on bumpy trails, particularly on wet trails, and if you also wash your bike frequently, especially with a power washer, you might need to overhaul the clutch every month or six weeks. You'll know you need to overhaul it if the derailleur creaks upon suspension movement or when shifting to a larger cog.

If you want to overhaul the lower pivot or replace the jockey-wheel cage, skip to 5-40 when you get to step 4.

1. **Turn the clutch lever switch to "off."** It is atop the clutch housing (Figs. 5.4, 5.5, and 5.15).
2. **Remove the clutch cover.** Undo the three screws with a 2mm hex key and pull off the cover (Fig. 5.58). Keep track of the gasket; it may stay on the derailleur or come off with the cover.
3. **Pull the clutch straight off.** It's the circular band with two ears. The cam unit, which has the cam and the clutch-adjustment screw on it, will come up off its rectangular shaft along with the clutch's ears, which run through it (Fig. 5.59).

5.59 Inset view of cam unit installed on clutch ears with cam in correct orientation

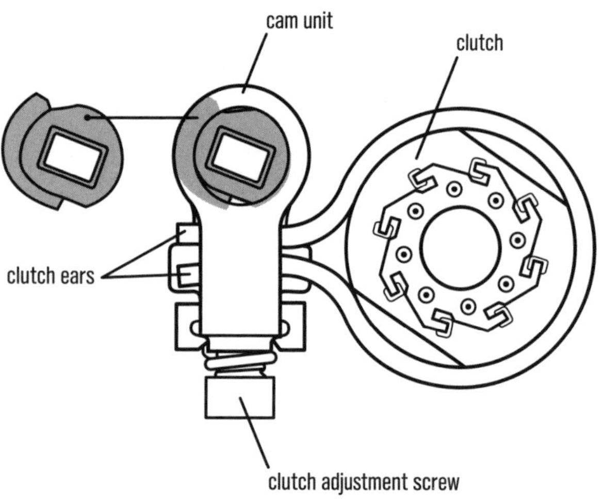

4. **Clean everything.** With a rag, wipe off the round shaft, the rectangular shaft, the inside and outside of the clutch, and the cam.
5. **Grease the outside of the clutch.** The best grease for this is Shimano's Nexus internal-gear-hub grease; it can handle the heat and will be more durable than most other types of grease (the clutch gets hot with rear-suspension movement on bumpy trails). Don't grease the clutch's central hole or the round shaft; if you get grease on either part, the clutch cannot produce sufficient friction to do its job.
6. **Push the clutch and the cam unit back on their shafts.** Rotate the cam so when its rectangular hole lines up with the rectangular shaft, its lobe is pointed away from the clutch (Fig. 5.59). If you have the cam pointed the other way, tightening the clutch adjustment screw won't squeeze the clutch's ears toward each other and cinch down on the round shaft and restrict the jockey-wheel cage's rotation. The trick to getting the clutch to go on fully is to pull forward on the jockey wheel cage as you push the clutch onto the shaft.
7. **Replace the clutch cover.** Make sure the gasket is in its groove to seal the edge of the cover from the derailleur body. Replace the three screws.
8. **Adjust the clutch as in 5-2h.**

5-37

OVERHAULING THE REAR DERAILLEUR

Except for the jockey wheels, clutch, and upper and lower main pivots, most rear derailleurs are not designed to be disassembled. If the pivot springs seem to be operating effectively, all you need to do is overhaul the jockey wheels (5-34) and clean and lubricate the parallelogram and spring as described in 5-38.

5-38

PERFORMING A MINOR WIPE AND LUBE

1. **Clean the derailleur as well as you can with a rag. Be sure to get in between the parallelogram plates.**
2. **Drip chain lube on both ends of every pivot pin.**

3. **Grease the spring contacts.** If there is a clothes-pin-type spring in the parallelogram (as opposed to a coil spring running diagonally from one corner of the parallelogram to the other), put a dab of grease where the spring end slides along the underside of the outer parallelogram plate.

5-39

OVERHAULING THE UPPER PIVOT

LEVEL 2

CAUTION: *Don't do this job unless you absolutely must in order to rehabilitate a poorly functioning rear derailleur. The strong spring resists your best intentions at reassembly, and you may not be able to get it back together properly, even with a second set of hands.*

1. **Remove the rear derailleur.** This usually takes a 5mm hex key to unscrew it from the frame and to disconnect the cable.
2. **With a screwdriver, pry the circlip** (Fig. 5.60) **off the threaded end of the mounting bolt.** Don't lose the bolt; it tends to fly when it comes off.
3. **Pull the mounting bolt and the upper pivot spring out of the derailleur.**
4. **Clean and dry the parts with or without the use of a solvent.**
5. **Grease the parts liberally and reassemble them.**
6. **Engage the spring ends into their holes.** If there are several holes, and you don't know which one the spring was in before, try the middle one. (If the derailleur does not keep tension on the chain well enough, you can later try another hole that increases the spring tension.)
7. **Squeeze the parts together, and replace the circlip with pliers.**

5.60 Rear-derailleur pivots

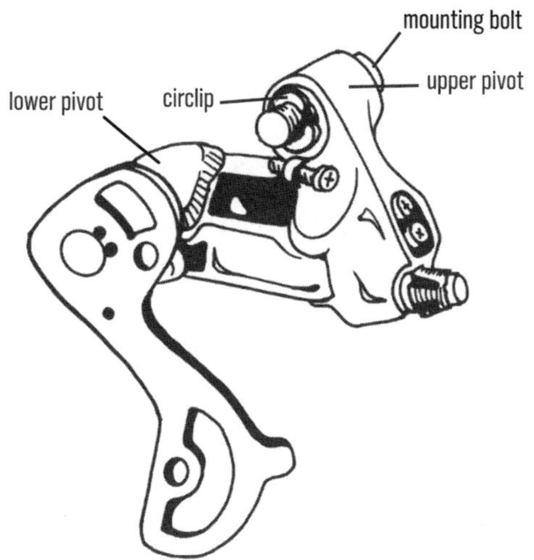

5-40

OVERHAULING THE LOWER PIVOT; REPLACING JOCKEY-WHEEL CAGE; ADJUSTING SPRING TENSION

LEVEL 2

If shifting is sluggish in both directions but the cable is moving freely (5-6 to 5-15), and the cable tension is correct (5-2d), try adjusting the lower pivot spring by turning the b-screw (Fig. 5.7) until the chain gap is correct (5-2f).

On some rear derailleurs without a clutch, chain retention and proximity of the guide pulley to the cogs can be further improved by increasing the lower pivot spring tension after disassembling the lower pivot. With the same procedure, you can also replace a bent jockey-wheel cage or increase the capacity of the rear derailleur by substituting a longer cage for a shorter one. And, of course, you can just grease the pivot parts to restore free action.

a. For recent Shimano rear derailleurs with a clutch (a.k.a., "chain stabilizer")

1. **Perform the clutch service in 5-36 up to step 4.** Then continue here with the next step. Unlike as with the clutch service alone, however, you need to remove the derailleur from the bike before continuing.
2. **Remove the jockey-wheel-cage stopscrew** (Fig. 5.13)**.** Allow the jockey-wheel cage to rotate around and release the tension from the spring.
3. **Unscrew the round shaft with a 4mm hex key** (Fig. 5.58)**.** Remove the jockey-wheel cage.
4. **Clean and grease the jockey-wheel-cage axle and the spring.**
5. **Install the spring.** Line up the spring ends with the notch inside the pivot housing and with the hole in the jockey-wheel cage it came out of (usually the more clockwise hole; see Fig. 5.58).
6. **Push the jockey-wheel cage into place.**

7. **Tighten the round shaft into the cage axle.**
8. **Replace the stopscrew** (Fig. 5.13). First rotate the jockey-wheel cage counterclockwise to wind its spring until the stopscrew's hole comes around the bottom of the pivot; then screw in the stopscrew.
9. **Continue on from step 4 in the clutch service in 5-36 above.**

b. For rear derailleurs without a clutch mechanism

1. **Remove the derailleur from the bike.** This usually takes a 5mm hex key to unscrew it from the frame and to disconnect the cable.
2. **Open the derailleur knuckle.** Shimano derailleurs can be divided into two types: ones that have a setscrew on the side of the lower pivot and ones that do not. If yours has a setscrew (Fig. 5.11), remove it using a 2mm hex key, and pull the jockey cage away from the derailleur.

   If the derailleur has no setscrew, find and unscrew the tall cage stopscrew on the derailleur cage (Fig. 5.13); it is located near the upper jockey wheel. This screw is designed to maintain tension on the lower pivot spring and prevent the cage from springing all the way around. Once the screw is removed, slowly guide the cage around until the spring tension is relieved. Remove the upper jockey wheel, and unscrew the pivot bolt from the back with a hex key, usually 5mm, sometimes 6mm (Fig. 5.14). Be sure to hold the jockey-wheel cage to keep it from twisting.
3. **Determine in which hole the spring end has been placed, and then remove the spring.**
4. **Clean and dry the bolt and the spring with a rag.** A solvent may be used if necessary.
5. **Grease all parts liberally.**
6. **Replace the spring ends in their holes in the derailleur body and jockey-wheel cage** (Fig. 5.12). Put the spring in the adjacent hole if you want to increase its tension (see 5-2g).
7. **Close the derailleur knuckle.** If the derailleur has a setscrew, push the assembly together, wind the spring, and replace the setscrew. If the derailleur does not take a setscrew, wind the jockey-wheel cage back around, screw it all back together with the pivot bolt, and replace the stopscrew.

5-41

OVERHAULING THE PARALLELOGRAM

Very few derailleurs can be completely disassembled. Those that can (Mavic, first-generation SRAM, and 2002 high-end SRAM) have removable pins holding them together. Circlips on the pin ends can be popped off with a screwdriver, after which the pins can be pulled out. Disassemble the derailleur carefully in a box so that the circlips do not fly away, and make note of where each part belongs. Clean all parts, grease them, and reassemble.

5-42

REPLACING STOCK BOLTS WITH LIGHTWEIGHT VERSIONS

Lightweight aluminum and titanium derailleur bolts used to be common replacement items for many derailleurs, but as the bolt shapes on modern derailleurs change so rapidly with each model year, making them has ceased to be a profitable business for small machine shops. Removing and replacing jockey-wheel bolts is simple, as long as you keep all of the jockey-wheel parts together and put the inner cage plate back on the way it was. Upper and lower pivot bolts are replaced following the instructions in 5-39 and 5-40.

TROUBLESHOOTING REAR-DERAILLEUR AND RIGHT-HAND SHIFTER PROBLEMS

Once you have made the adjustments outlined previously, the drivetrain should be quiet and should stay in gear, even if you turn the crank backward. If you cannot fine-tune the adjustment so that each click with the right shifter results in a clean, quick shift, you need to check some of the following possibilities. For skipping- and jumping-chain problems, see also the troubleshooting section at the end of Chapter 4.

5-43

DERAILLEUR TROUBLESHOOTING CHART

Ensure that the rear derailleur's low-gear limit screw prevents it from going into the spokes before test riding.

TABLE 5.1 — TROUBLESHOOTING SHIFTING PROBLEMS

REAR DERAILLEUR	
Chain jams between small cog and frame	Tighten cable (5-2e, 5-2d).
	Turn high-gear limit screw clockwise (5-2b).
Rear derailleur touches spokes	Turn low-gear limit screw clockwise (5-2c).
	Check derailleur-hanger alignment (17-5).
Chain falls between spokes and large cog	Turn low-gear limit screw clockwise (5-2c).
Chain won't go onto large cog	Tighten cable (5-2e, 5-2d).
	Turn low-gear limit screw counterclockwise (5-2c).
	Chain is too short (4-8)—replace with a longer one.
	Derailleur cage too short; replace cage (5-40) or derailleur.
Shifting sluggish to a larger cog	Tighten cable (5-2e).
	Turn off Shimano clutch (5-2h).
	Loosen Shimano clutch (5-2h).
Shifting sluggish to a smaller cog	Loosen cable (5-2e).
	Lubricate or replace cable (5-6–5-19).
	Lubricate return spring (5-37).
Shifting sluggish both directions	Replace or lubricate cable (5-6–5-19).
	Adjust b-screw (5-2f).
Derailleur creaks on downshifts or suspension movement	Overhaul derailleur clutch (5-36).
FRONT DERAILLEUR	
Chain falls off to outside	Check derailleur position (5-5a).
	Turn high-gear limit screw clockwise (5-5d).
Chain falls off to inside	Check derailleur position (5-5a).
	Tighten cable (5-5e).
	Turn low-gear limit screw clockwise (5-5c).
	Install inner stop (5-51).
Chain rubs inner cage plate in low gear	Check derailleur position (5-5a).
	Turn low-gear limit screw counterclockwise (5-5c).
	Loosen cable (5-5e).
Chain rubs outer cage plate in high gear	Check derailleur position (5-5a).
	Turn high-gear limit screw counterclockwise (5-5d).
	Tighten cable (5-5e).
Derailleur hits crankarm	Check derailleur position (5-5a).
	Turn high-gear limit screw clockwise (5-5d).
Shifting sluggish to big chainring	Check derailleur position (5-5a).
	Tighten cable (5-5e).
Shifting sluggish to small chainring	Check derailleur position (5-5a).
	Loosen cable (5-5e).
	Turn low-gear limit screw counterclockwise (5-5c).
	Replace or lubricate cable (5-6–5-19).

Make small adjustments each time before rechecking shifting. Turn cable barrel adjusters one click (or one-eighth turn) each time. Turn limit screws one-eighth turn each time.

If there are multiple options listed for a given problem in Table 5.1, perform them in order; if the first fix doesn't work, try the next one.

5-44

SHIFTER COMPATIBILITY

To work properly, the shifter must be compatible with the derailleur and cogs. This is especially important if any of these parts are not original equipment on the bike. It should be obvious that a 7-speed shifter will not work on an 8-, 9-, 10-, 11- , or 12-speed cassette, but components of different brands for the same number of speeds often will not work together either. If the shifter is a different brand from the derailleur, be certain that they are designed to work together. The most common example of this is SRAM's Grip Shift, some models of which are specifically designed to work with a Shimano rear derailleur (SRAM got its start making twist shifters to retrofit onto Shimano-equipped bicycles). Otherwise, only SRAM rear derailleurs work with SRAM shifters.

Generally, frictional shifters will work with anything, albeit slowly and inefficiently, since the user controls how the derailleur lines up, but indexed (click) shifters dictate the derailleur and cog spacing. Shifting performance will be optimized by using components that were all designed to work together as a system—same brand, model, and number of speeds for the shifters, front and rear derailleur, cogs, chain, and, to a lesser extent, the crank. Other combinations may work, but it can take considerable trial and error to find compatibility among different brands.

If the shifter and derailleur are incompatible, you will need to change one of them. If the quality of the parts is equal, I suggest replacing the less costly item (usually the shifter).

The rear derailleur's shift-activation ratio—the amount of lateral movement of the rear derailleur divided by the amount of cable pull required to generate that amount of lateral movement (i.e., the number of millimeters of lateral displacement of the rear derailleur per millimeter of cable pull)—is built into the derailleur. And the cable pull—the amount the cable moves with each click—is built into the shifter. The cable-pull per shift multiplied by derailleur's shift-activation ratio is equal to the distance the derailleur's chain cage moves laterally with each shift. To shift properly, this must be equal to the distance from the center of one rear cog to the center of the next (the cog pitch). Here's the equation:

TABLE 5.2 — REAR SHIFTING SPECIFICATIONS

BRAND AND NUMBER OF SPEEDS	CABLE PULL (mm)	SHIFT RATIO	COG PITCH (mm)
Shimano 6 Road/Mountain	3.2	1.7	5.5
Shimano 7 Road/Mountain	2.9	1.7	5
SRAM (1:1) 7 Mountain	4.5	1.1	5
Shimano 8 Road/Mountain	2.8	1.7	4.8
SRAM (1:1) 8 Mountain	4.3	1.1	4.8
Shimano 9 Road/Mountain	2.5	1.7	4.35
SRAM (1:1) 9 Mountain	4	1.1	4.35
Shimano 10 Mountain	3.4	1.2	3.95
SRAM (Exact Actuation) 10 Road/Mountain	3.1	1.3	3.95
Shimano 11 Mountain	3.6	1.1	3.9
SRAM (X-Actuation) 11 Mountain	3.5	1.12	3.9
SRAM (X-Actuation) 12 Mountain	3.45	1.01	3.5

Cable pull × Derailleur shift-activation ratio = Cog pitch

Cog pitch is equal to the thickness of a middle cog plus the thickness of one of the spacers separating it from the adjacent cogs on either side. Cog pitch decreases as the number of cogs increases. Shift-activation ratios, cable pull, and cog pitches vary not only with the number of rear cogs but also from brand to brand, and even sometimes from model to model within a brand.

Table 5.2 shows cable pull (in millimeters), shift-activation ratio, and cog pitch (in millimeters) based on the number of speeds for the two major component manufacturers.

5-45

STICKY CABLES

Check the derailleur cables to confirm that they run smoothly through the housing. Sticky cable movement will cause sluggish shifting. Lubricate the cable by smearing it with chain lube or a specific lubricant that came with the shifters (5-19).

If lubricating the cable does not help, replace the cable and housing (see 5-7 to 5-17).

5-46

BENT REAR-DERAILLEUR HANGER

A bent hanger will hold the derailleur crooked and bedevil shifting. Most newer bikes have replaceable hangers. Replaceable hangers are intentionally sacrificial, made to bend or break in a crash, saving your derailleur. It's a good idea, therefore, to keep a spare on hand, and to carry it with you when riding. Instructions for straightening the hanger are in 17-5.

5-47

BENT REAR-DERAILLEUR CAGE

A bent derailleur cage will hold the jockey wheels at an angle to the cogs. A mildly bent cage can be straightened by hand. Eyeball the crank for vertical reference.

5-48

LOOSE PIVOTS (WORN-OUT REAR DERAILLEUR)

A loose and floppy rear derailleur will not shift well. Replace it.

TROUBLESHOOTING FRONT-DERAILLEUR AND LEFT-HAND SHIFTER PROBLEMS

5-49

CHAIN SUCK

For chain suck problems, refer to the troubleshooting section at the end of Chapter 4.

5-50

CHAINLINE

Your bike probably has chainline problems if (1) the chain falls off to the inside no matter how much you adjust the low-gear limit screw, cable tension, and derailleur position; (2) there is chain rub, noise, or auto-shift problems in mild cross-gears that are not corrected with derailleur adjustments; or (3) the front derailleur cannot move the chain onto the large chainring even if the outer limit screw is backed all the way out.

NOTE: *If problem 2 occurs intermittently on a full-suspension bike (for example, when you are sitting but not when you are standing), it could be that as the rear suspension compresses, the chain is no longer running in the groove formed for it in the front-derailleur cage. Instead, the chain is hitting the forward ridge above the groove. You may have to put up with this problem because your alternatives are to either tighten the suspension (i.e., increase the spring rate) so that it does not move as much, or to get a derailleur without the internal cage shaping (which won't shift as fast).*

Chainline is the imaginary line connecting the center of the middle chainring with the middle of the cogset (Fig. 5.61). In theory, this line should be straight and parallel to the vertical plane of the bicycle. Even owners of new bikes may find poor chainlines, owing to mismatched cranks and bottom brackets.

The chainline is adjusted by moving or replacing the bottom bracket to move the cranks left or right. You can roughly check the chainline by placing a long straightedge against the middle chainring and back to the rear cogs; it ideally would come out in the center of the rear cogs. But the tire-clearance issues of most mountain bikes, particularly on full-suspension models with large tires or fat bikes, make this ideal unreasonable to expect; the chainrings have simply got to be farther out than the cogs or the bike designer cannot fit everything in. And as the rear hub width has changed on mountain bikes, so has the chainring position. The Boost 148mm rear hub-width standard for full-suspension bikes has resulted in a consequent increase in bottom bracket spindle length. And for fat bikes, whose rear hub width is approaching 200mm, the cranks have become appropriately far apart as well.

Continue to the next section for a precise chainline-measurement method.

5-51

PRECISE CHAINLINE MEASUREMENT

LEVEL 2

You will need a caliper (Fig. 1.4). The position of the middle chainring, as measured from the center of the seat tube to the center of the middle chainring, is often called the chainline, although this is only the front end point of the line.

1. **Find the front end point of the chainline** (CL_F in Fig. 5.61) **as follows:**
   (a) Measure from the left side of the down tube to the outside of the large chainring (d_1 in Fig. 5.61). (Do not measure from the seat tube, as it may be oval where it meets the bottom-bracket shell.)
   (b) Measure the distance from the right side of the down tube to the inside of the inner chainring (d_2 in Fig. 5.61).
   (c) Add these two measurements and divide the sum by two:

$$CL_F = (d_1 + d_2) \div 2$$

5.61 Measuring chainline

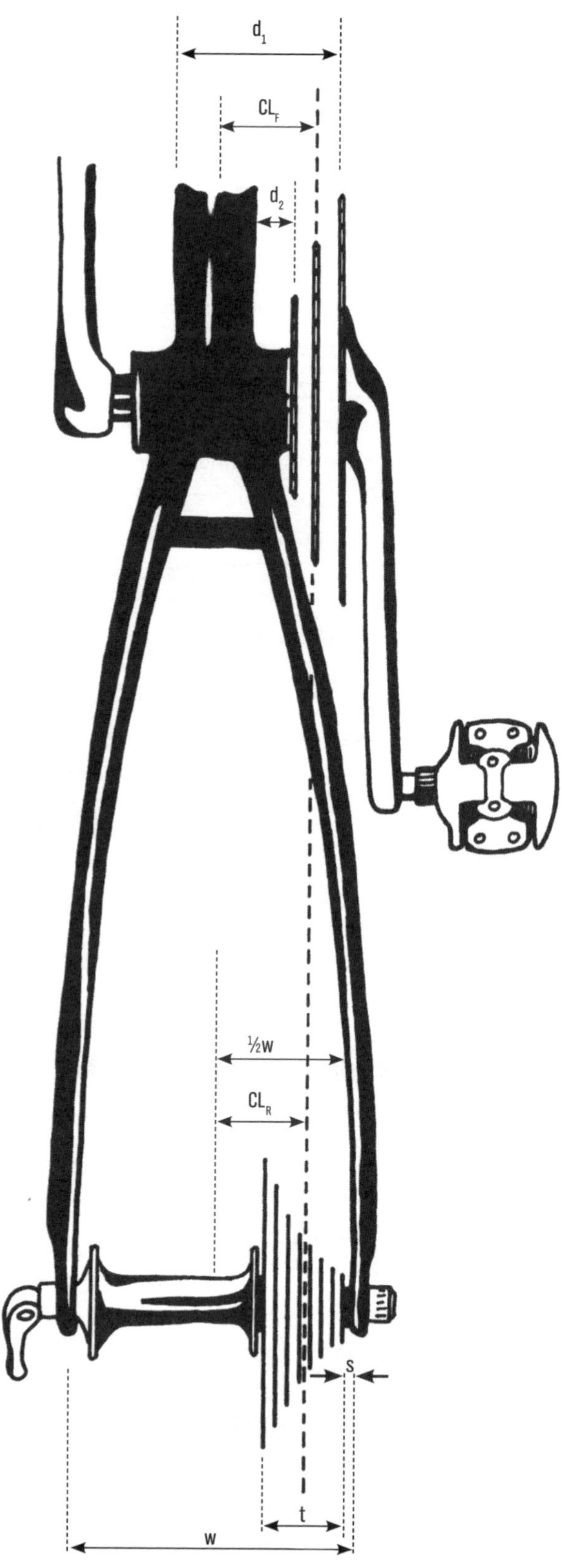

2. **Find the rear end point of the chainline** (CL_R in Fig. 5.61). This is the distance from the center of the plane of the bicycle to the center of the cogset.
   (a) Measure the thickness of the cog stack, end to end (t in Fig. 5.61).
   (b) Measure the space between the face of the smallest cog and the inside face of the dropout (s in Fig. 5.61).
   (c) Measure the length of the axle from dropout to dropout (w in Fig. 5.61). This dimension is also called the "axle overlock dimension," referring to the distance from locknut face to locknut face on either end. Generally, on mountain bikes with quick-release rear hubs since 1989 or so, this will be 135mm. Mountain bikes with through-axles generally have either 12×142mm or 12×148mm (Boost) rear axles. Fat bikes tend to have 170mm or 190mm rear axle lengths.
   (d) Subtract one-half of the thickness of the cog stack and the distance from the inside face of the right rear dropout from one-half of the rear-axle length:

$$\mathbf{CL_R = w/2 - t/2 - s}$$

3. **See if $CL_F = CL_R$.** If it does, the chainline is perfect. This, however, almost never occurs on a mountain bike, because of considerations about chainstay clearance of the tire and the chainrings, prevention of chain rub on large chainrings in cross-gears, and inward movement range of the front derailleur. For example, Shimano specifies a "chainline" (meaning CL_F, the front end point of the chainline) as 47.5mm for bikes with a 68mm-wide bottom-bracket shell and 50mm for 73mm-wide shells (both of these specified dimensions are plus or minus 1mm). CL_R, the rear end point of the chainline, on the other hand, usually comes out around 44.5mm for a cassette on a hub with a 135mm quick-release axle.

   Shimano's specifications, then, are primarily intended to avoid chainring rub on chainstays, not to promote ideal shifting. It designs its derailleurs to be optimized for a front chainline a few millimeters wider than the rear chainline.

5.62 Third Eye Chain Watcher

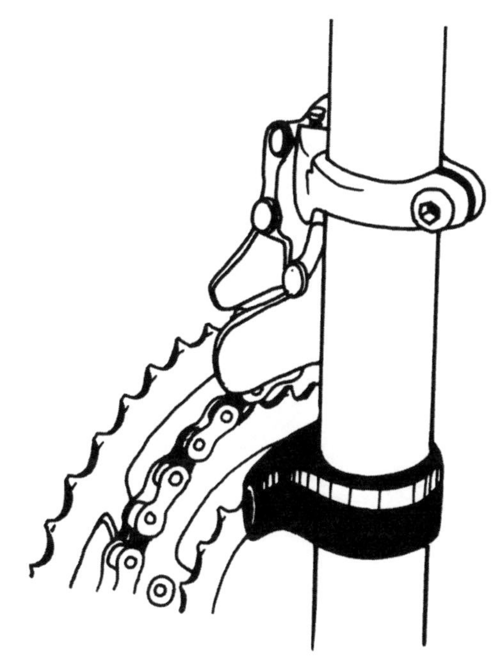

4. **To improve the chainline, move the chainrings.** There is little or nothing you can do with the rear cog position. The chainrings are moved by using a different bottom bracket, by exchanging the bottom-bracket spindle for a longer or shorter one, or by moving the bottom bracket right or left (bottom-bracket installation and overhaul are covered in Chapter 11).

NOTE: *The chainline can be off if the frame is out of alignment (see 17-13). If that's the case, it is probably not something you can fix yourself.*

5. **If the chain is still dropping on shifts, or if you don't want to mess with the chainline, install an inner chain stop.** Try a Third Eye Chain Watcher (Fig. 5.62), a Jump Stop, or a Deda Elementi Dog Fang. All three clamp around the seat tube next to the inner chainring. Clamp one on and adjust the position so that it nudges the chain back on when it tries to fall off to the inside.

It is exciting to discover electrons and figure out the equations that govern their movement; it is boring to use those principles to design electric can openers.

—NEAL STEPHENSON

6

ELECTRONIC SHIFTING SYSTEMS

TOOLS

Flat-blade and Phillips screwdrivers, small and medium
Torx T8 and T30 keys
Shimano TL-EW02 Di2 plug-in tool

OPTIONAL

Park Tool IR-1 or IR-1.2 Internal Cable Routing Kit
Snapring pliers
Shimano TL-FDM905 Di2 front-derailleur-cage wrench (to replace the front-derailleur cage, which is available as a separate part)
32mm headset wrench

Although you can shift modern cable-actuated derailleurs quickly and precisely with relatively light pressure from your thumbs, electrically actuated shifting—or electronic shifting, as the bike industry calls it—is even faster and more powerful, and some shifts are possible that simply cannot be made with cable-actuated derailleurs. Greatly improved shift quality with Shimano Di2 electronic derailleurs, especially in front, allows shifting under full power with little chance of dropping the chain. Furthermore, electronic derailleurs prevent shifts that cable shifters allow, no matter how ill advised; for instance, Di2 won't let the rider shift to small-small gear combinations that would allow the chain to be completely slack and hence vulnerable to jumping off when hitting bumps.

For example, a cable-actuated front derailleur can't upshift to the big chainring while sprinting out of the saddle over the crest of a hill and gaining speed. To make that shift, the rider would have to ease off the pedals, thus losing momentum and being overgeared once the shift finally happens. But an electronic system will make that shift every time. Another example: While braking hard on a rough, steep descent, you can shift from high gear all the way to low gear in anticipation of a steep uphill turn by simply holding down an electronic shift button. Try making that full-range shift while simultaneously braking hard and judiciously with a cable-actuated derailleur!

Shimano Di2 can also grant the simplicity of a "one-by" system with a single shifter without losing the gear range of a 2×11 or 3×11 drivetrain. You simply ask for an upshift or downshift with the right (rear) shifter, and the system fulfills your request by shifting either just the rear derailleur, or both the front and rear derailleurs. Shimano calls this one-hand shifting "Synchro Shift," and you can toggle between manual mode and two different Synchro modes on the fly.

In addition to saving weight (by eliminating one shifter if you don't care to ever have the manual option) and shifting more precisely and with less effort, Synchro allows shifting with one hand. It doesn't matter which hand, either; you can customize which shifter does what with a tablet or smartphone and a wireless app or a PC and a wired interface.

Synchro shifting is not just for injured, lazy, or gram-counting riders; it makes smarter shifts than we can often be trusted to do. It ensures that the gear-ratio steps that we are taking are consistent and thus makes maximal use of the gears we have. For instance, with a 2×11 Di2 drivetrain, we get 13 unique, evenly spaced gear ratios and nothing else; the racer benefits from this at least as much as the lazy, inexperienced, or inattentive rider would. A cross-country racer will be faster with Synchro Shift not only by making consistently good shifts without ever jamming or dropping the chain even when brain cells are oxygen-starved, but also by minimizing chain drag, since the bigger the chainring and cog that the chain is engaged on, the lower the chain resistance (because the chain bends less). Synchro shifting on a 2×11 in either of the preprogrammed options, dubbed S1 and S2, will keep the chain on the big chainring while downshifting through all 11 cogs from the smallest to the largest rear cog; only then will it finally shift to the inner chainring at the same time that it drops the chain back down to the second-biggest rear cog. And then the 13th and final gear combination it uses is small front to biggest rear.

Di2 can also control the electronic damping-control motors in Fox iRD front and rear shocks with the Di2 shifters and using the Di2 battery. The suspension settings are displayed on the same screen as the shifting information.

In short, electronic shifting offers us a new world of better shifting and lower weight with less thinking about what gear we're in. And, on the Shimano digital display (Fig. 6.2), namely the SC-MT800 (XT) or SC-M9050 or SC-M9051 (both XTR, the latter being Bluetooth-compatible), or on an ANT+ cycling computer paired with the system, we can see what gear we're in, how much battery charge we have left, what shifting program we're using (Synchro S1 or S2 or Manual), and get a beep alert when we're at the end of the gear range or two warning beeps in Synchro shift when we're about to experience a front shift.

6-1

OPERATING ELECTRONIC SHIFTERS

The operation of electronic shift levers is analogous to what we're accustomed to with cable-actuated shifters (5-1). That said, Di2 now allows you to program the shift switches to perform shifts differently than the way they're set up out of the box. Furthermore, mountain bike Di2 components can be mixed and matched with road Di2 components. For instance, mountain-bike Di2 derailleurs coupled with road drop-bar Di2 levers and a mountain-bike double crank will provide a much lower gear range on a road bike. Or, a mountain-bike Di2 rear derailleur coupled with a road drop-bar Di2 lever on a one-by (single-chainring, 1×11) bike will allow use of a much bigger rear cassette on a cyclocross bike. And of course if you are setting up your mountain bike with a drop bar (as an adventure bike, for example), Di2 simplifies the process.

a. Rear Shimano Di2 Electronic Thumb Shifter, XT or XTR

In the default configuration, to go to a larger cog, push the bigger, lower, more outboard paddle (Fig. 6.1) on the

6.1 Shift switches on Shimano XT/XTR Di2 electronic right lever

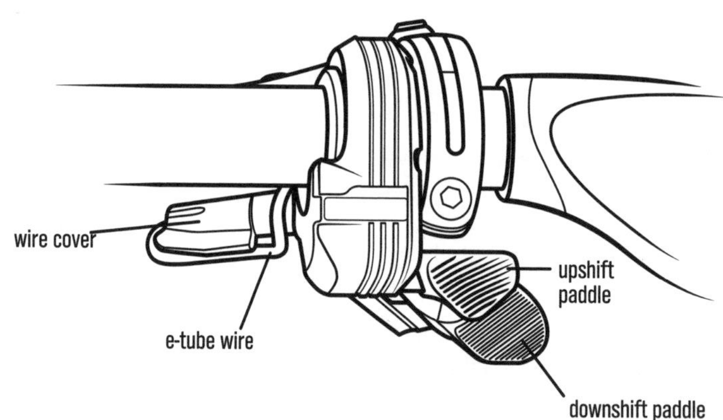

right shifter. To go to a smaller cog, push the smaller, upper, more inboard paddle.

When you get to the lowest or the highest gear, as well as each time you try and shift further in that direction once you've reached the end of the gears, Di2 gives you the bad news by emitting a single beep from the digital display unit. At that point, you're just going to have to get by with the gear you're in. A paired ANT+ cycling computer (see 6-10) will emit the same warning beep.

b. Front Shimano Di2 Electronic Thumb Shifter, XT or XTR

In the default configuration, to go to a bigger chainring, push the bigger, lower, more outboard paddle (Fig. 6.1) on the left shifter. To go to a smaller chainring, push the smaller, upper, more inboard paddle.

Note that, even in Manual mode, the front derailleur sometimes moves itself after a rear shift; as you shift through the rear cogs, the Di2 front derailleur automatically moves over to avoid chain rub at two different points when shifting through the rear cassette in either direction.

c. Di2 Battery Level

To view the battery-level indicator on the digital display (Fig. 6.2), click any shift paddle; this lights up the display and shows the battery level. If you don't have the display, press and hold a shift paddle until the indicator LED on Junction A (Fig. 6.3) lights up; green indicates full, flashing green is half full, red is quarter full, and flashing red is near empty. If you have the system paired with a cycling computer, the battery level will show up there, too.

When the battery runs out of charge, the front derailleur stops working first, then the rear derailleur. They stop at the last gear positions they were in.

d. Shift Mode

To switch from Manual (M) shifting mode to Synchro Shift (either S1 or S2), double-click the mode button. Each double click toggles to the next setting option, and it is shown on the digital display (Fig. 6.2) or on a paired ANT+ cycling computer.

Manual mode means the front derailleur only shifts when you push one of the buttons on the left shifter.

6.2 Di2 digital handlebar display, bottom view

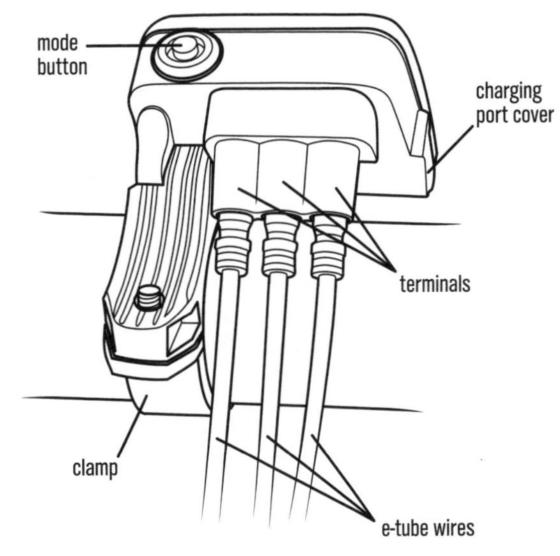

6.3 Di2 Junction A

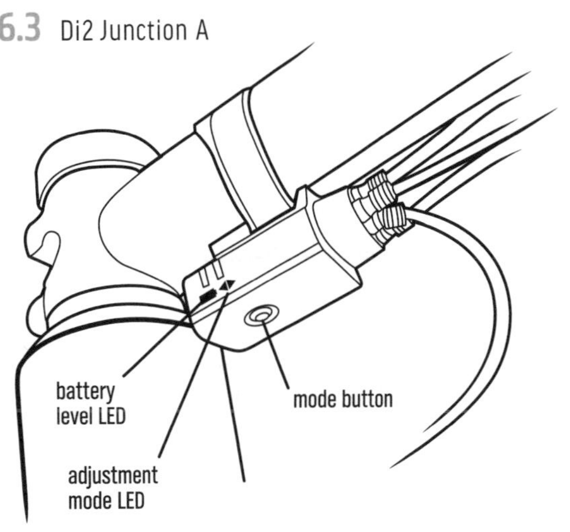

With either S1 or S2, you only need to shift the right shifter; the system will automatically shift the front derailleur at certain points in either shifting direction, accompanied by a simultaneous shift of the rear derailleur back in the other direction to make the gear step commensurate with single steps of the rear cassette.

In Synchro Shift mode, the front shift will not come as a surprise if you don't have ear buds plugged into your head; the digital display unit will emit a quick double beep to warn you when you've reached the gear on the rear cassette at which a front shift will come on the next push of the right shifter. A paired ANT+ cycling computer that is Synchro-Shift compatible (see 6-10) will do the same thing. The double beep is distinct from the single beep that tells you, "Sorry, buddy, you have no more gears left. Deal with it." when you reach or try to surpass either the lowest or the highest gear in the range.

NOTE: *If you make changes to any of the hardware or firmware or shift programming, the system will revert to Manual mode until you again double-click the mode button to select a Synchro-Shift mode.*

6-2

BATTERY CHARGING

With a Shimano internal battery, plug the charger wire (Fig. 6.27) into the port hidden under the square cover on the side of the digital display (Fig. 6.2) or of Junction A (Fig. 6.3). To charge an external Di2 battery, remove it from the bike (Fig. 6.8) and plug it into its charging pod. Charge time for Di2 batteries is approximately 1.5 hours.

Charge the battery fully before using it the first time. You can charge the battery with any amount of charge left in it. Leaving the battery uncharged for extended periods can damage it, and storing it with at least a half charge is preferable. When storing for extended time, recharge it periodically.

6-3

ADJUSTING ELECTRONIC DERAILLEURS ON THE FLY

You can adjust electronic front and rear derailleurs while you are riding (or stopped). This is useful for fine-tuning or if you change to a different rear wheel.

Adjust the rear derailleur:

1. **Push the mode button to go into adjustment mode.** Click and hold down the mode button on the digital display (Fig. 6.2) until the "R" of the F/R adjustment indicator flashes. If you don't have the display, press and hold the mode button on Junction A (Fig. 6.3) until the red LED with the "+-" icon lights up; adjustment mode will also show up on a paired cycling computer.
2. **Click the appropriate shift switch.** For instance, if the rear derailleur is slow to shift the chain from a bigger to a smaller cog, click the upshift switch (the smaller, upper paddle) on the right lever once or twice. Each click moves the derailleur an amount similar to what mechanical rear derailleur moves in response to turning a cable barrel adjuster a single notch (5-2d or 5-5e). The display will show how many positive or negative steps the derailleur has moved; there are 32 total steps, 16 in each direction.
3. **Exit adjustment mode.** Push the mode button again to turn off adjustment mode. The display will return to the standard screen, or the red LED on Junction A will turn off.
4. **Shift through the gears.** If shifting is still not ideal, repeat steps 1–4.

Adjust the front derailleur:

1. **Shift to the lowest gear.** The chain should be on the smallest chainring and largest rear cog.
2. **Push the mode button to go into adjustment mode.** Click and hold down the mode button on the digital display (Fig. 6.2) until the "R" of the F/R adjustment indicator flashes. If you don't have the display, press and hold the mode button on Junction A (Fig. 6.3) until the red LED with the "+-" icon lights up; adjustment mode will also show up on a paired cycling computer.
3. **Click the appropriate shift switch.** Click either the upper or lower paddle until the inner cage plate just barely clears the chain without rubbing. The upper paddle moves it farther from the chain, and the lower paddle moves it closer to the chain. Each click moves the derailleur a similar amount to a single notch of a cable barrel adjuster on a mechanical system (5-2d or 5-5e). The display will show how many positive or negative steps the derailleur has moved over; there are 32 total steps, 16 in each direction.
4. **Exit adjustment mode.** Push the mode button again to turn off adjustment mode; the display will return to the standard screen, or the red LED on Junction A will turn off.
5. **Shift through the gears.** If the chain rubs the inner cage plate, or if the chain does not shift up to the big chainring, repeat steps 1–5.

INSTALLING ELECTRONIC SHIFTING SYSTEMS

Since electronic systems are so different from cable-operated shifting systems, you may be wary of installing one of these the first time. However, you might find it to be quicker, if not easier, than installing a cable-shift system.

6-4

BATTERY INSTALLATION

Your first decision is what kind of a battery to use. If your bike's frame won't allow internal wires, an external Shimano Di2 battery is your only choice. If your bike accepts internal routing and you don't have a dropper post (Chapter 14), then you might as well choose a seatpost battery.

With internal routing and a dropper post (which of course has no room inside for a battery), you could use an external battery, or you could conceivably wrap a cylindrical Di2 battery in bubble wrap, attach a zip-tie loop to the top of it to grab it by, and shove it down inside the seat tube. However, if you're using a stealth dropper post (one with the cable or hydraulic hose running internally up the seat tube to the seatpost), you won't want to, or possibly even be able to, house the battery within the seat tube. Instead, you can install it inside the steering tube.

Also, if you are going to use a Bluetooth connection to a smartphone or tablet or an ANT+ connection to a cycling computer, you need a battery with enough memory. New Di2 batteries are Bluetooth compatible; for internal mount, the cylindrical BT-DN110 (Fig. 6.4) replaces the identically shaped SM-BTR2; for external mount, use either the blocky BT-DN100 (Fig. 6.7) replacing the SM-BTR1, or the cylindrical BT-DN110 installed inside the cylindrical SM-BTC1 bottle-cage-mounted battery case.

a. Shimano Di2 seatpost battery installation

The Di2 cylindrical battery (Fig. 6.4) is very slim. There are a number of ways to secure the battery in the seatpost. Here are two: Shimano/PRO's method and Ritchey's method.

6.4 Prying an e-tube wire out of a Di2 battery terminal with a TL-EW02 tool

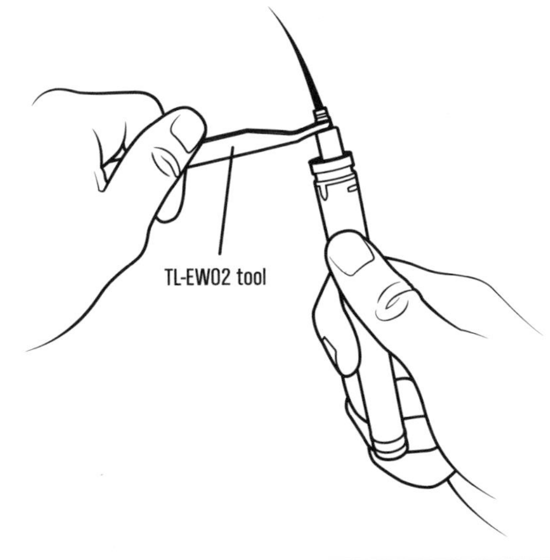

Before installing the battery into the seatpost, attach an e-tube wire of the appropriate length (long enough to extend a few inches past the bottom bracket when the seatpost is installed at the correct height) to the battery. Using the prongs on the TL-EW02 tool, pry out the plug protecting the battery connector (Fig. 6.4). To install, slip the end of the wire into the slender cylindrical end of the plastic TL-EW02 tool (Fig. 6.22) until the base of the connector hits the stop inside the tool. Push the wire into the junction port at the end of the battery with the TL-EW02 tool until it clicks into place.

Shimano/PRO

1. **Glue the Di2 seatpost sleeve into the bottom of the seatpost.** Epoxy or any other glue for metal is fine. Make sure the sleeve is the right size for the seatpost. If you have a PRO-brand (a Shimano subsidiary) Di2 seatpost, this sleeve will already be in place.
2. **Insert the battery.** Slide it in, terminal end down, after you've put the two included clamshell adapter pieces around it at that end (Fig. 6.5).
3. **Secure the battery with the snapring.** Sandwich the wave washer between the pair of flat washers, and push the washer sandwich in against the end of the battery (Fig. 6.5). Squeeze the snapring together with snapring pliers (Fig. 1.2), and slide it

6.5 Installing the sleeve in the seatpost and a Di2 SM-BTR2 or BT-DN110 battery into the sleeve

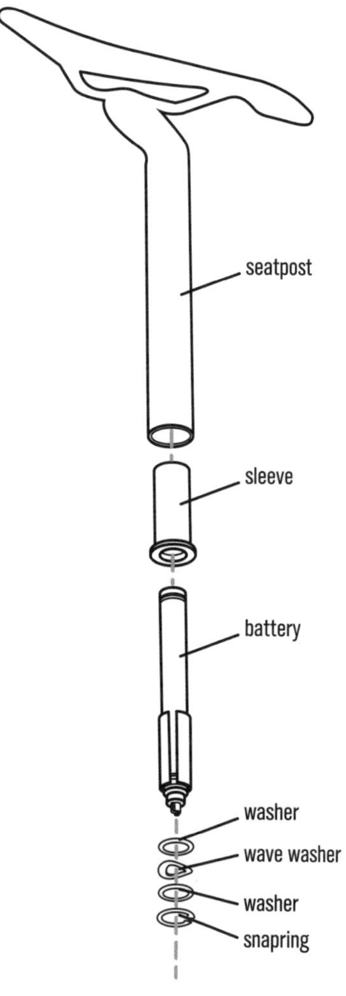

6.6 Installing Shimano Di2 internal battery into seatpost with Ritchey adapter

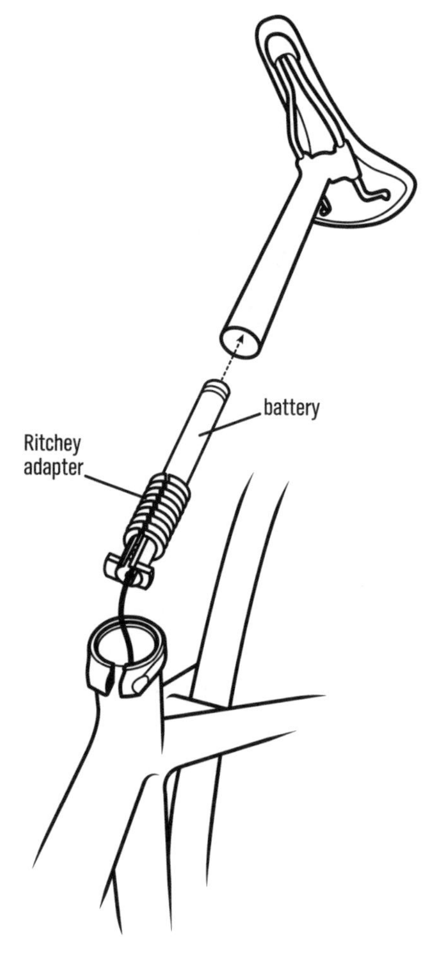

into the seatpost sleeve so that it snaps into the groove at the end of the seatpost sleeve.

4. **Drop the wire from the battery down the seat tube.** You may need to fish it through to the bottom bracket as in 6-5.
5. **Install the seatpost.** Pull the wire until it sticks out of the bottom-bracket shell a few inches.

Ritchey

1. **Ensure that the adapter is the correct size for the seatpost.**
2. **Put the rubber clamshell Ritchey battery adapter onto the battery.** With the adapter's ears at the wire end, nestle the battery into each half of the adapter (Fig. 6.6). There are grooves and notches to line up with corresponding protrusions on the battery.
3. **Push the battery and adapter up into the seatpost.** Keep pushing until the ears are at the end of the seatpost (grab these to remove the battery). The adapter's soft ribs will grip the inside of the seatpost and keep the battery secure. If the seatpost is cut off at an angle, don't worry about it; just shove the adapter and battery in as far as you can. It will stay in place just fine.
4. **Drop the wire from the battery down the seat tube.** You may need to fish it through to the bottom bracket as in 6-5.
5. **Install the seatpost.** Pull the wire so it sticks out of the bottom-bracket shell a few inches.

NOTE: *3T's internal Di2 battery mount fits in a 31.6mm-diameter seatpost; its installation is a combination of the above Shimano/PRO and Ritchey methods. You* *install the battery with washers and a snapring inside the 3T sleeve just like into the Shimano/PRO sleeve (Fig. 6.5). And then you push the 3T sleeve, which has prongs on the side to grip the inside of the seatpost, up into the seatpost as you would with the Ritchey adapter.*

b. Shimano Di2 steering-tube battery installation

You will need a Shimano cylindrical (BT-DN110 or SM-BTR2) battery, a PRO PRAC0085 two-piece, ribbed clamshell Di2 internal steering-tube battery holder, and a foam handlebar grip. Select a battery wire long enough to reach from under the fork crown to the display or to Junction A.

I. PRO Di2 stem and Di2 handlebar

If you have a PRO Di2 stem and Di2 handlebar, the battery wire can come out of the top of the steering tube and enter the stem through a hole just in front of the steering tube.

1. **Run the wires out to the shifters.** With the handlebar free, drop each Di2 wire into the oval hole at the center of the bar, run it out the hole near the end of the handlebar (see 6.6a for internal-routing tips), and double it back down the groove. Slide on the shifter and plug in the wire, wrapping it over the groove in the wire cap (Fig. 6.1). With two shifters, you'll also want to plug in a Junction B (Fig. 6.22) that will live inside the bar (Fig. 6.13).
2. **Install the stem.** Screw in the PRO Di2 stem's adjuster ring fully. Slide it onto the steering tube. The steerer should end 2mm below the top edge of the stem, so cut it (12-7, step 7d), or put spacers below. Line up the stem; tighten the upper clamp bolt. With a 32mm headset wrench on the stem's adjuster ring, adjust the headset. Tighten both clamp bolts to the correct torque.
3. **Install the battery.** Plug in the battery wire. Put the PRAC0085 clamshell halves around the battery (Fig. 6.7). Push the battery holder into the steerer. Hook its ears on the top edge facing right and left. The stem's top cap will snap over them. To pull out the battery holder, use its wire loop, plugging the ends into the holes at the bases of the ears. Install a Junction B (Fig. 6.22) inside the stem (Fig. 6.13). Additional wires run down the groove between the halves of the battery holder. Wires to the bike can drop out of the bottom of the steerer or over the top edge through the stem clamp groove (break off the rear tab on the stem top cap to clear the wire).
4. **Install the handlebar.** Plug a wire from the handlebar into the Junction B inside the stem. Run another wire from this Junction B out the slot in the handlebar clamp to the display (or to Junction A).

II. Standard stem and bar

With a standard stem and bar, you can install the battery upside down and secure it with a zip tie.

1. **Remove the front wheel.** Set it aside.
2. **Remove the stem, leaving it clamped to the handlebar.** Strap the handlebar to the fork.
3. **Knock the star nut out of the steering tube.** Using a long wooden dowel, tap the star nut down through the steerer from top to bottom with a hammer.

6.7 Installing SM-BTR2 or BT-DN110 battery inside steering tube

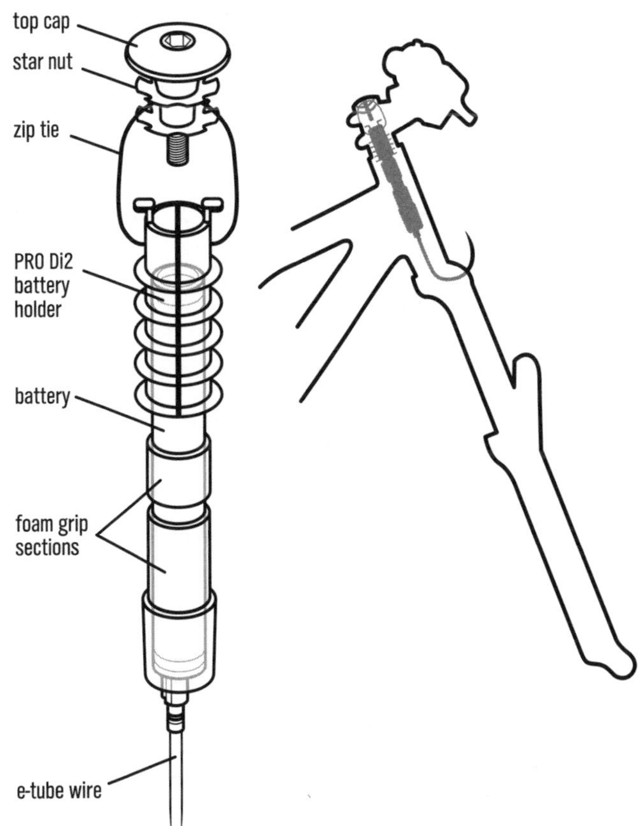

4. **Cut two, inch-long sections of a foam grip.** Slide one onto the battery about one-third of the way from the terminal, and slide the other over the terminal so it overhangs by a half-inch.
5. **Assemble the internal battery holder** above the upper piece of handlebar grip. The holder's ears will be at the top (non-terminal end) of the battery. Run a long, thin zip-tie through the hole at the base of each ear.
6. **Install the battery into the steerer** (Fig. 6.7).
7. **Loop the zip-tie around a lower prong of a new star nut.** Leave its tail end poking through the upper teeth of the star nut.
8. **Install the star nut** (Fig. 12.15).
9. **Remove the star-nut installation tool and pull on the zip-tie.** Grip the zip-tie's tail with needle-nose pliers and pull slack through. Trim with side cutters. At this point, the battery assembly should be hanging from a zip-tie under the star nut, held in place by friction with the Pro Di2 battery holder and padded with rings of foam handlebar grip. The battery terminal should be visible from the underside of the crown.
10. **Cut out a circular piece of flattened foam grip the size of the inside of the grip.** Poke a small hole in this circular plug, and push one end of the e-tube wire through the hole.
11. **Plug the wire into the battery.**
12. **Glue in the plug.** With flexible glue (Gorilla Glue or similar), bond it into the end of the grip section extending past the end of the battery inside the steering tube.
13. **Plug the battery wire into the digital display or into Junction A.**
14. **Reinstall the stem and wheel.**

c. Shimano Di2 external battery installation

There are two general types of Di2 external battery holders. One is the original type, in which the rectangular battery snaps into a bracket with a lever closure (Fig. 6.8). The bracket is attached either to a long strap that mounts under a water bottle cage, or a shorter one designed to bolt to dedicated, threaded Di2 battery mounts. The second holder is a thin, cylindrical battery case that mounts alongside a water bottle cage and accepts the thin, rod-shaped battery (Fig. 6.9).

If your frame accepts internal wiring, I don't see the point in using an external battery. Given that you can install the battery inside the seatpost or the fork steerer, and it's simple to recharge the battery through the port on the digital display or Junction A, why have a battery hanging out in the elements? Just sayin'. . . .

Original-style Di2 external battery installation

1. **Loosely bolt the battery bracket to the frame mounts.** For internal wiring through a hole in the frame between the mounts, temporarily mount it using only a single, partially screwed-in bolt at the top of the mount to allow access to the hole in the frame for wires connected to the battery. Frames may have various battery-mounting locations. For bikes without a specific battery mount, bolt it underneath the bottle cage on the down tube. Use the long mounting plate for a water-bottle mount; use the short mount for bolting to dedicated Di2 mounts below the bottle cage or under the down tube or left chainstay. With a Di2 bracket under the bottle mount, ensure that there

6.8 Removing and installing Shimano Di2 SM-BTR1 or BT-DN100 external battery: open lever, press release button, pull battery out; reverse to install

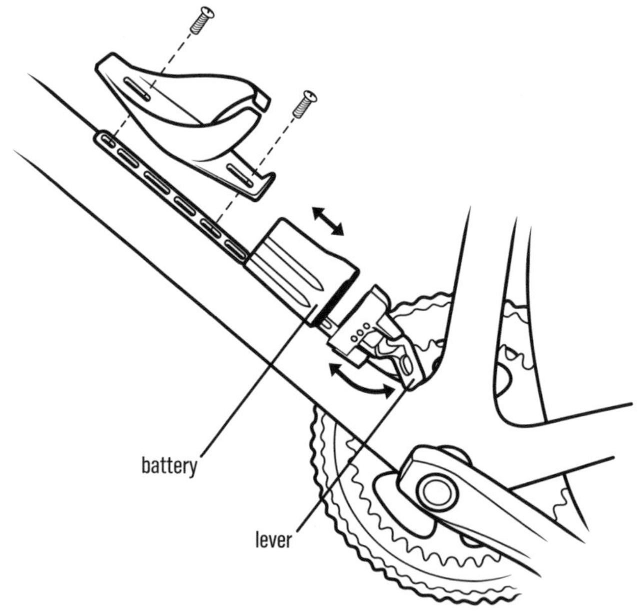

6.9 Mounting BTR2 or BT-DN110 battery with external battery case

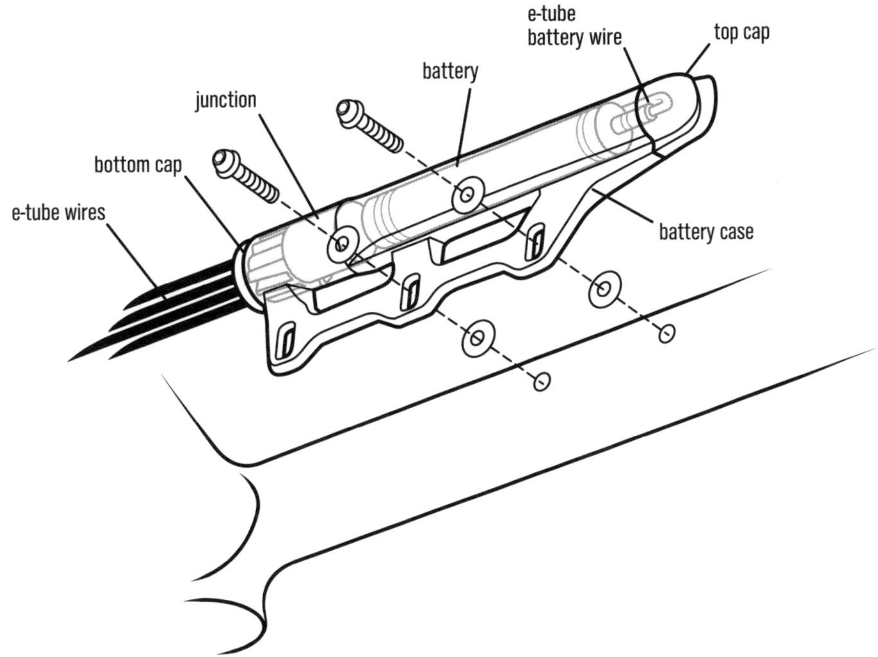

is at least 108mm from the base of the bracket to the bottom of the bottle cage (Fig. 6.8); this allows enough room to slide the battery up to remove it before it hits the bottle.

2. **Push the wire(s) into the hole in the frame.** Skip to Step 3 for external routing. Guide the wire(s) out one end of the bottom-bracket shell. The easiest way is with a magnetic guide tool—a magnet on the end of a shift cable (Fig. 6.16), but you can easily run these down through the seat tube with a stiff wire from the top or by pulling them down with a wire you run up from the bottom bracket.
3. **Tighten the battery to the mounts.** Use the supplied spacers between the mounts and the frame.

Low-profile external Di2 battery installation in battery case

Shimano SM-BTR2 or BT-DN110 cylindrical battery within SM-BTC1 battery case (Fig. 6.9).

1. **Bolt the battery case under the water bottle cage.** With the narrow end up, bolt through the forward or rearward pair of the battery case's three holes.
2. **Plug the e-tube wires into the battery case junction.** Use the TL-EW02 Di2 plug-in tool to snap each wire into place. The junction will accept as many as six wires and eliminates the need for a lower Junction B in the system. Plug in wires to the derailleurs, to the display or Junction A, and, if applicable, to the front and rear Fox iRD shocks; this would fill all six ports. Plug the battery wire into the port with the notch on it. Install dummy plugs into any open ports with the TL-EW02.
3. **Run the battery wire.** Wrap the battery wire back around the junction and down its side groove, into the battery case, and up to the terminal end of the battery case.
4. **Cover the ports with the junction spacer, and push the junction into place.** Run the wires (other than the battery wire) out the slots in the spacer. Line up the point of the teardrop shape on the junction with the corresponding shape in the end of the battery case.
5. **Install the junction cap.** Run the e-tube wires out of the hole in the cap, line up its teardrop shape with the corresponding shape in the end of the battery case, and snap the cap into place. Secure it with the fixing screw on the side.
6. **Wrap the protective sheet around the battery.** Remove the backing paper from the protective sheet and wrap it around the battery one turn; the

protective sheet's notch should cup one of the tab windows at the terminal end of the battery (Fig. 6.9).

7. **Install the O-ring into the groove at the non-terminal end of the battery.** It's packaged with the battery case.
8. **Slide the battery into the case.** Push the non-terminal end into the case until only the terminal is left sticking up out of the top end of the case. Leave free the battery wire coming up from the lower, junction end, of the case.
9. **Plug the battery wire into the battery terminal.** As always, use the TL-EW02 Di2 plug-in tool to push the wire's head in until it clicks.
10. **Install the winged battery cap.** It shields the terminal end of the battery. Secure it with the fixing screw on the side.

6-5

INSTALLING ELECTRONIC DERAILLEURS AND SHIFTERS

Make sure that you have the correct front derailleur adapter to fit the seat tube or the front derailleur mount on the seat tube or swingarm. Mountain bike Di2 front derailleurs come in two types: double and triple. They have the same mounting groove on the back to accept either of two mating band clamps (high clamp or low clamp), or one of two adapters for direct mounts (D or E).

Beware: The motors that drive electronic derailleurs are extremely powerful; if a button on the left shifter is pushed when your finger is between the derailleur and the large chainring, you will regret having had it there and will let loose some colorful language. For your own safety, rotate the cranks when shifting, and disconnect the battery before installing electronic components.

1. **Install the rear derailleur** (Fig. 6.10). Follow the instructions in 5-2 for cable derailleurs.
2. **Loosely mount the front derailleur** (Fig. 6.11). Set its cage above the height where the top of the chainring will be so that it won't interfere when you install the crank. Make sure you have the right adapter—Type D, Type E, high clamp or low clamp (Fig. 6.11)—and that you affix it properly (with the supplied bolt) to the derailleur.

6.10 Installing Shimano XTR Di2 rear derailleur

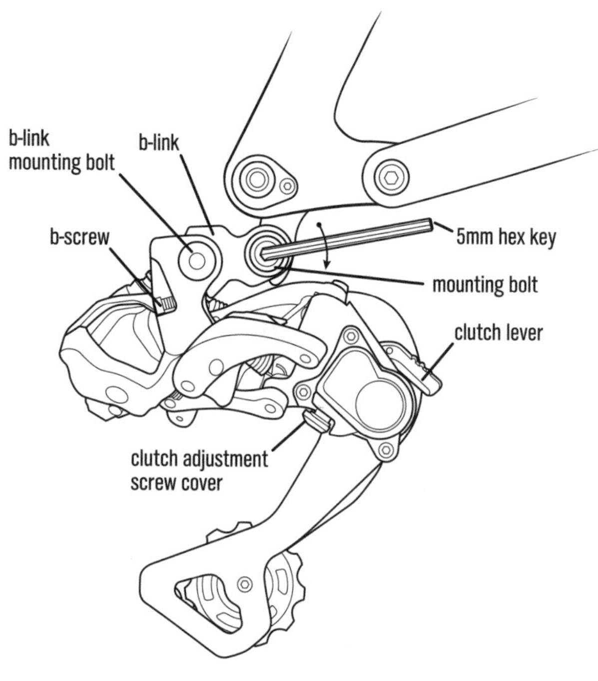

6.11 Shimano Di2 front derailleur and adapters

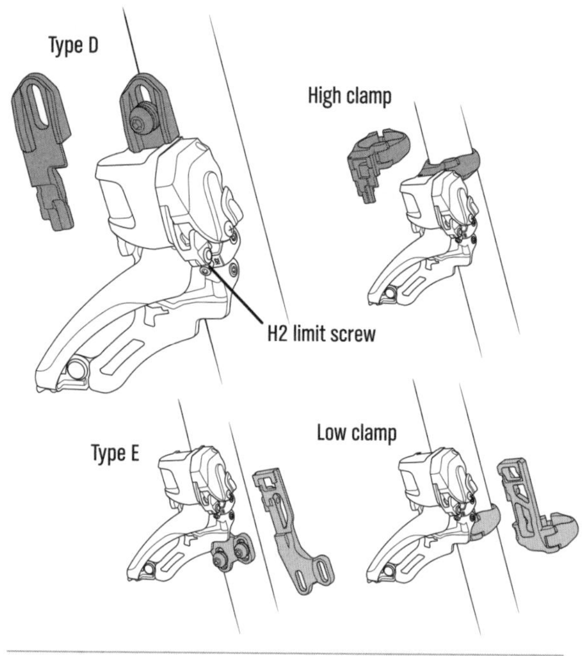

3. **Install the shifters.** Before tightening the clamping bolt (Fig. 6.1), set the shifter band laterally and rotationally where you want it. There are little set-screws (Fig. 6.25) on the back of each shift paddle so you can adjust them inboard and outboard independently.

4. **Install the digital display or Di2 upper junction.**
   (a) **Bolt the digital display** (Fig. 6.2) **to the handlebar,** or:
   (b) **Secure Di2 Junction A to the stem with the supplied strap** (Fig. 6.3). (You can also mount it to a Bar-Fly mount integrated with a headset spacer.)
      (i) **Wrap the strap over the stem shaft.**
      (ii) **Hook a slot on each strap end over a hook on either side of the Junction A mount.** Orient the mount's short end forward.
      (iii) **Trim the ends of the strap.**
      (iv) **Slide Junction A in from the side until it clicks.** The wire ports will point forward.

      To remove Junction A, push the little release tab on the rear of the mount and slide the junction box out from the side.

6-6

WIRE INSTALLATION

Shimano e-tube Di2 wires are universal, featuring the same plug-in connector at both ends. These wires come in a wide array of lengths, starting at 150mm and running long enough to wire a tandem. They plug into any component in the system interchangeably.

Use the following wiring schematics (Figs. 6.12 to 6.15) to determine how you're going to wire up your bike. Determine your wire lengths to reach from the component to the bottom-bracket shell with an inch or two of slack; they will connect to Junction B there. If you're going to use a D-Fly (Fig. 6.28) inline transceiver (see 6-10) somewhere in the system to transmit data wirelessly to your cycling computer, or to a smartphone or tablet via the E-Tube Project app, you'll need an additional wire.

6.12 Wiring diagram with battery inside seatpost and digital display

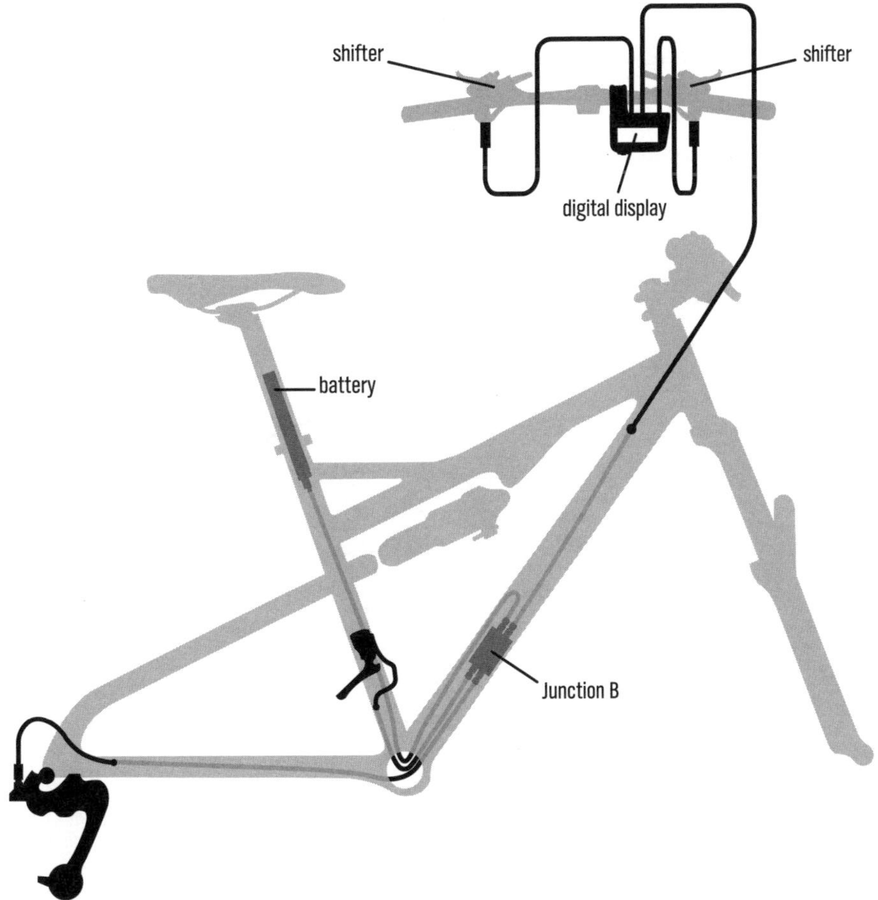

6.13 Wiring diagram with battery inside steering tube, internally routed handlebar, front and rear Fox iRD suspension controls, and digital display

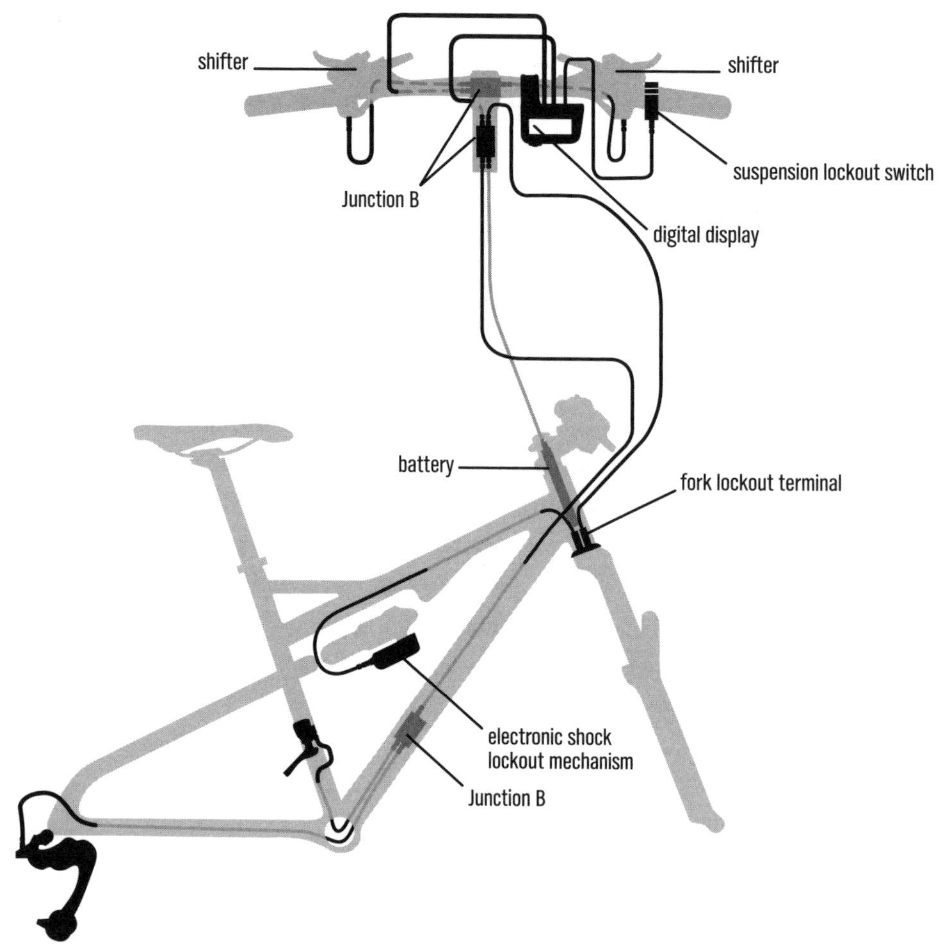

6.14 Wiring diagram with external battery case, front and rear Fox iRD suspension controls, and digital display

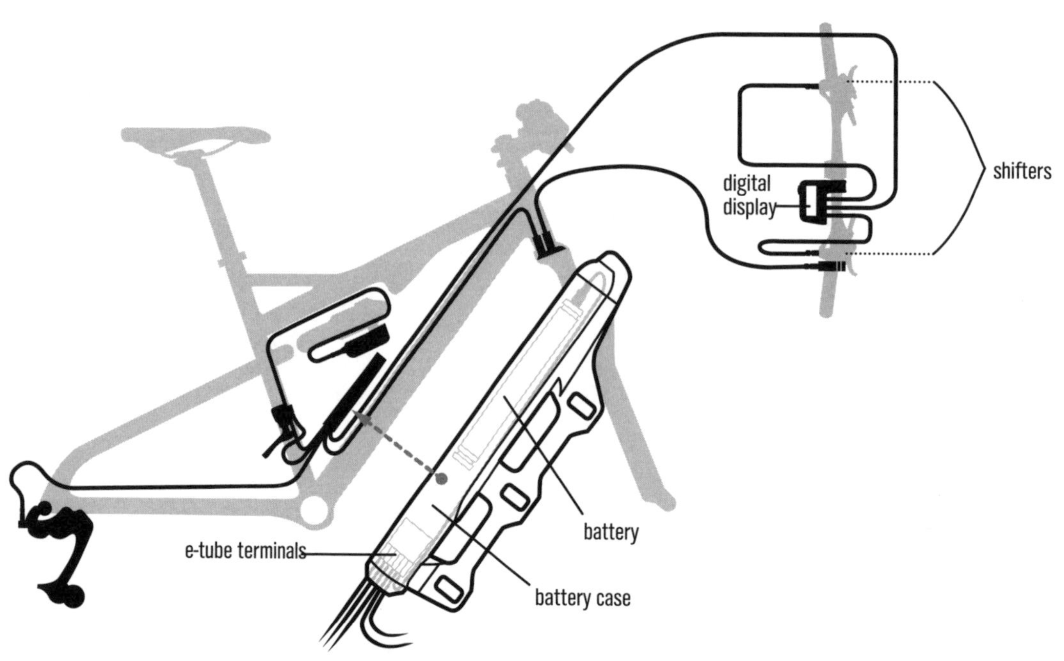

6.15 Wiring diagram with external clip-in battery, D-Fly Bluetooth transceiver, and Junction A

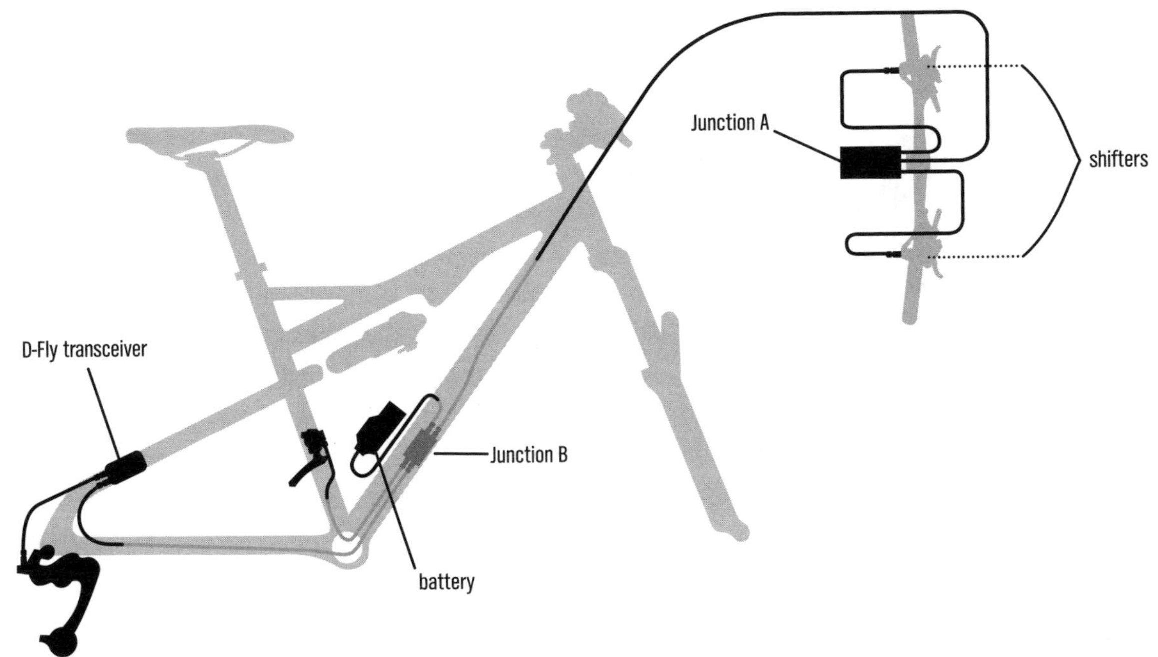

a. Di2 internal routing

The Park Tool IR-1 (or newer IR-1.2) magnetic internal wire-routing kit (Fig. 6.16) works fantastically well to feed wires through a frame, so these instructions explain how to route the wires using it. Otherwise, here are four other methods you can try for internally wiring a bike. Use one of these:

- Push a stiff wire with a hook bent in the end into a hole in the frame at the opposite end until it pops out of the hole at your end. Tape the hook to the connector on the Di2 e-tube wire, and pull the Di2 wire through and out the opposite hole with it.
- Push a shift cable in from the other end, tape the e-tube wire to it, and pull it back through. If you can't get the tip of the shift cable to come out the other end in the first place, try:
    - (a) Putting a little bend in the cable tip,
    - (b) Twisting it with a little bend in the tip when it has been inserted to the depth that it should reach the hole, or,
    - (c) Pulling the tip of the cable along through the frame tube with a strong magnet dragged along the outside.
- Slide a 4mm shift-cable-housing ferrule onto the end of a shift cable with the ferrule's open end toward the head of the cable. Slide it all of the way to the head until the cable head is down in the bottom of the ferrule. Snip the end off of a cable crimp cap (Fig. 5.26), open the cut end with a pick, and slide it onto the cable all the way to the head end, up against the ferrule. With the chopped-off cap against the ferrule, crimp the cap in place; you now have a shift cable with the head nested down inside of a ferrule secured in place. Now shove the cable in through a hole in the frame, leading with the bare cable tip, and use the tips in step 2 above to get it to come through. Push an e-tube wire head into the ferrule; it should fit tightly (it's best to test the ferrule size on an e-tube wire before starting this process); you can further crimp the ferrule down on the e-tube head to hold it. Now pull the cable back out through the frame, towing the e-tube wire with it.
- Using the crevice tool on a vacuum cleaner hose, vacuum a piece of string or dental floss from one end of the frame out the hole in the other end. (Secure the other end so you don't suck the entire piece of string through.) Tip the bike so the hole you're sucking out of is at the bottom. Tie or tape the end of an e-tube wire to the string, and pull it through. Better yet, cut the tip off a cable end cap and slip it onto

6.16 Park IR-1.2 internal wire-routing kit

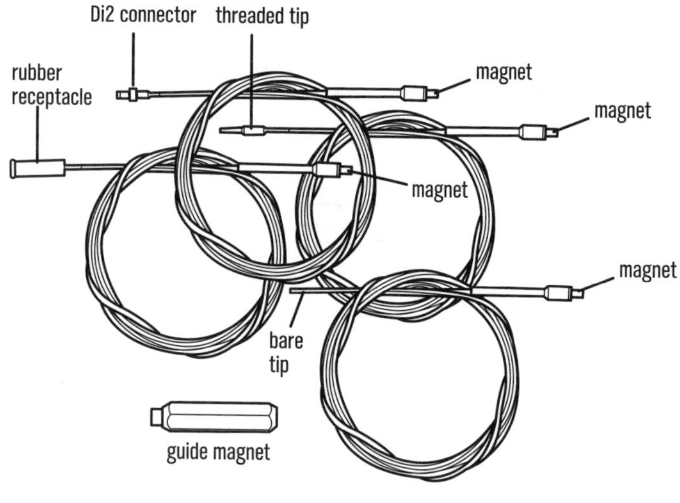

the end of the string. Slip the end of a shift cable—ideally the end of the Park IR-1.2 fishing wire—into the other open end of the cap and crimp it onto the cable and the string to hold them together. Pull that cable through with the string, and then use the Park magnet system; or, for a standard shift cable, use the method above in option c to get the Di2 wire through.

In terms of the order to do things, you can follow the wire-routing instructions listed below, substituting one of the above methods for the Park internal-wiring kit.

NOTE: *Even if the bike frame has internal-routing holes for shift cables, you will still have to route the electrical wires externally (see 6-6b below) if you can't feed the electric wires through, or if the tube openings into the bottom-bracket shell are not large enough to accommodate the wires and junction B (Figs. 6.22 to 6.24).*

1. **Plug an e-tube wire of the appropriate length into the rear derailleur.** The connection port is on the top back of the derailleur, just below and behind the mounting bolt (Fig. 6.10). As always, push it in with the TL-EW02 to ensure that the wire fully snaps into the port on the derailleur.
2. **Magnet end first, push the Park IR-1 lead cable with the rubber adapter on the other end in through the derailleur-wire hole near the dropout.** If you have the newer IR-1.2 kit, substitute the cable with the aluminum end (labeled "Di2 connector" in Fig. 6.16). Plug that end into the Shimano e-tube wire; it fits through smaller frame holes than does the rubber adapter.
3. **Shove the lead cable forward through the chainstay until the magnet appears at the bottom-bracket shell.** On a hardtail frame, you want the magnet to appear in the bottom-bracket shell. On a full-suspension frame, the magnet should pop out of a hole in the chainstay somewhere behind the bottom-bracket shell. If the lead magnet on the end of the cable gets hung up and doesn't come out in the bottom-bracket shell, do one of three things:
   (a) Using the same method as in Fig. 6.20, drag the lead cable's magnet along through the chainstay to its exit hole with the big magnet on the outside of the chainstay.
   (b) Shove the magnet end of the bare-tip Park fishing cable into the hole in the chainstay at the bottom bracket end and stick it to the magnet on the end of the lead cable already inside the chainstay; pull on the fishing cable until both magnets pop out of the hole.
   (c) Pull out the lead cable and instead push in the Park bare-tip fishing cable. Slip this cable in through the hole near the dropout, bare tip first. Using the big magnet on the outside of the chainstay, drag that bare cable tip through the chainstay and out the hole. Then stick the fishing cable's magnet (still hanging out of the hole by the dropout) to the magnet of the lead wire with either the rubber receptacle or the aluminum Di2 connector on its opposite end, and pull the fishing cable to drag the magnet of the lead cable out through the hole.
4. **Plug the free end of the rear derailleur e-tube wire into the Park lead cable.** Push the e-tube wire's connector into the rubber receptacle or onto the aluminum Di2 adapter on the Park lead cable (Fig. 6.17).
5. **Pull the lead cable through.** Pull until a few inches of the e-tube wire emerge from the bottom-bracket shell or from the hole in the chainstay near the bottom bracket. Pull the lead cable's rubber adapter or its aluminum Di2 adapter off the e-tube wire.

6.17 Pulling the rear derailleur e-tube wire through a chainstay

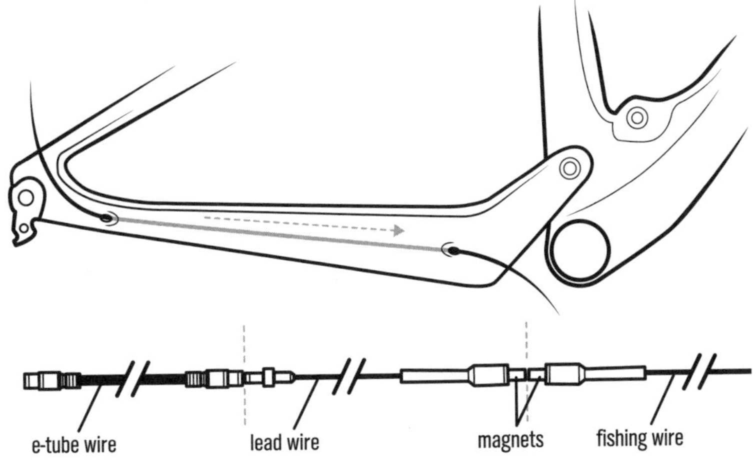

6.18 Installing front-derailleur e-tube wire into plug cover

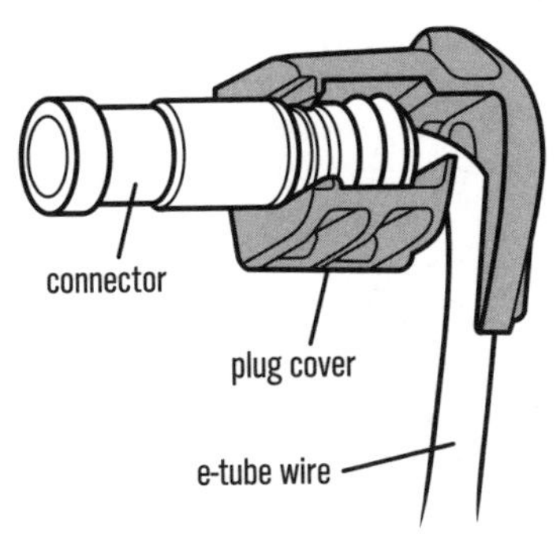

6.19 Installing e-tube wire into front derailleur by aligning plug cover with receptacle and pushing in until it clicks

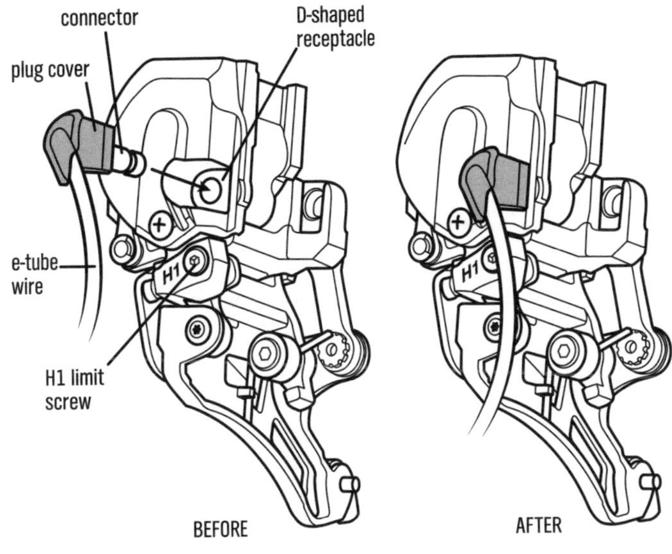

6. **Plug a wire into the front derailleur.** Install one end of a short e-tube wire into the front-derailleur plug cover, which bends the wire at 90 degrees right above its connector (Fig. 6.18). Align the D-shaped plug cover with the port receptacle on the inboard front of the derailleur and push in the plug cover with the connector inside until it fully snaps into the port (Fig. 6.19). (To remove it, pull the plug cover off; it will come off the wire by itself. Then pry the connector out with the fork end of the TL-EW02 [Fig. 6.4].)
7. **Push the other end of the e-tube wire into the hole in the seat tube near the front derailleur.** It should easily pop out into the bottom-bracket shell. If it doesn't, perform steps 2–6 with this front derailleur wire as well.
8. **Plug a wire into the digital display or Junction A.** Pick an e-tube wire that will reach all of the way to the bottom bracket. You can plug it into any of the three ports on the digital display or Junction A; the other two ports are for wires to the shifters.
9. **Magnet end first, push the Park Di2 lead cable in through the hole in the frame near the head tube.** Use either the lead cable with an aluminum e-tube-connector end or the one with the rubber-receptacle end. The hole can be on the top tube, head tube, or down tube. To get around a corner from the top tube or head tube, use the big magnet on the outside of the frame to guide the magnet end of the fishing cable (Fig. 6.20).

6.20 Guiding Park magnetic fishing cable from the head tube into the down tube with external magnet

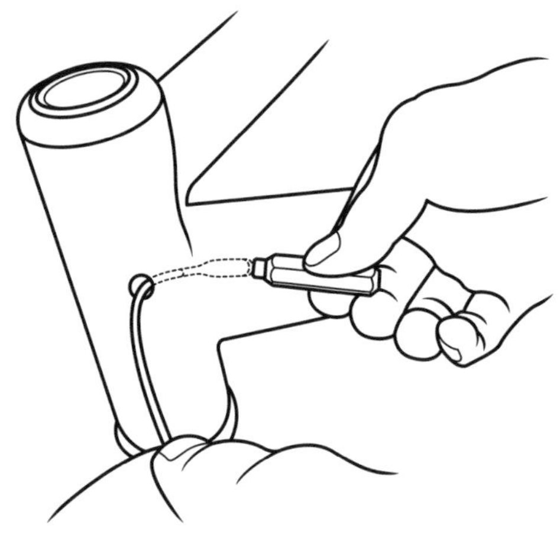

6.21 Pulling an e-tube wire into the bottom-bracket shell

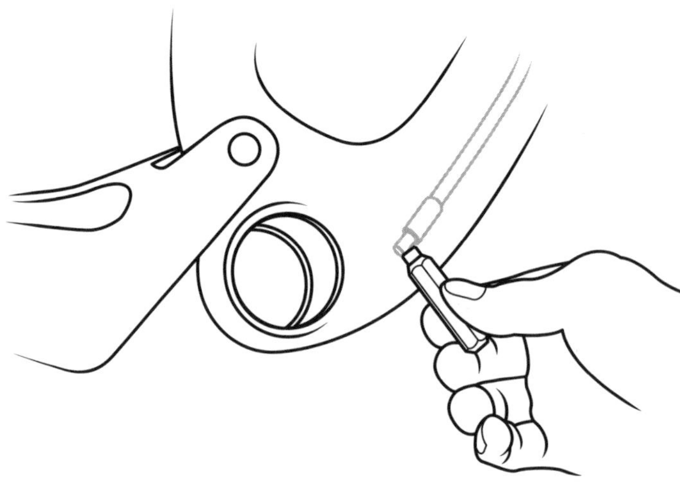

10. **Push on the cable until the magnet end appears at the bottom bracket.** On most frames, the cable should appear in the bottom-bracket shell. If you can't get the wire to pop out, try one of the other options listed in step 3.
11. **Based on battery location, determine the routing of the battery wire.** The e-tube wire from a battery inside the fork steering tube (Fig. 6.13) will come out from under the fork crown or up through the top cap, go into the frame near the head tube and down the down tube to pop out in the bottom-bracket shell (Fig. 6.21) or out of a hole ahead of the bottom-bracket shell. The e-tube wire from a seatpost-mounted battery (Fig. 6.12) will go down the seat tube, through the bottom-bracket shell (Fig. 6.22), and up into the down tube (Fig. 6.23). With an external battery (Fig. 6.8 or 6.9) and internal wiring, its e-tube wire will go into the down tube via a hole near the bottle bosses to pop out in the bottom-bracket shell.
12. **Shove the battery wire or the magnet end of the Park Di2 lead cable into the frame based on the routing you determined in step 11.** With the seatpost battery and perhaps even with the external battery, you may be able to just drop in the battery wire itself and have it pop out in the bottom-bracket shell (Fig. 6.22). If using the Park lead cable, use the IR-1 cable with the rubber receptacle

6.22 Connecting Di2 wires at the lower junction box (Junction B) at the bottom bracket

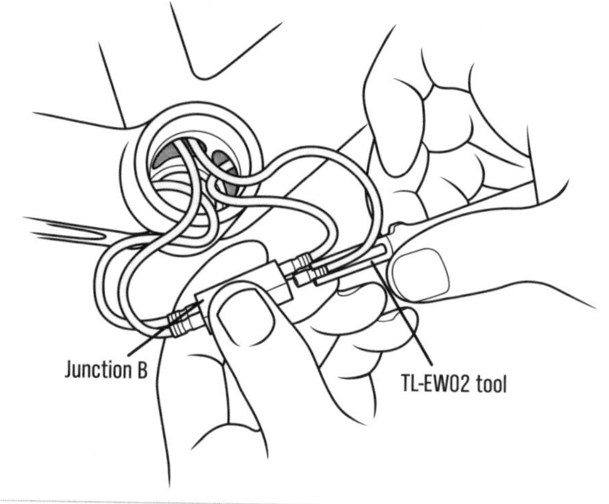

on the end (Fig. 6.21) or the IR-1.2 cable with the aluminum Di2 adapter on the end (Fig. 6.17). Push it through until it appears in the bottom-bracket shell (Fig. 6.22). If the lead magnet on the end of the cable gets hung up and doesn't come out of the hole or into the bottom-bracket shell, try one or more of the options in step 3 until it does.

13. **Pull the ends of the wires out of the bottom-bracket shell and plug all of them into Junction B** (Fig. 6.22). With some full-suspension frames, the rear derailleur wire may have to come around the bottom-bracket shell and into the hole in the underside of the down tube ahead of the bottom

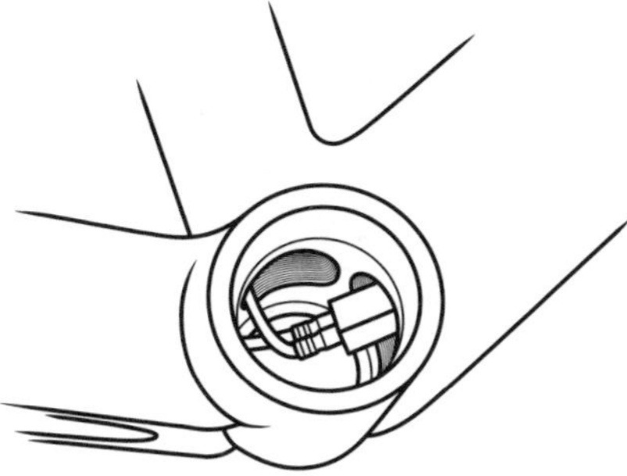

6.23 Shove the lower junction box (Junction B) up into the down tube.

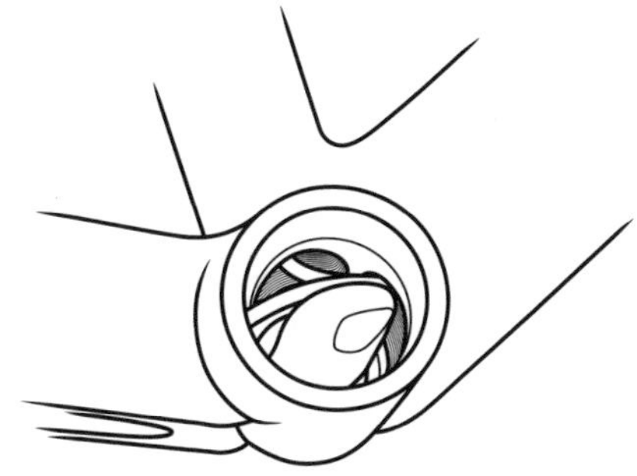

6.24 Ensure that the e-tube wires don't interfere with the bottom bracket.

bracket (Fig. 6.21). Use the TL-EW02 tool to fully insert them until they click (Fig. 6.22). Install a dummy plug into any unused ports on Junction B.

14. **Shove Junction B up into the down tube** (Fig. 6.23). The only wires visible inside the bottom-bracket shell should be the wires going to the front and rear derailleurs (Fig. 6.24), and, in the case of a seatpost battery, a battery wire going up into the seat tube. If applicable, replace the cover (Fig. 6.21) over the hole under the down tube.
15. **Check operation of the derailleurs.** Once you install the bottom bracket, you won't be able to get at the internal wires. To avoid any unpleasant surprises that will require removing the crank and bottom bracket to undo something you could have fixed before they went in, click the shifters and make sure that the derailleurs operate properly.
16. **Ensure that the front and rear derailleur wires don't interfere with the bottom bracket.**
17. **Install the bottom bracket (Chapter 11).**

b. Di2 external routing

1. **If you have the rectangular battery** (Fig. 6.8), **plug the battery wire into Junction B.** In this case, it will be the SM-JC40 Di2 Junction B that bolts on under the bottom-bracket shell. Use the TL-EW02 tool to push the wire's connector in until you hear a click. If you have the external cylindrical battery case (Fig. 6.9), skip to step 2.
2. **Plug e-tube wires into the components.** Select the wire lengths based on your frame and the configurations shown in Fig. 6.14 or 6.15. With the TL-EW02, plug them into: (1) the digital display or Junction A (Figs. 6.2, 6.3) at the handlebar or stem; (2) the rear derailleur (Fig. 6.10); and (3), if applicable, the front and rear Fox iRD shocks. On the front derailleur, first put the end of the wire into the plug cover that comes with the front derailleur (Fig. 6.18); it bends the wire at 90 degrees right above the plastic connector. Align the D-shaped plug cover with the port receptacle on the inboard front of the derailleur and push in the plug cover with the plug inside until it fully snaps into the port (Fig. 6.19).
3. **Plug these wires into the battery case or the SM-JC40 bolt-on Junction B.** Push them in with the TL-EW02 tool until they click. Cover and cap the wires to the battery case as explained in 6-4c. Route the rear derailleur wire along the underside of the chainstay so the chain can't drop on it.
4. **Provisionally tape the wires to the frame.** Later, clean the area with alcohol and then cover the wires with the supplied adhesive cover strips. With the battery case, you are done.
5. **Take up any wire slack in the bolt-on SM-JC40 Di2 Junction B's looping pegs (under the cover).** You can wind 120mm of slack into it.
6. **Bolt the SM-JC40 Junction B to the threaded cable-guide hole under the bottom-bracket shell.** Torque is 1.5–2 N-m.

c. Connect Di2 wires to shifters

1. **Run a wire from the digital display or from Junction A** (Figs. 6.2, 6.3) **to each shifter.**

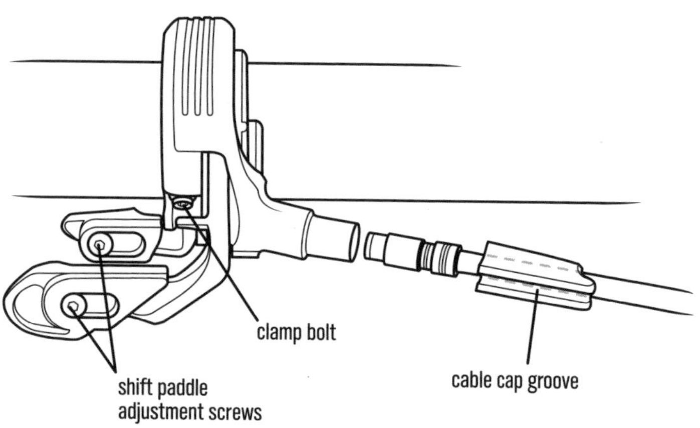

6.25 Run the wire through the shifter cable cap

6.26 Installing sealing grommet on wire entry into frame

2. **Pull the cable cap off the shifter port.** Slide the wire through the cap (Fig. 6.25).
3. **Plug the wire end into the shifter port.** Push it in with the TL-EW02 tool until it snaps in with a click.
4. **Push the cable cap back onto the shifter port.** The groove in the cable cap (Fig. 6.25) is for doubling the wire back down if the handlebar has internal wire routing.

6-7

COMPLETE COMPONENT INSTALLATION

1. **Install the bottom bracket plastic sleeve and bearings** (Chapter 11). Ensure that both derailleur wires (and the battery wire for a seatpost battery) pass above the sleeve (Fig. 6.24), and the bearing or bearing cup threads do not impinge on the rear derailleur wire.
2. **Install the cranks** (Chapter 11).
3. **Align the front derailleur as in 5-5a.** Skip to step 5 if you have an internal battery.
4. **If applicable, install the additional screw(s) in the external battery mount and tighten it down.** To keep the battery from bouncing around, the rectangular, bottle-mount-type Di2 battery mount (Fig. 6.8) requires either a zip-tie around the down tube at its tail or a third bolt hole in a dedicated threaded hole in the frame.
5. **Test the system.** Press the lever switches and ensure that the derailleurs move.
6. **Install the chain as in Chapter 4.**
7. **Clean up the wiring.** Tape or zip-tie the shifter wires to the handlebar. If bridging gaps, zip-tie or tape loose wires to each other or to brake hoses.
8. **On internally routed wires, slip a rubber sealing grommet around each wire where it protrudes from the frame.** Push the grommet into the hole (Fig. 6.26) until it engages properly and creates a seal. The grommets come in different shapes and sizes depending on frame hole shape and wire size. The standard frame hole size is 8mm × 7mm, with the hole for the battery wires from the original-style external battery (Fig. 6.8) being 14mm × 7.7mm, but Shimano also offers 7mm round grommets. Additionally, frame manufacturers using non-standard hole sizes will generally supply grommets with the frame. Hardware stores carry various rubber grommets if you need an odd size.

6-8

ADJUSTING SHIMANO DI2 ELECTRONIC DERAILLEURS AND SHIFTERS

a. Di2 rear derailleur

1. **Shift to the fifth largest rear cog.** See 6-1a for shifting instructions.
2. **Hold the mode button down on the digital display or on Junction A** (Figs. 6.2, 6.3). Hold it until the digital display shows “F R 0” with the “R 0” flashing, or until the LED on Junction A adjacent the

"+-" lights up red. This will switch the system into adjustment mode.

3. **Tap the right-hand upshift or downshift switch until the jockey wheel lines up straight under the cog** (Figs. 5.4 and 5.5). Listen for chain noise while turning the cranks. Shimano's suggested method is to tap the downshift switch (bigger, lower paddle) until the chain makes noise against the next largest cog. Then tap the upshift switch (smaller, upper paddle) five times to center the guide pulley under the cog. The screen displays how many positive or negative steps you have moved the derailleur over from its initial "0" position. There are 32 total adjustment steps, 16 in each direction, and each step is approximately equal to turning a derailleur barrel adjuster one notch.
4. **Revert out of adjustment mode.** Push the mode button on the digital display (Fig. 6.2) or on Junction A (Fig. 6.3) until the display goes back to the standard screen or Junction A's LED turns off.
5. **Rotate the cranks and shift through all of the cogs.** Check for silent operation. If not shifting perfectly, repeat steps 1–4.
6. **Shift to the largest cog.**
7. **Tighten the low-gear limit screw.** This is the outboard screw (Fig. 6.10); the limit screws are labeled "H" and "L." Tighten it with a 2mm hex key until it touches the link and ensures that the derailleur cannot shift into the spokes. If you tighten it too much, it either won't shift to the largest cog or it will continue to run the motor, wearing down the battery rapidly; back the low-gear limit screw off appropriately if that happens.
8. **Shift to the smallest cog.**
9. **Tighten the high-gear limit screw.** This is the inboard screw (Fig. 6.10). Tighten it with a 2mm hex key until it just touches the link.
10. **Set the b-screw as in 5-2f.** By turning the crank backward in both the front small chainring/biggest rear cog and front small/rear small combinations, check that the guide pulley (upper jockey wheel) is as close to the cog as possible without bumping up and down as the chain moves. On a middle cog, the chain gap distance between the top of the upper jockey wheel and the bottom of the cog (Fig. 5.6) should be 5–6mm if the largest cog is 42T or smaller and 8–9mm if it is bigger than 42T. With a 2mm hex key, turn the b-screw to achieve this.
11. **Adjust clutch if need be.** If you want more chain retention on bumps, tighten the clutch. If you want quicker shifting when the clutch lever switch is on, loosen the clutch. Instructions are in 5-2h. The clutch needs frequent overhaul if used in wet conditions or washed frequently; see 5-36.

NOTE: *In Manual (M) mode, the rear derailleur will not shift to the smallest rear cog while the chain is on the inner chainring. And in Synchro mode (S1 or S2), which you toggle to with subsequent double pushes of the mode button on the digital display or on Junction A, the rear derailleur will only go to the smallest cog after the derailleurs have made a double shift (near the middle of the cassette, the front derailleur will shift to a larger chainring and the rear derailleur will step back up to a larger cog). Use M mode for the above check of the b-screw setting; i.e., look at the chain gap at the smallest rear cog the system allows while the chain is on the inner chainring.*

b. Di2 front derailleur

1. **Shift to the largest chainring and largest rear cog.** See 6-1b for shifting instructions.
2. **With a 2mm hex key, loosen the front derailleur stroke-fixing screw.** It's on the front derailleur's leading edge, on the piece between the cage and the derailleur body and facing forward, directly below where the e-tube wire enters the body; it is labeled "H1" (Fig. 6.19).
3. **With a 2mm hex key, turn the high-gear limit screw.** It is on the side of the derailleur, toward the front, just below the cage pivot linkage. It is around the corner from the stroke-fixing screw and facing out; it is labeled "H2" (Fig. 6.11). Adjust this limit screw to bring the inner cage plate within 0.5mm from the chain. Counterintuitively, rotation direction is clockwise to move the cage outward; counterclockwise moves it inward.
4. **With a 2mm hex key, tighten the front derailleur stroke-fixing (H1) screw.** This holds the adjustment.

5. **Shift to the inner chainring.** The chain should still be on the largest rear cog; "1" should be displayed on the digital screen.
6. **Hold the mode button down on the digital display or on Junction A** (Figs. 6.2, 6.3). Hold it until the digital display shows "F R 0" with the "R 0" flashing or until the LED on Junction A adjacent the "+-" lights up red. The system is now in adjustment mode.
7. **Adjust the inner front-derailleur cage plate position to be 0.5mm from the chain.** Click the left upshift switch (bigger, lower paddle) to move the derailleur cage one step closer to the chain, or push the downshift switch (upper, smaller paddle) to move it away from the chain. There are 32 adjustment steps, 16 in each direction, and each step is approximately equal to one notch of a derailleur barrel adjuster.
8. **Exit adjustment mode.** Push the mode button on the digital display or on Junction A until the screen shows the standard display or the Junction A LED turns off.

The Di2 front derailleur shifts quickly, but in two steps; it pushes hard and fast to derail the chain without pushing so far that it overshifts the chainring, and then it carefully lines up the cage over the chain depending on which cog it is on in the rear.

On either chainring, the front derailleur automatically trims its position twice in either direction when the chain is shifted from one extreme of the cogset to the other. If you set the front derailleur position and trim adjustment properly, this auto-trim feature will eliminate or at least minimize chain rub in cross-gears.

6-9

OTHER ELECTRICAL DETAILS

a. Battery care

1. Do not wet the battery or charger terminals.
2. Do not subject the battery to temperatures above 140°F (60°C).
3. Do not cover the battery or charger when charging.
4. Do not allow metal objects or wet objects to connect across the battery or charger terminals.
5. Do not use if the battery shows signs of leakage.
6. Don't touch any leaking battery acid or get it on your clothing.
7. Use an AC charger only with AC voltages within the range printed on the unit.
8. Do not use the charger with electrical transformers designed for overseas use.

If the red ERROR indicator light on the battery charger (Fig. 6.27) flashes (it is marked with an icon of a battery with an exclamation point over it), the ambient temperature could be too high or low for charging, or there could be a problem with the battery or connections.

If the charger's orange CHARGE indicator flashes (it is marked with an icon of a battery with a lightning bolt over it), the current from the AC adapter plugged into the wall or the PC or USB hub connected to the USB cord could be too low. The charger needs at least 1 amp DC, so get an adapter that will deliver that, or don't connect to a USB hub, as that could be putting out too little current.

If the orange CHARGE indicator does not light up or goes out quickly, the battery may be fully charged. Check the battery level on the digital display, on Junction A, or on a paired cycling computer.

If there is a fault in the battery, charger, or system that prevents charging, either the red ERROR indicator light or the orange CHARGE indicator light will flash.

b. Sealing out the elements

Although Di2 components are designed to be waterproof and able to withstand wet weather and all sorts of muddy and wet mountain-bike riding conditions,

6.27 Shimano SM-BCR2 USB battery charger for cylindrical Di2 batteries

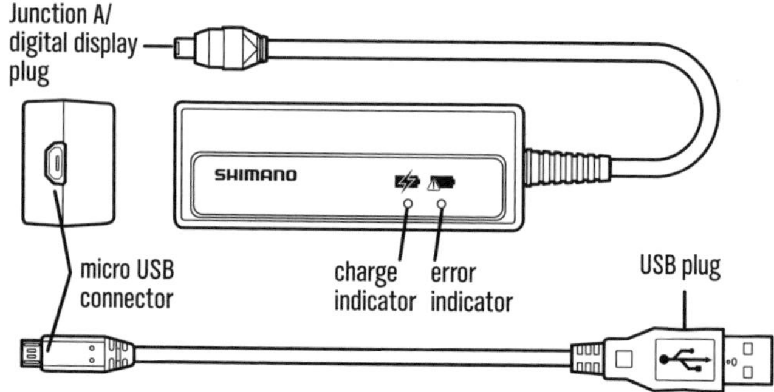

avoid submerging them in water beyond the occasional creek crossing. When washing the bike, ensure that all rubber sealing grommets (Fig. 6.26) are installed at wire-entry points and external electrical connections are fully snapped in place to properly seal out water. Avoid using solvents on the electrical components, and do not blast them directly with high-pressure spray, as in a car wash.

Avoid lubricating the rear derailleur links, as some lubricants can damage the O-rings that seal the electronic components inside. If you remove the cover from the rear derailleur clutch mechanism (Fig. 5.15), do not lubricate its central pivot pin, as the clutch will have insufficient friction to function properly. You will likely be removing that cover, because if you use your bike in wet conditions or wash it a lot, you will need to overhaul the clutch periodically (see 5-36).

Secure external electrical wires with tape or zipties to prevent snagging them. Di2 wires are completely interchangeable and can be obtained in different lengths.

Always connect and disconnect Di2 e-tube wires with the Shimano TL-EW02 tool; make sure the connection snaps together with an audible click to ensure a proper water seal.

c. Crash protection

If a Di2 rear derailleur is hit in a crash, it protects itself by uncoupling the motor from the mechanical shaft. You'll know this has happened when it won't shift properly. To re-couple a Di2 motor with its derailleur link, hold down the mode button on the digital display or on Junction A for five seconds or more; on the digital display, "P" will flash on the screen.

While holding down the mode button, the derailleur will not only re-couple with the motor, but it will also shift, and it will keep shifting as long as you hold the mode button down, until it reaches the last gear in the direction it's going. Because the derailleur has just been in a crash, the derailleur hanger on the dropout, as well as the b-link connecting the mounting bolt to the derailleur, could be bent inward; therefore, be aware of the danger of the derailleur shifting inward toward the largest cog. If the pedals are turning, and the bike is rolling along, this could shift the derailleur right into the spokes and tear it to shreds. This is exactly the sort of wallet-busting scenario the crash-protection de-coupling feature and the weak derailleur hanger and b-link are supposed to protect you from.

So, MAKE SURE that you check and, if need be, replace or straighten the derailleur hanger and b-link (you do carry spares, right?) before activating the re-coupling feature by holding down the mode button. And certainly avoid riding, or at least avoid using the largest few rear cogs, until you've had a chance to replace the b-link and the derailleur hanger if they're bent.

6-10

DI2 CONNECTIVITY

With any of the computers that are now integral parts of our daily lives, keeping their firmware updated is critical to having ongoing functionality, and Di2 is no exception. We also know that just performing firmware updates will only get us so far with an old computer; eventually, the latest programs will not run on it without some upgrade of the actual hardware. That, too, applies to Di2.

The fact that Di2 derailleurs and shifters are not wireless means that they are largely unaffected by wireless-connection problems. If the phone app bombs in the middle of a firmware update, or if the connection breaks between the Di2 system and an ANT+ cycling computer, the derailleurs probably will still work.

a. Hardware required for wired and wireless connectivity

Until 2016, downloading Di2 firmware, reconfiguring shifting, or performing diagnostic checks of the system all required a wired connection to a Windows PC (and not to a Mac; those remain unsupported). Now, however, Shimano's wireless E-Tube Project app (free on the App Store or on Google Play) for iPhones, Android phones, and tablet computers affords much of the same functionality. The smartphone app allows you to customize shifting and update firmware; the tablet app offers those functions and can also check for system errors and perform the system pre-set. Furthermore, the Di2 system can interact with many Garmin and other ANT/ANT+ cycling computers, displaying battery level and current gear combination, as well as let you switch between shifting modes.

You must have all of the appropriate hardware to get wireless connectivity. Fortunately, except for the first two generations of road Di2, all Shimano Di2 systems are backward-compatible, so you can upgrade some items to get this functionality while still using your existing shifters and derailleurs. Furthermore, all XT Di2 components are interchangeable with XTR Di2.

To enable Bluetooth and ANT+ compatibility, the battery must be either the BT-DN110 (cylindrical, for installation inside a seatpost, steering tube, or Shimano battery case—Figs. 6.6, 6.7, and 6.9) or the BT-DN100 original-shape external battery (Fig. 6.8); the older versions of these batteries (SM-BTR2 and SM-BTR1) have insufficient memory for Bluetooth connectivity. You also need a transceiver in the system; this can be either a digital display (Fig. 6.2) with a Bluetooth chip, namely the XTR SC-M9051 (not the SC-M9050) or the XT SC-MT800, or a D-Fly inline unit (Fig. 6.28)—either the EW-WU101 (both ports on the same end) or the EW-WU111 (one port on either end). With a D-Fly inline transceiver, you also need an additional short e-tube wire.

6.28 Inline D-Fly transceivers for Bluetooth connectivity of Di2 system

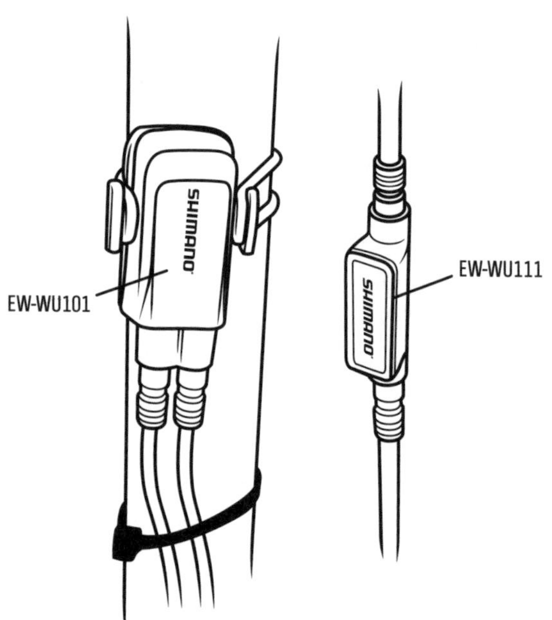

If you want to update, check or reconfigure a Di2 system and don't want the expense or trouble of obtaining wireless connectivity, you can download the E-Tube Project software from bike.shimano.com and plug a Windows computer into the bike's Di2 system with Shimano's SM-PCE1 PC interface device and accompanying USB cable. It plugs right into the charging port on Junction A or on the digital display. You can then update firmware, diagnose and correct problems in the system, and customize shifting options on the computer screen. Without the interface device, you can still perform firmware updates using the E-Tube Project software by plugging into the Junction A charger port with the USB charger (Fig. 6.27) connected to a Windows PC.

b. ANT+ connection with a cycling computer

What's possible in communication between cycling computers and Di2 is rapidly changing. As of 2017, there are two levels of compatibility with Di2: either just displaying battery level and current gear combination on the computer, or displaying Synchro Shift setting (in addition to battery percentage and gearing).

The instructions below apply to connecting popular Garmin computers to Di2; other Di2-enabled ANT+ computers have similar features. Recent Garmin models, such as the Edge 510, 520, 810, 820, and 1000, the Forerunner 735, 920, and 935, and the Fenix 5, can display Di2 shifting status and battery percentage. Garmins that also support Shimano's Di2 Synchro Shift capability are limited as of this writing to the more recent Edge 520, 820, and 1000 models. Note that Di2 connectivity is not supported on older Garmin models without updated Garmin firmware.

Setting up Di2 display on Garmin

1. **Select the icon of a crossed wrench and screwdriver on the home screen.** The settings menu will appear. On some older models, you can also access settings from any training page by touching the screen; an overlay will come up with the settings icon, as well as the home, left and right scroll, navigation search, and wireless connection icons (Fig. 6.29). To select a Garmin icon, touch it on the screen.

6.29 Garmin 810 with icon overlay over a training page

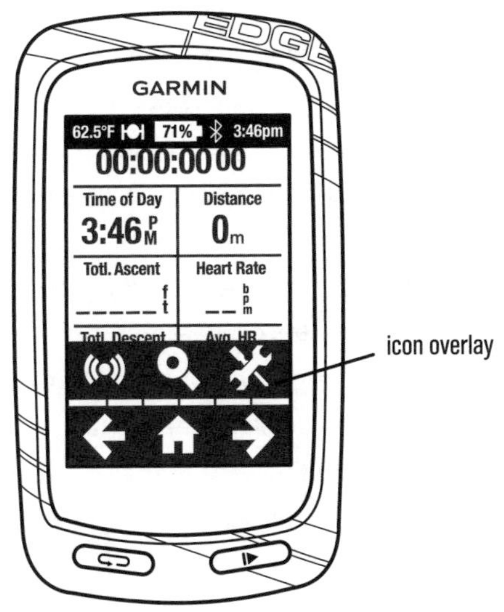

2. **Go to the ANT+ sensor list from the settings menu.** Depending on computer model, either select (a) "Bike Profiles" and select a bike stored in memory, or (b) "Sensors." Alternatively, on some older models, if you brought up the overlay on a training page (Fig. 6.29) instead of initially going to the home screen, get to the "Bike Sensors" list by selecting the wireless-connection icon (it looks like a dot broadcasting to the left and right). On some newer models, you can directly access the sensor list from any page by pulling down with your finger from the top of the screen; it pulls down a curtain with backlight and connection options; select "Connecting to Sensors."
3. **Select "Di2" on the sensors list.** If "Di2" is not one of the icons displayed, then the Garmin firmware is too old. Download "Garmin Express" online and create a login; from there, update the firmware. With updated Garmin firmware, return to step 1.
4. **Select "Enable" on the "Shimano Di2" screen.**
5. **Select "Search" or "Connect."** Once connected, it will display "Shimano Di2 found." For the Garmin to find your Di2 system, unplug and re-plug both wires to the D-Fly transceiver or all three wires to the digital display. If it doesn't connect, push the mode button for about a second on the digital Di2 handlebar display until a flashing "c" appears, or Junction A until the two LEDs flash red and green. If the "c" doesn't appear on the SC-M9051 or SC-MT800 digital display (i.e., if it passes on to adjustment mode or post-crash re-coupling mode), you must first update the SC-M9051 or SC-MT800 firmware with a wired connection to a PC (many of these units were shipped with the Bluetooth chip but without the firmware to make it work). If it still doesn't find Di2, restart the Garmin and try pairing it again. Once the Garmin is paired with Di2, continue with step 6 to create a screen on which to view the Di2 functions while riding.
6. **Select the settings icon.** It's the wrench/screwdriver icon. Again, find it by: (a) returning to the home page, or (b) pulling up the overlay on a training page (Fig. 6.29).
7. **Select "Activity Profiles."**
8. **Select one of the profiles.**
9. **Select "Training Pages" or "Data Screens."**
10. **Select a page that is currently shown as "Off."**
11. **Select "Enable" or "Enabled."** This will turn the page on.
12. **Increase, decrease, or maintain the number of fields on the page.** Select "+" or "-" to change the number of fields, then select the check mark in the lower right corner. To leave the number of fields unchanged, simply touch the check mark.
13. **Change fields to display Di2 information.** Touch any field to open the "Select a (Data Field) Category" page, scroll (using the arrows) to "Gears" and select it; this will bring up options like "Di2 Battery Level," "Front Gear," "Gear Ratio," "Gears," and "Rear Gear." With newer models, "Di2 Shift Mode," "Gear Battery," and "Gear Combo" will also appear. Fill as many fields as you want with Di2 fields. I recommend having at least two fields, namely "Gears," which shows graphically which front and rear gears the chain is on, and battery percentage. With newer Garmins, also create a "Di2 Shift Mode" field to show whether you are in M (Manual), S1 or S2 (pre-programmed Synchro Shift modes), or in a custom Synchro Shift mode (which you create with a smartphone or tablet in the E-Tube Project app or on a Windows PC connected via the SM-PCE1 interface device).

6.30 Garmin 1000 Di2 training page

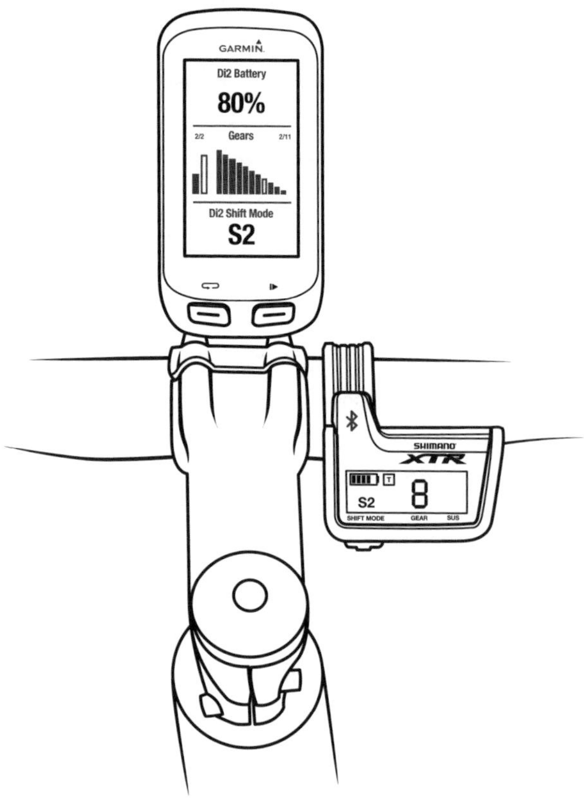

14. **Select the check mark at the bottom of the Garmin screen.** You're done. You now have a screen you can loop to when riding that shows what's going on with the Di2 system. If (when) the Garmin doesn't find the Di2 system when you turn it on to head out for a ride, you will have to go through steps 1–5 to pair them.

c. Using the E-Tube Project Wireless App

As with ANT+ cycling computers, you'll need either a Bluetooth-enabled digital display, or a D-Fly inline unit (either an EW-WU101 or EW-WU111) wired into your system.

1. **Download E-Tube Project and open it.** Find it for smartphones and tablets on iTunes or Google Play.
2. **Select the Bluetooth LE connection box and push the Di2 mode button on the bike's digital display or Junction A.** Push the mode button for perhaps a second—not long enough to bring up adjustment mode. A flashing "c" should appear on the digital Di2 handlebar display, or the two LEDs on Junction A will alternate flashing green and red. If the "c" doesn't appear on a SC-M9051 or SC-MT800 digital display (i.e., if it passes on to adjustment mode—see 6-8—or post-crash re-coupling mode—see 6-9c), you must first update the SC-M9051 or SC-MT800 firmware with a wired connection to a PC (many of these units were shipped with the Bluetooth chip but without the firmware to make it work). Once the connection is made, a box with the part number of the Di2 transceiver will pop up on the phone or tablet screen; select it.
3. **Update the firmware of the Di2 transceiver.** This will start automatically. During the process, it will ask you to change the default "000000" ID code of the Di2 transceiver; go ahead and do this. It requests "half-width alphanumeric characters," whatever those are; just pick another 6-digit number. At some point, the firmware update will stop before it's complete and will give you an error message. Don't worry; it's not you. It's also possible that you will be able to connect this time, and you won't get the error message until you try to connect the next time. In that case, you won't have to do the following step until you log onto the app next time. By the way, the tutorial on the app is not a video, with the short green bar showing how far it has (not) loaded; rather scroll to the right to view it page by page.
4. **Delete the Di2 transceiver on the tablet or smartphone.** Yes, really. If you don't do this, the app will try to connect to your Di2 transceiver with the old "000000" password and will display a spinning wheel of death without connecting. So select "Bluetooth" in your phone or tablet settings menu. Your Di2 transceiver's part number should appear on the list of your devices; click on its information box and select "Forget This Device."
5. **Select the Bluetooth LE connection box in the app.** It will display a box with the part number of your Di2 transceiver; select that for the app to connect to. Now you should be connected again, with your new password.
6. **Update firmware for all components.** Ones that are up to date will be grayed out and will say "latest."

7. **Play as you wish.** Now you can customize which shift buttons do what and which Synchro Shift protocols you can toggle between and how and at what points the derailleurs perform double shifts.
8. **Disconnect the app.** This is THE MOST IMPORTANT STEP, and it's not obvious in the app. Click on the three little bars in the upper right of the phone screen to bring up the main menu, and select "Bluetooth Disconnected." Right? It should say, "Disconnect Bluetooth" or words to that effect, but it doesn't. This will disconnect your phone or tablet and save whatever changes you have made during your session. If you don't do this, your Di2 components will continue to either be connected or to continue trying to connect to your phone or tablet. The symptom will be that the derailleurs will not respond to the shift buttons, and the charger will flash a fault light when plugged in; meanwhile, the Di2 battery will drain rapidly (the Bluetooth LE connection draws a lot of power), while not being able to be charged.

NOTE: *A similar possibility for neglecting the most important step (saving changes and disconnecting) exists with the wired PC connection to E-Tube Project. In the main menu, the last rectangular option box was always "Complete Setup" in years past. But now with the advent of the wireless connectivity, an additional "Bluetooth LE" option box has appeared in the column; this has pushed the "Complete Setup" box to the next page, and you have to know to scroll to it. Now you know.*

On a system without a digital display, once the firmware is updated so that the bike's Di2 system is fully Bluetooth-enabled, it is Synchro-Shift enabled as well. Junction A not only controls which shifting mode the bike is in (with double-clicks of the mode button), but now also displays it.

When you hold down a shift button (or a pair of them) to display the battery charge status, the battery LED first shows the battery status (see 6-1c), and then both LEDs show the shifting mode. If the battery LED glows green and the "+-" LED glows red, with neither of them flashing, the bike is in Manual shift mode. Two blinks of these green and red LEDs means S1 shift mode is operational, and three green/red blinks mean the bike is in S2 shifting mode.

A paired ANT+ cycling computer that is recent enough to support Synchro Shift will also display the shift mode. And, in the phone/tablet app or in the wired PC software, you can put custom shifting patterns of your design into those S1 and S2 slots.

6-11

ELECTRONIC DERAILLEUR TROUBLESHOOTING CHART

Ensure that the rear derailleur's low-gear limit screw (Fig. 6.10) prevents it from going into the spokes (6-8a) before test riding.

Make small adjustments and recheck shifting frequently. Press a shifter switch only once or twice each time while in adjustment mode (6-8a). Turn a limit screw one-eighth turn each time.

If there are multiple options listed for a given problem, perform them in order; if the first fix doesn't work, try the next one.

Turn to the next page for Table 6.1: Electronic Derailleur Troubleshooting Chart.

TABLE 6.1 — ELECTRONIC DERAILLEUR TROUBLESHOOTING CHART

REAR DERAILLEUR	
Nothing happens when push shift switch	Charge battery (6-2). Check wire connections. Disconnect from E-Tube Project app (6-10c).
Chain jams between small cog and frame	Click rear downshift switch while in adjustment mode (6-8a). Turn high-gear limit screw clockwise (6-8a).
Rear derailleur touches spokes	Click rear upshift switch while in adjustment mode (6-8a). Turn low-gear limit screw clockwise (6-8a). Check derailleur-hanger alignment (17-5).
Chain falls between spokes and large cog	Turn low-gear limit screw clockwise (6-8a). Click rear upshift switch while in adjustment mode (6-8a).
Chain won't go onto large cog	Click rear downshift switch once while in adjustment mode (6-8a). Turn low-gear limit screw counterclockwise (6-8a). Chain too short (4-8)—replace with longer chain. Derailleur cage too short—replace cage (5-40) or derailleur.
Shifting sluggish to a larger cog	Click rear downshift switch while in adjustment mode (6-8a). Turn off Shimano clutch (5-2h). Loosen Shimano clutch (5-2h).
Shifting sluggish to a smaller cog	Click rear upshift switch while in adjustment mode (6-8a). Shifting sluggish both directions. Adjust chain gap with b-screw (5-2f).
Derailleur creaks on downshifts or suspension movement	Overhaul derailleur clutch (5-36).
FRONT DERAILLEUR	
Nothing happens when push shift switch	Charge battery (6-2). Check wire connections. Disconnect from E-Tube Project app (6-10c).
Chain falls off to outside	Turn H2 adjustment screw counterclockwise (6-8b, steps 2–4)
Chain falls off to inside	In lowest gear and adjustment mode, step derailleur outward one step by clicking upshift paddle once (6-8b, steps 6–8). Install inner stop (5-51).
Chain rubs inner cage plate in low gear	Check derailleur position (5-5a). In lowest gear and adjustment mode, step derailleur inward one step by clicking downshift paddle once (6-8b, steps 6–8).
Chain rubs outer cage plate in high gear	Check derailleur position (5-5a). Turn top (H2) limit screw clockwise (6-8b, steps 2–4).
Chain rubs a cage plate in cross-gear	Click appropriate front shift paddle in adjustment mode (6-8b, steps 6–8).
Derailleur hits crankarm	Check derailleur position (5-5a). Turn top (H2) limit screw counterclockwise (6-8b, steps 2–4).
Shifting sluggish to big chainring	Turn the top (H2) adjustment screw clockwise (6-8b, steps 2–4). Shifting sluggish to small chainring. In lowest gear and adjustment mode, step derailleur inward one step by clicking downshift paddle once (6-8b, steps 6–8).

The planet's spinning a thousand miles an hour around this gigantic nuclear explosion while these people roll these machines with rubber tires . . .

—JOE ROGAN

7

TIRES

TIRES, TUBES, AND TUBELESS TIRES

TOOLS

Tire pump
Tire levers
Tube patch kit

OPTIONAL

Tire sealant
Soft brush
Air compressor
Small Phillips screwdriver
Valve-core remover
Large syringe (for sealant)

Tires provide grip and traction for propulsion and steering, and the air pressure in them is your first line of suspension. On most mountain bikes, inner tubes keep the air inside the tires, and the rest depend on an airtight tire and rim to keep air in their tubeless tires.

This chapter addresses how to replace or repair a tire or inner tube, how to install and maintain tubeless tires, and how to determine tire pressure. Tubular (i.e., "sew-up") mountain bike tires are much too rare to cover in this book, so please consult *Zinn and the Art of Road Bike Maintenance* for instructions on working with tubular tires.

7.1 Schrader valve

7.2 Presta valve

7-1

REMOVING A STANDARD TIRE AND TUBE

NOTE: *If you have tubeless tires, skip to 7-2.*

1. **Remove the wheel** (see 2-2 through 2-6). If the wheel is on a Cannondale Lefty one-legged fork, you can skip this step, because you can change the tire while the wheel is on the bike!
2. **If the tire is not already flat, deflate it.**

(a) To deflate a Schrader valve (the kind of valve you would find on a car tire), push down on the valve pin with something thin enough to fit in that won't break off, such as a pen cap or a paper clip (Fig. 7.1).

PRO TIP — VALVE EXTENDERS

VALVE EXTENDERS ARE MADE FOR Presta valves, and they solve the problem of using an inner tube with a short valve or a short tubeless valve on a deep-section rim that requires a long valve. The simplest type of valve extender is simply a thin tube like a drinking straw with threads inside one end to screw onto the cap threads of the valve (Fig. 7.3A). Unscrew the little nut atop the valve hard against the end of the threads so that it always stays open and does not screw closed on its own. If the valve has some problems—an imperfect seal, a bent rod—it can leak when the nut is not tightened down, making this type of valve extender problematic.

To deflate tires that have simple drinking-straw-type valve extenders, you need to insert a thin rod (a spoke is perfect) down into the valve extender to release the air.

To install drinking-straw-type valve extenders so that they seal properly and allow easy inflation, you need to unscrew the little nut on the Presta valve until it is against the mashed threads at the top of the valve shaft (they are mashed to keep the nut from unscrewing completely). Back the nut firmly into these mashed threads with a pair of pliers so that it stays unscrewed and does not tighten back down against the valve stem from the vibration of riding, thus preventing air from going in when you pump it. Wrap a turn of Teflon pipe thread tape around the top threads on the valve stem before screwing on the valve extender; if you do not, air will leak out during pumping and the pressure gauge on your pump will not give an accurate reading of tire pressure. Tighten the valve extender onto the valve with a pair of pliers.

The best alternative, if you have inner tubes or tubeless valves with removable Presta valve cores, is to get valve extenders for them that are threaded at the base with the same thread as on the base of a valve core. They will have threads inside the other end to accept a valve core (Fig. 7.3B). Deflating or inflating the tire with one of these valve extenders is no different from deflating or inflating any tire with a standard Presta valve, making them the valve extender of choice.

7.3A Drinking-straw-type valve extender

7.3B Removable-core valve extender (showing separate Presta valve core)

7.3C Topeak/Spinergy valve extender

To install one of these valve extenders, unscrew the valve core (counterclockwise) with a Presta valve-core wrench or adjustable wrench and remove it. Screw the valve extender into the valve body where the core was; tighten it firmly with pliers or a wrench on its wrench flats. Screw the valve core into the valve extender, and tighten it with a Presta valve-core wrench or adjustable wrench.

Some valve extenders (Topeak, Spinergy), now going out of style with the easier access to tubes and valves with removable valve cores, have a thin knurled knob on top with a shaft running all the way down to grab the nut atop the valve (Fig. 7.3C). They actually allow you to tighten or loosen the valve nut with the extender in place, thus behaving for all intents and purposes like a standard Presta valve. With this type of extender, unless it has a rubber seal at its base, you should also wrap a turn or two of Teflon pipe thread tape around the top threads on the valve stem before screwing on the valve extender; if you do not, air will leak out during pumping and the pressure gauge on your pump will not give an accurate reading of tire pressure. Tighten the valve extender onto the valve stem with a pair of pliers.

(b) Presta valves are thinner and have a small threaded rod with a tiny nut on the end. To let air out, unscrew the little nut a few turns and push down on the thin rod (Fig. 7.2). To seal, tighten the little nut down again (with your fingers only!); leave it tightened for riding.

NOTE: *If your bike has deep-section rims (often carbon) that are too tall for a normal valve stem, you will require longer valves or "valve extenders"—thin, threaded tubes that screw onto the Presta valve stems (Figs. 7.3A, 7.3B, and 7.3C)—so that you can inflate and deflate the tire. See the Pro Tip on valve extenders for more information.*

3. **Push up on the sidewall with your thumbs to lift it over the rim wall.** Start adjacent to the valve stem; see the Pro Tip below on tire removal and installation for why. If you can push the tire bead off the rim with your thumbs without using tire levers, by all means do so, for there is less chance of damaging the tube and the tire.
4. **If you can't get the tire off with your hands alone, insert a tire lever, scoop side up, between the rim sidewall and the tire until you catch the edge of the tire bead.** Again, this is most easily done adjacent to the valve stem. Be careful not to catch any of the tube under the lever.
5. **Pry down on the lever until the tire bead is pulled out over the rim** (Fig. 7.4). If the lever has a hook on the other end, hook it onto the nearest spoke. Otherwise, keep holding it down.
6. **Place the next lever a few inches away.** Do the same thing with it (Fig. 7.4).
7. **If needed, place a third lever a few inches farther on.** Pry the bead out, and continue sliding this lever around the tire, pulling the bead out as you go (Fig. 7.5) until one bead is off. Some people slide their fingers around under the bead, but beware of cutting your fingers on sharp tire beads.

NOTE: *There are various "quick" tire levers on the market that require the use of only one lever. But if the tire is really stubborn, the tried-and-true three-lever method outlined here may be the one to resort to.*

7.4 Removing the tire with levers

PRO TIP — TIRE REMOVAL AND INSTALLATION

TIRE REMOVAL IS EASIEST IF YOU START near the valve stem, and tire installation is similarly best accomplished by finishing at the valve stem. That way, the beads of the deflated tire can fall into the dropped center of the rim on the opposite side of the wheel, making it effectively a smaller-circumference rim. If you instead try to push the tire bead off (or on) the rim opposite the valve stem, the circumference on which the bead is resting is larger, because the valve stem is forcing the tire beads to stay up on their bead-seat rim ledges on the opposite side of the rim from where you are working (Fig. 7.7 shows the tire beads, rim ledges, and valley).

Starting at the valve stem, and letting the opposite tire bead fall into the rim valley, greatly improves your odds of mounting and removing a tubeless tire without tools, which is important because prying a tubeless tire on or off with tire levers can damage the thin sealing flap along the tire bead.

When mounting a tire and tube, finishing at the valve stem, and then pushing up on the valve (7-6), minimizes the chances of getting a bit of inner tube stuck under the tire bead, which, upon inflation, can blow the tire off the rim. It can temporarily deafen you if it happens while pumping, or it can strand you with a torn-open tube if it happens when you're out riding.

7.5 Pulling out bead with third lever

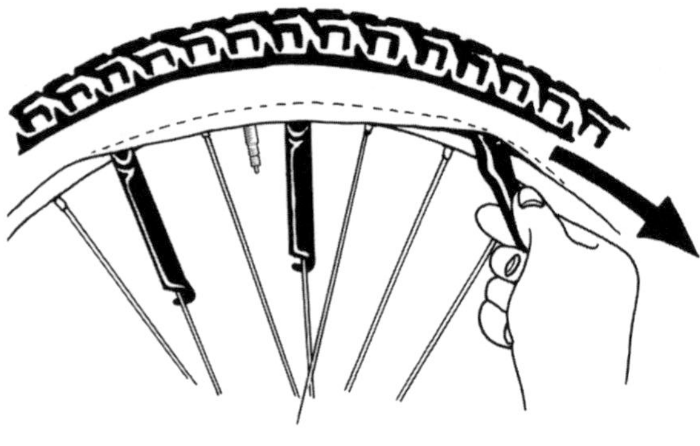

7.6 Removing the inner tube

8. **Pull the tube out** (Fig. 7.6). If you are patching or replacing the tube, you do not need to remove the other side of the tire from the rim. If you are replacing the tire, the other bead should come off easily with your fingers, starting at the valve stem. If it does not, use the tire levers as outlined previously.

7-2

REMOVING A TUBELESS TIRE

If your wheels have tubeless tires (often labeled UST, for "Universal Standard Tubeless"), remove and install the tires with only your hands if at all possible, as tire levers can damage the sealing flap that extends beyond the bead of tubeless tires (Fig. 7.7). If you are planning to patch the tire, you must find the leak before removing it from the rim (7-3).

1. **Remove the wheel** (see 2-2 to 2-6). If the wheel is on a Cannondale Lefty one-legged fork, you need not remove the wheel!
2. **If the tire is not already flat, deflate it at the valve.** Have the wheel vertical and the valve at 5- or 7-o'clock to prevent sealant from squirting out.
   (a) Tubeless tire valves screw into the rim with rubber seals around them. They can be either Schrader valves (the kind of valve you would find on a car tire), Presta valves (Fig. 1.1B), or both. Original Mavic tubeless valves are both; unscrew and remove the outer, Schrader-size externally threaded tube to make this valve a Presta valve.
   (b) To deflate a Schrader valve, push down on the valve pin with something thin enough to fit in that won't break off, such as a pen cap or a paper clip (Fig. 7.1).
   (c) To let air out of a Presta valve, unscrew the little nut a few turns and push down on the thin rod (Fig. 7.2). To seal, tighten the little nut down again (with your fingers only!). Leave it tightened for riding.
3. **Push inward on the tire bead all the way around with your thumbs.** This pops the bead off the bead-seat ledge, and off the "hump" of a UST rim (Fig. 7.7), so that it'll fall into the dropped center of the rim. A well-fit tubeless fat-bike tire may require you to stand on the tire sidewall with the wheel lying on the floor to dislodge the tire bead.

7.7 Cross-section of UST tubeless tire and rim

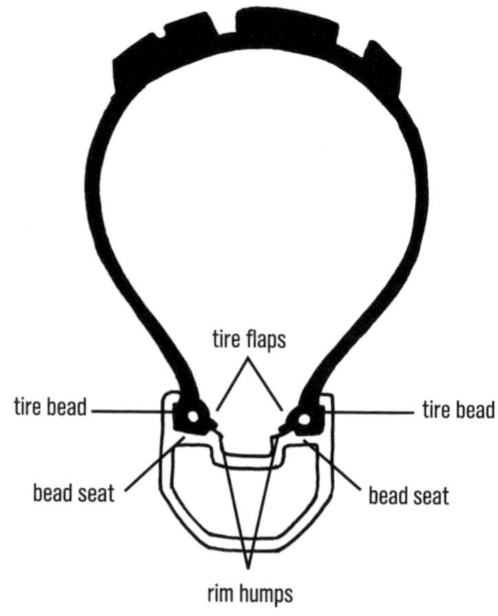

4. **Starting adjacent to the valve stem, push the tire off the rim with your thumbs.** The Pro Tip on tire removal and installation in 7-1 explains why you start at the valve stem. To preclude damaging the tire's sealing flap, avoid using tire levers unless you won't be reusing the tire.
5. **Dispose of the sealant pooled up in the bottom of the tire.**

7-3

FINDING LEAKS

Keep in mind that you can patch only small holes. If the hole is bigger than the eraser end of a pencil, a round patch is not likely to work. A slit of up to an inch or so can be repaired with a long oval patch. To find a leak in a tube, you need to remove it from the tire. To find a leak in a tubeless tire, you must find the leak with the procedure below before removing the tire from the rim to patch it.

1. **If the leak location is not obvious, put some air in the tube or tire.** For a tube, inflate it until it is two to three times larger than its deflated size. Be careful. It will explode if you put too much air in. For a tubeless tire, leave the tire on the rim and inflate it to 25–50 psi, then continue with steps 2 and 3.
2. **Listen/feel for air coming out.** Mark the leak(s).
3. **If you cannot find the leak, submerge the tube or tire in water.** Look for air bubbling out (Fig. 7.8), and mark the spot(s). With a tubeless tire, you can also spread soapy water all over it with a soft bottle brush. Dish soap in water works great and is more effective at locating the leak (due to longevity of the soap bubbles) than submerging the tire and wheel in a big tub or pond.

7.8 Checking for a puncture

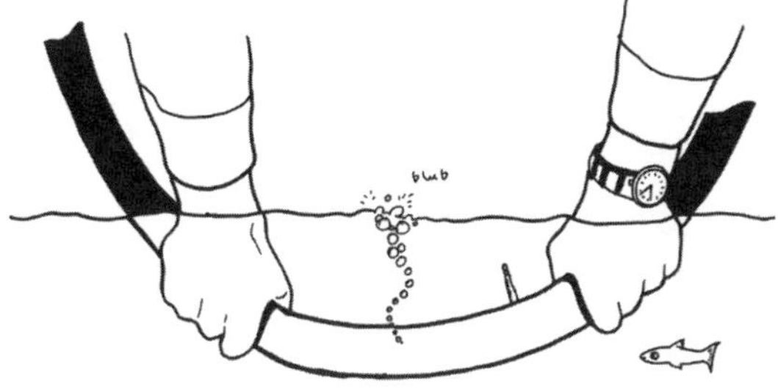

7-4

PATCHING A LEAK WITH STANDARD PATCHES

NOTE: *The following instructions apply to patching a tube, but the procedure is the same for patching a tubeless tire, except that you patch the inside of the tire versus the outside of a tube. Also, you can always stick a tube inside a tubeless tire if you don't want to deal with patching the tire. As for repairing a tubeless tire where there are a lot of cacti or thorns, it is arduous and next to impossible to find and patch all of the holes. Rather than throw the (expensive) tire out, put some tire sealant inside it (7-7 or 7-9).*

1. **Dry the tube thoroughly near the hole.**
2. **Rough up and clean the surface within about a 1-inch radius around the hole.** Use a small piece of sandpaper (usually supplied with the patch kit). Do not touch the sanded area, and don't rough up the tube with one of those little metal "cheese graters" that come with some patch kits. They tend to do to your tube what they do to cheese.
3. **Use a high-quality patch kit for bicycle tires.** It will have thin, usually orange, gummy edges surrounding a black rubber patch in the center.
4. **Apply patch cement over an area centered on the hole** (Fig. 7.9). Apply a thin, smooth layer to an area that is bigger than the patch.

7.9 Smearing the patch glue

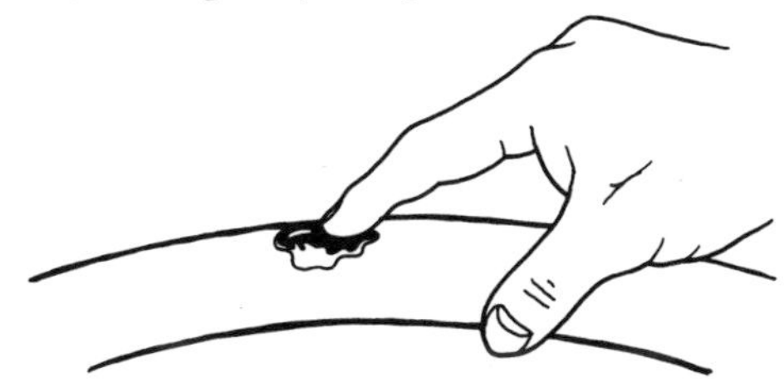

7.10 Removing the cellophane backing

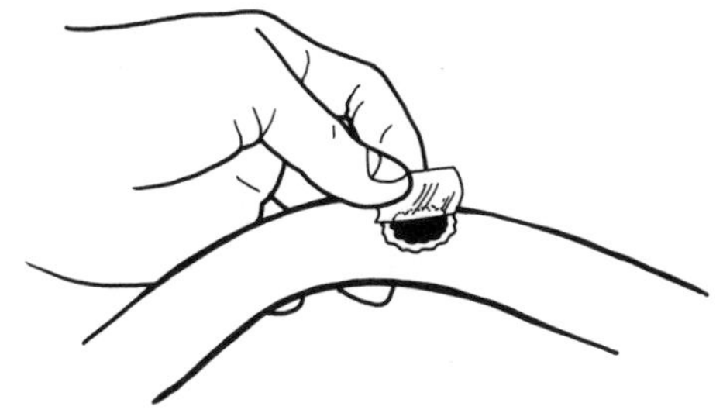

PRO TIP — TIRE SIZE

IDEAL TIRE SIZE IS RELATED TO terrain and riding style. Some of this is obvious: Bumpier surfaces and higher speeds demand bigger tires. Climbing, especially on smooth surfaces, rewards light weight, and smaller tires can be lighter. However, rolling resistance on anything but smooth roads will generally be improved with a larger, softer tire.

Thing is, it's not always simple to figure out what tire size you're actually getting from the dimensions listed on the tire or on the box. The good news is that there is an international standard numbering system for bicycle tires and rims that is straightforward and understandable. Formerly called "E.T.R.T.O." (European Tyre and Rim Technical Organisation), it has become "ISO" (International Organization for Standardization). The first number in the ISO code is the width of the tire or the inner width of the rim between the sidewall hooks ("W" in Fig. 7.11). The second number determines what rim the tire will fit on and is the rim's "bead seat diameter" ("BSD" in Fig. 7.11), which is the diameter of the shelf that the tire bead rests on. So a 700 × 23 road tire would be a 23–622, a 29 × 2.35 mountain bike tire would be a 61–622, and a 28 × 1 road tire would be a 25–622. (Yes, 700C, "28-inch" road, and 29-inch mountain bike tires all take the same rim diameter.)

The bad news is that, even though tires often have the ISO code molded into them, tires and rims are not listed in catalogs this way.

Numbers like 26-inch or 700C may have once referred to the diameter of the tire in inches or millimeters, but with time, different tire widths for each given rim size came along. These tires consequently had different diameters from the original upon which the number was based, but the 26-inch or 700C designation was carried forward anyway, followed by a second number indicating the tire width in inches, such as 26 × 2.35.

For mountain bikes, you will generally want to buy only tires with a decimal tire width (26 × 2.35, 27.5 × 3.0, 29 × 2.2, etc.). Avoid tires with the width indicated as a fraction, because equivalent numerical values do not equate to equivalent tire sizes. Even though, for example, 1½ is numerically equivalent to 1.5, the rim size will almost always be different for tires listed this way. For instance, 26 × 1.0 through 26 × 2.5 (with two ignorable exceptions) are tire sizes that fit on 26-inch mountain bike rims, which have a BSD of 559mm (so a 26 × 1.0 is an ISO 25–559). But a 26 × 1 is a 650C (ISO 571) triathlon size, 26 × 1¼ and 26 × 1⅜ are both ISO 597, 26 × 1½ is 650B (now called 27.5, ISO 584), and 26 × 1¾ is an ISO 571 Schwinn cruiser size (yes, same BSD as a 26-inch triathlon tire).

7.11 Inner rim width (W) and bead seat diameter (BSD)

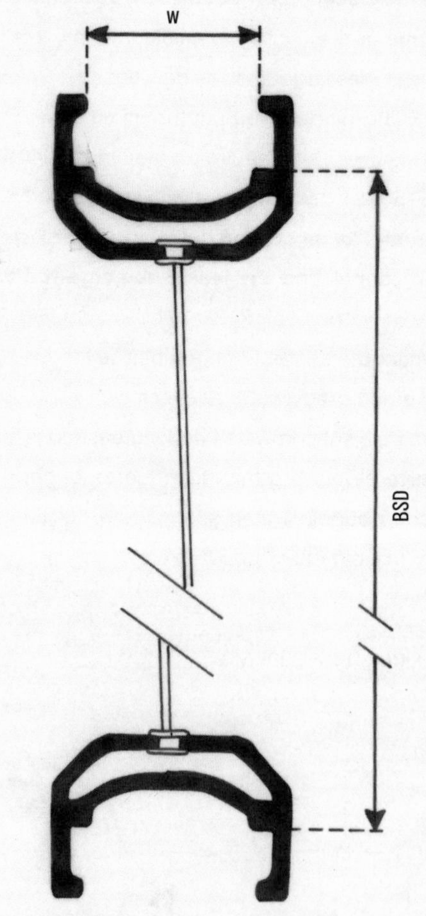

5. **Let the glue dry.** Wait until there are no more shiny, wet spots (5–10 minutes).
6. **Remove the foil backing from the gummy underside of the patch.** Don't remove the cellophane top cover.
7. **Stick the patch over the hole, and push it down in place.** Make sure that all the gummy edges are stuck down. That's it!
8. **Although there is no need to do so, you may remove the cellophane top covering.** Be careful not to peel off the edges of the patch when removing the cellophane (Fig. 7.10). If the cellophane atop the patch is scored, fold the patch so that the cellophane will split at the scored cuts. Peel outward from the center, and avoid pulling the newly adhered patch away from the tube.

7-5

PATCHING A LEAK WITH GLUELESS PATCHES

There are a number of adhesive-backed patches on the market that do not require cement to stick them on. Most often, you simply need to clean the area around the hole with an alcohol pad supplied with the patch. Let the alcohol dry, peel the backing off, and stick on the patch. The advantage of glueless patches is that they are very fast to use, take little room in a pack, and free you from the experience of opening your patch kit only to discover that the glue tube has dried up.

On the downside, I have not found any glueless patches that stick nearly as well as the standard type. With a standard patch installed on a tube, you can inflate the tube to look for more leaks without having it in the tire. If you do that with a glueless patch, it usually lifts the patch enough to start it leaking. You must install it in the tire and on the rim before putting air in it after patching. And don't expect the glueless patch to be a permanent fix, as you can with a glued-on standard patch.

7-6

INSTALLING A TUBE AND TIRE

Feel around the inside of the tire to see if there is anything sticking through that can puncture the tube. This is best done by sliding a rag all the way around the inside of the tire. The rag will catch on anything sharp and will keep your fingers from being cut by whatever is stuck in the tire.

7.12 The rim strip protects the inner tube from the ends of the spokes

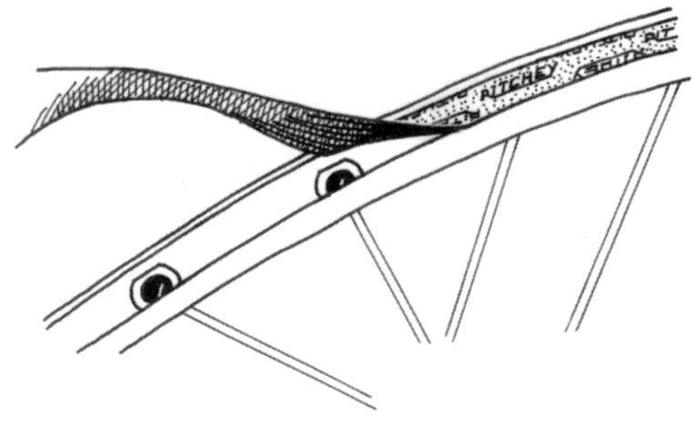

1. **Check for tire wear.** Replace any tire that has worn-out or cracked tread or areas (inside or out) where the tread-casing fibers appear to be cut or frayed.
2. **Examine the rim tape.** Check that the rim tape, which covers the spoke holes (Fig. 7.12), is in place and that there are no spokes or anything else sticking up that can puncture the tube. Replace the rim tape if necessary. With an asymmetrically drilled rim, make sure the adhesive and/or the fit of the rim tape is very good. If the rim tape slides over even a little bit, it can expose the edge of one of the offset spoke holes and puncture the tube.
3. **By hand, push one bead of the tire onto the rim.** Check the direction of the tire rotation, if marked, and orient the tire label or pressure rating to be at the valve stem.
4. **Optional: Smear talcum powder around the inside of the tire.** This prevents the tube and tire from adhering to each other. Wear a mask and don't inhale talcum powder; it's bad for your lungs.
5. **Put just enough air in the tube to give it shape.** Close the valve if it is a Presta.
6. **Push the valve through the valve hole in the rim.**
7. **Push the tube up inside the tire all the way around.**
8. **Starting at the side opposite the valve stem, push the tire bead onto the rim with your thumbs.** See the Pro Tip on tire removal and installation in 7-1 for the reason. Be sure that the tube doesn't get pinched between the tire bead and the rim; having some air in the tube will minimize this issue.

PRO TIP — TIRE DIRECTION

TIRE DIRECTION MAKES A DIFFERENCE for technical riding. On the front, you generally want the concave or V-shaped scooping edges of the tread blocks at the bottom of the tire pointed forward for braking (so they point back at the top; see Fig. 2.2), whereas on the rear, you want the scooping edges oriented backward for propulsion traction. Some tires have an arrow indicating rotation direction for use on either the front or the rear. If not, hold the tire up above your head and look at the tread as the ground sees it. Consider which way the wheel is rotating and what happens during braking and driving. The best way to orient the tread will then be apparent.

7.13 Installing a tire by hand

7.14 Finishing installation at the valve

7.15 Seating the tube by pushing up on the valve

9. **Complete the tire bead installation.** With your thumbs, work around the rim in both directions toward the valve stem (Fig. 7.13). Finish from both sides at the valve (Fig. 7.14). You can usually install a mountain bike tire without tools; if you cannot, first try deflating the tube when you have gotten as far around as you can with your hands. You should now be able to push the tire on the last bit, as deflating the tube will allow the bead on the far side, opposite the valve stem, to drop into the lower center of the rim. If this does not allow you to complete the mounting by hand, use tire levers to pry the tire bead on, but make sure you don't catch any of the tube under the edge of the tire bead. Finish at the valve.
10. **Push up on the valve stem.** This pulls up any nearby folds of the tube stuck under the tire bead (Fig. 7.15) when you pushed the last bit of bead onto the rim. You may have to manipulate the tire so that all the tube is tucked within it.
11. **Inspect for any part of the tube that might be protruding from under the edge of the tire bead.** Go all the way around the rim. If you have a fold of the tube under the edge of the bead, it can blow the tire off the rim either when you inflate it or while you are riding. It will sound like a gun blast and will leave you with an unpatchable tube.
12. **Pump up the tire.** Generally, 35–45 psi is a good amount on a standard tire (see the Pro Tip on tire pressure for additional information). Much more, and the ride gets harsh. Much less (except on very fat tires; see sidebar) and you run the risk of a pinch flat, or "snakebite."

PRO TIP — LOW TIRE PRESSURE

YOU CAN GET A SMOOTHER RIDE, better traction across hillsides, and lower rolling resistance on rough terrain by using low tire pressure (under 30 psi). That's one of the primary benefits of tubeless tires, since you can't pinch-flat a tubeless tire (there is no tube to pinch!), although you can dent the rim. Begin by experimenting with tire pressures, and don't be afraid to try even lower than 20 psi, even with 2.2-inch-wide tubeless tires—you might discover that you like it!

When you reduce the tire pressure, you also reduce the stress the air pressure creates on the tire casing. Because hoop stress in a cylinder increases with diameter, you cannot run a fat tire at the same high pressure as a skinny tire. This hoop stress, which you can feel as hardness of the tire, is proportional to the tire diameter. That is, if you have a tire that is twice the diameter of another one (twice as fat), and you run them both at the same pressure, the fatter tire will have double the hoop stress, and, consequently, it will feel much harder than the skinnier tire. (If you are familiar with vectors, you can see why this would be: Newton's Third Law requires that there be equal and opposite vector components of the tension in the tire casing balancing the outward force vectors created by the tire pressure. The fatter the tire, the flatter the tire profile, and hence the smaller the percentage of a given tension vector of the tire casing is pointed radially inward; therefore, at a given tire pressure, the fatter the tire the higher the casing tension.)

High hoop stress can lead to blowing the tire off the rim as well as premature failure of the tire casing, not to mention a jarring ride and high rolling resistance. Where this is particularly an issue is with fat-bike tires (4-inch width or greater), as riders often can't get their heads around running them at super-low pressures of between 2 psi and 8 psi. But you are asking for big trouble if you run a fat-bike tire at 20 psi or higher. Also be aware that, the wider the rim, the wider the mounted tire becomes, further increasing the hoop stress at a given pressure.

Again, hoop stress is proportional to the cross-sectional area of the tire and rim and hence to the cross-sectional diameter of the inflated tire. For instance, if you were to increase a given tire diameter from 2.2 inches to 2.35 inches, either by getting a wider model of the same tire or by getting a wider rim, the stress on the tire casing and the rim walls (which you feel as hardness of the tire) would go up if inflated to the same pressure. To maintain the same stress and hardness of the inflated tire, choose the pressure in the wider tire by multiplying the pressure in the narrower tire by 2.2/2.35. So, if you were running the 2.2 tire at 27 psi (1.86 bar), you'd run the 2.35 tire at 25.3 psi (1.74 bar). With a 5-inch fat-bike tire of the same thickness, you'd multiply by 2.2/5, and the equivalent pressure would be 11.9 psi (0.82 bar).

WARNING: *Never* exceed 20 psi on any fat rim (60–100mm wide), even when trying to seat a tubeless tire, and never have more than 12 psi in fat-bike tires for any length of time. Due to the from the high pressure, the tires will stretch and the rim walls will flex outward; weak rims will fail once the tires stretch enough.

When a fat-bike tire blows off a rim, it can be extremely dangerous. Think about the number of pumps you put into a fat tire to reach a given pressure; that's pure stored energy—enough to lift a truck. If your tubeless fat-bike tire were to explode at 20 psi or more, the sealant will penetrate your skin like a tattoo, your ears will ring for hours, and your eyes will not have time to blink before the sealant hits your corneas. And there is no gain to higher pressure; tests have demonstrated that effective rolling resistance doesn't change measurably above 12 psi for a 170-pound rider on a fat-bike tire.

On snow, a fat-bike tire at 2 psi (0.14 bar) leaves a flat track, reducing the irritation among skiers sharing the trail, and it rolls with less resistance than at double, triple, or quadruple the pressure. However, keeping a tubeless tire from burping—and thus having to deal with the problem in cold temperatures—at 2 psi requires special measures. See the note on tubeless fat-bike tires at the end of section 7-7.

7-7

INSTALLING A TUBELESS TIRE

The time involved in mounting the following categories of tubeless tire/rim combinations increases as you go down the list, and the probability of ultimate success decreases:

- Installing a UST tubeless tire on a UST tubeless rim
- Installing a TLR tubeless-ready tire on a UST tubeless rim
- Installing a UST tubeless or TLR tubeless-ready tire on a TCS tubeless-conversion rim
- Installing a UST tubeless or TLR tubeless-ready tire on a rim not designated for tubeless conversion but converted to tubeless anyway
- Installing a standard tire as tubeless on a UST tubeless or TCS tubeless-conversion rim
- Installing a standard tire as tubeless on a rim not designated for tubeless conversion but converted to tubeless anyway
- Installing a fat-bike tire tubeless to run it problem-free at 2 psi in the cold

"UST" stands for Universal Standard Tubeless and refers specifically to a two-part rim/tire system (Fig. 7.7). Mavic, Michelin, and Hutchinson developed the UST system together, and Shimano became part of the UST partnership after its inception in 1999. To get the UST tubeless designation, a tire must hold air without sealant inside. This is a high bar, so "tubeless ready" ("TLR") and "tubeless conversion system" (TCS) came along, which make no guarantees and require sealant to fill air porosity in the tire and along the rim/tire interface.

A UST tire has a thick rubber layer coating its interior and a rubber flap along its bead (Fig. 7.7) to ensure an airtight seal on a UST rim (although using sealant is still a good idea with a UST system to seal punctures). A tubeless-ready tire (different manufacturers may use different designations than "TLR," but the idea is the same) will have the same bead-sealing flap as a UST tire, but it will have less rubber coating the inside, rendering it lighter and more supple. In order to hold air, it requires sealant to fill the tiny holes that otherwise would let air bleed out of the sidewalls and along the bead if not used on a UST rim. Similarly, some standard tube-type MTB tires can be mounted tubeless, but they require sealant to fill orifices in the sidewalls as well as to seal all along the tire bead.

A UST rim will have ridges ("humps") along the inboard edges of the bead ledges (Fig. 7.7) and a rim valley of specified depth and width, and it will not have any spoke holes penetrating the rim bed. In addition to sealing the bead flaps, the bead-ledge humps lend more security in keeping the tire on the rim in case of a sudden loss of pressure (due to a cut casing or "burping" the tire on a hard corner). The hump also reduces the chances of burping air out of the tire in the first place.

Many standard rims can be sealed with tubeless rim tape to be used tubeless, but not all rims will work. The rim valley has to be wide and deep enough to allow both tire beads to drop into it and not so deep that the tire won't pop up out of the valley to seat on the rim ledges when given a blast of compressed air. The rim will have to be taped with sealing tape or with a tubeless rim strip to seal off the spoke holes. Rims marked TCS or a similar designation for tubeless conversion will generally work well.

Be aware that any rim without the UST ridges (humps) along the medial edges of the rim ledges will be more susceptible to burping on hard, bumpy corners; the side impact breaks the bead seal for an instant and lets a burp of air escape. The pressure drops so much with each burp that subsequent side impacts don't need to be nearly as high to burp more air, so continuing to corner hard will result in a flat tire. For safety reasons, pump the tire back to acceptable pressure after a burp.

To prevent damage to the tire's bead-sealing flap, avoid using tire levers for mounting and removal, if at all possible. Starting and finishing at the valve stem make this easier (see the Pro Tip on tire removal and installation in 7-1).

1. **Evaluate your tools and materials.** If the valve has a removable core (Fig. 7.16) and you have either a sealant syringe (Fig. 7.18) or sealant in a squeeze bottle with a tip for a valve stem, you can add sealant after the tire is seated (this is preferable, especially for adding more sealant at a later date); otherwise, you will be pouring sealant into the open

tire before you fully install it. The sealants to use are thin ones that coagulate upon air blowing past them—not viscous (often brightly colored, glycol-based) sealants with chopped fibers in them to plug holes. A UST tire requires less sealant than a TLR tire, which in turn requires less than a standard tube-type tire you're setting up as tubeless.

2. **Check if you have a UST rim or will need a tubeless rim strip.** Looking down on the top of a UST rim, you won't see any spoke holes, and you will see a "hump" along the medial edge of each bead ledge (Fig. 7.7). Check that the rim edges are not dented, which could allow air to escape at those spots. If you have a UST rim, skip to step 8.
3. **Remove the existing rim strip.** If the rim already has a rim strip for an inner tube, remove it (Fig. 7.12) so that the tubeless rim tape can stick and seal. The exception is a fat-bike rim (60–100mm wide) with big weight-saving holes between spokes. With a rim like that, you will need to put the sealing tape over the rim strip (you should still remove, clean, and reinstall it), because the sealing tape alone lacks the structural integrity to constrain the air pressure at those big holes.
4. **Clean the rim.** Leave no residue to prevent the sealing tape from sticking.
5. **Choose a tubeless sealing tape of the appropriate width for the rim.** It should be wide enough to butt against the rim walls, and thus cover the spoke holes and rim ledges, and not so wide as to run up the walls. Fat-bike rims require two overlapping strips of tape.
6. **Tape the rim.** Line up the tape so that it is smooth. Start before the valve hole, pass over it, go around and overlap the hole, and cut the tape. Lacking tubeless tape, you can instead wrap two layers of fiberglass-reinforced strapping tape around the rim, completely covering the rim holes, then cover it with two layers of electrical tape to seal it, overlapping the strapping-tape edges. On fat-bike rims, lay the first strip of tubeless tape down along one side, along the rim wall; pull it very tight as you go, because one edge of the tape will be going around a bigger diameter than the other edge, and you don't want to leave wrinkles that won't seal with the tape overlapping it when you lay down the strip along the other rim wall.
7. **Punch through the tape at the valve hole.** Use a small Phillips screwdriver or an awl. *Do not* use a knife to cut the hole, as you can get leakage around the slit, and it can further split down the rim tape.
8. **Install the valve.** Pick a tubeless valve of the appropriate length. Ideally, get one with a removable valve core (Fig. 7.16); it will have wrench flats on the tiny valve-cap threads at the tip. If the valve's rubber foot is asymmetrical, align its long length with the rim valley. With a fat-bike rim (60–100mm wide), make sure you have a valve for a single-wall rim. If the rubber base has too long of a taper or the threads on the valve stem don't go down to the rubber base, tightening the valve collar will not pull the rubber base fully up into the valve hole so that it seats into such a thin rim.
9. **Install the valve collar.** If there is a rubber O-ring included with the valve, install that first. Tighten the collar finger-tight against the rim, so you can remove it on the trail if you puncture and need to put in an inner tube. Put a drop or two of sealant in the rim bed around the base of the valve.
10. **Determine tire rotation direction.** Orient the tire based on the arrows on the tire sidewall, or see the Pro Tip on tire direction.

7.16 Tubeless valves for standard, double-wall rims (left) and for single-wall (fat bike) rims (right)

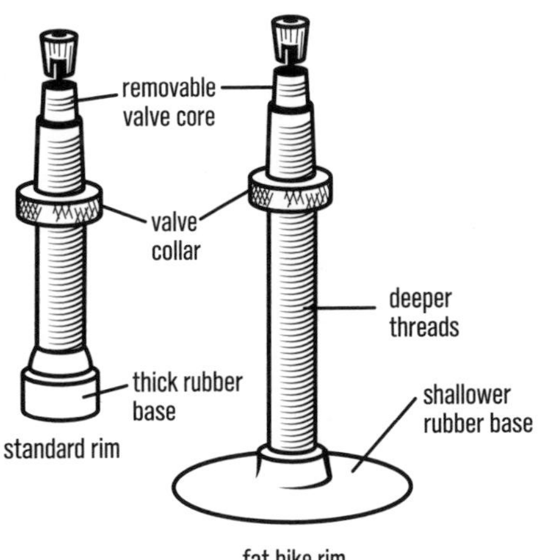

11. **Wet the edges of the tire.** Put dish soap in the water to facilitate sealing the tire as well as to help find leaks.
12. **Push one bead of the tire onto the rim.** Locate the brand label or recommended tire pressure imprint near the valve. Starting opposite the valve, push one bead of the tire onto the rim with your thumbs and fingers only, finishing at the valve (Fig. 7.13).
13. **Push the other bead onto the rim by hand.** Again, start on the side of the rim opposite the valve and finish at the valve (for why, see the Pro Tip on tire removal and installation in 7-1). If you have a removable valve core and a sealant syringe or squeeze bottle with a tip that fits into a valve, install the tire completely (Fig. 7.14). Otherwise, stop just before you get the last 6–8 inches (15–20cm) of tire bead onto the rim, and skip to step 17.
14. **Remove the valve core.** You need a valve-core remover tool. If the core has dried sealant on it, clean it with your fingers. If it is damaged, get a new one.
15. **Inflate the tire.** This only applies with the valve core removed. Blow up the tire to its max pressure until you hear the beads pop up onto the rim ledges. This is most easily done with an air compressor (Fig. 1.4) and a Presta chuck or adapter. Another great method is to use an air tank you pump up first with a floor pump; this can be a separate, water-bottle-size, carry-along unit, or a tubeless-inflation chamber integrated into the floor pump. Lacking those, if you are working with a UST tire and rim, you should be able to seat it with a floor pump, and you may be able to seat a TLR tire this way as well. With a fat-bike tire, if the bead is too far from the rim to seat, first move it over there by slipping a plastic tire lever under the bead and sliding it along the bead, pulling the bead over along the rim wall.
16. **Deflate the tire.** Simply pull the air chuck off the valve stem; without a valve core in it, the air will rush out.
17. **Add sealant.** Shake first. A good amount for most tires is ¼ cup (2 fl. oz., or 60mL); fat tires obviously require much more, as will lightweight tube-type tires; a suggested amount may be printed on the tire packaging. With a removable valve core, squirt a measured amount from a syringe or squeeze bottle in through the valve stem. Without a removable valve core, pour or squirt tire sealant into the open tire at the bottom (Fig. 7.17), rotate the open section to the top, push the last of the bead on the rim, and skip to step 19.

7.17 Putting tire sealant in a tubeless tire

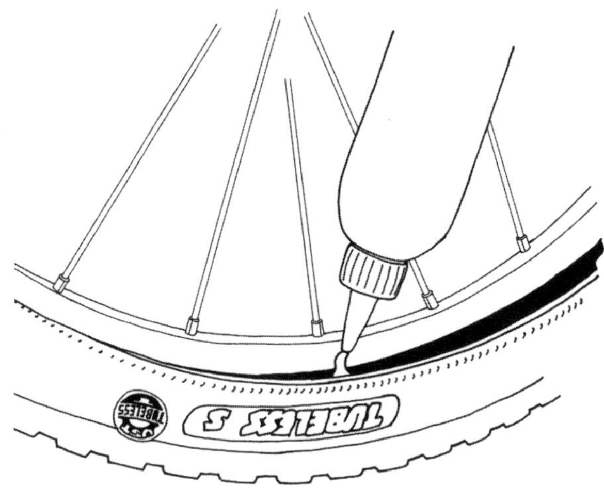

18. **Install the valve core.**
19. **Inflate the tire.** Unless you already did so in step 15, put air in as fast as possible until you hear the tire seat. See the inflation options in step 15. Inflate to the maximum pressure on the tire label.
20. **Check that the tire is correctly seated.** Look at the tire's mold line all of the way around along the rim edge; it should be a constant distance from the rim. If it's hidden in spots, the tire is not seated. If you can flex it with your fingers so that the mold line pops up out of the rim and lines up, you're good. If not, deflate it (with the valve at 5- or 7-o'clock to keep sealant from getting in it), lubricate the bead at the low point with soapy water, and re-inflate. Keep at it until you get it seated properly.
21. **Spin the wheel.**
22. **Shake the wheel.** Swing it up and down, moving your hands around the rim, to coat the entire inside of the tire with sealant.
23. **Coat the tire sidewalls with soapsuds.** A cup (0.25 liter) of water with a squirt of dish soap in it is a good quantity and dilution.

24. **Move sealant to any leaks.** See where air bubbles appear in the soapsuds and tip the wheel so that point is down. Hold the wheel in that position until air bubbles cease to appear there. Repeat at any other points where air bubbles appear.
25. **Pump up (or deflate) to your desired riding pressure.** Go ride your bike if it's holding air, or leave the wheel resting horizontally across the top of a bucket so that the sealant sits in the sidewalls and continues to fill holes.

NOTE: *Fat-bike tires (4-inch to 5-inch width) roll on snow with less rolling resistance and leave a flatter track less objectionable to skiers at 2 psi (0.14 bar) than at higher pressures. Tubeless saves a lot of weight over fat tubes, which can weigh over a pound, but 2 psi leaves them vulnerable to burping air when cornering, and you don't want to have to remove a valve core from a flat rim and install a tube in wintertime temperatures.*

ENTER FATTY STRIPPERS: *Flat, 150mm-wide, thin latex loops you stretch over the rim. Push the valve through the hole in the fatty stripper first, then, in step 8 above, install it into the valve hole. Stretch the fatty stripper flat over the rim strip and/or tubeless tape. Continue with the next steps, installing the tire on top of the latex strip, leaving its edges hanging over the rim. In step 17, use a latex-based sealant and continue with the rest of the steps. The latex sealant, the latex strip, and the tire will eventually all bond together for an airtight seal, forming, essentially, a tubular tire that will not burp air on hard corners at 2 psi. If you so desire, you can trim the excess fatty stripper hanging over the rim; it takes a sharp knife while stretching the latex strip, but it tends to tear more than it cuts cleanly. Sometimes beautification takes time.*

7-8

PATCHING TIRE CASING (SIDEWALL) WITH A STANDARD TIRE AND TUBE

Unless it is an emergency, do not try to patch a sidewall! If the casing is cut, it's best to get a new tire because patching the tire casing is dangerous. No matter what you use as a patch, the tube will find a way to bulge out of the patched hole, and when it does, the tube will burst and the tire will go flat immediately. Furthermore, the casing is compromised and can explode. Imagine coming down a steep descent and suddenly your front tire goes completely flat—you get the picture. In emergency situations, you can put layers of nonstretchable material, such as a dollar bill, an empty energy bar wrapper (or two), or even a short section of the exploded tube (double thickness is better) between the tube and tire (see 3-3d and Fig. 3.2).

7-9

TIRE SEALANTS

Tire sealants can virtually eliminate flat tires caused by tread punctures, but they do not fix sidewall cuts, pinch flats, or rim-side punctures. Tire sealants used to be made of green goo full of chopped fibers. These have largely been superseded by thinner sealants that have a liquid latex or similar base. These sealants can come as a liquid in a bottle, like Stan's NoTubes, Caffélatex, Orange Seal, or Schwalbe Doc Blue (Fig. 7.18), or as an aerosol, like Vittoria Pit Stop or Hutchinson Fast'Air (Fig. 7.19).

NOTE: *You can also purchase inner tubes with sealant already inside.*

ALSO NOTE: *The inner tube must be of proper size. If it is too small (for instance, a 26 × 1.325 to 1.5-inch tube inside a 26 × 2.0-inch tire) and stretches when inflated in the tire, any holes will stretch open and won't seal.*

a. Putting tire sealant in a tubeless tire

The best-case scenario is if you have a removable valve core and a sealant syringe (Fig. 7.18) or sealant in a squeeze bottle with a nozzle that fits a valve stem. See steps 14–25 of section 7-7 above.

Without a removable valve core, simply pop the tire off one side of the rim (7-2), shake the bottle, and squirt 2–4 oz. (¼–½ cup) or so of sealant into the tire (Fig. 7.17). Push the tire bead back onto the rim (7-7) and inflate. See steps 17–25 of section 7-7 above. This method works okay with glycol-based chopped-fiber sealants and better with modern, thin, quick-drying ones.

Some thin sealants can be injected via a syringe hose right through the valve (with the valve core in place), but Stan's and chopped-fiber sealants will clog the valve.

7.18 Liquid tire sealants

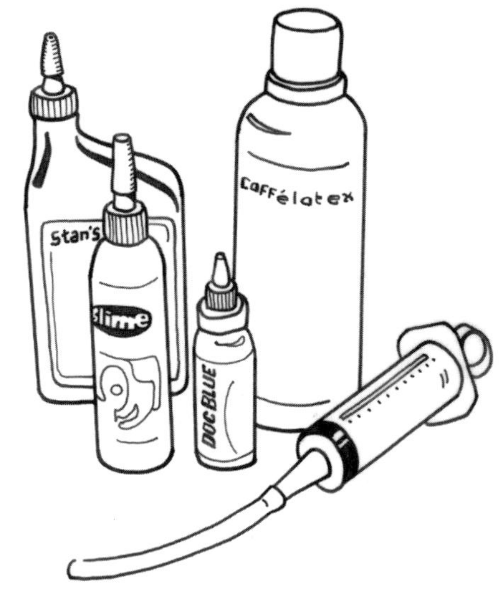

7.19 Aerosol latex sealants

b. Installing aerosol sealant after a puncture

This is a great on-trail fix for any tire, tubed or tubeless.

1. **Deflate the tire.** It probably already is, or you wouldn't be doing this.
2. **Screw the end of the aerosol sealant can's nozzle or hose** (Fig. 7.19) **onto the valve.** Open the Presta valve (Fig. 7.2), or remove the Schrader valve (Fig. 7.1) cap first.
3. **Deploy the full contents of the can into the tire.** This will inflate the tire and fill it with sealant.
4. **Rotate the wheel until the hole is at the bottom.** Hold the wheel that way until the sealant has filled the hole and no more air is escaping. Go ride, or spin the wheel for a while to further spread the sealant around in the tube.

Alternatively, the old-school way to seal a slow leak in an inner tube, or to add some puncture resistance to it, is to pour a can of evaporated milk into a pump you don't care about and pump it in through the valve. This works quite well for tiny leaks, but if you get a blowout after the milk has been in there awhile, boy, does it ever stink!

c. Maintaining sealant-filled tires

Inflating or deflating

Before opening the valve, have the stem at 4 o'clock, and wait a minute for the sealant to drain away; if you don't, sealant will leak out when the valve is depressed, eventually clogging the valve. Even if you're careful about this, over time the valve may still become clogged with sealant. In that case:

1. **Remove the valve core.** Use a valve-core remover or an adjustable wrench.
2. **Push a spoke or similar stiff wire into the valve stem to clear it.**
3. **Pick dried sealant off the valve core.**
4. **Re-install the valve core.**

Sealing punctures

1. **If you find that the tire has gone flat, pump it up and ride it a bit to see if it seals.**
2. **If you get numerous punctures, you may need to pump repeatedly and ride before the tube seals up.**
3. **Remove embedded nails and other foreign objects.** Rotate the wheel so that the puncture is at the bottom so the sealant will flow there and seal the hole.
4. **If it won't seal, establish whether it is even a sealable hole.**

Pinch flats, caused by pinching the inner tube between the tire and rim, are nearly impossible to seal because one of the two "snakebite" holes is on the rim side, and centrifugal force pushes the sealant to the tire side. This also applies to a puncture caused by a protruding spoke head, or anything that causes a hole on the rim side. You will need to replace the tube.

Sidewall gashes won't seal; the tire needs to be replaced.

I'm just sitting here watching the wheels go round and round.

—JOHN LENNON

8

WHEELS

RIMS, SPOKES, HUBS, CASSETTES, AND FREEWHEELS

TOOLS

Spoke wrench
13mm, 14mm, 15mm, 16mm, 17mm, 20mm, 21mm, 22mm, 23mm, 24mm, 28mm cone wrenches
17mm open-end wrench (or an adjustable wrench)
Flat-blade screwdriver
2mm, 5mm, 6mm, 10mm, 14mm, 15mm, 17mm hex keys
Socket set, including 14mm and 15mm sockets
Grease
Oil
Pedro's Vise Whip or chain whip
Pin spanner—adjustable or specific to the hub
Cassette lockring remover
Large adjustable wrench
Freewheel remover (if your wheel does not have a freehub)

OPTIONAL

5mm and 5.5mm Y-spoke wrench for internal nipples
Soft hammer
Wooden hammer
Fine-tip grease gun
Citrus solvent
13⁄16-inch or 21mm socket and handle
Rohloff HG-IG-Check cog-wear indicator gauge

Wheels on mountain bikes are generally strung together with spokes. The hub is at the center, and its bearings allow the wheel to turn freely around an axle. The rim is supported and aligned by the tension on the spokes. On many bikes, the rim serves as both support for the tire and a braking surface.

On the rear wheel, a freewheel or cassette freehub (rear hub with a built-in freewheel) allows the wheel to spin while coasting and engages when forward force is applied to the pedals (Fig. 8.1).

Wheelbuilding is covered in Chapter 15. This chapter addresses how to maintain rims, true wheels, overhaul hubs and freehubs, maintain and replace freewheels, and maintain and replace cogs. Have at it.

RIMS AND SPOKES

8-1

CHECKING RIM CONDITION

LEVEL 2

You never want a rim to fail while riding; the consequences can be severe. To ensure that this won't happen to you, replace any rim with a significant defect. Chapter 15 explains all of the steps required to rebuild a wheel with a new rim, should you decide to do it yourself.

Check over the rims for cracks, particularly at the spoke holes, the valve hole, and the seam (opposite the valve hole). If you find a crack, replace the rim.

If your bike has rim brakes, inspect the rims for a wear indicator (which might be tiny). Modern rims have markings to let you know when the brakes have worn the rim sidewalls too thin. If the sidewalls become too thin, the tire can force the sidewalls out and push them open like a limp taco shell, causing braking and tire-retention issues. Rim-wear indicators can consist of small holes drilled partway into the brake track that, once gone, indicate that too much sidewall material has been worn off. A different type of indicator that can be found on some newer rims consists of a dark spot underneath the surface that only appears once the wear limit has been reached. Mavic rims have a dark hole that appears when the rim is deeply worn; the hole is directly opposite the valve hole and is identified by a sticker.

8.1 The whole thing

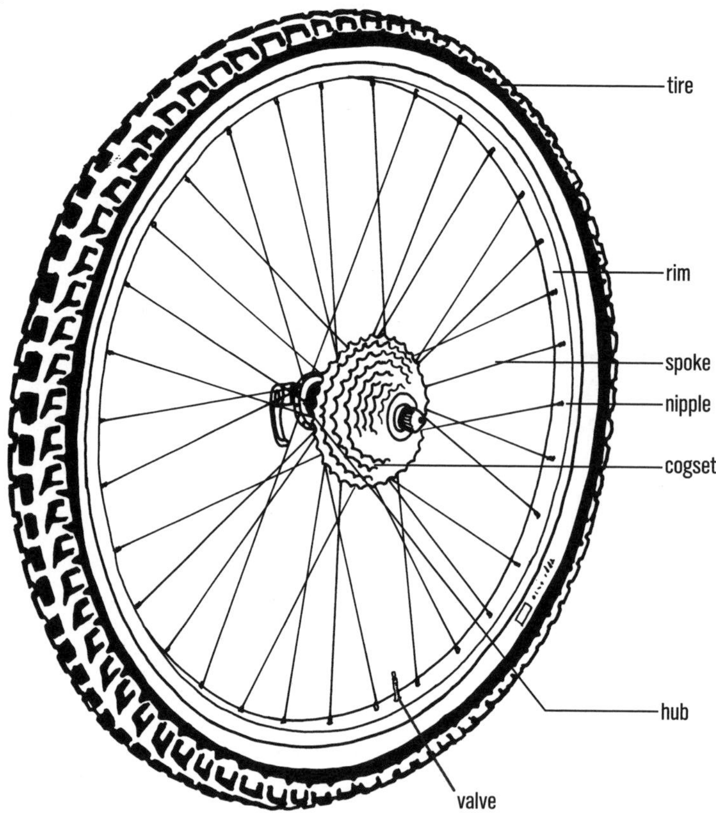

8-2

TRUING A WHEEL

For more information on truing wheels, see section 15-4 in the wheelbuilding chapter.

If a wheel has a wobble caused by loose spokes, you can straighten it out by adjusting the tension on the spokes. An extreme bend in the rim itself cannot be fixed by spoke truing alone, because the spoke tension on the two sides of the wheel will be so uneven that the wheel will rapidly fall apart. Instead, you need to replace the rim.

To adjust spoke tension, you'll need a spoke wrench of the right size for the spoke nipples—they come in different sizes (as well as shapes), and you will wreck the nipples if you use a spoke wrench that is too large.

1. **Check for broken or loose spokes in the wheel.** Feel for any spokes that are so loose that they flop around. If there is a broken spoke, follow the replacement procedure in 8-3. If there is a single loose spoke, check to see that the rim is not dented or cracked in that area. If the rim is damaged, I recommend replacing it (Chapter 15). If the rim looks okay, mark the loose spoke with a piece of tape, and tighten it up with the spoke wrench until it seems to be at the same tension as adjacent spokes on the same side of the wheel (pluck the spoke and listen to the tone). Then follow the truing procedure here.
2. **Check the hub-bearing adjustment.** Grab the rim while the wheel is on the bike and flex it side to side. If the bearings are loose, the wheel will clunk side to side. Tighten the hub before you true the wheel, or else the wheel will behave erratically. Follow the hub-adjustment procedure in 8-6d, steps 1–4.
3. **Put the wheel in a truing stand if you have one.** Otherwise, leave the wheel on the bike and suspend the bike in a bike stand or from the ceiling, or turn it upside down on the handlebar and saddle.
4. **Adjust the truing stand feeler so that it scrapes the rim at the biggest wobble.** If you do not have a stand and are truing the wheel on the bike,

push one of the brake pads over to serve as a trueness indicator.

5. **Where the rim scrapes, tighten the spoke or spokes that come to the rim from the opposite side of the hub, and loosen the spoke or spokes that come from the same side of the hub as the rim scrapes** (Figs. 8.2 and 8.3). This approach will pull the rim away from the feeler or brake pad. When correcting a wheel that is laterally out of true (wobbles side to side), always adjust spokes in pairs: one spoke coming from one side of the wheel, the other from the opposite side.

NOTE ON DIRECTION TO TURN THE NIPPLES: *Tightening spokes is similar to opening a jar upside down. With the jar right side up, turning the lid counterclockwise opens the jar, but this reverses when you turn the jar upside down (try it and see). Spoke nipples are just like the lid on that upside-down jar. In other words, when the nipples are at the bottom of the rim, counterclockwise tightens and clockwise loosens (Figs. 8.2 and 8.3). The opposite is true when the nipples you are turning are at the top of the wheel. It may take you a few attempts before you catch on. If you temporarily make the wheel worse, simply reverse what you have done and start over.*

It is best to tighten and loosen by small amounts (about one-quarter turn at a time), decreasing the amount you turn the spoke nipples as you move away from the spot where the rim scrapes the hardest. If the wobble gets worse, then you are turning the spokes in the wrong direction.

NOTE ON TWISTED SPOKES: *As you turn the nipple on a tight spoke, particularly a thin one or a flat, aero one, the spoke will tend to twist. To avoid this, when you tighten or loosen a spoke, twist it in the correction direction, and then twist back half as far. This will be enough to unwind most spokes. Aero spokes or superlight spokes may still have a twist in them. With aero spokes, it is best to prevent the spoke from twisting as you turn the nipple. For flat steel spokes, DT Swiss makes a red plastic spoke wrench with a long, conical groove down the spoke side to fit an L-shaped steel tool to keep the spoke from twisting (Fig. 1.4). The long part of the L-shaped tool is slotted for the aero spoke to keep it from twisting as you turn the nipple, and its conical exterior fits in the groove in the spoke wrench. For large, flat, aluminum spokes, you'll need a slotted tool from the wheel manufacturer to hold them. With superlight round spokes, you won't really be able to tell whether the spokes have unwound, other than the wheel may ping when you first ride it as the spokes unwind. You may need to retrue the wheel once they have settled in.*

NOTE ON INTERNAL NIPPLES: *Some deep-section carbon rims have nipples inside the rim because the spoke holes through the rim wall toward the hub are not big enough for a nipple to fit through. In this case,*

8.2 Lateral truing if rim scrapes on left

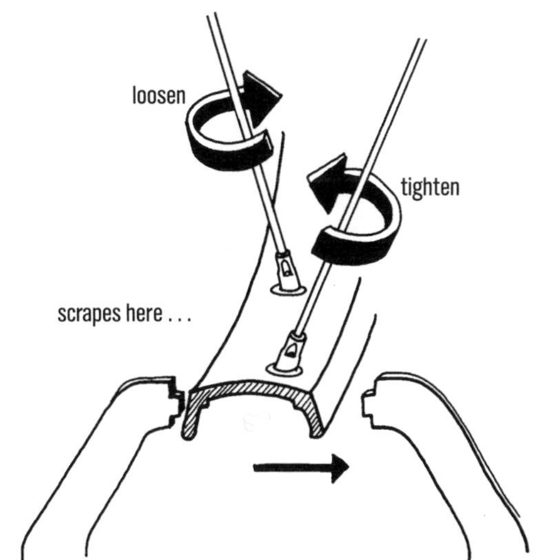

8.3 Lateral truing if rim scrapes on right

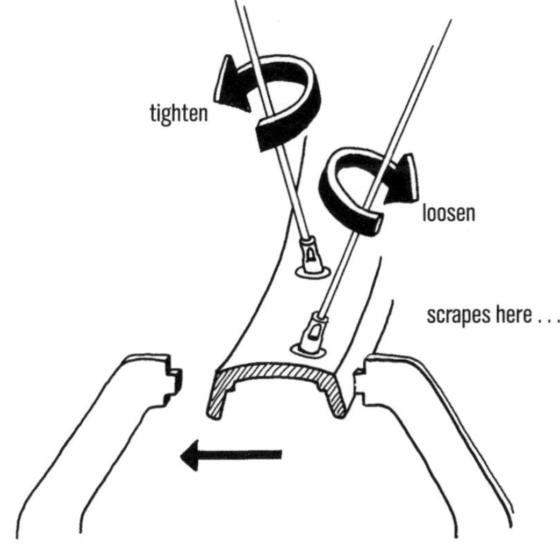

remove the tire and rim strip, then reach down into the spoke holes with the correct-size socket. The simplest tool for this is either a Y-wrench with 5mm and 5.5mm hex sockets and a square-drive socket (for upside-down standard nipples inside the rim) on the ends of its three long arms (Fig. 1.4) or a specialty wrench for the particular brand of internal spoke nipples.

NOTE ON NIPPLES AT THE HUB: *Some wheels have the spoke nipples at the hub and not at the rim. You need a special spoke wrench (it should come with the wheels) to get at them. You must be particularly careful about the rotation direction to make sure you are tightening or loosening as you intend (see the preceding discussion about jar lids).*

1. **As the rim moves toward center, readjust the truing-stand feeler (or the brake pad).** Again, set it to scrape the most out-of-true spot on the wheel.
2. **Check the wobble first on one side of the wheel and then on the other.** Adjust spokes accordingly so that you don't end up pulling the whole wheel off-center by chasing wobbles on only one side. As the wheel gets closer to true, you will need to decrease the amount you turn the spokes to avoid overcorrecting.
3. **Accept a certain amount of wobble, especially if truing in the bike.** The in-the-bike method of wheel truing is not very accurate and is not at all suited for making a wheel absolutely true. If you have access to a wheel-dishing tool, check to make sure that the wheel is centered (15-5). And if you notice wide variations in spoke tension on the same side of the wheel or the wheel hopping up and down as it turns, remove the tire and check the radial trueness (wheel roundness, 15-4b). If there is a big dent or flat spot in the rim, you may need to replace it; the wheel will rapidly lose trueness if the spoke tension is uneven due to the rim being D-shaped.

8-3

REPLACING A BROKEN SPOKE

1. **Get a new spoke of the same length and thickness as the spoke you are replacing.**

8.4 Weaving in a new spoke

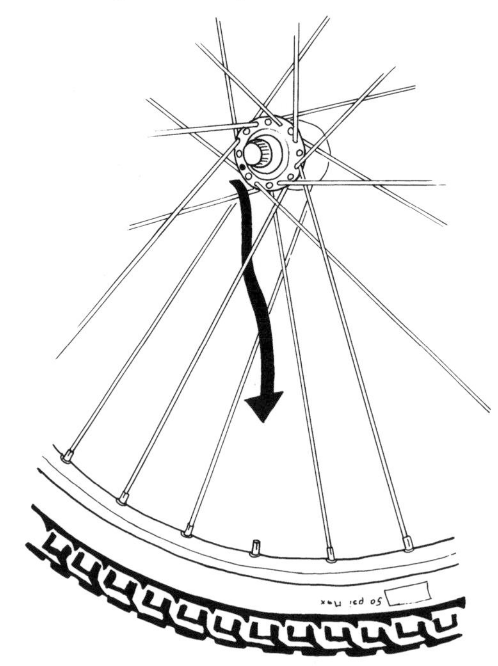

NOTE: *The spokes on the front wheel are usually not the same length as the spokes on the rear. Also, the spokes on the drive side of the rear wheel are almost always shorter than those on the other side. Same goes for a disc-brake side of a front wheel.*

2. **Thread the spoke through the spoke hole in the hub flange.** If the broken spoke is on the drive side of the rear wheel, you will need to remove the cassette cogs or the freewheel to get at the hub flange (8-11 and 8-12). If the spoke is adjacent to the disc-brake rotor, remove the rotor (9-13a).
3. **Weave the new spoke in with the other spokes just as it was before** (Fig. 8.4). It may take some bending to get it in place.
4. **If the nipple is in good shape, thread the spoke into the same nipple.** Otherwise, use a new nipple. You'll need to remove the tire, tube, and rim strip to install one.
5. **Mark (with tape) and tighten the new spoke.** Tighten it up about as snugly as the neighboring spokes on that side of the wheel are tightened. If the wheel was true before the spoke broke, tighten just this one spoke until the wheel is true again.
6. **Follow the steps for truing a wheel as outlined in 8-2.**

HUBS

8-4

SERVICING HUBS

LEVEL 2

Hubs should turn smoothly and noiselessly. If they are regularly maintained (and of decent quality to start with), you can expect them still to be running smoothly when you are ready to give up on the rest of your bike.

There are two general types of hubs: the sealed-bearing (or cartridge-bearing) type (Fig. 8.5) and the standard cup-and-cone type (Figs. 8.6 and 8.7). Modern through-axle hubs usually have cartridge bearings, but Shimano hubs are always loose-bearing, whether they have through-axles (Fig. 8.7) or quick-release axles. All hubs have a hub shell that contains the axle and bearings and is connected to the rim with spokes.

Loose ball bearings roll along smooth bearing surfaces (bearing races) that consist of a pair of bearing cups in the hub shell and a pair of conical nuts called "cones" that thread onto the axle (Fig. 8.6). On each side of the hub, the cone presses the balls against the bearing race inside the cup as the cup turns. In high-quality hubs, the bearing-race surfaces on the cone and cup have been polished to minimize friction. The operation

8.5 Quick-release front hub with cartridge bearings

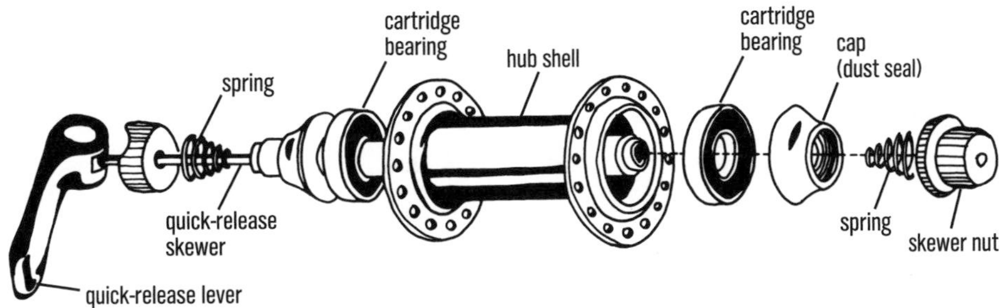

8.6 Quick-release front hub with loose ball bearings

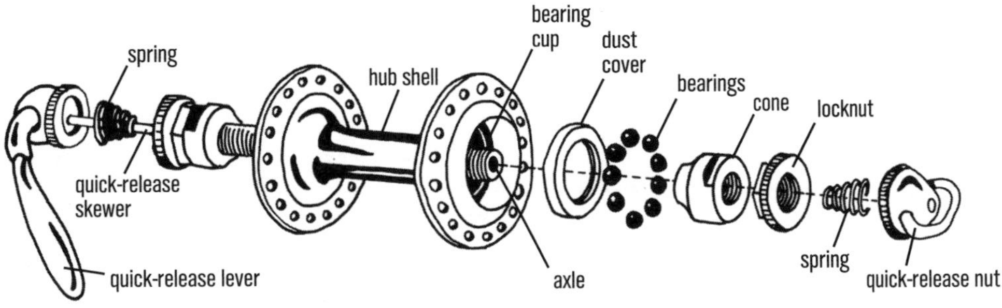

8.7 Shimano through-axle front hub with loose bearings

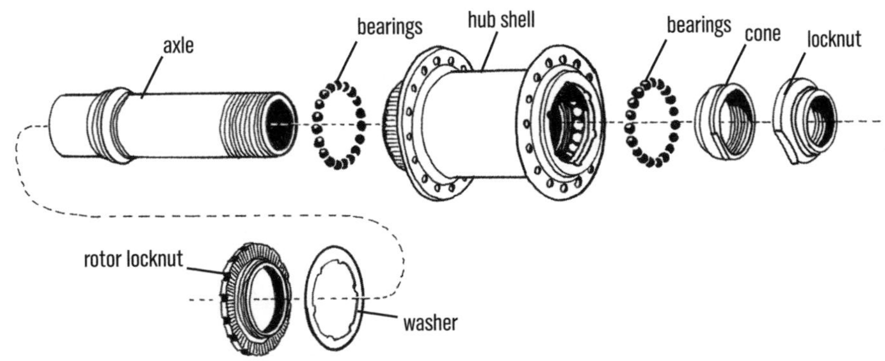

of the hub depends on the smoothness and lubrication of the cones, balls, and cups. Outboard of the cones are one or more spacers (or washers), followed by threaded locknuts that tighten down against the cones and spacers to keep the hub in proper adjustment. The rear hub will have more spacers on both sides, especially on the drive side (Fig. 8.25).

The term "sealed-bearing hub" is a bit of a misnomer, because many cup-and-cone hubs offer better protection against dirt and water than some sealed-bearing hubs. The term "cartridge-bearing hub" is more accurate, because the distinguishing feature of these hubs is that the bearings, races, and cones are all assembled as a cartridge unit at a bearing factory and then plugged into a hub shell that is machined to accept the cartridge bearing. Cartridge-bearing front hubs have two bearings, one on either end of the hub shell (Fig. 8.5). Rear hubs (Fig. 8.23) have at least two and come with additional bearing cartridges to stabilize the rear freehub body (the part onto which the rear cogs are attached), assuming it is not a hub for a thread-on freewheel as shown in Figure 8.25.

Cartridge-bearing hubs can have any number of axle-assembly types. Some have a threaded axle with locknuts quite similar to a loose-bearing hub. Much more common on mountain bikes are aluminum axles, often very large in diameter with correspondingly large bearings (the large diameter gives them rigidity). Their end caps usually snap on, screw on, or are held on with setscrews or circlips. The large-diameter axles and bearings are meant to prevent independent movement of the legs of suspension forks or rear-suspension assemblies.

Through-axles have largely displaced quick-release axles on the hubs of performance mountain bikes. Due to their high degree of stiffness and ability to minimize the independent movement of the fork legs, as well as their much lower chance of coming free of the fork with improper assembly, large-diameter (15mm or 20mm) through-axles have become standard on front hubs (see 2-5 and 2-10). On rear wheels, 12mm through-axles have become standard. These days, one end of the through-axle threads into an insert in the dropout; on older through-axle forks, the fork ends clamp around the ends of the axle.

By changing the axle end caps, many quick-release hubs can be easily converted to accept through-axles.

Hub axles have grown in length over time, becoming progressively longer as the number of rear cogs has increased and as tires have become wider.

8-5

REMOVING HUB FROM BIKE

1. **Remove the wheel from the bike** (2-2 to 2-7).
2. **Remove the quick-release skewer or other hardware that holds the wheel onto the bike.**

8-6

OVERHAULING LOOSE-BEARING HUB, FRONT OR REAR

These instructions apply to quick-release hubs as well as to loose-bearing through-axle hubs (i.e., Shimano). All Shimano (loose-bearing) through-axle hubs have a second, tubular axle with threaded cones and locknuts through which the through-axle passes to mount the wheel.

Take some time to evaluate the hub's condition before disassembling it. Doing so will help you isolate problems. Spin the hub while holding the axle, and turn the axle while holding the hub. Does it turn roughly? Is the axle bent or broken? Wobble the axle side to side. Is the bearing adjustment loose?

NOTE: *Some hubs have large rubber seals covering the axle nuts. These seals can squeal hideously, even though the inside workings of the hub are in good shape. The squeal can be caused by dust in the seal or by the seal not being seated properly against the hub face. Pull the seal off by squeezing and yanking it. Brush it off, put it back into its mating grooves on the hub, and the noise will probably be gone. If the seal is made of metal, lubricate the edges with lightweight bicycle grease.*

a. Disassembly

1. **Set the wheel flat on a table or workbench.**
2. **Slip a cone wrench of the appropriate size onto the wrench flats on one of the cones.** Size is usually

8.8 Shimano quick-release rear hub; the cone takes a 17mm cone wrench, and the locknut/end cap takes a 5mm hex key

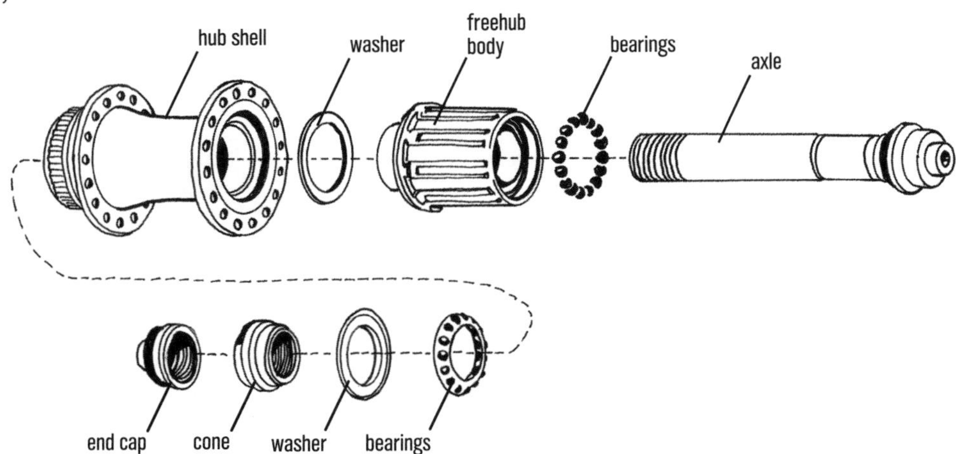

13mm, 14mm, 15mm, or 17mm with a quick-release axle. With a Shimano through-axle hub, the cone wrenches required can be, depending on front/rear and model, 17mm, 20mm, 22mm, 23mm, 24mm, or 28mm. On a rear wheel, work on the nondrive side.

3. **Put an appropriately sized wrench on the locknut on the same side.** This can be a box wrench, cone wrench, or adjustable wrench and can include any of the sizes listed in step 2. Alternatively, the hub might require a 5mm hex key inside the end of the combination locknut/end cap (Fig. 8.8)
4. **While holding the cone with the cone wrench, loosen the locknut** (Figs. 8.9 and 8.10). This procedure may take considerable force, as these are often fastened together very tightly to maintain the hub's adjustment. Make sure that you are unscrewing the locknut counterclockwise ("lefty loosey, righty tighty," as shown in Figs. 8.9 and 8.10).
5. **Hold the axle in place as you unscrew the locknut.** As soon as the locknut loosens, move the cone wrench from the cone on top to the cone on the opposite end of the axle, if there is one on the other end; otherwise, hold the other end with your fingers. The locknut will generally unscrew with your fingers; use a wrench if necessary to get past any damaged or dirty threads.
6. **Slide off any spacers.** If they will not slide off, the cone will push them off when you unscrew it. Note that some spacers have a small tooth or key that

8.9 Loosening or tightening a locknut on a quick-release hub

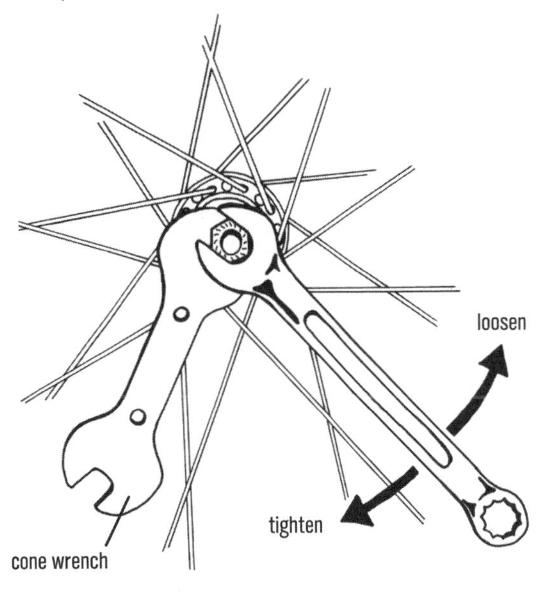

8.10 Loosening or tightening a locknut on a Shimano through-axle front hub

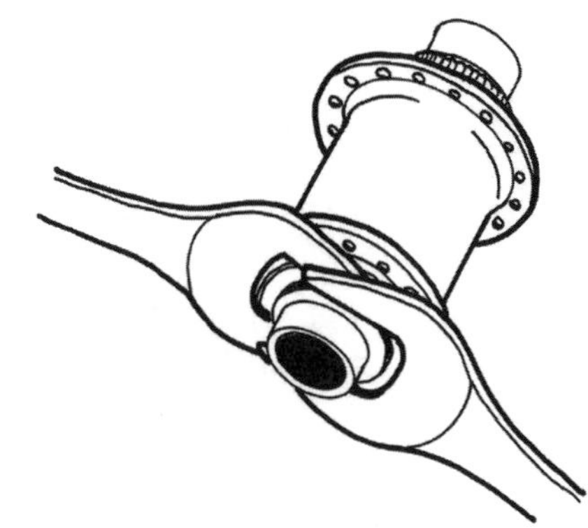

corresponds to a groove along the axle; make sure it's lined up before you unscrew the cone, or you'll damage the threads.

7. **Unscrew the cone from the axle.** Ideally, do this with your fingers, but if the threads are damaged or the keyed washer is stuck, you may need to hold the opposite cone with a wrench and use a wrench on this cone. If the cone is a split type with a pinch bolt to clamp it onto the axle, loosen the pinch bolt first.
8. **Keep track of the various parts.** Line up nuts, spacers, and cones on your workbench in the order in which they were removed (Figs. 8.6, 8.7, and 8.8), perhaps even with a wire tie through them, to guide you when reassembling the hub.
9. **Flip the wheel over.** To catch any bearings that might fall out, place a rag underneath the wheel and put your hand over the end of the hub from which you removed the nuts and spacers.
10. **Pull the axle up and out.** Be careful not to lose any bearings that might fall out of the hub or that might stick to the axle (Fig. 8.11). Leave the cone, spacers, and locknut all tightened together on the opposite end of the axle from the one you disassembled. If you are replacing a bent or broken axle, measure the amount of axle sticking out beyond the locknut, and put the cone, spacers, and locknut on the new axle identically.
11. **Remove all the ball bearings from both sides of the hub.** They may stick to a screwdriver with a coating of grease on the tip, or you can push them down through the center of the hub and out the other side with a screwdriver. Tweezers or a small magnet might also be useful for removing bearings. Put the bearings in a cup, a jar lid, or the like. Count the bearings, and make sure you have the same number from each side. Standard bearing counts and sizes are: nine ¼-inch bearings per side on a rear quick-release hub, and 10 ³⁄₁₆-inch bearings per side in a front QR hub.
12. **Gently pop off the seals that are pressed into either end of the hub shell.** This may require a screwdriver (Fig. 8.12) or just your fingers. Be careful not to deform the seals; leave them in if you can't pop them out without damage. If they are not removed, it is tedious, but not impossible, to clean the dirty grease out of their concave interiors with a rag and a thin screwdriver.

8.11 Installing or removing axle in a Shimano through-axle front hub

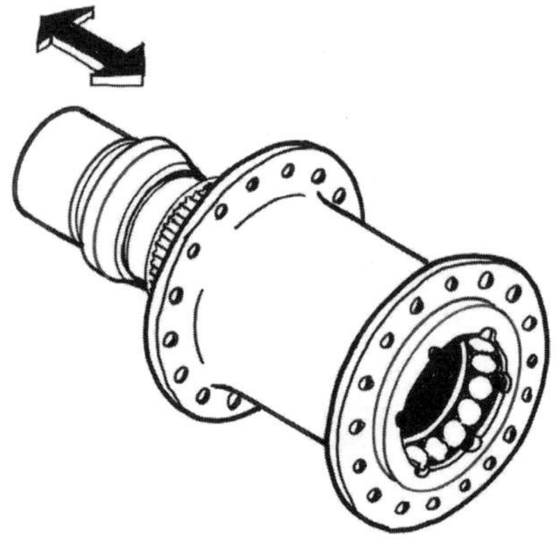

8.12 Removing a dust cap

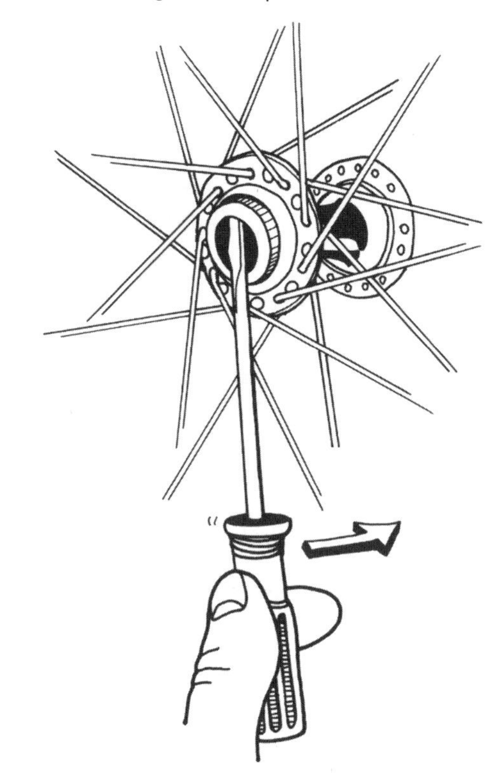

b. Cleaning

1. **Wipe the hub shell clean with a rag.** Remove all dirt and grease from the bearing surfaces. Using

a screwdriver, push a rag through the axle hole in the hub and spin the wheel around it to clean out grease or dirt. Wipe off the outer faces of the shell. Finish with a very clean rag on the bearing surfaces. They should shine and be completely free of dirt or grease. If you let the hub go too long between overhauls, the grease may have solidified and glazed over so completely that you will need a solvent to remove it. If you are working on a rear hub, take this opportunity to lubricate the freehub body (8-14.)

2. **Wipe the axle, nuts, and cones with a rag.** Clean the cones well with a clean rag. Again, a solvent may be required if the grease has solidified. Get any dirt out of the threads on the disassembled axle end; if you do not, the cone will push any dirt that remains into the hub upon reassembly, negating your work.
3. **Wipe the grease and dirt off the seals.** A rag over the end of a screwdriver is sometimes useful for getting inside. Again, glaze-hard grease may have to be removed with a solvent. Keep the solvent out of the freehub body.
4. **Wipe off the bearings.** Rub all of them together between two rags. This may be sufficient to clean them completely, but small specks of dirt can still adhere to them, so I prefer to take the next step as well.
5. **Superclean and polish the bearings.** (If you are overhauling low-quality hubs, skip to step 6.) I prefer to wash the ball bearings in a plugged sink with an abrasive soap such as Lava, rubbing them between my hands as if I were washing my palms. This really gets them shining, unless they are caked with glaze-hard grease. Make sure you have plugged the sink drain! This method has the added advantage of cleaning your hands for the assembly step. It is silly to contaminate your superclean parts with dirty hands. If there is hardened glaze on the bearings, soak them in a solvent. If that does not remove the glaze, buy new bearings; be sure to buy the right size (take one of the old ones with you to the bike shop).
6. **Dry all bearings and any other wet parts.** Inspect the bearings and bearing surfaces carefully. If any of the bearings has pits or gouges, replace all of the bearings. The same goes for the cones. A lack of sheen or a patina on either balls or cones indicates wear and is cause for replacement. Most bike shops stock replacement cones. If the bearing cups in the hub shell are pitted, the only thing you can do is buy a new hub. Regular maintenance and proper adjustment can prevent pitted bearing races.

NOTE: *Using new ball bearings when overhauling standard cup-and-cone hubs ensures round, smooth bearings. However, do not avoid performing an overhaul just because you don't have new ball bearings. Inspect the balls carefully. If there is even the slightest hint of uneven wear or pitting on the balls, cups, or cones, discard the bearings and complete the overhaul with new bearings and perhaps even new cones.*

c. Assembly and lubrication

1. **Press in the seals or dust covers on both ends of the hub shell.**
2. **Smear grease with your clean finger into the bearing race on one end of the hub shell.** I like using light-colored or clear grease so that I can see if it gets dirty, but any bike grease will do. Grease not only lubricates the bearings; it also forms a barrier to dirt and water. Use enough grease to cover the balls halfway. Too much grease will slow the hub by packing around the axle.
3. **Stick half of the ball bearings into the grease.** Make sure you put in the same number of bearings that came out. Distribute them uniformly around the bearing race.
4. **Smear grease on the cone that is still attached to the axle, and slide the axle into the hub shell.** Lift the wheel up a bit (30-degree angle), so that you can push the axle in until the cone slides into position and keeps all the bearings in place. On rear hubs it is important to replace the axle and cone assembly into the same side of the hub from which it was removed because of the asymmetrical spacing between sides due to the cogs.

5. **Turn the wheel over.** Hold the axle pushed inward with one hand to secure the bearings (Fig. 8.13).
6. **Smear grease into the bearing race that is now facing up.** Lift the wheel and allow the axle to slide down just enough so that it is not sticking up past the bearing race. Make sure no bearings fall out of the bottom. If the race and bearings are properly greased and the axle remains in the hub shell, they are not likely to fall out.
7. **Place the remaining bearings uniformly around in the grease.** The top end of the axle should still be below the bearing race. Make sure you have inserted the correct number of bearings.

8.13 Push inward on the axle and flip the wheel over

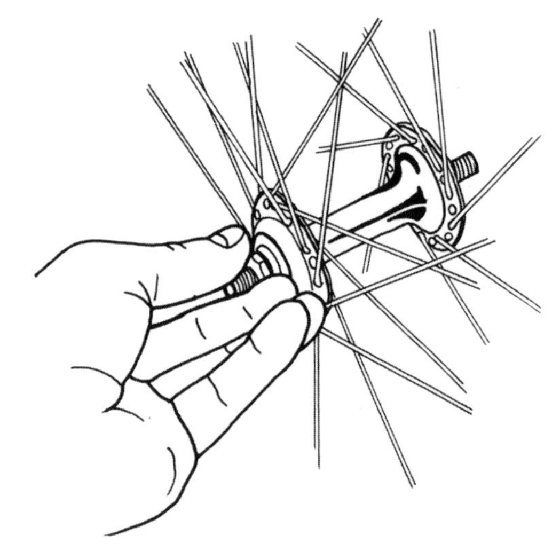

8.14 Set the axle on the workbench to keep the cone in contact with the bearings

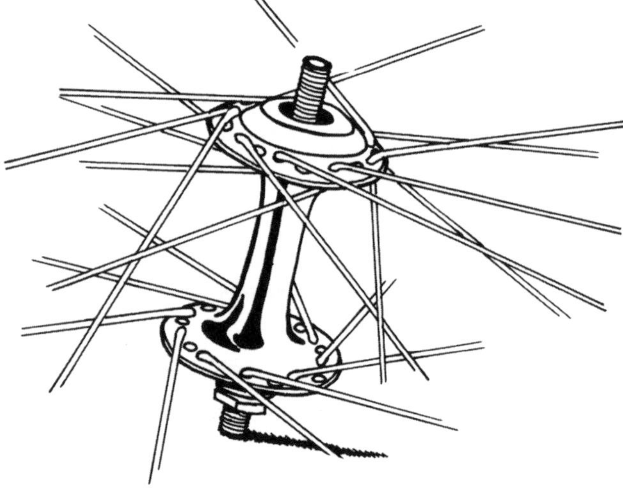

8. **Slide the axle back into place** (Fig. 8.11).
9. **Set the wheel down on the axle end.** Lay the wheel flat on the table so it rests on the lower axle end, seating the cone against the bearings (Fig. 8.14).
10. **Apply a film of grease to the top cone's bearing surface.**
11. **Screw the top cone down into place.** Use your fingers. Seat it so that it just kisses the bearings.
12. **In correct order, slide on any washers and spacers.** Watch for those washers with the little tooth or key that fits into a lengthwise groove in the axle.
13. **Use your fingers to screw on the locknut.** Note that the two sides of the locknut are not the same. If you are unsure about which way the locknut goes back on, check the orientation of the locknut that is on the opposite end of the axle (this locknut was not removed during this overhaul and should therefore be in the correct orientation). As a general rule on quick-release hubs, the rough surface of the locknut faces out so that it can get a better bite into the dropout. On through-axle hubs, the thin edge is to the outside.

d. Hub adjustment

1. **Tighten the cone until it lightly contacts the bearings.** The axle should turn smoothly without any roughness or grinding, and there should be a small amount of lateral play. Tighten the locknut until it is snug against the cone.
2. **Place the cone wrench into the flats of the hub cone.** Tighten the locknut with another wrench (Figs. 8.15 and 8.16). To hold the adjustment, tighten the locknut as much as you can against the cone and spacers. Be aware that you can ruin the hub if you accidentally tighten the cone against the bearings instead of against the locknut. If the cone is a split type with a pinch bolt to clamp it onto the axle, tighten the pinch bolt.
3. **If the adjustment is off, loosen the locknut while holding the cone with the cone wrench.** If the hub is too tight, unscrew the cone a bit. If the hub is too loose, screw the cone in a bit.
4. **Repeat steps 1–3 until the hub adjustment feels right.** It should have a slight amount of end play so

8.15 Tightening or loosening a locknut

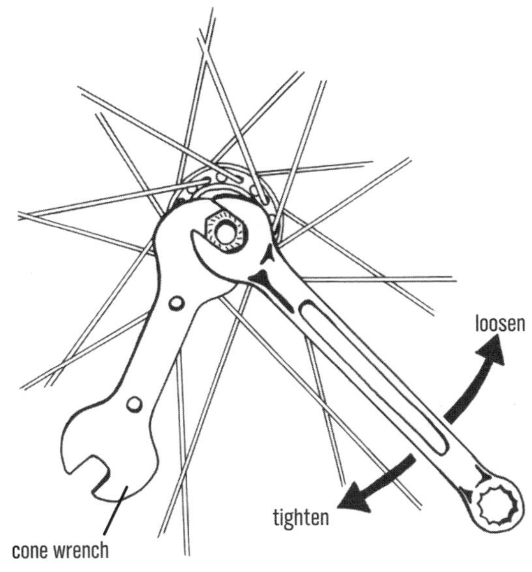

8.16 Tightening or loosening a locknut on a Shimano through-axle front hub

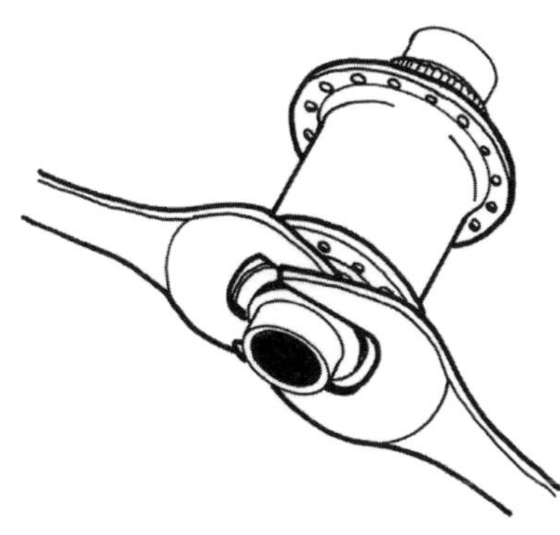

that the pressure of the quick-release skewer will compress it to a perfect adjustment. If the hub is a nutted type (Fig. 2.7), you want to adjust it without any play in it. Tighten the locknut firmly against the cone to hold the adjustment.

NOTE: *You may find that tightening the locknut against the cone suddenly turns your Mona Lisa–perfect hub adjustment into something less beautiful. If it is too tight, back off both cones (by using a cone wrench on either side of the hub, each on one cone) a fraction of a turn. If too loose, tighten both locknuts a bit. If still off, you might have to loosen one side and go back to step 1. It's rare that I get a hub adjustment perfectly dialed in on the first try, so expect to tinker with the adjustment a bit before it's right.*

5. **Put the skewer back into the hub.** Skip this step with a through-axle hub. Make sure that the conical springs have their narrow ends to the inside (see Figs. 8.6 and 8.25.)
6. **Install the wheel in the bike.** Tighten the skewer (Fig. 2.8), through-axle (Fig. 2.5), or axle nuts (Fig. 2.7). Check that the wheel spins well without any side play at the rim. If it needs readjustment (too tight or too loose), go back to step 3.

Congratulate yourself on a job well done! Hub overhaul is a delicate job, and it makes a significant difference in the longevity and performance of your bike.

8-7

OVERHAULING CARTRIDGE-BEARING HUBS

These instructions apply to both through-axle and quick-release hubs.

Evaluate the hub's condition before disassembling it. Spin the hub while holding the axle, and turn the axle while holding the hub. Does it turn roughly or squeal or squeak? Wobble the axle side to side. Is there side play?

Compared to inexpensive hubs with loose bearings, cartridge-bearing hubs (Figs. 8.5 and 8.22) generally need less maintenance, though some hubs have better seals than others. However, if you ride through water above the hubs or direct a high-pressure sprayer into them, you can expect water and dirt to get through the seal no matter how good it may be. If the ball bearings inside the cartridges get wet or start making noise, they should be overhauled or replaced. Maintenance generally consists of replacing the cartridge bearings as units, but if done regularly enough, the cartridge bearing can re-packed with grease and given new life.

Unless you have a few days before you need to use your wheel, have some new bearings ready on standby to put in; even if you're not planning to replace them, you may find that you have to, once you get inside the hub. If you can easily get at the bearing cover seal, you may not need to remove the bearings from the hubs to grease them. See step e below.

There are many types of cartridge-bearing hubs, and it is outside the scope of this book to explain in detail how to disassemble every one of them, but the instructions below cover the vast majority of them (and rear Mavic hub overhaul is in 8-14d). It is usually not too hard to figure out how to take apart any hub, whether it is a through-axle hub or a quick-release hub, but there is often an important order to it that may not be obvious.

For a Mavic rear quick-release hub, skip to 8-14d.

For a DT Swiss rear hub, skip to 8-14e.

a. Remove the end cap(s)

Most cartridge-bearing hubs generally have axle end caps (Fig. 8.5) that butt up against the bearings and must be removed to remove the axle. You should be able to use one of the following approaches:

- Pull or pry off the end caps if they don't have a hex hole inside or wrench flats outside; they are often held on by a rubber O-ring engaging a groove at the end of a quick-release axle and inside the end cap or inside the bore of a through-axle hub and around the edge of the end cap. A good yank will get one off. With a quick-release hub, you may need an axle-clamp tool (1-4, Fig. 1.4) clamped in a vise to grab the tip of the end cap securely enough to pull it off by pulling up on the wheel. With a through-axle hub, you may be able to pry each end cap off with your finger inside or by sticking the end of the through-axle into the end cap and tipping it back and forth until the cap pops out.
- Unscrew the end caps on a quick-release hub that has hex holes inside; use a 5mm hex key inserted into the central 5mm hex holes in either axle end cap; generally only one cap will come off this way. You may find a receptacle for an 8mm or 10mm hex key in the bore of the axle after one end cap is off; insert that hex key and, with a 5mm hex key still in the opposite end cap or with your fingers on a smooth, conical end cap, unscrew that cap (Fig. 8.35).
- Unscrew the caps on either axle end with cone wrenches engaged on wrench flats on each cap (Fig. 8.5); generally only one cap will come off this way.
- On Mavic, unscrew the axle with a 5mm hex key while holding the other end with either a pin tool in the adjuster ring (front) or a 5mm hex key (front) or 10mm hex key (rear; Fig. 8.35) in the bore of the axle, often after pulling off the end cap on the adjuster end.
- Loosen a tiny setscrew pinching each collar-type end cap around the quick-release axle. One collar may be threaded and will unscrew once the setscrew is loosened, and the other (unthreaded) will slide off. Push out the axle bearing and the aluminum bearing cover by pushing on the threaded end of the axle. The rear hub may have a collar with a tiny setscrew; remove the screw and push the axle out of the drive side, and then pull off the freehub body. You can remove the bearings on each end of the freehub body and both ends of the front and rear hub shells with light finger pressure.
- Yank the rear cassette straight off by hand (DT Swiss); the freehub body will come off with the cogs and will take off the end cap with it (Fig. 8.36).
- Slide the end caps off after loosening a setscrew on the side of each cap.
- Loosen three radial setscrews (with a tiny—2mm—hex key) on the axle collar (on the nondrive side on a rear hub), which sometimes must be accessed through a tiny hole in the end of the hub shell (rotate the axle to line each setscrew up under the hole in succession). Once the collar is loose, pull out the end cap, and then pull the entire axle assembly (including the freehub on a rear hub) out from the opposite side.
- If you see hex flats inside a quick-release axle's bore, try sticking a hex key in there (except if there are also setscrews; see above). If both axle ends have them, put a 5mm hex key in each end and unscrew the end caps. If only one end fits a hex key, the other end must first pull off to reveal a hex hole into which you can insert a second hex key to work against the first to unscrew the axle (Fig. 8.35).
- Some (generally very old) cartridge-bearing hubs have threaded axles with locknuts that tighten against each other to hold the bearing adjustment, much like a loose-bearing hub (Figs. 8.6, 8.25). You remove the axle just as you would for a loose-bearing hub (8-6).

b. Remove the axle

With a front through-axle hub, skip to step d. Be aware that in a rear through-axle hub, you still will need to do this step, because there will often be a second, tubu-

8.17 Tapping out an axle (and a cartridge bearing)

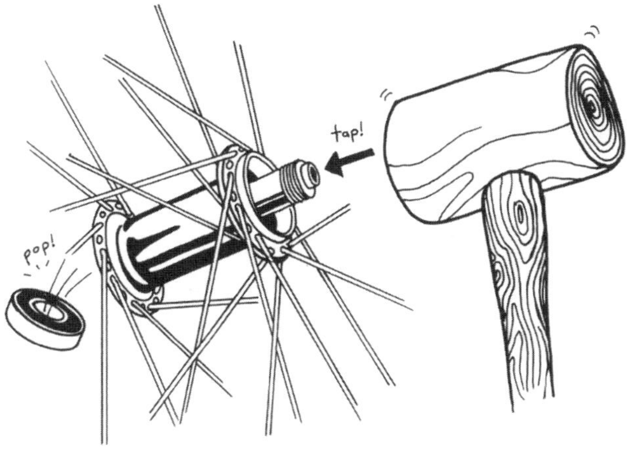

8.18 Unscrewing the axle of a Stan's through-axle rear hub

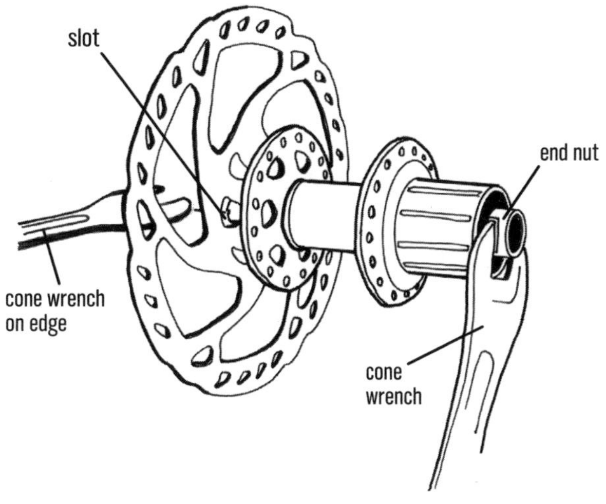

lar axle to keep the freehub lined up surrounding the through-axle that holds the wheel in.

A bearing and/or the freehub body may come out with the axle. Disassemble it over a clean area, and be ready to catch pawls, springs, or ratchet rings (Figs. 8.35 to 8.37). One of these methods should work to get the axle out:

- Pull on the axle.
- Some axles have a threaded collar under the end cap that unscrews by hand or is fixed in place by a pinch bolt; loosen the pinch bolt, put a hex key in the opposite end of the axle, and unscrew the collar; now pull out the axle.
- On a rear hub, try pulling on the freehub body (or on the cogs, if you have not removed them); the axle may come out with the freehub body.
- Tap the axle out from the end whose cap came off; it will sometimes drive the bearing on the opposite side out with it (Fig. 8.17). There will sometimes be an enlarged flange with wrench flats behind that bearing; if yours has one, put an open-end wrench on it and a cone wrench on or a hex key in the end cap, and unscrew the end cap to free the bearing.
- See step d for more axle types that drive a bearing out with them.
- On a through-axle hub, if you see slots on the end of the tubular axle, put the edge of a cone wrench (Fig. 8.18) in the slots to unscrew the axle. Hold the flats on the drive-side end nut as in Fig. 8.18, or, if the axle is unscrewing from a threaded collar, hold that with an adjustable or specialty pin tool (Mavic).

c. Remove the freehub body (rear hub only)

- Carefully pull the freehub body off the hub while turning it counterclockwise (Fig. 8.37); the axle may come out with the freehub body. Sometimes the freehub body comes off when you push the axle out.
- If there are spacers under the freehub body (Fig. 8.8), note their orientation. Keep them, clean them, and remember to put them back in under the freehub on installation, or the freehub body will likely bind up.

d. Remove the bearing(s)

If the bearings are neither frozen, rusted, nor making horrible noises, you may be able to re-pack them with grease, rather than replace them. And if you can easily get at the bearing cover seal, you may not need to remove the bearings to grease them. See step e below.

If there is a tubular spacer separating the bearings, which there generally is in through-axle hubs and often is in quick-release hubs, move it over when pushing out the first bearing so you can get your bearing drift against the back of the opposite bearing. Pull the spacer out after the first bearing comes out. Clean it and remember to put it in again after you put in the first bearing.

If you are replacing the bearings rather than trying to re-pack them with grease, you don't need to worry much about damaging the bearings when pushing them

8.19 Tapping out a cartridge bearing with a long hex key

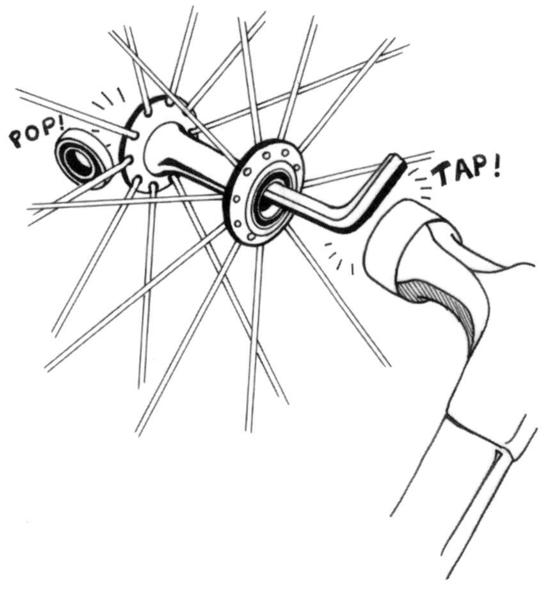

out. Setting the wheel flat atop a bucket is a good way to hold it and keep track of flying parts when you knock the bearings out.

- If you have a blind-hole bearing puller, that is the ideal tool for this job. Slip the collet of the correct size inside the bearing, turn it to expand it so its flanges hook on the back of the bearing, and slide the sliding weight against its stop to pull the bearing out.
- On a through-axle front hub, you should be able to get behind the opposite (bottom, if the wheel is sitting flat) bearing with a drift (a solid rod with a flat end), a big hex key (Fig. 8.19), a big flat-blade screwdriver, or a punch. Set the implement in at an angle against the back of the opposite bearing and tap the end of it with a hammer as you work its tip around the circumference of the bearing to gradually work it out of its nest in the hub shell. Push the spacer between the bearings off to the side so you can get at the back of the opposite bearing.
- Some axles have a shoulder against the inner face of the bearing that allows you to push it out. In this case, smack the end of the axle with a soft hammer (Fig. 8.17) to drive the bearing out (or if the axle end is recessed, with a rod, hex key, or drift punch just smaller in diameter than the end of the axle). Or you can set the end of the axle down on the workbench and push down on the hub or the spokes emanating from it. If it's too stubborn to just tap out while you're holding the wheel, first stand the hub up, with the bearing to be removed facing down, on a bucket, two wood blocks, or padded vise jaws separated by enough space to allow the bearing to pass between them as it comes out. Unless the fit is very tight, driving the bearing out with the axle usually does not damage a bearing. On a DT Swiss rear hub (Fig. 8.36), you can get the nondrive bearing out this way, but you cannot remove the drive-side bearing without first unscrewing and removing the large threaded ring with inwardly radiating teeth that engage the freehub star ratchets. This requires a special DT Swiss tool held in a very solid vise.

e. Overhaul (or replace) the bearings

Cartridge bearings are vulnerable to lateral stress; if you had to remove them to get at them and had to use a lot of force to pound them out, or if they are rusted, pitted, or otherwise damaged, they will need to be replaced. If they're worth servicing, here's how you can overhaul them.

1. **Remove the bearing covers.** The most common type of hub bearing is also the simplest to access; it has a rubber-coated metal seal on each side. Gently pop off the rubber-coated bearing seal by sliding a razor blade (Fig. 8.20) or box-cutter knife blade under the edge and prying it up. Avoid tearing or cutting the rubber edges of the seal, but if you bend the seal, you can easily flatten it out again; it is soft aluminum. If the bearing seals are steel instead, you may find a thin, flat circlip (C-shaped retaining ring) around the outer edge that you can remove; slip

8.20 Removing the bearing seal

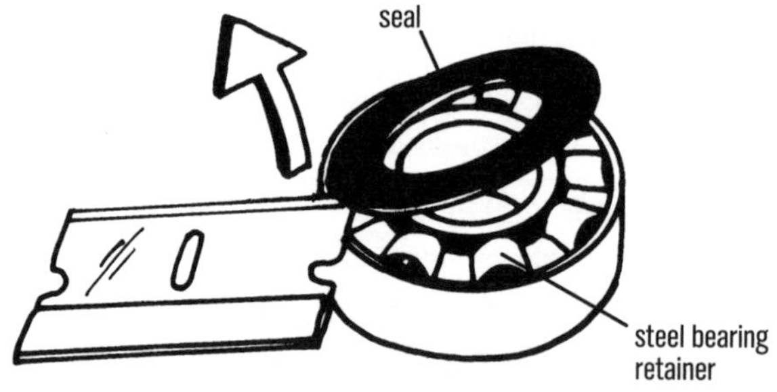

a thin blade under one tip of the circlip and work around to pop it out. The bearing cover should come right out with some help from a razor blade. If the bearing has steel seals and no circlip, you probably cannot remove them without damaging them (you would remove them by pounding a sharp awl into the edge of the seal and prying up); it's best to buy new bearings.

2. **Clean out the bearing.** Squirt citrus-based solvent into the bearing (wear rubber gloves and protective glasses) to wash out the grease, water, and dirt. Scrub with a clean toothbrush. Don't brush your teeth again with this toothbrush.
3. **Blow out the bearing with compressed air.** This evaporates and blows out the solvent and may blow out some grit as well.
4. **Pack the bearing with grease.** A general-purpose grease is generally fine for this, especially one designed for durability under wet conditions. If this is a ceramic bearing, you probably want to use a grease specific for those. A fine-tip grease gun or syringe is a nice tool for this but you can also use your finger.
5. **Snap the bearing covers back on.** Replace the bearing if it doesn't turn smoothly. If you really want to rescue this bearing (for instance, if it's an expensive ceramic bearing whose steel bearing races have rusted), you can disassemble it completely, polish the bearing races and bearings, and bring it close to new. This is possible only if it has a plastic bearing retainer inside; forget it if the bearing retainers are metal (i.e., when the seals are off, you would be looking at a bumpy silver metal ring on either side as in Fig. 8.15, rather than a smooth plastic one). See 12-13 for how to disassemble a cartridge bearing.

f. Install and adjust the bearings

The closest thing the average home mechanic will have to a proper tool for pressing in a cartridge bearing is a quick-release skewer and a socket wrench or the old bearing. The ideal tool is a hub bearing press (Fig. 1.4), which has a central threaded shaft and discs of different sizes to fit various bearing sizes; tightening the wing nut on the end with the proper pair of discs against the bearings drives them in evenly.

NOTE: *Installing a bearing with a hammer can damage it and prevent it from spinning smoothly.*

1. **Grease the bearing.** Put a layer of grease around the outside of the new bearing and in the bearing seats in the hub shell.
2. **Get the bearing started.** Place it in proper alignment where it's going to go in. Lightly tap it in a bit with the wooden butt of a hammer handle to get it started.
3. **Press in the first bearing.** Place a socket whose outside diameter (OD) is just slightly smaller than the bearing's OD against the new bearing, or place the old bearing atop the new bearing (but note that if the bearing seat is deeper than the bearing, the old bearing can get stuck). It's generally better to install one bearing at a time to ensure that it's going in straight, especially because you may have only one socket of the right size anyway. Using large washers or something of the sort to protect the bearing seat on the opposite end, install the skewer (without the springs on it) and tighten it until the bearing is fully pressed in place (Fig. 8.21). If you're very careful to not slip and to keep the bearing lined up straight, you can alternatively

8.21 Pressing the bearing in with the skewer and a socket

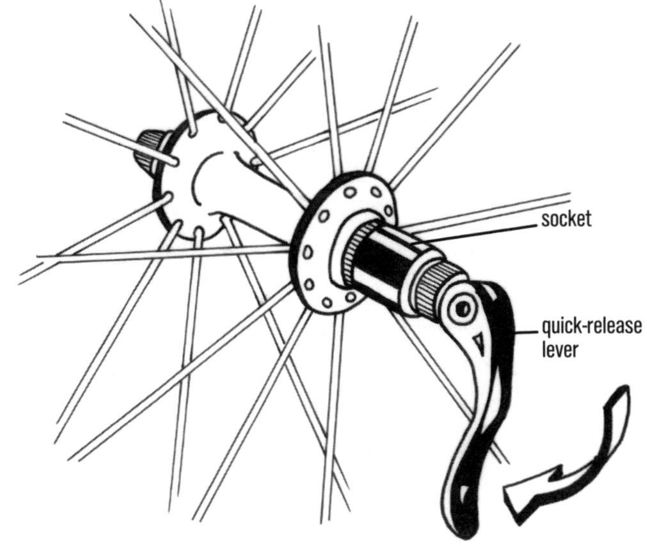

tap on the socket atop the bearing with a wooden hammer or hammer handle. If the bearing is hard to press in, check that it's going in straight; if it isn't, knock it out again and start over.

When you install ceramic cartridge bearings or other high-quality bearings, it is worth thinking ahead of time about maintenance when you are determining the orientation of the bearings. You may as well keep it rolling smoothly, since you don't want to be making this investment frequently. Ceramic bearings for hubs will most likely be hybrid ceramic bearings, so even though their ceramic balls cannot rust and are more than twice as hard as steel balls, their races are still steel and can rust. (Full-ceramic bearings are generally not used in hubs, as their brittle ceramic outer races can crack when pressed into a tight hole.)

Cartridge bearings have bearing retainers (Fig. 8.20) that separate the balls from each other. This design reduces friction by preventing neighboring balls, whose adjacent sides are turning in opposite directions, from rubbing against each other. The bearing retainer may be plastic and asymmetrical, so when you remove the bearing seals (with a razor blade slipped under the edge), you'll see the balls on one side and you'll see only the plastic retainer from the other side. Proper maintenance requires cleaning out the bearing and slathering new grease into it, so before you press the bearings in, determine on which side the balls are visible and make sure that side faces outboard. That way, you can pry off the bearing covers and clean and grease them easily. If you were to completely disassemble and overhaul this bearing instead, you would need to remove it from the hub to get at the other side of the bearing retainer.

If the bearing has a symmetrical retainer (usually steel; Fig. 8.20) concealing the balls on both sides, the best you can do to clean them out is to squirt in solvent followed by compressed air. Disassembling the bearing is only possible with a plastic, asymmetrical bearing retainer. Removing a steel bearing retainer (Fig. 8.15) ruins it, and you won't be able to reassemble the bearing.

4. **Replace the sleeve between the bearings.** The sleeve separates the bearings the proper distance in the hub shell so they won't bind when clamped into the frame or fork.
5. **On a through-axle hub, press in the second bearing.** Follow the same procedure as in steps 1–3 above. When you press in the second bearing with the socket and skewer, you'll also need another, bigger socket or a thick, giant washer for the other end of the hub; the washer should be large enough to clear the outside of the bearing. Place the washer or bigger socket with its open end over the bearing you just pressed in; this allows the skewer to push against the hub shell, rather than against the end of the axle. Run the skewer through the axle, both bearings, and the sockets, and tighten it. Skip to section g.
6. **On a quick-release hub, reassemble the hub axle, second bearing, and end caps the reverse of the way they came apart.** Install the axle (put a bit of grease on it first) and press in the other bearing, generally using the same method you did with the first bearing. When you press in the second bearing with the socket, you'll also need another, bigger socket for the other end of the hub that is large enough to clear the outside of the bearing. Place the bigger socket with its open end over the bearing you just pressed in; this allows the skewer to push against the hub shell, rather than against the end of the axle. Run the skewer through the axle, both bearings, and the sockets, and tighten it. If one bearing was originally tightened between a shoulder on the axle and a thread-on end cap, tighten the new bearing onto the axle that way first, and then tap the axle assembly in. On many hubs, you can also press in the outer bearing in the freehub body with a socket and the skewer once the freehub body is in place on the hub (Fig. 8.22).
7. **Check the bearing adjustment.** Sometimes the bearings will be out of alignment slightly after installation, making the hub noticeably hard to turn. A light tap on either end of the axle with a soft hammer will sometimes free them. If the hub has threaded end caps and you find that tightening them binds up the hub, then you'll want to put Loctite on the threads and tighten them only enough to make the hub adjustment perfect once the wheel is tightened with the skewer into the

8.22 Pressing the bearing into the freehub

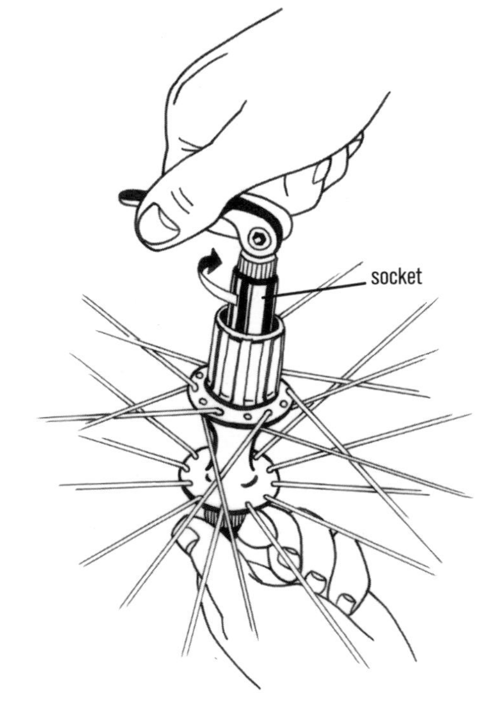

frame or fork; the threadlock compound will prevent the end caps from loosening.

Many cartridge-bearing hubs have no adjustment; a tubular spacer between the inner races of the two bearings keeps them at the proper separation, and the end caps push against the other side of the inner bearing races. Other systems have no spacer between the bearings; the bearings just sit in a pocket in each end of the hub, and pressure from the outside has to be adjusted to remove axle end play without side-loading the bearings.

On Mavic and Hadley hubs, you adjust the bearing with a pin spanner on the adjuster ring once the hub is tightened into the frame or fork with the quick-release skewer.

On White Industries hubs, there are three setscrews pointing radially inward through a sliding collar on one end of the hub. Each setscrew is at 120 degrees from the next, and on some hubs must be accessed through a hole in the hub shell that must be rotated over each setscrew. Loosen the three setscrews with a 2mm hex key and slide the collar inward against the bearing to remove the end play. If the adjustment is still loose, try rotating the collar on the axle end cap first; it may be that the setscrews keep going back into the indentations they made in the collar before and thus prevent changing the adjustment unless you rotate them to a new area of the collar.

g. Install the end caps on a through-axle hub

Replace the end caps the way you removed them. If they were hard to pry off with the end of the through-axle, they may not snap on easily; use your bearing drift and a wooden hammer to tap around the edges of the end cap to gradually push it into place.

8-8

UPGRADING BEARINGS

Following the instructions in 8-6 or 8-7 (depending on whether the hub is loose bearing or cartridge bearing), you can replace old bearings with supersmooth ceramic bearings, higher-grade steel bearings, or new bearings of the same type. The grade of the bearing rises with increasing smoothness, hardness, roundness, and uniformity in size of the balls and smoothness and hardness of the races. Ceramic bearings are generally higher-grade than even the best steel ones, because the ceramic balls are harder, smoother, rounder, and more uniform in size than steel balls. Expect to pay a lot for ceramic bearings (and even for high-grade steel bearings). Ceramic balls should run smoother and last longer, because they cannot rust and are less sensitive to lubrication.

Hybrid ceramic bearings have ceramic balls and steel races; full-ceramic bearings have ceramic races as well as ceramic balls and are the most expensive. Full-ceramic bearings are generally installed at the factory because the ceramic race cannot be pressed in (it has no flexibility and may crack). Instead, the seat in the part into which it fits must be larger so that the bearing fits in without pressure, and then it must be glued in.

8-9

WORKING WITH GREASE GUARD HUBS

Wilderness Trail Bikes, SunTour, and others have made high-end hubs, some labeled Grease Guard, that have small grease ports that accept a small-tipped grease gun. Injecting grease into these ports forces it through the bearings from the inside out, squeezing old grease out the outer end. Grease injection does not eliminate the need for overhauling the hubs; it merely extends the amount of time between overhauls. Furthermore, these systems are only as good as you are about using them.

FREEHUBS, FREEWHEELS, AND COGS

Freehubs and freewheels are freewheeling mechanisms, meaning that they allow the rear wheel to turn forward freely while the pedals are not turning.

A freehub is an integral part of the rear hub. The cogs slide onto the freehub body, engaging longitudinal grooves, or splines (Fig. 8.23). A freehub can also be called a cassette hub; the group of cogs is the cassette.

SRAM's XD freehub body (Fig. 8.24A) has only short splines on its inboard end, and they are all the same width. It has threads around its waist rather than splines. It is narrower at its outboard end to allow the use of a 10-tooth small cog, rather than an 11-tooth, which is the smallest attainable with splined freehubs. The XD cassette cogset comes as one piece, rather than having a separate lockring and some loose cogs (Fig. 8.24B). When you tighten the integrated lockring on a SRAM XD cassette, you turn an internally threaded sleeve inside onto the threads around the waist of the XD freehub body. It accepts a standard, splined cassette lockring remover for removal and installation (Fig. 8.24C).

A freewheel is a separate unit with the cogs attached to it. The entire freewheel threads onto the drive side of the rear hub (Fig. 8.25). Interchanging cogs

8.23 Quick-release freehub with cartridge bearings and a cassette cogset

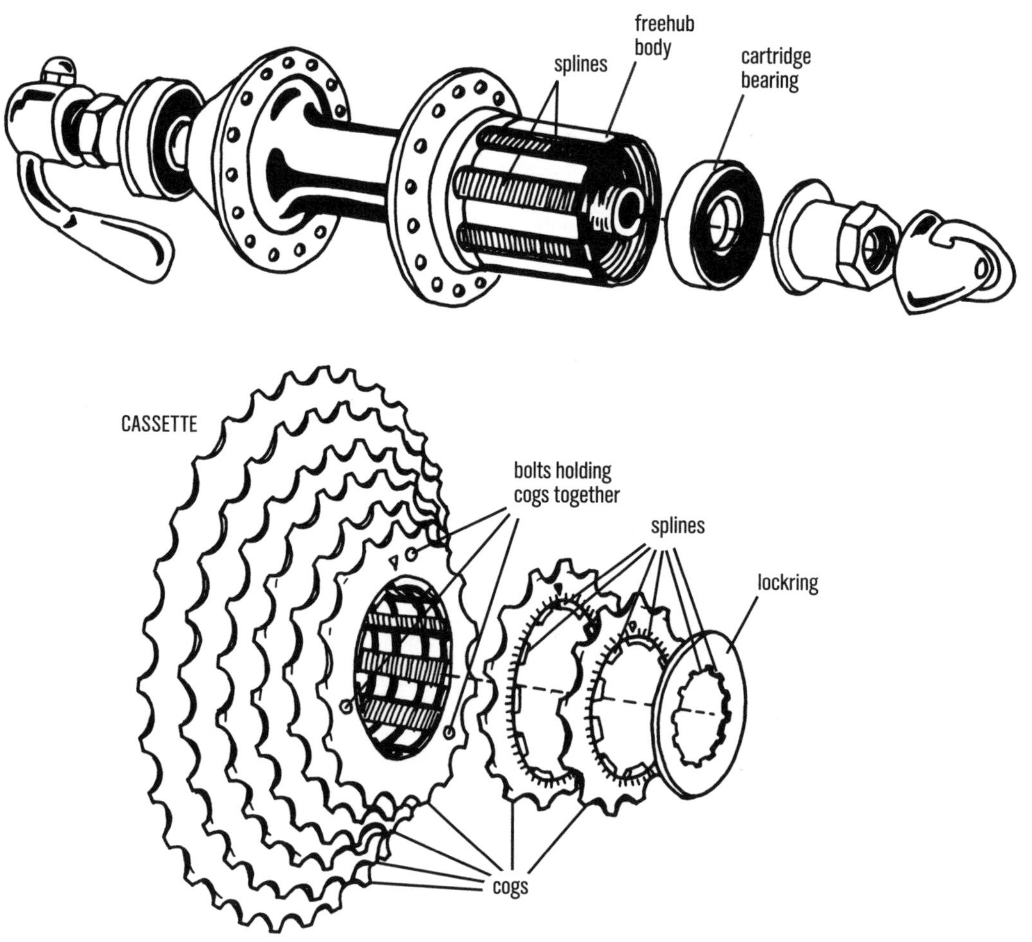

8.24A SRAM XD freehub body allows the use of a 10-tooth small cog and a standard cassette lockring remover

8.24B SRAM 10-50 cassette

8.24C Cassette lockring remover

8.25 Threaded quick-release rear hub with loose ball bearings and a freewheel

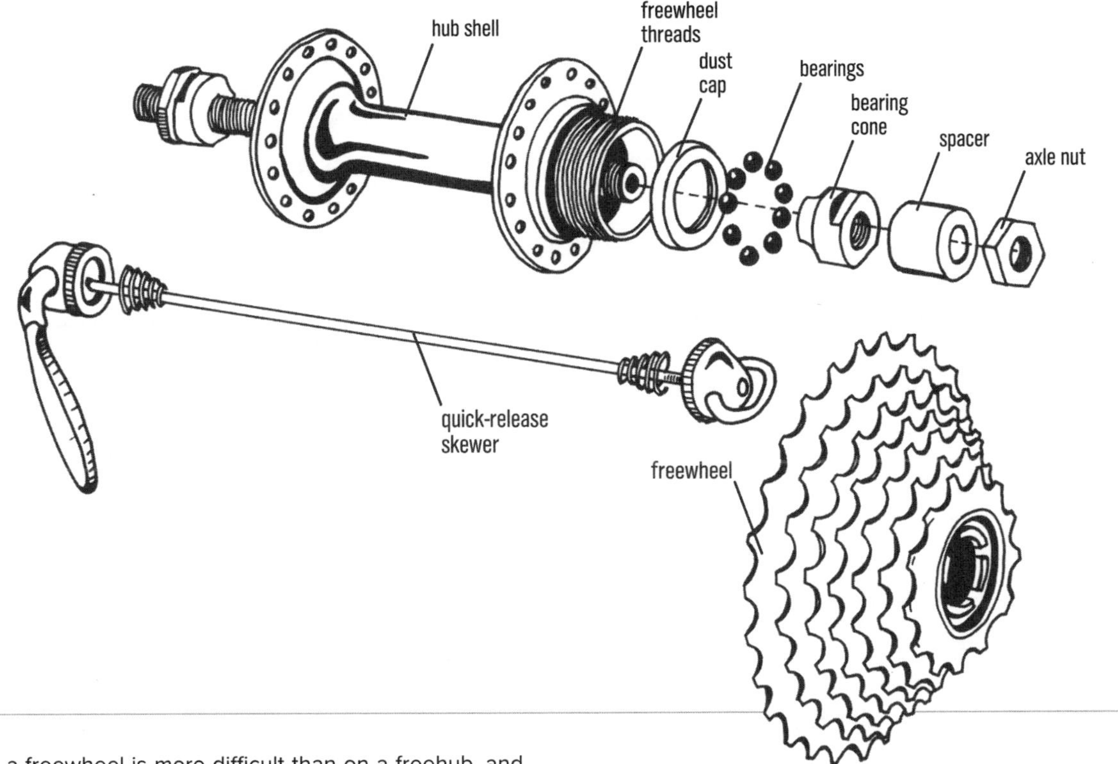

on a freewheel is more difficult than on a freehub, and a freewheel does not support the drive side end of the hub axle, two reasons that thread-on freewheels have fallen out of fashion. Freewheels can be removed by using a tool made to fit the specific freewheel. Entire freewheels with different gear combinations can be changed in this way.

Freehubs and freewheels usually rely on a series of spring-loaded pawls that engage internal teeth when pressure is applied to the pedals but allow the bike to freewheel when the rider is coasting. The pawls riding over the teeth as they rotate make the familiar clicking noise you hear when coasting.

Many freehubs can be lubricated without removing them from the hub. Changing gear combinations is accomplished by removing the cogs from the freehub body and putting on different ones (8-11).

8-10

CLEANING REAR COGS

The quickest, albeit perfunctory, way to clean the rear cogs is to slide a rag or Gear Floss string back and forth between each pair of cogs (Fig. 8.26). The other way is to remove them (see 8-11) and wipe them off with a rag or immerse them in solvent.

8-11

CHANGING CASSETTE COGS

Removal for SRAM and Shimano is the same; installation is similar, with only minor differences.

1. **Assemble your tools.** You will need a Pedro's Vise Whip or chain whip, a cassette-lockring remover (Fig. 1.2), a wrench (adjustable or open) to fit the remover, and the cog(s) you want to install. The Vise Whip allows you to clamp the cog and lift and control the entire wheel with it, whereas the dangling chain whip will fall off if the freehub spins back, and you can tear some bark off your knuckles on the cog teeth when a tight lockring finally lets loose and the chain whip falls off the backward-spinning cassette. Note that some very old cassettes from the early 1980s have a threaded smallest cog instead of a lockring. These require two Vise Whips or chain whips (or one of each) and no lockring remover.
2. **Remove the quick-release skewer from the hub axle** (Fig. 8.25).
3. **Clamp the Vise Whip** (Fig. 8.27) **or wrap the chain whip** (Fig. 8.28) **around a cog at least two up from the smallest cog.** Wrap a chain whip in the drive direction (Fig. 8.28). In place of the Vise Whip or chain whip, you can substitute a Pedro's cog wrench, which has two pins to hold the first cog; it will be labeled "11T" on one side and "12T" on the other side, because the pins will be located differently depending on the size of the first cog (11- or 12-tooth); it won't work on a SRAM cassette with a 10-tooth first cog, or a cassette starting with bigger than a 12T cog.

8.26 Cleaning the cogs

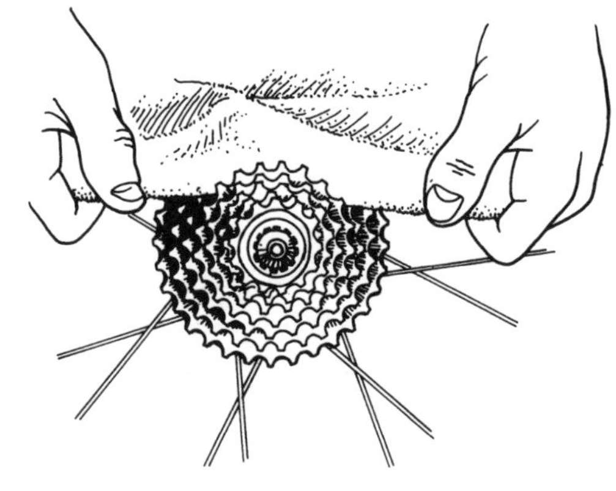

8.27 Removing a cogset with a Pedro's Vise Whip

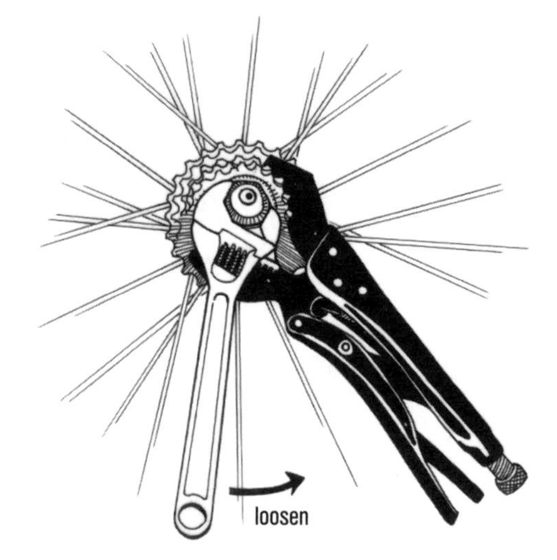

8.28 Removing a cogset with a chain whip

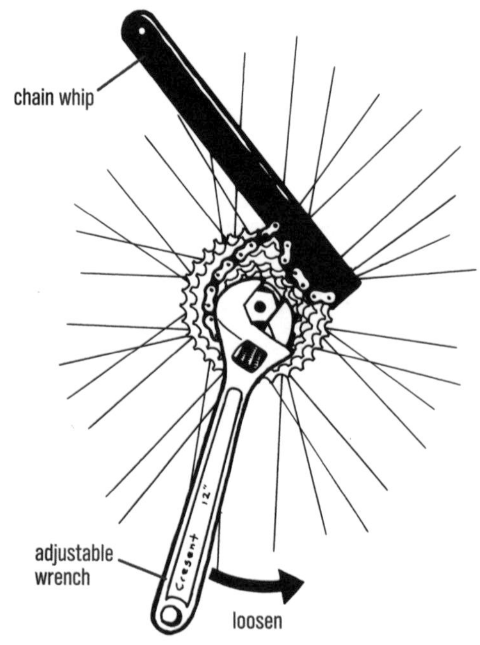

4. **Insert the splined lockring remover** (Fig. 8.24) **into the lockring.** The lockring is the threaded flange with a splined bore holding the smallest cog in place. Unscrew the lockring in a counterclockwise direction while holding the Vise Whip or chain whip to keep the cassette from turning (Figs. 8.27 and 8.28). If the lockring is so tight that the tool pops out and damages it, install the quick-release skewer without its springs through the hub and tool and tighten it against the lockring to hold the removal tool in place. Loosen the lockring a fraction of a turn, remove the skewer, and unscrew the lockring the rest of the way. If the lockring suddenly comes free, be careful if you're using a chain whip; it will fall off the cog when the cassette suddenly spins backward, and you can tear skin off the knuckles of the hand holding the chain whip. This is the reason for the Pedro's Vise Whip or cog wrench.
5. **Pull the cogs straight off.** Some cassette cogsets are comprised of single cogs separated by loose spacers, some cogsets are bolted together, and some cogsets are a combination of both.
6. **Clean the cogs with a rag or a toothbrush.** Use a solvent if necessary, and don't put the toothbrush back by the bathroom sink!
7. **Inspect the cogs for wear.** If the teeth are hook-shaped or the chain can be lifted off them when it is wrapped around the cog under tension, they may be ripe for replacement. If you have access to Rohloff's HG-IG-Check cog-wear indicator gauge (Fig. 1.4), wrap its measurement chain around the cog (while the cogset is on the freehub) and pull on the handle. If the last chain roller on the tool hooks on the tooth and resists your flipping it in and out of the tooth pocket while the tool handle is under pressure, or worse, if the entire measurement chain except the first roller slides easily away from the cog teeth while the handle is under pressure, the cog is worn out. The tool works only on cogs up to 21 teeth.
8. **Replace the cogs.**
   (a) If you are replacing the entire cogset, just slide the new one on. Usually, one spline is wider than the others to ensure proper alignment (Fig. 8.29), so clock the cassette and any loose cogs accordingly so they drop on. A SRAM XD cassette requires no certain clocking, since the XD splines are all symmetrical; the same is true with original, lockring-free Shimano cassettes and freehubs from the early to mid-1980s.

8.29 Large spline vs. large spleen

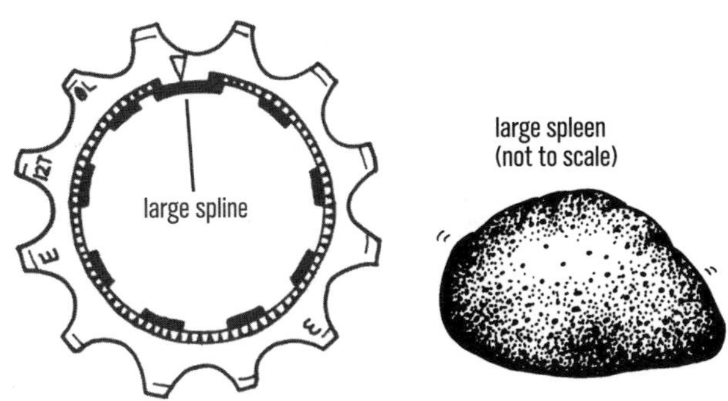

   (b) If you are installing a cassette with 9 or more cogs, see the notes under step 9.
   (c) If you are replacing some individual cogs within the cogset, be certain that they are of the same type and model. For example, not all 16-tooth Shimano cogs are alike. Most cogs have shifting ramps, differentially shaped teeth, and other asymmetries. They differ with model as well as with sizes of the adjacent cogs, so you need to buy one for the exact location and model. Install them in decreasing numerical sequence with the numbers facing out.

NOTE: *Some bolt-together cogsets are held together by three long bolts (Fig. 8.23) and can be disassembled for cleaning and then reinstalled onto the freehub as individual cogs to facilitate future cog changes and cleaning (the bolts are there only to speed bike assembly at the factory). But high-end cogsets usually have lightweight splined aluminum carriers onto which the cogs are mounted. These cannot be disassembled, and you must replace the entire carrier with cogs assembled onto it.*

9. **When all the cogs are on, tighten the lockring with the lockring remover and wrench.** (If you have the old, early-to-mid-1980s freehub type with the thread-on first cog, screw that on tight instead,

using a chain whip.) Make sure that all of the cogs are seated and can't wobble side to side, which would indicate that the first or second cog is sitting against the ends of the splines. All cogs are integrated into a SRAM XD cassette, so there are no separate parts; ensure that it is lined up straight so that the fine threads inside don't cross-thread onto the freehub body. If the cogs are loose after you've tightened the lockring, loosen the lockring, line up the first and second cog until they fall into place, and tighten the lockring again. Make sure that the lockring is compatible with both the freehub and the cogset; these can vary depending on manufacturer and size of the first cog.

NOTE ON COMPATIBILITY: *These instructions for removing and replacing cogs apply to 12-, 11-, 10-, 9-, 8-, 7-, and 6-speed cogsets. But the freehub bodies vary, so make sure you use, for example, only an 8-speed cogset on a freehub body designed for eight cogs. SRAM 11- and 12-speed XD freehub bodies are the same.*

NOTE ON 11-TOOTH COGS: *Some 8- and 9-speed freehub bodies will not accept 11-tooth cogs (for example, 1992–1994 XTR freehub bodies will not accept 11–28 or 11–30 cogsets). To accept the small 11-tooth cog, the splines of current, Shimano-compatible freehub bodies stop about 2mm before the outer end of the freehub body. If you are motivated to do so, you can grind the last 2mm of splines off an old-style 8-speed freehub so that it will accept an 11-tooth cog. But be aware that the steel is very hard, and you may need a rotary grindstone to do the job.*

8-12

CHANGING FREEWHEELS

LEVEL 3

If the bike has a freewheel (Fig. 8.25) and you want to switch it with another one, follow this procedure. Replacing individual cogs on an existing freewheel is beyond the scope of this book and is rarely done these days because of the unavailability of spare parts.

1. **Obtain the appropriate freewheel remover for your freewheel, a big adjustable wrench to fit it, and the freewheel you are replacing it with.**
2. **Remove the quick-release skewer, and take the springs off it.**
3. **Place the freewheel remover into the end of the freewheel so that the notches or splines engage.** Slide the skewer back in from the nondrive side, and thread the skewer nut back on, tightening it against the freewheel remover to keep it from popping out of its notches.
4. **Loosen the freewheel counterclockwise.** Put the big adjustable wrench onto the flats of the freewheel remover. It may take considerable force to free the freewheel, and you may even need to put a large pipe on the end of the wrench for more leverage. Have the tire on the ground for traction as you apply pressure to the wrench. Once the freewheel pops loose, stop and loosen the skewer nut before continuing to unscrew the freewheel; otherwise, you will snap the skewer in two as you continue to unscrew the freewheel.
5. **Loosen the skewer nut a bit, and unscrew the freewheel a bit more.** Repeat until the freewheel is loose enough that there is no longer any danger of having the freewheel remover pop out of the notches it engages.
6. **Remove the skewer and spin off the freewheel.**
7. **Clean and grease the threads on the hub and on the new freewheel.**
8. **Thread on the new freewheel by hand.** Be careful starting the freewheel onto the fine, aluminum threads. You can tighten it using a chain whip or the freewheel remover and a wrench, but it will tighten itself the first time you ride the bike.
9. **Replace the skewer with the narrow ends of its conical springs facing inward.**

8-13

REPLACING/FLIPPING ROHLOFF SPEEDHUB COG

LEVEL 3

To remove a cog from a Rohloff SpeedHub, you will need the special Rohloff driver (sprocket remover tool). If the cog is worn and you're replacing the chain, you can simply flip it over and double its life, because it is symmetrical.

1. **Remove the quick-release skewer and take the springs off it.**
2. **Place the Rohloff driver into the center of the cog so that its four prongs engage the two notches in the end of the hub.** Slide the skewer back in from the nondrive side, and thread the skewer nut back on, tightening it against the freewheel remover to prevent the remover from popping out of its notches.
3. **Clamp the cog with a Vise Whip** (Fig. 8.27), **or wrap a chain whip around it in the direction opposite the pedaling direction** (the reverse direction of Fig. 8.28).
4. **Holding the cog remover (driver) with a 24mm wrench or an adjustable wrench, unscrew the cog counterclockwise with the Vise Whip or chain whip.**
5. **Pull the cog off, leaving the driver in place.**
6. **Using the driver to line up the cog with the threads, screw on the flipped-over or new cog.** Tighten it (clockwise) with the Vise Whip or chain whip. Now you can remove the driver and replace the quick-release skewer, along with its springs.
7. **Check that the cog flanges are pristine.** If the flange of the cog that recesses into the hub shell is damaged, it could in turn damage the rubber seal on the end of the hub, causing oil to leak out of the hub.

8-14

LUBRICATING FREEHUB MECHANISMS

Most people ignore their freehubs, even if they maintain the rest of their bicycle well. In addition, the simple method employed in some bike shops of dunking freehubs in a tub of solvent has the net negative effect of pulling contaminants deep within the freehub, which then lets solvent seep out during riding, thereby contaminating lubricants in the freehub and wheel bearings. This bad practice, and the seeming impenetrable black-box nature of the freehub, could explain why freehubs are one of the most often purchased replacement parts in bike shops. But a little simple lubrication on a regular basis can prevent their demise.

Many freehubs can be adequately cleaned and lubricated simply by dripping chain lube into them once the axle assembly is out (8-14a explains how). But this method allows you to drip in only a thin lubricant, which will not protect or hold up as long as a thicker formulation will. Also, a thin lubricant will get into the wheel bearings, contaminating and thinning the grease on them. See 8-14b for instructions for improved injection of cleaning solvents as well as thicker, more protective lubricants into a Shimano freehub, and 8-14c for completely overhauling a Shimano freehub.

Some high-end freehubs have grease-injection holes on the freehub body that accept a fine-tip grease gun. With these, you remove the cogs to get at the hole, and then clean that hole before inserting the grease gun tip (otherwise you will push a plug of dirt right into the freehub mechanism). Rather than using bearing grease in the grease gun, inject a thinner lube into it, such as outboard-motor gear oil. Using one of these will keep grease from thickening inside the freehub and sticking to the pawls in cold temperatures, preventing engagement of the freehub. In most freehubs, the pawl springs are very light, and it does not take much in the way of sticky, cold-thickened lube to stop them from pushing the pawls radially outward. Believe me, it's a good way to end up on your nose when you apply power to the pedal and the freehub slips.

If the freehub has teeth on the faces of the hub shell and freehub (DT Swiss/Hügi freehubs), you can just drip oil into the crease between the freehub and the hub shell as you turn the freehub counterclockwise, but it makes no sense to do this because they pull apart easily for proper lubrication; see 8-14e. And for Mavic and all manner of freehubs with radial teeth inside of the hub shell, see sections 8-14d, f, and g.

a. Basic freehub oiling

LEVEL 2

1. **Disassemble the hub-axle assembly as in 8-6 or 8-7.** Standard freehubs are Shimano or similar, with tiny, loose bearings inside, and a better method for lubricating them, described in the next section, is to remove them first. However, some lubrication is way better than nothing, so at least do this much.

2. **Wipe clean the inside of the drive-side bearing surface.**
3. **Drip in chain lube between the bearing surface and the freehub body as you spin the freehub counterclockwise.** Set the wheel on a flat surface with the freehub pointed up toward you. You will hear the clicking noise of the freehub pawls smoothing out as lubricant reaches them. Keep it flowing until old black oil flows out of the other end of the freehub.
4. **Wipe off the excess lube, and continue with the hub overhaul.**

b. Shimano freehub lubrication without disassembly

LEVEL 2

This method lets you avoid removing the easily damaged freehub-body dust seal, but it doesn't provide enough pressure to ensure that the gear oil gets in everywhere, the way complete overhaul (below) does. Still, the following method is better than nothing.

1. **Disassemble the hub-axle assembly as described in 8-6 and 8-7.**
2. **Remove the freehub body.** Unscrew the internal freehub-fixing bolt with a 10mm hex key (Fig. 8.30) on quick-release hubs and the entire freehub body with a 14mm or 15mm hex key on through-axle hubs (Fig. 8.31A). Holding the hex key in a securely mounted bench vise and turning the wheel as in Fig. 8.31B may be necessary if it's really tight.
3. **Remove the rubber seal at the back.**
4. **Close off the bottom of the freehub body.** Use a big rubber stopper.
5. **Completely flush out the freehub.** Pour solvent into the outer opening, spinning the mechanism and thereby letting contaminants run out. Repeat until clean. Don't dunk the freehub in a solvent tank; this will pull dirt into it.
6. **Lubricate the freehub.** Squirt in a quantity of outboard-motor gear lube, and then park the body on paper towels and let the excess drain off.
7. **Replace the rubber seal on the backside of the freehub body.**
8. **Tighten the freehub body into the hub shell.** Use the big hex key down the center of the freehub

8.30 Removing/installing the bolt and freehub body on a Shimano quick-release hub

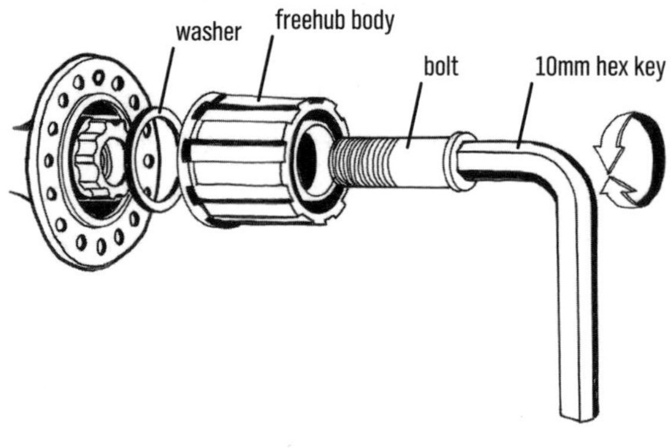

8.31A Removing/installing a Shimano threaded freehub body on a through-axle hub with a 14mm or 15mm hex key

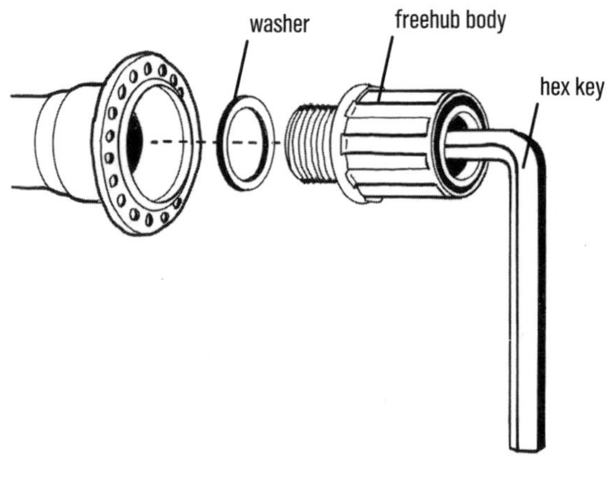

8.31B Use a vise to hold the hex key securely

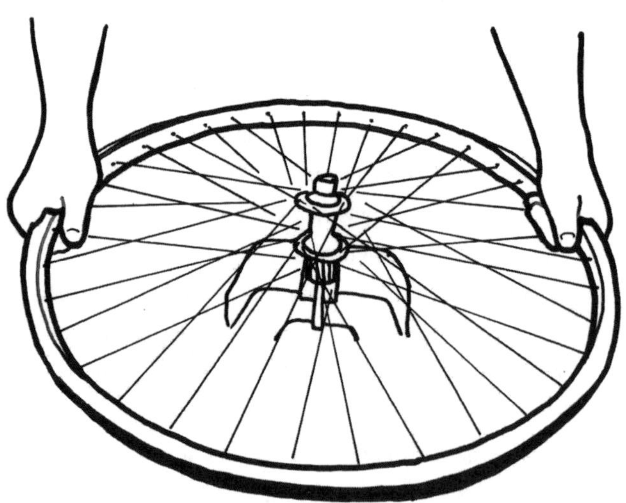

again. Don't forget any washers that may have been under it (Fig. 8.30).

9. **Overhaul the hub and replace the axle assembly (8-6).**

c. Complete overhaul of Shimano freehub

LEVEL 3

If the above two methods don't get your Shimano freehub working the way you want it to, and you're more committed to this freehub than to buying a new one and screwing it onto your hub, then you can completely overhaul it. You will have to make a tool for it, however, because that does not seem to be a tool anyone sells, although it is possible that some two-tooth freewheel removers might be the right diameter and could be used with minor modification.

NOTE: *I illustrate this method photographically in my* Mountain Bike Performance Handbook.

1. **Make your freehub disassembly tool.** Grind down the end of a socket that is about 28mm in outside diameter at its open end—a 13⁄16-inch or 21mm socket is the correct size—to form two diametrically opposed teeth about 5–6mm tall and 3.6–3.7mm wide. Do this on a bench grinder and form sharp corners on either side of the teeth with a file. The socket steel will be tough, so you can expect this operation to be a bit tedious.
2. **Disassemble the hub-axle assembly as in 8-6 or 8-7.**
3. **Remove the freehub body.** Depending on hub model, unscrew the freehub body with a 10mm, 14mm, or 15mm hex key (Figs. 8.30 and 8.31). Some have an internal freehub-fixing bolt, and others have a threaded root to the pawl carrier that screws directly into the hub shell. If it's too hard to turn, clamp the hex key in a vise so it's sticking straight up, push the wheel down until the hex key engages, and turn the entire wheel counterclockwise as shown in Fig. 8.31.
4. **Remove the freehub dust cover.** If you have a blind-hole bearing puller, that is the ideal implement for this job and will prevent damage to the dust cover. Otherwise, carefully tap out the dust cover from the backside with the screwdriver down the center of the freehub body against the backside of the dust cover. Lightly tap the handle with a hammer as you work your way around.

NOTE: *Shimano freehub dust caps are made of stamped sheet metal, and they may be ruined upon removal. Shimano does not sell them separately, so do your best to avoid bending or kinking the cap.*

5. **Set the freehub body back in place in the hub shell.** Set the wheel flat on the workbench. The lobed teeth (or the threads in a through-axle hub) will hold the freehub body in place.
6. **Unscrew the bearing race.** Put the two teeth of your socket tool into the two notches in the bearing race (Fig. 8.32) and turn it *clockwise* with the socket wrench handle—it's left-hand threaded! Hold the socket and freehub down firmly while you turn it, because it's very tight, and you risk damaging your tool's teeth and the bearing race's notches.
7. **Pull the race/cone out.** It might be easiest to pull off the freehub shell and push the newly freed pawl carrier down and out with your finger. The hub-bearing race is a cone on the backside for a row of 25 tiny (1⁄8-inch) freehub bearings sitting atop the pawl carrier; these will probably all come clattering out, so be prepared. So will the other 25 tiny ball bearings on the other end of the pawl carrier. Place

8.32 Unscrew the left-hand threaded bearing cup in a clockwise direction with the teeth of a custom tool made from a socket engaging the notches in the cup

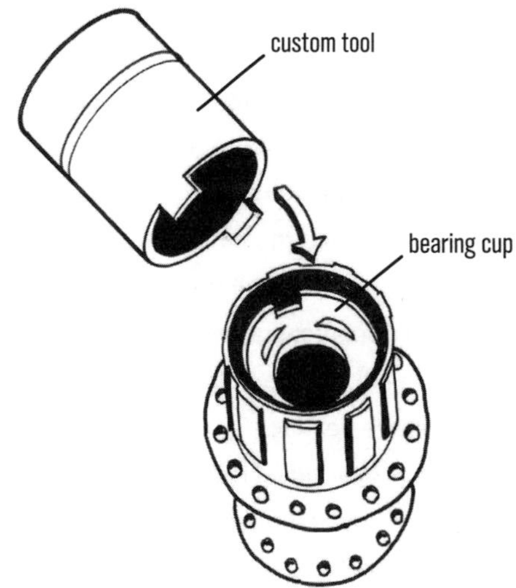

8.33 Shimano freehub body

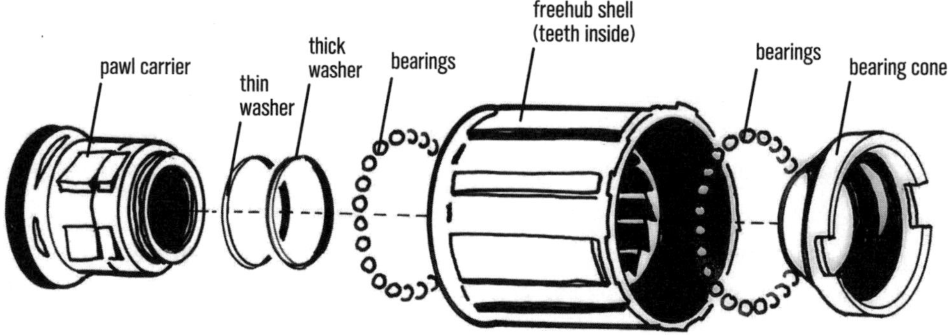

a soft, clean rag on your workbench to slow down the bearings as they race to their hiding places in the far corners of your shop.

8. **Clean everything.** Clean the pawl carrier, the pawls, and all of the bearing surfaces (Fig. 8.33); solvent will help. Keep track of any washers inside and note where they go. Clean the ball bearings unless you're replacing them; in that case, get 50 ⅛-inch balls.
9. **Judiciously apply grease.** Grease only the pawl hinges (Fig. 8.34); don't get grease under the pawls, which could stick them down and prevent the free-hub from locking up for forward pedaling. Grease the teeth inside the freehub shell, and grease the bearing races inside the shell on either end of the teeth, on the base of the pawl carrier, and on the cone on the bottom of the hub bearing race. Apply just enough grease to stick the bearings in place but not enough to coat the entire area and perhaps under the pawls.
10. **Carefully install the bottom bearings.** Set the pawl carrier in place standing straight up, engaged into the hub shell, and line the balls up around the base of the pawl carrier; tweezers are a good tool for this.
11. **Replace the washers.** There are usually two or three that go atop the pawl carrier.
12. **Drop the freehub shell down onto the pawl carrier.** Turn it counterclockwise to get the pawls to go into the teeth and the bearing race in the shell to sit down on the bearings.
13. **Install top bearings.** Line up another 25 of the ⅛-inch balls around the bearing race along the top of the shell's teeth.

8.34 Grease the pawl hinges only

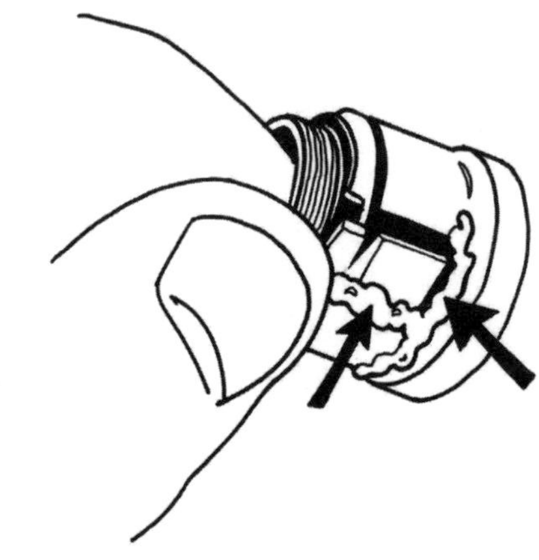

14. **Screw in the cup/cone.** Use your homemade, toothed socket again to tighten the hub bearing race with the two notches into the freehub shell (*counterclockwise!*). Tighten it very tightly.
15. **Replace the dust cover.** Carefully tap it back in by lightly tapping the screwdriver handle with a hammer as you work its tip around the groove in the dust cover. Just tap it in until it hits a soft stop; don't tap it down far enough to later impinge on the hub bearings sitting in the bearing race underneath it.
16. **Replace the rubber seal on the backside of the freehub body.**
17. **Bolt the freehub body into the hub shell.** Tighten it with a 10mm, 14mm, or 15mm hex key down the center of the freehub (Figs. 8.30 and 8.31).
18. **Check the operation.** If it spins well and locks up well, pat yourself on the back.
19. **Overhaul the hub and replace the axle assembly (8-6).**

d. Mavic freehub lubrication

1. **With the wheel upright, remove the axle end cap(s).** Depending on model, whether it's a quick-release hub or through-axle hub, this usually involves pulling the non-drive-side dust cap straight off.
2. **Remove the axle.** With a QR hub, using two hex keys, one in either end, loosen counterclockwise, unscrew, and remove (Fig. 8.35). Depending on the model, this takes two 5mm hex keys or one 10mm and one 5mm, or it can require unscrewing external locknuts from external threads on a steel axle with cone wrenches. With a Mavic through-axle hub, hold the nondrive axle end with a 14mm wrench and unscrew the drive-side end cap with a 17mm cone wrench. To get the axle out, unscrew it from the adjuster ring on the nondrive end while holding it with a Mavic or adjustable pin tool; you can now push the axle out. The freehub on Mavic through-axle hubs has radial pawls around its base and teeth inside the hub shell as in Fig. 8.35, so follow steps 5 and 6 in 8-14f before replacing the axle the reverse of how you got it out.
3. **Turn the wheel on its side, freehub up.** Place the wheel on a clean surface where you can catch—or at least see—any pawls or pawl springs that might fly away.
4. **Rotate the freehub body slowly counterclockwise as you pull up on it, and remove it** (Fig. 8.35). Slide the freehub up a little, find where the pawls are, and hold them with your fingers before pulling the freehub completely off.
5. **Pull off the outer rubber seal.**
6. **Clean the pawls, springs, seal, and hub shell.**
7. **Replace the seal, square side in.** Do not use grease on the seal; it's outside the freehub body, and lubricating it will only attract dirt.
8. **Replace the springs and pawls.**

8.35 Removing Mavic quick-release rear axle and freehub

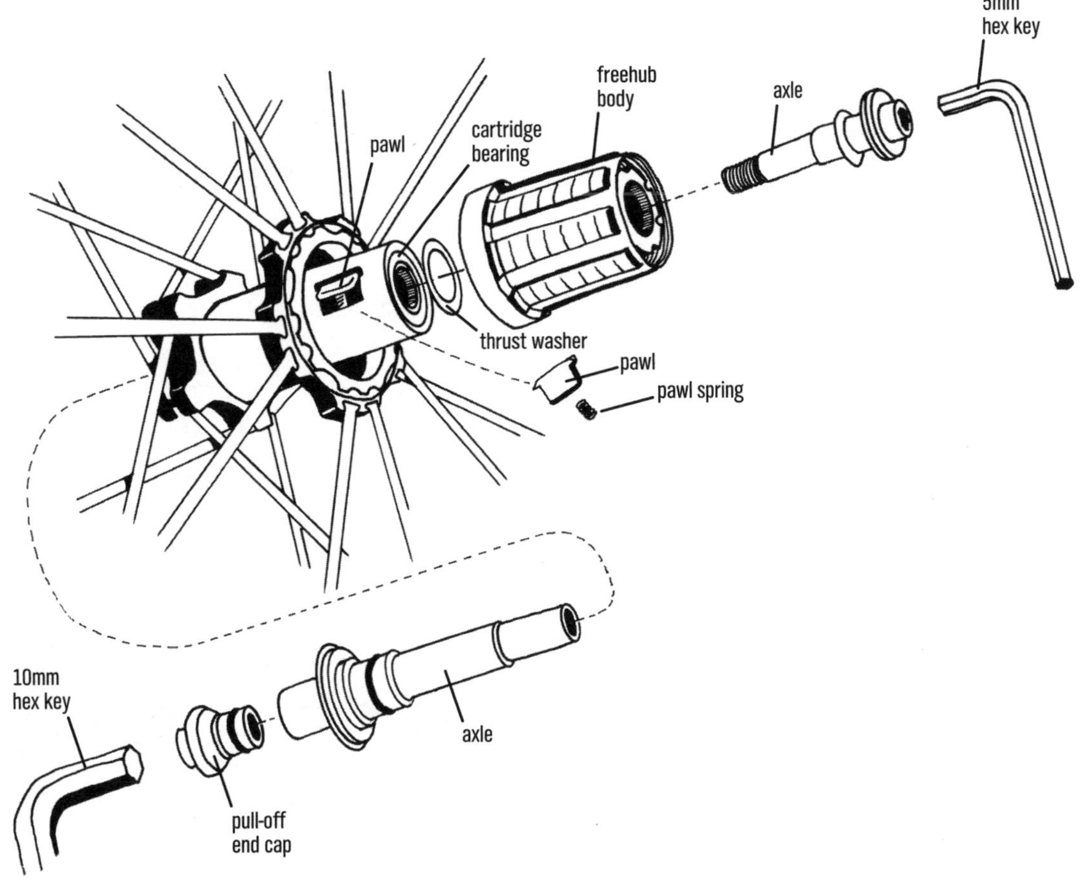

9. **Lubricate the freehub.** Put 10 or so drops of 10-weight mineral oil (Mavic recommends its M40122 mineral oil or Phil Wood Tenacious Oil) on the plastic bushing and the ratchet teeth in the freehub.
10. **Reinstall the freehub body.** Turn it counterclockwise while holding the pawls down with your fingers.
11. **Replace the axle.** Simple.

e. DT Swiss star-ratchet freehub lubrication

LEVEL 1 DT Swiss and DT-Hügi high-end star-ratchet freehubs pull apart easily for cleaning and lubrication.

1. **Remove the quick-release skewer.**
2. **Lay the wheel on its side, cogs up; grasp the cogset and pull up.** Hold the wheel down. The freehub body will come off (Fig. 8.36), bringing the axle end cap with it.
3. **Clean and grease the spring, both star-shaped ratchets, and the teeth that engage on the freehub body and hub shell.**
4. **Push the freehub and end cap back on, and replace the skewer.**

   That's it!

f. Lubricating freehubs with three radial pawls mounted on cartridge-bearing hubs

LEVEL 1 The freehub body is essentially the same as on the Fulcrum hubs mentioned below, but they exist on myriad brands of cartridge-bearing hubs. The hub shell has radial teeth pointing inwardly on its drive end, and the base of the freehub body slips down inside and has three or more pawls that flip outward to engage the teeth (Fig. 8.37). Often, there is a circular spring around all three pawls that flips them outward; it also generally keeps them from flying away when you pull the freehub body off. This is a nice, clean design, but it is always possible that the freehub body instead has a single tiny coil spring behind each pawl, and these parts definitely can go flying when you pull the freehub body off unless you put a twist tie around them as mentioned below (8-14g).

1. **Access the freehub body by first removing the axle as in 8-7.**

8.36 DT Swiss star-ratchet freehub removal and lubrication

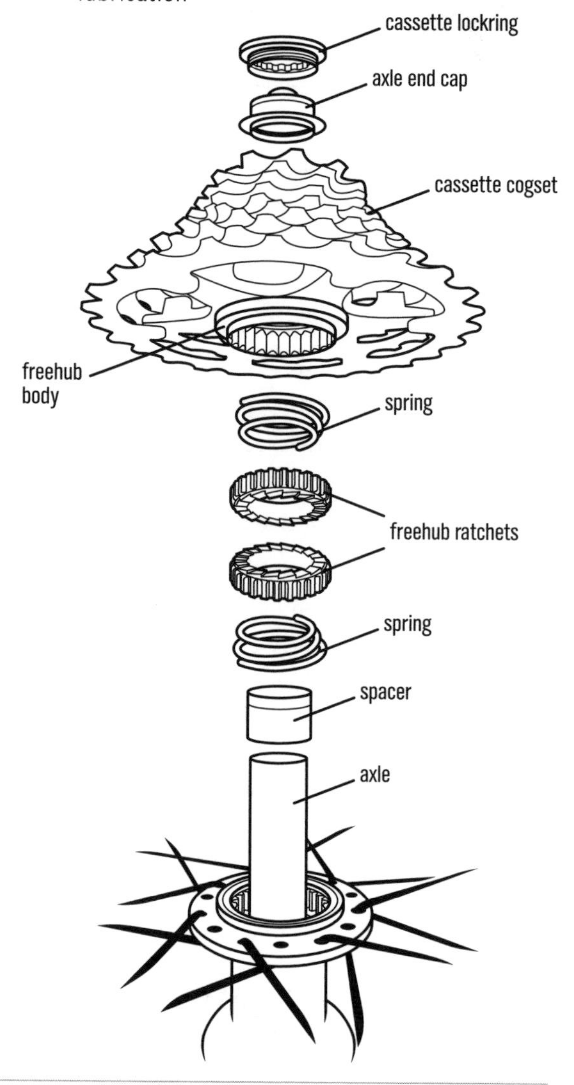

8.37 Radial-pawl freehub removal and lubrication

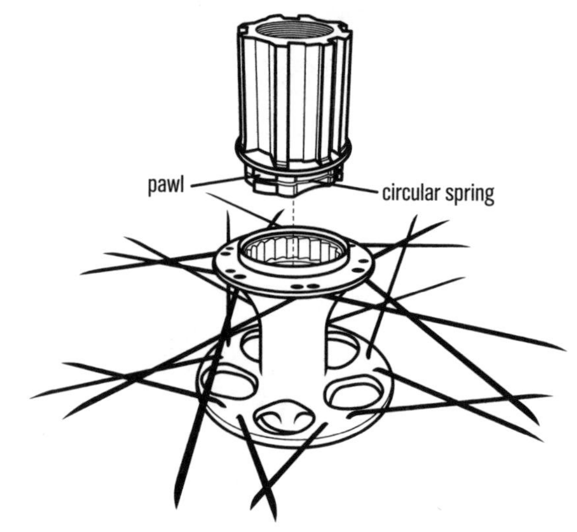

2. **Pull the freehub straight off.** Turn it counterclockwise as you pull. A single, circular wire spring will generally be wrapped around all three pawls (it fits in a groove in the freehub body and the pawls; Fig. 8.37). Nonetheless, exercise care that nothing goes flying. If there were spacers under it, note how they were oriented; clean them and set them aside for later reinstallation. If the pawls are not retained by a wire spring, be ready to collect them and the coil springs beneath them.
3. **If the freehub turns roughly, drive the bearings out of the freehub body.** If it spins smoothly, skip to step 5. You may need to remove a snapring with snapring pliers (Fig. 1.2) from the drive-side end before you can remove the bearings. They can sometimes be pulled out with your fingers, or you can drive both of them out the drive-side end onto a wooden bench with a 14mm or 15mm socket. You may also be able to drive them out one at a time with the axle, with a screwdriver or with a hex key of appropriate diameter. If there are spacers between the bearings, keep track of them and their orientation in the freehub.
4. **Press in the bearings.** If you have a bearing press, use that. Otherwise, use the method shown in Figures 8.21 and 8.22. If there were spacers between the bearings, put them in before putting in the second bearing. If there was a snapring securing the bearing, replace that.
5. **Clean and lightly grease the pawls and the radial teeth inside the hub shell.** Don't get grease under the pawls. A good alternative to light grease is heavy oil like Phil Wood Tenacious Oil or outboard-motor gear oil on the pawls and teeth.
6. **Slide the freehub body back in while slowly turning it counterclockwise.** If there were spacers behind the freehub body, put them in first, in the orientation they were originally in. Push inward on each pawl with a pencil tip as you turn and push in on the freehub body, until all three are engaged and the freehub body drops into place.
7. **Reassemble the hub as in 8-7.**

g. Fulcrum freehub lubrication

1. **On high-end Fulcrum quick-release rear hubs, remove the skewer.**
2. **Unscrew the locknut.** Insert a 5mm hex key into the drive-side axle, and put a 17mm open-end or box wrench on the drive-side locknut. While holding the 5mm hex key, unscrew the locknut *clockwise* (in other words, it's left-hand threaded, and you need to unscrew it in the opposite direction from what you would expect). Older Fulcrum models have a little setscrew on the 17mm locknut that you loosen with a 2mm hex key in order to unscrew the locknut.
3. **Pull the freehub straight off.** Older models have individual coil springs under each of the three pawls. These can go flying, and they are hard to clean and to insert back into the hub shell. When removing the freehub body, wrap a twist tie around the three pawls as they expose themselves from the hub shell as you pull; if you don't do this, the three pawls and the three springs will fly away. Newer Fulcrum models have a single circular wire spring wrapped around all three pawls (Fig. 8.37); it fits in a groove in the freehub body as well as one cut across the flanks of each pawl. With these, you can pull the freehub off with abandon, and nothing will fly away.
4. **Clean and grease the three pawls and the radial teeth inside the hub shell.**
5. **Slide the freehub back in while slowly turning it backward.** Push inward on each pawl with a pencil tip as you do this until all three are engaged and the freehub body drops into place. Again, older Fulcrum models require the use of a twist tie to hold the pawls in place as you push it in. Pull the twist tie off after the pawls are inside the hub shell and before the freehub is pushed all the way in.
6. **Tighten the locknut.** While holding the drive-side end of the axle with a 5mm hex key inside its bore, tighten the locknut *counterclockwise*—remember, it's left-hand threaded! This locknut generally tightens with a 17mm cone wrench.

8-15

LUBRICATING THREAD-ON FREEWHEELS

1. **Wipe dirt off the face of the fixed part of the freewheel surrounding the axle.**
2. **Drip lubricant into the crease between the fixed and moving parts of the freewheel as you spin the cogs in a counterclockwise direction.** Set the wheel flat with the cogs facing up toward you. You will hear the clicking noise inside become smoother as you get lubricant in there. Be sure to keep the lubricant going in until the old, dirty oil flows out the backside around the hub flange.
3. **Wipe off the excess oil.**

9

DISC BRAKES

If you think about it long enough you will find yourself going round and round . . . until you finally reach only one possible, rational, intelligent conclusion. The law of gravity and gravity itself did not exist before Isaac Newton. No other conclusion makes sense.

—ROBERT M. PIRSIG, *ZEN AND THE ART OF MOTORCYCLE MAINTENANCE*

TOOLS

- 2.5mm, 3mm, 4mm, 5mm, and 6mm hex keys
- 7mm, 8mm, 9mm, and 10mm box-end wrenches
- 9mm and 10mm open-end wrenches
- Small adjustable wrench
- Pliers
- Grease
- Screwdrivers, flat-head and Phillips
- Torx T10 and T25 wrenches
- Cable-housing cutter
- Sharp knife
- Isopropyl alcohol
- Plastic tire lever
- Brake pad spacer
- Brake bleed block
- Drywall-sanding screen
- Bleed kit for your hydraulic brake
- Hydraulic fluid specific to your brake

OPTIONAL

- Rotor-alignment tool (Park DT-3 and/or dial indicator)
- Rotor-tuning fork(s)
- Plastic grooved blocks for clamping hydraulic hose

CONTINUED ON P. 184

Disc brakes, which squeeze the pads against a hub-mounted disc (or "rotor"), stay much cleaner than rim brakes, as mud and water are thrown away from them by the tires, rather than into them, resulting in little drop-off in performance in wet conditions. Additionally, the rim does not heat up during braking, so disc brakes can offer consistent performance over a wide range of conditions.

Disc brakes can be hydraulic (activated by pressure applied to one end of a column of non-compressible liquid) or mechanical (a.k.a. cable actuated, by tension on cables). On cable-actuated disc brakes, the brake levers are the same as on rim brakes, as are the procedures for hookup to the levers and calipers, cable routing along or through the frame, and maintaining the cables and housings, so refer to the early sections of Chapter 10 for all of those operations. Installing disc brakes correctly is a must, so follow the directions in this chapter carefully. Once properly installed, discs require less maintenance than rim brakes because the tire is not dragging mud into them. There is no need to be intimidated by disc brakes; although they are small and enclosed and therefore somewhat mysterious, they are really quite simple.

This chapter focuses on the disc-brake calipers, both cable-actuated and hydraulic, and on hydraulic levers and hoses.

There's no need to release a disc brake to remove a wheel. Just drop the wheel right out. The disc will fall straight out of the caliper (other than in a couple of old designs for frames and forks without built-in disc-brake mounts).

NOTE: *Never squeeze the lever on a hydraulic disc brake without a disc or a spacer between the pads, as you can push a caliper piston too far out. For bike travel with the wheel out, insert a spacer between the pads—either one that came with the brake or a chunk of corrugated cardboard you cut for the purpose.*

9-1

CHECKING AND REPLACING DISC-BRAKE PADS

LEVEL 1

Disc-brake pads are less easy to see than rim-brake pads, so you should be even more diligent about checking them for wear. The friction material needs to be at least the thickness of a dime (about 1.2mm); with most

TOOLS, CONT.

Air compressor
Hose cutter
Hacksaw
Piston-bore-plug removal tool specific to your one-piece hydraulic disc caliper
Full set of Torx wrenches, up to T30
Digital caliper or micrometer
Park IR-1 or IR-1.2 internal-routing kit
Q-tips

pads this means that the entire pad and backing plate together should be greater than 2.5mm. Some brakes supply a pad-wear gauge; Magura, for instance, has a 4.1mm-thick "finger" sticking off the plastic pad spacer that comes with the brakes. Apply the brakes against the rotor by squeezing the lever, and the finger should fit between the steel ears of the pads. If it doesn't, the distance between the pad backing plates is too small, indicating that the pad material has worn off and new pads are in order.

Rotors wear, too. If over 0.2mm has been removed from the rotor thickness, it is due for replacement. The thickness change is so small that it must be measured with a very accurate gauge, like a digital caliper or a micrometer.

The wheel must be off for most disc-brake pads to be removed, but some top-loading pads can be pulled out from the top of the caliper while the wheel is in the bike. On cable-actuated disc brakes, unscrew the inboard pad-adjuster knob or bolt until it stops to spread the pads as far apart as possible before pulling the pads out.

Many pads require that you remove a cotter pin or bolt before pulling the pads out (Fig. 9.1). The cotter pin may be a threaded bolt, a pin with a retaining clip holding it in, or both (Fig. 9.24). Catch the spreader spring, too.

On cotterless pads, with the wheel off, grab a tab on the pad (Figs. 9.2 and 9.3) with your fingers or needle-nose pliers and pull it toward the center of the caliper slot and out. You have to do this from underneath the caliper, so it's often not easy to get to. The pads need space to pull inward and out, one at a time. With hydraulic brakes, you may need to first slip a plastic pad spacer, tire lever, or flat-head screwdriver between the pads and carefully rock it back and forth to separate them without damaging them; on cable-actuated brakes (Fig. 9.2), you may need to back out the pad adjuster knob.

The best way to clean the pads is to sand them, face down, on a flat piece of drywall-sanding screen lying on a flat surface so that the material removed from the surface of the pad falls through the screen, rather than being rubbed back onto the face of the pad. You can also clean the pads with isopropyl alcohol or a dry, oil-free rag, or rub the pads against each other. Once

9.1 Removing top-loading brake pads

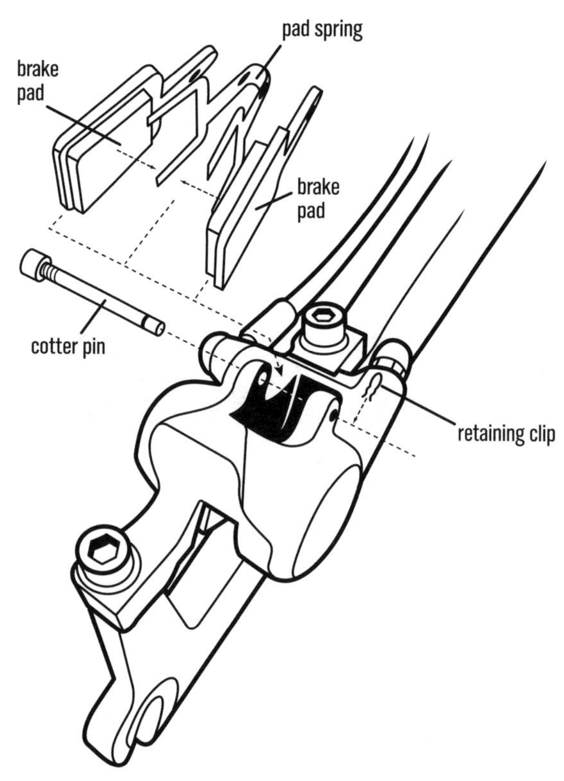

9.2 Removing cotterless brake pads

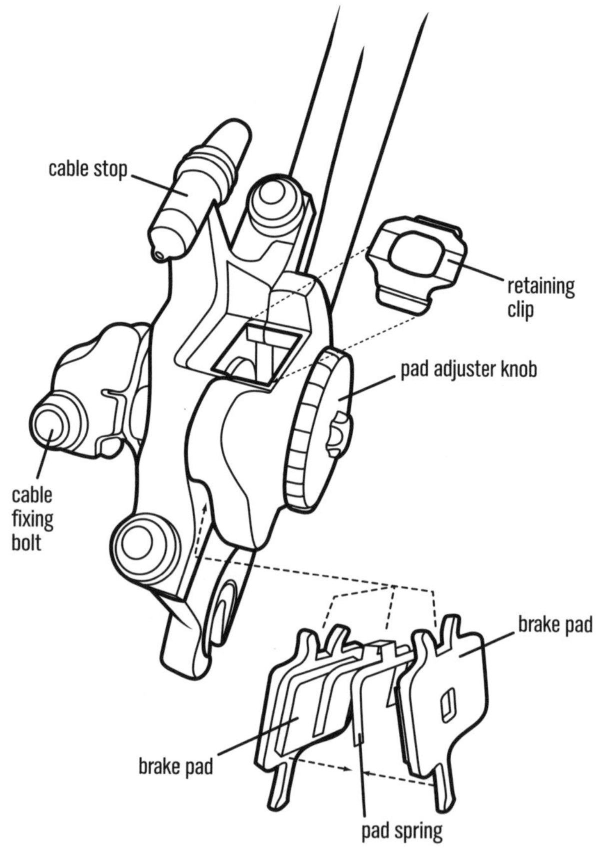

9.3 Changing Hayes cotterless brake pads (the caliper is shown apart only for the purpose of illustration; the pads clip in and out without dismantling the caliper)

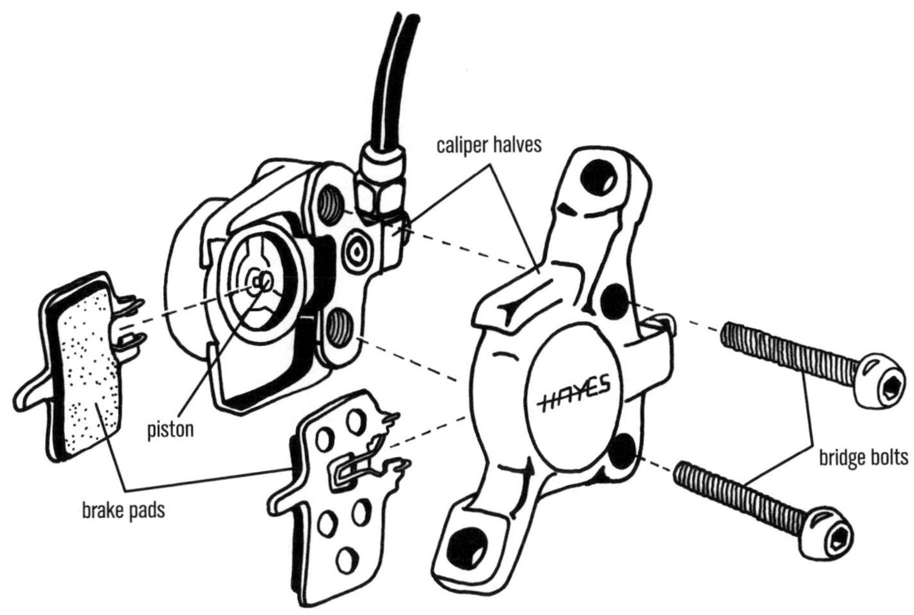

clean, check for pad wear, scoring, or glazing—anything that could damage the rotor or endanger braking effectiveness. Brakes fade when the pads and rotors get too hot, after which blue discoloration of the rotor and glazing of the pads occur, indicating that resins holding the pad material have broken down and recrystallized on the surface. Glazed pads must be discarded.

Although it's not mandatory, you'll reduce the potential for introducing dirt into a hydraulic caliper if, before replacing the pads, you clean around the pistons with a Q-tip soaked in brake fluid for that brake.

Install the new pads the way the old ones came out, noting that the left and right pads may differ; it should be obvious if you try to put a pad in the wrong side.

Cotterless pads usually snap back in—either onto a nub on the piston with a wire catch (as on Hayes, Fig. 9.3) or magnetically (Fig. 9.2). If the pads are not symmetrical and you reverse them, they may not snap back into place because the piston is often offset from the center of the cutaway for the pad. If there is a spring-steel pad spreader clip between cotterless pads (Fig. 9.2), it may go in after the pads are inserted, or it may go in like a sandwich with the pads.

On cottered pads, the ears on the pads may not line up with the cotter hole if reversed (Fig. 9.24). On cottered pads that have a little butterfly-shaped spring-steel piece that pushes the pads apart (Figs. 9.1 and 9.24), make a sandwich of the new pads and the butterfly spring and push them back in together. Then push in or screw in the cotter pin and replace its circlip if it has one.

9-2

SELECTING AND BURNING-IN DISC-BRAKE PADS

Make sure you buy pads meant for the exact make and model of the brake; there are myriad shapes of disc-brake pads, and they're not interchangeable.

When buying pads, you may have a choice of pad compounds, and your choice should be based on the type of riding you do. Metallic pads deal better with the heat generated by high-speed descents, whereas resin pads give better modulation, and hence control, of wheel speed. Resin pads also wear out faster than metallic pads, especially in wet conditions.

New pads need to be burnished (or burned in or bedded in) with repeated braking before they reach full braking power. If you instead just slam on the brakes when the pads are new, you can damage them so that they won't reach full power, and they may squeal mercilessly to remind you on every ride.

Burn in the pads by braking firmly and evenly without letting the brake get too hot. It's best to do this with one brake at a time, rather than with both brakes. Every manufacturer has a different procedure, but all say that it takes somewhere in the range of 20 to 40 stops to bed in the pad. I recommend that you start by applying the brake 10 times to bring the speed down from about 10–12 mph to walking speed. Then increase the speed to 15–18 mph and brake to walking speed 10 more times. This works the heat cycle evenly over the rotor and reduces the potential for squeal problems.

Don't bring the bike to a dead stop during the bed-in process.

9-3

PUSHING DISC-BRAKE PISTONS BACK IN WHEN PADS RUB

a. Push the pistons back

Pistons on hydraulic brakes sometimes get pushed so far out that they drag on the rotor or won't even let the rotor back in when the wheel is out. This can easily happen if you pull the lever when there is no rotor or spacer between the pads. You will have to push the pistons back in, and on some brakes this is best done with the pads out, whereas on others it is best done with the pads in.

If the wheel is out and the rotor will not go in, you will first have to push the pads and pistons back by jamming in the plastic pad spacer (which you should have installed when you pulled the lever when the wheel was out; it would have prevented this from happening). Once you have some space between the pads, you might as well try pushing the pistons back with the pads in. Using a plastic tire lever or a flat-head screwdriver, carefully (so that you don't gouge the pads) tip the tire lever or screwdriver back and forth until there is enough space between the pads to allow the rotor to turn without rubbing.

On Hayes and Stroker brakes, to get the pistons fully back in place, remove the pads (9-1; Fig. 9.3) first. Carefully push the pistons back in with the box end of a wrench, with the size depending on brake model (8mm for Stroker, 9mm for El Camino, and 10mm for HFX). Avoid pressing on and damaging the pin sticking out of the piston (Fig. 9.3); this engages the wire catch on the back of the pad and holds it in. Replace the pads.

b. Clean the pistons

Sometimes, under normal usage, the brake doesn't retract the pads fully, and they rub. This indicates contamination, and you'll have to clean around the piston to get this to stop.

On most hydraulic disc brakes, each piston is retracted after application by a square-cross-section O-ring seal surrounding the waist of the piston (Fig. 9.4). This square seal or quad ring sits in a groove running around the bore of the piston cylinder; you can see it in cross-section in Figure 9.5. When the lever is squeezed, fluid is forced in behind the piston, the piston moves

9.4 Piston and square seal

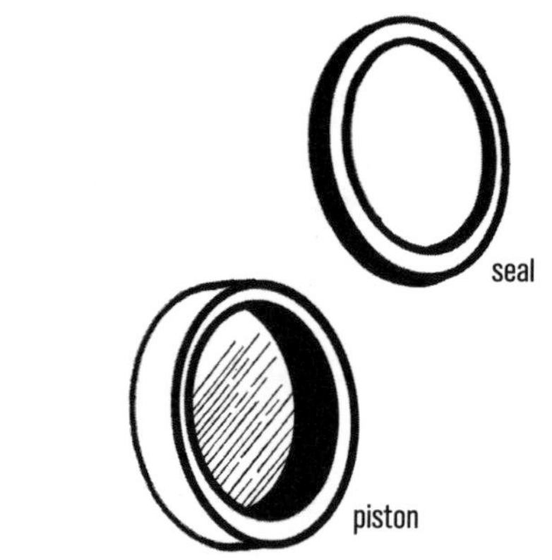

9.5 Caliper cylinder wall, cutaway view

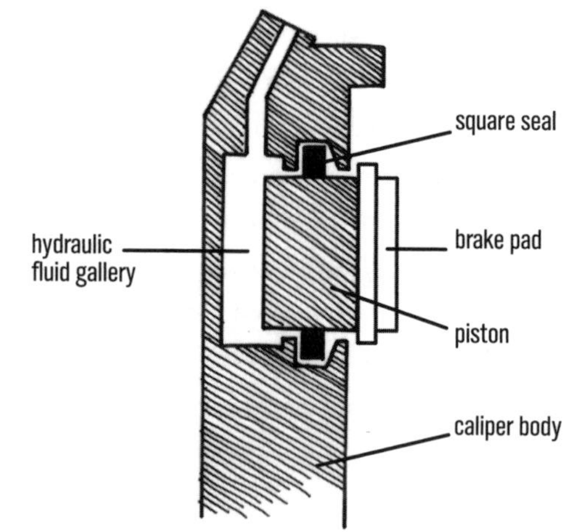

outward, and the square seal twists out into the tapered section of the groove shown in Figure 9.5. When the lever is released, hydraulic pressure is relieved and the square seal untwists back to its original configuration, bringing the piston back with it, as long as the seal is not leaking due to contamination or damage.

If dirt is present, it can inhibit piston retraction, either by breaking the seal or by creating more friction around the sides of the piston than the square seal can overcome. In this case, simply forcing the pistons back into their bores exacerbates the problem by pushing even more dirt under or against the square seal. So clean around the piston and lubricate it.

If you have this problem of pad spacing from the rotor being reduced to almost nothing while riding, remove the pads (9-1). While holding one piston in place with a plastic tire lever or a box-end wrench, carefully squeeze the lever to push its mating piston out a bit more to expose more of it for lubrication. Using a cotton swab soaked in hydraulic fluid of the type that's in your brake, wipe off any grime from around the piston and lubricate it the same way with clean hydraulic fluid.

Carefully prying against the opposite side of the caliper, push the piston back into its bore with the plastic tire lever or box-end wrench. Repeat the procedure to clean and lube the other piston, push it back in as well, and replace the pads.

If you use the wrong implement to push directly on the piston, you can crack it.

Also, if the piston comes out too far and you cannot push it back in far enough against fluid pressure, open the bleed screw slightly while pushing the piston back, and then close the bleed screw immediately to prevent the entry of air.

9-4

INSTALLING AND ADJUSTING DISC BRAKES

LEVEL 2

Simply stated, you just bolt the rotor to the hub, tighten the lever onto the handlebar (see 10-8 for instructions on this), bolt the caliper to the mounts on the frame or fork, and tie down or internally route the hose or cable. But the space between the pads and rotor is small, and the speed of accurate mounting depends on you, the brake, and the type of mount the brake accepts.

The two types of caliper mounts built into mountain-bike frames and forks are International Standard (IS) mounts (Fig. 9.11) and post mounts (Fig. 9.9). IS mounts are drilled transversely (toward the wheel) and are not threaded, whereas post mounts are threaded directly into the frame or fork, parallel to the plane of the bicycle. IS mounts, front or rear, are 51mm apart. Since the 2000 model year, the post-mount standard for frames and forks is 74mm. Older post mounts are less standardized, so be aware of this if you intend to put an old brake on a new frame, or vice versa. Original Hayes and Manitou fork post mounts were 68.8mm apart. For a brief time, Hayes and Manitou adopted 70mm front post-mount caliper spacing but abandoned it within a season. Rear chainstay post mounts and early seatstay post mounts were 21.5mm apart. (Confused yet?)

CAUTION: *Unless you want to do a lot of fiddling and chasing of obsolete parts, don't buy pre-2000 disc brakes on eBay or anywhere else—chances are they will not work with current forks and hubs. In 1997 or thereabouts, many disc-brake and suspension-fork makers agreed on the IS mount of 51mm front and rear, but the agreement did not cover rotor mounting, and manufacturers were all over the map with rotor-mounting systems. So rotors made prior to 2000 other than Hayes will generally not fit on current disc-brake hubs.*

In 2000, fork and disc-brake makers adopted the IS 2000 standard, which also incorporated the six-bolt rotor-mounting pattern (Fig. 9.6) that Hayes had established. And although old Hayes rotors will work on current wheels, Hayes's post-mount caliper dimensions, as you read previously, were all over the place until 2000, so you would at minimum need an obsolete adapter to put a pre-2000 Hayes brake on a fork with IS mounts, and it would not work at all with a current post-mount fork. Also, the Hayes rear chainstay post-mount dimensions have been abandoned.

The good news is that since 2000, the forks, frames, brakes, and rotors of the major manufacturers are completely cross-compatible. This compatibility includes a second rotor-mounting standard, Shimano's Center Lock rotor mount (Figs. 9.7 and 9.8).

Avoid touching the rotor's braking surface and getting grease or oil on it or the pads. If brake performance ever drops off or squealing occurs, clean the rotor and pads with alcohol; if that doesn't fix it, sand the brake pads facedown on a flat piece of drywall-sanding screen. And for obvious reasons, never touch a rotor that's hot after heavy braking.

Never squeeze the lever without a disc or another spacer between the pads, as you can push a piston all the way out. For bike travel with the wheel out, insert a spacer between the pads—either one that came with the brake or a chunk of corrugated cardboard you cut for the purpose.

After installation, follow the pad bed-in procedure in 9-2 to get full brake performance.

a. Rotor installation and removal

Installation

At the turn of this millennium, Hayes's six-bolt pattern (Fig. 9.6) became the standard. However, beginning in 2003, Shimano's brake rotors adopted a splined hub attachment, called Center Lock (Figs. 9.7 and 9.8), which has become a second standard. On a Center Lock brake, the splined aluminum spider adapter riveted to the steel rotor slips onto the splines of the hub, and a single lockring holds it in place. Other manufacturers offer hubs compatible with these rotors, and adapters are available to allow mounting six-hole rotors on Center Lock hubs. (Incidentally, by design, the rotor should still be positioned in the same place relative to the axle end in either case, so a wheel with a Center Lock rotor should work in a brake set up for a bolt-on rotor of the same diameter.)

Six-bolt rotor

1. **Loosely bolt the rotor to the hub flange** (Fig. 9.6). The logo on the rotor should face outward so that the rotor turns in the proper direction.
2. **Gradually snug the bolts, alternately tightening opposing bolts rather than adjacent bolts.** A T25 Torx wrench (like a hex key but with a star-shaped end) is usually required for this. Torque for rotor bolts ranges from 18 in-lbs (2 N-m) for some manufacturers to 55 in-lbs (6 N-m) for others.

9.6 Installing/removing 6-bolt rotor

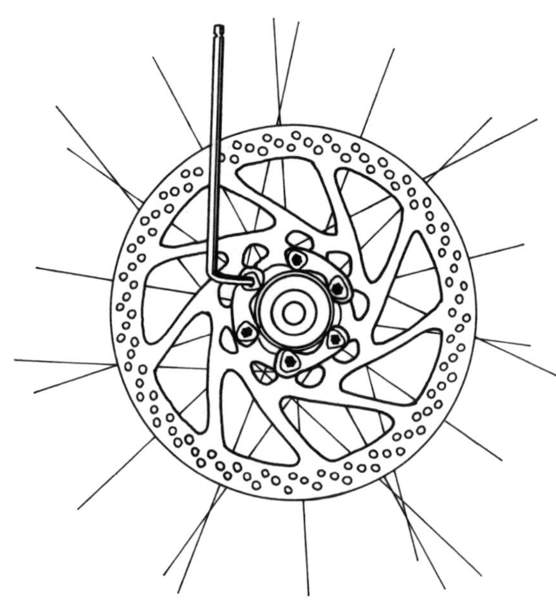

NOTE: *Some older disc-brake hubs have no rotor-mounting flange and require a 6-bolt adapter; the rotor is bolted to the adapter.*

Center Lock splined rotor

1. **Slip the rotor splines over the hub splines.** The logo should be facing you.
2. **Thread on the rotor-securing lockring.** Depending on model, tighten it with either a splined lockring-remover tool for rear gear cassettes (Fig. 9.7), or with splined bottom-bracket wrench for a bigger lockring (Fig. 9.8). If you have a torque wrench that fits the lockring tool, tighten it to 40 N-m (350 in-lbs).

Six-hole rotor on Center Lock hub

You need a Center Lock adapter to attach a 6-hole rotor to a Center Lock hub.

1. **Slide the adapter's flange onto the hub splines.** If it has six pins, those should be facing you.
2. **Place the rotor onto the adapter.** With a DT Swiss adapter (much quicker), you will drop its six holes onto the six pins. There is another type of adapter that also requires no bolts—with it, after you drop the rotor onto the aluminum 6-hole adapter, you drop a steel ring with six pins pointing down through the rotor holes and into the six holes of the adapter.

9.7 Installing/removing Shimano Center Lock splined rotor onto the splines of a Center Lock hub with a cassette lockring tool

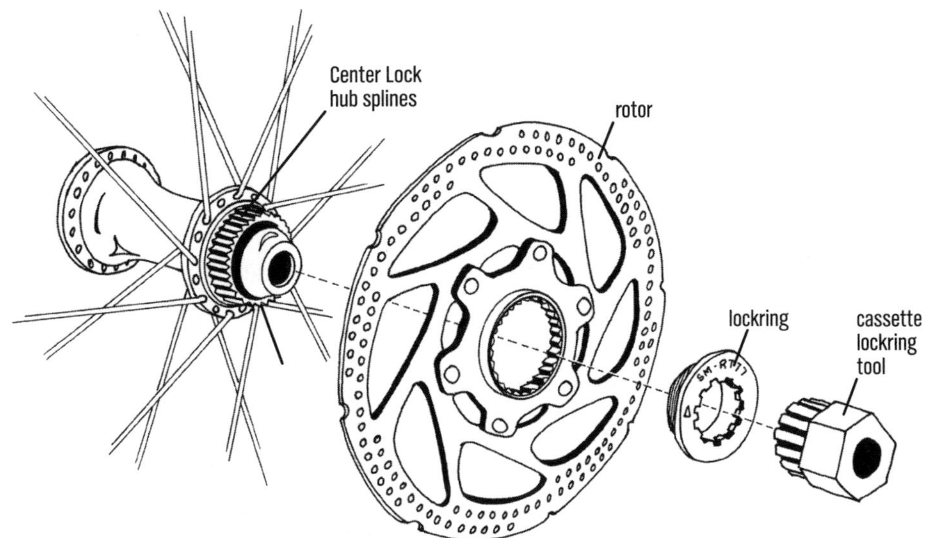

9.8 Installing/removing Center Lock splined rotor onto the splines of a Center Lock hub with a splined bottom-bracket wrench

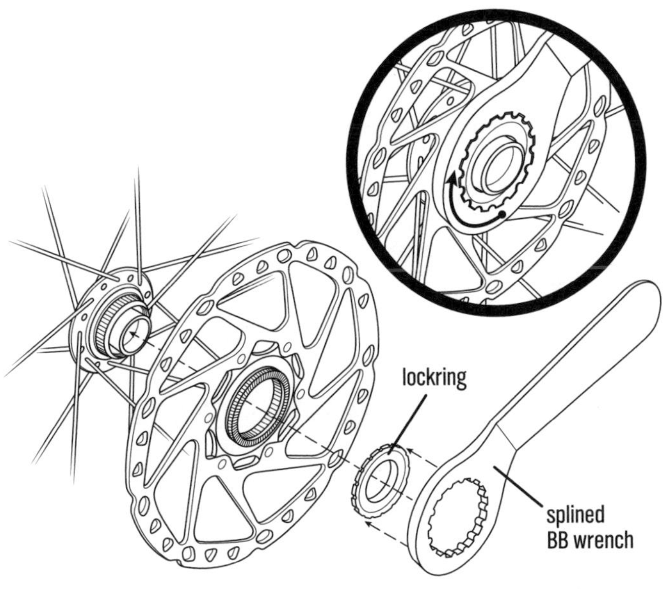

With the Shimano adapter, bolt the rotor onto the adapter's threaded holes with six bolts and a Torx T25 wrench. The rotor's logo should face you.

3. **Secure the rotor with the lockring.** Depending on lockring, use a cassette lockring tool (Fig. 9.7) or use an external-bottom-bracket splined wrench (Fig. 9.8); tighten to 40 N-m.

Removal

Multiple-bolt rotor

When removing a bolt-on disc rotor, you must loosen all the screws a fraction of a turn before unscrewing any of them. On braking, the rotor may rotate relative to the hub a bit and lean against one side of each screw. If you remove one screw while the others are still tight, the rotor hole's wall will still be pressed against the side of the screw, and the screw threads will be damaged. You will then wreck the threads in the hub when you force the damaged screw back in. So loosen them all slightly first, and then unscrew completely.

If you have an old Shimano 6-bolt rotor with the plates bent up around the triangular bolt heads (Fig. 9.6), bend the plate edges back down before unscrewing the bolts.

Center Lock rotor

Unscrew the lockring with a cassette lockring-removal tool (Fig. 9.7) or an external-bottom-bracket cup tool (Fig. 9.8), and pull the rotor off.

b. Post-mount disc-brake caliper installation

The beauty of post-mount brakes (Fig. 9.9) is that the brake can be moved laterally without the necessity of installing shims (thin washers) between the caliper and the mounts (as must be done with an IS brake on IS mounts; see section c below). This advantage applies even if you are mounting an IS brake via an adapter to a post-mount frame or fork, or vice versa.

1a. **If you are using a post-mount caliper on an IS frame or fork, tighten the correct adapter bracket to the IS mounts first** (Fig. 9.10). Torque is usually 6–8 N-m.

1b. **If you are using an IS caliper on a post-mount frame or fork, tighten the correct adapter bracket to the caliper first.** Torque is usually 6–8 N-m.

2. **Loosely bolt the post-mount caliper to the post mounts on the fork** (Fig. 9.9)**, frame, or adapter bracket** (Fig. 9.10)**.** If the brake has concave and convex washers, keep them in the same order as in Figure 9.9.

NOTE: *If you are installing a hydraulic caliper that is not connected to its lever, skip to section e to cut the hose to length, then skip to 9-6 to fill it with fluid and bleed it, and then come back here and begin with step 3.*

3. **Install the wheel.** The caliper slot will be over the rotor, and the caliper will have some lateral freedom of movement.

4. **While squeezing the brake lever, shake the caliper to get it to find its natural position over the rotor, and tighten the mounting bolts.** Alternate tightening each bolt a bit at a time to avoid twisting the caliper.

With a cable-actuated brake (Fig. 9.9), first turn the adjuster screw on the wheel side clockwise a few clicks to bring the inboard pad closer to the rotor. Even if there is a knob on the adjustment screw, it is easier on the knuckles to use the Torx T25 key or hex key it requires through the spokes to turn the screw. Then turn the outboard pad adjustment screw (if present; Fig. 9.9) clockwise until it stops (otherwise, pull the cable tight by pulling the lever), which will cause the pads to squeeze the rotor from both sides. Now tighten the mounting bolts to the correct torque setting. Ultimately, you want the spacing even on both sides or a third of the gap between pads and rotor to be on the inboard side and two-thirds of it to be on the outboard side, so after hooking up the cable (10-6) and tightening the cable anchor bolt to torque,

9.9 Mounting a cable-actuated post-mount brake caliper on fork post mounts

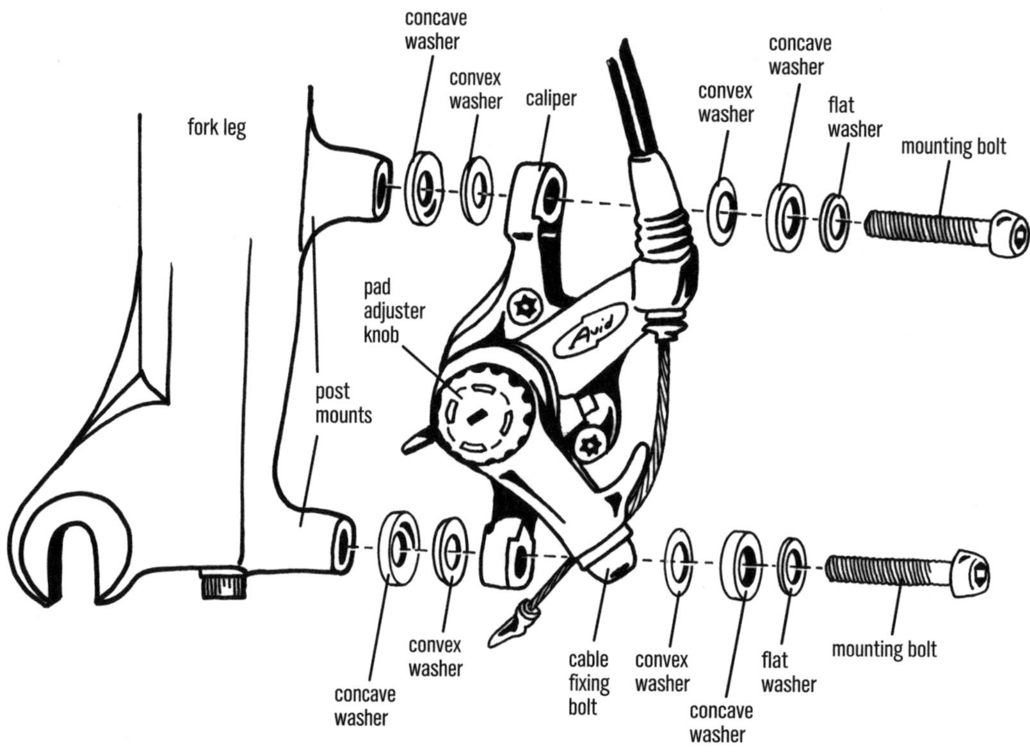

9.10 Mounting a hydraulic post-mount caliper on IS mounts

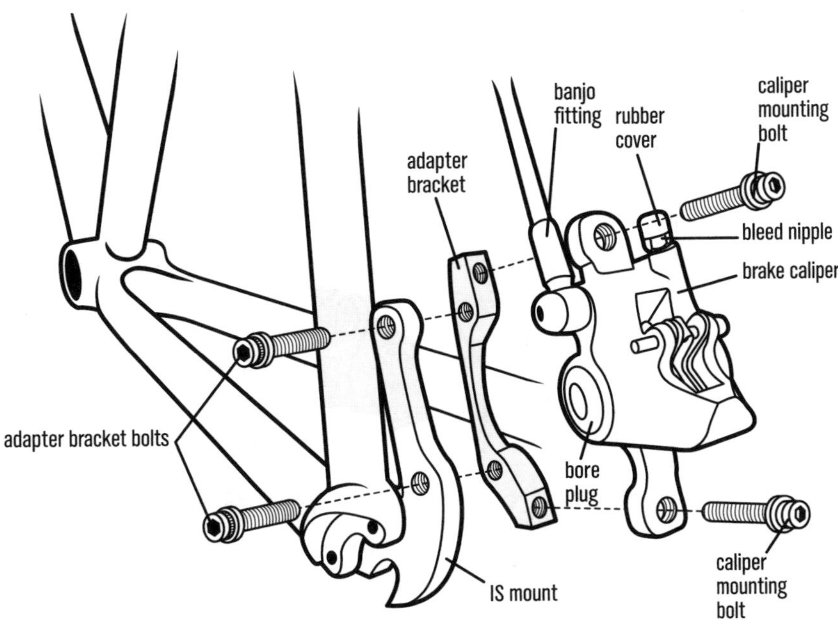

back out the two knobs (or the cable tension and inboard knob) appropriately to achieve this. Be aware that, while hydraulic disc-brake pads self-adjust to maintain the same spacing to the rotor as they wear, cable-actuated ones do not. As the pads wear, adjust the pads of brakes with dual adjuster screws (Figs. 9.2 and 9.9) farther toward the rotor by turning in (clockwise) both adjuster screws at the caliper, rather than by tightening the cable at the fixing bolt or barrel adjuster. On brakes without an outboard adjuster screw, turn in the inboard adjuster screw and tighten the cable at one of the barrel adjusters.

5. **Spin the wheel to check for brake rub.** If you hear rubbing, peer through the gap between the rotor and the pads with a white background for contrast. Note which pad (or worse, which side of the caliper slot) is rubbing. Loosen the bolts again, and slip a business card or two between the rubbing pad and the rotor.
6. **Repeat steps 4 and 5 until the rotor spins without rub.** If you're desperate, just loosen the bolts (maybe one at a time and then push the caliper with only one bolt tight), eyeball the gap, push the caliper as you see fit, and tighten while holding the caliper.
7. If the rotor is bent, straighten it. See 9-5 on rotor truing.

Shimano supplies little plastic clips to snap over the bolts around the knurled part of their head to prevent the bolts from unscrewing.

NOTE ON CENTERING BRAKES WITH ONLY ONE MOVING PAD: *Some hydraulic disc brakes and almost all cable-actuated disc brakes work by flexing the rotor toward a fixed pad. The above procedure will work for centering hydraulic ones, provided you first tighten the fixed-pad adjuster screw about a half turn. Once the caliper is bolted in place, back out the fixed-pad adjuster screw so the pad just barely clears the rotor.*

c. IS disc-brake caliper installation onto IS mounts

IS brake calipers have two transverse-drilled threaded bolt holes and attach directly to IS mounts (Fig. 9.11).

NOTE: *If you are using a caliper adapter bracket to mount a post-mount brake caliper to an IS fork or frame, first bolt the adapter to the frame or fork mounts, and then follow the directions in section c for installing a caliper onto post mounts. If you are using a rotor that is a different size from the rotor that the brake came with, you will need an adapter.*

ANOTHER NOTE: *If you are installing a caliper that is not connected to its lever, skip to section e to cut the hose*

9.11 Mounting a hydraulic IS brake caliper on IS mounts

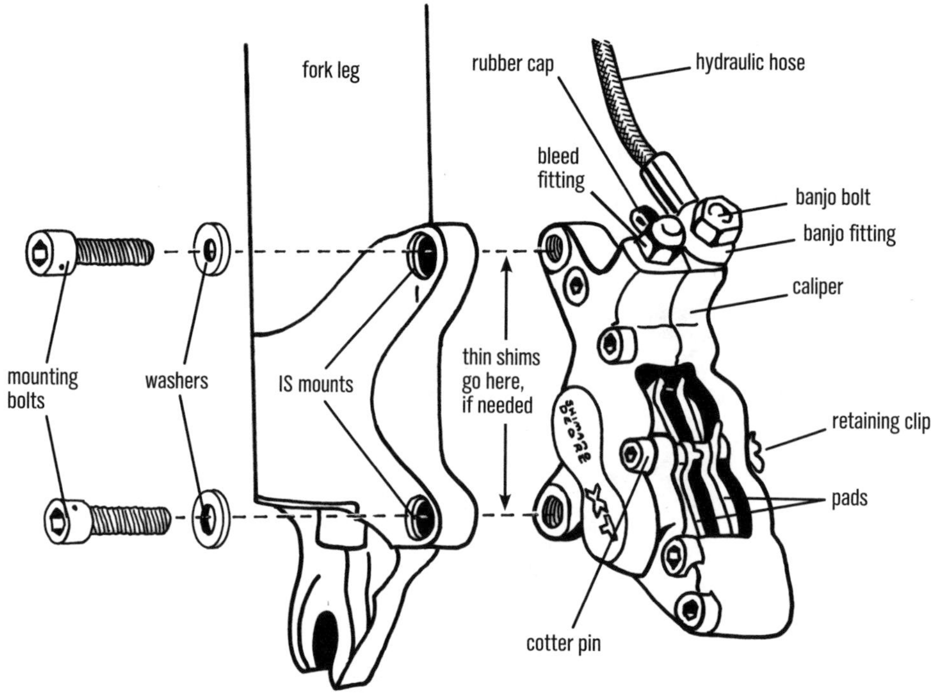

to length, then skip to 9-5a to fill it with fluid and bleed it, and then come back here and begin with step 1.

1. **Install the wheel on the bike.**
2. **Slip the caliper over the rotor and up against the frame or fork mounts.**
3. **Loosely install the mounting bolts** (Fig. 9.11).
4. **Pull the brake lever to squeeze the pads against the rotor.**
5. **Measure the gap between the caliper and the mounting tab at each bolt while squeezing the brake lever.**
6. **Make a stack of shim washers adding up to that measurement.**
7. **Unscrew the mounting bolts, and slide each stack of shim washers on its bolt between the caliper and mount tab as indicated in Figure 9.11.**
8. **Tighten the bolts.** It's best to apply a counterclockwise torque on the caliper with your hand while tightening the bolts. This will take out the play so that the rotor won't twist the caliper on its mounts on the first hard brake application. Bolt-tightening torque varies from 53 to 110 in-lbs, depending on brand.
9. **Spin the wheel.** The pads should not rub. If they do, add or remove shims until the rubbing has been eliminated.
10. **If the rotor is wobbly, straighten it.** See 9-7 for instructions.

NOTE: *Some IS brake calipers have only one moving pad; they flex the rotor toward the stationary inboard pad. The stationary inboard pad is adjusted independently with a 5mm hex key or with a thumbscrew so that it just clears the rotor without rubbing.*

Some IS brakes use a floating caliper in which the entire caliper moves as the pad (or pads) on the outboard side push(es) against the rotor and pull(s) the stationary pad (or pads) over to the rotor. It is almost impossible to eliminate pad rub with these brakes. The same is true with a floating rotor: The caliper is fixed, but the rotor slides laterally on plastic bushings. These old designs are now rare.

d. Hookup of cable-actuated disc brakes

Route the cable housing to the brake following the procedures in 10-6 for cable installation. Secure the housing with zip-ties where there are no cable stops or hose/housing guides. Push the cable through the housing stop on the caliper, and tighten it under the cable-fixing bolt.

e. Cutting hydraulic disc-brake hoses to length

When you route the hose to the brake, curve it smoothly without kinks and without large loops that can catch on things, and make sure the hose is not so short that it is tight across spans where it is vulnerable to damage, particularly when the suspension moves.

If the frame has internal hose routing, do not cut the hose to length until after you have routed it through the frame—this reduces the likelihood that you will have to bleed the brake. You can use Park's IR-1 or IR-1.2 internal wire guide tool as in Figure 5.38; one of the guide wires has a conical auger tip you can screw into the end of the hose to pull it through, following the internal routing instructions in 5-15. Or, if it's a new frame with temporary guide tubes running through it, you can slide a cable through the guide tube and then through the hydraulic hose. Pull the temporary tube out while pushing the hydraulic hose in with the cable still bridging them so the hose can follow; you will have to bleed the brake, as the cable will have displaced some fluid.

If the frame has external disc-brake hose guides, use those. Otherwise, secure the hose to the frame or fork with zip-ties, tape, guides that clip or screw into cable guides, or adhesive-backed hose guides.

If the hose is too long and one end of it has a permanent end crimped to it, do not cut that end. Generally, on an end you can cut, there will be a brass, olive-shaped ring around the end of the hose (Fig. 9.12); a sleeve nut squeezes the olive to seal against leaks. The brass olive will need to be replaced after you cut the hose. Do not cut hydraulic hoses sheathed in braided stainless-steel wire; these must be purchased in the correct length for the frame.

Whenever you cut a hose, keep the end of the hose as well as the connection port on the lever or caliper pointed up to avoid losing fluid and hence creating an air bubble once the hose is connected.

1. **Remove the wheel and the brake pads** (9-1). You do not want to get brake fluid on the rotor or the pads.

9.12 Hose sealing system: the sleeve nut squeezes the brass olive, which squeezes the hose supported by the barbed fitting, which either pushes or screws into the end of the hose.

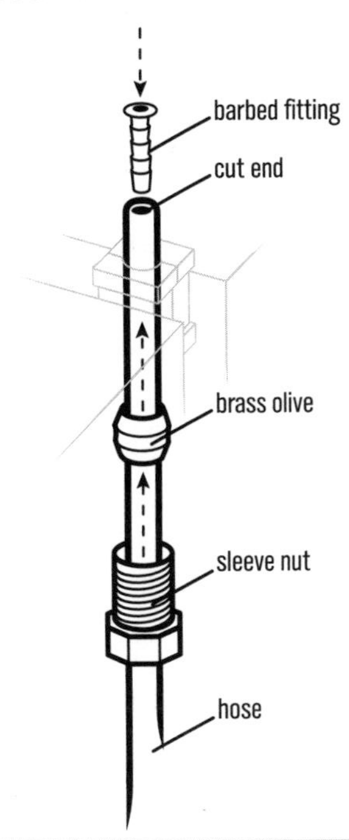

9.13 Inline hose connection at caliper

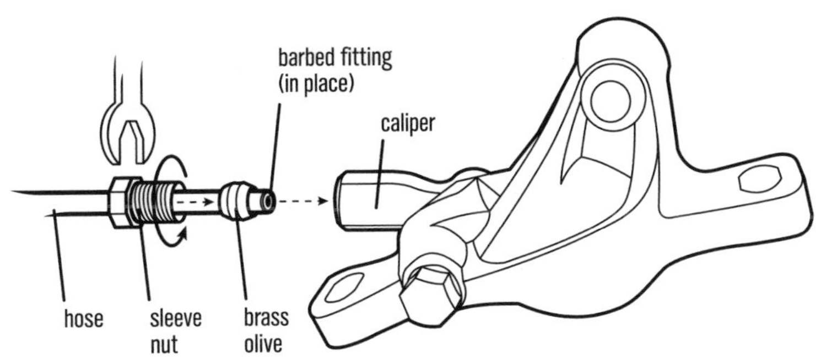

2. **Disconnect the hose from the fittings.** Do this at an inline connection at either the lever (Fig. 9.15) or the caliper (Fig. 9.13), but not at an end that has a crimped fitting, like at a banjo fitting (Fig. 9.14). Unscrew the sleeve nut (Fig. 9.13), which is often covered by a plastic or rubber cover; slide that up the hose for access (Fig. 9.15).
3. **Gently pull the hose straight off.** Be very careful, because on some brakes you may break a thin, barbed nipple that runs up into the tube. As part of the sealing system, the thin barbed nipple extends up inside the hose under the brass olive ring, and

9.14 Banjo hose connection at caliper

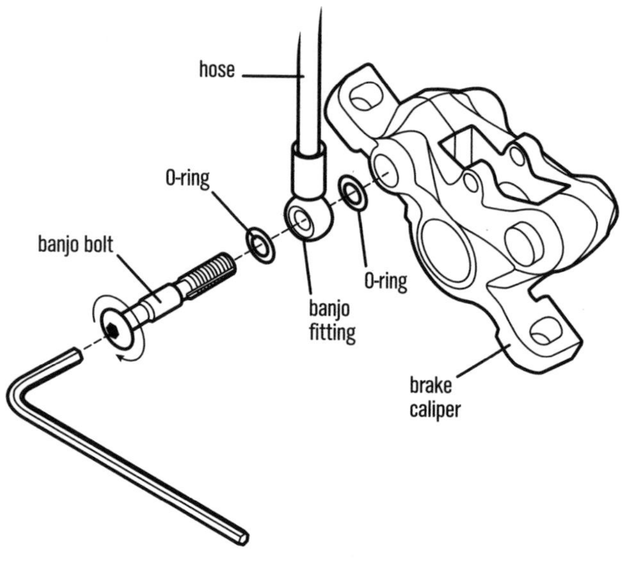

the olive, compressed by the sleeve nut, tightens the hose around it. This barbed nipple is generally a separate piece that slides up or screws into the hose as in Figure 9.12. On some brakes this barbed nipple is part of the lever, and if you bend the hose sideways while you pull on it, you can break it off. Instead, pull the hose straight off.

4. **If the hose did not pull off easily, carefully cut the brass olive open with a hacksaw.** Peel it away from the hose, and pull the hose off without damaging the nipple. This applies in the case of old Hayes levers that have an integrated barbed nipple inside the threaded hole.

5. **Slide the sleeve nut and rubber cover up the hose beyond where you plan to cut.**

6. **Cut the hose to length with a sharp knife, making a clean, perpendicular cut.** Many brakes come with a pair of grooved plastic blocks with which you can clamp the hose in a vise while cutting. Ideally, use a hose-cutter tool. Again, tip the hose end up to prevent loss of fluid and consequent entry of air into the system.

7. **Slip the new brass olive over the end of the hose.** The hose nut and rubber nut cover (Fig. 9.15) should be further up the hose already.

8. **Install the barbed fitting.** If the brake has a separate barbed fitting, tap it into the hose while it is still held between the plastic grooved blocks. Or, if it is threaded externally and has a Torx hole in the end, screw it into the hose with a Torx key.

9. **Tighten the sleeve nut (use a maximum torque of 40 in-lbs).** Slide the plastic or rubber nut cover back into place over the nut.

10. **Smooth the hose routing.** Seek the smoothest, least vulnerable path from lever to caliper. Some brakes allow adjustment of the angle of the banjo at the caliper. Loosen the banjo bolt (Fig. 9.14) a quarter turn or so (unscrew it too much and you could lose oil and/or allow air in), rotate the banjo to get the hose path you want, and tighten the banjo bolt.

11. **Confirm that the hose routing works.** Make sure that the hose does not get yanked when you turn

9.15 Shimano XT Dual Control master cylinder, exploded view

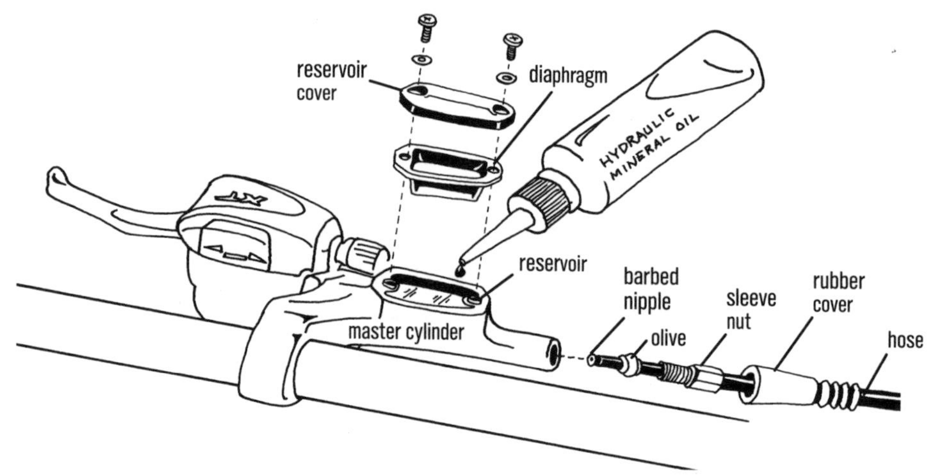

the handlebar or when the suspension moves up and down.

12. **Skip to 9-6 for bleeding/filling.** However, if the brake was previously connected and properly bled, and you think you might have prevented the entry of air into the system by keeping the hose end and connection port for it both pointed up, then install the pads and wheel and squeeze the lever. If the lever feels firm, you're done. If it doesn't, there is air in the line, and you must bleed the air from the system (9-6).

f. Lever reach, lever pull, and pad spacing

The reach adjustment may consist of a small setscrew under or on the front of the lever blade, perhaps under a dust cover, or it may be a knob on the lever blade. This may or may not be independent of any lever pull and pad spacing adjustments.

Lever pull and pad spacing are closely related, as the closer the pads are to the rotor, the less pull it takes to stop. But the pads will rub if too close. Adjustments exist on some brakes but not on others.

Lever pull and reach with cable-actuated brakes can be adjusted by the cable barrel adjuster and reach adjustment screw on the lever (see 10-9 for details). Pad spacing (which affects lever pull) on cable-actuated disc brakes can be adjusted with a knob or screw on the wheel side of the caliper. Same goes with hydraulic brakes with only one moving piston.

Some hydraulic brakes have an adjuster screw or knob on the lever that pushes the master-cylinder piston, thereby changing the amount of lever travel required. Some have a knob on the lever or reservoir to adjust the pad contact point in the lever's travel while riding. Some brakes have a knob to adjust leverage that actually moves the lever pivot in and out via a cam mechanism; changing leverage changes lever pull.

On brakes with both a reach adjustment and a pad position/application adjustment, don't adjust the reach to the closest position as well as the pad position to its maximum firmness. This could result in insufficient lever movement to achieve the necessary brake power to safely control your bike.

9-5

TRUING DISC-BRAKE ROTORS

Even when you get a caliper perfectly centered over a rotor (disc), the brake can squeal and howl if the rotor gets bent. The spacing between brake pads and rotor is so tight—around 0.015 inch (0.4mm)—that there is little room for rotor wobble. Bicycle rotors are thin and relatively unprotected; they can be bent by rocks thrown up while you are riding, when you are in a crash, or when you are packing your wheel in a car or a bike bag. They can also warp from heat buildup on a long, steep descent on a hot day. One way or another, the rotors will likely get bent eventually, so you need to be able to straighten them.

If a rotor is really potato-chipped, you may first need to remove it from the hub and pound it as flat as you can with a hammer on an anvil. Then you can proceed with any of the methods detailed here.

a. Eyeballing the rotor in the caliper

You can often do an adequate job by eye to at least minimize brake-pad rub, but be forewarned that this approach requires patience; it can be hard to tell on which pad the rotor is rubbing as the gap is so small. Place a piece of white paper on the floor or the wall, below or level with the caliper, so that you can see the space between the rotor and the pads. Slowly turn the wheel, marking with a felt-tip pen where the disc rubs on each pad. Carefully bend the disc into alignment with your fingers, rechecking it constantly by spinning it again through the brake.

b. Rigging up a pointer

A rotor-truing gauge like the Park DT-3 that clamps to a truing stand makes this job a snap.

A truing method almost as cheap as eyeballing and more accurate (and one that requires as much patience) is to attach a pointer to the frame or fork in such a way that you can adjust it to graze the rotor as you would a feeler on a truing stand. This pointer may be made of wire bent around the caliper. You can also remove the caliper and screw a piece of metal with a hole in it onto one of the caliper-mounting tabs. Bend the rotor away from the pointer where it touches as you rotate the wheel. The pointer must be mounted securely for this method to work; otherwise, you can make mistakes, thinking the disc is bent one way when it is actually bent the other.

c. Truing a rotor with a dial indicator

If you straighten rotors a lot, eyeballing the rotor or using a rigged-up pointer will soon drive you mad. Happily, a dial indicator tool (Fig. 9.16) shows the lateral position of the rotor within 0.001 inch (0.025mm), so you can get the rotor as straight as it was when it was brand new. The Park DT-3i.2 clamps onto the DT-3 on a truing stand.

Set the dial indicator foot against the rotor. Rotate the indicator face cover so that the needle is on zero; wherever the needle indicates the greatest deflection in either direction, bend the rotor back, continually rechecking it with the dial indicator.

You can bend the rotor back with your thumbs. Better yet are rotor tuning forks (Fig. 9.16), which slip onto the rotor and provide leverage to precisely bend the rotor. You can use a single fork and do a good job. To more precisely locate your bend, stabilize the rotor in position as in Figure 9.16 by using three forks: use two forks, one on either side of the bent spot, and then bend the rotor with the third fork to eliminate the warped spot.

9.16 Truing a bent rotor with a dial indicator and rotor-tuning forks

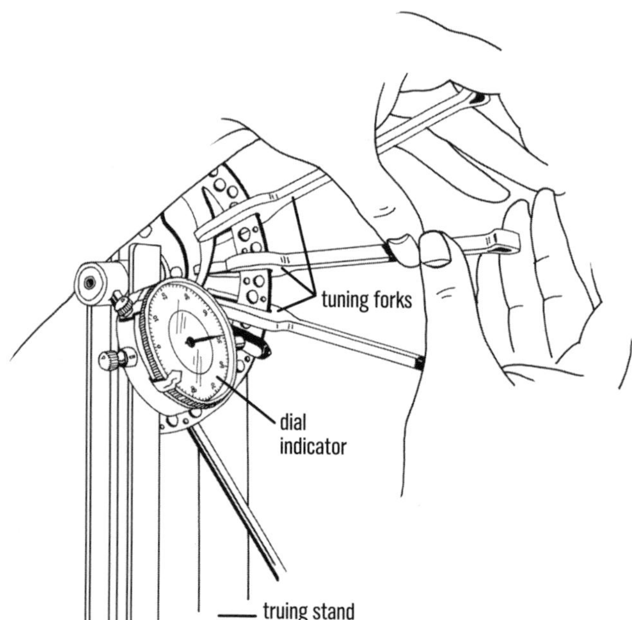

9-6

BLEEDING (OR FILLING) HYDRAULIC DISC BRAKES

LEVEL 2

Brakes must be bled whenever they have air in the system. The symptom is a lever that is not firm when pulled and/or that becomes more firm with repeated pumping. Separately, given enough usage in dirty conditions, dirt can get past the seals and contaminate the fluid, so flushing the old fluid out with new fluid will improve performance.

Bleeding brakes is not difficult, but it does require patience and extreme cleanliness. Work in a well-lighted area with clean tools. Place a clean sheet on the floor below the bike to catch any dropped parts. Observe the notes and cautions regarding DOT fluid and exercise care at all times.

The procedure for filling an empty brake system is the same as for bleeding one. In general, you move fluid down through the system by filling the reservoir at the top and forcing fluid down with the lever or by sucking from a syringe at the caliper, or you force fluid up from a syringe or squeeze bottle through the caliper to the lever reservoir, or you do a combination of both. Air bubbles float up to the top of the fluid to the reservoir.

It is unrealistic to include complete bleeding instructions here for every brake. With one of the following four methods, you should be able to bleed almost any brake, but unless you have one of the brakes specifically described here, it is preferable to follow specific instructions that come with a brake's bleed kit.

With all brakes:

1. **Remove the wheel.**
2. **Remove the brake pads** (9-1).
3. **Install a spacer block between the pistons** (Fig. 9.17). The spacer allows you to apply hydraulic pressure while keeping the pistons pushed back in their bores. Many brakes come with a bleed spacer.

IMPORTANT—PAD PROTECTION: *Avoid getting fluid on the pads, which will ruin them. Replace pads contami-*

9.17 Installing bleed block into caliper in place of brake pads

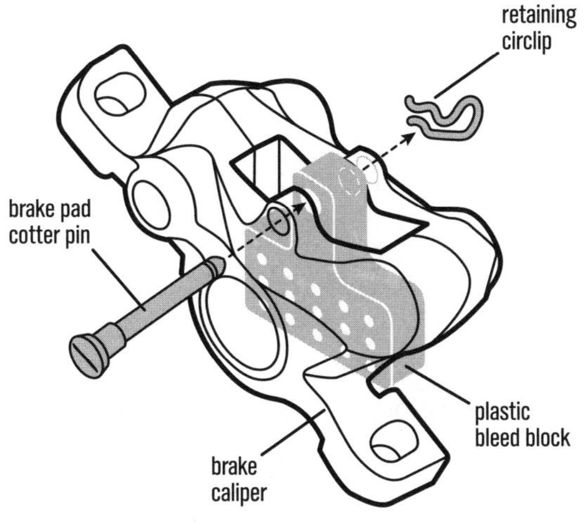

nated by brake fluid, and clean rotors contaminated by brake fluid with rubbing alcohol.

IMPORTANT: FLUID TYPE: *Use the recommended brake fluid for your brake. Some systems use mineral oil, and some use DOT (automotive) brake fluid (DOT stands for Department of Transportation). Do not interchange mineral oil and DOT fluid in a brake; doing so will ruin the seals inside.*

Not all mineral oil is the same (viscosity, purity, and boiling point vary), nor is all DOT fluid the same. DOT has a standardized numbering system. The higher the DOT number, the higher the boiling point, and your brake was designed to operate in a certain temperature range with a DOT fluid for that range. If you were to use DOT 3 fluid, for instance, in a brake designed for DOT 5.1, you might be without brakes when you need them the most—when they get really hot under heavy braking (see following warning).

WARNING: BOILING FLUID: *With brakes using DOT fluid, add fluid only from a container that has never been opened before. DOT fluid absorbs water, and the more water it has absorbed, the lower its boiling point. Opening the container for a short time can be enough to bring the boiling point down significantly (if you leave a full glass of DOT fluid out overnight in a humid area, it will overflow the glass by morning!).*

Why is the boiling point of the fluid important? A hydraulic brake works because liquids are essentially noncompressible, so pushing on a piston at one end of a column of liquid (at the lever) can push a piston just as forcefully at the other end of the column of liquid (at the caliper). Gases, in contrast, are compressible; that's why you have compressed air in your tires and front and rear shocks. But if hydraulic fluid boils, gas bubbles form in the hydraulic lines, and pulling the lever will only compress the gas; it won't push the caliper pistons.

Vapor lock occurs when the caliper gets so hot that the fluid inside boils. Let go of the lever so it can cool enough to brake for the next turn. Once vapor lock happens, you need to replace the fluid in a DOT-fluid brake with new DOT fluid. With mineral oil, you need only let the brake cool down, because oil doesn't absorb water.

IMPORTANT: *DOT fluid can dissolve paint, so wipe it off wherever it drips on the bike, and rinse immediately with water or isopropyl alcohol.*

a. Modern Shimano brake bleed

You will need a Shimano bleed kit for this procedure, which includes Shimano brake oil, a syringe, clear tube, oil funnel, and plug.

1. **Rotate the brake lever to horizontal.** Snug up the lever clamp bolt to keep it level through this procedure. Secure the bike in a position such that the hose trends downward the entire way to the caliper. You can unbolt the caliper so that it hangs by the hose.
2. **Remove the lever bleed screw** (Fig. 9.18).

9.18 Bleed funnel installed in a Shimano lever bleed port

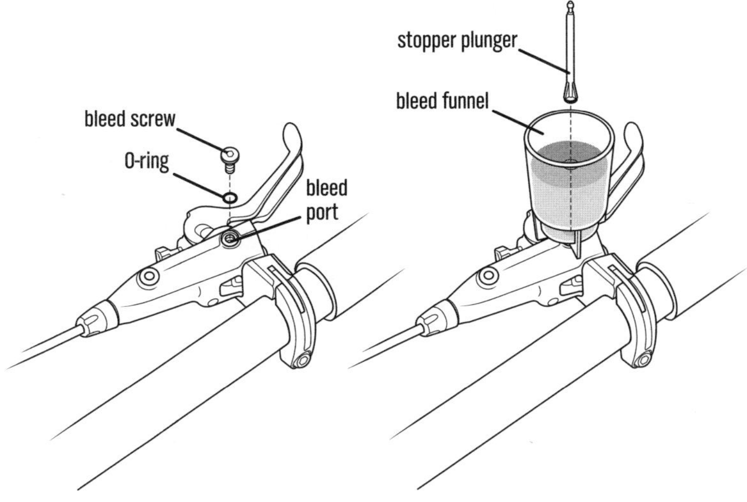

9.19 Bleed syringe connected to a Shimano caliper bleed nipple

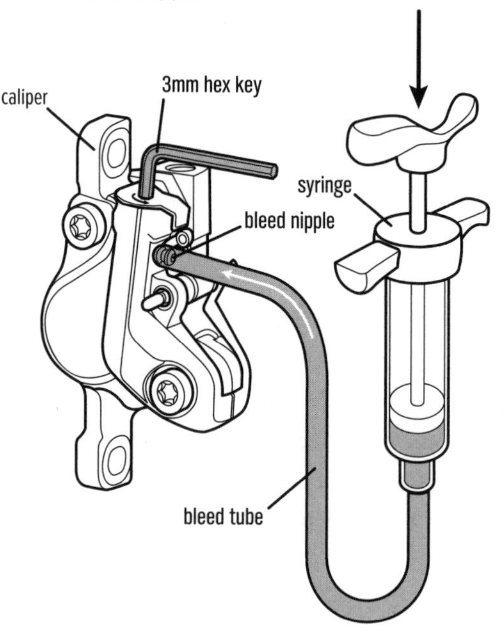

9.20 Catch bag connected to Shimano caliper bleed nipple

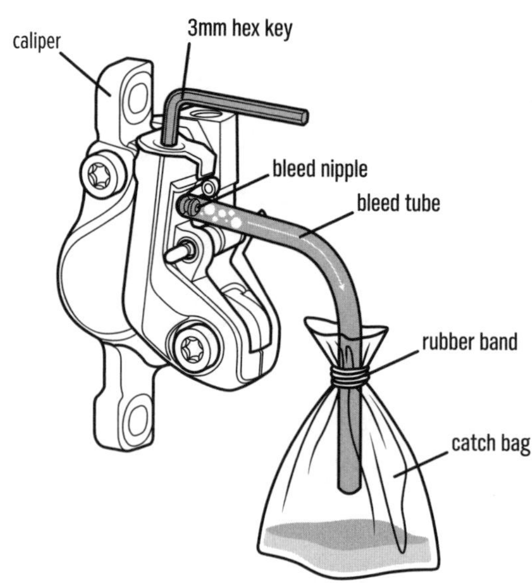

3. **Install the bleed funnel.** Thread the funnel's tip into the bleed hole atop the lever (Fig. 9.18).
4. **Fill the syringe with oil.** Install the clear bleed tube on it, and suck oil out of a bottle of Shimano brake oil.
5. **Attach the syringe to the caliper bleed nipple.** Flip open the rubber cap covering the bleed nipple, and connect the tube to the bleed nipple (Fig. 9.19).
6. **Slightly open the caliper bleed screw.** It is above the bleed nipple; loosen it one-eighth turn with a 3mm hex key (Fig. 9.19).
7. **Push oil up to the lever.** Push the syringe plunger (Fig. 9.19), watching oil (and air bubbles) appear in the funnel (Fig. 9.18). Continue to push new oil up until you see no more air bubbles coming up.
8. **Gently tighten the caliper bleed screw.** This prevents oil or air from going in or out of the bleed nipple.
9. **Pull the syringe off of the clear tube.** Cover with a rag.
10. **Put the catch bag on the tube.** Rubber-band the plastic bag on (Fig. 9.20).
11. **Allow oil to gravity-flow into the bag.** Again, loosen the caliper bleed screw one-eighth turn with a 3mm hex key (Fig. 9.20). Some air bubbles will accompany the oil running down from the funnel into the bag. Until air bubbles cease to appear in the bag, keep adding oil to the funnel; if you allow the funnel's throat to be exposed, you will let air into the lever.
12. **Gently tighten the caliper bleed screw.**
13. **Squeeze air out of the caliper.** Hold the brake lever depressed while repeatedly opening and closing the caliper bleed screw for about a half second each time. Air bubbles released from the caliper will appear in the bag. Repeat a few times until you're convinced no more air bubbles will appear.
14. **Gently tighten the caliper bleed screw.**
15. **Pump the brake lever.** Air bubbles will pop up in the oil in the funnel. When you stop seeing bubbles, pull the brake lever as far as it will go; the lever should feel firm.
16. **Rotate the lever up and down and pressurize the closed system.** While making sure that there is sufficient oil in the funnel to always keep its throat covered, tilt the brake lever first up 30 degrees or so, and then down 30 degrees or so. In each position, pump the lever, looking for air bubbles appearing in the funnel. Air can hide in a corner right by the bleed port, and this step is key to getting it out.

Repeat until you're convinced no more air bubbles will appear. Gently plunge with the funnel stopper (shown in Fig. 9.18) to get a few more tiny bubbles out. Squeeze the lever fully; feel if it is any firmer.

17. **Plug the funnel with the stopper.** Check that its O-ring is in good condition so oil won't leak out.
18. **Remove the funnel and close the bleed port.** Keep the stopper in the funnel's throat while you remove it. Replace and tighten the bleed screw while oil is domed up atop the bleed hole. Wipe off oil drips from the caliper, lever, and hose.
19. **Check the system.** Remove the bleed block, replace the pads, install the wheel, and check that the brake operates properly. Pat yourself on the back if it does.
20. **Zip-tie the lever around the handlebar overnight to check for fluid leaks.** If the zip-tie is still tight around the lever and the handlebar grip the next morning, there are no leaks.

b. Vacuum-bleeding Avid and SRAM brakes

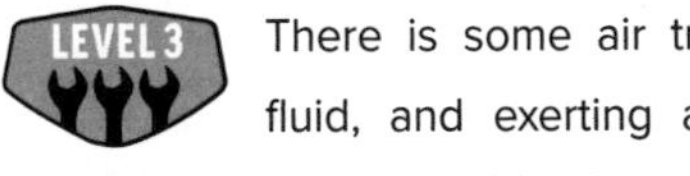

There is some air trapped in any DOT fluid, and exerting a vacuum over the fluid, as in the following bleed method, can draw some of that air out. The SRAM bleed kit is required for this operation and has a pair of syringes with screw-on fittings and hose clamps.

1. **Draw fluid up into the two syringes.** Use Avid/SRAM DOT Fluid, ideally from a previously unopened container (see note on boiled fluid above); DO NOT use any fluid that is not DOT 4 or DOT 5.1. Fill one syringe half full, and fill the other syringe one-quarter full.
2. **Expel the air from the fluid in both syringes.** Holding a rag over the end of the syringe hose, point each syringe up, push the plunger, and expel any air bubbles.
3. **Draw dissolved gas bubbles out of each syringe.** Close the clamp on the hose and pull the plunger so tiny champagne-like bubbles appear in the fluid and get larger the harder you pull (Fig. 9.21). Don't pull hard enough to suck air in past the plunger or past the clamp on the hose. While still pulling the plunger, repeatedly flick your finger against the syringe to free any bubbles sticking to the sides and encourage them to float to the top. Open the clamp and expel collected air as in step 2. Repeat a few times, but don't try to remove all of the bubbles, because there will be no end to it.

9.21 De-gassing DOT fluid

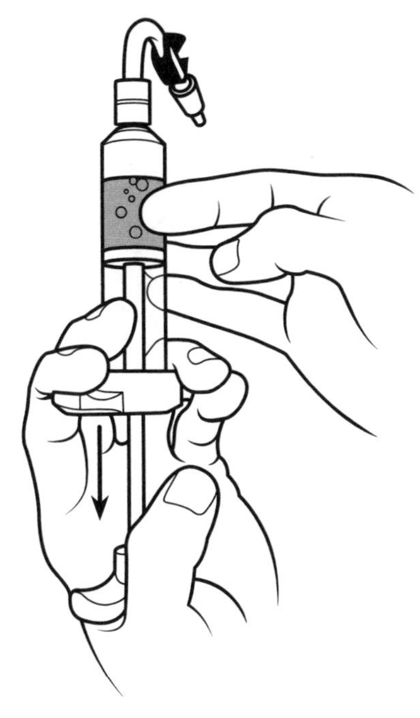

4. **If the brake lever has a pad-contact knob, turn it all the way to the "out" position.** You will be turning the opposite direction of the inscribed arrow. Also, turn the lever-reach adjustment out to maximum reach.
5. **Screw the syringes onto the bleed ports.** Using a Torx T10 key, remove the bleed screw from the center of the banjo bolt on the side of the caliper, then screw in the half-full, degassed syringe. Remove the T10 bleed screw from the top of the lever body, and screw in the quarter-full, degassed syringe. Wipe away any fluid that may have dripped before it gets on paint or skin.
6. **Push fluid up from the caliper.** Holding both syringes upright, push the caliper syringe plunger (Fig. 9.22) to force half of its fluid up through the system to the lever syringe, along with any air bubbles. If the fluid coming out at the lever is dark (dirty), refill and de-gas the syringes again, and push half of the caliper syringe fluid up again, ensuring that

9.22 Pushing and pulling on the plunger of the SRAM bleed syringe on the caliper

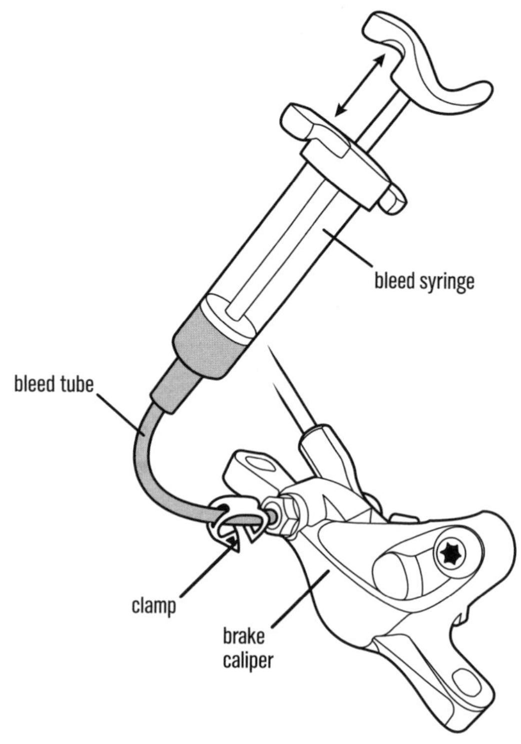

9.23 Pushing and pulling on the SRAM bleed syringe on the lever

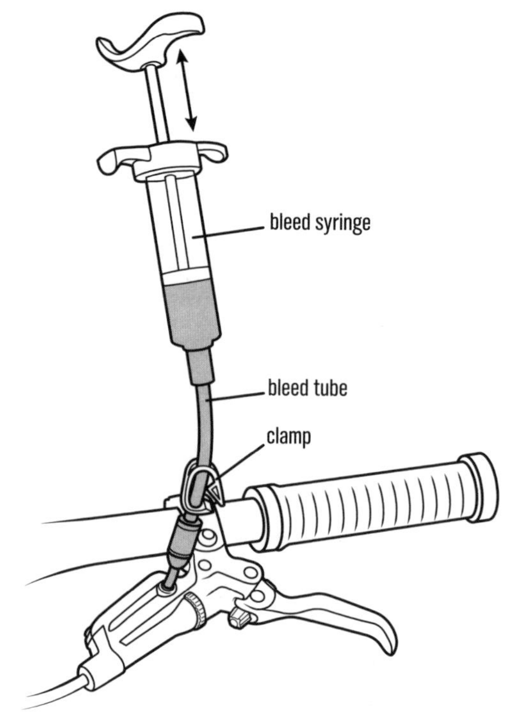

the fluid appearing in the lever syringe is clean. Close the clamp on the lever syringe.

7. **Pull the brake lever to the handlebar and secure it with a strap or rubber band.** This closes the hole connecting the reservoir and the master cylinder.
8. **Suck the air out of the caliper.** Without pulling so hard that you pull air past the plunger seal, pull a vacuum on the caliper syringe's plunger with the syringe upright (Fig. 9.22), drawing air bubbles up through the fluid in the syringe. Push and pull the syringe plunger a few times until bubbles stop coming up the syringe hose.
9. **Holding the brake lever to the bar, remove the band around it.**
10. **Push the caliper syringe plunger in while slowly allowing the brake lever to open.** Close the caliper syringe clamp.
11. **Remove the caliper syringe and replace the caliper bleed screw.** Don't overtighten the tiny T10 bleed screw.
12. **Bleed the lever.** Open the lever syringe clamp, turn the syringe upright, and firmly pull on its plunger (Fig. 9.23). This sucks air out of the lever. Gently push the plunger to pressurize the system. To free bubbles, pull and release the brake lever, allowing it to snap back 10 times, followed by firmly pulling on the syringe plunger again. Repeat until large bubbles cease to appear in the syringe hose. Lightly push the syringe plunger to pressurize the system, and close the syringe clamp.
13. **Remove the syringe and replace the lever bleed screw.** Don't overtighten the tiny T10 bleed screw.
14. **Clean DOT fluid off the lever, caliper, and bike.** Wipe it first, then spray it with water or isopropyl alcohol and wipe again.
15. **Check the system.** Squeeze the levers hard and look for fluid appearing around joints.
16. **Zip-tie the lever around the handlebar overnight to check for fluid leaks.** If the zip-tie is still tight around the lever and the handlebar grip the next morning, there are no leaks.
17. **Return pad-contact and reach adjustments to previous settings.** Do in reverse what you did in step 4; tweak these adjustments to your preference.

c. Bleeding older Shimano brakes with a screw-on reservoir cover

LEVEL 3 One way to bleed brakes is to add extra fluid to the reservoir at the master cylinder (the lever) and squeeze old fluid and air out at the caliper as well as allow bubbles to come up to and out of the fluid in the reservoir. This method works for older Shimano brakes and will work for others with a lever reservoir that has a removable cover.

1. **Level the lever.** Turn the handlebar and the lever so that the reservoir is horizontal, and ensure that the hose is trending downward the entire way to the caliper. You can unbolt the caliper so that it hangs by the hose.
2. **Remove the reservoir cover and diaphragm** (Fig. 9.15). Tiny screws secure it; make sure your screwdriver or wrench is completely engaged in each one.

IMPORTANT: *Make sure you remove the rubber diaphragm under the cover! You would not be the first one to think you were looking at the inside of the reservoir and not the top of the diaphragm, wondering why the fluid you keep adding does not disappear and why the brakes don't tighten up!*

3. **Add fluid to the lever reservoir** (Fig. 9.15). Use fluid specified by the brake manufacturer.
4. **Completely remove the rubber cap from the fitting's nipple** (Fig. 9.11).
5. **Put a box-end wrench on the bleed fitting.** Size will vary with application. If you don't have a box-end wrench, a standard open-end wrench will do, but be mindful not to round off the nipple hex as you work.
6. **Put a clear tube on the caliper bleed-fitting nipple** (Fig. 9.24) **leading into a bottle or a plastic bag.** It is preferable to put a box-end wrench on the bleed fitting (as in Fig. 9.20 rather than as in Fig. 9.24) before pushing the hose onto the nipple—the wrench then stays in place as you open and close the fitting. Make sure you have a spacer (you can make one out of a chunk of wood) between the pistons (Figs. 9.17 and 9.24).

9.24 Shimano XT two-piston disc-brake caliper with spacer block and bleeding apparatus, exploded view

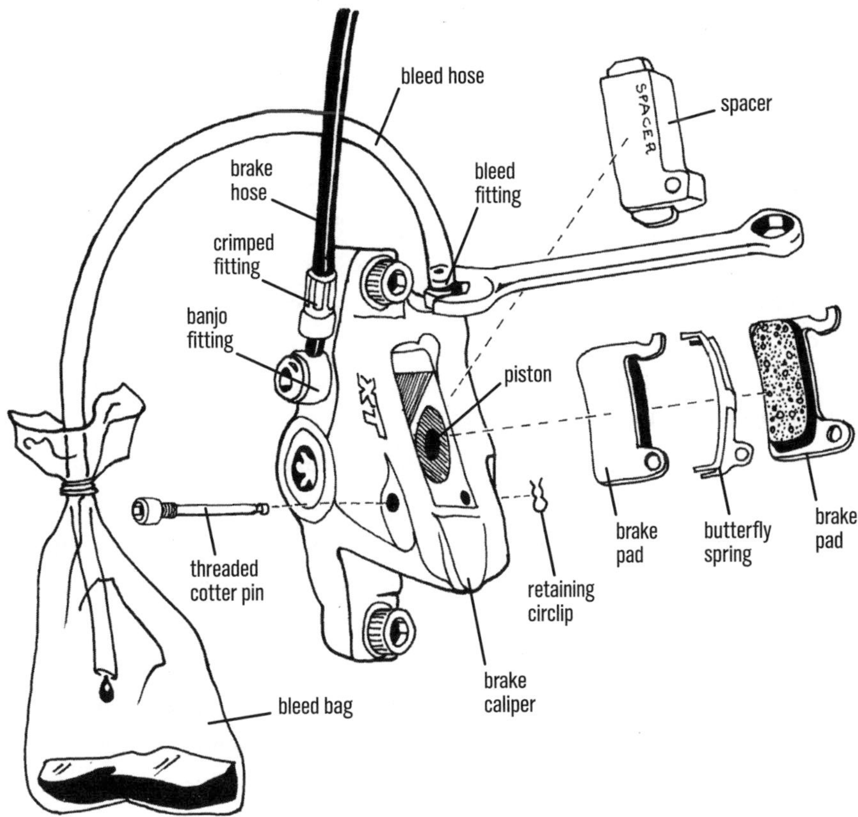

7. **Open the caliper bleed fitting one-eighth turn.**
8. **Squeeze the lever repeatedly.** Tighten the bleed fitting before releasing the lever each time. This pushes fluid in and air out.
9. **Use both of these techniques to be sure that you remove all the air:**
   (a) Push air bubbles out of the caliper: Squeeze the lever with the bleed nipple open, tighten the nipple, release the lever, open the bleed nipple, squeeze the lever, tighten the nipple, release the lever, and repeat, keeping the reservoir topped up, until no more bubbles appear in the bleed tube.
   (b) Repeatedly squeeze the brake lever with the bleed nipple closed, making sure that you keep the fluid level in the reservoir topped up. While you are squeezing, air bubbles should rise through the port into the reservoir.
10. **While squeezing the brake lever, open and close the bleed nipple in rapid succession for about a half second each time.** This optional procedure can release trapped air bubbles from the caliper. Repeat two or three times (refilling as needed at the reservoir), and finish by tightening the bleed nipple.
11. **Squeeze the lever fully.**
   (a) If the lever feels firm, as it should, skip to step 12.
   (b) If the lever still does not feel firm, squeeze and hold it while you shake the hose and the caliper, and tap on the hose and the caliper with the plastic head of a screwdriver to free any stuck air bubbles; again, the hose needs to be trending up the entire way from caliper to lever for the bubbles to go to the lever. When they have all been removed, the lever should feel firm. Continue with step 12.
12. **Refill the reservoir to the top** (Fig. 9.15). Be careful not to overfill.
13. **Replace the diaphragm, cover, and cover screws** (Fig. 9.15).
14. **Install the pads and wheel.** Check brake function.
15. **Retighten the brake lever to the handlebar in its normal riding position.**
16. **Clean off brake fluid.** Use a rag and water or rubbing alcohol.
17. **Zip-tie the lever around the handlebar overnight to check for fluid leaks.** If the zip-tie is still tight around the lever and the handlebar grip the next morning, there are no leaks.

d. Bleeding older Hayes, Stroker, and other brakes up from the caliper

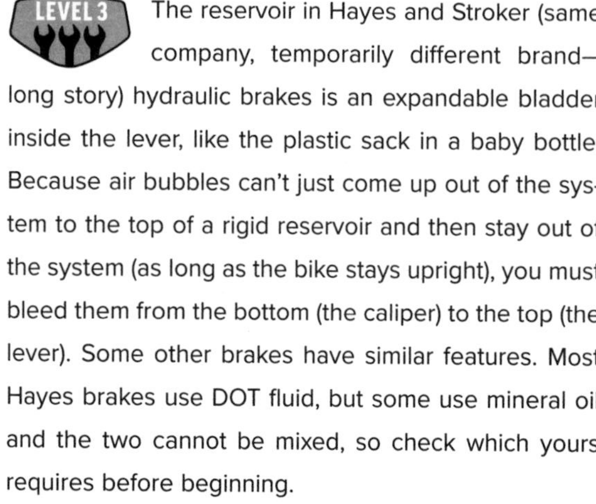

The reservoir in Hayes and Stroker (same company, temporarily different brand—long story) hydraulic brakes is an expandable bladder inside the lever, like the plastic sack in a baby bottle. Because air bubbles can't just come up out of the system to the top of a rigid reservoir and then stay out of the system (as long as the bike stays upright), you must bleed them from the bottom (the caliper) to the top (the lever). Some other brakes have similar features. Most Hayes brakes use DOT fluid, but some use mineral oil and the two cannot be mixed, so check which yours requires before beginning.

1. **Orient the lever so that the bleed screw is at the top.** Mount the bike in a stand, turn the handlebar, and rotate the lever on the handlebar such that the lever is the highest point in the system. Turn the lever until the bleed screw—at the base (both sides) of El Caminos and Strokers and at the front of the lever on HFX—points straight up (Fig. 9.25).
2. **Push the pistons fully back into their cylinders.** Carefully push the pistons back in with the box end of a wrench (use an 8mm for Stroker, 9mm for El Camino, and 10mm for HFX). Avoid damaging the pin sticking out of the piston (Fig. 9.3). Put in a bleed block.
3. **Completely remove the rubber cap from the caliper bleed fitting.** Pull its leash off, too.
4. **Push a short section of clear tube onto the tip of the squeeze bottle or syringe.** A squeeze bottle comes with an old Hayes bleed kit; push the tube past the ridge on the tapered tip so that it stays on. You can also use a syringe (Fig. 9.26).
5. **Fill the squeeze bottle or syringe with brake fluid.** For most Hayes models, use DOT 4 brake fluid

9.25 Catching fluid being bled from a Hayes HFX Mag brake

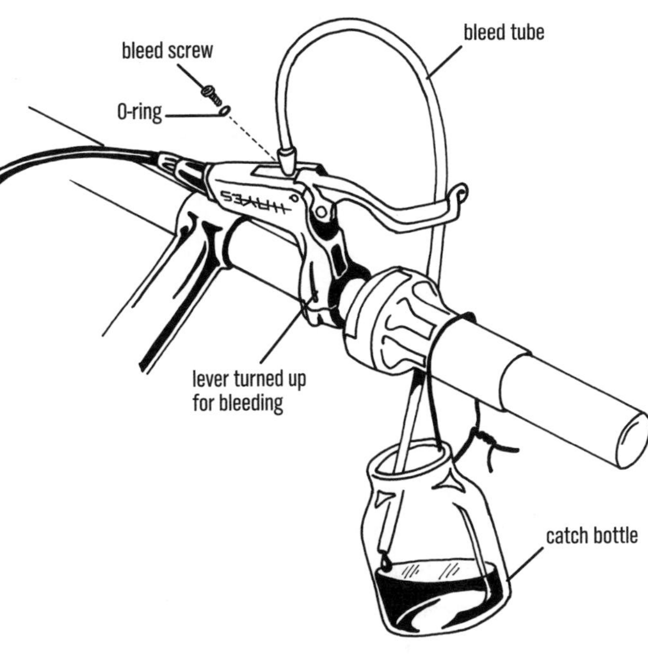

9.26 Bleeding a brake from the caliper with a syringe

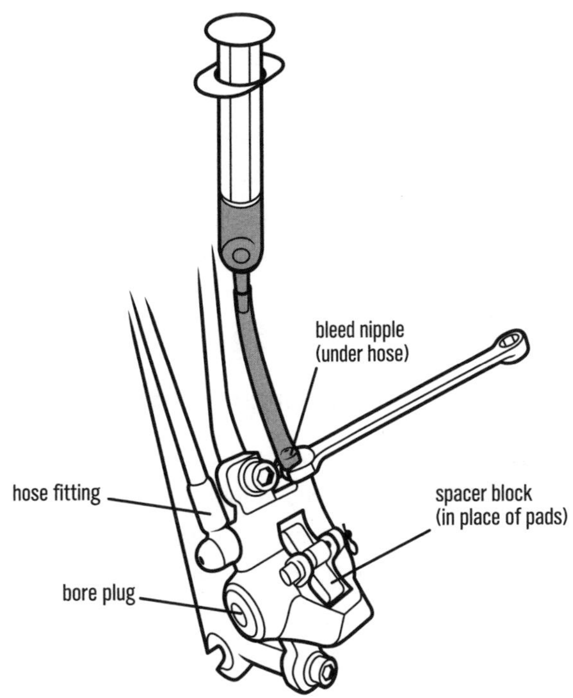

from a previously unopened container (see previous warning about boiled fluid).

6. **Push the other end of the tube over the caliper bleed fitting.** See the bleed fitting in Figure 9.28 and how the hose goes on in Figure 9.26.
7. **Connect the longer bleed tube to the lever bleed hole.** The old Hayes/Stroker bleed kit has a longer clear tube with a selection of plastic and metal fittings for various Hayes and Stroker models. Plug the proper fitting into the tube. Remove the screw or plug from the lever bleed hole, and stick the fitting into it (Fig. 9.25).
8. **Hang a container from the handlebar.** Wire or zip-tie it on, and direct the other end of the tube into it (Fig. 9.25).
9. **With the caliper bleed fitting closed, squeeze the fluid bottle repeatedly.** The bottle or syringe should be pointed straight down; keep it that way throughout the following steps. Trapped air bubbles in the tube come up into the bottle.
10. **Loosen the bleed fitting on the caliper one-fourth turn, and squeeze new fluid in for a count of five.** If the tube is popping off the caliper bleed nipple, put an "olive"—the barrel-shaped brass ring seal (Figs. 9.12, 9.13, and 9.15)—on the bottle's bleed tube before you stick it on the nipple. Push the olive down over the tube and nipple so that the tube won't pull off.
11. **Let off for about three counts (until the squeeze bottle returns to its natural shape).** This draws air out of the caliper and up into the bottle.
12. **Squeeze for five counts, let off for three.** Repeat until no more bubbles come out of the caliper.
13. **Squeeze firmly on the bottle.** Keep squeezing until clean fluid without bubbles comes out of the tube at the lever.
14. **While still squeezing the bottle, quickly pull the lever to the grip and release.** Look for air bubbles coming out of the bleed tube at the lever. Repeat until no more air emerges.
15. **While still squeezing, close the caliper bleed fitting.** Don't overtighten the fitting—it is small, and you need to tighten it only enough to create a seal.
16. **Remove the tubes from the caliper and lever.**
17. **Replace the lever bleed screw or plug.**
18. **Replace the rubber cap on the caliper bleed fitting.**
19. **Install the pads** (9-1) **and wheel, and pump the lever.** The lever should feel firm, and it should not

come back to the grip. Repeat the bleed if the lever feels spongy.

20. **Clean off DOT fluid.** Use a rag and water or rubbing alcohol.
21. **Check for fluid leaks by putting a zip-tie around the lever overnight.** If the zip-tie is still tight around the lever and the handlebar grip the next day, the system is completely sealed.

9-7

OVERHAULING DISC BRAKES

LEVEL 3

Regular bleeding, combined with cleaning around the pistons, keeps dirt out of the system and lengthens the life or the time between overhauls of hydraulic brakes. Some manufacturers offer factory service only on their calipers, and others offer neither tools, parts, nor factory service; if bleeding and exterior cleaning aren't enough to maintain satisfactory operation, you have to buy a new caliper. The latter is the case with Shimano brakes, so it is worthwhile to create a regular schedule of bleeding and cleaning the outside of the pistons.

On many disc brakes, overhaul is relatively simple, as long as you have an air compressor to get the pistons out of the caliper. There are two kinds of hydraulic calipers: clamshell models whose two pieces bolt together, and single-piece calipers. Buy new seals for the part of the brake you are overhauling before you start; if the parts are not available, that indicates that the manufacturer does not support overhauling them and may require factory service or purchase of a new caliper or lever. A speck of dirt or hair in a hydraulic disc brake can cause a leak, so work in a clean area with clean methods.

a. Overhauling a clamshell hydraulic caliper

1. **Remove the caliper from the bike.** See Figures 9.9 to 9.11.
2. **Remove the brake pads** (9-1).
3. **Disconnect the hose.** Plan on replacing the brass olive ring seal on an in-line hose connection (see 9-4e and Figs. 9.12, 9.13, and 9.15). If your brake instead has a banjo fitting (Fig. 9.14), so-named because the head of it looks like a . . . you guessed it, unscrew the hollow bolt holding it on (Fig. 9.14), but don't disconnect the hose from the banjo.
4. **Remove the bridge bolts** (Figs. 9.27 and 9.28) **holding the caliper clamshell halves together.**
5. **Remove the piston(s).** This is best done by blowing compressed air into the fluid-transfer hole while plugging either the bleed hole or the fluid-entry hole with your finger (Figs. 9.27 and 9.28), depending on which piston you are removing. Be careful not to get hit with fluid or parts. Wear safety glasses, and cover the piston with your hand so that no springs or other parts fly away.
6. **Dig the piston seals out of their grooves in the cylinder bores.** Use a fingernail or a toothpick to avoid scratching the cylinder walls. You'll find there are relatively few parts inside the caliper—an object that you might assume is much more com-

9.27 Clamshell hydraulic caliper, exploded view

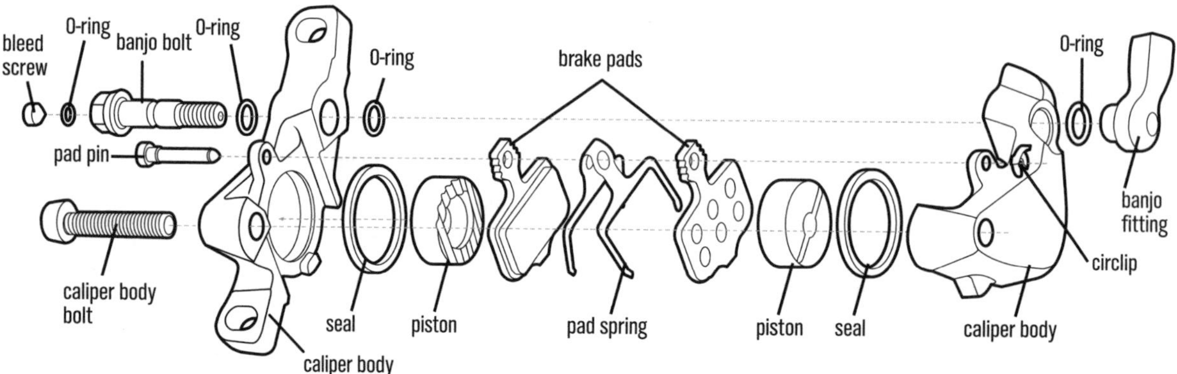

9.28 Hayes clamshell hydraulic caliper, exploded view

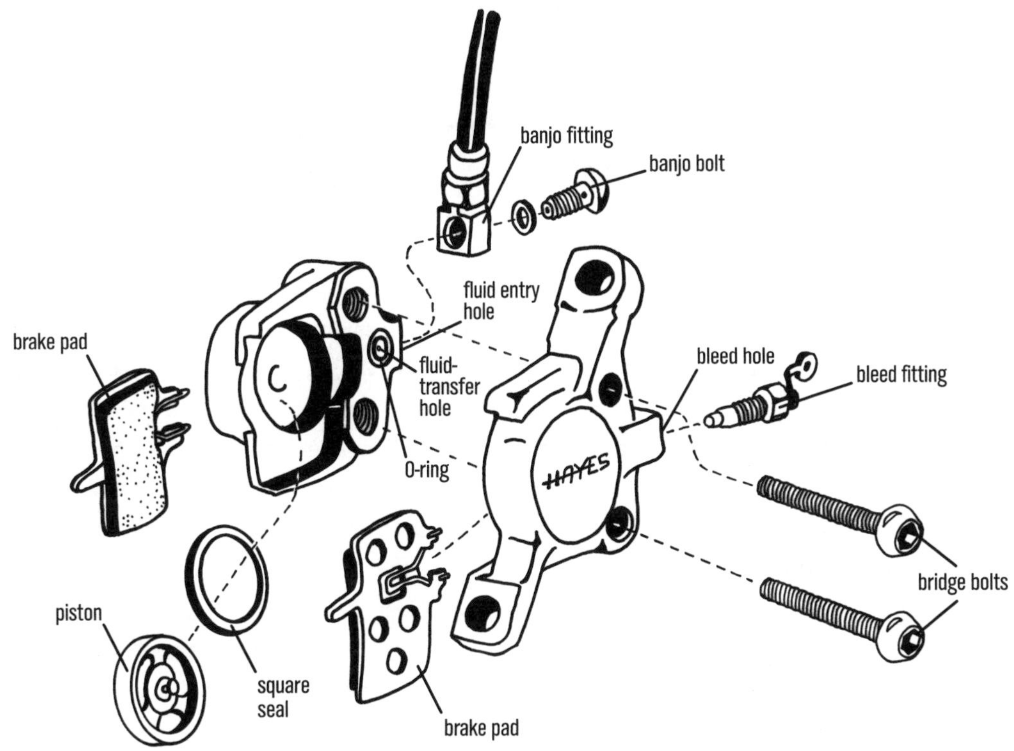

plicated. There are usually just a couple of pistons and a few seals (Figs. 9.27 and 9.28).

7. **Clean all parts carefully with isopropyl alcohol.** Inspect. Replace any cracked or scratched parts.
8. **With compressed air, clean out the caliper-seal grooves and the bleeder hole.** Wear safety glasses. Check that the seal grooves are completely clean.
9. **Let the parts dry.** Compressed air is humid, which contaminates DOT fluid.
10. **Lubricate the pistons and new seals with brake fluid.**
11. **Put all of the parts back together in the way that you found them.**
12. **Bolt the caliper together to the recommended torque.** Check Appendix D or the brake manual for torque specs.
13. **Reinstall the hose.** Use a new olive and barbed fitting for an in-line connection (Figs. 9.12 and 9.13), and new O-rings for a banjo connection (Fig. 9.14). Failure to use new seals will lead to leaks.
14. **Install and center the caliper** (9-4b or 9-4c).
15. **Bleed the system** (9-6).

b. Overhauling a one-piece hydraulic caliper

Do not start this procedure if you have not obtained new seals and the proper tool for the bore plug(s). If the brake manufacturer doesn't sell them, they could be making it clear that they are not permitting you to do this procedure.

1. **Remove the caliper from the bike.** See Figures 9.9 to 9.11.
2. **Remove the brake pads** (9-1).
3. **Disconnect the hose.** If it has a banjo fitting (Figs. 9.26 to 9.29), unscrew the hollow bolt holding it on (Fig. 9.14), but don't pull the hose off.
4. **Remove the bore plugs** (Fig. 9.29). You will need a tool specific to the brake caliper for this.
5. **Push the outer piston out.** Reach in with your finger through the open bore to push the piston into the rotor gap so that it will fall out.
6. **Blow the inner piston out.** Blow with compressed air through the fluid-entry (banjo) hole to push the piston into the rotor gap so that it will fall out.
7. **Clean all parts carefully with isopropyl alcohol.** Inspect. Replace any cracked or scratched parts.

8. **With compressed air, clean out the caliper-seal grooves and the bleeder hole.** Wear safety glasses. Check that the seal grooves are completely clean.
9. **Let the parts dry.** Compressed air is humid, which contaminates DOT fluid.
10. **Lubricate the pistons and new seals with brake fluid.**
11. **Install the new square seals in the cylinder grooves.**
12. **Install the inner piston.** Slide the piston up into the rotor gap and push it in place with your finger through the open bore.
13. **Install the outer piston.** Do this the same way as you did in step 12.
14. **Install new bore plug seals** (Fig. 9.29).
15. **Tighten in the bore plugs.**
16. **Reinstall the hose.** Use a new olive and barbed fitting for an in-line connection (Figs. 9.12 and 9.13), and new O-rings for a banjo connection (Fig. 9.14).
17. **Install and center the caliper** (9-4b or 9-4c).
18. **Bleed the system** (9-6).

c. Overhauling the lever (master cylinder)

Levers rarely need to be overhauled if you bleed the brakes regularly. They are high above the dirt and grime, so any dirt getting into the piston has to travel up from the caliper, which ain't easy.

Generally, you need to remove the lever from the handlebar (Fig. 9.30), disconnect the hose (Fig. 9.15), and remove the lever blade from the housing. Carefully keep unscrewing things or removing circlips and pins until the lever comes apart, noting the order of the parts as you progress. Don't try to remove the rubber seals from the piston. On some brakes (Hayes HFX, for instance; Fig. 9.25), the cylinder is separate from the lever body rather than being machined into it. In that case, remove the cylinder assembly. It is a cartridge that you replace as a unit; do not take out the little piston and seals.

Clean the parts with isopropyl alcohol and inspect them, replace any worn ones, lubricate the parts with brake fluid, and put them back together. Install the lever blade, mount the unit onto the handlebar, connect the hose, and bleed the system (9-5).

9.29 One-piece hydraulic caliper, exploded view

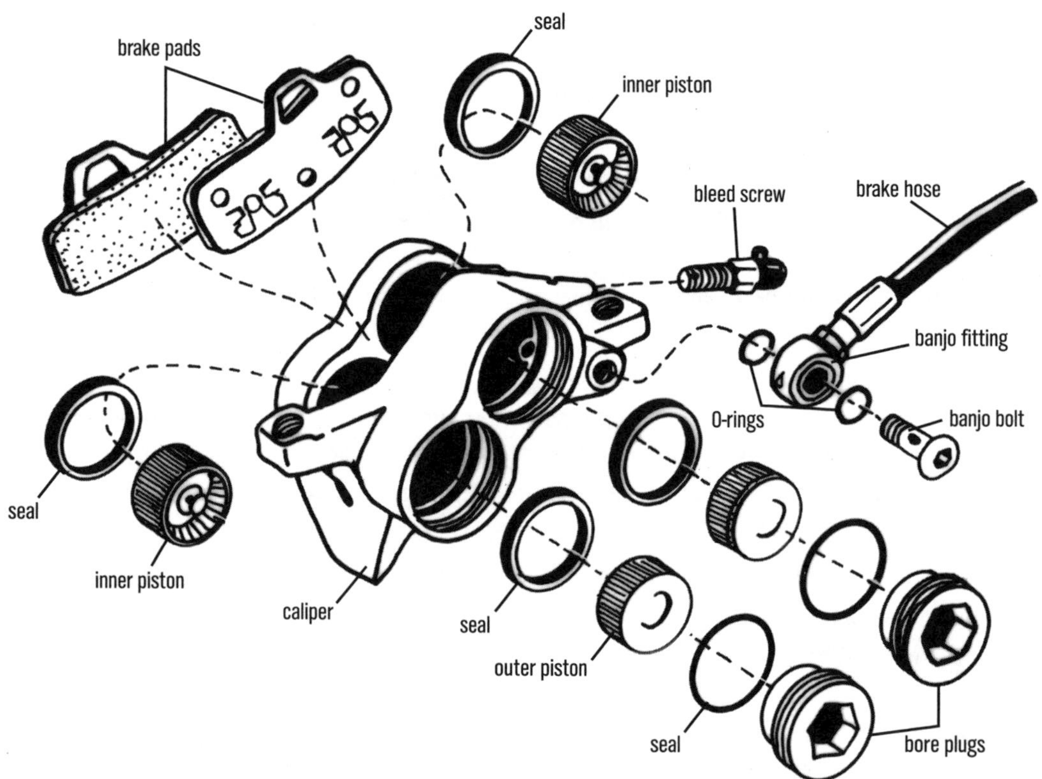

9.30 Removing and installing a hydraulic master cylinder (i.e., lever) with a two-piece handlebar clamp

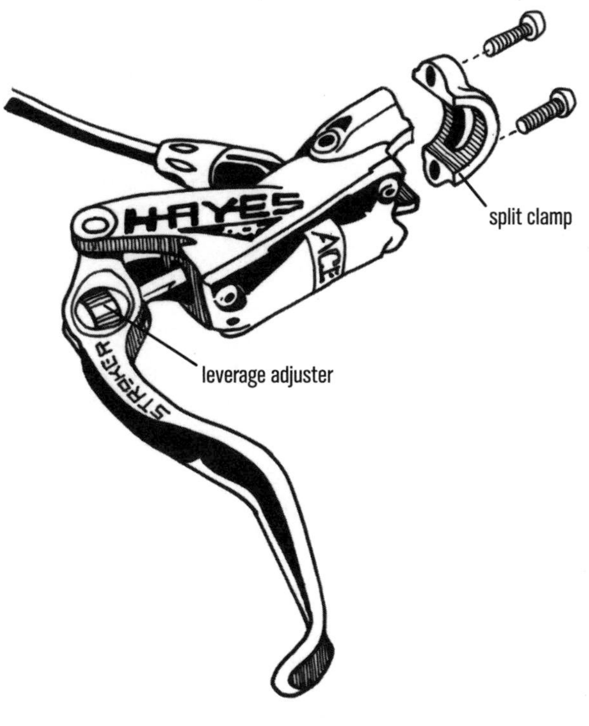

d. Overhauling mechanical disc-brake calipers

Mechanical disc brakes (Fig. 9.9) usually push the pistons by means of a number of ball bearings rolling in a nautilus-shaped track. Methods to disassemble them vary, and the how-to is usually not in the accompanying instruction manual. But as long as you are careful, keep the parts in order, and don't lose any, it is not particularly complicated to take mechanical disc brakes apart and put them back together.

Remove the pads, after which the pistons will usually come out one at a time through the rotor slot. If you can do it this way, you will avoid taking the entire brake apart, which is fine, as the pistons are probably all you need to clean anyway. For example, on Avid, to get out either piston, turn its red plastic screw (Fig. 9.9) clockwise until it pops out. The wheel-side piston is threaded on the outside and has a flat rectangular bar sticking out from the back to engage the plastic knurled disc. Once this piston is out, snap out the red knob—it just pops out with little prongs to hold it in when you replace it. After cleaning the inboard piston and its threaded receptacle, grease the threads and replace that piston and the red knob (don't grease the knob). The outboard piston has a rod sticking straight out from the back. A ring of spring steel around the piston holds it in a hole in the drive mechanism. Don't grease any part of this piston; it doesn't need grease, and grease will only attract dirt. The pistons are magnetic, so expect a bit of a hassle getting them lined up and back in; they will be attracted to each other. Drive in the outboard piston with the inboard piston (turn the red knob). Installing the pads and a spacer in between may help.

If you want to get at the ball-bearing mechanism, start by unbolting the arm. Simply clean and grease everything, and put it back in the way you found it. Be careful, but don't be intimidated.

TROUBLESHOOTING

The first thing to check with any brake is that it stops the bike!

1. **While the bike is stationary, pull each lever.** See that it firmly engages the brake while the lever is still at least a finger's width away from the handlebar grip. If not, skip to 10-2 to 10-4 for tensioning cable, 9-4 to 9-6 for adjusting and bleeding hydraulic disc brakes or adjusting mechanical disc brakes.
2. **Moving at 10 mph or so, apply each brake one at a time.** By itself, the rear brake should be able to lock up the rear wheel and skid the tire, and the front brake alone should come on hard enough that it will cause the bike to pitch forward. Careful. Don't overdo it and hurt yourself testing your brakes!

If you can't stop the bike quickly, you must make some adjustments and perhaps do a little cleaning.

See the next page for Table 9.1: Troubleshooting Disc Brakes.

TABLE 9.1 — TROUBLESHOOTING DISC BRAKES

SQUEALING	
Pads glazed or improperly bedded-in	Replace pads and bed-in properly.
Squealing or howling disc brake due to vibration	Replace or true a bent disc rotor.
	Tighten loose rotor.
	Center caliper to stop pad rub.
	Try different brake pads.
	Tighten the caliper-mounting bolts to torque spec.
	Try a different wheel.
	Check for play in suspension pivots.
	With a hardtail, stiffen the brake mount. To isolate the problem, duct-tape in a dowel-rod strut triangulating the brake-mount area of the seatstay against the chainstay. If the squealing stops, get a new frame or have a framebuilder or carbon-repair shop install a permanent reinforcement.
Squealing due to oil or road de-icer splashed on rotor and pads	Clean rotor with rubbing alcohol.
	Check for oil leak at caliper.
	Reveal a clean layer of the pad by sanding it facedown on drywall-sanding screen so contaminants fall away through the screen.
	Use no solvent other than rubbing alcohol on disc-brake pads.
LOW POWER	
Flexing of cable-actuated brake arm or lever	Install new brakes, but try eliminating other factors first to see if braking power improves.
Stretching of cable	Replace cable and housing. Always use compressionless housing.
Compression of brake housing	Replace cable and housing. Always use compressionless housing.
Insufficient coefficient of friction between the pads and disc	Try different pads; replace worn pads. Organic pads offer higher power; metallic (sintered) pads offer greater durability.
Oil and grime on the discs and pads	Clean rotors with rubbing alcohol; see "Squealing" for pad-cleaning instructions.
Hydraulic brake system has a leak or has air in it	Check for leaks; bleed the system.
TOO MUCH LEVER TRAVEL	
WITH HYDRAULIC BRAKE	
Air in system	Bleed brake.
Pad-contact adjustment out too far	Adjust pad-contact point.
Excessive pad spacing on single-piston caliper	Tighten fixed-pad-adjustment screw on caliper.
WITH CABLE BRAKE	
Cable too long	Tighten cable.
Brake pads worn	Replace pads.
Excessive pad spacing	Tighten pad-adjustment knobs on caliper.

PAD DRAG	
PADS RUB ROTOR BECAUSE OF OFF-CENTER CALIPER, OVERLY TIGHT BRAKES, OR UNTRUE ROTOR	
One pad rubs all the way around the rotor	Center the caliper.
Both pads rub all the way around on a cable-actuated brake	Loosen the cable and/or back out pad-adjustment knobs on caliper.
Both pads rub all the way around on a hydraulic brake	Push pistons back into their bores by prying pads apart with a plastic lever or by removing pads and using the head of a box-end wrench to push pistons back in.
Rotor wobbles back and forth against pad(s)	True the rotor (9-5).
SLOW RETURN	
Cable sticking	Lubricate cable; thaw and dry it if frozen.
Hydraulic pistons sticking	Clean around hydraulic pistons and seals: Remove pads. Hold one piston in with a plastic tire lever and squeeze the lever to push other piston out; clean around it with Q-tip soaked in correct brake fluid for the brake system. Push that piston in with box end of small box wrench or tire lever. Clean other piston the same way.
LOOSE CALIPER	
Caliper rattles or clunks	Tighten mounting bolts on brake caliper and on caliper adapter.
GRINDING NOISE	
Sounds like there is sand on the pads	Clean rotors with rubbing alcohol; see “Squealing” for pad-cleaning instructions.
SOFT LEVER THAT PUMPS UP AND GETS FIRMER (HYDRAULIC BRAKES ONLY)	
Air in hydraulic brake system	Check for leaks; bleed the system.

I prefer a bike to a horse.
The brakes are more easily checked.
—LAMBERT JEFFRIES

10

RIM BRAKES

CABLES, LEVERS, AND CALIPERS

TOOLS

- 2.5mm, 3mm, 4mm, 5mm, and 6mm hex keys
- 9mm, 10mm, and 13mm open-end wrenches
- Small adjustable wrench
- Pliers
- Grease
- Flat-head and Phillips screwdrivers
- Torx T10, T25 wrenches
- Cable-housing cutter

In the formative years of mountain biking, by far the most common brake was the cable-actuated center-pull cantilever (Figs. 10.16 to 10.32). But by the late-1990s, sidepull-cantilever brakes, otherwise known as V-brakes or linear-pull brakes (Fig. 10.11), almost completely eliminated standard cantilevers except on bikes using road bike brake levers, such as cyclocross bikes. And now, disc brakes (Chapter 9) have replaced V-brakes as the standard on mountain bikes.

V-brakes offer more flexibility than cantilevers: They can be used on rear-suspension frames without added complexity because they do not require a cable hanger, and parallel-push V-brake designs allow swaps among wheels with different rim widths without pad readjustment. V-brakes are also more powerful than cantilevers because their arms are longer, and the direct cable pull from one arm to the other is more efficient than yanking up on a straddle cable tying the two arms together, the way that cantilevers operate. Adjusting V-brakes is also quicker and simpler than adjusting cantilevers.

Still, there are a lot of old-style center-pull cantilever brakes out there; hence, working on them is thoroughly covered in this chapter. They're light and simple, they offer good mud clearance, and, above all, they stop your bike. Like V-brakes, cantilevers pivot on bosses attached to the frame and fork.

There are several other mountain bike brake options as well, most of which are only found on bikes from the last century. Linkage brakes that mount on the cantilever bosses offer the same advantage of operating without a cable stop as V-brakes and disc brakes. Hydraulic rim brakes (Fig. 10.33) that mount on cantilever bosses were long the choice of observed trials riders because of their high stopping power and ability to lock up the wheel. Roller-cam brakes (Fig. 10.40) and U-brakes (Fig. 10.39) also mount on pivot posts attached to the frame and fork, but the posts for these brakes are positioned farther from the hub than those used for standard cantilevers and V-brakes.

Rim brakes are light and simple. The drawbacks are that they heat the rims, and they can fade from overheating on a long, steep descent. That heat can even burst the tire. Rim brakes also become less powerful when wet or contaminated with dirt. But if you have them on your bike, this chapter will help you get the most out of them and keep them working properly for years to come.

10-1

RELEASING BRAKES TO REMOVE A WHEEL

- **V-brakes** (Fig. 10.11): Hold the brake-arm link while pulling back and up on the cable noodle until it comes out of the slotted hole in the link (see Fig. 2.1). You can also hold the pads against the rim and pull the cable noodle back and up to release it from the link, but this approach requires more force.
- **Cantilevers** (Fig. 10.22) **and U-brakes** (Fig. 10.39): Hold the pads against the rim, and pull the head of the straddle cable out of the hook at the end of one brake arm (see Fig. 2.2).
- **Magura hydraulic rim brakes** (Fig. 10.33): If the brake has a stiffening arch over the wheel (Fig. 10.33), pull its left end off the bolt head it slips over. If the brake has a quick-release lever on one side, open it and pull the brake bracket off the brake boss (Fig. 10.34). If there is no quick-release, unscrew the mounting bolt on one side to pull the brake bracket off.
- **Other types:** See the section on your particular brake.

CABLES AND HOUSINGS

Cables transfer braking force from the levers to the wheels on nonhydraulic brakes, and proper installation and maintenance of the cables are critical to good brake performance. If there is excess friction in the cable system, the brakes will not work properly, no matter how well the brakes, calipers, and levers are adjusted. Each cable should move freely; replace any cable that has any broken or rusty strands.

10-2

ADJUSTING CABLE TENSION

As brake pads wear and cables stretch, the cables need to be shortened to keep the pads in close proximity to the rims. The barrel adjuster on the brake lever (Fig. 10.1), through which the cable passes, offers easy adjustment. The cable should be tight enough that the lever cannot be pulled all the way to the grip, yet loose enough that the brake pads (assuming they are centered and the wheels are true) are not dragging on the rims.

10.1 Adjusting cable tension

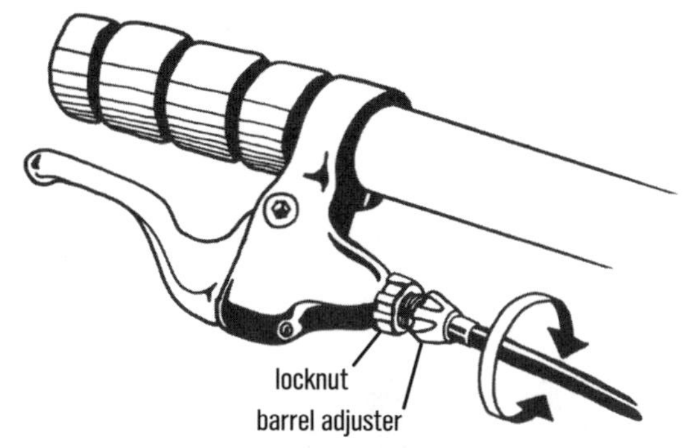

10-3

INCREASING CABLE TENSION

1. **Back out the barrel adjuster by loosening its locknut and turning it counterclockwise** (Fig. 10.1). (Determine clockwise versus counterclockwise rotation direction of the barrel adjuster from the perspective of the end where the cable housing enters.) Some barrel adjusters have no locknut (Fig. 10.10); just turn them and they will hold their adjustment by friction.
2. **Adjust the cable tension so that the brake lever does not hit the grip when the brake is applied fully.**
3. **Lock in the tension by tightening the locknut against the lever body while holding the barrel adjuster.** Again, some levers do not have a locknut on the barrel adjuster and stay in place without it (Fig. 10.10).
4. **If you need to take up more slack than the barrel adjuster allows, tighten the cable at the brake.** Screw the barrel adjuster most of the way in so as to leave some adjustment in the system for brake setup and cable stretch over time. Loosen the bolt clamping the cable at the brake. Check the cable for wear. If there are any frayed strands, replace the cable (see 10-6). Otherwise, pull the cable tight and retighten the clamping bolt. Use the lever barrel adjuster to fine-tune the pad clearance at the rim.

10-4

REDUCING CABLE TENSION

1. **Back out the locknut on the barrel adjuster** (Fig. 10.1) **a few turns (counterclockwise).** Determine clockwise versus counterclockwise rotation direction of the barrel adjuster from the perspective of the end where the cable housing enters.
2. **Turn the barrel adjuster clockwise.** As you turn it, the locknut will go with it and tighten up against the lever body, so keep unscrewing the locknut (Fig. 10.1) as you screw in the barrel adjuster. Stop when the brake pads are properly spaced from the rim.
3. **Tighten the locknut clockwise against the lever body to lock in the adjustment** (Fig. 10.1).
4. **Double-check that the cable is tight enough that the lever cannot be squeezed all the way to the grip.** Check pad alignment with the braking surface and adjust as needed (see 10-10b and 10-16 for V-brakes and cantilevers, respectively).

10-5

MAINTAINING CABLES

1. **If the cable is frayed or kinked or has any broken strands, replace the cable** (see 10-6).
2. **If the cable is not sliding well, lubricate it.** Use molybdenum disulfide grease if you have it; otherwise, try a chain lubricant. White lithium-based greases are not suitable; they can eventually gum up cables and restrict movement.
3. **To lubricate, open the brake** (via the cable quick-release as you do when removing a wheel; see 10-1).
4. **Pull each section of cable housing out of each slotted cable stop.** If your bike does not have slotted cable stops, you will have to detach the cable at the caliper and pull it completely out.
5. **Slide the housing up the cable, and rub lubricant with your fingers on the cable section that was inside the housing.**
6. **Pop the housing ends back into place.**
7. **If the cable still sticks, replace it.** You should probably replace the cable housing at the same time.

10-6

INSTALLING CABLES

As with chains and derailleur cables, brake-cable replacement is a maintenance operation, not a repair operation; don't wait until a cable breaks or seizes to replace it. Purchase good-quality cables with lined housings. Most brake-cable housing is spiral-wrapped to prevent splitting under braking pressure (see Fig. 5.26). Teflon-lined housing reduces friction and is a must on a mountain bike unless you're using a cable-housing system with a separate cable sheath, such as Nokon housing, which you assemble onto the sheath out of separate snap-together segments.

NOTE: *When you install a new cable, it is a good idea to replace the housing as well, even if you don't think it needs to be replaced. Daily riding in particularly dirty conditions may demand the replacement of cables and housings every few months.*

Except in cases in which manufacturers supply lubricants with their cables and housings to be applied during their installation, in my opinion it is usually best not to use a lubricant on new cables and lined housing. Lubricant can attract dirt and even gum up inside the housing.

1. **Remove the old cable.**
2. **Cut the housing sections long enough that they do not make any sharp bends.** If you are replacing existing housings, look at where they bend before removing them. If the bends are smooth and do not bind when the wheel is turned or the suspension moves, cut the new housings to the same lengths. If you see that binding has been occurring, cut each new segment longer than you think necessary and then trim it back to give the smoothest path possible for the cable, without the cable tension being affected by turning the handlebar or by movement of the rear swingarm on a full-suspension bike. Use a cutter specifically designed for cutting housings, or a sharp side-cutter. Cut the housing carefully by applying the cutting blade to a space between the housing coils; this will create a clean cut that will allow the cable to pass through smoothly.

3. **After cutting, make sure the end faces of the housing are open and cut off straight.** If not, flatten the end with a file, a grinding wheel, or a clipper. Eliminate any sharp points that might cut into the cable or impede it.
4. **If the end of the Teflon housing liner is mashed shut after cutting, open it up with a sharp object.** Use a nail or a toothpick, or push a length of cable through from the opposite end to open up the closed-off liner end.
5. **Slip a ferrule over each housing end for support** (see Fig. 5.26).
6. **Decide which hand you want to control which brake.** USA standard is that the right hand controls the rear brake, but you can set up your bike as you prefer.
7. **Tighten the adjusting barrel to within one turn of being screwed all the way in.**
8. **Rotate the barrel adjuster and locknut so their slots line up with slots on the lever and lever body** (Fig. 10.2).
9. **Insert the round head of the cable into the lever's cable hook** (Figs. 10.2, 10.8, and 10.10).
10. **Pull the cable into the lined-up slots on the barrel adjuster and locknut.**
11. **Turn the barrel adjuster so that the slots are offset to prevent the cable from slipping back out.** If you have an old-style lever that is not slotted, you will have to feed the entire length of the cable from the cable hook out through the hole in the lever body.
12. **Slide the rear brake cable through the housing sections, and then route the cable and housing from the brake lever to the brake, sliding the housing and cable into the slot in each stop** (Fig. 10.3). If you don't have slotted stops, you will have to feed the cable through the hole in each cable stop. See 5-15 for internal cable routing.
13. **For the front brake, continue with step 14. For the rear brake, skip to step 15.**
14. **Terminate the housing for the front brake.** With a V-brake, terminate the housing in the top of the noodle guide tube (Fig. 10.11 or 10.12). With a cantilever brake and a suspension fork, terminate the front-brake housing at the stop on the fork brace (Fig. 10.4). For cantilevers without suspension, you may have a cable stop that is integral to the stem or on a hanger above the headset (Figs. 10.5 and 10.6); if the brake cable passes through either a cable stop hanging from the stem or a stem through-hole, I recommend bypassing it, as either requires readjustment of the front brake with every change in stem height. Instead, use a cable hanger with a collar that slips around the stem or steering tube above the headset (Fig. 10.5) or one that slips into the headset stack between locknuts (Fig. 10.6).
15. **Attach the cable to the brake.** Pull the cable taut and tighten the cable-clamping bolt (see the section on your type of brake). Pull the lever as hard as you can, and squeeze it repeatedly for about a minute to stretch the new cable.

10.2 Cable installation at the brake lever (note that slots are lined up in lever body, barrel adjuster, and locknut)

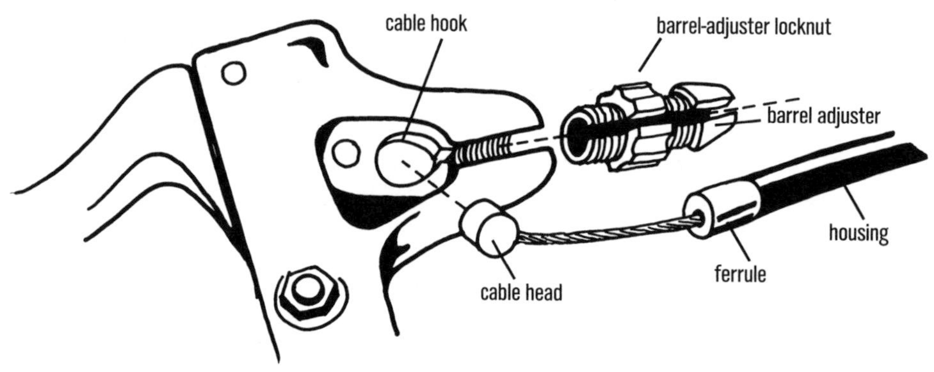

10.3 Slotted cable stop

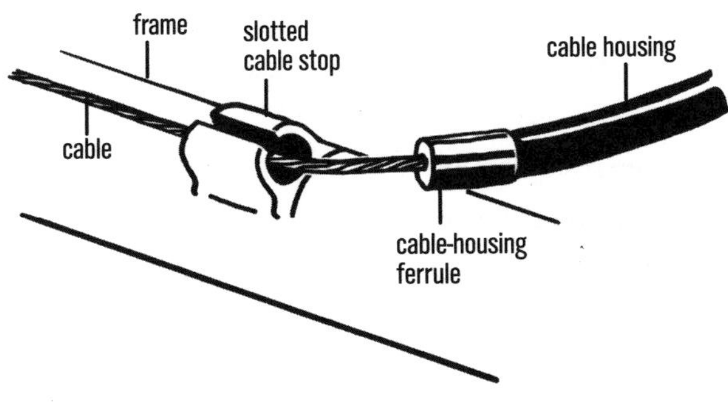

16. **Adjust cable tension with the lever barrel adjuster** (as described in 10-2 to 10-4).
17. **Cut off cable ends about 2½ inches (65mm) past the cable anchor bolts.**
18. **Crimp end caps on all exposed cable ends to prevent fraying** (Fig. 5.45), **and bend the extra to the side.**
19. **Follow the adjustment procedure for your brake, if need be.**

NOTE: *Once the cable has been properly installed, the lever should snap back quickly when released. If it does not, recheck the cables and housings for free movement and sharp bends. Release the cable at the brake, and check the levers for free movement. With the cable still loose, check that the brake pads do not drag on the tire as they return to the neutral position, make sure the brake arms rotate freely on their pivot bosses, and check that the brake-arm return springs pull the pads away from the rims.*

10.4 Suspension-fork cable hanger

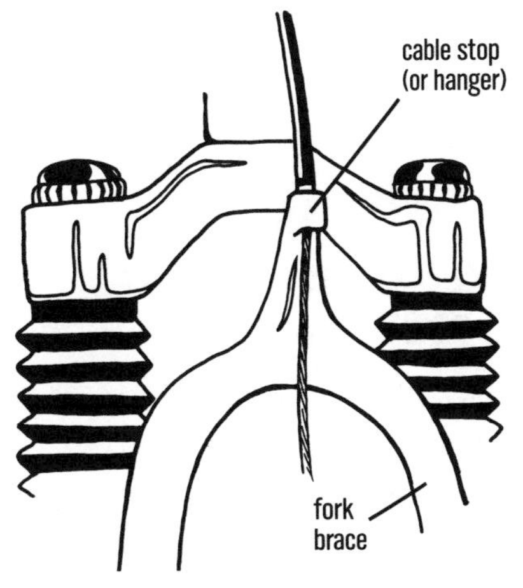

10.5 Stem-clamp cable hanger

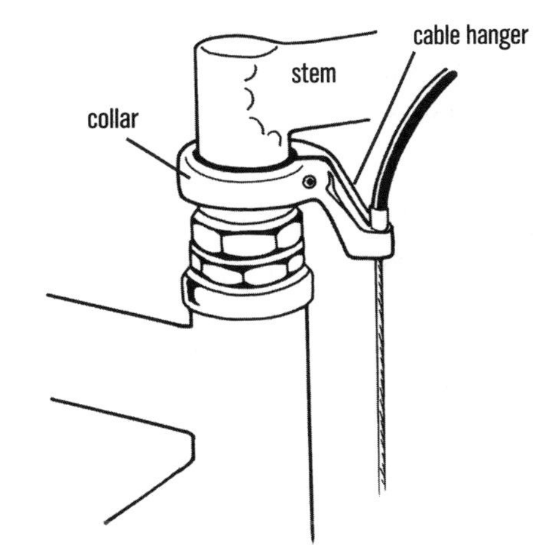

10.6 Headset cable hanger

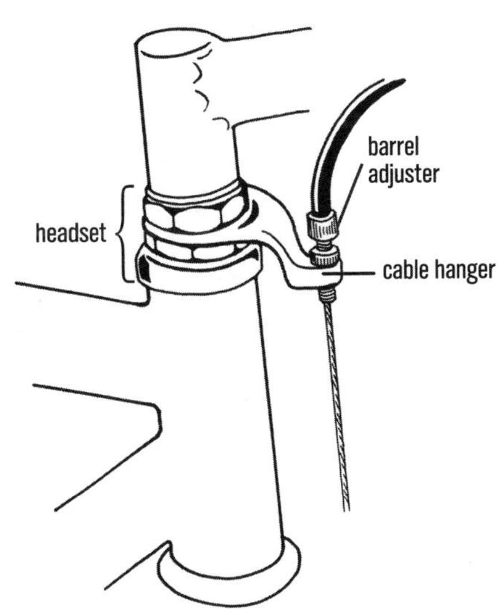

BRAKE LEVERS

The levers must operate smoothly and be positioned so that you can easily reach them while riding.

10-7

LUBRICATING AND SERVICING LEVERS

1. **Lubricate all pivot points in the lever with grease or oil.**
2. **Check return-spring function for levers that have them.**

3. **Make sure that the lever or lever body is not bent in a way that hinders movement.**
4. **Check for stress cracks; if you find any, replace the lever.**

10-8

REMOVING, INSTALLING, AND POSITIONING LEVERS

LEVEL 1

Brake levers mount on the handlebar inboard of the grip and bar end. They are also mounted inboard of twist shifters and usually outboard of thumb shifters (Fig. 10.7). Dual-lever trigger shifters on separate band clamps can mount inboard or outboard of the brake lever, depending on personal preference. Integrated systems include both brake lever and shifter in a single unit (Fig. 10.10).

1. **If a bar-end hand grip is installed, remove it by loosening the mounting bolt and sliding the bar end off the bar.**
2. **Remove the rubber handlebar grip.** Lift the edges on both ends, squirt rubbing alcohol or water underneath, and twist the grip until it becomes free and slides off. For bolt-on grips, simply loosen the pinch bolts and slide them off. See 12-1 for other methods.
3. **If a twist shifter is installed, remove it by loosening the mounting bolt and sliding the shifter off the bar** (Fig. 10.7).
4. **Loosen the brake lever's mounting bolt with a hex key, and slide the lever off** (Fig. 10.7).
5. **Slide the new lever on, and replace the other parts in the order in which they were installed.** If necessary, slide on grips using rubbing alcohol (it dries quickly) as a lubricant; water works, too, but the grips will slip for a while.
6. **Make certain the levers do not extend beyond the ends of the handlebar.** Rotate them and slide them inward to your preferred location. Also see the Pro Tip on optimizing V-brake performance.
7. **Tighten all mounting bolts on levers, shifters, and bar ends.**

NOTE: *If your bike has a carbon-fiber handlebar, you may need to use lower-than-normal torque on the mounting bolts to prevent damage to the handlebar. Consult the handlebar or bicycle manufacturer regarding this detail.*

10.7 Brake lever position relative to shifter

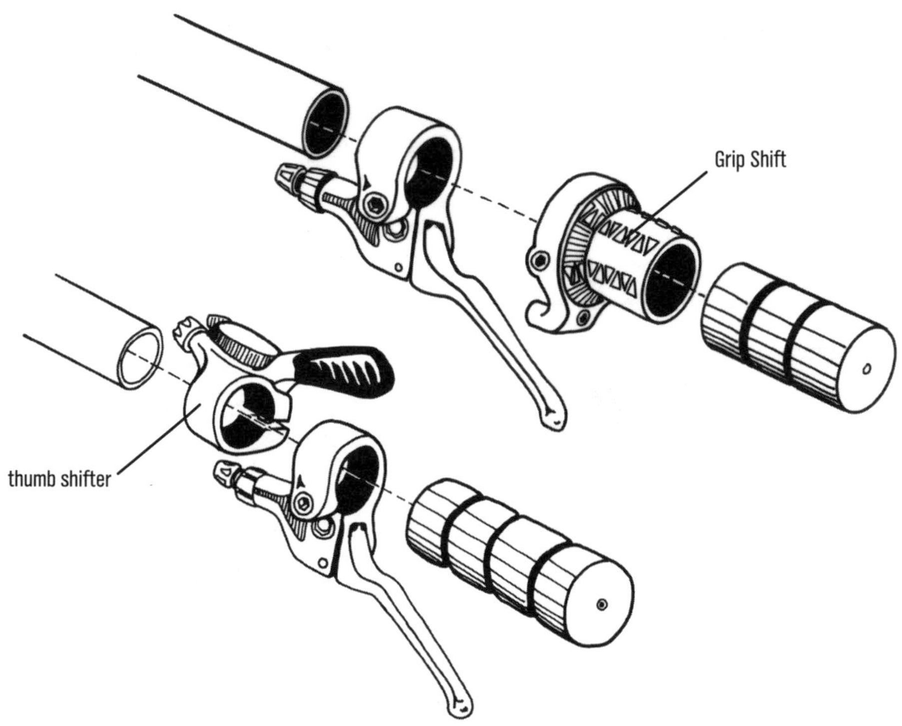

10-9

ADJUSTING REACH AND LEVERAGE

Some levers have a reach-adjustment setscrew; it may be on the lever body near the barrel adjuster (Figs. 10.8 to 10.9) or on top of (Fig. 9.30) or under the lever. If you have small hands, you may want to tighten the reach-adjustment setscrews so that you can reach the levers more easily.

Some brakes also have a leverage adjustment (Figs. 10.8 and 10.10), which moves the cable end in or out relative to the lever pivot. The closer the cable passes by the pivot, the higher the leverage becomes, but the less cable the lever pulls, and vice versa. To start with, set the leverage adjustment at the position that offers the weakest leverage (sometimes demarcated with an "L" on the lever), where the cable head or cable path is farthest from the pivot. Only increase the leverage if you become very confident in using the brakes.

On more expensive brake levers, a long screw performs the leverage adjustment (Fig. 10.10). On some Shimano and SRAM levers, you can adjust leverage by installing, relocating, or removing a series of inserts. On some Shimano LX, DX, and M600 levers, adjust leverage by loosening a small bolt on the upper face of the lever arm with a 3mm hex key, sliding the leverage adjuster up or down, and retightening the bolt (Fig. 10.8). The ends of the adjustment range are generally clearly marked with an "L" for lowest leverage (cable path farthest from the lever pivot) and an "H" for highest leverage (cable path closest to the lever pivot). These Shimano (LX, DX, and M600) and SRAM levers have a hook to hold the cable end far out along the lever (Fig. 10.8); the cable passes over a trough whose position away from the pivot determines the leverage. On other levers, a rotating notched eccentric disc adjusts the cable-head position relative to the pivot. Again, analogous to sitting on a teeter-totter, leverage is increased (and amount of cable pulled is reduced) if the cable head or cable path is closer to the lever pivot, and vice versa.

NOTE: *The levers for V-brakes are initially set up with intentionally low leverage (and correspondingly high cable pull), because of the high leverage of the long brake arms. If you use a lever from a cantilever brake*

10.8 Shimano brake lever for simple V-brakes

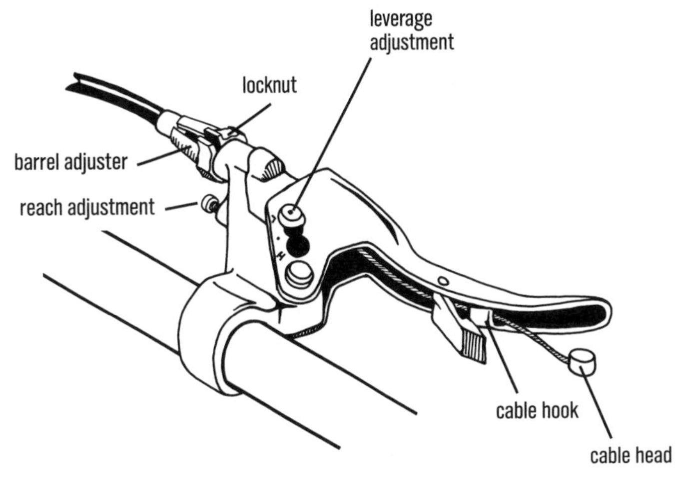

10.9 Brake reach adjustment

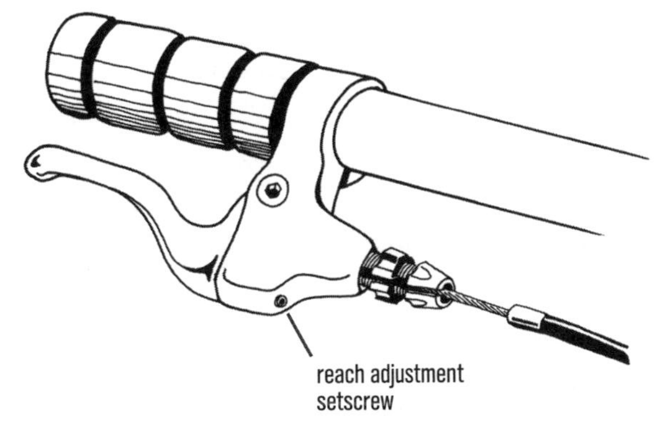

10.10 Rapidfire integrated shift/brake levers (XTR shown)

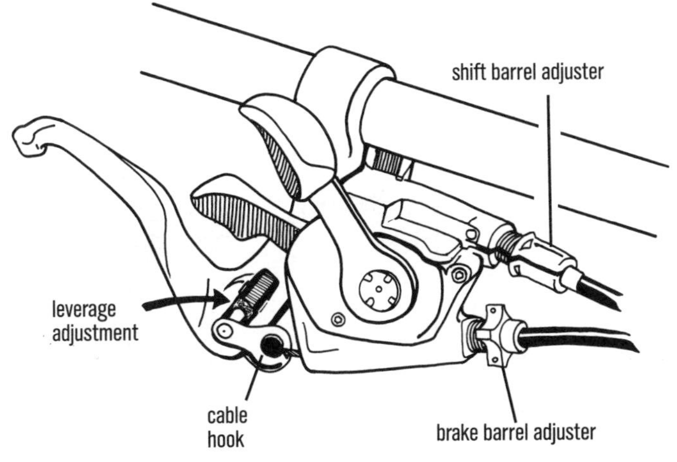

with a V-brake, you have more leverage and can end up on your nose. Always start with V-brake levers adjusted to the lowest leverage (cable passing farthest from the lever pivot), and increase from that setting if you wish.

V-BRAKES

As you can see in Figures 10.11 to 10.12, V-brakes (a.k.a. sidepull cantilevers) have tall, nearly vertical arms, a horizontal cable-hook link on top of one arm, and a cable clamp on top of the other. A curved aluminum guide pipe (noodle) hooks into the horizontal link, taking the cable from the end of the housing out through the link, and directing it toward the cable clamp on the opposite arm. V-brake pads usually have threaded posts. Some V-brakes have parallel-push linkages (Fig. 10.11), which move the brake pads horizontally rather than in an arc around the brake boss the way a cantilever moves. Simple V-brake designs mount the pad directly to the arm so that it moves in an arc (Fig. 10.12).

Because of their long arms, V-brakes are extremely powerful and can be grabby if used with a center-pull cantilever brake lever; it is important that you use levers that are designed for use with V-brakes (see 10-9 regarding leverage).

10.11 Shimano parallel-push V-brake

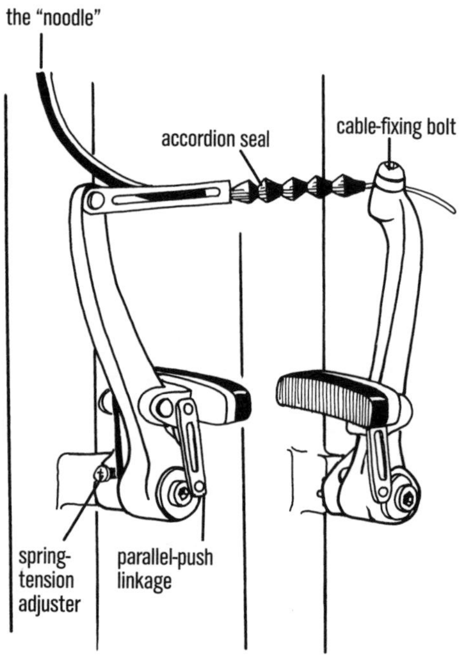

10.12 Simple V-brake (a.k.a. "sidepull cantilever brake")

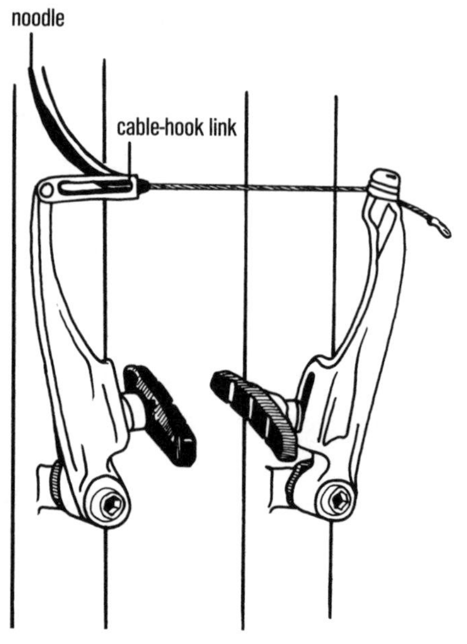

10.13 Finalizing pad-to-rim adjustment

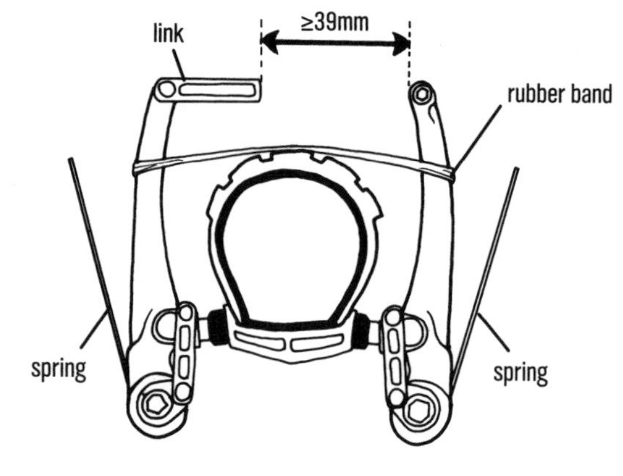

10-10

INSTALLING AND ADJUSTING V-BRAKES

a. V-brake mounting

1. **Grease the brake bosses on the frame or fork.** A light smear does it.
2. **Slide each brake arm on, inserting the spring pin into the center hole of the boss.** You may need to pull outward on the return spring (the tall vertical wire—see Fig. 10.13) to get the pin to line up with the center hole.

3. **Tighten the brake bolt with its washer into the boss.** You want this bolt to be snug, but if you overtighten it, you may mushroom the end of the brake boss so that the brake arm does not pivot freely. Don't overdo it; see the torque table in Appendix D.

b. V-brake pad adjustment

These instructions apply to brake pads with threaded posts. For V-brakes with unthreaded pad posts, follow the pad-adjustment procedure for cantilever brakes in 10-16, coupled with the pad offset described in step 3 here.

1. **Roughly adjust each pad.** Loosen the pad nut, push the arm toward the rim, and tighten the pad nut with the pad flat against the rim.
2. **Determine the proper amount of pad offset from the brake arms.** While holding the pads against the rim, measure the space between the end of the link and the inside edge of the opposite brake arm (Fig. 10.13); this length should be at least 39mm. If it is less than 39mm, the end of the noodle may hit the opposite arm when the brake is applied, particularly as the pads wear. Obviously, this placement would prevent the brakes from grabbing the rims, which is not what you have in mind when you apply the brakes.
3. **Set the pad offset.** Ensure that the length shown in Figure 10.13 is ≥39mm by interchanging concave washers of various thicknesses nesting over convex washers on either side of the mounting tab (Fig. 10.14). By interchanging taller and shorter concave washers from one side of the mounting tab to the other, you set the pad offset so that (a) the space between the end of the link and the inside edge of the opposite brake arm (Fig. 10.13) is at least 39mm, and (b) the top of each brake arm is a little outside of vertical relative to the brake mounting bolt when the brake is applied (i.e., the arms are approximately parallel).
4. **Finalize the pad-to-rim adjustment.** On brakes with vertical return springs such as in Figure 10.13, flip the springs off their retention pins and connect the tops of the arms with a rubber band to lightly hold the pads against the rim. Otherwise, hold the pad against the rim or put a rubber band around the brake lever after you have connected the cable.
5. **Loosen the pad anchor nut, and then tighten the pad-fixing nut with the pad held flat against the rim.** The pad's top edge should be about 1mm below the edge of the rim. Applying toe-in to the pads (Fig. 10.21), so that the front end of the pad hits the rim before the rear end, is not necessary in

10.14 V-brake pad-holder assembly

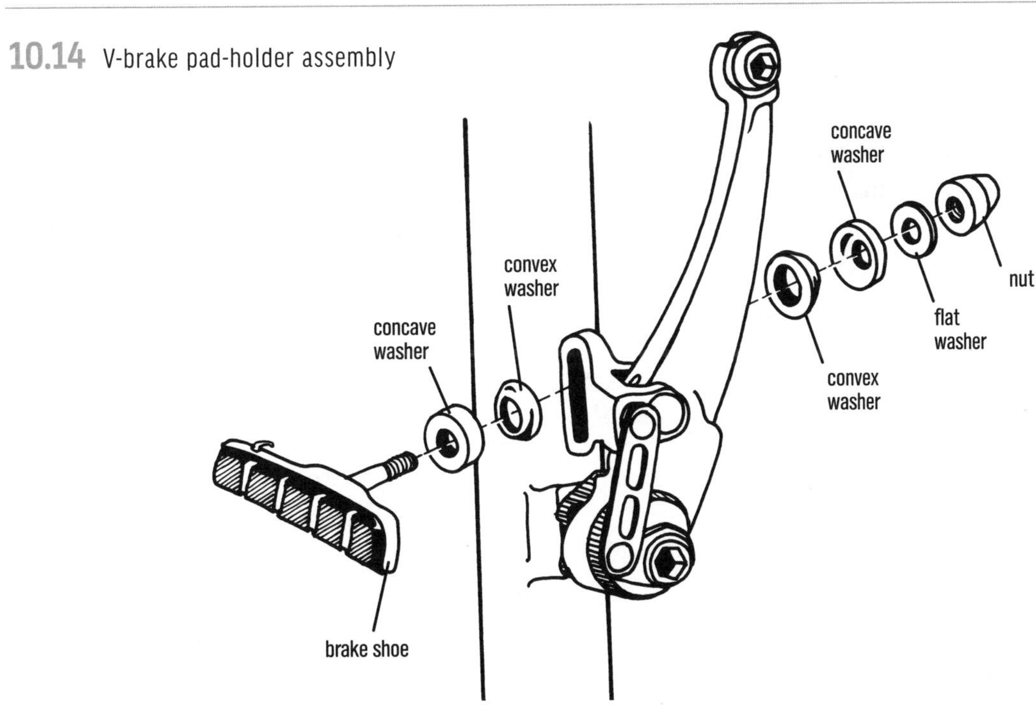

10.15 Pad replacement on V-brakes

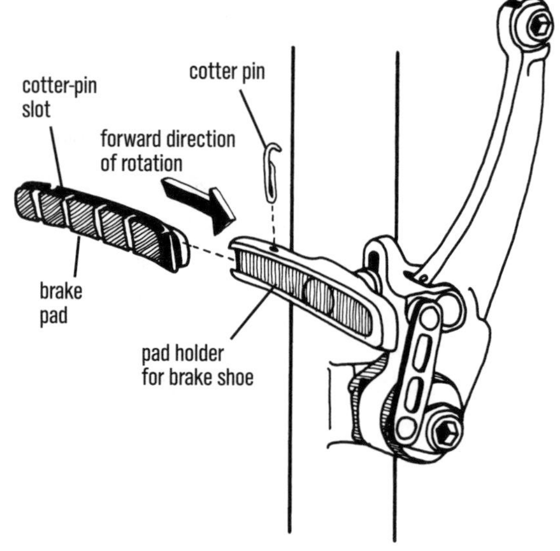

many cases, but it is recommended if the brakes squeal. To get just a bit of toe-in, slip a paper clip between the tail of the brake pad and the rim, and then hold the pad against the rim and tighten it down.

6. **Rehook the return springs behind the retention pins.**

NOTE: *On high-end parallel-push V-brakes, the linkage attached to the pad-mounting bracket keeps the pad moving horizontally as it contacts and leaves the rim surface (Figs. 10.11 and 10.13 to 10.15). When interchanging wheels with these brakes, there is usually no need to adjust the pads if the rim width varies; the only necessary adjustment is to the cable length.*

c. Threading of the cable to the brake through the curved alloy guide pipe (the noodle)

Of the two supplied noodles for the rear brake, pick the one whose curvature and length best fit the frame for a smooth cable path. Bend the noodle if need be. Hook the head of the noodle into the notch in the horizontal link.

1. **Slip the rubber accordion-like dust boot onto the cable, big end first.** Push it over the tip of the noodle (Fig. 10.11).
2. **Connect the cable to the anchor bolt on the opposite arm.**
3. **Set the cable length so that there is 1–1.5mm of space between each pad and the rim.** Tighten the cable anchor bolt with the lever barrel adjuster screwed out one turn. Make sure the wheel is centered in the frame or fork.

d. V-brake centering and/or spring-tension adjustment

- Some V-brakes use a vertical return spring (Fig. 10.13) adjusted by a screw at the mounting pivot on each arm (Fig. 10.11); turn the screw clockwise to move the arm farther from the rim, and vice versa. A quick way to increase spring tension or center the brakes on the trail without fooling with the screws is to bend one or both vertical springs outward after pulling them off the retention pins on the back of the arms as in Figure 10.13, and rehook them behind the retention pins. Hold the spring with your thumb near its base while bending it outward at the top to avoid breaking the plastic adjuster housing.
- Dia-Compe 747s use a spring adjuster cam rotated by a 5mm hex key; turn the cam toward the imprinted "H" or "L" for more or less spring tension.
- Avid Arch Supremes have an innovative (and very quick) way to set the spring balance. While lightly squeezing the lever so that the pads touch the rim, loosen and retighten the plastic knob at the top of the arch. The W-shaped spring passes through the knob and hooks on the arms; it automatically finds its balance point when the knob is loosened.

If the V-brake springs do not adjust with any of these three methods, look at spring-tension adjustment for cantilever brakes, 10-18, because any of the spring configurations used in cantilevers could be built into a V-brake.

IMPORTANT: *After these adjustments, squeeze the brake lever hard a number of times to stretch the cable and make sure it does not slip at the anchor bolt.*

NOTE: *If a brake arm does not turn freely on the pivot boss, the boss may be damaged. Bulging or mushroomed bosses can be filed and sanded smaller; bent or broken ones must be replaced (with luck they are the bolt-on type; otherwise new ones must be welded on).*

ANOTHER NOTE: *Parallel-push V-brakes (Fig. 10.11) often do not hold their pad-centering adjustment and may rub the rim after a ride that they began centered. This is because any bit of grit in any of the*

PRO TIP — OPTIMIZING V-BRAKES

TO GET MAXIMUM PERFORMANCE from V-brakes with a minimum of application effort, you can increase the leverage at the brake lever. But if you do so, you must also position the lever so that you can reach it only with your forefinger; otherwise you may grab too much brake and do an endo (i.e., go arse over teakettle).

Even though a lot of people want a hard feel to the brakes, the harder the brakes feel, the less power there is. The hard feel indicates less mechanical advantage; it feels hard because you are doing all of the work! A softer feel indicates that you have more leverage. You will do less work and stop the bike more easily.

If you set up the brake levers for high leverage, you will lose some pad travel. You will be moving the cable hook (or the cable path) closer to the lever pivot (10-10), usually by turning in a leverage-adjustment screw (Fig. 10.8), by removing some inserts under the cable hook, or by moving the position of an adjuster screw (Fig. 10.10).

Move the levers inboard on the handlebar so that the tip of the lever is under your forefinger. Hold the handlebar with three fingers (and your thumb), and pull the lever with your forefinger. Make sure you pull on the end of the lever, because that is where the leverage is. You will find that you can grip the handlebar better, and your arms will stay more relaxed when braking. In addition, it will be comfortable to simply rest your forefingers on the levers so that you will be ready to brake at any time.

Note that it is easy to move the brake lever inboard far enough with twist shifters and with Shimano integrated brake and shift levers, but it may not be easy with trigger shift levers that have a separate band clamp. The band clamp usually goes inboard of the brake lever, and it may prevent the lever from moving inboard enough for a rider with large hands (and bar ends taking up some handlebar real estate) to get unimpeded one-finger braking. The shifter band clamp may hit the bulge or curve of the handlebar and stop before it has moved inward enough that the brake lever clears the second finger. And even though wider handlebars are back in fashion, the bends in a riser bar may be too far outboard to allow the shifter and the brake lever to move inboard as far as you might wish. Try putting the trigger shifter outboard of the brake lever, and see if you can get the function and finger clearance you desire.

numerous pivots will change the return friction on one side relative to the other side, especially as the pivots break in.

10-11

REPLACING V-BRAKE PADS IN PAD HOLDERS

It is only on high-end brakes that you will find removable pads that slide in and out of permanent pad holders.

1. **Remove the pad-securing pin or screw.** With a pair of pliers, remove the cotter pin from the top of the pad holder (Fig. 10.15). Or, if a screw from the side holds the pad, unscrew it so only a few threads are engaged.
2. **Slide the old pad out of its groove in the pad holder** (Fig. 10.15).
3. **Slide in the new pad.** Pay attention to the "R" and "L" markings for right and left and the forward direction arrow, if present.
4. **Replace the cotter pin or screw, and check that the pad is secure in the holder.**

NOTE: *Pads meant for straight grooves are not interchangeable with most pad holders that have a curved groove. The pads are flexible enough that they can be jammed into each other's holders in a pinch, but the outer curvature of the pad will no longer match that of the rim.*

10-12

REPLACING A PAD ON V-BRAKES WITH A ONE-PIECE PAD AND THREADED POST

1. **Note how the washers are stacked on the pad post** (Fig. 10.14).

2. **Unscrew the shoe anchor nut, and remove the old pad and post from the arm.**
3. **Replace the concave and convex washers as they were and bolt the new pad to the arm.** The convex washers are placed on either side of the brake arm with flat sides facing each other (Fig. 10.14). The concave washers are placed adjacent to the convex washers so that the concave and convex surfaces meet and allow angular adjustability of the pad.
4. **Follow the pad-adjustment procedure in 10-10b.**

10-13

REPLACING A PAD ON V-BRAKES WITH UNTHREADED PAD POSTS

Follow pad replacement and adjustment procedures for cantilever brakes, 10-15 and 10-16.

CANTILEVER BRAKES

10-14

INSTALLING CANTILEVER BRAKES

LEVEL 1

Make sure to install the brakes with all of the parts in their original order. Return springs are not interchangeable from left to right and will often be of different colors to distinguish them.

1. **Grease the brake pivot bosses** (Fig. 10.16). A light coat is all you need.
2. **Install the bushings, if applicable.** Each arm may have a separate inner sleeve bushing; if so, slip one onto each brake post.
3. **Determine the type of return-spring anchor.** If the brake arms have either no spring-tension adjustment or a setscrew on the side of one of the arms for adjusting spring tension, continue to step 4. Such brakes anchor the spring in a hole at the base of the brake post. If the brake arms instead have a large nut for adjusting spring tension behind or in front of the brake arm (Fig. 10.16), skip to step 7.
4. **Slip the brake arm and return spring onto the brake post.** Insert the lower end of the spring into the hole at the base of the post (if there are three holes, try the center hole first; use a higher hole to make the brake snappier, a necessity with lower-quality or old brakes). Ensure that the top end of the spring is inserted into its hole in the brake arm as well.
5. **Install and tighten the mounting bolt into the cantilever boss.** Make the bolt snug but do not overtighten; a bolt that's too tight may mushroom the boss and cause the brake arm to stick.
6. **Skip the next three steps.** They are only for brake arms with tension-adjustment nuts.

10.16 Cantilever brake assembly

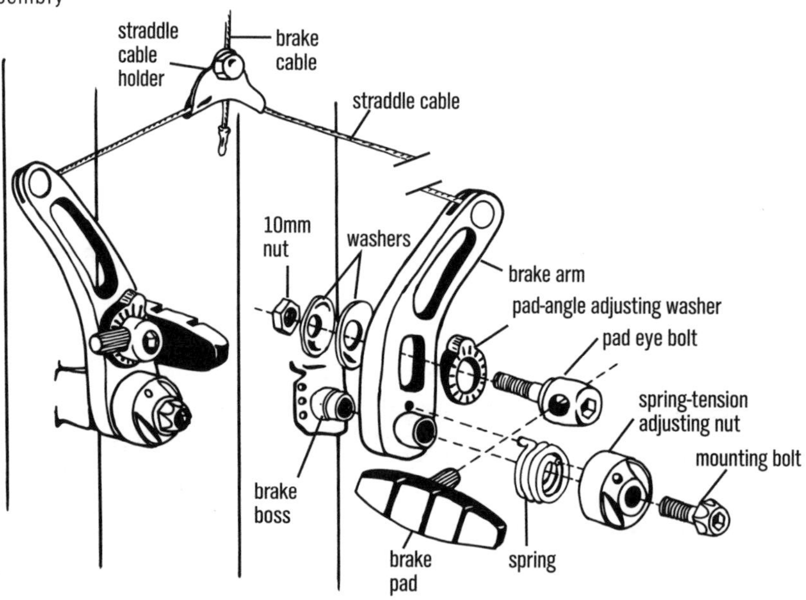

7. **Slip the brake arm onto the cantilever boss.**
8. **Install the spring.** One end inserts into the hole in the brake arm and the other inserts into the hole in the adjusting nut.
9. **Install and tighten the mounting bolt while holding the adjusting nut.** Make the bolt snug but do not overtighten; a bolt that's too tight may mushroom the boss and cause the brake arm to stick.

10-15

REPLACING CANTILEVER PADS

1. **Remove the old pad.** Loosen the bolt that clamps the pad post and pull the pad free.
2. **Install the new pad.** Most cantilevers rely on an eye bolt with an enlarged head and a hole through it to accept the pad post (Fig. 10.16). Some cantilevers instead have a slotted clamp with a hole for the pad post (Fig. 10.17). A few cantilevers use a threaded pad post that passes through a slot in the brake arm (Fig. 10.18), as on a V-brake.

10.17 Cylindrical clamp cantilever brake

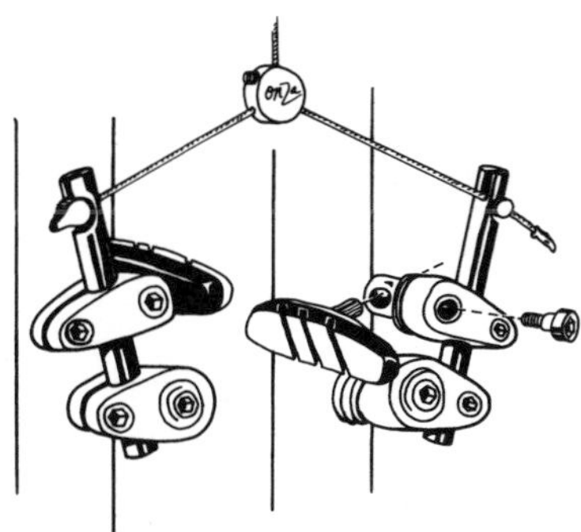

10.18 Cantilever brake with threaded pad posts

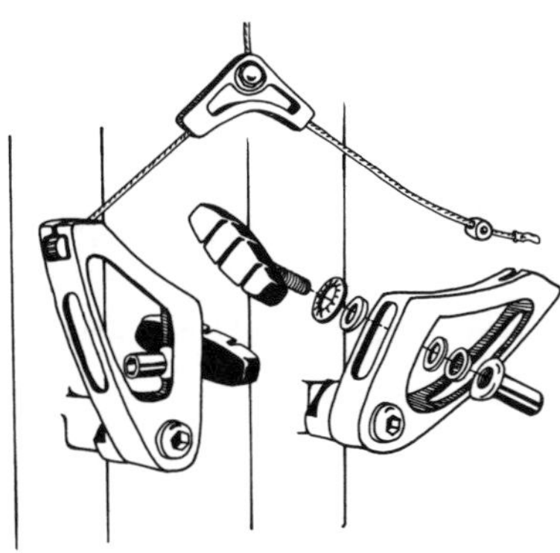

10-16

ADJUSTING CANTILEVER PADS

There are five separate adjustments that must be made for each pad (labeled "a" through "e" in Figs. 10.19 to 10.21):

(a) Offset distance of the pad from the brake arm (extension of the pad post) (a, Fig. 10.19)
(b) Vertical pad height (b, Fig. 10.20)
(c) Pad swing in the vertical plane for mating with the rim's sidewall angle (c, Fig. 10.19)
(d) Pad twist to align the length of the pad with the rim's curvature (d, Fig. 10.20)
(e) Pad swing in the horizontal plane to set toe-in (e, Fig. 10.21)

Cantilevers that feature a cylindrical brake arm are by far the easiest to adjust (Fig. 10.17). Pad adjustment is simple because the pad is held to the cylinder with a clamp that offers almost full range of motion. Other cantilevers employ a single-pad eye bolt to hold all five adjustments (the eye bolt and washers are exploded in Fig. 10.14 and are seen from above in Fig. 10.21). It requires a bit of manual dexterity to hold all five adjustments simultaneously while tightening the bolt.

With all types of cantilevers:

1. **Lubricate the pad anchor threads.**
2. **Set the pad offset** (a, Fig. 10.19). Loosen the pad-clamping bolt, and slide the post in or out of the clamping hole. The farther the pad is extended away from the brake arm, the greater the angle of the brake arm from the plane of the wheel. A benefit of this is that leverage is increased (see straddle-cable angle in Figs. 10.24 and 10.25). The drawbacks are that the brake feels less firm because less force is required to pull the lever, and clearance between the rider's heels and the rear brake arms is reduced, an issue with small frames. A good initial position is with the post clamped in the center of its length.

NOTE: *With threaded-post pads (Fig. 10.18), set the pad offset with washers between the brake arm and pad.*

3. **Roughly adjust the vertical pad height** (b, Fig. 10.20). Slide the pad-clamping mechanism up and down in the brake-arm slot. With cylindrical-clamp

10.19 Distance of pad to fixing bolt (a) and angle against rim (c)

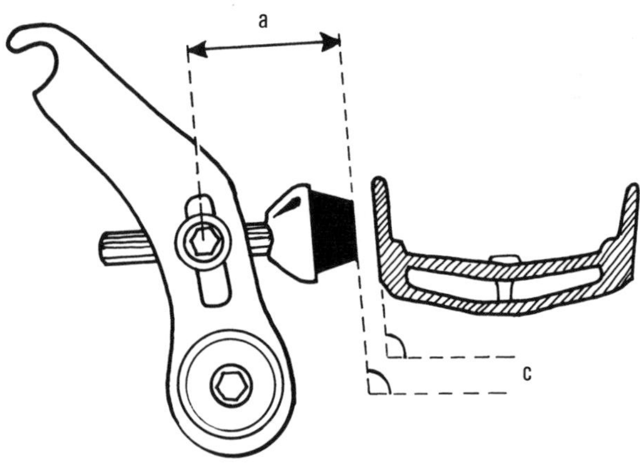

10.20 Up and down (b) and twist (d)

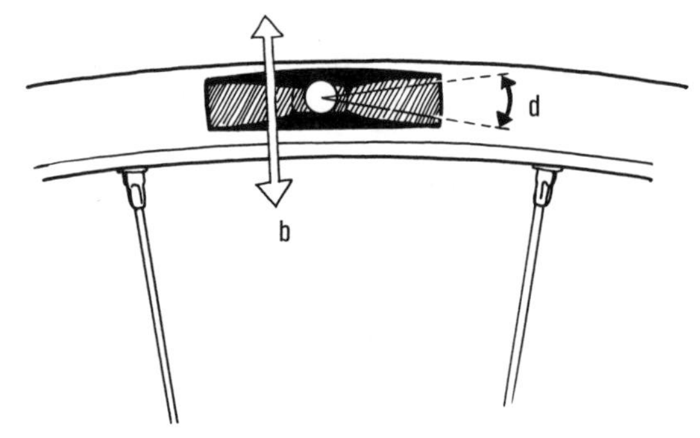

10.21 Brake pad toe-in (e)

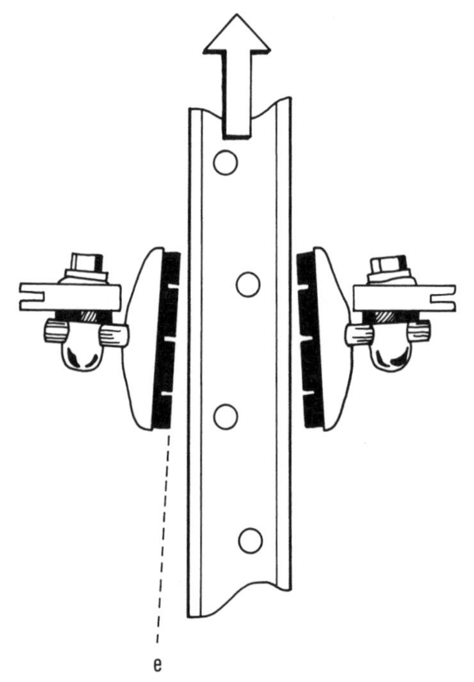

brakes (Fig. 10.17), loosen the bolt clamping the pad holder to the brake arm, and snug the bolt back up once the rough adjustment is reached. With all other types, leave the pad bolt just loose enough so that you can move the pad easily, and continue.

4. **Adjust pad swing in the vertical plane** (c, Fig. 10.19). Make the face of the pad meet the rim flat with its top edge 1–2mm below the top of the rim. Fine-tune this adjustment by simultaneously sliding the pad up or down while rotating it to meet the rim flat.
5. **Adjust the pad twist** (d, Fig. 10.20). Make sure the top edge of the pad is parallel to the top of the rim. Long pads require precision with this adjustment. With cylindrical-clamp brakes (Fig. 10.17), the pad-securing bolt may now be tightened.
6. **Finally, adjust the pad toe-in** (e, Fig. 10.21). The pad should either be adjusted flat to the rim or toed-in so that when the forward end of the pad touches the rim, the rear end of it is 1–2mm away from the rim.

If the pad is toed-out, the heel of it will catch and tend to chatter, making for a grabby feel and squealing noise. If the brake arms are not stiff, or if they fit loosely on the cantilever boss, the same thing will happen when flat. Toe-in is a must with flimsy brake arms and will need to be adjusted frequently to keep them quiet as the pads wear.

NOTE: *If the bike shudders under hard braking, add more toe-in. If that doesn't fix the shudder and the bike does not have a suspension fork, there's too much flex in the fork steering tube, thus pulling the cable tighter when the pads grab the rim. Replacing the front cantilever brake with a V-brake (Figs. 10.11 to 10.12) will fix this problem. The shudder will also probably disappear if you use a cable hanger attached to the fork, rather than a cable hanger above the headset as in Figures 10.5 and 10.6.*

On cylindrical-arm brakes with two anchor bolts (Fig. 10.17), the toe-in is adjusted by loosening the bolt that holds the vertical height adjustment of the pad. Because you have already tightened the other bolt that holds the pad in place, simply loosen this second bolt and swing the pad horizontally until you arrive at your preferred toe-in or flatness setting. Tighten the bolt again, and you are done with pad adjustment.

With any brake using a single bolt to hold the pad as well as control its rotation, you now have the tricky task of holding all the adjustments you have made and simultaneously tightening the nut. Most eye-bolt systems are tightened with a 10mm wrench on the nut on the back of the brake while the front is held with a 5mm hex key. Help from someone else to either hold or tighten is useful here. Probably the trickiest brake to adjust has a toothed or deeply notched washer between the head of the eye bolt and a flat brake arm (Fig. 10.16). The adjusting washer is thinner on one edge than on the other, so rotating it (by means of the tooth or notch) toes the pad in or out. With this type, you must hold all the pad adjustments as you turn this washer and then keep it and the pad in place as you tighten the nut. It's not an easy job, and the adjustment changes as you tighten the bolt.

The other common type has a slotted brake arm with a convex or concave shape, and cupped washers separate the eye-bolt head and nut from the brake arm (Fig. 10.22). The concave-against-convex surfaces allow the pad to swivel, and tightening the bolt secures everything. Again, you may not get it on the first try. Threaded posts (Fig. 10.18) also employ such washers.

NOTE: *Some curved-face brakes do not hold their toe-in adjustment well; you may need to sand the brake-arm faces and washers to create more friction between them.*

Brakes with a cylindrical arm and a clamp secured only by the pad eye bolt are adjusted functionally in the same way as the curved-face ones with cupped washers. A rare but simple-to-adjust type (Campagnolo, Fig. 10.23) has a ball joint at each pad eye bolt.

10.22 Curved-face cantilever brake

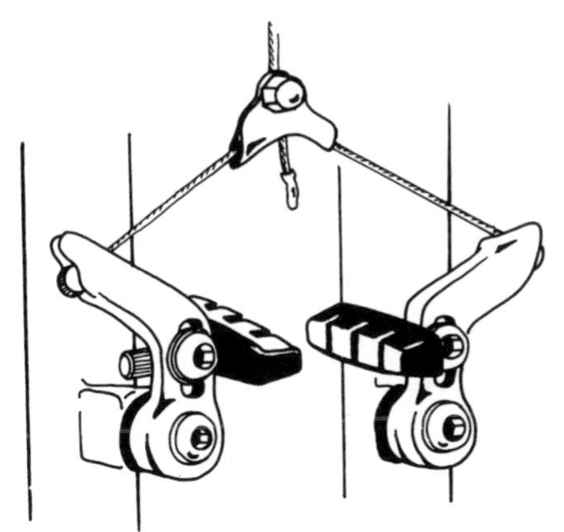

10.23 Ball-joint cantilever brake

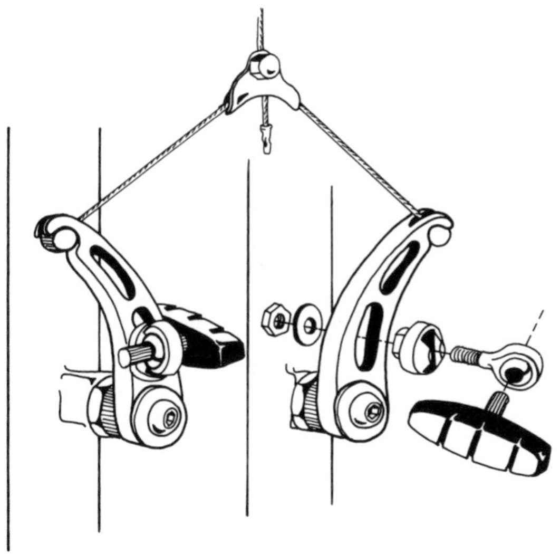

10-17

ADJUSTING STRADDLE CABLE

The straddle cable should be set so that it pulls on the brake arms in such a way as to provide optimal braking. This is not always the adjustment that produces the highest leverage, for sometimes brake feel (i.e., stiffness when pulling the lever) is improved when leverage is reduced because you are doing more of the work. In general, I recommend setting the straddle cable for high leverage and reducing it from there to improve lever feel.

With any lever arm, the mechanical advantage is highest when the force is applied at right angles to the lever arm. For general purposes, set the straddle cable so that it pulls as close to 90 degrees to the brake arm as possible (Fig. 10.24). An esoteric and more precise argument is that once the pad hits the rim, the actual lever arm is the line from the face of the pad to the cable attachment point on top of the arm (because the pad, not the brake boss, now becomes the fulcrum). If you set the straddle cable at 90 degrees from this line, the leverage is maximized (Fig. 10.25). Pull the cable to the desired position with pliers, and tighten the nut on the straddle-cable hanger with a wrench (Fig. 10.26).

10.24 Cable angle when open

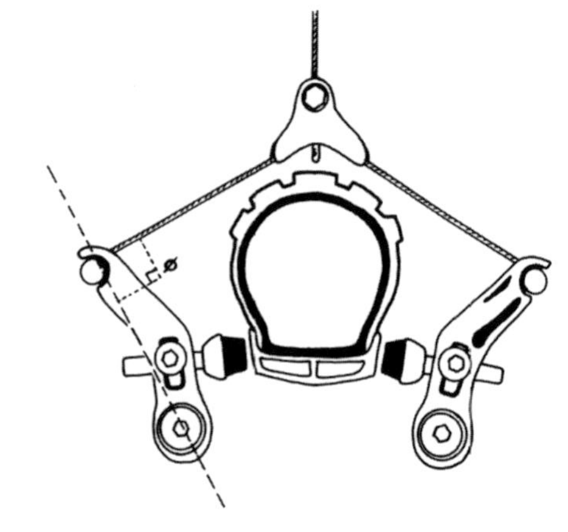

10.25 Cable angle when closed

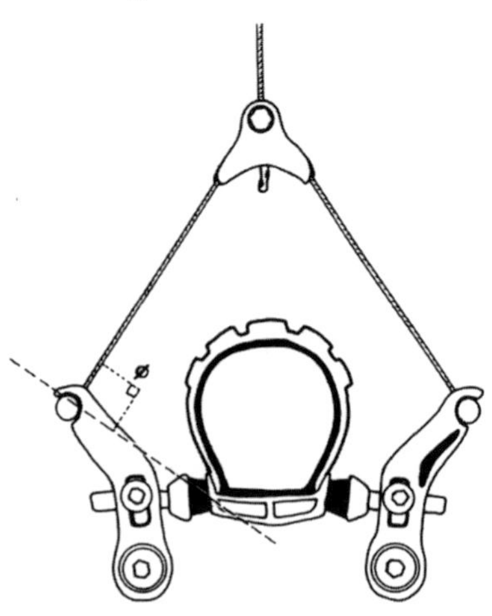

10.26 Tightening straddle-cable holder to brake cable

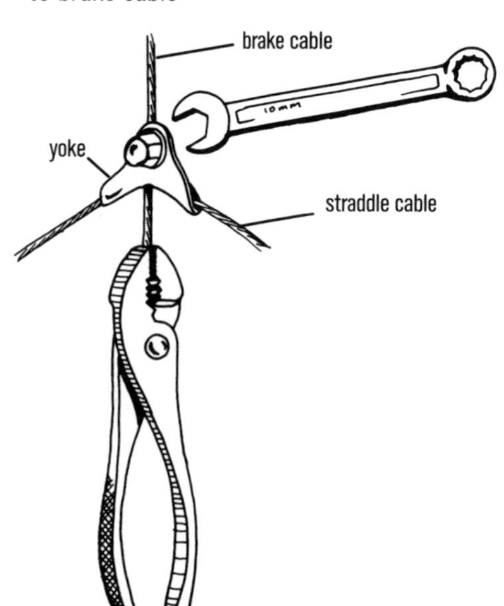

With low-profile brake arms, a 90-degree straddle-cable angle results in a short straddle cable set very low and close to the tire. Make sure that you allow at least an inch of clearance over the tire to prevent mud, or a bulge in the tire, from engaging the brake.

The straddle cable has a metal blob on at least one end (Figs. 10.27 to 10.29). The blob fits into the slot atop one brake arm and acts as a quick-release for the brake. The other end of the straddle cable is fixed to the opposite brake arm.

On Shimano cantilevers built since 1988, the brake cable connects directly to the cable clamp on one brake arm, and a link wire hooks to the other arm. On post-1993 Shimano cantilevers, the cable passes through

10.27–10.29 Straddle cables

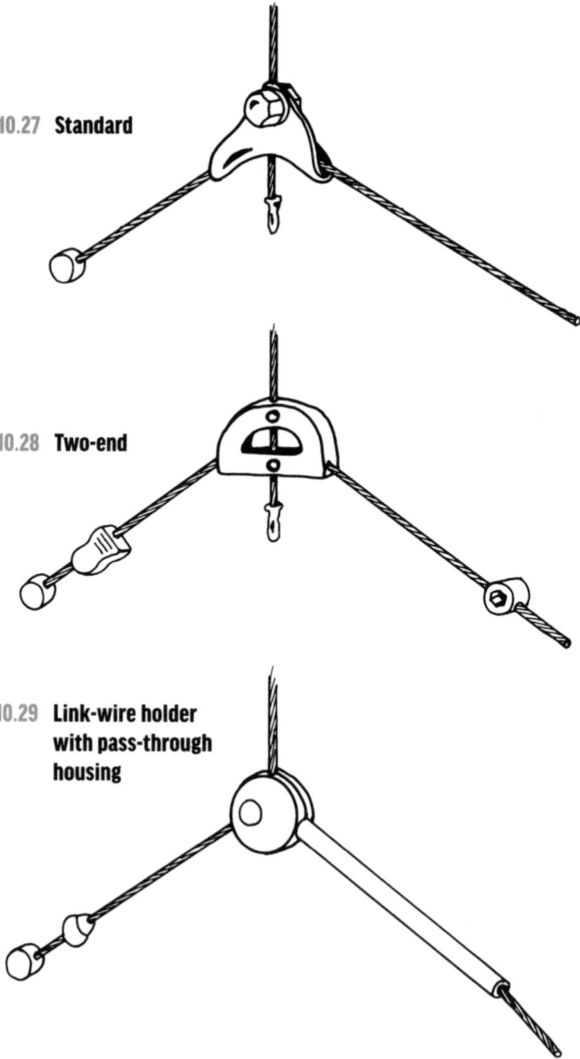

a link-wire holder accommodating not only a link wire but also a fixed length of cable housing (Fig. 10.29). The brake cable passes directly through the link-wire holder and housing segment to the cable clamp on the brake arm. The mechanic has no choice of straddle-cable settings; the setting is predetermined.

From 1988 to 1993, Shimano brakes did not have the housing segment on the link-wire holder; instead, the holder was clamped to the brake cable. For this type, simply set the cable length from the link-wire holder to the brake arm the same on both sides.

Some brakes do not have a cable clamp on either brake arm; both arms are slotted to accept the blob on the end of a straddle cable or link wire. In this case, a small cylindrical clamp forms a second blob on the end of the straddle cable (Fig. 10.28), or a link-wire holder that holds two separate link wires is used.

With any straddle cable, after its length is set, the position of the straddle-cable holder is set by loosening the bolt (Fig. 10.26) or setscrews (Fig. 10.28) that hold it onto the end of the brake cable and sliding it up on the brake cable. Tighten it in place. It is set properly when the brake engages quickly and the lever cannot be pulled closer than a finger's width from the handlebar. Some cable slack can be taken up with the barrel adjuster on the brake lever (Fig. 10.1).

The lateral position of the straddle-cable holder can be changed with setscrews as well. The holder should generally be centered on the straddle cable, but sometimes the brake cable pulls asymmetrically as it comes around the seat tube. In these cases, the straddle-cable holder may need to be offset for the brakes to work (Fig. 10.30).

10-18

ADJUSTING SPRING TENSION

The spring-tension adjustment centers the brake pads about the rim and also determines the return spring force. There is only one adjustment to make on brakes with a single setscrew on the side of one brake arm. Turn the screw (Fig. 10.31) until the brakes are centered and the pads hit the rim simultaneously when applied. Higher spring tensions can be achieved by moving the spring to a higher hole in the brake boss.

Some brakes rely on a large tensioning nut in front of (Fig. 10.16) or behind the brake arm and do not use the holes in the brake bosses as anchors. On these, the tensioning nuts may be turned on both arms to get the combination of return force and centering you prefer. You must loosen the mounting bolt while holding the tensioning nut with a wrench. Turn the nut to the desired

10.30 An offset straddle-cable stop requires an offset straddle hanger

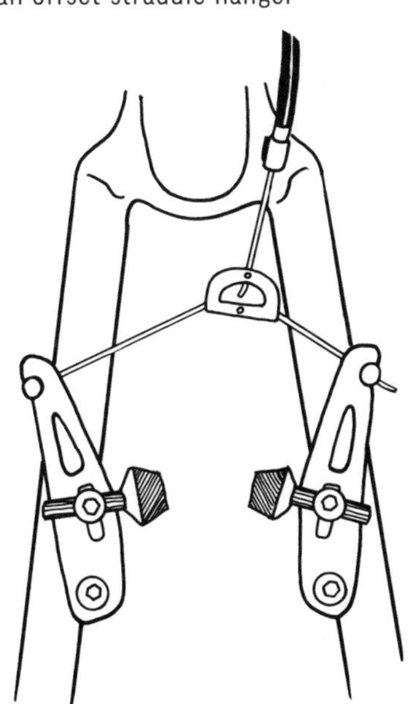

10.31 Adjusting return-spring tension with a setscrew

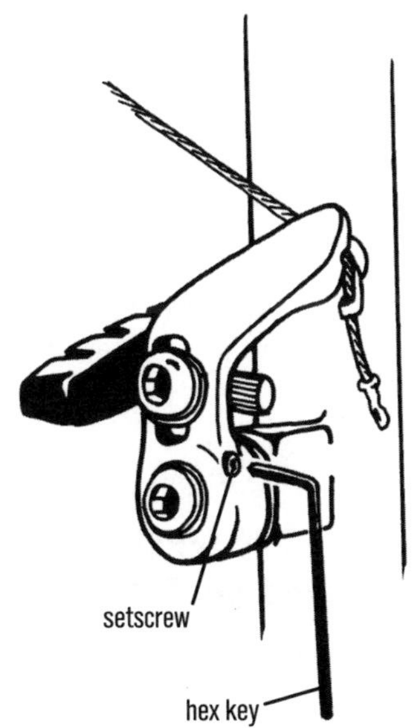

10.32 Adjusting return-spring tension with tensioning nut

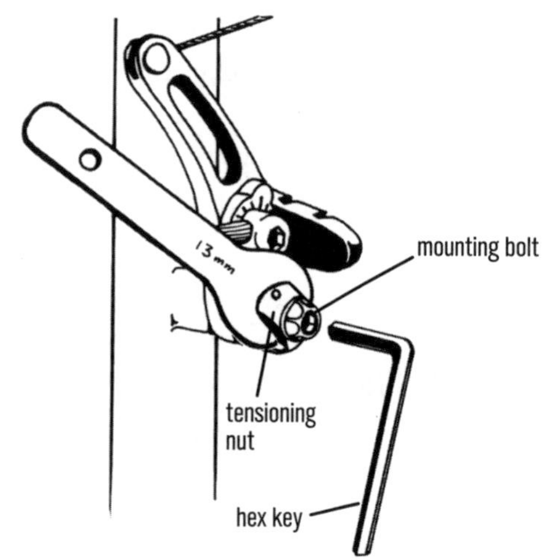

10.33 Magura hydraulic brake

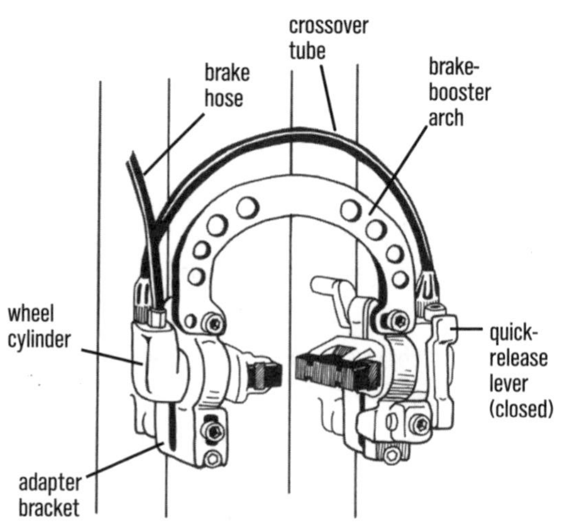

tension, and while holding it in place with the wrench, tighten the mounting bolt again (Fig. 10.32).

On really old brakes without a tension adjustment, centering is accomplished by removing the brake arm and moving the spring to another hole on the boss. It is a rough adjustment at best, and some bosses do not have more than a single hole. When this adjustment fails, you can twist the arm on the boss to tighten or loosen the spring a bit. That, of course, is an even rougher adjustment.

NOTE: *If the brake arms do not rotate easily on the brake post, there is too much friction. Remove the brake, and check that the post is not bent or split, in which case a new one needs to be screwed in or welded on. If not bent, the post is probably too fat to slide freely inside the brake arm, owing to either paint on it or bulging or mushrooming of the post because of overtightening of the brake mounting bolt. In this case, if it's the replaceable type, screw a new post into the frame or fork. Otherwise, file and sand around the post to reduce its diameter. File and sand uniformly, only a little at a time; avoid making the post too thin.*

10-19

LUBRICATING AND SERVICING CANTILEVER BRAKES

The only lubrication necessary for cantilever brakes is on the cables, levers, and brake arms. This should be performed whenever braking feels sticky. Lever and cable lubrication is covered in 10-5. Cantilevers can be lubricated by removing them, cleaning and greasing the pivots, and replacing them (10-14).

HYDRAULIC RIM BRAKES

This section applies to brakes that are fully hydraulic and are mounted on the cantilever bosses. The most common type is Magura (Fig. 10.33), and these instructions, while possibly applicable to others, focus on Maguras.

The Magura brake has great stopping power, is practically maintenance-free, and is simple to adjust. The system is completely sealed from dirt, and there are no cables and housings to wear. Pads are replaced simply by pulling them out by hand and pushing new ones in. A screw on the lever adjusts pad-to-rim spacing as easily as turning a barrel adjuster on a cable-actuating lever.

10-20

INSTALLING AND ADJUSTING MAGURA RIM-BRAKE CALIPERS ONTO CANTILEVER-BRAKE BOSSES

1. **Snap a C-shaped plastic ring around each wheel cylinder.** The ring is supplied with the brakes.
2. **Assemble the adapter brackets around the plastic ring on each wheel cylinder.** Use the supplied

10.34 Magura brake-pad installation and quick-release operation

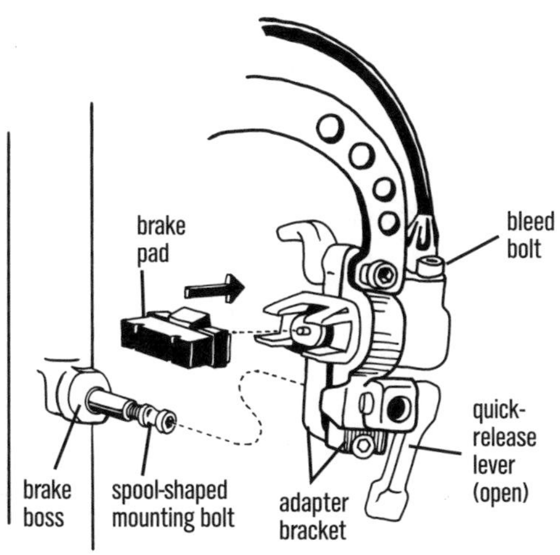

10.35 Installing and/or adjusting elbow

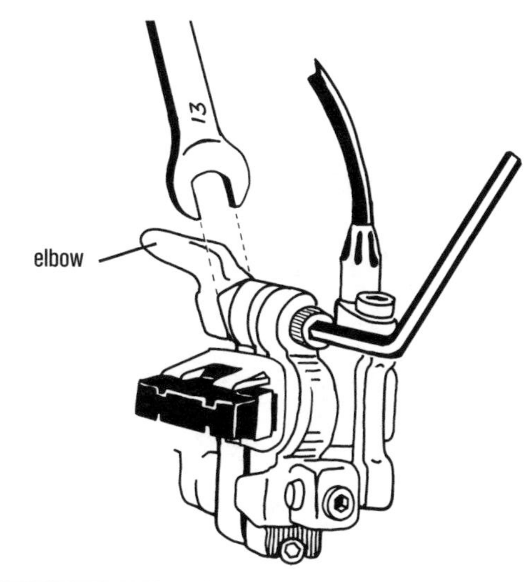

4mm bolts, installing the L-shaped elbow behind the top bracket hole (Fig. 10.35). Right and left wheel cylinders are normally determined by orienting the crossover hose connecting the cylinders toward the bike (Fig. 10.33)—that is, the hose from the lever and the bleed hole are away from the fork or seatstay.

NOTE: *The adapter brackets (Fig. 10.33) are asymmetrical and can be reversed from left to right to move the slave cylinder closer to the rim or vice versa. Normal mounting is with the bracket imprinted with Magura on the right when facing the brake.*

3. **Slide the included D-shaped washer onto the cantilever boss.** The flat side of the washer is on top. A thick washer with a setscrew must be used with some suspension forks to clear the fork brace.
4. **Bolt the adapter bracket to the cantilever boss with a bolt and washer.** If mounting a bracket with the quick-release feature, first screw the mounting bolt with the spool-shaped head a few turns into the brake boss. Slide the QR unit over the mounting slot on the adapter bracket (Fig. 10.34). Push the bracket onto the brake boss so that the spool-shaped mounting bolt head comes through the hole in the quick-release unit. Flip the QR lever up to its closed position. Tighten the mounting bolt with a 5mm hex key. You can now remove this side of the brake by merely flipping the quick-release lever down and pulling the bracket straight off (Fig. 10.34). To install, push the bracket onto the boss so that the bolt head sticks out through the hole in the QR unit, and flip the lever up.

NOTE: *The QR unit is installed properly when the quick-release lever's pivot pin is above the spool-shaped mounting bolt, not below it.*

5. **Set the pad position.** Slide the adapter bracket up or down to adjust the height of the wheel cylinder. Loosening the mounting bolt and the other bolt (or bolts) holding the bracket together (Fig. 10.35) allows the cylinder to be slid in or out and rotated. Using these adjustments in tandem, set the pad-to-rim contact so that the pad hits the rim flat about 2mm below its upper edge. See to it that the pad holders do not drag on the tire. When retracted, the pads should sit 2–3mm away from the rim.
6. **Position the elbows** (Fig. 10.35). They support the brake and simplify repositioning after removal. Loosen the bolt above the wheel cylinder. With a 13mm open-end wrench, rotate the elbow until it contacts the inner side of the seatstay or fork leg, and tighten the bolt (Fig. 10.35). Aftermarket elbows are available to better fit certain suspension forks.
7. **If you have the "brake booster" arch** (Fig. 10.33), **loosely bolt it onto the right bracket adapter.** Its oval mounting hole goes on the right. Swing the arch over the wheel, and slide it laterally until the booster's bottom left hole lines up with, and slips over, the bolt head on the left bracket. Tighten the top bolt on the right bracket to fix the booster

10.36 Installing hydraulic brake hose

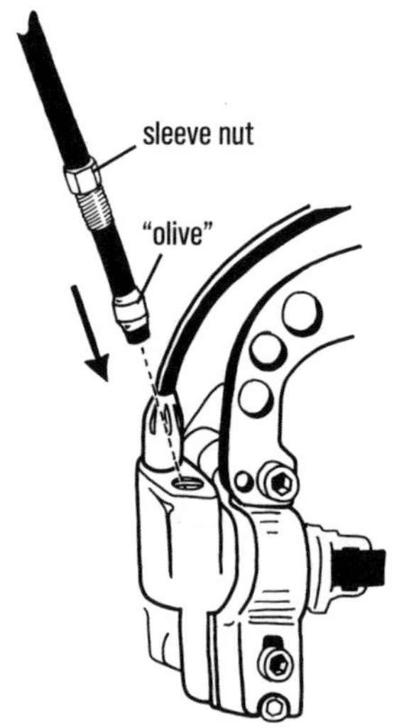

in place. When releasing the brake, pull the left side of the booster off the bolt head, and leave it attached to the right cylinder when pulling it off.

8. **If the hose is not connected, attach it** (Fig. 10.36). See 9-4e for hydraulic hose routing and cutting.
9. **If the brake is not firm when you pull the lever, bleed it.** See 10-22.
10. **Fine-tune pad-to-rim spacing by turning the adjustment screw on the lever.** Some systems take a 2mm hex key; the screw is under the lever. Later systems have a finger-operated knob on the front of the lever.

10-21

REPLACING THE MAGURA PADS

Pad replacement is very simple:

1. **Remove the wheel.**
2. **Grab the pad and pull it straight out** (Fig. 10.34).
3. **Push the new pad in.** Pay attention to the rotation-indicator arrow on it.

That's it!

10-22

BLEEDING MAGURA RIM BRAKES

LEVEL 2

A spongy feeling at the lever indicates the need to bleed air from the lines. Bleeding is a maintenance operation that also clears out dirt that has crept in and should be done every 500 miles or so. If the oil inside comes out clean, you will know you can wait longer next time. There is rarely a need to bleed due to air bubbles, because air does not get in unless the system is opened or gets damaged. That said, any air bubbles will rise toward the highest point in the system, so they should be up in the lever after any ride, meaning that it will not take much new fluid to drive them out.

1. **Back out the lever adjustment screw(s).** Later models have a 2mm microadjustment screw under the lever and a 2mm reach-adjustment screw on top of the lever. Newer models have a single knob on the front of the lever.
2. **Remove the bleed bolt on the right brake wheel cylinder** (Fig. 10.34).

IMPORTANT: *Never squeeze the lever while the system is open; fluid will squirt out.*

3. **Fill the Magura syringe with Magura brake fluid.** The syringe has a tube with a barbed fitting on the end (Fig. 10.37).
4. **Invert the syringe, and push any air up and out with the plunger.**
5. **Screw the fitting into the bleed hole on the wheel cylinder** (Fig. 10.37). (In a pinch, a squirt bottle of fluid can be used instead of the syringe, but it is easy to allow air in this way. You will need another person to hold the bottle tip tightly in the bleed hole and squeeze the bottle while you open and close the lever bleed bolt.)
6. **Tip the bike or rotate the lever on the handlebar so that the lever bleed bolt is at its highest point.**
7. **Remove the bleed bolt from the lever.**
8. **Screw in a piece of tubing with a barbed fitting into the lever bleed hole.** Have the tube drain into a bottle of fluid hanging from the handlebar (Fig. 9.25).
9. **Push fluid into the wheel cylinder with the syringe** (Fig. 10.37). Let it push fluid out of the lever and into the catch bottle.
10. **Reinstall the bleed bolt at the lever and tighten it.** Make sure the bolt and washer are free of grit.
11. **Remove the syringe or bottle from the wheel cylinder.** Leave a dome of fluid bulging from the hole.

12. **Put the bolt back in and tighten it.** Again, the bolt and washer must be free of grit.
13. **Squeeze the lever.** If the bubbles have been driven out, the brake will no longer feel spongy. Repeat until the sponginess is gone. You may need to bleed the system again after riding; the bubbles will have collected at the lever.
14. **Adjust the pad spacing and lever reach to your liking.** Use the two 2mm bolts or the single knob on the lever.

10.37 Bleeding and/or filling Magura brake hoses

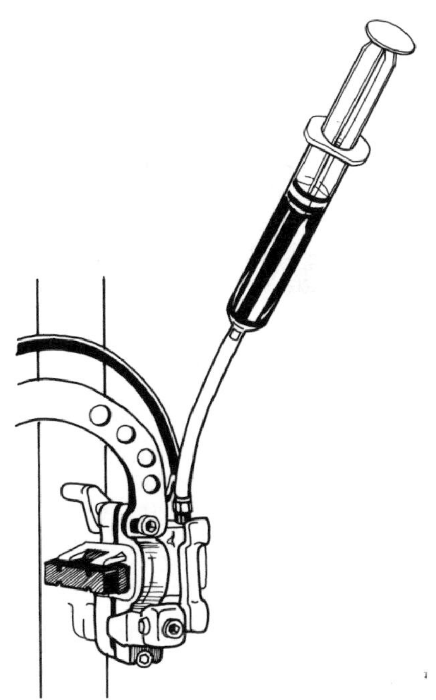

LINKAGE BRAKES

There are so many vastly different linkage brakes (Fig. 10.38) besides V-brakes floating around on older bikes that including them all in detail is not possible here. Linkage brakes are often quite similar to cantilevers or V-brakes and are adjusted, centered, and mounted in much the same way. If in doubt, refer to specific instructions from the manufacturer.

U-BRAKES

10-23

INSTALLING U-BRAKES

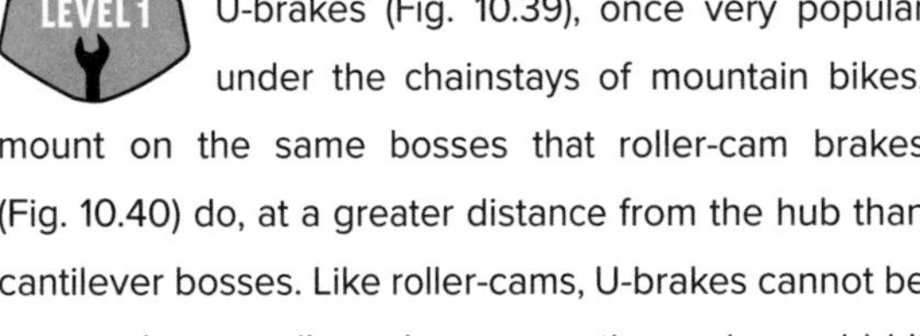

U-brakes (Fig. 10.39), once very popular under the chainstays of mountain bikes, mount on the same bosses that roller-cam brakes (Fig. 10.40) do, at a greater distance from the hub than cantilever bosses. Like roller-cams, U-brakes cannot be mounted on cantilever bosses, as the pads would hit the tire rather than the rim.

1. **Grease the pivots.** Use a light coat.
2. **Slide the arms onto the pivots.**
3. **Screw in the mounting bolts.**
4. **Tighten the straddle-cable yoke to the brake cable** (Fig. 10.26).
5. **Attach the straddle cable to the cable clamp on one arm.**

10.38 Linkage brake

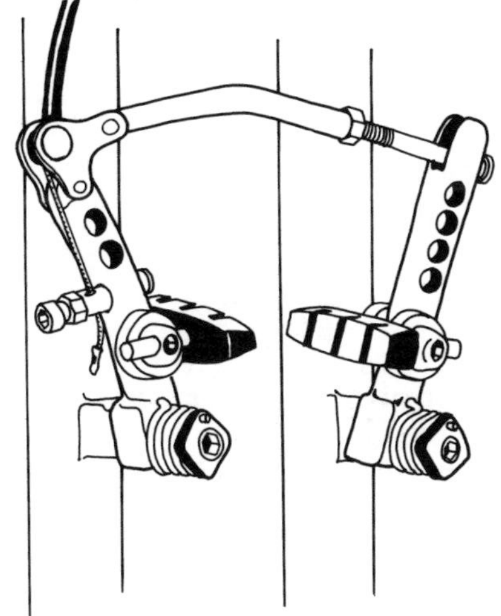

10.39 U-brake (usually under chainstays)

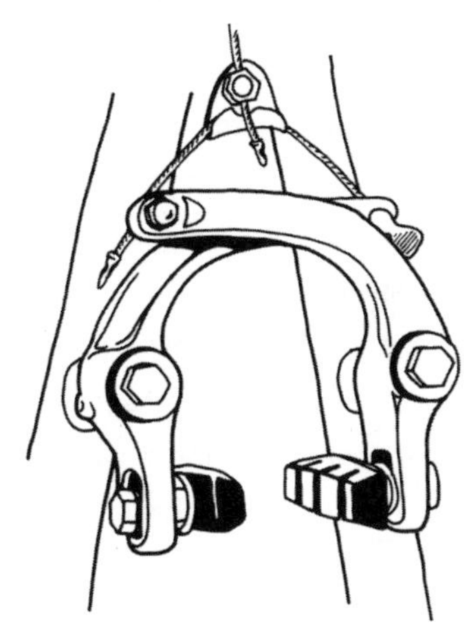

10.40 Roller-cam brake

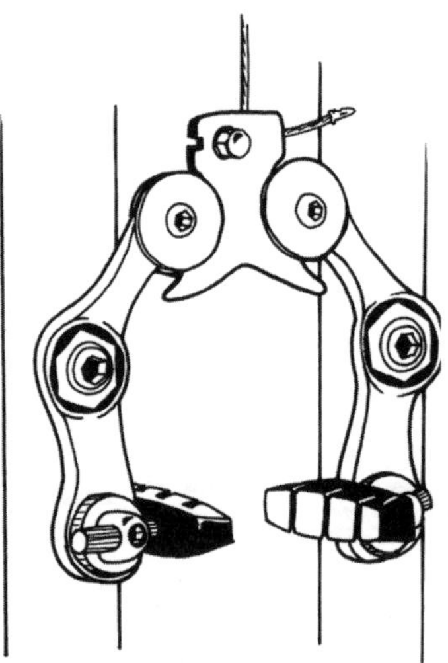

6. **With a chainstay-mounted U-brake, set the position of the straddle-cable yoke the easy way.** Loosen the straddle-cable bolt and squeeze the lever to the grip after slipping the yoke up against the bottom-bracket cable guide. Then tighten the bolt. This is the highest position on the cable and allows for the longest possible straddle cable.

10-24

ADJUSTING U-BRAKES

1. **Set the position of the rear straddle-cable yoke as outlined in step 6 in the previous section.** On a front brake, set the position about 2 inches (5cm) above the brake.
2. **Tighten the straddle cable while pulling the brake cable tight with pliers** (Fig. 10.26). Make sure you have tightened the anchor bolt enough that the cable does not slip.
3. **Check that you cannot pull the lever closer than a finger's width from the grip.** Tension the cable as needed with the straddle-cable yoke or the lever barrel adjuster (Fig. 10.1).
4. **Set the spring tension.** Release the straddle cable, loosen the mounting bolt, and swing the pad away from the rim; then tighten the mounting bolt. Center the brake by setting the spring tension on one arm first, followed by the other arm in the same fashion. If the brake has a small hex-head setscrew on the side of one arm, use it to make fine spring-tension adjustments.

10-25

REPLACING AND POSITIONING U-BRAKE PADS

U-brakes rely on brake pads with threaded posts. Install them with the original spacers in their original orientation. The pads should hit the center of the braking surface and should have a small amount of toe-in. There is no adjustment for spacing from the brake arm. Hold the pad in place with your hand while tightening the nut with a wrench. As the pads wear, they tend to slide up on the rim and hit the tire, so check this adjustment frequently. You should also regularly clear hardened mud from inside the brake arms; it can build up here on U-brakes and abrade the tire sidewalls.

ROLLER-CAM BRAKES

10-26

REMOVING AND INSTALLING ROLLER-CAM BRAKES

Roller-cam brakes (Fig. 10.40) mount on U-brake bosses attached to the fork and either the chainstays or the seatstays. These brakes are mounted farther from the hub than cantilevers or V-brakes, and, like U-brakes, they will not work on standard cantilever bosses.

Roller-cams are removed by first pulling the cam plate out from between the rollers on the ends of the arms. Remove the mounting bolts, and pull the arms off the bosses.

Installation is performed in reverse: Grease the bosses and the inside of the pivots as well as the edges of the cam plate and the mounting bolts.

10-27

ADJUSTING ROLLER-CAMS

1. **Check that the pulleys spin freely.** Loosen them with a 5mm hex key on the front and an open-end wrench on the back. The pulleys should rest on the

narrow part of the cam (Fig. 10.40), which gives the greatest mechanical advantage when the brakes are applied. Change pad spacing by changing the location of the cam on the cable.

2. **Center the brake.** Use a 17mm wrench on the nut surrounding the mounting bolt to adjust spring tension. Loosen the mounting bolt, make small adjustments to the 17mm nut, and tighten the mounting bolt again.
3. **Tighten the cam onto the cable.**

10-28

REPLACING AND POSITIONING ROLLER-CAM PADS

The pad eye bolt is held on the front with a 5mm hex key. The bolt in the back is adjusted by using a 10mm open wrench. The pads should be toed-in slightly. As the pads wear, they tend to slide up the rim and rub the tire, so check this adjustment periodically.

TROUBLESHOOTING BRAKES

The first thing to check with any brake is that it stops the bike!

1. **While the bike is stationary, pull each lever.** See that it firmly engages the brake while the lever is still at least a finger's width away from the handlebar grip. If not, skip to 10-2 to 10-4 for tensioning cable, or 10-20 to 10-22 for adjusting and bleeding hydraulic rim brakes.
2. **Moving at 10 mph or so, apply each brake one at a time.** By itself, the rear brake should be able to lock up the rear wheel and skid the tire, and the front brake alone should come on hard enough that it will cause the bike to pitch forward. Careful. Don't overdo it and hurt yourself testing your brakes!

   If you can't stop the bike quickly, you must make some adjustments and perhaps do a little cleaning. See the troubleshooting table on the next page for specific advice.

10.41 Removing dirt or embedded aluminum from brake pad

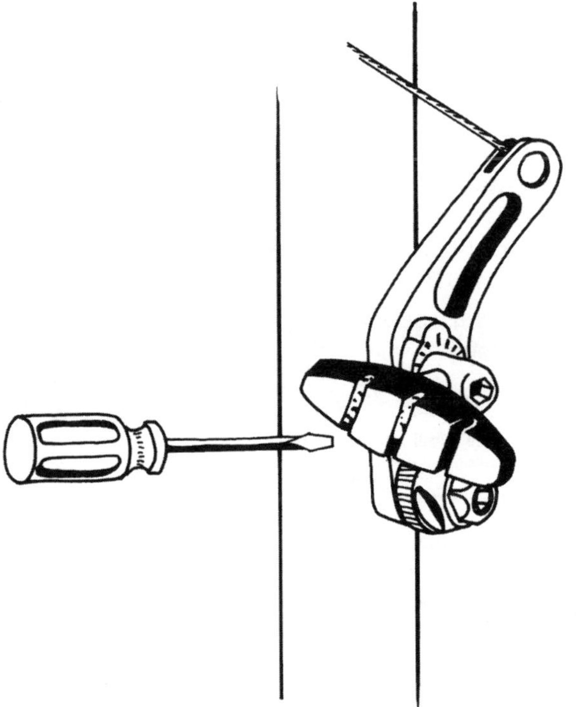

10.42 Worn brake pad

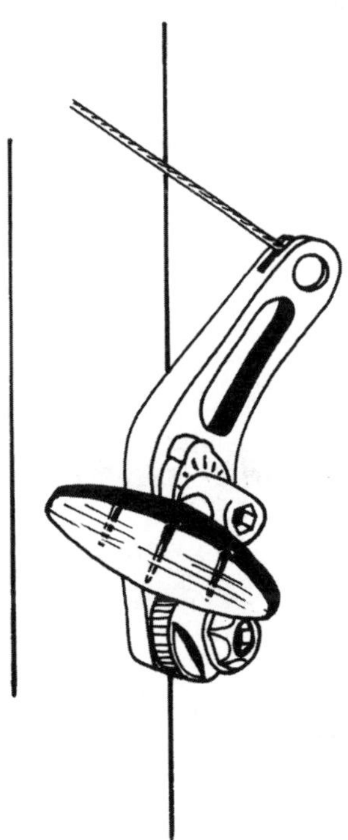

TABLE 10.1 — TROUBLESHOOTING RIM BRAKES

SQUEALING	
Pads glazed	Replace pads; sand surface of pad; dig embedded dirt or aluminum out of pads (Fig. 10.41).
Pads worn (Fig. 10.42)	Replace pads.
No pad toe-in with flexy brake arms	Toe pads in (Fig. 10.21).
Squealing or howling due to vibration	Tighten brake calipers on pivot bosses if loose.
Front cantilever shudder	Get V-brake or fork-mounted cable stop (Note in 10-16).
LOW POWER	
Sticking cable	Lubricate cable; replace cable and housing.
Stretching of cable	Tighten cable; replace cable and housing.
Frayed cable	Replace cable and housing.
Compression of brake housing	Replace cable and housing.
Insufficient friction between pads and rim	Clean pads and rims; try different pads.
Oil and grime on the rim and pads	Clean rims and pads; sand pads.
Hydraulic rim brake has air in it	Check for leaks; bleed the system.
Poor pad alignment	Align pads with rim (Figs. 10.13 and 10.19 to 10.21).
High straddle-cable angle	Set angle of brake arms and straddle cable close to 90 degrees (10-17, Figs. 10.24 and 10.25).
Flexing seatstays or fork legs	Install horseshoe-shaped "brake booster" bridging mounting bolts for V-brakes, cantilevers, or hydraulic rim brakes (Fig. 10.33).
Flexing of brake arms	Install new brakes; eliminate other factors first.
TOO MUCH LEVER TRAVEL	
Cable too long	Tighten cable (10-2 to 10-4).
Brake pads worn	Replace pads.
Poor pad alignment	Align pads with rim (Figs. 10.13 and 10.19 to 10.21).
Hydraulic rim brake has air in it.	Check for leaks; bleed the system (10-22).
See "Low Power" above	
PAD DRAG	
One pad rubs	Adjust spring balance on brake arms; lubricate arm on brake pivot boss if sticking; true wheel if only hitting at certain points (8-2).
Both pads rub	Loosen cable; increase lever reach (10-9); adjust pads so they're closer to brake arms; true wheel if only hitting at certain points.
SLOW OR INCOMPLETE LEVER RETURN	
Cable sticking	Lubricate cable; thaw and dry it if frozen; replace cable and housing.
Sharp bends in cable housing	Replace cable and housing, using longer housing with smooth curves.
Frayed cable	Replace cable and housing.
Lip on top edge of pad	Slice or sand off lip and align pads with rim (Figs. 10.13 and 10.19 to 10.21).
LOOSE BRAKE ARM	
Brake arm rattles or clunks	Tighten mounting bolt on brake arm.
GRINDING NOISE	
Sounds like there is sand on the pads	Replace pads; dig embedded dirt or aluminum out of pads (Fig. 10.41).
SOFT LEVER THAT GETS FIRMER WHEN PUMPED (HYDRAULIC BRAKES ONLY)	
Air in hydraulic brake system	Check for leaks; bleed the system.

If you don't have time to do it right the first time, you must have time to do it over again.

—ANONYMOUS

11

CRANKS AND BOTTOM BRACKETS

The crankset, your bike's power center, consists of a pair of crankarms, a bottom bracket (a.k.a. BB), and chainrings (Fig. 11.1), all attached by chainring bolts and crank bolt(s). A traditional bottom bracket is a spindle supported by bearings that thread into the BB shell. However, on modern, integrated-spindle (a.k.a. two-piece) cranksets (Fig. 11.2), it's less clear where the bottom bracket ends and the crankarms begin, because the spindle is integrated as a single piece with the right crankarm. The bearings are thus usually called the bottom bracket on a two-piece crank.

CRANKARMS AND CHAINRINGS

11-1

REMOVING AND INSTALLING CRANKS

Most modern cranks are easy to remove with a single hex key. The two-piece (Fig. 11.2) design means that only the left arm needs to come off; the spindle pulls out of the bearings along with the right crankarm.

NOTE: *"Right" refers to the drive side of the bike, and "left" refers to the non-drive side.*

To remove a traditional three-piece crankset (Fig. 11.1), you must remove the crank bolt and the washer under it, and then you will need a separate tool to pull the crank off. You will need either a thin-wall 14mm, 15mm, or 16mm socket wrench or a large 8mm hex key in order to remove the crank bolt. You will generally need a crank puller (Fig. 11.6) to take off the crankarms.

Removal

a. Integrated-spindle cranks with two pinch bolts on the left arm

1. **Unscrew the cap from the left arm completely.** Shimano models require a special splined tool; others require a hex key.
2. **Loosen the two pinch bolts holding the arm onto the spindle** (Fig. 11.3). Use a 5mm hex key.
3. **Pull off the left arm.**
4. **Pull the right arm (and attached spindle) straight out.** You may need to tap the end of the spindle with a rubber mallet to get it started. If there is a bearing seal stuck on the spindle, leave it there; when you install the crank, the seal will go back against the bearing.

TOOLS

- Torx T30 wrench or socket driver
- 14mm socket wrench
- 3/8-inch drive ratchet or torque wrench
- 3/8-inch drive extension
- Chainring nut tool
- External-bearing bottom-bracket wrench or socket
- Shimano TL-FC16 splined left arm cap installation tool or equivalent
- Internal bottom-bracket splined socket with large opening for ISIS/Octalink
- Snapring pliers
- Crank puller with ends for square-taper and splined spindles
- Pin spanner (or adjustable pin tool)
- Toothed lockring spanner
- Adjustable wrench
- Pliers
- Flat hand file
- Razor blade or box cutter
- Grease
- Rubber mallet

CONTINUED ON P. 236

TOOLS, CONT.

OPTIONAL

- Shimano TL-FC35 XTR crank-removal tool
- Shimano TL-FC17 XTR bearing-preload adjustment tool
- Headset press
- Notched external-bearing bottom-bracket wrenches or sockets (or step-down inserts for wrenches or sockets) in all sizes: 39mm, 41mm, 44mm, 46mm, and 48.5mm
- Park BBT-30.4 bearing removal tool/installation bushings for 30mm-spindle bottom brackets
- Park BBT-90.3 bearing removal tool/installation bushings for 24mm-spindle bottom brackets
- 15mm and 16mm socket wrenches
- Dust cap pin tool
- Threadlock compound
- External-bearing bottom-bracket wrench or socket for all sizes of 16-notch and 12-notch cups—39mm, 41mm, 44mm, 46mm, and 48.5mm
- Rounded-nose cylindrical collet expander
- Bearing puller

11.1 Square-taper crankset, exploded view

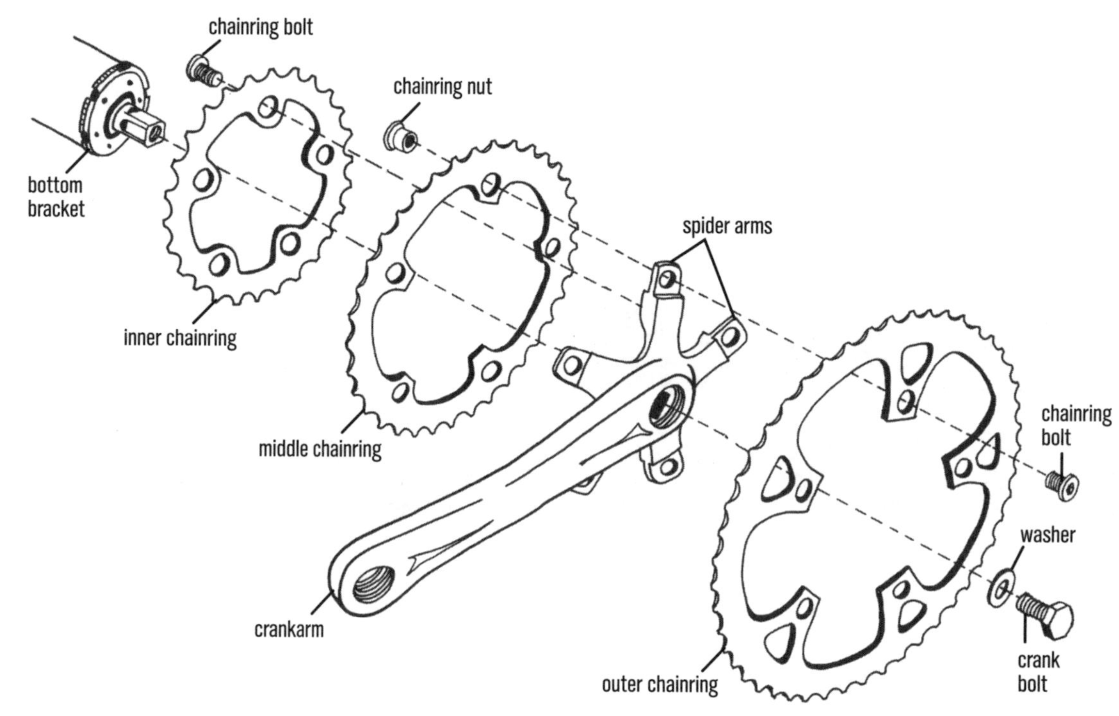

11.2 Integrated spindle (a.k.a. two-piece) crankset, exploded view (2003–2005 Shimano XTR FC-M960 shown); the three spacers are to fit a 68mm-long bottom-bracket shell—only one spacer (drive side) is required with a 73mm shell

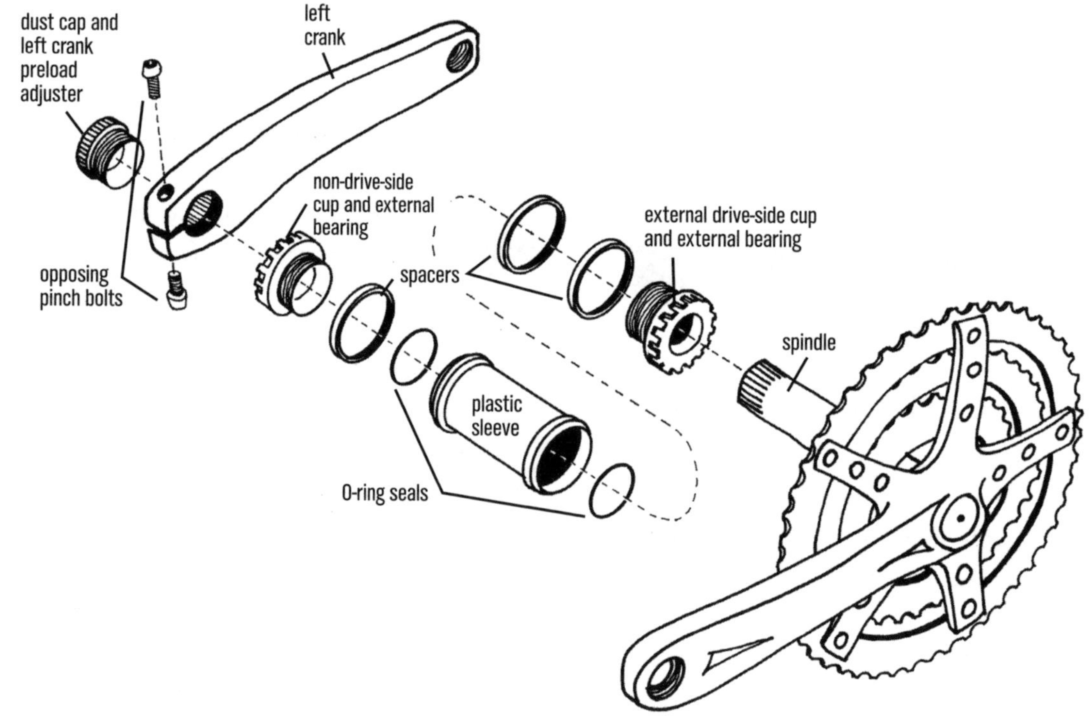

11.3 Removing and installing left two-bolt crankarm (Shimano HollowTech II shown)

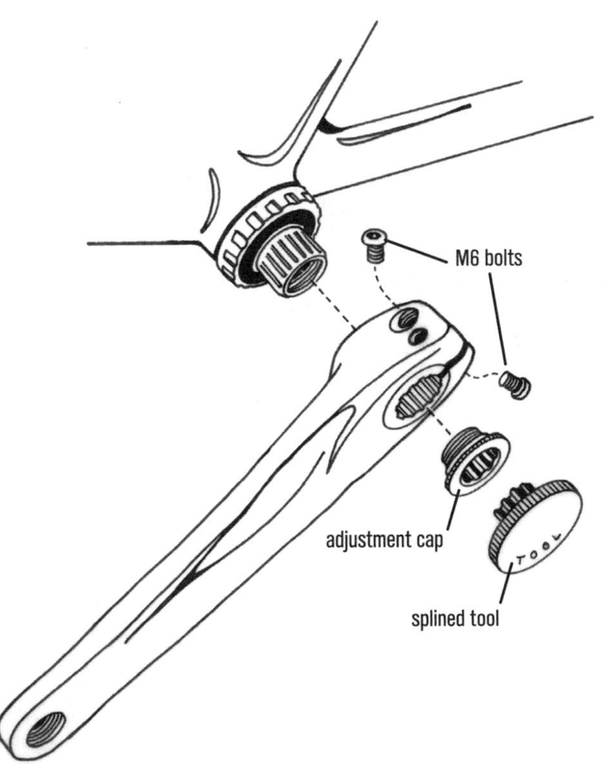

b. Integrated-spindle cranks with a single crank bolt

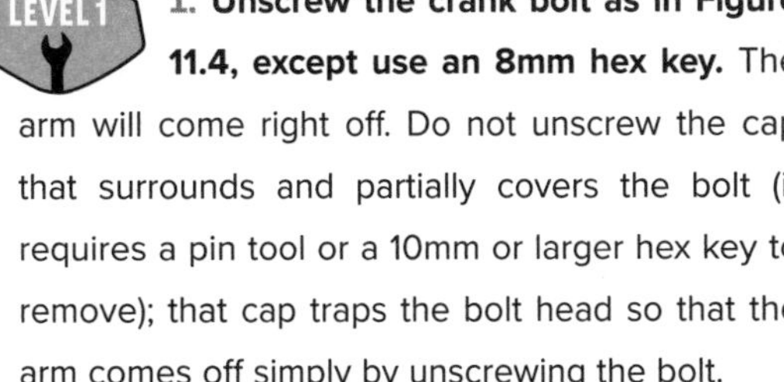

1. **Unscrew the crank bolt as in Figure 11.4, except use an 8mm hex key.** The arm will come right off. Do not unscrew the cap that surrounds and partially covers the bolt (it requires a pin tool or a 10mm or larger hex key to remove); that cap traps the bolt head so that the arm comes off simply by unscrewing the bolt.

NOTE: *The 2007–2010 Shimano XTR (FC-M970) cranks (Fig. 11.5) don't have this self-extractor cap and require a separate tool for crank extraction. Unscrew the plastic retainer ring around the bolt head—it unscrews clockwise(!) because it has left-hand threads. Thread the TL-FC35 tool shown in Figure 11.5 into the arm (counterclockwise) to hold the bolt head in; tighten it in as far as it will go, at least 3.5 turns. Now unscrew the bolt; it will push the arm off.*

2. **Pull the right arm (and attached spindle) straight out.** Tap the end of the spindle with a mallet if it's stuck. If a bearing seal comes off with the arm, you can clean it in place or pull it off and put it back on the bearing.

11.4 Removing and installing a crank bolt

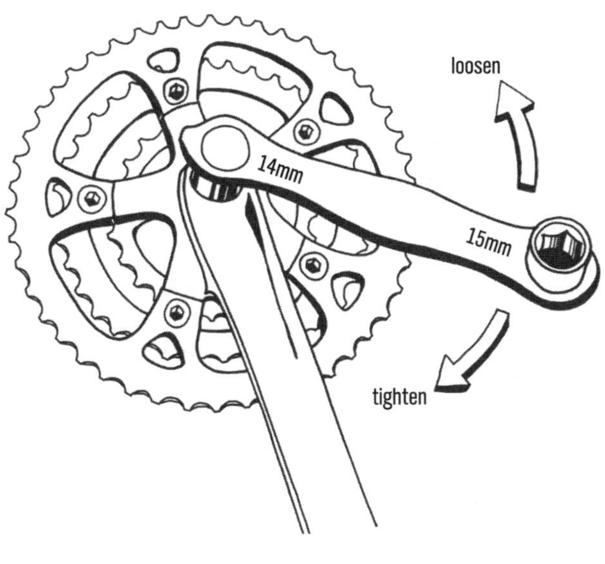

NOTE: *Race Face X-Type cranks are the opposite of other integrated-spindle cranks; the left arm is fixed to the spindle. The drive-side arm comes off with an 8mm hex key.*

c. Three-piece cranks (square taper, Shimano Octalink, and ISIS)

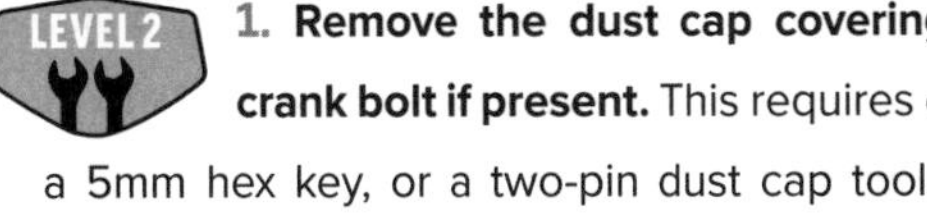

1. **Remove the dust cap covering the crank bolt if present.** This requires either a 5mm hex key, or a two-pin dust cap tool, or a screwdriver.
2. **Remove the crank bolt with the appropriate wrench** (Fig. 11.4). Make sure that you extract the washer (Fig. 11.1) with the bolt. If you leave it in, you will not be able to pull the crank off. If it remains in the crank, fish it out with a thin screwdriver or a pick.

NOTE: *Some cranks that accept a hex key in the crank bolt are self-extracting and don't require a crank puller (Fig. 11.6). The crank bolt is held down by a retaining ring threaded into the crank; as the bolt is unscrewed, it pushes on the ring and pushes the crank off.*

Square taper, Octalink, and ISIS are three different bottom-bracket and crankarm interface standards. Square-taper bottom-bracket spindles are square on the end (Figs. 11.1, 11.20, 11.23, and 11.24) and fit into a square hole in the crankarm. The spindle ends are tapered to tighten into the crank as the arm is pushed

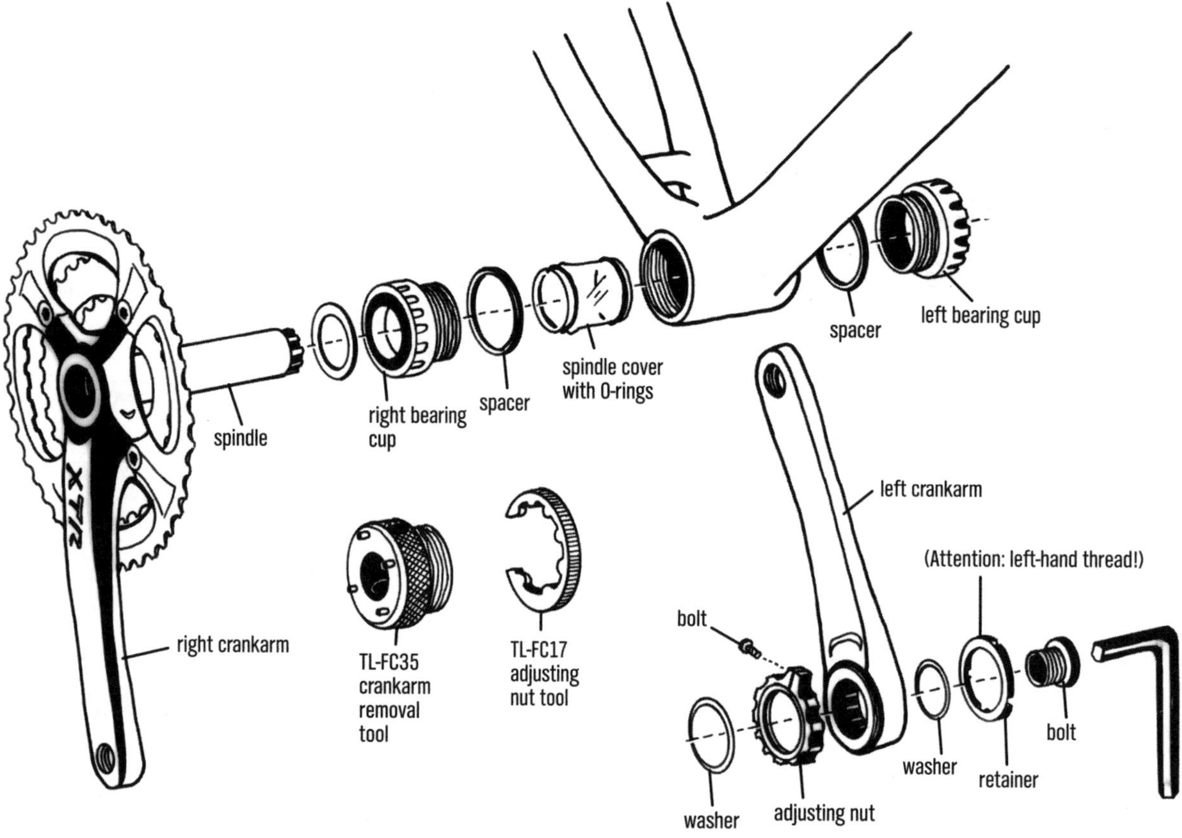

11.5 Removing and installing a Shimano 2007–2010 XTR FC-M970 crankset

into the spindle. ISIS (Fig. 11.20) and Shimano Octalink (Fig. 11.21) are hollow spindles (a.k.a. pipe spindles) with longitudinal splines on the ends.

3. **Unscrew the center push bolt of the crank puller** (Fig. 11.6) **so that its tip is flush with the face of the tool.** Make sure the flat end of the push bolt is the right size for the bottom bracket; the push bolt end is much smaller for a square-taper spindle than for an ISIS or a Shimano Octalink splined spindle.
4. **Thread the crank puller into the hole in the crankarm.** Be sure that you thread it in (by hand) as far as it can go; otherwise, you will not engage sufficient crank threads when you tighten the push bolt, and you will damage the threads. Future crank removal depends on those threads being in good condition.
5. **Tighten the push bolt clockwise.** Keep turning (Fig. 11.6) until the crankarm pulls off the spindle.
6. **Unscrew the puller from the crankarm.**

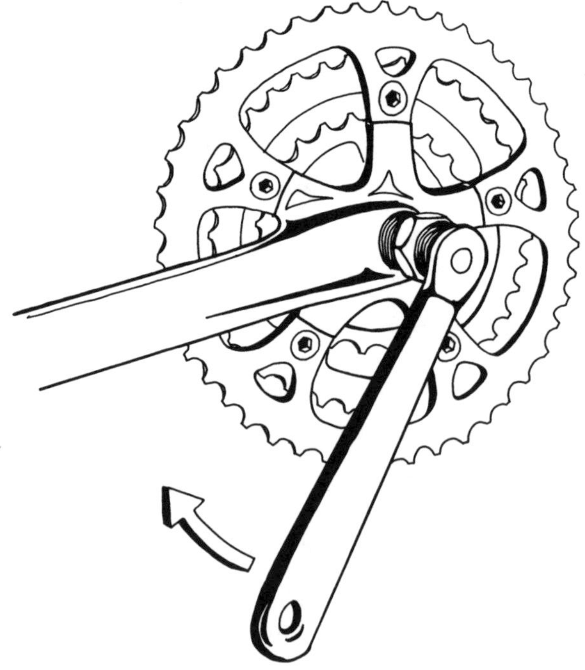

11.6 Using a crank puller

Installation

a. Integrated-spindle cranks with two pinch bolts on the left arm

1. **Grease the spindle tip and the bore of each bearing.**
2. **Push the spindle (which is attached to the right crankarm) in through the bearings from the drive side.**
3. **Slide the left arm onto the end of the spindle.** Check that the crank is at 180 degrees from the right arm.
4. **Gently tighten the left-side adjustment cap.** Use the special plastic splined cap tool for Shimano or a hex key for FSA (Fig. 11.3). Torque is not tight—0.4–0.7 N-m—just enough to pull the right and left cranks over against the bottom-bracket cups.
5. **Tighten the two opposing (greased) pinch bolts** (Fig. 11.3). Using a 5mm hex key, alternately tighten each bolt one-quarter turn at a time. Torque is 10–15 N-m.
6. **Recheck the torque after one ride as the crank may settle in and the bolt will need retightening.**

b. Integrated-spindle cranks with a single crank bolt

1. **Grease the spindle tip and the bore of each bearing.**
2. **Push the spindle (which is attached to the right crankarm) in through the bearings from the drive side.**
3. **Slide the left arm onto the end of the spindle.** Check that the crank is at 180 degrees from the right arm.
4. **Tighten the crank bolt with an 8mm hex key.** Torque is high; see Appendix D.
5. **Recheck the torque after one ride as the crank may settle in and the bolt will need retightening.** Periodically check the torque from then on.

NOTE: *These instructions also apply to Race Face X-Type cranks except with left and right interchanged, because the left Race Face crankarm is integrated with the spindle.*

ANOTHER NOTE: *Installation of 2007–2010 Shimano XTR (FC-M970; Fig. 11.5) follows the previous instructions with these exceptions: Before installing the left crankarm, tighten the adjuster nut (adjustment ring) to eliminate the gap between it and the crankarm. Install the arm. Adjust the bearing play by turning the adjustment ring with the TL-FC17, and then tighten its pinch bolt with a 2.5mm hex key. Screw in the crankarm retainer ring counterclockwise with the TL-FC35. When new, the crankarm retainer ring is already installed; check that it is tight with the TL-FC35.*

c. Three-piece cranks (square taper, ISIS, and Octalink)

1. **Slide the crankarm onto the bottom-bracket spindle.** With square-taper spindles, clean all grease from both parts. Grease may allow the soft aluminum crank to slide too far onto the spindle and deform the square hole in the crank. With an ISIS or a Shimano Octalink splined spindle, however, do grease the parts. With ISIS and Octalink cranks you must be careful to line up the crank splines with those on the spindle before tightening the crank bolt.
2. **Install the crank bolt.** Apply grease to the threads and tighten (Fig. 11.4). Apply titanium-specific anti-seize compound for titanium spindles and for titanium crank bolts. If you have aluminum or titanium crank bolts, first tighten the cranks with the greased steel bolt to the specified torque; then replace the steel bolt with the lightweight bolt, and tighten it to spec.

NOTE: *Here is where a torque wrench comes in handy. Tighten the bolt to about 32–49 N-m (300–435 in-lbs), and as high as 59 N-m for some steel oversized bolts in ISIS spindles (see Appendix D). If you're not using a torque wrench, make sure the bolt is really tight, but don't muscle it until your veins pop.*

3. **Replace the dust cap.**
4. **Check front-derailleur adjustment** (see Chapters 5 and 6). Removing and reinstalling the right crankarm could affect chainring position and hence shifting.
5. **Recheck the torque after one ride as the crank may settle in and the bolt will need retightening.** Periodically check the torque from then on.

11-2

MAINTAINING CHAINRINGS

The chainring teeth should be checked periodically for wear, the chainring bolts should be checked periodically for tightness, and the chainrings themselves should be checked for trueness—watch them as they spin past the front derailleur.

Chainrings for "one by" drivetrains (i.e., 1×11 or 1×12, having a single front chainring and no front derailleur) generally have, for improved chain retention, much taller teeth than chainrings on double and triple cranks. Also for improved chain retention, the teeth alternate between fat and thin; the fat teeth fit into the space in the chain between outer link plates, and the thin teeth go into the narrower space between inner link plates. Make sure the chain is put onto the ring properly, and check that there is no damage to the teeth that would cause the chain to lift off the chainring.

1. **Wipe the chainring down and inspect each tooth.** The teeth should be straight and uniform in size and shape. If the teeth are hook-shaped (Fig. 11.7), the chainring needs to be replaced. The chain should be replaced as well (see Chapter 4), because a worn chain rapidly wears the teeth into a hook shape.

CAUTION: *Don't be deceived by the seemingly erratic tooth shapes designed to facilitate shifting; you can tell the pattern is intentional if it repeats regularly. Shifting ramps located on the inner side (Fig. 11.8) that are meant to speed chain movement between the rings often look like cracks on cheaper chainrings, because they are pressed into the ring rather than being a separate piece riveted on, as on better chainrings.*

NOTE: *Another wear-evaluation method is to lift the chain from the top of the chainring; the greater the wear of either part, the farther the chain separates. If it lifts more than one tooth, the chain, and perhaps the chainring, need to be replaced.*

2. **Remove minor gouges and small burrs in the chainring teeth with a file.**
3. **If an individual tooth is bent, try bending it back carefully with a pair of pliers or an adjustable wrench.** It will likely break off; heed the message and buy a new chainring.
4. **While slowly turning the crank, watch where the chain exits the bottom of the chainring.** See if any of the teeth are reluctant to let go of the chain. That can cause chain suck. Locate any offending teeth, and see if you can correct the problem with pliers and a file. If the teeth are really chewed up or cannot be improved, the chainring should be replaced.

11.7 Worn chainring teeth

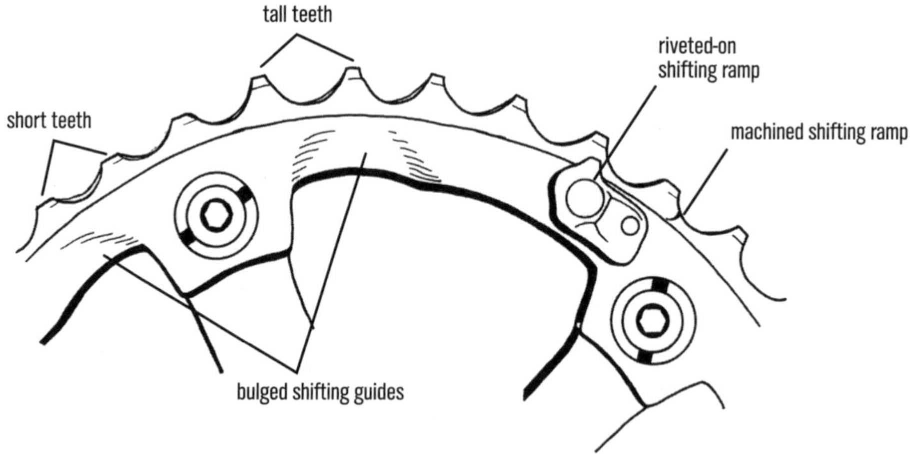

11.8 Chainring shifting ramps and asymmetrical teeth

11-3

TIGHTENING CHAINRING BOLTS

Check that the bolts are tight by turning them clockwise while holding the nut on the backside from turning. Most chainring bolts take either a 5mm hex key or a Torx T30 key on the bolt and a two-pronged chainring-nut tool (Fig. 11.9) or a 6mm hex key on the nut. Grease or thread-lock compound on the bolts can prevent creaking.

11-4

FIXING WARPED CHAINRINGS

Looking down from above, turn the crank slowly and see whether the chainrings wobble back and forth relative to the plane of the front derailleur. If they do, make sure there is no play in the bottom bracket by adjusting the bottom bracket (11-10, step 15). It is normal to have a small amount of chainring and crank flex when you pedal hard, but excessive wobbling will compromise shifting. Small, localized bends can be straightened with an adjustable wrench (Fig. 11.10). If the chainring is really bent, replace it.

11-5

REPAIRING BENT CRANKARM SPIDERS

If you installed a new chainring and are still seeing serious back-and-forth wobble, chances are good that the spider arms (Fig. 11.1) on the crank are bent. If not, then it has to be a bent spindle. Either way, if the crank is new, this is a warranty item, so return it to your bike shop for replacement.

11-6

REPLACING A CHAINRING

a. Replacing either of the two largest chainrings

LEVEL 1

1. **Unscrew the chainring bolts** (Fig. 11.9). You'll need either a 5mm hex key or a Torx T30 wrench. Hold the nut on the backside if it turns. The nut will take either a pronged chainring-nut tool (Fig. 11.9) or a thin screwdriver in a pinch, or a 6mm hex key.

11.9 Removing and installing chainring bolts

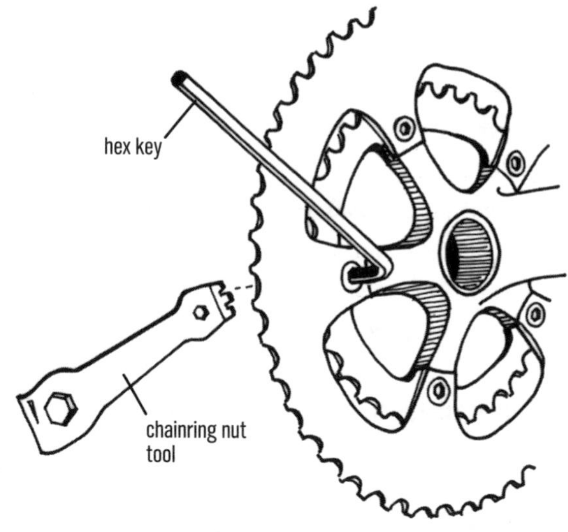

11.10 Straightening a bent chainring

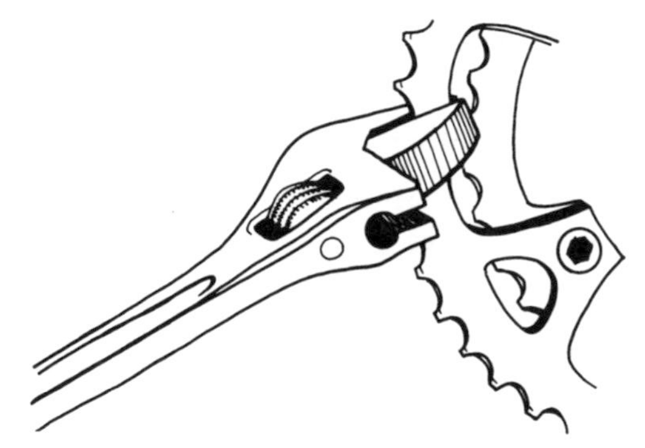

11.11 Outer and middle chainrings

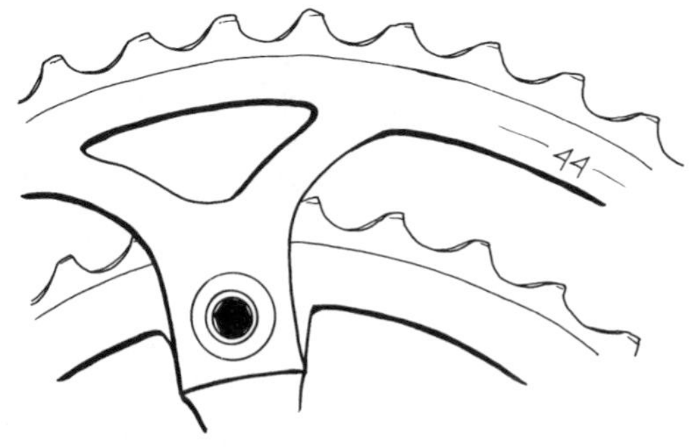

2. **Install the new chainrings.** The outer chainring has a protruding pin meant to keep the chain from falling down between it and the crankarm. Make sure this pin lines up with the crankarm and faces the arm. The middle and inner rings each have a little bump protruding radially inward that is also to line

up under the crankarm. The chainring bolt holes will have recesses for the heads of the chainring bolts and nuts; make sure that these recesses face away from the spider arm tabs. If the chainrings are inverted or clocked wrong relative to the crank, the shift ramps (Fig. 11.8) will not work.

3. **Lubricate the bolts and tighten them** (Fig. 11.9).

NOTE: *Whenever you change the outer chainring size, you must reposition the front derailleur for proper chainring clearance, as described in Chapters 5 and 6.*

b. Replacing the innermost chainring (a.k.a. "granny gear")

1. **Pull off the crankarm** (11-1).
2. **Remove the bolts securing the chainring.** These bolts are threaded directly into the crankarm (Fig. 11.1) and take either a 5mm hex key or a Torx T30 wrench.
3. **Install the new chainring.** Its little bump, which protrudes radially inward, lines up under the crankarm. Make sure that the chainring is oriented so that the recesses for the heads of the chainring bolts receive those parts and are not facing inward toward the spider.
4. **Lube and tighten the bolts.**

NOTE: *Some cranks do not accept separate chainrings; the chainrings come on and off as a set, or they are not removable. The 1996–2002 Shimano XTR and 1997–2003 XT cranks rely on a slip-on spider system that spins off all three chainrings from the crankarm as a single unit (Fig. 11.12). After removing a circlip (by prying it off with a screwdriver), use a special lockring tool to loosen the chainring-spider-securing lockring; a female-threaded tool that goes on the crank bolt holds the lockring tool in place (Fig. 11.12). Once the spider is off, you can interchange chainrings within the set or simply pop on a whole new set.*

Economical cranks often have chainrings riveted to the crank or riveted to each other and bolted to the crank as a unit. In either case, if you want to replace a chainring, you must replace either the entire crank or the chainring set.

5. **Replace the crankarm** (11-1).

Go ride your bike.

11.12 Removing Shimano 1996-2002 XTR FC-M950/952 chainrings

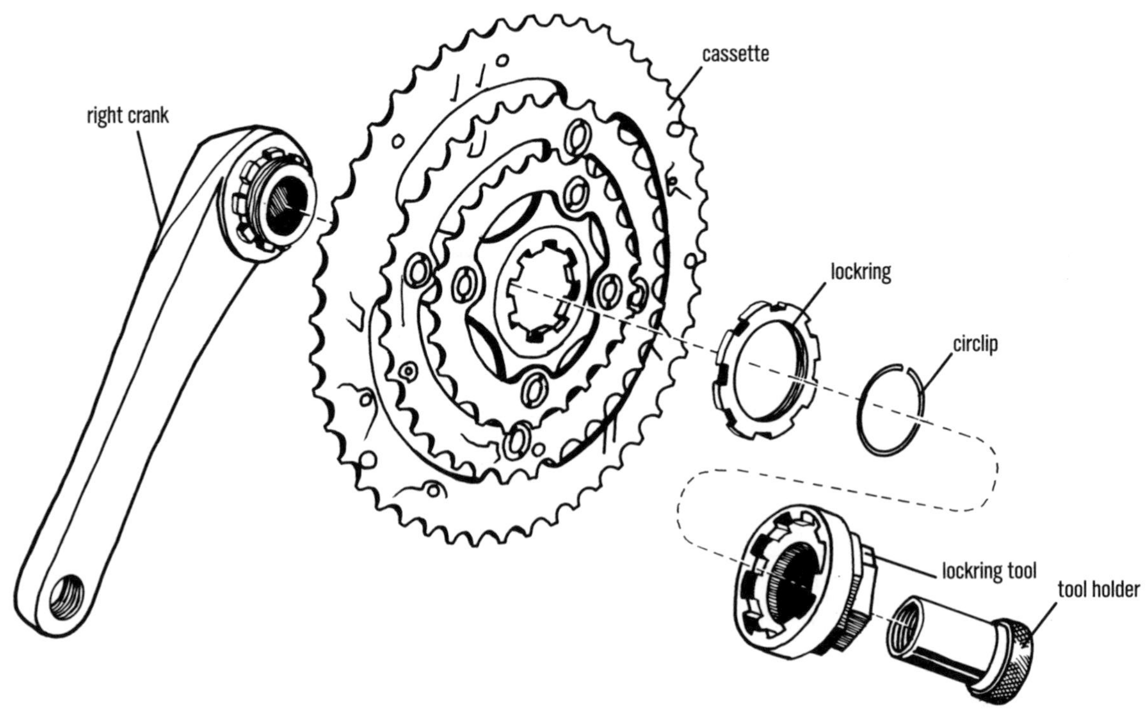

BOTTOM BRACKETS

LEVEL 2

Traditionally, mountain-bike bottom brackets threaded into the frame's bottom-bracket shell. Originally, these were square-taper bottom brackets (Fig. 11.13), and the cranks were separate from the bottom-bracket spindle (three-piece cranks). External-bearing bottom brackets (Fig. 11.14) for two-piece (integrated-spindle) cranks (Fig. 11.2) then took over.

Recently, however, the trend has been toward frames with unthreaded bottom-bracket shells. There are a number of different widths and diameters of these.

a. Threadless bottom brackets

Eliminating bottom-bracket threads is convenient for factory assembly and may allow the use of full-carbon bottom-bracket shells, so threadless bottom brackets with press-in bearings (Figs. 11.15 to 11.17) abound. However, there is no standard for width or diameter of the frame's bottom-bracket shell, and fat bikes with

11.13 Bottom-bracket assembly

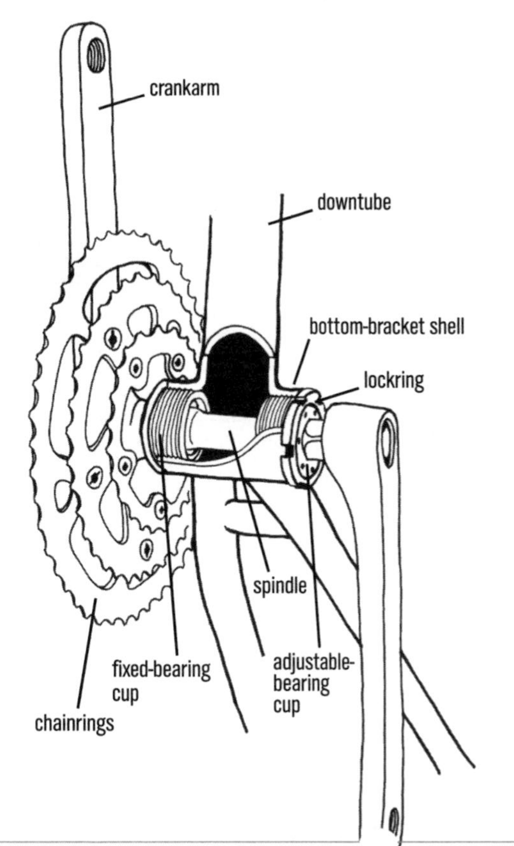

11.14 Removing and installing external-bearing cups; note left-hand threads on drive side, whether the threads are standard BSA or larger T47 size. The 2.5mm spacer on each side is for a SRAM/Truvativ GXP crank on a 68mm bottom-bracket shell; no spacers are to be used with a GXP crank on a 73mm bottom-bracket shell. One drive-side spacer is standard with a Shimano or FSA 24mm-spindle crank on a 73mm bottom-bracket shell.

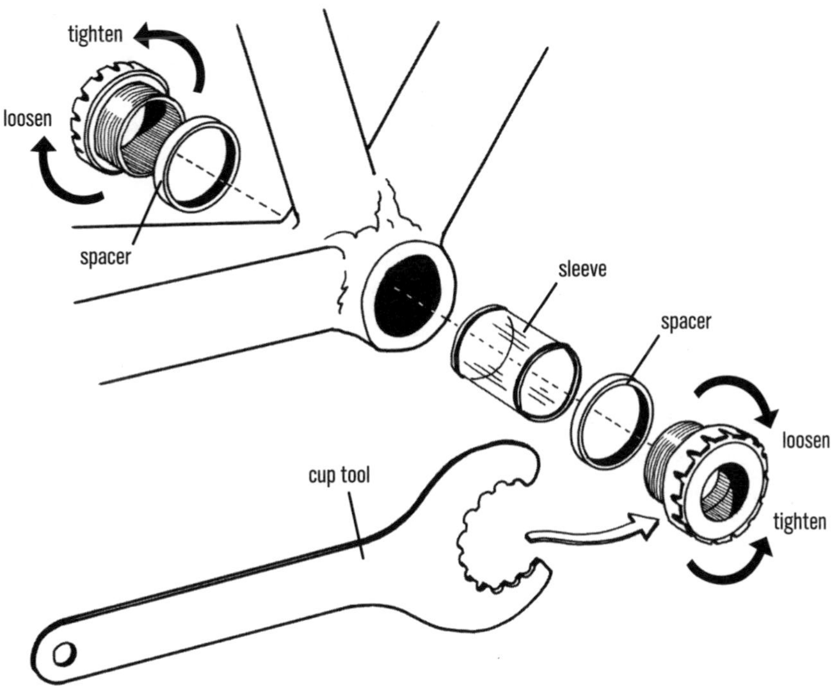

11.15 BB30 crankset, exploded view

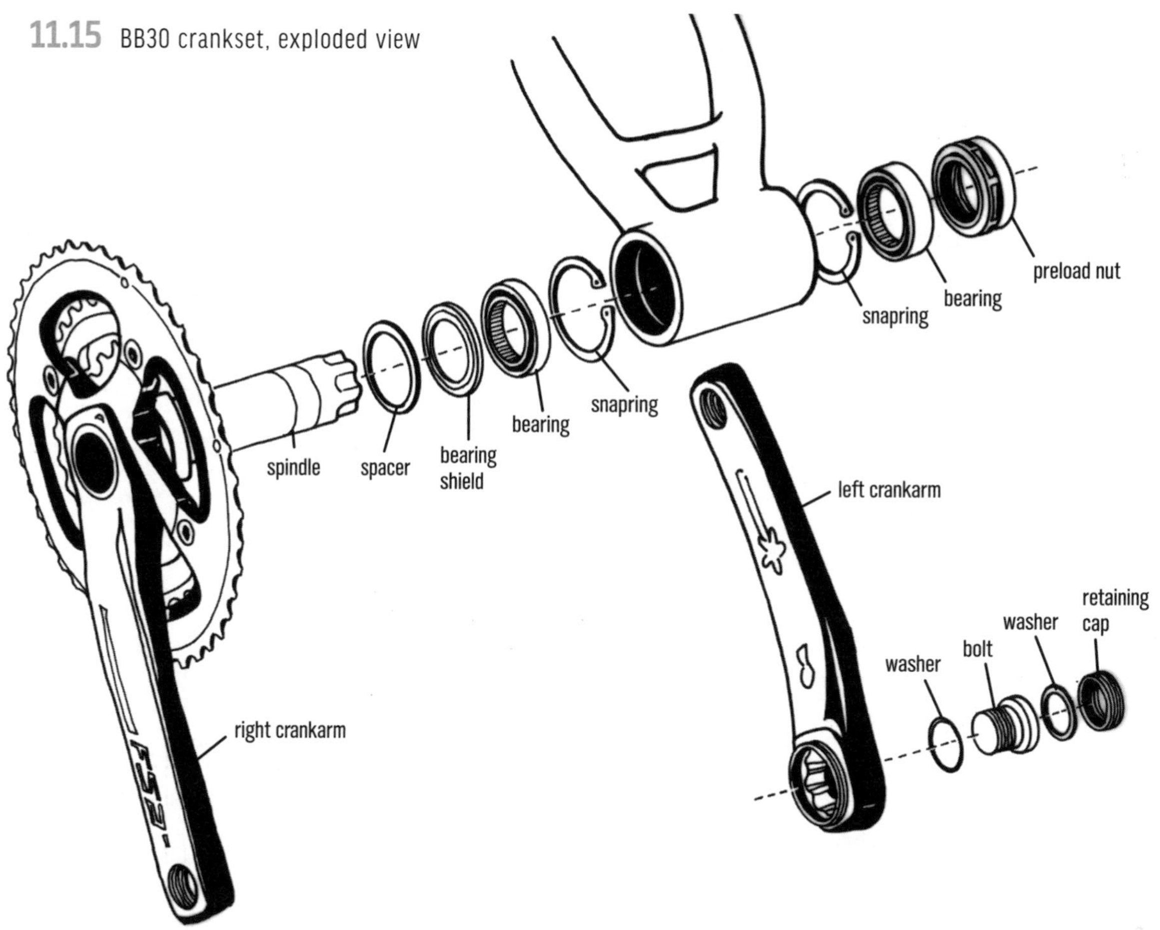

11.16 PressFit 30 (PF30) or BB392 EVO bottom bracket; a PF24 bottom bracket also looks identical, but with 24mm ID bearings, rather than 30mm ID

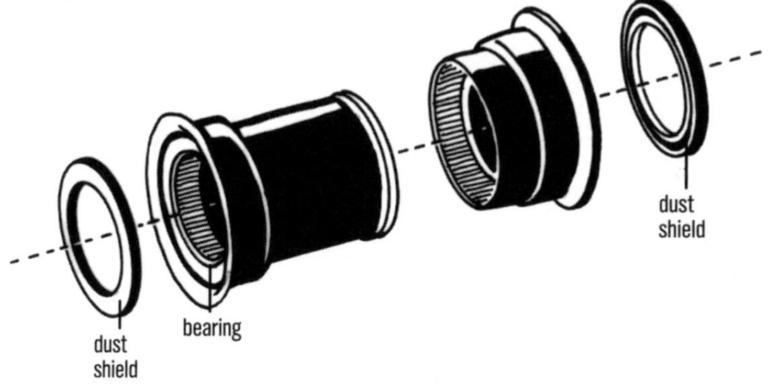

100mm- and 121mm-wide shells make such a standard yet more elusive. Apart from some early designs for three-piece cranks, threadless bottom brackets take integrated-spindle cranks with either 24mm or 30mm spindles.

A bottom bracket pressed into an unthreaded shell, if the fit is not perfect, can create annoying creaking while pedaling. Some aftermarket bottom-bracket makers (Wheels Mfg., Enduro Bearings, Praxis Works, etc.) have largely solved this creaking problem by making bottom brackets for threadless shells that thread together—one aluminum cup has female thread and the other has male thread (Figs. 11.18 and 11.19), and they meet in the middle. Threading the two cups tightly to each other and against the faces of the shell can prevent the movement of the cups in the shell that causes creaking.

11.17 Trek BB95 with Truvativ GXP crankset, exploded view

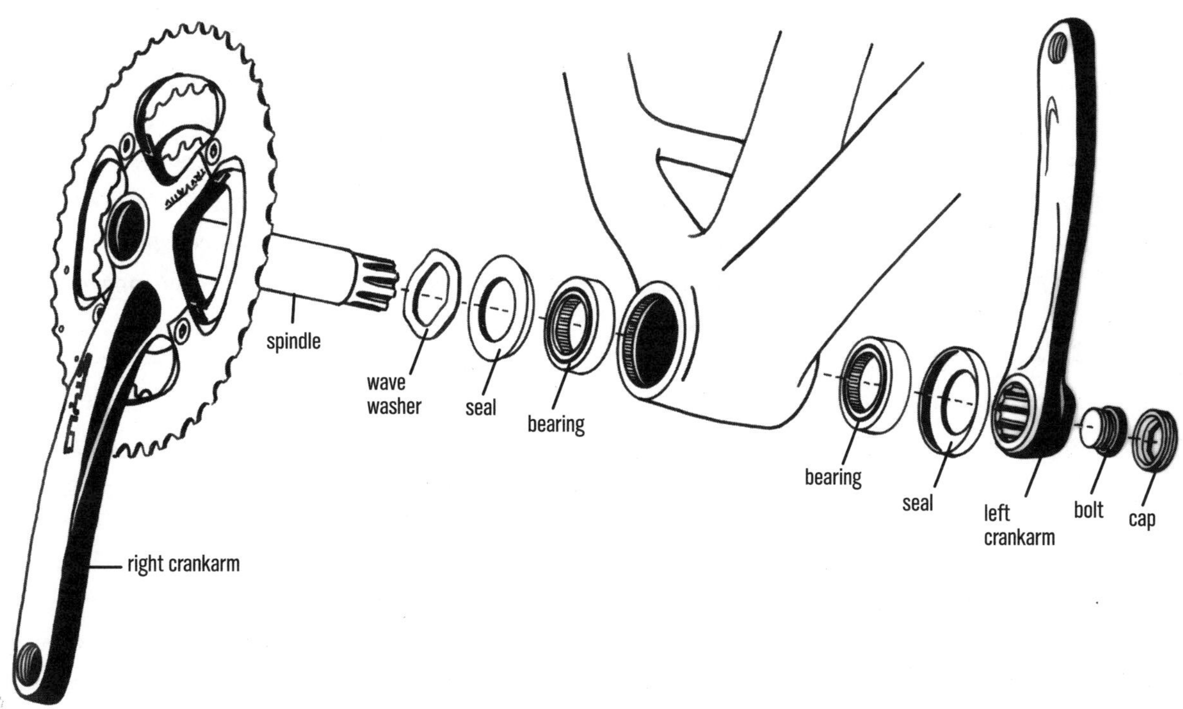

11.18 Wheels Mfg. thread-together bottom bracket with external bearings for a 24mm spindle for BB30 or PF30 bottom-bracket shells

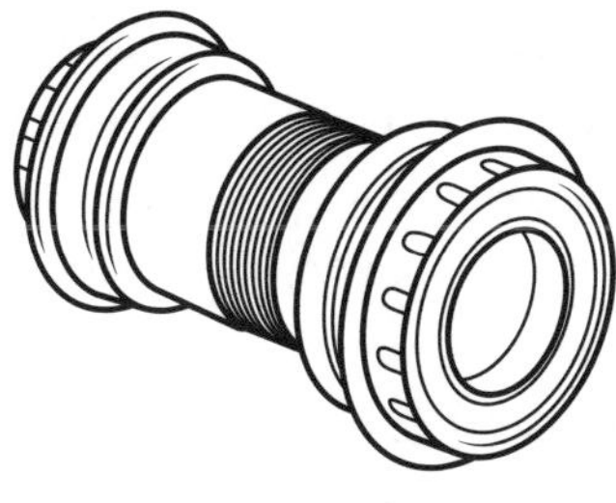

11.19 Wheels Mfg. thread-together bottom bracket with internal bearings and a thin, notched flange to replace press-in bottom bracket for 30mm spindles in PF30 or BB392 EVO, for 24mm spindles in BB92 (PF24), and in longer versions for fat bikes in both spindle sizes with 100mm- and 121mm-long bottom-bracket shells

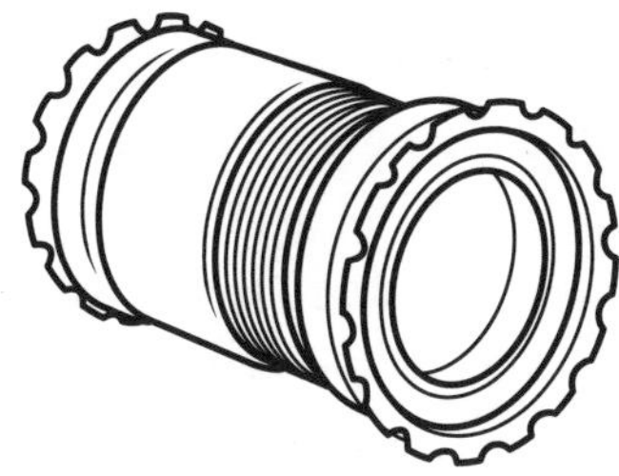

b. BB30, PF30, PF121, and BB392 EVO

BB30, PressFit 30 (a.k.a PF30), PF121, and BB392 EVO cranks all have 30mm-diameter bottom-bracket spindles that fit inside 30mm inside diameter (ID) bearings.

On BB30 mountain bikes, the bearings press into a 68mm- or 73mm-wide, 42mm-ID bottom-bracket shell and are prevented from going in deeper by snaprings seated in grooves in the inner diameter of the shell (Fig. 11.15).

PF30 bearings are the same size as BB30 bearings (42mm outside diameter [OD], 30mm ID), but they are housed inside 46mm-diameter plastic sleeves (Fig. 11.16) that press straight into the 46mm-ID bottom-bracket shell without any need for snaprings (aluminum bearing cups are available aftermarket). PF30 mountain bikes have 73mm-wide bottom-bracket shells, while PF121 fat bikes have 121mm-wide shells with the same inside diameter. Thread-together aluminum bottom brackets for PF30 bottom-bracket shells (Figs. 11.18 and 11.19) offer a more secure (and creak-free) fit than a press fit. These come in external-bearing types for cranks with 24mm integrated spindles (Fig. 11.18) that install with a 44mm-diameter notched bottom bracket wrench (also available in 42mm-OD size for BB30 shells). They also are available in internal-bearing models for 30mm spindles (Fig. 11.19) and require a 48.5mm-diameter notched bottom bracket wrench to engage their thin, notched flanges.

BB392 EVO bottom brackets are essentially the same as PF30 bottom brackets; the difference is that the BB392 EVO bottom-bracket shell is 92mm wide (86mm for road BB386 EVO), rather than 73mm wide (or 68mm for road PF30). The key difference is the length of the 30mm-diameter spindle; a BB392 EVO spindle is almost 2cm longer than a BB30 crank spindle. This takes away the ankle-clearance advantage of a BB30 crank, but it offers much greater versatility in terms of bikes the crank will fit on. A BB392 EVO crank will fit on almost any bike other than a fat bike, including bikes with traditional threaded shells; the extra spindle length makes room for outboard bearings on a 73mm-long shell, thus allowing a 30mm spindle to be used with a standard (34mm-ID) threaded bottom-bracket shell.

When the BB392 EVO crank is used with a BB392 EVO bottom-bracket shell, an option for a press-fit bottom bracket is a thread-together aluminum bottom bracket with internal, 30mm-ID bearings (Fig. 11.19).

c. BB92 and BB95

A number of road and mountain bike press-fit bottom bracket systems, called variously BB86, BB90, BB92, and BB95, are named for the width of the bottom-bracket shell. BB86 and BB92 are often called the "Shimano press-fit system" for road and mountain bikes, respectively, even though other crank manufacturers make bottom brackets for this standard as well. They are also sometimes called PF41, because they press fit inside a 41mm ID bottom-bracket shell. Since other press-fit standards instead call out the diameter of the spindle, rather than the ID of the shell, for simplicity and symmetry with PF30, I'll instead call BB86 and BB92 PressFit 24, or PF24, because the spindle diameter is 24mm. PF24 bottom brackets accept standard, 24mm integrated-spindle cranks (Fig. 11.2).

A BB86 road shell is 86.5mm wide, whereas a BB92 mountain bike shell is 91.5mm wide. Again, both bottom-bracket shells have a 41mm ID. Like PF30 (Fig. 11.16), the bearings for BB86 and BB92 are incorporated into plastic sleeves that press into the bottom-bracket shell (aluminum bearing cups are available aftermarket). On PF24, however, the bearings have a 37mm OD, and the sleeves have a 41mm OD. Each sleeve's shoulder is 1.75mm wide, creating a 90mm overall width for road BB86 (86.5mm + 1.75mm + 1.75mm) and a 95mm overall width for mountain BB92, making the width from bearing face to bearing face exactly the same as on the Trek systems. Wheels Mfg., among others, makes thread-together aluminum bottom brackets for PF86 and BB92 (PF24) bottom-bracket shells (Fig. 11.19); these accept 24mm spindles. Since the wider bottom-bracket shell doesn't allow room for external bearings with a standard 24mm integrated-spindle crank, the bearings sit within the shell, and there is a thin, notched flange outside of the shell for a notched, external-bearing bottom bracket wrench.

BB90 and BB95 are Trek's slip-fit bearing systems for road and mountain bikes, respectively. The road bottom-bracket shell is 90mm wide, whereas the mountain bottom-bracket shell is 95mm wide. Both BB90 and BB95 shells have a 37mm ID for the bearing seat, which is molded directly into the frame (Fig. 11.17). The 37mm OD × 24mm ID bearing is the same that you would find inside the threaded cup of any external-bearing system, and it is compatible with any of the standard external-bearing/integrated-24mm-spindle cranksets (Fig. 11.2). Trek supplies bearing sets for all 24mm-spindle cranksets, as do many aftermarket bottom bracket manufacturers, and the bearings slip into place with finger pressure alone.

d. Threaded bottom brackets

ISO (a.k.a. English or BSA) threads are standard on threaded mountain bike bottom brackets. The bottom-bracket cups are usually engraved "1.370 × 24," denoting a 1.370-inch thread diameter and a thread pitch of 24 threads per inch. It is important to remember that left-hand threads on the drive side are standard on English/ISO bottom brackets. In other words, you tighten the right-hand (drive-side) cup by turning it counterclockwise (Fig. 11.29). Meanwhile, the threads on the non-drive cup are right-hand threads and are therefore tightened clockwise.

ISO-threaded mountain bike bottom-bracket shells come in two widths: 68mm (older) and 73mm; threaded shells for fat bikes are 100mm or wider. Bottom brackets for three-piece cranks generally have stamped or printed numbers such as “73–118” (meaning that the shell is 73mm wide and the spindle is 118mm long). Many external-bearing bottom brackets are designed to work for both shell widths by means of 2.5mm spacers added on either side between the bearing cups and the shell faces.

A recent, larger threaded bottom bracket standard called T47 has M47 × 1 threads. T47 bottom-bracket shells come in lengths from 68mm (road) to 121mm (fat bike). The large ID of the shell allows the use of a 30mm spindle with the bearings internal to a threaded shell. Even short-spindle, high-ankle-clearance BB30 cranks will work on 68mm (road) and 73mm (MTB) T47 threaded shells, while BB392 EVO cranks require either a 92mm T47 shell and an internal-bearing bottom bracket, or they require external-bearing cups on a 73mm shell. Like BSA threads, T47 threads are left-hand on the drive side and right-hand on the non-drive side.

A thread-together bottom bracket for a (threadless) PF30 or BB392 EVO bottom-bracket shell (Fig. 11.19) can also be used in a T47 bottom-bracket shell.

e. Two-piece threaded cranksets

With the bearing cups of integrated-spindle, external-bearing cranksets (Fig. 11.2) being external to the bottom-bracket shell (Fig. 11.14), the bearings and spindle can be far larger (and hence stiffer) than on three-piece cranksets (whose bearings are contained within the 34mm-ID threaded bottom-bracket shell). The external bearings can also be right up against the crankarms (Fig. 11.3), adding more stiffness.

For integrated-spindle bottom brackets to work properly, threads on both sides of the frame's bottom-bracket shell must be aligned, and the end faces of the shell must be parallel. If you are installing an expensive bottom bracket and have any doubts about the frame, it is a good idea to have the bottom-bracket shell tapped (threaded) and faced (ends cut parallel) by a qualified shop possessing the proper tools. These tools are pictured in Figure 1.4. This procedure will improve durability and freedom of movement and will reduce the likelihood of creaking while pedaling.

f. Three-piece threaded cranksets

For three-piece cranks, the most common type of bottom bracket is probably the cartridge type (Fig. 11.20 or 11.21); it has cups that accept the splined removal tool shown in Figure 11.21. These bottom brackets can have a square-taper spindle (Fig. 11.20), an ISIS splined spindle (Fig. 11.21), or a Shimano Octalink splined spindle (Fig. 11.22).

Although most Octalink bottom brackets are the cartridge type, the first generation of Shimano XTR Octalink bottom brackets had four sets of loose, adjustable, and overhaulable bearings: two sets of tiny ball bearings and two thin sets of needle bearings (Fig. 11.22).

NOTE: *Shimano has two noninterchangeable Octalink standards: Octalink 1 and Octalink 2. Octalink 1 (Fig. 11.22), found only on XTR and on some Shimano road*

11.20 Shimano cartridge, square-taper bottom bracket

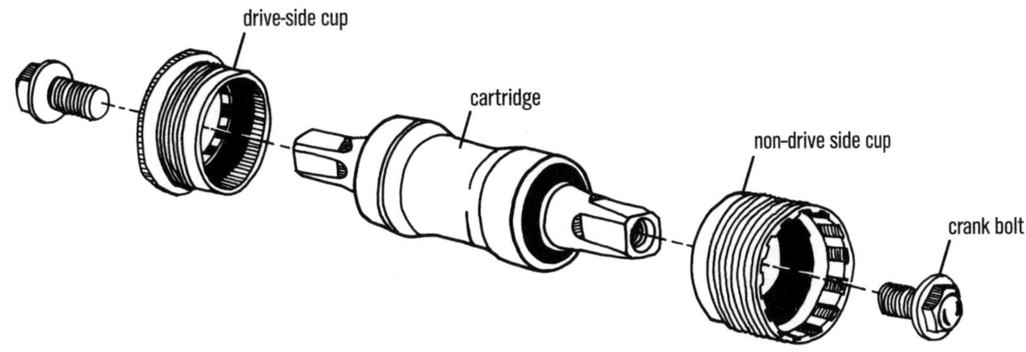

11.21–11.24 Threaded, internal-bearing bottom brackets

11.21 ISIS cartridge bottom bracket

11.22 Shimano XTR Octalink pipe spindle bottom bracket with loose bearings

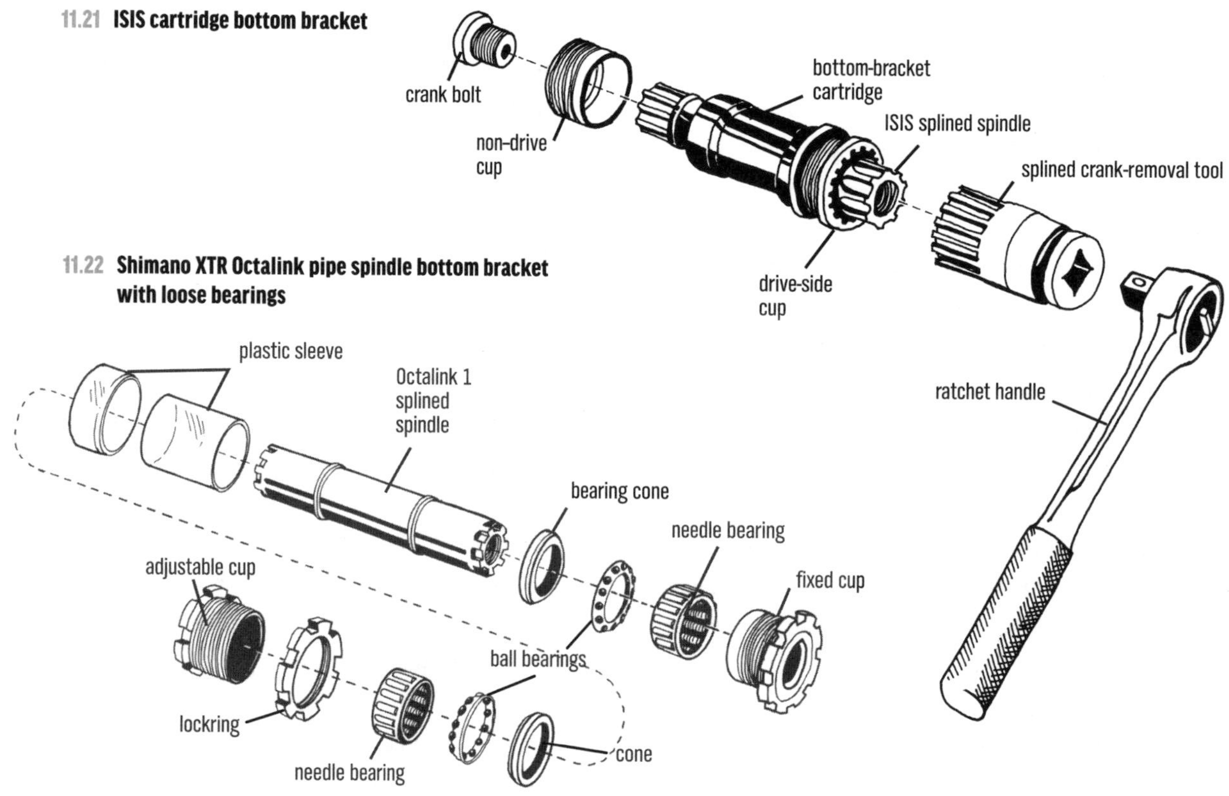

11.23 Standard cup-and-cone, loose ball bearing, square-taper bottom bracket

11.24 Adjustable cartridge bearing, square-taper bottom bracket

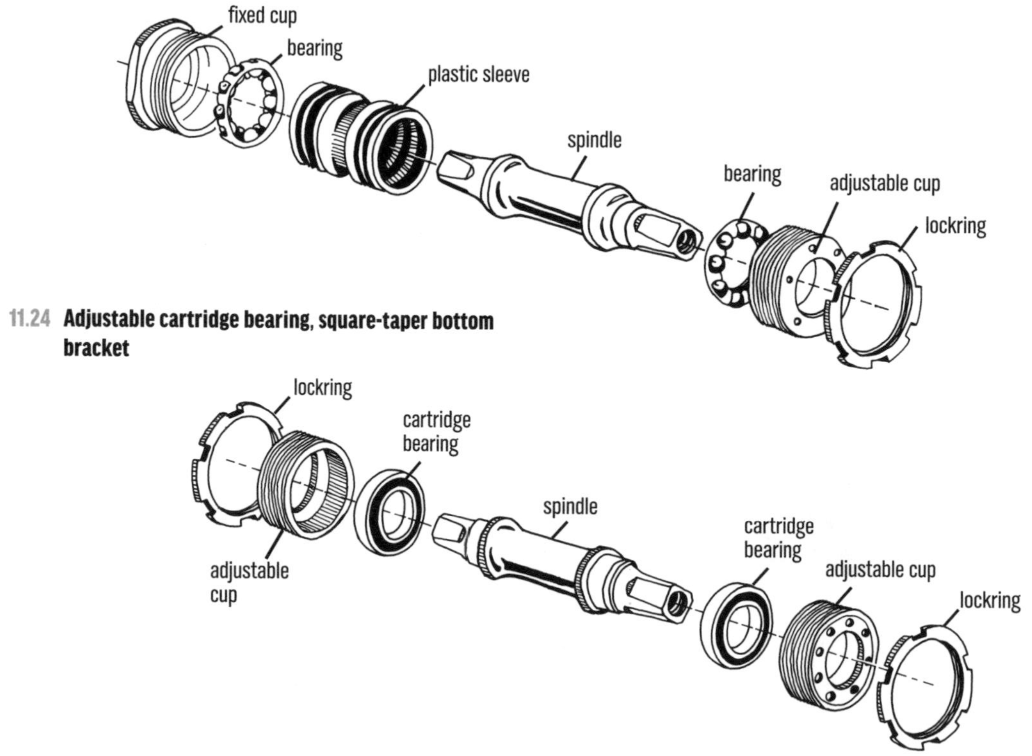

models, has eight (hence "Octa") spline ridges that are 2.2mm wide separated by 5mm-long valleys. Octalink 2 works with Shimano XT, LX, and Deore cranks, and its eight spline ridges are 2.8mm wide, while its valleys are 9mm long.

The most common mountain bike bottom bracket prior to the 1990s was the square-taper cup-and-cone style with loose ball bearings (Fig. 11.23). As discussed previously regarding external-bearing two-piece cranks, the bottom-bracket shell for cup-and-cone bottom brackets must be threaded concentrically, and the faces of the shell must be faced parallel to each other and perpendicular to the spindle. Otherwise, the bearings will drag and wear excessively.

Another older bottom-bracket type has cartridge bearings secured by an adjustable cup and lockring at either end (Fig. 11.24).

Some square-taper bottom brackets do not thread into the bottom-bracket shell. One type, found on old Fisher, Klein, and Fat Chance frames, uses cartridge bearings held into an unthreaded bottom-bracket shell by snaprings in machined grooves similar to BB30 (Fig. 11.27). Very early Merlin frames use a variation of this; the bearings are a press-fit into the shell and secured with Loctite.

11-7

INSTALLING A THREADED EXTERNAL-BEARING BOTTOM BRACKET

1. **Add spacers as required.** This is only necessary in a few circumstances.
   (a) 68mm bottom-bracket shell: On Shimano, FSA Mega Exo, and Race Face X-Type, slip two of the three supplied 2.5mm spacers over the thread of the right cup and the remaining one on the thread of the left cup (Fig. 11.2); on SRAM/Truvativ GXP, put one spacer on the right cup and one on the left cup (Fig. 11.14).
   (b) 73mm bottom-bracket shell: With Shimano and FSA, slip one spacer on the right cup and none on the left; use no spacers for SRAM/Truvativ GXP.
   (c) E-type front derailleur (Fig. 5.20): Instead of placing a spacer on the right cup, slide on the front-derailleur bracket. Because of the added spacing, this derailleur will not work with a SRAM/Truvativ GXP crankset except on a 68mm bottom-bracket shell.

NOTE: *A removable plastic sleeve (Fig. 11.14) keeps contamination away from the backside of the bearings. Keep the sleeve on the right cup when installing the cup.*

2. **Grease the threads.**
3. **Start the cups by hand.** Turn the right (drive-side) cup counterclockwise and the left cup clockwise.
4. **Tighten the cups.** Use the notched tool designed to fit the notched cup (Fig. 11.14). Torque is high (35–50 N-m), so yank on the tool pretty hard without a torque wrench.
5. **Install the spindle and crankarms.** Follow the instructions in 11-1a, b, or c, depending on type.

11-8

INSTALLING A THREADLESS BOTTOM BRACKET

Threadless, press-in bottom-bracket installation is similar to threadless headset installation (12-22).

a. PF30 (30mm spindle), BB392 EVO (30mm spindle), BB92 (PF24, 24mm spindle), and PF121 (30mm spindle)

1. **Clean and grease inside the ends of the bottom-bracket shell.**
2. **Press or screw in the bearing cups.** For press-in cups, using a headset press, press in each cup with the bearing inside, one at a time (Fig. 11.25). Ideally, use bushings in the bearings like the ones that come with the Park BBT-30.4 tool for PF30, PF121, and BB392 or with the BBT-90.3 tool for BB92 (PF24); in a pinch, the flat faces of the headset press will also work. Ensure that both bearing cups are seated fully. Place the manufacturer-supplied seals over them.

   Alternatively, if you are using an aluminum thread-together bottom bracket, slide the two cups in, one

11.25 Pressing in BB92 (PF24), BB30, or PF30 cups with a headset press

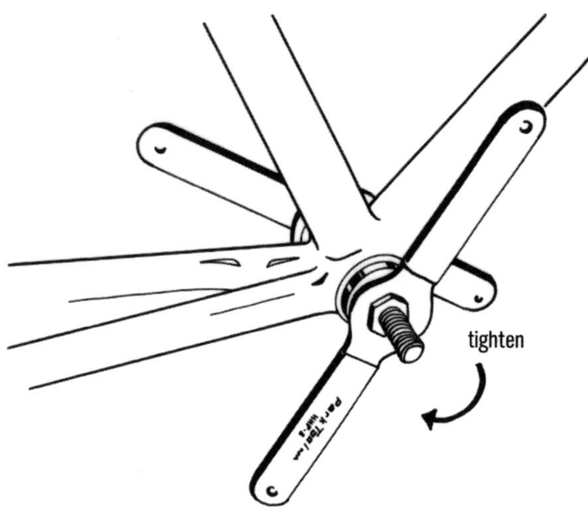

11.26 Pressing in a Trek BB95 bearing by hand

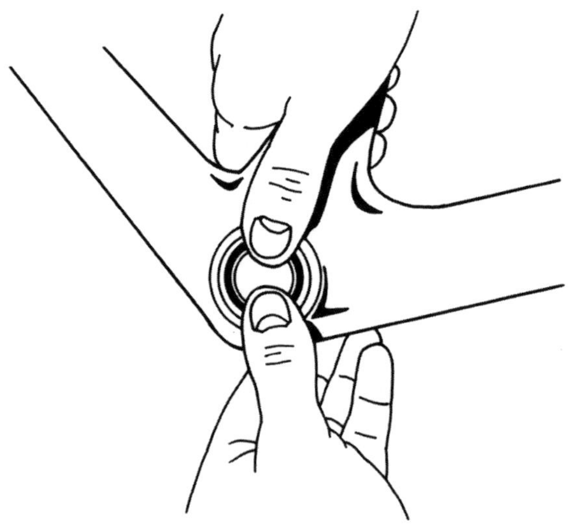

from each end, and screw them together where they meet in the middle; it doesn't matter which cup goes in which end. Tighten them with a 16-notch bottom-bracket wrench (48.5mm-diameter for PF30 and BB392, and 44mm for BB92/PF24).

3. **Install the spindle and crankarms.** Follow the instructions in 11-1a, b, or c, depending on type.

b. BB95 (Trek frame with a 95mm-wide bottom-bracket shell, 24mm integrated spindle)

1. **Clean and grease inside the ends of the bottom-bracket shell.**
2. **Press in the bearings and install the spindle and crankarms.**
   (a) For Shimano HollowTech II, FSA Mega Exo, and other non-GXP cranksets, simply push the bearings in by hand (Fig. 11.26) and place the Trek-supplied seals over them (Fig. 11.17). Install the spindle and crankarms, following the instructions in 11-1a, b, or c, depending on type.
   (b) For Truvativ GXP, push the bearings in by hand (Fig. 11.26), making sure that you put the one with the smaller bore (22mm rather than 24mm) in the left side with its protruding inner bearing-ring lip facing inward. Place the flat, rubbery bearing seal against the right bearing, and slip the wavy washer onto the crankarm (Fig. 11.17). Push in the (greased) spindle from the drive side, place the metal bearing cover over the spindle and left bearing, and engage the left arm on the (greased) spindle splines. Tighten the crank bolt to 50–57 N-m.

NOTE: *Trek does have a plastic bearing-installation tool for its BB90/BB95 frames, but it's generally not a necessity. Lacking that, if they won't go in by hand, use a headset press and the bushings from the Park BBT-90.3 tool to press them in.*

c. BB30 (30mm spindle)

1. **Clear the bottom-bracket shell of any metal chips and other detritus.**
2. **Grease the contact surfaces.** Apply a thin layer of grease to the snapring grooves and surfaces outboard of them and to the snaprings themselves.
3. **Insert the snaprings.** If the snapring has a hole (eye) on either end, push the tips of the snapring pliers into the holes, squeeze the handles to reduce the snapring's diameter, and install the snapring into the groove inside one end of the bottom-bracket shell (Fig. 11.27). If the snapring does not have holes, compress the ring by pushing it into the shell, push the ring's square-cut end to the groove in the shell so that the ring edge drops into the groove, and work around, pushing the rest of the ring into place. Check that the snapring is fully engaged in the groove all the way around. Repeat for the other

11.27 Installing a snapring in a BB30 shell

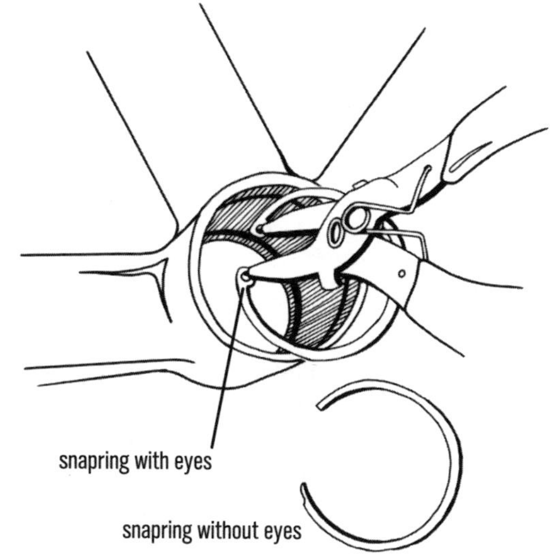

snapring in the other end. (To remove a snapring with holes on the ends, compress it with snapring pliers and pull it out. To remove a snapring without eyes, slip a screwdriver blade under the pointed end of the snapring, push the blade inward beyond the groove, and keep pushing on the screwdriver blade, working from that end toward the square-cut end until the snapring is free.)

4. **Press in the bearings.** Beg, borrow, or buy tool bushings for BB30 bearings (FSA, Park, Wheels Mfg., and Cannondale sell such bushings). With a headset press, press in both bearings simultaneously or one bearing at a time (Fig. 11.25) depending on the bushings you have. Obviously, you are pressing both bearings in until they stop at the circlip. Without a headset press, if you're careful and have BB30 tool bushings, you should be able to press the bearings in straight using a large bench vise.
5. **Grease the backside of each bearing.** This protects it from water trapped in the shell.
6. **Place the supplied aluminum bearing shields against the bearings.** Their machined grooves face inward toward the bearings.
7. **Push the spindle in from the drive side.** Lightly grease the spindle bearing areas, splined end, and threads first. You'll probably need a rubber mallet, because BB30 spindles are designed to be a light interference, or friction-based, fit.
8. **If supplied, put a wavy washer on the left end of the spindle.** It goes between the bearing shield and the crankarm and takes out lateral play.
9. **Push on the left crankarm and tighten the crank bolt.** Ideally, use a torque wrench (8mm or 10mm hex driver) to torque spec (torque is often less than with external-bearing cranks; see Appendix D for torque specs).
10. **Check the wavy washer.** It should be compressed somewhat, but not flattened. If it is not slightly compressed, remove the crank, add as many spacers against the bearing shields as are required to fill the space so that the wavy washer will be slightly flattened when the crankarm is on, and reinstall the crankarm.
11. **If the bearings fit loosely or creak, they need Loctite.** Remove them (11-13), smear a thin layer of Loctite 609 retaining compound where they sit inside the bottom-bracket shell, and reinstall them. Allow the Loctite to cure as specified in the directions or, if you do not have the instruction sheet, at least 24 hours.

d. Square-taper press-in

Small-diameter unthreaded bottom-bracket shells with snapring grooves can be found on Fisher, Klein, and Fat Chance frames from the 1980s. Two snaprings retain the bearings in the shell, and the spindle has a square taper. If the shell is bored to the correct diameter, you can probably seat the cartridge bearings by hand. If not, you can press them in with a vise or a headset press, using a large socket or other flat-ended cylindrical object as a drift pushing against the bearing, as long as whatever you use is just slightly smaller in diameter than the OD of the bearing. The inboard bearing stops are usually shoulders or snaprings on either end of the axle.

If the bearings will go in by hand, install one snapring with snapring pliers into the groove in one end of the shell. Push the entire assembly of axle and two bearings in from the other side of the bottom-bracket shell. Install the other snapring, and you're done.

If you can't press them in by hand, press one bearing in as far as the snapring groove by using one vise jaw or headset-press face against the shell face, the other against first the bearing, until it reaches the shell, and then the drift to push the bearing into the bottom-bracket shell's bore. Install the snapring. Install the spindle, and push the other bearing in the same way until you can install the other snapring.

11-9

INSTALLING A THREADED CARTRIDGE BOTTOM BRACKET

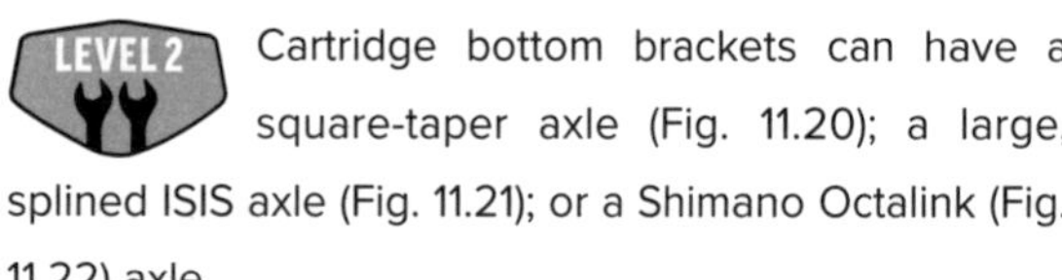

LEVEL 2 Cartridge bottom brackets can have a square-taper axle (Fig. 11.20); a large, splined ISIS axle (Fig. 11.21); or a Shimano Octalink (Fig. 11.22) axle.

1. **Thread the left cup (clockwise) in three to four turns by hand.** It has no lip on it, it goes into the non-drive side, and it's right-hand threaded.
2. **Slide the cartridge into the bottom-bracket shell.** Pay attention to the right and left markings on the cartridge. The cup with the raised lip and left-hand thread is the drive-side cup (the cup shown on the left in Fig. 11.20 and on the right in Fig. 11.21).
   (a) Add a 2.5mm spacer. If you have a 68mm-width bottom-bracket shell (yes, measure it!) and are using a bottom bracket designed for both 73mm and 68mm bottom-bracket shells, add at least one spacer on the drive side.
   (b) An E-type front derailleur's bracket mount, shown in Figure 5.20, slips between the shell and the cup lip. Some ISIS bottom brackets will accept only an E-type front derailleur with a 68mm shell, because there is room only for a single spacer with 68mm and there is no spacer room with a 73mm. Some Shimano bottom brackets are designed to accept an E-type front derailleur on either a 68mm or a 73mm shell; with these, you install either two 2.5mm spacers or one spacer and the E-type bracket between the right cup and a 68mm shell. With a 73mm shell, you install either the E-type front-derailleur bracket or a single 2.5mm spacer against the shell.
3. **Tighten the right (drive-side) cup.** Tighten it counterclockwise with a splined cup installation tool

11.28 Removing and installing non-drive-side cup on cartridge bottom bracket

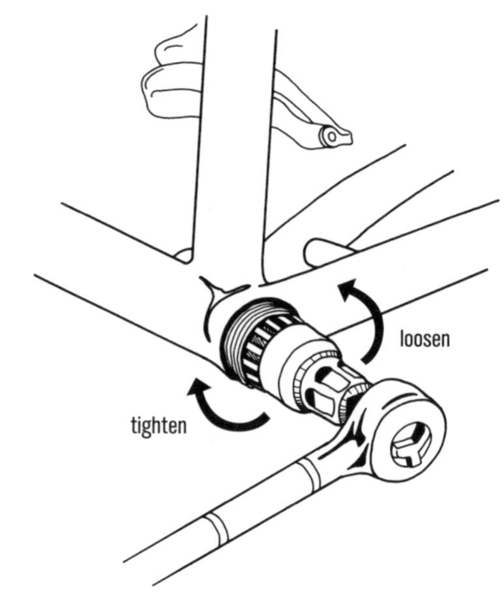

like the one depicted in Fig. 11.21; tighten with an open-end wrench, a torque wrench, or a standard socket wrench (as shown in Fig. 11.28, except on the drive side) until the lip seats against the face of the bottom-bracket shell. Recommended torque is high (see Appendix D).

4. **Tighten the left (non-drive) cup.** With the same tool, tighten the cup clockwise against the cartridge (Fig. 11.28) to the same torque as the right cup. There is no adjustment of the bearings to be done.
5. **Install the crankarms.** Follow the instructions in 11–1c, under Installation. Check that the ends of the cranks are centered about the chainstays; if not, remove the cranks and bottom bracket, add or remove spacers as needed, and go back to step 1 above.

11-10

INSTALLING CUP-AND-CONE BOTTOM BRACKETS

LEVEL 2 In cup-and-cone bottom brackets (Figs. 11.22 and 11.23), ball bearings ride between cone-shaped bearing surfaces on the spindle and cup-shaped races in the threaded cups. The fixed cup (the left-hand cup in Figs. 11.13 and 11.23 and the right-hand cup in Fig. 11.22) has a lip on it and fits on the right

(drive) side of the bike (Fig. 11.29). The adjustable, opposite, cup is held in position by a lockring threaded onto the cup against the bottom-bracket shell. A retaining cage usually holds the individual ball bearings together.

Again, it is a good idea to have the bottom-bracket shell tapped (threaded) and faced (ends cut parallel) by a qualified shop possessing the proper tools (Fig. 1.4).

1. **Unless you have a shop fixed-cup tool, have a shop install the fixed cup for you.** The shop tool ensures that the cup goes in straight and tight. The tool pictured in Figure 11.29 can be used in a pinch, but it may let the cup go in crooked and will slip off before you get it really tight. The fixed cup must be very tight (see Appendix D for torque) so that it does not vibrate loose. Remember that English-threaded fixed cups are tightened counterclockwise (Fig. 11.29).
2. **Wipe the inside surface of both cups with a clean rag.**
3. **Put a medium layer of clean grease on the bearing surfaces.** The balls should be half covered; any more is wasted and attracts dirt.
4. **Wipe the axle with a clean rag.**
5. **Figure out which end of the axle is toward the drive side. The** drive side may be marked with an "R," or choose the longer end (when measured from the bearing surface). If there is writing on the spindle, it will usually read right side up for a rider on the bike. If there is no marking and no length difference, the spindle orientation is irrelevant.
6. **Slide one set of bearings onto the drive-side end of the axle** (Fig. 11.30). Make sure you orient the retainer cage correctly. The balls, rather than the retainer cage, should rest against the axle-bearing surfaces. Because there are two types of retainer cages with opposite designs, you need to be careful to avoid binding, as well as smashing, the cages. If you're still confused, there is one easy test: If the retainer cage is installed correctly, the axle will turn smoothly; if the retainer is installed the wrong way around, it won't. If the bottom bracket has loose ball bearings with no retainer cage, stick them into the greased cup. Most rely on nine balls; you can confirm that you are using the correct number by inserting and removing the axle and checking to make sure that they are evenly distributed in the grease and are neither piled up (indicating you have too many) nor gapped for more balls. The XTR loose-bearing bottom bracket in Figure 11.22 requires the installation of a bearing ring, a set of ball bearings, and a needle-bearing set on the spindle.

11.29 Drive-side fixed cup; note left-hand threads

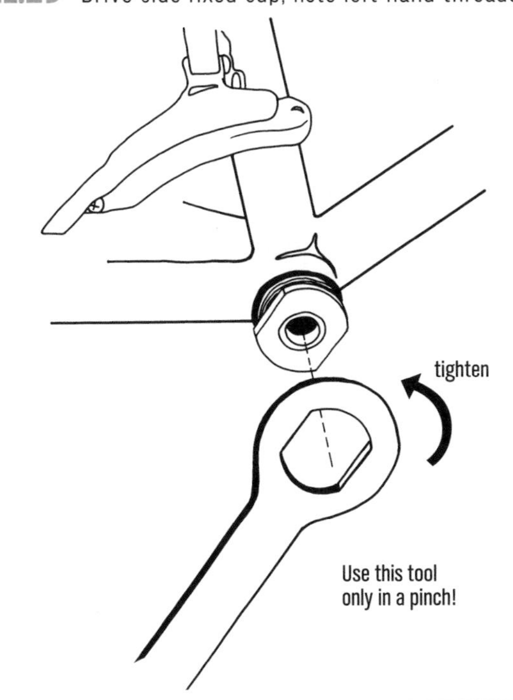

11.30 Placing the axle in the bottom-bracket shell

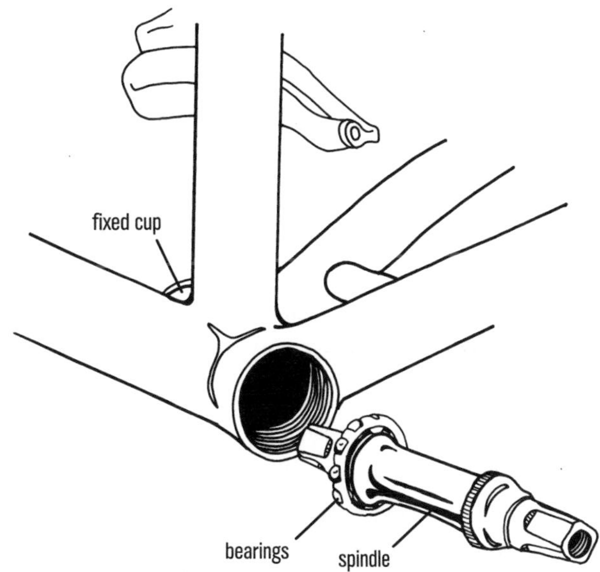

7. **Slide the axle into the bottom-bracket shell.** Push the bearings into the fixed cup (Fig. 11.30). Use your pinkie to stabilize the end of the axle from the fixed-cup end as you slide it in.
8. **Insert the protective sleeve.** The plastic sleeve shown in Figures 11.22 and 11.23 keeps dirt and rust from falling from the frame tubes into the bearings, so if you don't have a sleeve, get one. Push it against the inside edge of the fixed cup.
9. **Place the second bearing set into the greased adjustable cup.** If there is a bearing retainer, make sure it is properly oriented. On an XTR loose-bearing BB (Fig. 11.22), install the bearing ring, ball bearings, and needle bearings on the spindle as you did on the other end.
10. **Screw in the adjustable cup.** Leaving the lockring off, turn the cup clockwise, ensuring that it is going in straight over the axle. Screw the cup in as far as you can by hand, ideally all the way until the bearings seat between the axle and cup.
11. **Locate the appropriate tool for tightening the adjustable cup.** Most cups have two holes that accept the pins of an adjustable cup wrench called a pin spanner (Fig. 11.31) or a splined end for a tool like the one in Figure 11.21. Another type of adjustable cup has two flats on which you may use an adjustable wrench. The XTR loose-bearing BB in Figure 11.22 requires a toothed lockring spanner or the lockring tool shown in Figure 11.12.
12. **Carefully tighten the adjustable cup against the bearings.** Do not overtighten. Turn the axle periodically with your fingers to ensure that it moves freely. If it binds up, you have gone too far; back it off a bit. Overtightening can cause the bearings to dent the bearing surfaces of the cups, and the axle will never turn smoothly again.
13. **Screw the lockring onto the adjustable cup by hand.**
14. **Tighten the lockring against the bottom-bracket shell.** Use a lockring spanner that fits the lockring while holding the adjustable cup in place (Fig. 11.31). If you turn the bicycle upside down, you can pull down harder on the wrenches.
15. **Check for play.** As you snug the lockring against the bottom-bracket shell, wiggle the bottom-bracket spindle periodically. Because the lockring pulls the cup out of the shell minutely, it loosens the adjustment. The spindle should turn smoothly without free play in the bearings. I recommend installing and tightening the drive-side crankarm onto the drive end of the spindle (Fig. 11.4) at this time so that you can push the crank from side to side to check for free play. Adjust the cup so that the axle play is just barely eliminated.

11.31 Tightening the lockring

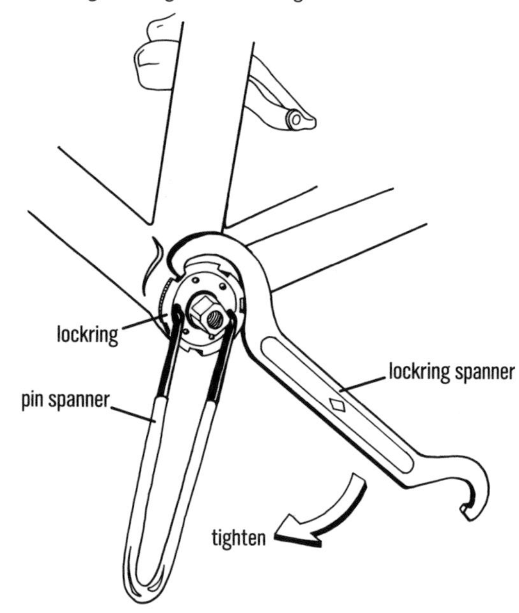

16. **While holding the cup, fully tighten the lockring** (Fig. 11.31). To prevent the bottom bracket from coming out of adjustment during a ride, tightening it as tightly as you can is about right (recommended torque is in Appendix D). You may have to repeat this step a time or two until you get the adjustment just right.

11-11

INSTALLING OTHER TYPES OF BOTTOM BRACKETS

LEVEL 2

On cartridge-bearing bottom brackets with adjustable cups on both ends (Fig. 11.24), simply install the drive-side cup and lockring, slide the cartridge bearing in (if it is not already pressed into the cup), slip the spindle in, and then install the other bearing, cup, and lockring. Tighten each lockring while holding the adjustable cup in place with a pin spanner (Fig. 11.31); the cup could also take a splined cup tool like the one in Figure 11.12 or 11.28. Adjust for free play as described in 11-10, steps 12–16.

(a) The advantage of having two adjustable cups is that you can center the cartridge by moving it from side to side in the bottom-bracket shell. If the chainrings end up too close or too far away from the frame (see Fig. 5.61 and 5-50), you can move one cup in and one out to shift the position of the entire cartridge.

(b) Sometimes cartridge-bearing bottom brackets bind up a bit during adjustment and installation. A light tap on each end of the axle usually frees them.

OVERHAULING THE BOTTOM BRACKET

A bottom-bracket overhaul consists of cleaning or replacing the bearings, cleaning the axle and bearing surfaces, and regreasing them. With any type, both crankarms must be removed (11-1).

11-12

OVERHAULING INTEGRATED-SPINDLE BOTTOM BRACKETS

1. **Remove the crankarms** (11-1). This requires loosening pinch bolts or a big crank bolt.
2. **Remove the rubber cover seal covering the bearing.** To get this seal off to reveal the bearing, slip a blade under the edge and pry it up (Fig. 11.32), possibly working around the outside with a thin screwdriver for FSA and Shimano cover seals that extend inside the bearing bore. In many cases, the drive-side cover seal stays stuck on the crank spindle and pulls off with the right crank.
3. **Pry out the circular seal between the bearing's inner and outer rings.** With a blade, get under the edge of the seal and pry it off (Fig. 11.33).
4. **Clean the bearing.** At a minimum, wipe it out if it's not dirty inside. Otherwise, with solvent and a clean toothbrush, scrub dirty grease out of the bearing.

11.32 Pry bearing seal from external bearing cup with a box-cutter blade

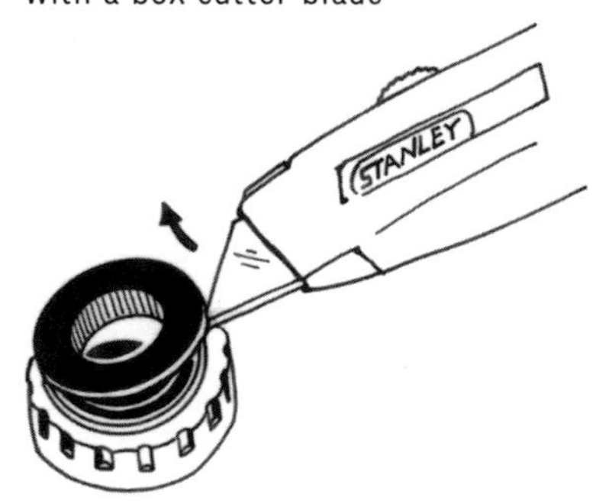

11.33 Pry bearing seal from cartridge bearing with a box-cutter blade

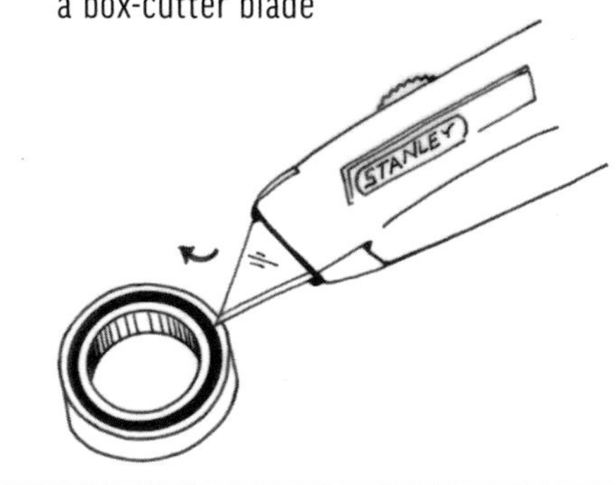

5. **Blow the bearing out with compressed air.** Wear safety glasses. If you don't have an air compressor, dry it as best you can.
6. **Repack the bearing with clean grease.**
7. **Replace the bearing seal and the bottom-bracket cover seal.**
8. **Reinstall the crankarms** (11-1).

You can't get at the other side of the bearing without removing it from the cup with a special bearing puller (see 11-13), but do the best you can to flush the bearing out and dry it out afterward.

If the bearings still do not turn well after this or are gritty, they will have to be replaced; see 11-14.

11-13

REPLACING BEARINGS IN INTEGRATED-SPINDLE BOTTOM BRACKETS

LEVEL 3

You can replace the bearings with the same type, or you can make the axle spin better than ever by replacing the bearings with upgraded steel ones or with the ultimate: ceramic bearings. Ceramic bearings are expensive, but they deliver such a performance advantage that no top pro cyclist will ride without them. Ceramic ball bearings are lighter, smoother, 2.5 times rounder, 2.5 times harder, and 50 percent stiffer than steel balls and are less affected by heat.

a. Interchanging the bearings inside threaded external-bearing cups

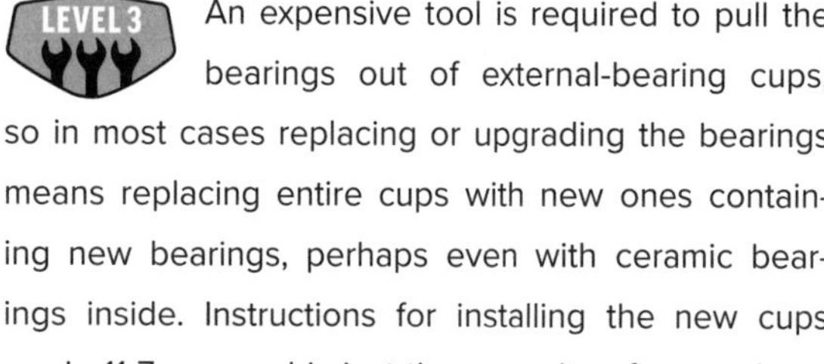

An expensive tool is required to pull the bearings out of external-bearing cups, so in most cases replacing or upgrading the bearings means replacing entire cups with new ones containing new bearings, perhaps even with ceramic bearings inside. Instructions for installing the new cups are in 11-7; removal is just the opposite after you have removed the cranks as in 11-1.

If you have the required tool and want to upgrade the bearings inside external-bearing cups, first select the bearings from your bike shop or a source such as WheelsMfg.com, EnduroForkSeals.com, CeramicSpeed .com, or BocaBearings.com. Make sure the bearings are correct for the crankset. Even though the spindle is 24mm, not all manufacturers use bearings with a 24mm ID. Stock FSA and Shimano bearings are 25mm ID; a thin plastic shim integral with the outer bearing cover (making the cover shaped like a top hat) brings the bore down to the 24mm diameter of the spindle (it's visible in Fig. 11.32). If you are simply replacing the bearings and reusing the stock seals, get 25mm ID bearings. Otherwise, Enduro's Shimano/FSA replacement kit uses a 24mm ID bearing and a thin outer silicone cover seal instead of the stock one.

1. **Remove the crankarms** (11-1).
2. **Unscrew the cups from the frame** (11-7, Fig. 11.14). Remember, the drive-side cup will be left-hand-threaded.
3. **Remove (or not) the bearing cover seals.** Slip a razor blade under the edge (I mean the seals covering the face of the outboard cup, not the actual seals on the cartridge bearings) and pry it up (Fig. 11.32), possibly working around the outside with a thin screwdriver blade for FSA and Shimano cover seals that extend inside the bearing bore. You can leave them on, and when you push out the bearings in the remover tool, the bearings will push the cover seals off.
4. **Get a specialty tool for removing and installing bearings from external-bearing cups.** It must ensure complete bearing insertion and proper alignment. Enduro has a nice one (made by Sonny's Bike Tools), as does Phil Wood. The key to the external bottom-bracket bearing puller is a two-piece collet. These instructions are for the Enduro tool; the Phil Wood tool is slightly different.
5. **Drop the collet down into the bearing from the inboard side.**
6. **Push a rounded-nose cylindrical collet expander into it.** It will spread the collet inside the bearing bore so that its lips will catch the back of the inner bearing race. Many bearing cups have an internal shelf that would prevent you from pushing the bearing out if you did not have this collet.
7. **Put the bearing cup's outboard end facedown into the tool's cup holder.**
8. **Push the bearing out by applying pressure on the collet expander.** With the Enduro tool, you accomplish this by running a big bolt through the expander and tightening the bolt. This often takes considerable force, and especially in the case of SRAM/Truvativ GXP, the bearing finally comes free with a loud pop.
9. **Determine bearing orientation.** Think ahead about maintenance before determining the orientation of the bearings. Even though ceramic balls cannot rust and are more than twice as hard as steel balls, the steel races can rust, and they need maintenance. Ceramic cartridge bearings, like most steel cartridge bearings, have bearing retainers that separate the balls from each other. This reduces friction by preventing neighboring balls, whose adjacent sides are turning in opposite directions, from rubbing each other. The bearing retainer may be asymmetrical, so when you remove the bearing seals with a razor blade (Fig. 11.33), you'll see the balls on one side and you'll see only the retainer from the other side. Proper maintenance requires cleaning out the bearing and pumping new grease into it, so before you press the bearings in, determine which

side the bearing retainer is closed on and make sure that side faces inboard. That way, you can pry off the bearing covers and clean and grease them easily as in 11-12. SRAM/Truvativ recommends a service interval of 100 hours for the ceramic bearings in its XX crankset. Service interval will obviously depend on how you use the bike.

10. **Place the bearing on an insert that fits snugly in the bearing's ID.** One end of the insert is 24mm in diameter, and the other is 25mm.
11. **Place the insert and bearing inside the bore of the tool's cup holder, facing upward.**
12. **Place the bearing cup over the bearing, and put a support ring atop it.**
13. **Run the bolt through and tighten the bolt.** Tighten with an 8mm hex key until it hits a dead stop (Fig. 11.34).
14. **Check that the bearing is in as far as the old one was.** If not, flip the bearing cup over, with the insert still inside the bearing bore, and put the cup back into the tool's cup holder with the cup's threaded section down inside. Reinstall the bolt and tighten it until it stops. This second pressing step is always necessary with SRAM and Truvativ GXP cranks (24mm ID bearings), because the replacement bearing (a standard size, 7mm thick) is 1mm narrower than the (proprietary) bearing employed by SRAM/Truvativ.

11.34 Press new bearing into external-bearing cup with Enduro tool

15. **On a non-drive side SRAM/Truvativ GXP bearing, install its sleeve.** GXP cranks require an 11.5mm-wide, 1mm-thick sleeve that stops the non-drive shoulder of the GXP spindle and establishes the side-to-side position of the crankset. Use the special GXP insert in the tool to install the sleeve.
16. **Install the cups as in 11-7 and the crankarms as in 11-1.** Spin them.

b. Interchanging the bearings in threadless bottom brackets

I. PF30, BB392, BB92 (PF24), and PF121

These instructions apply to PressFit systems for both 30mm and 24mm spindles.

1. **Insert the bearing remover.** On PF24 (BB92), use the Park BBT-90.3 bearing removal tool, which is simply a slimmer version of a headset cup remover. As in 12-20, slide the remover tool in, narrow end first (Fig. 12.37), and pull it out from the other end until its wings snap outward behind the bearing. On PF30, BB392, and PF121, use the BBT-30.4 and follow the instructions for BB30, except that the tool pushing against the bearing will pop out the entire plastic (or aftermarket aluminum) cup with the bearing still inside.
2. **Push the old bearing out.** Smack the remover with a hammer (Fig. 12.38).
3. **Repeat for the other bearing.**
4. **Install the new bearings as in 11-8a.**

II. BB95

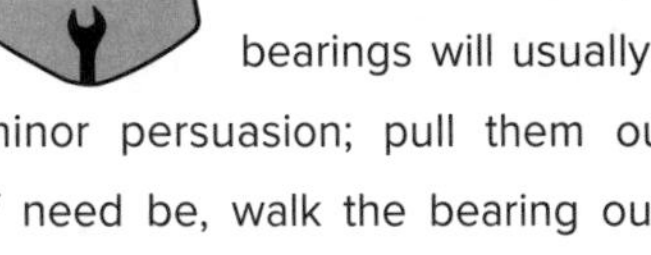

With Trek's BB95 (Fig. 11.17) system, the bearings will usually come out with only minor persuasion; pull them out with your finger. If need be, walk the bearing out a bit at a time by placing a big hex key against it from the opposite side and gently tapping it with a hammer as you move the hex tip around from side to side on the backside of the bearing.

Put the new bearings in as described in 11-8b.

III. BB30

LEVEL 2

With a BB30 crank (Fig. 11.15), you'll need a special bearing puller. Enduro has a removal/installation tool that works similarly to the tool described in 11-13a. As in 11-13a, you insert the collet and collet expander into the bearing and drive it out by tightening the tool's center bolt.

The following instructions, however, are for Park's simpler BBT-30.4 T-shaped BB30 bearing puller (actually, it's a pusher-outer).

1. **Angle the BBT-30.4's T-end in through one bearing, and push it straight in against the other bearing.** Make sure the tool does not hit the snapring's eyes (Fig. 11.27).
2. **Center the BBT-30.4 shaft with the dummy bearing insert in the near side.**
3. **Smack the handle with a hammer.** The bearing on the far side will pop out (Fig. 11.35).
4. **Install the new bearings.** See 11-8c.

If you don't have the correct tool, you'll probably want to have a shop do this to avoid mauling the inside of the bottom-bracket shell, but if you're careful, you can reach in against the backside of the opposite bearing (not against the snapring!) with a rod or big hex key that you tap with a hammer. Work a little bit on each side of the bearing, moving the end of the rod from side to side and around the bearing as you tap it to slowly walk the bearing out.

Install the new bearings as in 11-8c.

11.35 Push out BB30 bearing with Park BBT-30.4 tool

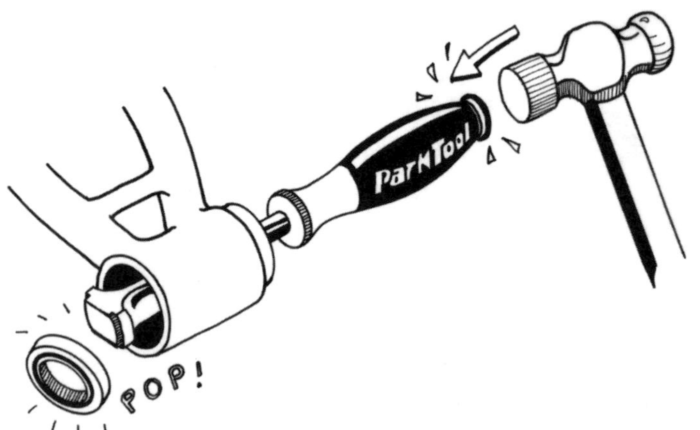

11-14

OVERHAULING CARTRIDGE BOTTOM BRACKETS (OR NOT!)

Standard cartridge bottom brackets (Figs. 11.20 and 11.21) are sealed units and cannot be overhauled. They must be replaced when they stop performing properly. Remove the cranks as in 11-1. Remove the bottom bracket by unscrewing the cups with the splined cup tool (Fig. 11.28), and install a new bottom bracket as directed in 11-9.

11-15

OVERHAULING CUP-AND-CONE BOTTOM BRACKETS

Cup-and-cone bottom brackets (Figs. 11.22 and 11.23) can be overhauled entirely from the non-drive side after you have removed the crankarms as described in 11-1.

1. **Remove the lockring.** Use a lockring spanner as in Figure 11.31, except the lockring spanner and the rotation direction will be reversed.
2. **Remove the adjustable cup.** Use a tool that fits the particular cup (usually a pin spanner like the one in Fig. 11.31).
3. **Leave the fixed cup in place.** Check that it is tight in the frame by putting a fixed-cup wrench on the cup and trying to tighten it counterclockwise (Fig. 11.29)—without letting it slip off and maul the cup.
4. **Clean the cups and axle with a rag.** There should be no need for a solvent unless the parts have a hard glaze on them.
5. **Clean the bearings with a citrus-based solvent.** Don't remove them from their retainer cages. A simple way to clean them is to drop the bearings in a plastic bottle, fill it with solvent, cap it, and shake it. A toothbrush may be required afterward, and a solvent tank is certainly handy if you have access to one. If the bearings are not shiny and in perfect shape, replace them. Balls with dull luster and/or rough spots or rust should be replaced.
6. **Wash the bearings in soap and water.** This removes solvent and remaining grit.

7. **Dry the bearings.** Towel them off and let them thoroughly air-dry. An air compressor is handy here.
8. **Install the bottom bracket as in 11-10.**
9. **Install the crankarms as in 11-1 and Figure 11.4.**

TROUBLESHOOTING CRANK AND BOTTOM-BRACKET NOISE

11-16

CREAKING NOISES

Mysterious creaking noises can drive you nuts. Just as you think you have your bike tuned to perfection, a little noise comes along to ruin your ride, and these annoying little creaks, pops, and groans can be a bear to locate. Pedaling-induced noises can originate from almost anything connected to the crankset, such as movement of the bottom bracket in the bottom-bracket shell, movement of the cleats on your shoes, movement of the crankarms on the bottom-bracket spindle, loose chainrings, or poorly adjusted bearings. Of course, noises can also originate from seemingly unrelated components, such as the seat, seatpost, front derailleur clamp, frame, wheels, or handlebar. Creaking occurs when parts that are supposed to be fixed together instead move against each other when large loads are applied to them. Insufficient tightening of fasteners or lack of grease between parts is often the cause.

Before spending hours overhauling the drivetrain, spend some time trying to isolate the source of the noise. Try different pedals, shoes, or wheels. Pedal out of the saddle, and pedal without flexing the handlebar. If the source of the creak turns out to be the saddle, seatpost, wheels, or handlebar, turn to the appropriate chapter for directions on correcting the problem.

If the creaking is in the crank area, do the following:

1. **Check to make sure that the chainring bolts are tight, and tighten them if they are not** (Fig. 11.9).
2. **If that step does not solve the problem, make certain that the crankarm bolts are tight** (Figs. 11.4 and 11.5). If they are not, the resulting movement between the crankarm and the bottom-bracket spindle is a likely source of noise. If the crank is of a different brand than the bottom bracket, check with the manufacturer or your local shop to make sure that they are recommended for use together. Incompatible cranks and spindles will never properly join.
3. **Replace creaking press-fit bottom bracket with a thread-together one.** PF30, PF24 (BB92), BB386, and other press-in bottom brackets will creak if the bottom-bracket shell is oversized or not round. Remove the cups (11-13b, section I). You can try slathering the exterior of the cups with grease and reinstalling them (11-8a), but that will probably be only a temporary fix. The best method to get silent performance out of a threadless bottom-bracket shell that is not to spec is to use a thread-together bottom bracket (Fig. 11.18 and 11.19).
4. **If the bearing fit is not ideal, apply Loctite.** Bearings may move inside BB95, BB30, and ancient threadless square-taper bottom bracket shells and creak, if the bottom-bracket shell is oversized or not round. Remove the bearings (11-13b), smear a thin layer of Loctite 609 retaining compound where they sit inside the bottom-bracket shell, and reinstall them.
5. **Grease bottom bracket threads.** Remove and grease the threads of threaded bottom bracket cups. Reinstall the bottom bracket as in 11-7, 11-19, or 11-10.
6. **Apply grease if the cartridge moves inside the cup.** Rust, and alternating wet and dry cycles, can break down the glue bond that holds the drive-side cup onto a cartridge bottom bracket (Figs. 11.20 and 11.21). If the cartridge can move within the cup, it will creak. Fix it by removing the bottom bracket (11-14 and Fig. 11.28) and slathering grease inside both cups and around the outside of the cartridge where the parts meet. Grease the cup threads and tighten the bottom bracket back into place (Fig. 11.28).
7. **If the creaking persists, recheck the adjustments and inspect the bottom bracket threading and facing.** The bottom bracket can creak owing to improper adjustment, lack of grease, cracked bearings, worn parts, or loose cups. All of these things require adjustment or overhaul via the procedures

outlined in 11-12 through 11-15. Many integrated-spindle designs (Fig. 11.2) as well as cup-and-cone bottom brackets (Figs. 11.22 and 11.23) are very sensitive to the bearings being out of parallel, and creaking may occur if the bottom-bracket shell is not perfectly tapped and faced. This is a job for a good bike shop.

8. **Now for the bad news.** If creaking persists, the problem could be rooted in the frame. Creaks can originate from cracks in and around the bottom-bracket shell, so be sure to check that area. The threads in the bottom-bracket shell could also be worn to the point that they allow the cups to move slightly. Neither of these is a good sign, unless, of course, you were hoping for an excuse to buy a new frame.

11-17

CLUNKING NOISES

1. **In the case of crankarm play, grab the crankarm and push on it from side to side.**
   (a) If there is play, tighten the crankarm bolt (Figs. 11.3 to 11.5; torque spec is in Appendix D).
   (b) If there is still crankarm play with a cup-and-cone bottom bracket (Figs. 11.13, 11.22, and 11.23) or a cartridge-bearing bottom bracket with a lockring on either side (Fig. 11.24), adjust the bottom-bracket spindle end play (11-10, steps 12–16).
   (c) If bottom-bracket adjustment does not eliminate crankarm play, or if the bike has a non-adjustable cartridge bottom bracket (Figs. 11.20 and 11.21), the bottom bracket is loose in the frame threads, and you should tighten it up. With a cup-and-cone bottom bracket, you can go back to 11-10 and start over, making sure that the fixed cup is very tight. Adjustable-cup lockrings need to be equally tight (Fig. 11.31) once the spindle end play is adjusted properly.

   The lockrings and fixed-cup flanges must be flush with the bottom-bracket shell all the way around; if they are not, the bottom bracket must be removed, and the bottom-bracket shell must be faced (cut parallel) by a shop equipped with a facing cutter.
   (d) If the crankarm play persists, the hole in the crankarm could be worn out. This is a common problem with square-taper bottom brackets (Figs. 11.20, 11.23, and 11.24) if the bike was ridden without the crankarm bolt properly tightened. You will have to buy another crankarm. Make sure you tighten this one to the correct torque spec (Appendix D) and keep it tight.
   (e) If the bottom-bracket fixed cup or lockring will not tighten completely or keeps coming loose, then either the bottom-bracket cups are stripped or undersized, or the frame's bottom-bracket shell threads are stripped or oversized. Either way, it's an expensive fix, especially the frame-replacement option! Get a second opinion if you reach this point.
2. **In the case of pedal end play, grab each pedal and wobble it to check for play.** If either is loose, see Chapter 13.

11-18

HARD-TO-TURN CRANKS

If the cranks are hard to turn, overhaul the bottom bracket (see 11-12 through 11-15)—unless you want to continue intensifying your workout or boosting the egos of your cycling companions. The bottom bracket may be shot and need to be replaced.

11-19

INNER CHAINRING DRAGS ON CHAINSTAY

If you hear noise because the inner chainring is dragging on the chainstay, then the bottom-bracket spindle is too short; you need a spacer between the bottom-bracket shell and the drive-side external-bearing cup (only applies to a 68mm-wide bottom-bracket shell); the square or ISIS hole in the crankarm is deformed (from riding with the crank bolt loose) so that the crank slides on too far; the frame is bent; or you have installed a larger inner chainring.

If the bottom-bracket spindle is too short, you need a new one of the correct length. If you have a 68mm-wide bottom-bracket shell with external bearings and don't have a spacer on the drive side, you need at least one 2.5mm spacer between the shell and the external-bearing cup (Shimano requires two drive-side spacers or one spacer along with the bracket of an E-type front derailleur); see 11-9, step 2. Shimano threaded, external-bearing cups for Hollowtech MTB cranks on a 73mm shell require a single, 2.5mm cup spacer on the drive side.

If the square or ISIS hole in the crank is badly deformed, you need a new crankarm; otherwise it will continue to loosen up and cause problems. If the frame is bent, refer to 17-13 to check on it and then decide what to do based on those instructions. If the chainring is too large, you need to get a smaller one.

With an adjustable cartridge-bearing bottom bracket with a lockring on either end (Fig. 11.24), the entire bottom bracket may be offset to the left. To move it to the right, loosen the lockrings, back out the right cup, screw the left cup in farther, adjust the end play, and retighten the lockrings (Fig. 11.31).

NOTE: *See 5-50 (Fig. 5.61) on chainline to establish proper crank-to-frame spacing. The chainring might not be rubbing the frame, but the crank is still too far inboard if the derailleur cannot move inward far enough to shift to the granny gear. The mechanism on a Shimano top-swing front derailleur in particular can hit the seat tube and stop before the chain drops onto the inner ring. The problem is compounded with a large-diameter seat tube. And if the bottom bracket is too long, you may have the opposite problem. The front derailleur will not reach the large chainring.*

The great thing in this world is not so much where you stand, as in what direction you are moving.

—OLIVER WENDELL HOLMES JR.

12

HANDLEBARS, STEMS, AND HEADSETS

TOOLS

Headset wrenches (two) sized to headset
Metric hex keys
Bike stand
Hammer
Screwdriver
Hacksaw
Flat file
Round file
Grease
Isopropyl alcohol
Scissors, tin snips, or knife

OPTIONAL

Air compressor with blow gun
Star-nut installation tool
Threadless saw guide
Carbon-specific hacksaw blade
Carbon assembly paste or spray
Channellock pliers
Securely mounted vise
Crown race slide punch
Crown race remover
Headset press
Headset cup remover
Small torque wrench
Head-tube reamer
Crown race facer
Syringe

You maintain or change your bike's direction largely by applying force to the handlebar. If everything works properly, variations in that pressure will result in your front wheel changing direction. Pretty basic, right? Right. But there is a somewhat complex series of parts between the handlebar and the wheel that makes that simple process possible. The parts of the steering system are illustrated in Figure 12.1. In this chapter, we cover most of that system by going over handlebars, stems, and headsets.

GRIPS

12-1

REMOVING THE GRIP

If the grip is shot, just cut it off with a knife. Otherwise, do the following:

1. **Remove the handlebar end plug.** It should just pull out. If you can't edge it out with your fingernail, use a thin screwdriver.
2. **If present, remove the bar end** (12-7).
3. **If present, loosen grip pinch bolts.** If there are tiny bolts on collars on each end of the grip (a lock-on grip), loosen the bolts and skip to step 6. Alternatively, if you have closed-end grips and an air compressor, you can blow them off by puncturing the end of one grip and blasting air into it with the end of the air blow gun pressed against the hole. Seal the end of the other grip if it also has a hole, and cover the end of the handlebar with your hand to blow off the other one.
4. **Roll back an edge of the grip** (Fig. 12.2).
5. **Squirt rubbing alcohol on the handlebar and the underside of the exposed grip.** Lacking alcohol, use water (Fig. 12.2). Flip the rolled-up edge back down, and repeat steps 4 and 5 on the other end of the grip.

PRO TIP — GRIP REMOVAL

A SYRINGE CAN BE USED to inject rubbing alcohol under the grip. The needle can be slipped under the grip from the end, and it can even be pushed through the grip. With alcohol underneath, the grip will slide off in seconds.

12.1 Steering assembly (shown without brakes for clarity)

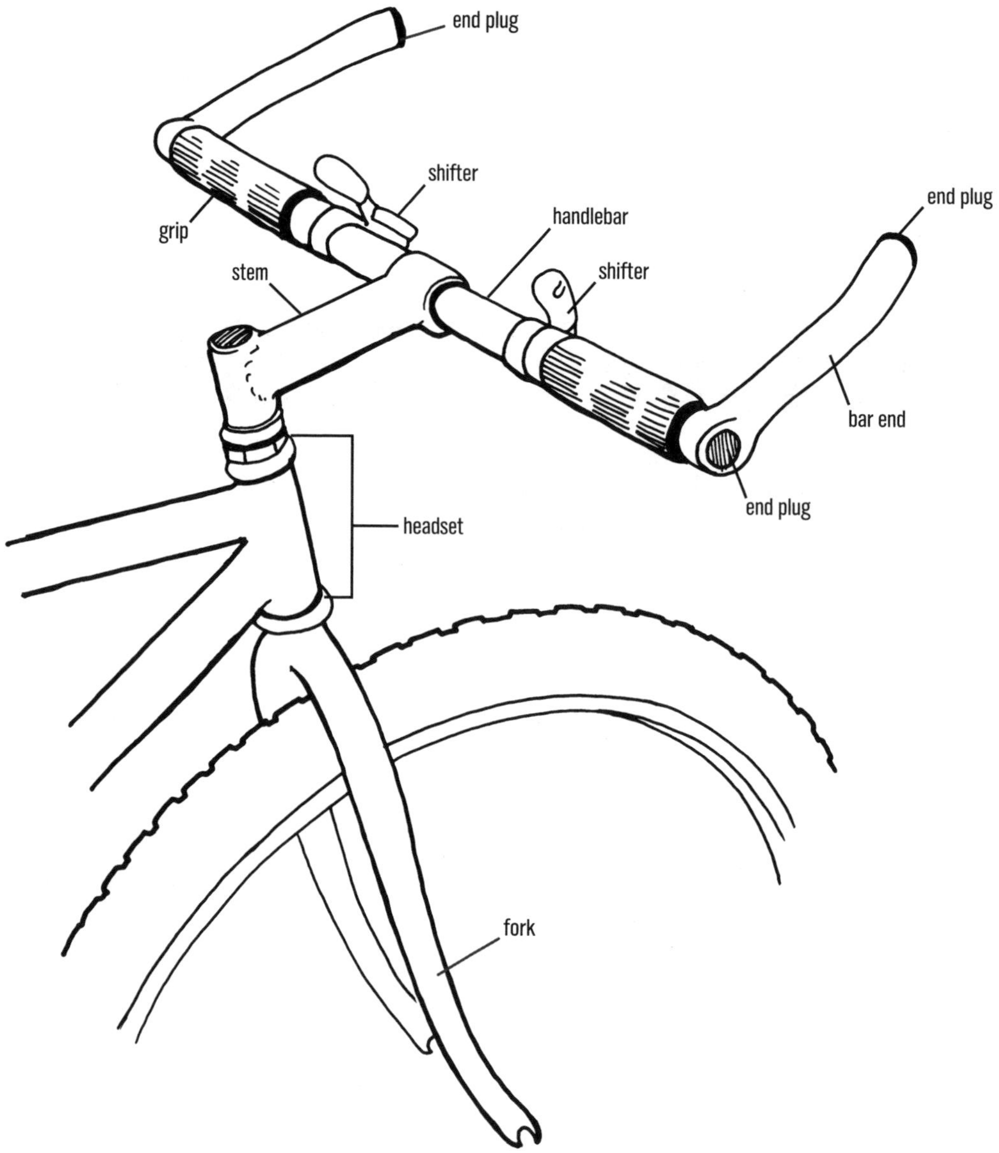

12.2 Dripping rubbing alcohol under a grip

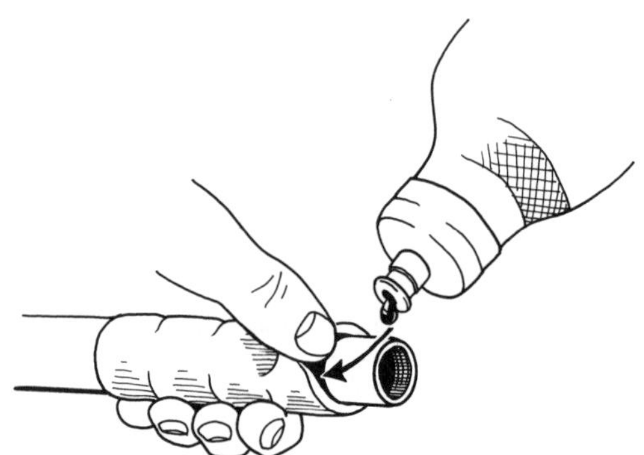

12.3 Grip removal

6. **Starting at the ends, twist the grip back and forth as you pull outward on it** (Fig. 12.3). The sections that are wet underneath will slip easily, and the dry middle section will start moving as the ends twist. Pull the grip off.

12-2

INSTALLING THE GRIP

1. **Squirt rubbing alcohol inside the grip.** This is not necessary with a lock-on grip; it will have a clamping collar with a binder bolt on either end of the grip and will just slide on with the bolts loose. Rubbing alcohol lubricates well and dries quickly. Water dries slowly, so the grip will slip for a few days. Hairspray and spray adhesives can be used to prevent grip slippage, but they are harmful to breathe, and they can set up permanently, thwarting subsequent removal and repositioning.
2. **Slide and twist the grip onto the handlebar.**

NOTE: *Some grips have a closed end. If you are going to use bar ends, you will need to cut off the closed end; you may also want to shorten the grip to adjust to your hand size or to adapt to a twist shifter. Some grips have a groove that marks where they are meant to be cut; you can use scissors (Fig. 12.4). If there's no groove, you can cut them anywhere you wish with scissors, tin snips, or a knife. If you have a thin, lightweight handlebar, you can easily cut off the end of the grip by hitting the end of the grip with a mallet after it is installed on the handlebar. The handlebar will cut a nice hole in the grip end like a cookie cutter!*

12.4 Trimming grip to accommodate bar end or Grip Shift

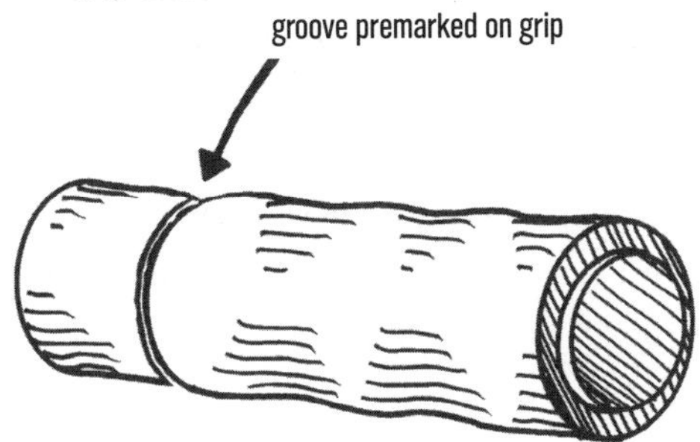

12.5 Clamp-type stem for threadless headset

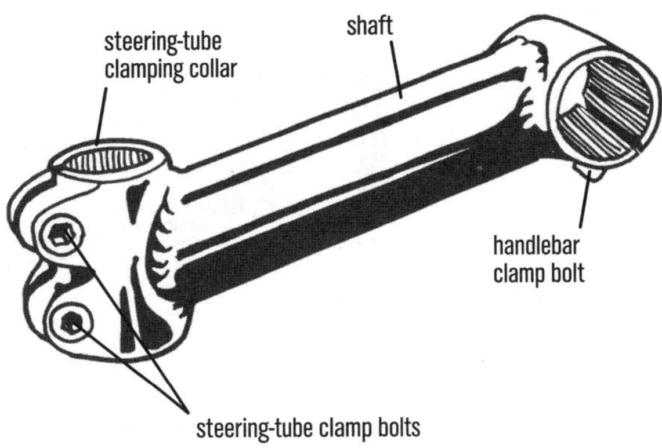

ANOTHER NOTE: *Grips used alongside Grip Shift and other twist shifters are shorter than standard grips, as part of the hand is sitting on the twist grip. Grips specifically designed for Grip Shift shifters are readily available in bike shops. If you can't find them, just cut the grips you have to the proper length.*

3. **If this is a lock-on grip, tighten the screws on the collar on each end.**

HANDLEBARS

12-3

REMOVING THE HANDLEBAR

1. **If you have a stem with a separate front faceplate, remove the faceplate.** The bar will drop out (Fig. 12.6).
2. **With a non-front-opening stem: Remove the bar ends and grips** (Fig. 12.7)**, followed by the brake levers and shifters.** Instructions for removal of bar ends and grips are in 12-1, for brake levers in Chapters 9 and 10, and for shifters in Chapters 5 and 6.
3. **Loosen or remove the stem's handlebar clamp bolt(s).** With a single-bolt stem (Fig. 12.5), just loosen that bolt. This bolt usually takes a 5mm hex key.
4. **Pull out the handlebar.**

12.6 Threadless headset and stem, cutaway view

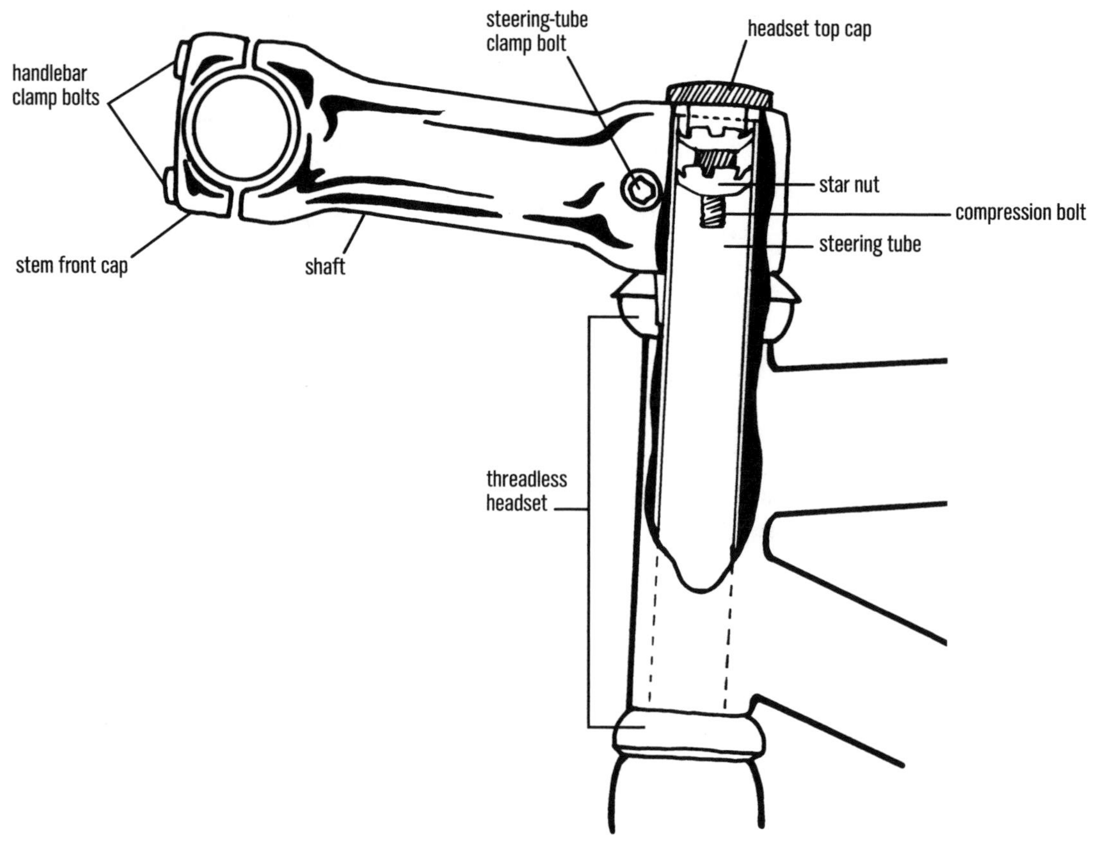

12-4

INSTALLING THE HANDLEBAR

1. **Loosely mount the handlebar in the stem clamp.** With a single-bolt stem, remove the stem clamp bolts, grease the threads, and reinstall the bolt. Grease the inside of the stem clamp, and grease the clamping area in the center of the handlebar. With a front-opening stem, place the front cap over the handlebar and replace the bolts.
2. **Twist the handlebar to the position that you find most comfortable.** Usually, this is with ends pointed up and back, but the exact position is a matter of personal preference.
3. **Tighten bolts that clamp the handlebar to the stem.** Tighten to the recommended torque (Appendix D). The torque applied is particularly important with expensive, lightweight stems and handlebars. You can pinch, and thereby weaken, a lightweight handlebar by overtightening, and the high-strength tubing will crack right next to the stem. Light stems come with small bolts with fine threads, and overtightening can strip the threads inside the stem. If you don't have a torque wrench and you do have a lightweight stem with small bolts (e.g., M5 or M6 bolts, which take 4mm and 5mm hex keys, respectively), use a short hex key so that you can't get much leverage. Proper torque is even more important with carbon-fiber handlebars.
4. **Check the gaps at the stem clamp.** Make sure that there is the same amount of space between the stem and either edge of the front plate on a front-opening stem. Any stem whose clamp gap(s) gets pinched until touching when tightened around the handlebar needs to be replaced, along with the handlebar. This is because overtightening has stretched the stem cap and deformed and weakened the handlebar.

12-5

MAINTAINING AND REPLACING THE HANDLEBAR

A bike cannot be controlled without a handlebar, so you never want one to break on you. Do not look at the handlebar on your bike as a permanent accessory. All aluminum handlebars will eventually fail. If titanium, steel, or carbon-fiber handlebars are repeatedly stressed above a certain level, they will eventually fail as well. What that level is depends on the particular handlebar. The trick is to replace it before it fails.

Keep the handlebar clean. Regularly inspect it for cracks, crash-induced bends, corrosion, and stressed areas. If you find any sign of wear or cracking, replace the handlebar. Never straighten a bent handlebar—replace it! If you crash hard on your bike, consider replacing the handlebar even if it looks fine. If you have had a crash and can see no problems with the handlebar, remove the bar ends and check whether it is bent at the bar-end edges. A carbon-fiber handlebar can be broken internally, and the damage may not be visible from the outside. If the bar has taken an extremely hard hit, it's a good idea to replace it rather than gamble on its integrity. This is especially true with lightweight handlebars; the high hardness of the materials used may prevent visible bending, but they may be so weakened that they will soon shear off.

Some manufacturers recommend replacing stems and handlebars every four years. If you rarely ride the bike, this is overkill. If you ride hard and ride often, every four years may not be frequent enough. Do what is appropriate for you, and be aware of the risks.

BAR ENDS

12-6

INSTALLING BAR ENDS

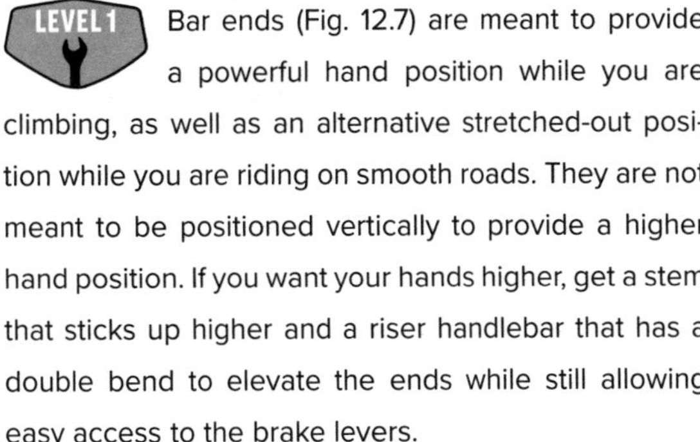

Bar ends (Fig. 12.7) are meant to provide a powerful hand position while you are climbing, as well as an alternative stretched-out position while you are riding on smooth roads. They are not meant to be positioned vertically to provide a higher hand position. If you want your hands higher, get a stem that sticks up higher and a riser handlebar that has a double bend to elevate the ends while still allowing easy access to the brake levers.

Now that short, flat handlebars (Fig. 12.1) have gone out of style, replaced by long riser bars, bar ends out at the ends of the bars (Fig. 12.7) have become obsolete. They not only often head off at weird angles due to the bend in the handlebar, but they are also too wide to properly support the shoulders. But inner bar ends (Fig. 12.8), placed inboard of the grip, do make a lot of sense on wide handlebars. They offer another hand position to relax the hands, and a more aerodynamic one at that, much like the position on a road bike atop the brake hoods.

1. **Slide the brake levers and shifters inward.** See Chapters 5 and 6, and 9 and 10 for moving shifters and brake levers, respectively.
2. **Outer bar ends: slide the grips inward.** This makes room at the end of the handlebar for the bar end. See 12-1 for instructions on moving the grips. Skip to step 4.

12.7 Grip and bar-end assembly, exploded view

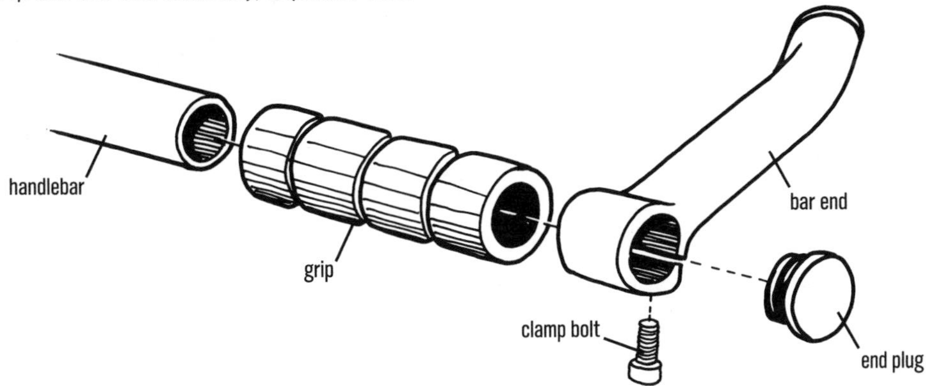

12.8 Inner bar ends

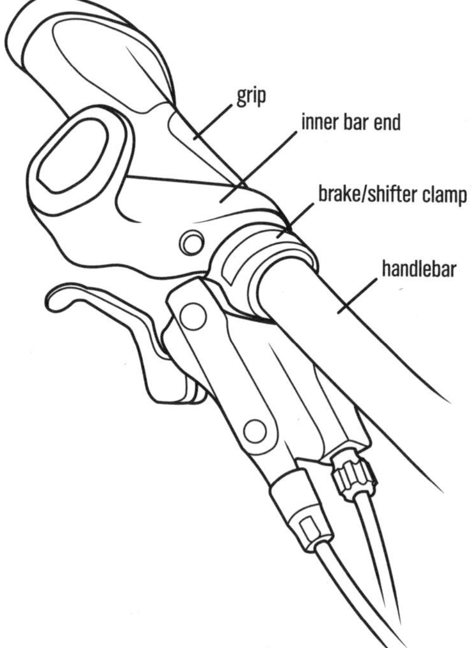

3. **Inner bar ends: remove the end plugs and grips.** See 12-1. This allows setting the inner bar ends between the grip and the brake lever.
4. **Loosen the bolt on the bar-end clamp.** This bolt usually accepts a 5mm hex key.
5. **Slide the bar end onto the handlebar** (Fig. 12.7).
6. **Tighten the clamp bolt just enough to hold the bar end in place.** Rotate the bar ends to the position you like (see Appendix C, C-3e, for position recommendations).
7. **Tighten the clamp bolt.** Make sure it is snug. (Recommended torque is in Appendix D.)
8. **With inner bar ends, install the grips and end plugs.** See 12-2.

NOTE: *The ends of some superlight handlebars can be damaged by bar ends and therefore come equipped with small, cylindrical, aluminum inserts that provide support under the bar ends when inserted at the ends of the handlebar. These handlebars cannot be shortened without forgoing the option of bar ends.*

SAFETY NOTE: *Bar ends, be they inner or outer, are not to be used on freeride, enduro, or downhill bikes. And in technical or dangerous riding conditions, hold the handlebar grip, not the bar end.*

12-7

REMOVING BAR ENDS

1. **Loosen the bolt on the bar-end clamp.** This bolt usually accepts a 5mm hex key. With inner bar ends, you'll also have to remove the grips and end plugs (12-1).
2. **Pull off the bar end** (Fig. 12.7).

STEMS

The approximately horizontal stem connects to the approximately vertical steering tube of the fork (which is either 1, 1⅛, or 1¼ inches in diameter) and clamps around the handlebar, which has one of two diameters: 25.4mm or 31.8mm. Stems come in two basic types: (1) for threadless steering tubes (Fig. 12.5) or (2) for threaded steering tubes (Fig. 12.10). Some stems have shock-absorbing mechanisms with pivots and springs to provide suspension (Fig. 12.11).

Stems for threadless steering tubes (Fig. 12.5) have a clamping collar for the fork steering tube. Because the steering tube has no threads, the top headset cup slides on and off. In this case, the stem plays a dual role: It clamps around the steering tube to connect the handlebar to the fork, and it also keeps the headset in proper adjustment by preventing the top headset cup from sliding up the steering tube (Figs. 12.5 and 12.13). If you have

12.9 Quill-type stem

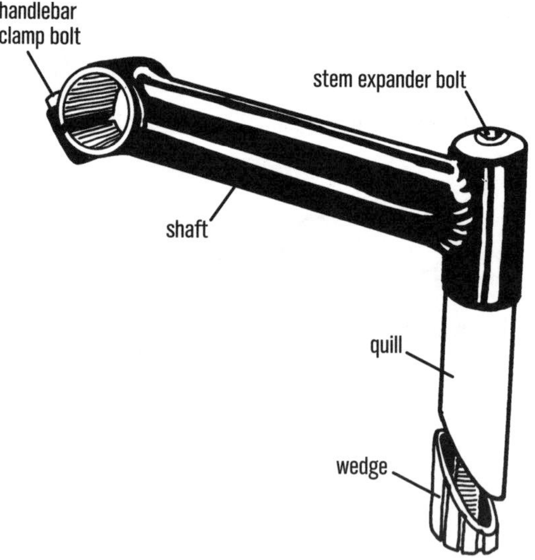

12.10 Threaded headset and quill stem, cutaway view

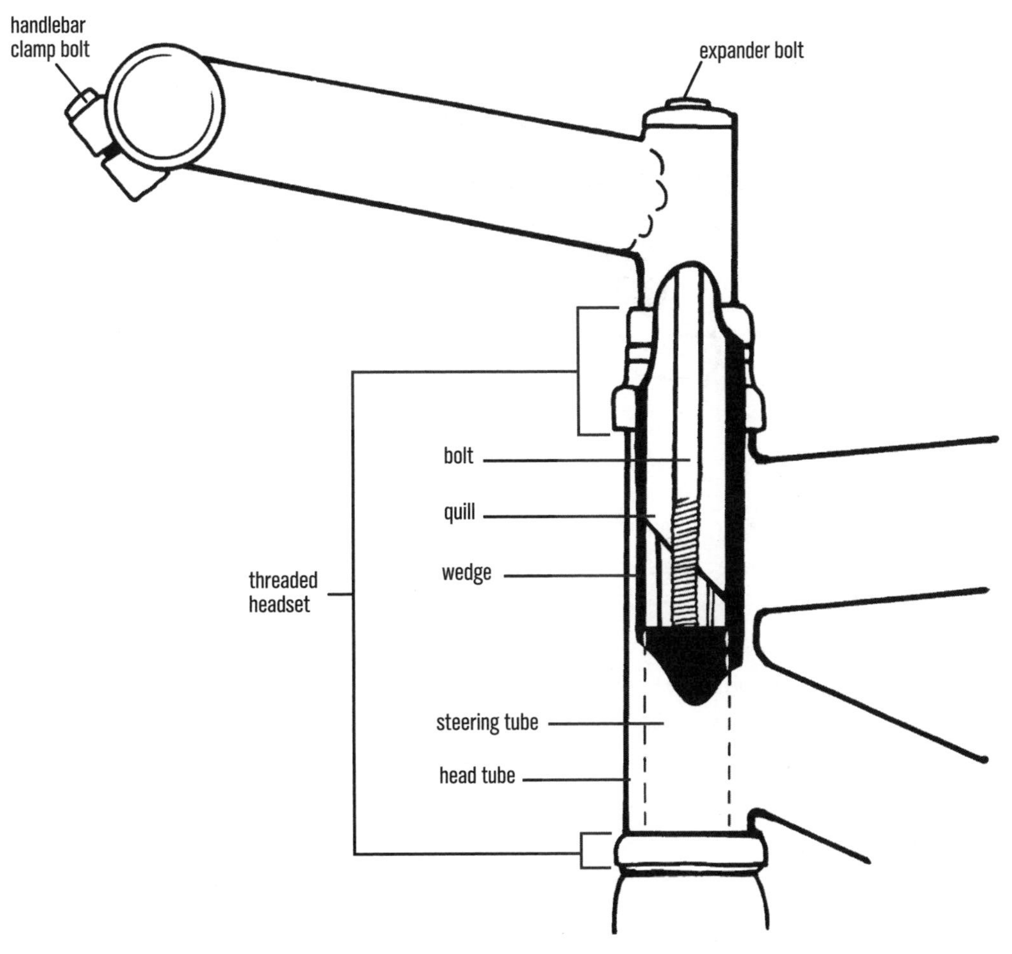

12.11 Suspension stem, quill type

12.12 Threadless headset cup held in place by stem

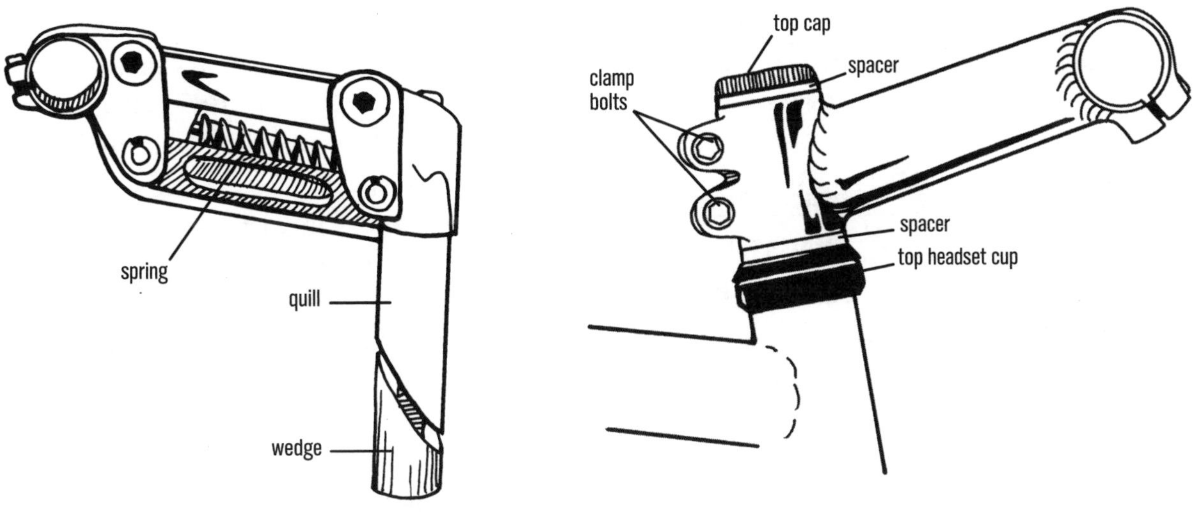

12.13 Loosening or tightening compression bolt on a threadless headset

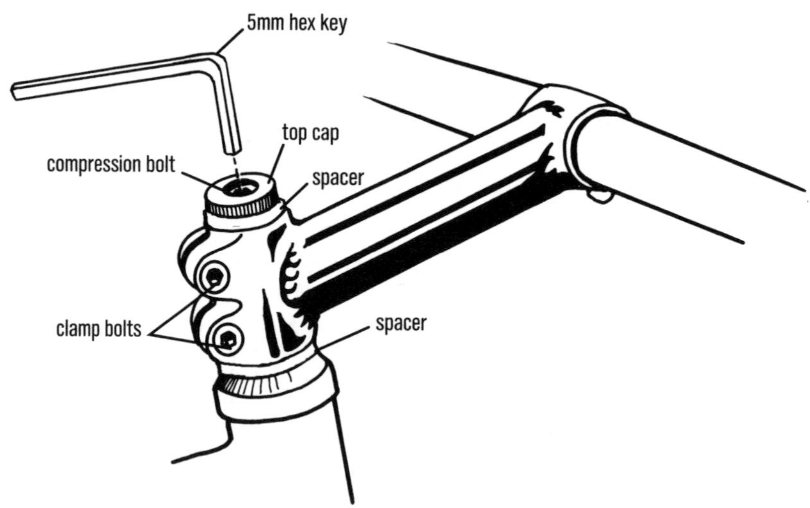

an old 1-inch-diameter steering tube (the old standard) and a stem for a 1⅛-inch steering tube (the current standard), you can get a slotted aluminum reduction bushing to allow the stem to be used with the steering tube.

Stems for threaded steering tubes (Fig. 12.9) were the standard until the mid-1990s. They have a vertical quill, which extends down into the steering tube of the fork and binds to the inside of the steering tube by means of a wedge-shaped plug pulled up by a long bolt that runs through the quill (Fig. 12.10).

Suspension stems used to be quite popular and were made for both threadless and threaded steering tubes. The Softride stem (Fig. 12.11) uses a parallelogram system with four pivots to prevent the handlebar from twisting as it moves up and down. Others have a single pivot around which the handlebar swings in an arc. The spring is usually a steel coil or an elastic polymer (elastomer). Some suspension stems also come with a hydraulic damper to control the speed of movement.

12-8

REMOVING A CLAMP-TYPE STEM FROM A THREADLESS STEERING TUBE

1. **Loosen the horizontal bolts clamping the stem around the steering tube.** Alternate loosening them a bit at a time.
2. **Remove the adjusting bolt in the top cap covering the stem clamp** (Fig. 12.13). Unscrew it with a 5mm hex key. Removing this bolt will allow the fork to fall out, so hold the fork as you unscrew the bolt.
3. **With the bike standing on the floor to keep the fork from falling out, pull the cap and the stem off the steering tube.** Leave the bike standing until you replace the stem, or slide the fork out of the frame, keeping track of all headset parts.
4. **If the stem will not budge, see 12-14a.**

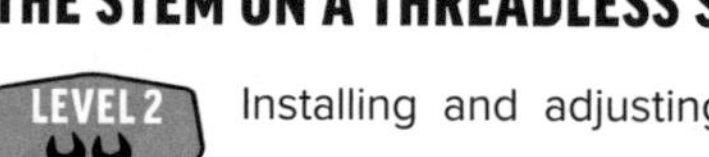

INSTALLING AND ADJUSTING THE HEIGHT OF THE STEM ON A THREADLESS STEERING TUBE

LEVEL 2

Installing and adjusting the height of a stem on a threadless fork are much more complicated than installing and adjusting the height of a standard stem in a threaded fork. On a threadless fork, the stem is an integral part of the headset (Fig. 12.6), so any change to the stem position alters the headset adjustment. That's why this procedure carries a level 2 designation.

1. **Stand the bike up on its wheels so that the fork does not fall out.**
2. **Grease the parts.** Grease the top end of the steering tube if it is steel or aluminum, but leave it dry if it is carbon fiber. Remove the stem clamp bolts, grease their threads, and then thread them back into place but do not tighten.
3. **Slide the stem onto the steering tube.**

4. **Set the stem height to the desired level.** If you want to place the stem in a position higher than directly on top of the headset, you must put spacers between the bottom of the stem clamp and the top piece of the headset. There must be contact—through direct contact or through spacers (and including the top crown on a double-crown fork)—between the headset and the stem. Otherwise, the headset will be loose.
5. **Check the steering-tube overlap.** To adjust the threadless headset, the top of the stem clamp (or, ideally, spacers placed above it) should overlap the top of the steering tube by 3–6mm (⅛–¼ inch) (Fig. 12.14). If it does, skip ahead to step 8.

NOTE: *I recommend always having one spacer above the stem, especially with a carbon steering tube. If you add the spacer, measure the 3–6mm overlap from the top of the spacer, not from the stem. That way, the entire stem clamp is clamped onto the steering tube, and there is no chance for the upper part of the clamp to pinch the end of the steering tube.*

6. **If the steering tube is too short, move spacers or replace the fork.** If the top spacer or the top of the stem clamp overlaps the top of the steering tube by more than 6mm (¼ inch), the steering tube is too short to set the stem height where you have it. If you have spacers below the stem, remove some until the edge of the top spacer (or the stem clamp) overlaps the top of the steering tube by 3–6mm. If you cannot lower the stem any farther, or do not want to, get either a fork with a longer steering tube, a stem with a shorter clamp, or a stem that is angled upward more to attain your desired handlebar height. The stem is cheaper and easier to replace than the fork.

   On many old suspension forks, you could simply replace the steering tube and fork-crown assembly with a longer one and bolt the existing fork legs into it. But nowadays, to get a longer steering tube, you must replace the fork.
7. **If the steering tube is too long, move spacers or cut the tube.** If the top of the steering tube sticks up above, or if it is less than 3mm (⅛ inch) below the edge of the top spacer (or the stem clamp), you have a choice.
   (a) If you want the option to raise the stem for a higher handlebar position, stack some headset spacers on top of the stem clamp until the spacers overlap the top edge of the steering tube by at least 3mm.
   (b) If, however, you are sure you will never want the stem any higher, then cut off the excess tube. First, mark the steering tube along the top edge of the spacer above the stem clamp or the stem clamp itself (if you are not heeding my advice to always have at least a single thin spacer above the stem). Remove the fork from the bike. Make another mark on the steering tube 3mm below the first mark. Place the steering tube into a padded vise or bike-stand clamp.
   (c) There is a star-shaped nut that is inserted inside the steering tube (Fig. 12.6). On an existing fork, the star nut will be in place; new forks will come with one to be installed. The bolt through the top cap screws into it to adjust the headset bearings. If the star nut is already inside the steering tube and it looks like the saw will hit it, move the star nut down—see step 8 for instructions to do that.
   (d) Make your cut straight. Wrap a piece of tape around the steering tube and cut along the edge of the tape. If you are not sure your cut will be straight, start it a little higher and file it down flat to the tape edge. If you really want to be safe, use a saw guide like Park Tool's

12.14 Minimum steering-tube length

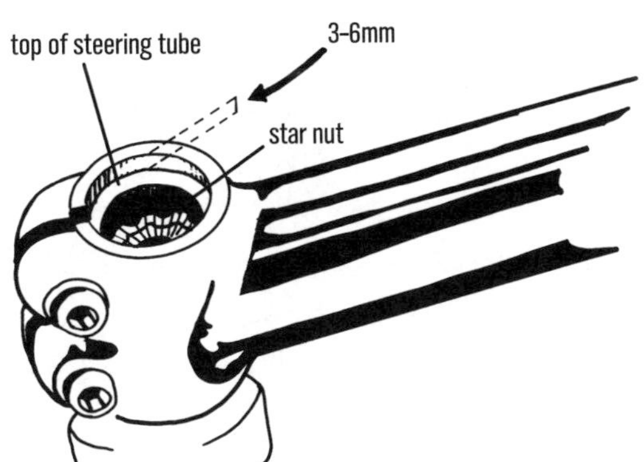

SG-6 or SG-8 specifically designed to help you make a straight cut. Measure twice and cut once; you can always shorten the steering tube a little more, but you cannot make it longer! Use a round file on the inside of the tube and a flat file on the outside to remove any metal burrs left by the hacksaw or cutter.

(e) Put the fork back in, replacing all of the headset parts in the way that they were originally installed (Figs. 12.19 to 12.21). Return to step 1.

NOTE: *If you are cutting a carbon-fiber steering tube, cut three-quarters of the way through and then turn the steering tube over and cut from the other side to meet your prior cut. This will prevent peeling away layers of carbon with the blade on the inside of the back cut. To anchor the top cap bolt in a carbon steering tube and to prevent the stem clamp from crushing the carbon steering tube, set an expander plug (Fig. 12.26) inside the steering tube under the stem clamp, instead of a star nut; tighten its expander bolt with a hex key.*

8. **Check that the edges of the star-shaped nut are at least 12mm below the top edge of the steering tube.** The headset top cap must not hit it when you tighten the adjusting bolt. If the nut is not in deep enough, drive it deeper into the steering tube with a star-nut installation tool (Fig. 12.15). The tool threads into the nut, and you hit it with a hammer until it stops; the star nut will now be set 15mm deep in the steering tube. If you do not have this tool, have a bike shop set the nut for you. If you insist on driving the star nut in without the proper tool, put the adjusting bolt through the top cap, thread it five turns into the star nut, and tap the top of the bolt with a mallet until the star nut is 15mm (⅝ inch) below the top of the steering tube. Use the top cap as a guide to keep it going in straight. It is easy to mangle the star nut if you do not tap it in straight.

NOTE: *Steering-tube wall thickness depends on whether it is steel or aluminum, its grade and heat treating, and the usage the fork is designed for. Therefore, the stock headset star nut may not fit and will bend when you try to install it. This is not a big problem, because star nuts of various sizes can be purchased separately. If the star nut goes in crookedly, take a long punch or rod,*

12.15 Setting the star nut in the steering tube

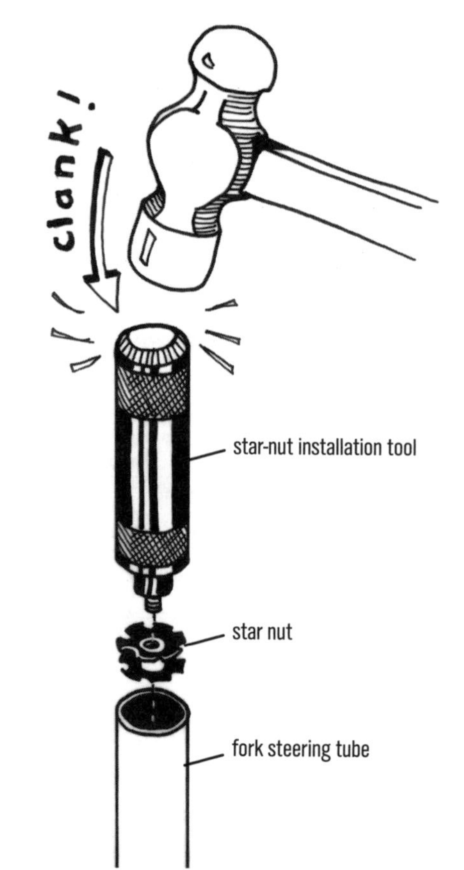

set it on top of the star nut, and drive it all the way out of the bottom of the steering tube.

Standard ID (inside diameter) at the top of a 1-inch steel steering tube is 22.2mm (⅞ inch), 25.4mm (1 inch) for a 1⅛-inch steel steering tube, and 28.6mm (1⅛ inches) for a 1¼-inch steel steering tube; the ID of aluminum steering tubes will be less. If a star nut was not packaged with the fork, and the one packaged with the headset does not fit, buy one of the correct size. Don't fuss around with trying to bend the legs of a star nut that doesn't fit.

9. **Install the headset top cap on the top of the stem clamp or spacer(s) above it.** Grease the threads of the top-cap adjusting bolt, and thread it into the star nut inside the steering tube (Fig. 12.13).
10. **Adjust the headset and stem before tightening the stem bolts.** Line the stem up straight with the front wheel. Headset adjustment is explained in 12-16.
11. **Tighten the stem bolts.** Tighten to the recommended torque (see Appendix D).

12.16 Freeing the stem wedge

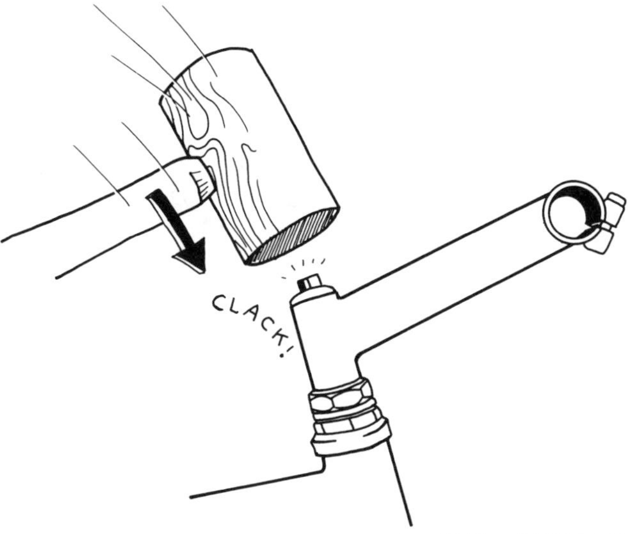

12-10

REMOVING A STANDARD QUILL-TYPE STEM FROM A THREADED FORK

1. **Loosen the stem anchor bolt on the top of the stem.** Unscrew it three turns or so. Most stem bolts take a 6mm hex key.
2. **Tap the top of the bolt down with a hammer** (Fig. 12.16). This disengages the wedge from the bottom of the quill.
3. **Pull the stem out of the steering tube.** If the stem will not budge, see 12-14b.

12-11

INSTALLING AND ADJUSTING THE HEIGHT ON A QUILL STEM IN A THREADED FORK

1. **Lube all contacting parts.** Grease the stem quill, the bolt threads, the outside of the wedge or conical plug (Fig. 12.9), and the inside of the steering tube.
2. **Tighten the bolt until it just pulls the wedge or plug into place.** Don't tighten the bolt enough to prevent the stem from inserting into the steering tube.
3. **Slip the stem quill into the steering tube** (Fig. 12.10) **to the depth you want.** Make sure the stem is inserted beyond its height-limit line.
4. **Line up the stem with the front wheel, and tighten the bolt.** It needs to be tight, but don't overdo it. You can overtighten the stem bolt to the point that you bulge out the steering tube on the fork, so be careful. Recommended torque is in Appendix D.

12-12

MAINTAINING AND REPLACING THE STEM

A bike cannot be controlled if the stem breaks, so make sure yours doesn't break. Because aluminum has no fatigue limit, all aluminum parts will eventually fail. Steel and titanium parts that are repeatedly stressed more than about one-half of their tensile strength will eventually fail as well. Stems and handlebars are not permanent accessories on your bike. Replace them before they fail on you.

Clean the stem regularly and examine it for corrosion, cracks, bends, and stressed areas. If you find any, replace the stem immediately. If you crash hard on your bike, especially hard enough to bend the handlebar, replace the stem and possibly the fork. It makes sense to err on the side of caution. Lightweight, expensive stems, in particular, need to be replaced after a violent impact, even if you see no visible signs of stress. The hard, thin material is not likely to bend, but it may be weakened.

Some manufacturers recommend replacing stems and handlebars every four years. If you rarely ride the bike, this is overkill. If you ride hard and ride often, every four years may not be frequent enough. Do what is appropriate for you, and be aware of the risks.

12-13

SETTING STEM AND HANDLEBAR POSITIONS

Complete treatment of this subject is in Appendix C-3. Here are some brief suggestions:

- At least initially, set the handlebar twist so that the bends in the handlebar are pointed up and back. I also recommend setting the bar ends, if installed, so that they are horizontal or tipped up between 5 and 15 degrees from horizontal (C-3e).
- Setting handlebar height and reach is very personal. Much depends on your physique, your flexibility, your

frame, your riding style, and a few other preferences. Again, this subject is covered in depth in Appendix C.

- Because I do not know anything about you personally, I will leave you with a few simple guidelines:
  - (a) If you climb a lot, you will want the handlebar lower and farther forward to keep weight on the front wheel on steep uphill trails.
  - (b) If you descend technical trails a lot, you will want the handlebar higher and with less forward reach.
  - (c) If you ride a lot on pavement, a low, stretched-out position is better aerodynamically and may be more comfortable. A low position means that the handlebar grips are about 7–12cm (2¾–4¾ inches) lower than your saddle. A stretched-out position would place your elbow at least 5cm (2 inches) in front of your knee at the top of the pedal stroke. Inner bar ends (Fig. 12.8) can improve comfort and aerodynamics.

12-14

REMOVING A STUCK STEM

LEVEL 3

A stem can get stuck onto, or into, the steering tube because of poor maintenance. Periodically regreasing the stem and steering tube will keep them sliding freely, and the grease will form a barrier against sweat and water getting in between the two. If the stem is really stuck, be careful; you may ruin the fork as well as the stem and headset while trying to get the stem out. In fact, you're better off having a shop work on removing the stem, unless you really know what you are doing and are willing to accept the risk of destroying a lot of expensive parts.

a. Removing a stuck stem from a threadless fork

1. **Remove the top cap** (Fig. 12.13) **and the bolts clamping the stem to the steering tube.**
2. **Spread the stem clamp.** Insert a coin into the slot between each bolt end and the opposing threadless half of the binder lug (Fig. 12.17). Turn the bolts around, install them from the opposite direction, and tighten each bolt against each coin so that the clamp slot opens wider.

12.17 Sticking a coin in the crack to spread the stem clamp

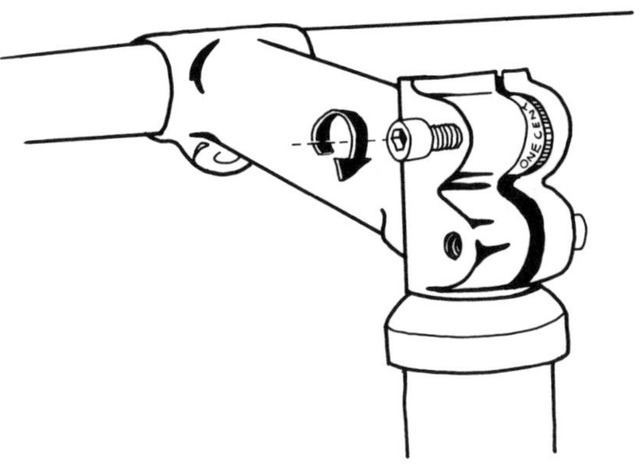

NOTE: *If the stem is the type that comes with a single bolt in the side of the stem shaft ahead of the steering tube (Fig. 12.6), loosen the bolt a few turns and tap it in with a hammer to free the wedge. Some penetrating oil, and perhaps some heat to expand it, may be required to free this type of stem from around the steering tube.*

3. **If the stem still will not come free,** pour ammonia, automotive antifreeze, or Coca-Cola in around the stem clamp to dissolve the aluminum oxide sticking the parts together and let the stem sit—overnight if need be. If it is still stuck, use thermal expansion and contraction to free the parts: Try heating the stem clamp with a hair dryer and discharging a tire inflation cartridge inside the steering tube. Still no luck? You may have to hold the crown in a vise, following instructions 8 to 10 in 12-14b on freeing a quill-type stem. Failing that, your last resort is to saw through the base of the stem clamp and the steering tube and replace the stem and fork.

b. Removing a stuck stem from a threaded fork

1. **Unscrew the expander bolt on top of the stem three turns or more.**
2. **Smack the bolt (or the hex key in the bolt) with a hammer** (Fig. 12.16). Do this until the bolt drops down, indicating it has completely disengaged the wedge.
3. **Grasping the front wheel between your knees, twist back and forth on the handlebar.** Don't use

all of your strength, because you can ruin a fork and a front wheel this way.

4. **If the stem didn't budge, squirt ammonia around it where it enters the headset.** Let the bike sit for several hours, adding ammonia every hour or so. (This assumes it's an aluminum stem; ammonia or Coca-Cola dissolves aluminum oxide, whereas penetrating oil dissolves steel rust.)
5. **Turn the bike over, and pour ammonia, automotive antifreeze, or Coca-Cola into the bottom of the fork steering tube.** Let it run down around the stem quill for several hours, and add more every hour or so. Note that penetrating oil works only on steel parts.
6. **Repeat step 3.** Still stuck? Continue with step 7.
7. **Discharge a tire inflation cartridge inside the quill shaft to chill and shrink it.** Remove the stem expander bolt to allow access.
8. **If the stem doesn't come free, use a heavy-duty vise.** Make sure it is solidly mounted.
9. **Remove the front wheel** (Chapter 2) **and the front brake** (Chapter 9 or 10).

12.18 Clamping the fork crown in a vise

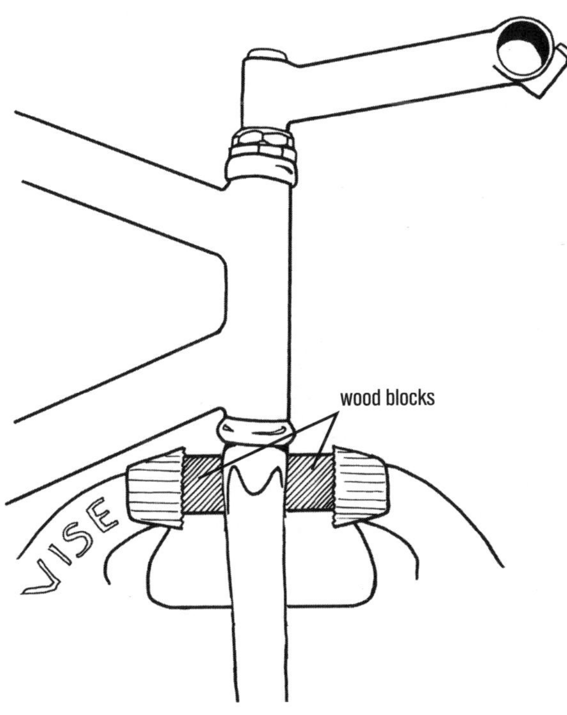

10. **Clamp the fork crown into the vise between wood blocks** (Fig. 12.18). If you have an old suspension fork with crown clamp bolts, remove the inner fork legs from the crown by loosening the crown bolts and yanking the legs out.
11. **Repeat steps 3 and 7.** If this doesn't work, you may need to saw off the stem just above the headset and have the bottom of the stem reamed out of the steering tube by a machine shop.

HEADSETS

HEADSET TYPES AND HOW TO DETERMINE FIT TO A GIVEN FRAME AND FORK

There are three basic types of threadless headsets: traditional (a.k.a. external or TR, Fig. 12.19), cupless internal (a.k.a. integrated or IS, Fig. 12.20), and press-in internal with lipped cups (a.k.a. semi-integrated, zero stack, low stack, or ZS, Fig. 12.21). Older mountain bikes have threaded steering tubes with threaded headsets (Fig. 12.22); these always have traditional external cups.

FORK MEASUREMENT

Headsets come in many sizes for mountain bikes, and the size refers to the diameter in inches of the fork steering tube. However, in these days of tapered steering tubes, you need to measure the diameter of the steering tube both at its base (crown race seat diameter) and at its top (stem clamp diameter). Standard mountain bike sizes are 1 inch, 1⅛ inches, 1¼ inches, 1½ inches, and a combination of two of these sizes. The diameter of the crown race seat will be 1–2mm larger than the headset size, whereas the steering-tube diameter at the top will be exactly the headset size. For instance, the steering-tube diameter of a fork for a 1⅛-inch headset will measure 28.6mm (1⅛ inches) at its top and approximately 30mm at its base; the steering-tube diameter of a fork for a 1½-inch headset will measure 38.1mm (1½ inches) at its top and approximately 39.8mm at its base; and the steering-tube diameter of a fork for a 1½ × 1⅛-inch "tapered" headset (Fig. 12.23) will measure 28.6mm at its top and approximately 39.8mm at its base.

12.19–12.22 Headsets

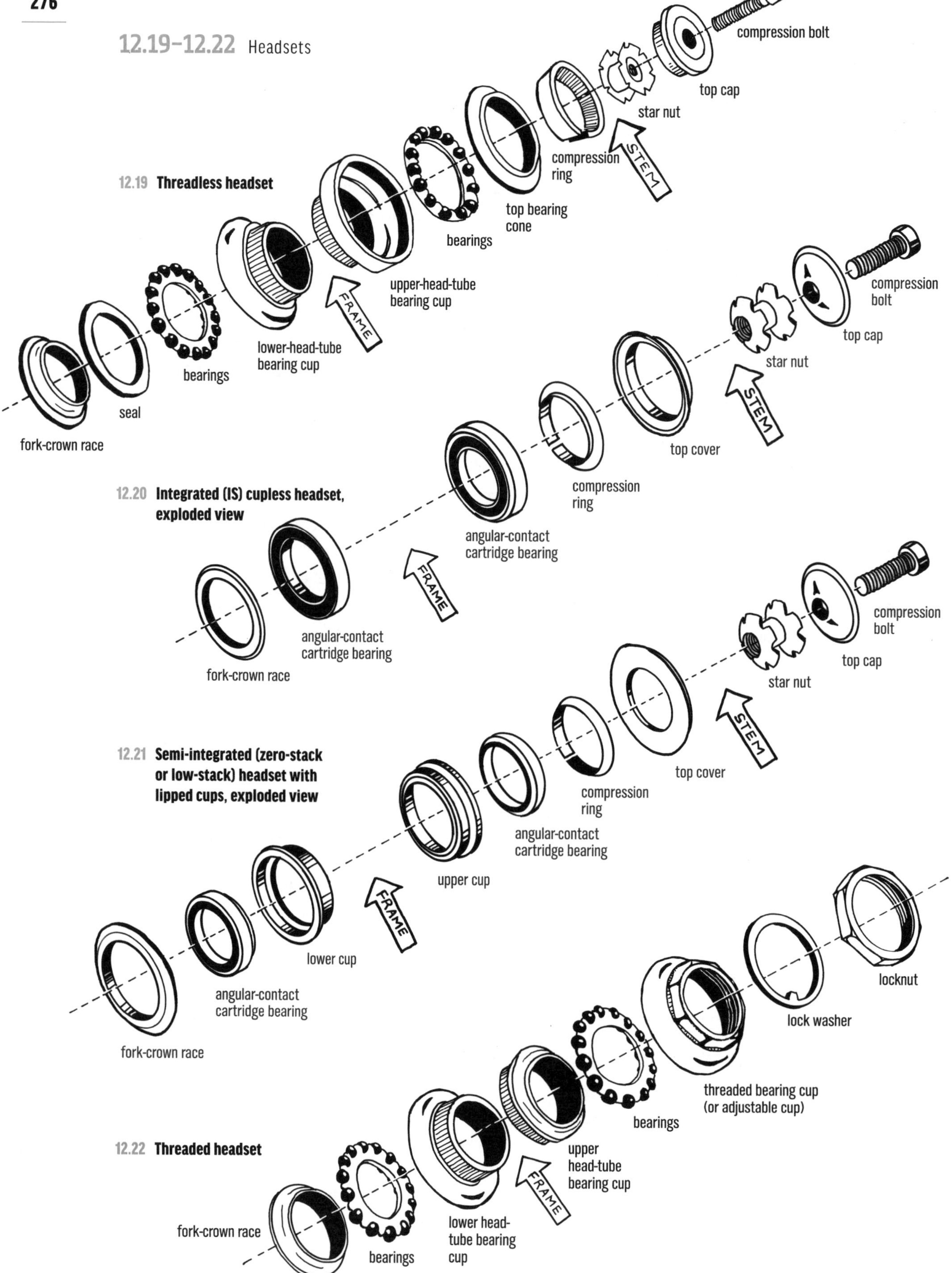

12.19 **Threadless headset**

12.20 **Integrated (IS) cupless headset, exploded view**

12.21 **Semi-integrated (zero-stack or low-stack) headset with lipped cups, exploded view**

12.22 **Threaded headset**

12.23 Trek/Gary Fisher E2 integrated cupless headset with tapered steering tube and differentially sized bearings

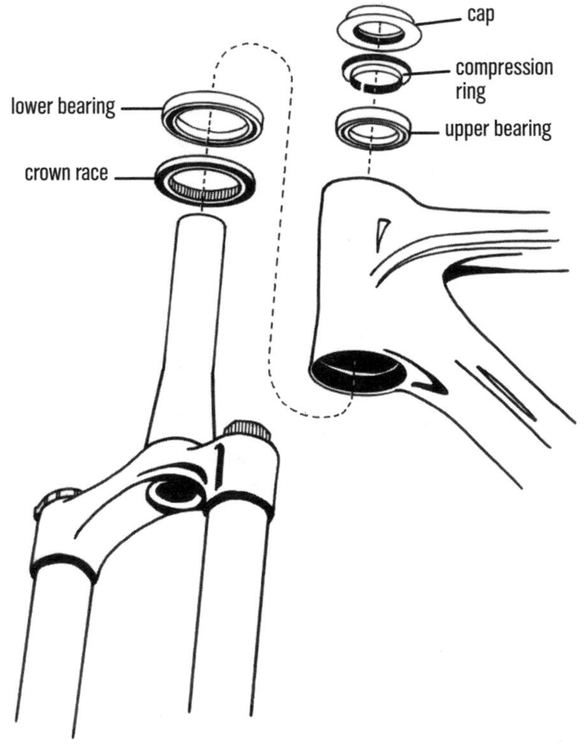

HEAD-TUBE MEASUREMENT

You determine the headset size as above from the diameter(s) of the fork steerer, and you determine the threadless headset type by examining and measuring the inside of the head tube (threaded headsets are simply determined by whether the fork has a threaded steerer or not). If the inside of the head tube is cylindrical and uniform in diameter (except for perhaps a small cut edge a few centimeters in where the reamer stopped), then it is intended for a press-in cup—either a traditional (TR) or semi-integrated (zero stack or low stack). If, however, there is a raised and chamfered (angled) surface ringing the inside of the head tube anywhere from 5mm to 25mm in, then it most likely takes an integrated (IS) headset.

Note that you can get reducer headsets to adapt a fork with a thinner steering tube or nontapered steering tube to a head tube that's meant for a steering tube that is either larger throughout or is tapered and larger only at its base.

THREADLESS HEADSETS

Threadless headset systems (Figs. 12.19 to 12.21) are lighter than threaded ones because they lack the stem quill, bolt, and wedge (Fig. 12.10) of a threaded headset (Fig. 12.22). The connection between the handlebar and the stem is also more rigid than the type of connection with an expanding stem wedge. Of course, fork manufacturers prefer threadless headsets because they do not have to thread their forks and they can offer a single steering-tube length; steering-tube diameter is now the only variable.

On a threadless headset, the top cup and a conical compression ring slide onto the steering tube (Figs. 12.19 to 12.21 and 12.23 to 12.25). The stem clamps around the top of the steering tube and above the compression ring. A star-shaped nut with two layers of spring-steel teeth sticking out from it fits into the steering tube and grabs the inner walls (Figs. 12.6 and 12.24). On a carbon-fiber steering tube, an expandable insert (Fig. 12.26) instead of a star nut serves the dual purpose of anchoring the top cap and protecting the steering tube from being crushed by the stem clamp.

12.24 Semi-integrated (zero-stack or low-stack headset system), cutaway view

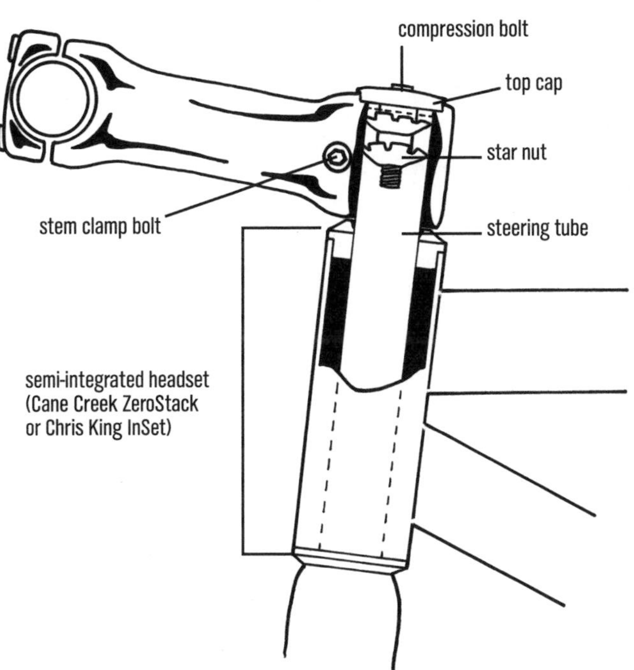

12.25 Integrated headset with drop-in bearing cups (Campagnolo shown), exploded view

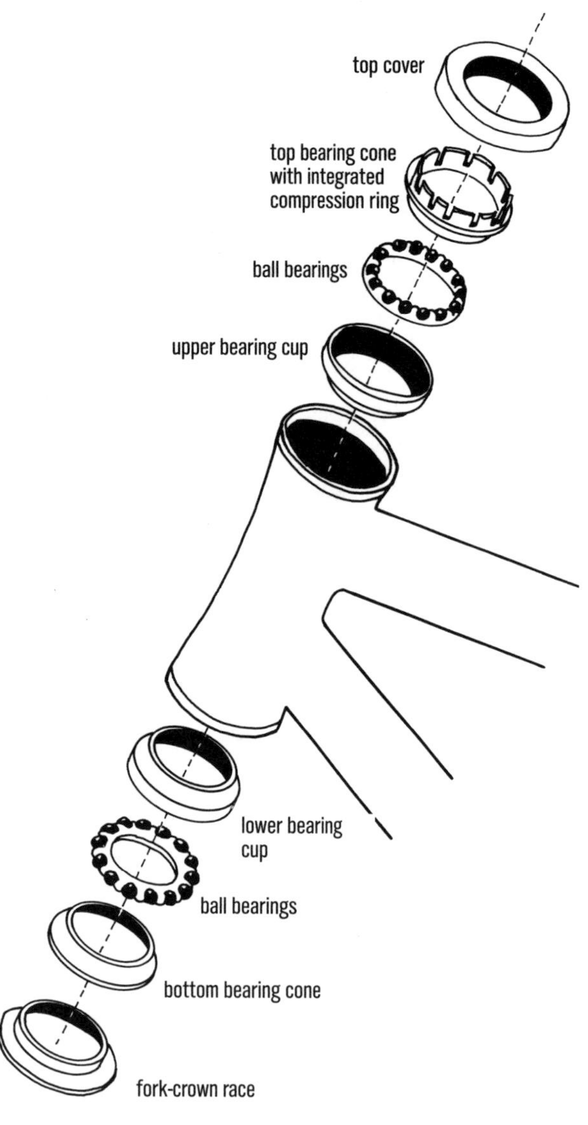

12.26 Inserting an expandable support/anchor plug into a carbon-fiber fork steering tube

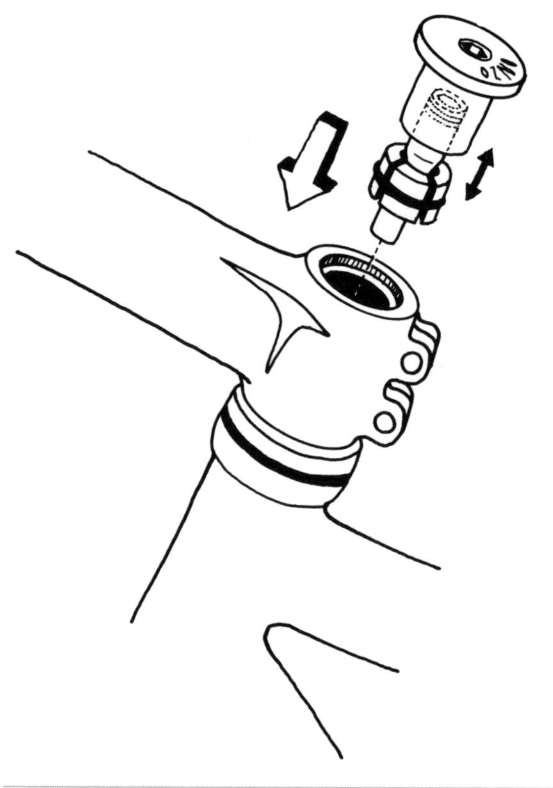

12.27 Needle bearings

A top cap sits atop the stem clamp and pushes it down by means of a long bolt threaded into the star nut to adjust the headset (Figs. 12.13 and 12.24). The stem clamped around the steering tube holds the headset in adjustment (Fig. 12.13).

The next generation of threadless headsets are internal, or integrated, headsets, concealed inside the frame's steering tube (Figs. 12.20, 12.21, and 12.23). Where traditional threaded and threadless headsets have bearing cups that are pressed into the head tube (Figs. 12.19 and 12.22), integrated headsets have bearings seated inside the head tube. The bearings either rest on a chamfered platform within the head tube itself (Figs. 12.20 and 12.25) or have semi-integrated press-in cups with thin flanges that extend out to the edges of the head tube (Figs. 12.21 and 12.24). Otherwise, the headset is identical to, and is adjusted in the same way as, the traditional threadless systems shown in Figures 12.6 and 12.19.

Prior to the 1990s, practically all headsets and steering tubes were threaded. The top bearing cup on a threaded headset has wrench flats, a toothed washer stacked on top of it, and a locknut that covers the top of the steering tube. That locknut tightens against the

12.28 Lower parts of cartridge-bearing headset

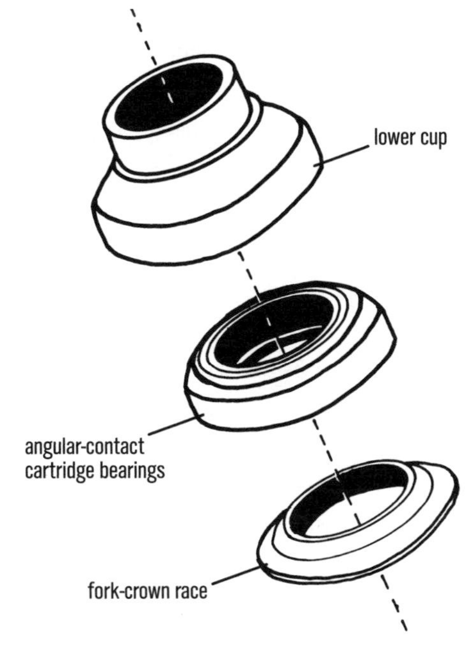

washer and top cup (Fig. 12.22). A brake-cable hanger (Fig. 10.6) and extra spacers may be included under the locknut.

Traditionally, headsets, whether threaded or threadless, used ball bearings held in some type of steel or plastic retainer or cage (Figs. 12.19, 12.22, and 12.25) so that you did not have to chase dozens of separate balls around when you worked on the bike. A variation on this design (Stronglight, some Ritchey models) has needle bearings held in conical plastic retainers (Fig. 12.27) riding on conical steel bearing surfaces.

Cartridge-bearing headsets are now the standard and generally employ angular-contact bearings (Fig. 12.28), because normal cartridge bearings cannot take the side (axial) forces encountered by a headset. Each bearing is a separate sealed, internally greased unit.

12-15

CHECKING HEADSET ADJUSTMENT

LEVEL 1

If the headset is too tight, the fork will be difficult to turn or at least will feel gritty when you do. If the headset is too loose, it will rattle or clunk while you ride. You might even notice some play in the fork as you apply the front brake.

1. **Check for headset looseness by applying the front brake and rocking the bike forward and back.** Try it with the front wheel pointed straight ahead and then with the wheel turned at 90 degrees to the bike. Feel for back-and-forth movement (or play) at the lower head cup with your other hand. If there is play, you need to adjust the headset because it is too loose.

NOTE: *This task is more complicated with a suspension fork and even with many rim brakes. There is always some side-to-side play in any suspension fork, as well as in many brakes; this makes it hard to isolate whether the play you feel is from the headset, the fork, or the brakes. You must feel each part as you rock the bike, and you may have to make some trial-and-error headset adjustments.*

If the headset is loose, skip to the appropriate adjustment section, 12-16 or 12-17.

2. **Check for headset tightness by turning the handlebar back and forth.** Feel for any binding or stiffness of movement. Also, check for a chunk-chunk-chunk feeling as the bar moves. This results from a pitted headset; the pits are indentations in the bearing surfaces made by overtightening the bearings, forcing the balls to make dents. If you feel this, you need a new headset (skip to 12-20). Lift the front wheel off the ground and lean the bike to one side and then the other to see how easily the headset turns and allows the front wheel to steer (be aware that cable housings can resist the rotation of the fork). Lift the bike by the saddle so that it is tipped down at an angle with both wheels off the ground. Turn the handlebar one way and let go of it; then repeat the other way. See if it returns to center quickly and smoothly on its own. If the headset does not turn easily on any of these steps, it is too tight, and you should skip to the appropriate adjustment section, 12-16 or 12-17.
3. **With a threaded headset, try to turn either the top nut (the locknut) or the threaded cup by hand.** They should be so tight against each other that they can be unscrewed only with wrenches. If you can tighten or loosen either part by hand, you need to adjust the headset; go to 12-17.

12-16

ADJUSTING A THREADLESS HEADSET

LEVEL 1 Adjusting a threadless headset, whether it is a modern integrated type (Figs. 12.20 to 12.25) or the traditional type (Figs. 12.6 and 12.19), is much easier than adjusting the threaded style. It is a level 1 procedure and usually takes only a hex key or two.

1. **Check the headset adjustment** (12-15). Determine whether the headset is too tight or too loose.
2. **Loosen the bolt(s) that clamp(s) the stem to the steering tube.**
3. **If the headset is too tight, loosen the bolt on the top cap.** Unscrew it about one-sixteenth of a turn with a 5mm hex key (Fig. 12.29). Recheck and repeat as necessary.
4. **If the headset is too loose, tighten the bolt on the top cap.** Tighten it about one-sixteenth of a turn with a 5mm hex key (Fig. 12.29). Be careful not to overtighten it, thereby pitting the headset. If you're using a torque wrench, 22 in-lbs is a recommended setting. Recheck the adjustment and tighten or loosen further as necessary.
   (a) If the cap does not move down and push the stem down, make sure the stem is not stuck to the steering tube. If it is, refer to 12-14a.
   (b) Another hindrance occurs if the conical compression ring shown in Figures 12.19 to 12.21 and 12.23 is stuck to the steering tube, preventing adjustment via the top cap bolt. With the stem off, gently tap the steering tube down with a mallet, and then push the fork back up to free the compression ring. Grease the ring and the steering tube and reassemble.
   (c) If neither the stem nor the compression ring is stuck, yet the cap still does not push the stem down, the steering tube may be so long that it is hitting the lip of the top cap and preventing the cap from pushing the stem down. The steering tube's top should be 3–6mm below the top of the top edge of the stem (Fig. 12.14) or of the spacers above the stem. If the steering tube is too long, add a spacer, or cut or file some off the top (see 12-9, step 7).
   (d) Another thing that can thwart adjustment is when the star nut is not installed deeply enough and the cap bottoms out on the star nut; it should be 12–15mm below the top of the steering tube. With metal steering tubes, tap the nut deeper with a star-nut installation tool (Fig. 12.15), or put the bolt through the top cap, thread it five turns into the star nut, and gently tap it in with a soft hammer while using the top cap to keep it going in straight (see 12-9, step 8). With carbon steering tubes, loosen the aluminum expander (Fig. 12.26) with a hex key, move it down, and retighten it.
   (e) With a double-crown fork (Fig. 16.18), you must loosen the clamp bolts on the upper crown. There are three of them—one clamps the steering tube, and the other two each clamp one upper fork tube (stanchion). If the headset is loose and these bolts are still clamped, tightening the headset compression bolt cannot push the stem and top crown down more. After the headset is adjusted properly, retighten the bolts to the torque specified by the fork maker.
   (f) Once you have fixed the cause of the adjustment problem, return to step 1.
5. **Tighten the stem clamp bolts.** Tighten to the recommended torque, given in Appendix D.
6. **Recheck the headset adjustment.** Repeat steps 2–4 if necessary.

12.29 Loosening or tightening the compression bolt on a threadless headset

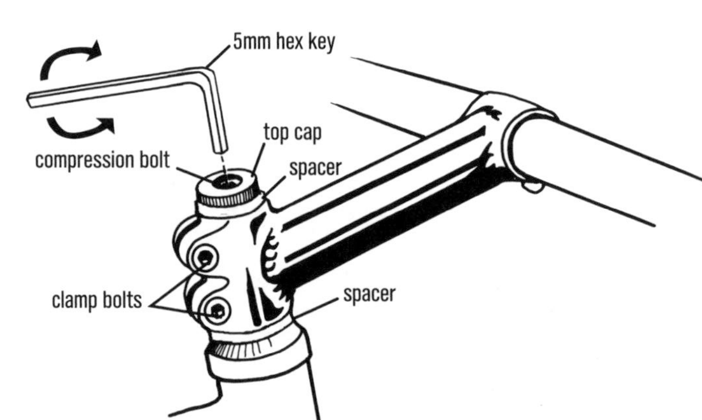

With some integrated headsets, you may need a thin shim washer under the bearing cover so that its edges do not drag and scrape on the top end of the head tube.

If the headset is adjusted properly, make sure that the stem is aligned straight with respect to the front wheel, and then go find something else to do, because you are done.

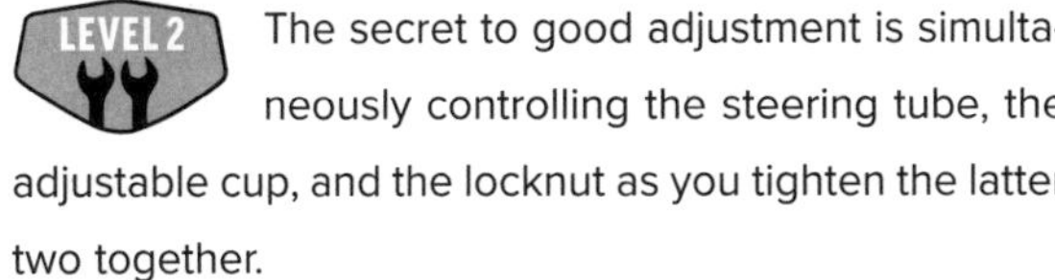

12-17

ADJUSTING A THREADED HEADSET

LEVEL 2 The secret to good adjustment is simultaneously controlling the steering tube, the adjustable cup, and the locknut as you tighten the latter two together.

NOTE: *Perform the adjustment with the stem installed. Not only does it give you something to hold to keep the fork from turning during the installation, but also there are slight differences in adjustment when the stem is in place as opposed to when it is not. Tightening the stem bolt can sometimes bulge the walls of the steering tube very slightly (Fig. 12.10), just enough for it to shorten the steering tube slightly and throw your original headset adjustment off.*

1. **Once you determine whether the headset is too loose or too tight, follow the steps outlined in 12-15.**
2. **Put a pair of headset wrenches on the headset's nuts.** Headset nuts come in a wide variety of sizes, so make sure you have the proper wrench sizes. Place the wrenches so that the top one is slightly offset to the left of the bottom wrench. That way you can squeeze them together to free the nut (Fig. 12.30).

NOTE: *People with small hands or a weak grip will need to hold each wrench at the end in order to get enough leverage.*

3. **Hold the lower wrench in place, and turn the top wrench counterclockwise to loosen the locknut.** It may take considerable force to break the locknut loose, because it needs to be installed very tightly to keep the headset from loosening.
4. **Depending on whether the headset was too loose or too tight, do one of the following:**

12.30 Loosening a headset locknut

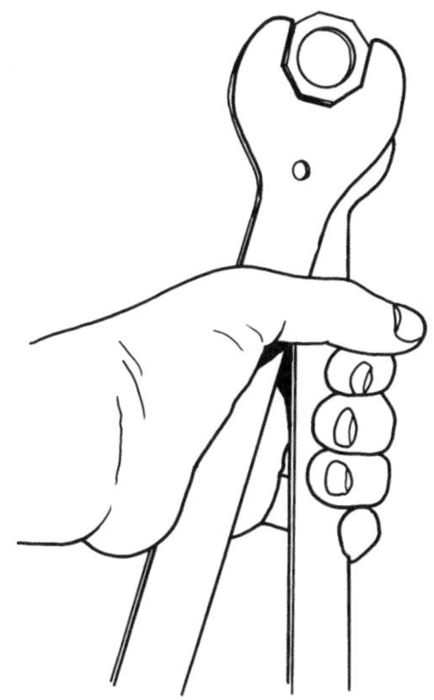

(a) If the headset was too loose, turn the lower nut (the threaded cup) clockwise about one-sixteenth of a turn while holding the stem with your other hand. Be very careful when tightening the cup; overtightening it can ruin the headset by pressing the bearings into the bearing surfaces and making little indentations. The headset then stops at the indentations rather than turning smoothly, a condition known as a pitted headset.

(b) If the headset was too tight, loosen the threaded cup counterclockwise one-sixteenth of a turn while holding the stem with your other hand. Loosen the cup until the bearings turn freely, but be sure not to loosen to the point that play develops.

5. **Holding the stem, tighten the locknut clockwise with a single wrench.** Make sure that the threaded cup does not turn while you tighten the locknut. If the cup does turn, either you are missing the toothed lock washer separating the cup and locknut (Fig. 12.22) or the washer you have is missing its tooth. In this case, remove the locknut and replace the toothed washer. Put it on the steering tube so that the tooth engages the longitudinal groove in the steering tube. Tighten the locknut again.

NOTE: *You can adjust a headset without a toothed washer by working both wrenches simultaneously,*

12.31 Tightening a headset locknut

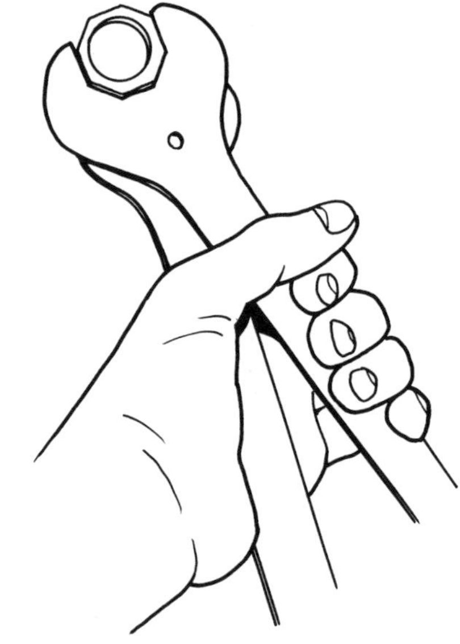

but it is trickier to adjust and often comes loose while riding. However, if the steering tube is cut too short to accept a washer in the headset, you may have to do without one.

6. **Check the headset adjustment again.** Repeat steps 4 and 5 until the headset is properly adjusted.
7. **With wrenches on both nuts, tighten the locknut (clockwise).** Snug it firmly against the washer(s) and threaded cup to hold the headset adjustment in place (Fig. 12.31).
8. **Check the headset adjustment again.** If it is off, follow steps 2–7 again. If it is adjusted properly, make sure the stem is aligned with the front wheel.

NOTE: *If you constantly get what you think to be the proper adjustment and then find it to be too loose after you tighten the locknut and threaded cup against each other, the steering tube may be too long, thereby causing the locknut to bottom out. Remove the stem and examine the inside of the steering tube. If the top end of the steering tube butts up against the top lip of the locknut, the steering tube is too long. Remove the locknut and add another spacer.*

If you don't want to add another spacer, file off 1 or 2mm of the steering tube. Be sure to remove any burrs both inside and out. Avoid leaving filings in the bearings or steering-tube threads. Replace the locknut and return to step 5.

12-18

OVERHAULING A THREADLESS HEADSET

These instructions apply to both integrated (Figs. 12.20, 12.21, and 12.23 plus 12.25) and traditional (Figs. 12.6 and 12.19) threadless headsets.

Like all other bike parts with bearings, headsets need periodic overhauls. If you use your bike regularly, you should probably overhaul a loose-bearing (Figs. 12.19 and 12.25) or needle-bearing (Fig. 12.27) headset once a year. Headsets with cartridge bearings (Figs. 12.20, 12.21, and 12.23) need less frequent maintenance; some angular-contact bearings can be disassembled and cleaned, and some cannot. With those that cannot, if a bearing fails, either replace the bearing or, if it has pressed-in bearings (Fig. 12.32), replace the entire cup (12-20).

1. **Mount the bike upside down in a work stand.** Alternatively, be ready to catch the fork when you remove the stem.
2. **Disconnect or remove the front brake** (Chapters 9 and 10), **unscrew the top cap bolt** (Fig. 12.29) **and stem clamp bolts, and remove the top cap and the stem.** If you have a double-crown fork (Fig. 16.21), loosen the bolts clamping the upper tubes (i.e., stanchions) and pull the top crown off as well.
3. **Remove the top headset cup, sliding the top cup, conical compression ring, and any other spacers above it off the steering tube** (Figs. 12.19 to 12.21, 12.23, and 12.25). It may take a tap with a mallet on the end of the steering tube, followed by pushing the fork back up, to free the compression ring.
4. **Pull the fork out of the frame.** The lower bearing and seal may come with it.
5. **Remove any bearing seals that are present.** Remember the position and orientation of each.
6. **Remove any bearings remaining in the head tube or cups.** Be careful not to lose any. Separate the top and bottom sets if they are of different sizes.
7. **Clean or replace the bearings:**
   (a) With either ball-bearing (Figs. 12.19 and 12.25) or needle-bearing (Fig. 12.27) headsets, put the bearings (leave the balls or needles in

their retainers) in a jar or old water bottle along with some citrus-based solvent. Shake. If the bearings from the top and bottom are of different sizes, keep them in separate containers to avoid confusion. Blot the bearings dry with a clean rag. Plug the sink, and wash the ball bearings (whether separate or held in retainers) in soap and water in your hands, just as if you were washing your palms by rubbing them together. Your hands will get clean for the assembly steps as well. Rinse bearings thoroughly and blot them dry. Air-dry completely.

(b) Some cartridge bearings (Fig. 12.28) can be pulled apart and cleaned. Hold the bearing over a container (to catch the balls) so that the beveled outer surface that fits into the cup faces down, and push up on the bearing's inner ring. The bearing should come apart—the inner ring will pop up and out with the bearings stuck to its outer surface. It may take a little rocking of the inner ring as you push up. If the bearing does not come apart, first pry off the plastic seal covering the bearings with a knife or razor blade (as shown in Fig. 11.32), and then try again. Wipe the bearings, bearing rings, and seals with a clean rag.

(c) If the bearings are the type that will not come apart, check to see if they turn smoothly. If they do not, buy new ones and continue to step 8.

8. **Wipe all of the bearing surfaces and areas the bearings might touch with clean rags.** Wipe the steering tube clean, and wipe the inside of the head tube clean with a rag over the end of a screwdriver.

9. **Inspect all of the bearing surfaces of loose-ball headsets for wear and pitting.** If you see pits (separate indentations made by bearings in the bearing surfaces), replace the headset; skip to 12-20.

10. **Grease all bearing surfaces.** If the headset uses cartridge bearings, apply grease conservatively. For the angular-contact cartridge bearing that you have disassembled, smear grease around the outside of the inner bearing ring, and stick the balls into the grease in the channel around the ring. With the outer ring sitting beveled side down on the table, push the inner ring (which has the balls attached) down into it (the internal bevel on the inner ring should be facing up, opposite the bevel on the outer ring). Snap the bearing seals back into place.

11. **Replace the lower bearing.** The bike should be upside down in the bike stand.

(a) For a cupless integrated headset (Figs. 12.20 and 12.23), set a bearing into the chamfered seat in the bottom of the head tube itself. For an integrated headset with drop-in cups (Fig. 12.25), place the lower cup into its chamfered seat in the bottom of the head tube.

(b) With loose-ball headsets, make sure you have the bearing retainer right side up so that only the balls and not the retainers contact the bearing surfaces (note the different upper cup styles and bearing orientations in Figs. 12.19, 12.22, and 12.25). If the retainer is upside down, it will contact one of the bearing surfaces so the headset will not turn well, and riding it that way will turn the retainer into jagged chunks of broken metal. To be safe, double- and triple-check the retainer placement by turning each cup pair and bearing in your hand before proceeding. Some loose-ball headsets have the bearings set up identically for both top and bottom (Fig. 12.22), where the top piece of each pair is a cup, the bottom piece is a cone, and the bearing retainer rides the same way in both sets. Many headsets, however, place both cups (and hence the bearing retainers) facing outward from the head tube (Figs. 12.19 and 12.25).

(c) If the ball bearings are loose with no bearing retainer, stick the balls into the grease in the cups one at a time, making sure that you replace in each cup the same number that you started out with.

(d) With angular-contact cartridge bearings, the beveled end faces into the cup (Fig. 12.28) or into the chamfered seat inside the head tube (Figs. 12.20 and 12.23).

12. **Reinstall any seals that you removed from the headset parts.**
13. **Slip the steering tube of the fork into the head tube so that the lower headset bearing seats properly** (Fig. 12.33).
14. **Slide the top cartridge bearing, cup, or cone onto the steering tube.** On loose-ball headsets, have the bearings already in the upper cup or on the upper cone. In the case of a cupless headset (Figs. 12.20 and 12.23), slide on the bearing alone. Keep the bike upside down at this point; it not only keeps the fork in place, but it also prevents grit from falling into the bearings as you put the cup on.
15. **Slide the (greased) compression ring and top cover onto the steering tube.** Ensure that the narrower end of the compression ring slides into the conical space in the top of the top cup (Figs. 12.20 and 12.23) or into the bearing of a cupless headset (Fig. 12.22). If you have a double-crown fork (Fig. 16.21), slide the top crown onto the steering tube and the upper fork tubes.

NOTE: *On a Campagnolo integrated headset (Fig. 12.25), which appears on some Cannondale mountain bikes, there is a plastic biconical compression ring inserted into the top bearing cone (Figs. 12.25 and 12.34), and like a normal split compression ring, it centers the top cone in the bearings. The upper edge of the plastic compression ring is notched like a turreted castle tower. Above this part comes the top bearing cover, whose inner edge is beveled for the turreted top conical edge of the plastic compression ring. To preload the bearings, you must first install the top bearing cover upside down (Fig. 12.34) and then push down on it. This preloads the bearings by pushing the top cone down. If you install the top bearing cover in its standard orientation before the top cone and plastic compression ring are slid down far enough to preload the bearings, the beveled inner edge of the top bearing cover will pinch the turreted upper conical edge of the plastic compression ring in place and not allow it to slide down farther. Once you have pushed the bearing cone down fully in this manner, flip the top bearing cover right side up and put it in place over the cone and compression ring (Fig. 12.35).*

16. **Replace the stem and spacers.** Slide on any spacers you had under the stem. Slide the stem on, and tighten one stem clamp bolt to hold it in place.
17. **Turn the bike right side up in the bike stand.**
18. **Check the steering-tube overlap.** The stem clamp or, ideally, the top spacer above the stem should extend 3–6mm above the top of the steering tube (Fig. 12.14).
    (a) If the overlap is correct, install the top cap on the top of the stem clamp and steering tube, and screw the bolt into the star nut inside the steering tube (Fig. 12.29).

12.32 Cartridge bearing pressed into headset cup

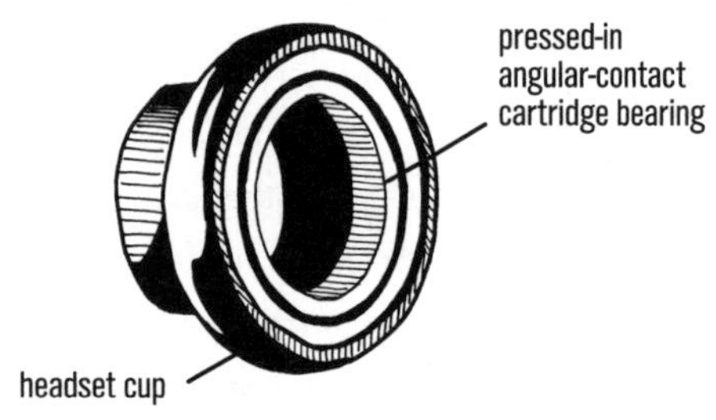

12.33 Setting the fork in the head tube to seat the bearings

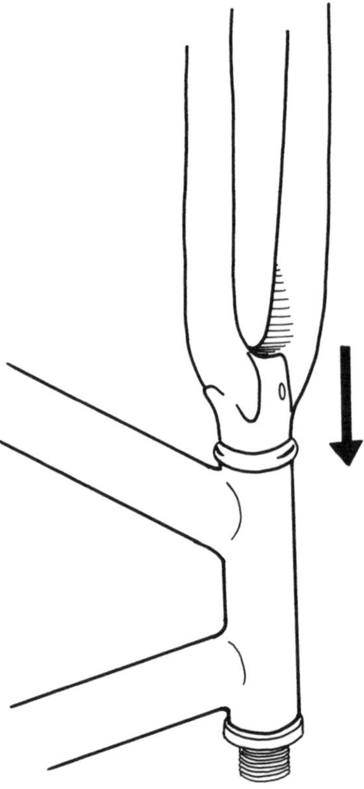

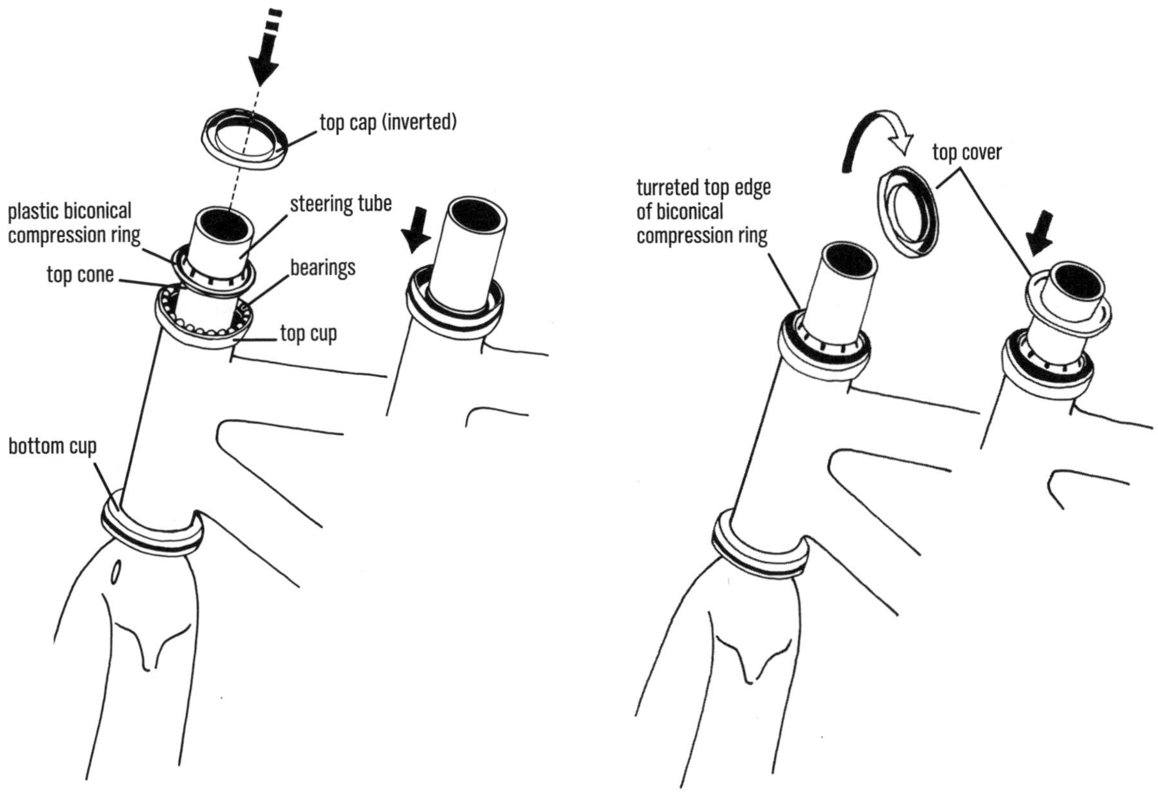

12.34 Seating Campagnolo threadless headset, part 1

12.35 Seating Campagnolo threadless headset, part 2

(b) If the steering tube is too long, add another spacer above the stem or remove the stem. Add a spacer under the stem or cut and/or file the steering tube shorter until the spacer above the stem overlaps it by 3–6mm.

(c) If the steering tube is too short, remove spacers from below the stem if there are any. If there are no spacers to remove, try a new stem with a shorter clamp. Install the top cap on the top of the stem clamp, and screw the bolt into the star nut (Fig. 12.29).

19. **Adjust the headset, following the instructions in 12-16.**

12-19

OVERHAULING A THREADED HEADSET

LEVEL 2

As with any other bike part with bearings, headsets need periodic overhauls. If you use your bike regularly, you should probably overhaul a loose-bearing (Fig. 12.22) or needle-bearing (Fig. 12.27) headset once a year. Headsets with cartridge bearings (Fig. 12.28) need less frequent overhaul; some angular-contact bearings can be disassembled and cleaned, and some cannot. With those that cannot, if a bearing fails, you either replace the bearing or, if it has press-in bearings (Fig. 12.32), you replace the entire cup (12-20).

1. **Disconnect the front brake** (Chapter 9 or 10).
2. **Remove the stem.** Loosen the stem bolt three turns, tap the bolt down with a hammer to free the wedge (Fig. 12.16), and pull the stem out.
3. **Remove the headset locknut.** Place one headset wrench on the locknut and one on the threaded cup. Loosen the locknut by squeezing the wrenches together, which will turn the locknut counterclockwise (Fig. 12.36). Unscrew the locknut from the steering tube.
4. **Mount the bike upside down in a work stand.** Alternatively, be ready to catch the fork when you remove the threaded cup.
5. **Remove the threaded cup.** The headset washer(s) will slide off the steering tube as you unscrew the top cup.
6. **Pull the fork out of the frame.**

7. **Remove any seals that surround the edges of the cups.** Remember their positions and orientations.
8. **Remove the bearings from the cups.** Be careful not to lose any. Separate the top and bottom sets of bearings if they are of different sizes.
9. **Clean or replace the bearings:**
   (a) With standard ball-bearing (Fig. 12.22) or needle-bearing (Fig. 12.27) headsets, put the bearings in a jar or old water bottle along with some citrus-based solvent. Shake. If the bearings from the top and bottom are of different sizes, keep them in separate containers to avoid confusion. Blot the bearings dry with a clean rag.
   (b) Some cartridge bearings (Fig. 12.28) can be pulled apart and cleaned; see 12-18, step 7b.
   (c) If the cartridge bearings will not come apart, check to see if they turn smoothly. If they do not, buy new ones and skip to step 12.
10. **Plug the sink, and wash the ball bearings in soap and water in your hands.** Wash as if you were washing your palms by rubbing them together. Do this with loose balls as well as with balls held in retainers. This also helps clean your hands for the assembly steps. Rinse bearings thoroughly and blot them dry. Let them air-dry completely.
11. **Wipe the bearing surfaces and other places the bearings might touch with clean rags.** Wipe the steering tube clean, and wipe the inside of the head tube clean with a rag over the end of a screwdriver.
12. **Inspect all the bearing surfaces of loose-ball headsets for wear and pitting.** If you see pits (separate indentations made by bearings in the bearing surfaces), you need to replace the headset. If that's the case, skip to 12-20.
13. **Grease all bearing surfaces.** A thin film will do, especially with cartridge bearings. For the angular-contact cartridge bearing that you have disassembled, see 12-18, step 10.
14. **Install the bearings in their cups.** The bike should be upside down in the bike stand.
   (a) Place a set of bearings in the top cup and a set in the cup on the lower end of the head tube.
   (b) With loose-ball headsets, make sure you have the bearing retainer right side up so that only the balls and not the retainers contact the bearing surfaces (note the different upper cup styles and bearing orientations in Figs. 12.19 and 12.22). If the retainer is upside down, it will contact one of the bearing surfaces so the headset will not turn well, and riding it that way will turn the retainer into jagged chunks of broken metal. To be safe, double- and triple-check the retainer placement by turning each cup pair in your hand with the bearing in between before proceeding. Some loose-ball headsets have the bearings set up identically for both top and bottom (Fig. 12.22), where the top piece of each pair is a cup, the bottom piece is a cone, and the bearing retainer rides the same way in both sets. Many headsets, however, place both cups (and hence the bearing retainers) facing outward from the head tube (Fig. 12.19). Also, watch for asymmetry in ball size; some headsets have smaller balls on top than on the bottom.

NOTE: *Stronglight and Ritchey needle-bearing headsets come with two pairs of separate conical steel rings. These are the bearing surfaces that sit on either*

12.36 Loosening a headset locknut

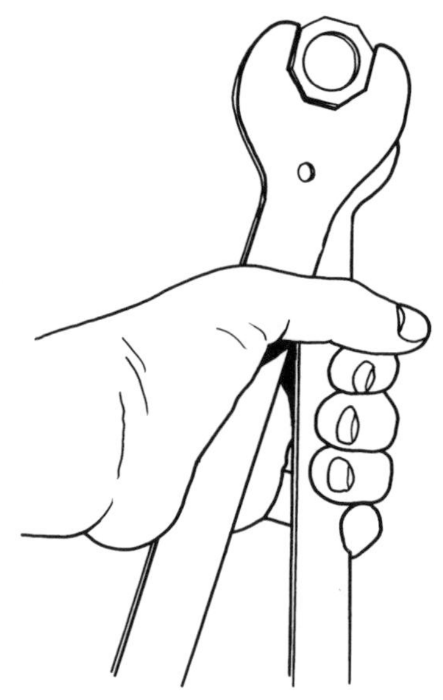

side of the needle bearings (Fig. 12.27). For each set of conical rings, you will find that one is smaller than the other. Place the smaller one on the bottom.

(c) If the ball bearings are loose with no bearing retainer, stick the balls into the grease in the cups one at a time, making sure that you replace in each cup the same number that you started with.

(d) With angular-contact cartridge bearings, the beveled end faces into the cup (Fig. 12.28).

15. **Reinstall any seals that you removed from the headset parts.**
16. **Slide the steering tube of the fork into the head tube.** Ensure that the lower headset bearing seats properly (Fig. 12.33).
17. **Screw the top cup, with the bearings in it, onto the steering tube.** Keeping the bike upside down at this point not only keeps the fork in place, but it also prevents grit from falling into the bearings as you thread the cup on.
18. **Slide on the toothed washer** (Fig. 12.22). Align the tooth in the groove going down the length of the steering-tube threads. If you have one, install the brake-cable hanger (Fig. 10.6) the same way. Screw on the locknut with your hand.
19. **Turn the bike over.**
20. **Grease the stem quill and insert it into the steering tube** (Fig. 12.10). Make certain that it is in deeper than the imprinted limit line.
21. **Line the stem up with the front wheel, and tighten the stem bolt.** See Appendix D for recommended torque.
22. **Adjust the headset as described in 12-17.**

12-20

REMOVING THE HEADSET

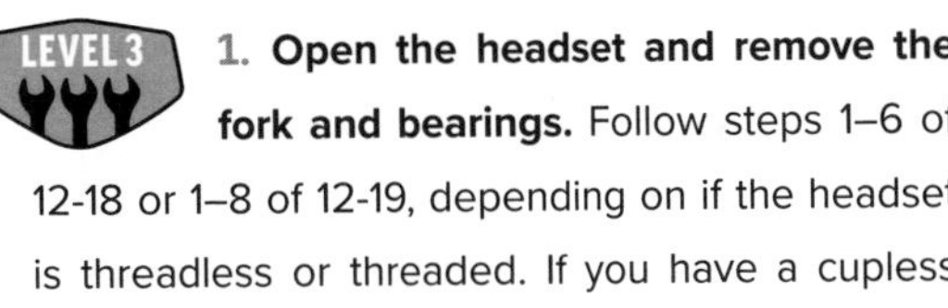

1. **Open the headset and remove the fork and bearings.** Follow steps 1–6 of 12-18 or 1–8 of 12-19, depending on if the headset is threadless or threaded. If you have a cupless integrated headset (Figs. 12.20, 12.23, and 12.25), skip to step 5.
2. **Slide the solid end of the headset-cup remover (a.k.a. "headset rocket") through the head tube** (Fig. 12.37). As you pull the headset-cup remover through the head tube, the splayed-out tangs on the opposite end of the tool pull through the cup and spread out.
3. **Strike the solid end of the cup remover with a hammer.** Smack it until you drive the cup out (Fig. 12.38). Be careful: The cup may go flying.

12.37 Inserting a cup-removal tool

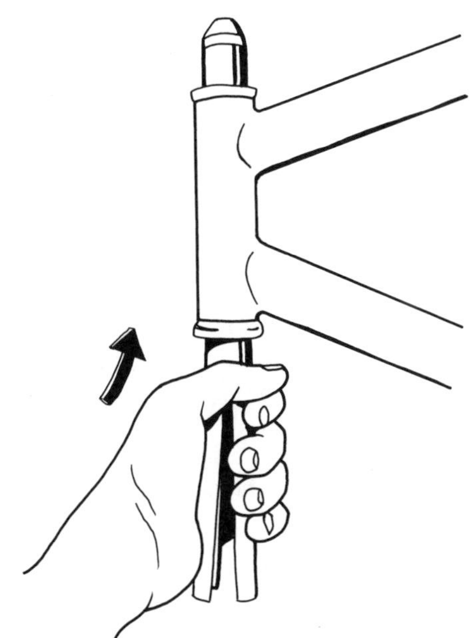

12.38 Removing a lower headset cup

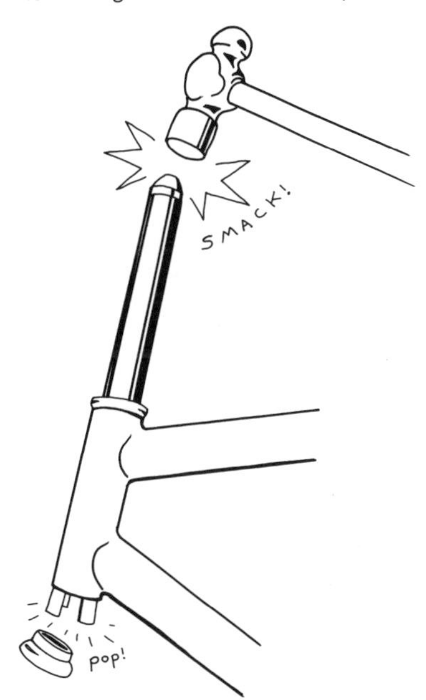

12.39 Removing a fork-crown race with a screwdriver

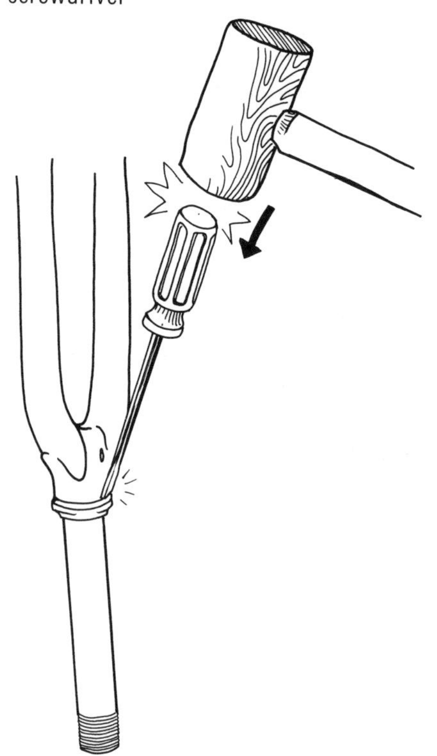

12.40 Removing a fork-crown race with a crown race remover

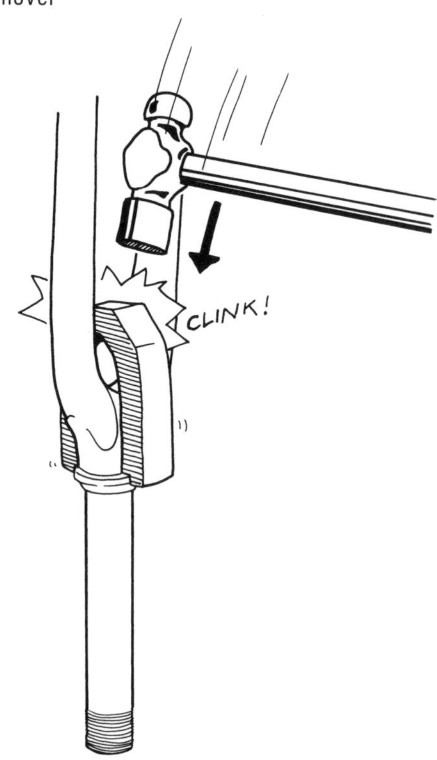

4. **Remove the other cup.** Slide the cup remover in through the opposite end of the head tube, and repeat steps 2 and 3 on the opposite end of the end tube.
5. **Remove the fork-crown race.**
   (a) With a suspension fork (or a rigid fork with a broad crown), getting the fork-crown race off can be a bear. Clamp the steering tube into a vise, or turn the fork upside down so that the top of the steering tube is sitting on the workbench. If you find a notch on the front and back of the fork crown under the fork-crown race, place the blade of a large screwdriver into the notch on one side of the crown so that it butts against the bottom of the headset fork-crown race. Tap the handle of the screwdriver with a hammer to drive the race up the steering tube a bit (Fig. 12.39). Move the screwdriver to the groove on the other side, and tap it again to move that side of the race up a bit. Continue in this way, alternately tapping either side of the race up the steering tube, bit by bit, until it gets past the enlarged section of the steering tube and slides off. With luck, the screwdriver will not be damaged, but be prepared to relegate it to use on similar jobs in the future.
   (b) If there is no notch and there is not enough to the fork-crown race's edge protruding over the crown to get a screwdriver against it, get the race to start moving up by using a sharp chisel or box-cutter blade. Tap the chisel or box-cutter blade with a hammer when its edge is between the race and the fork-crown edge. Once the crown race has moved up a bit, start using a broad screwdriver as previously described to "walk" the race off the steering tube's enlarged base.
   (c) If you are fortunate enough to have a Park Tool Universal crown-race remover tool (Fig. 1.4), use it! Using the finger-tightened screws at the bottom, tighten the blades in under the fork-crown race until they stop. Then tighten the top screw to slide the race up the steering tube.

12.41 Removing a fork-crown race with a vise

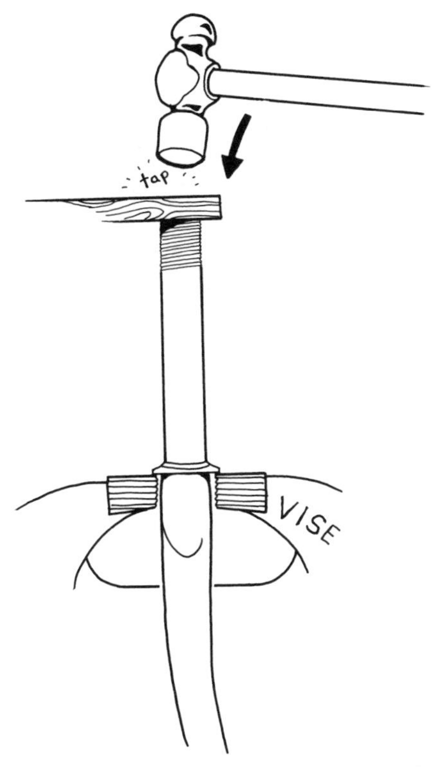

(d) If you have an old, narrow rigid fork, you can get the fork-crown race off as in (a), (b), or (c), but you can also do it more elegantly with an old-school crown-race remover (Fig. 12.40) or an appropriately-sized bench vise (Fig. 12.41). Stand the fork upside down on the top of the steering tube. Place the U-shaped crown-race remover so that it straddles the underside of the fork crown and its ledges engage the front and back edges of the fork-crown race. Smack the top of the crown-race remover with a hammer to knock the race off (Fig. 12.40).

(e) To remove the crown race with a bench vise, flip the brakes out of the way and slide the fork in, straddling the center shaft of the vise. Tighten the vise so that its faces lightly contact the front and back of the fork crown with the lower side of the fork-crown race sitting on top of the vise faces. Put a block of wood on the top of the steering tube to pad it. Strike the block with a hammer to drive the fork down and knock the race off (Fig. 12.41).

12-21

PREPARING A FRAME AND FORK PRIOR TO INSTALLING A HEADSET

LEVEL 3

You can install a new headset yourself if you have the necessary tools. Otherwise, get a shop to do it. You should also head to a well-equipped bike shop if you are working with a new frame that has not been properly prepared for the headset prior to installation. The job requires special tools to ream and square up the head tube faces, and not all shops have them.

If this is a frame with chamfered bearing seats for an integrated headset (Figs. 12.23 and 12.25), you shouldn't need to do anything except drop the bearings in, beveled end toward the head-tube seat, and follow the steps in 12-16.

With a headset with cups—whether they are semi-integrated (Figs. 12.21 and 12.24) or traditional (Figs. 12.19 and 12.22)—on a new frame (or on one on which headsets have become pitted in the past), make sure that the head tube has been reamed and faced. If not, you will need a bike shop equipped with the proper tools to do it for you. Reaming makes the head tube ends round inside and of the correct diameter for the headset cups to press in. Facing makes both ends of the head tube parallel so that the bearings can turn smoothly and uniformly. A head-tube reaming-and-facing tool is shown in Figure 1.4.

The fork-crown race needs to be the correct diameter to press onto the base of the steering tube. The crown-race seat on top of the fork crown must be faced in a way that places the race parallel to the head-tube cups. This is generally only a concern with rigid forks; suspension forks usually come with the tube properly machined to accept the fork-crown race.

The fork steering tube (threaded or threadless) must also be cut to the proper length. Remember, you can always go back and cut more off, but you can't add any, so be careful! You can wait until the headset (and stem, in the case of a threadless headset) is installed. Or you can figure out the length first.

The safest way to make sure you don't cut the steering tube too short is to install the headset first

(12-22). Determining the steering-tube length for an already installed threadless headset is detailed in 12-9. Once a threaded headset is installed, measure the excess length as shown in Figure 12.43 and add spacers that thickness, or remove the top nut and cut that much off the top. File off the burrs that the hacksaw left on the inside and outside edges of the steering-tube end.

12-22

INSTALLING A HEADSET

LEVEL 3

1. **Install the fork-crown race.** Slide the (greased inside) fork-crown race down on the fork steering tube until it hits the enlarged section at the bottom. Slide the crown-race slide punch up and down the steering tube, pounding the race down until it sits flat on top of its seat on the fork crown (Fig. 12.42). Some crown-race punches are longer and closed on the top and are meant to be hit with a hammer rather than to be slid up and down by hand. Check if installation is complete by holding the fork up against the light to see whether there are any gaps between the fork-crown race and the crown. For an integrated headset (Figs. 12.20, 12.23, and 12.25), you can skip the rest of this section and install and adjust the bearings as described in 12-18 and 12-16.

NOTE: *Extra-thin fork-crown races can easily be bent or broken by the crown-race punch. Chris King and Shimano both make support tools that sit over the race and distribute the impact from the punch.*

2. **Grease where the cups will go.** Put a thin layer of grease inside the ends of the head tube.
3. **Prepare the headset press and cups for installation.** By hand, place the headset cups into the ends of the head tube. Slide the headset-press shaft through the head tube. Press the button on the sliding end of the tool, and slide it onto the shaft until it bumps into one of the cups (Fig. 12.44). This same method, and often the same press, can be used for semi-integrated (Figs. 12.21 and 12.24) and traditional (Figs. 12.19 and 12.22) headsets. The press has separate inserts that may be appropriate to put into the cups before sliding the press into place.

12.42 Setting a fork-crown race

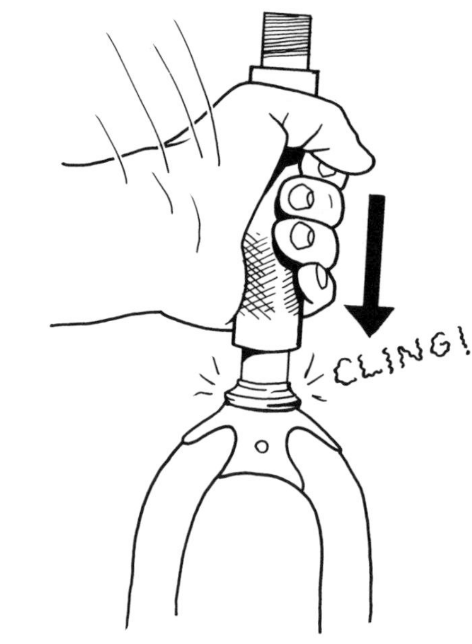

12.43 Measuring the amount of steering tube to cut

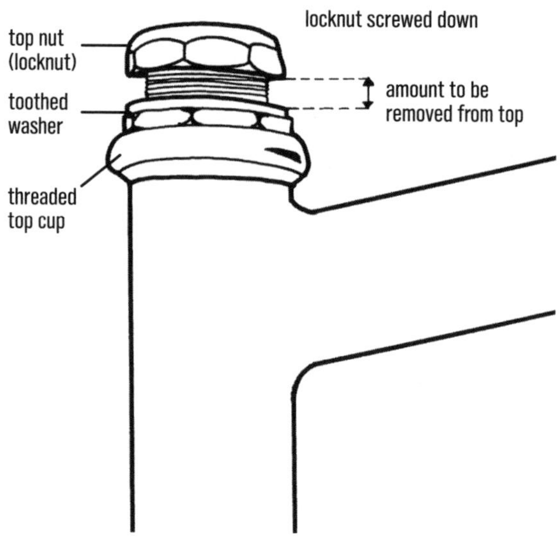

NOTE: *Make sure that the press and any installed press inserts make contact only with the outer cup flange or an inner shelf and not with the bearing seat, or the bearings may not roll smoothly.*

Find the nearest notch on the tool's shaft, and release the button to fix the detachable press end in place.

ANOTHER NOTE: *Some headsets have cartridge bearings that are permanently pressed into the cups (Fig. 12.32). If you use a headset press that pushes on the center of the cups, you will ruin the bearings. You need*

12.44 Pressing in headset cups with a headset press

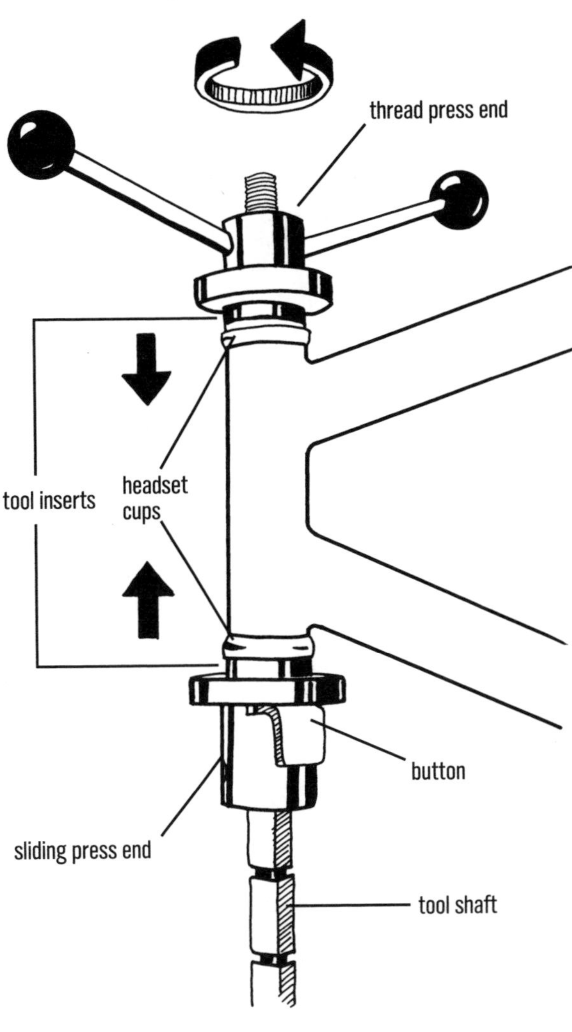

a press or an insert on the press that pushes the outer part of the cup and does not touch the bearings.

4. **Press in the cups.** Hold the lower end of the headset-press shaft with a wrench. That will keep the tool from turning as you press in the cups. Tighten the press by turning the handle on the top clockwise (Fig. 12.44). Keep tightening the tool until the cups are fully pressed into the ends of the head tube. Examine them carefully to make sure that they are pressed in square so that there are no gaps between the cups and the ends of the head tube.

NOTE: *You can easily crush thin headset cups with a flat-surface headset press when it is used without inserts, so be careful and stop when the cups reach the head tube.*

5. **Grease all bearing surfaces.** If you are using cartridge bearings (Fig. 12.25), a thin film will do.
6. **Assemble and adjust the headset.** Follow the instructions in 12-18 and 12-16 for a threadless headset and in 12-19 and 12-17 for a threaded one.

12-23

TROUBLESHOOTING STEM, HANDLEBAR, AND HEADSET PROBLEMS

a. Handlebar slips

Tighten the pinch bolts on the stem that holds the handlebar, but not beyond the maximum allowable torque (see Appendix D). With a front-opening stem, make sure that there is the same amount of space between the stem and the front plate on both edges of the front plate. With any stem, if the clamp closes against itself without holding the handlebar securely, be sure to check that the handlebar is not deformed or undersized and that the stem clamp is not cracked or stretched; replace any questionable parts (replace lightweight stems and bars if the clamp edges touch each other). You can slide a shim cut from a beer can between the stem and handlebar to hold it better if you have a heavy stem and handlebar, but don't try this with lightweight parts. Replacing parts is a safer option—there is always a reason that parts that are meant to fit together no longer do! With lightweight stems and handlebars, you cannot just keep tightening the small clamp bolts as you can the larger bolts on heavy stems because you will strip threads and/or cause handlebar and stem failures.

b. Creaking handlebar noises

Loosen the stem clamp first. Then grease the area of the handlebar that is clamped in the stem, slide the handlebar back in place, and tighten the stem bolt. Also, sanding the hard-anodized surface inside the stem clamp can eliminate creaking.

c. Slipping bar end

Tighten the bar-end clamp bolt. If you have to go beyond the specified torque, the handlebar or the bar end may be damaged and may need to be replaced.

d. Stem not pointed straight ahead

Loosen the bolt(s) securing the stem to the fork steering tube, align the stem with the front wheel, and tighten the stem bolt(s) again. With a threaded headset, the bolt you are interested in is a single vertical bolt on top of the stem. Loosen it about two turns, and tap the top of the bolt with a hammer to disengage the wedge on the other end from the bottom of the stem (Fig. 12.16). With a threadless headset, there are one, two, or three horizontal bolts pinching the stem around the steering tube (Figs. 12.5 and 12.6) that need to be loosened to turn the stem on the steering tube. Do not loosen the bolt on the top of the stem cap (Fig. 12.13); you'll have to readjust the headset if you do.

e. Fork and headset rattle

The headset is too loose. Adjust the headset (12-16 or 12-17).

f. Stem, bar, and fork assembly not turning smoothly

If it does not turn smoothly but instead stops in certain fixed positions, the headset is pitted and needs to be replaced. See 12-20 to 12-22.

g. Stem, bar, and fork assembly not turning freely

The headset is too tight. The front wheel should swing easily from side to side when you lean the bike or lift the front end. Adjust the headset (12-16 or 12-17, depending on type).

h. Stem stuck on or in fork steering tube

See 12-14.

To everything, turn, turn, turn . . .

—THE BYRDS, VIA PETE SEEGER

13

PEDALS

TOOLS

15mm pedal wrench
Small and large adjustable wrenches
2.5mm, 3mm, 4mm, 5mm, 6mm, and 8mm hex keys
Torx T25 key
Screwdriver
Pliers
Knife
Grease
Oil (chain lubricant)

FOR OVERHAULING PEDALS

7mm, 8mm, 9mm, 10mm, 17mm, and 18mm open-end wrenches
Shimano or Look splined pedal tool
Snapring pliers
13mm cone wrench
8mm, 9mm, 10mm, and 14mm sockets
Drift punch
Bench vise
Hammer

To best serve its purpose, a bicycle pedal needs only to be firmly attached to the crankarm and provide a stable platform for the shoe. A simple enough task, but you might be amazed at the different approaches that have been taken to achieve this goal. Of the two basic types of mountain bike pedals, the standard cage-type pedal, with or without a toeclip and strap (Fig. 13.1), is the simplest and cheapest. The clip-in pedal (Fig. 13.2), found on most mid- to high-end cross-country and trail mountain bikes, has a spring-loaded shoe-retention system, much like a ski binding. Clip-in pedals are also called "clipless" pedals because they have no toeclip.

Cage-type pedals are fairly common on lower-end bikes. They are relatively unintimidating for the novice rider, and the frame (or cage) that surrounds the pedal provides a large, stable platform (Fig. 13.1). For doing stunts and gravity-driven mountain biking, many riders prefer flat pedals to clip-in pedals, that is, standard pedals with a broad platform and often little steel pins sticking up to provide more traction with the shoe sole. Without a toeclip, the top and bottom of a standard or flat pedal are the same, and you can use just about any type of shoe. Using a toeclip

13.1 Cage-type pedal with toeclip and strap

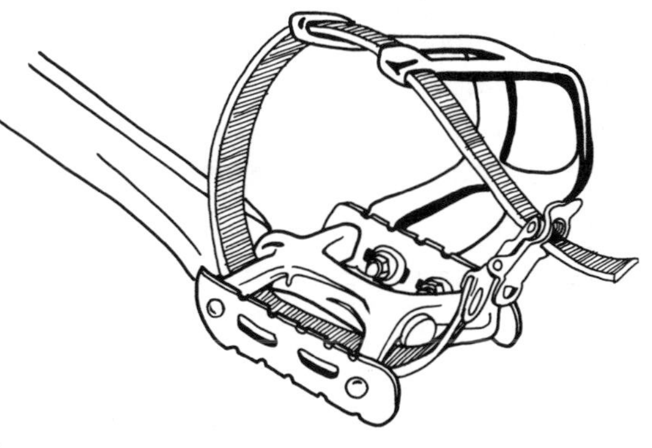

13.2 Clip-in pedal

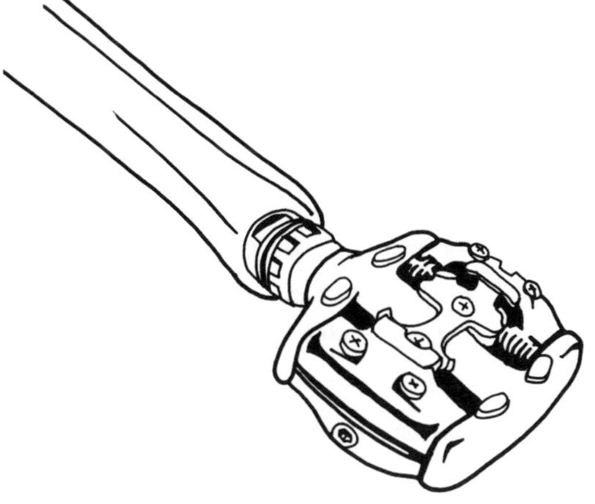

without a strap can keep your foot from sliding forward and still allow easy release in almost any direction.

The combination of a toe strap and a stiff-soled mountain bike shoe with aggressive tread works well to keep your foot on the pedal while you are riding even the roughest singletrack. When tightened, the strap allows you to pull up on the upward part of the pedal stroke, giving you more power and a more fluid pedal stroke. Of course, as you add clips and straps, the pedal becomes harder to enter and to exit, especially with boots or shoes with aggressive tread designs.

Clip-in models offer all the efficiency advantages of a good clip-and-strap combination, yet allow easier entry and exit from the pedal. These pedals are more expensive and require special shoes and accurate mounting of the cleats. Your choice of shoes is limited to stiff-sole models that accept cleats for your particular pedal. Once you have them dialed in, you will find that clip-in pedals waste less energy through flex and slippage and allow you to transfer more power directly to the pedals. This greater efficiency combined with low weight explains their universal acceptance among cross-country mountain bike racers.

As regards the shoe sole design, most clip-in mountain pedals are SPD compatible; SPD stands for Shimano Pedaling Dynamics. Shimano produced the first successful clip-in mountain pedal in the mid-1980s (Fig. 13.2) and set the shoe standards. SPD compatibility indicates that the cleat mounts with two side-by-side M5 screws, spaced 14mm apart, screwing into a movable, threaded cleat-mounting plate on a shoe with two longitudinal grooves in the sole (Fig. 13.6). However, SPD compatibility does not necessarily mean that one company's cleat will work with the pedal of another company. **NOTE:** *Some pedal cleats do work with other brands of pedals, but it's best to assume that they do not. Instead, stick with cleats and pedals of the same brand and type. Some cleats and pedals do not mix and match, and you might find yourself unable to disengage at a most inopportune moment.*

This chapter explains how to remove and replace pedals, how to mount the cleats and adjust the release tension with clip-in pedals, how to troubleshoot pedal problems, and how to overhaul and replace spindles on almost all mountain bike pedals. Incidentally, I use the terms "axle" and "spindle" interchangeably, as both terms are used by manufacturers and bike shops.

13-1

REMOVING AND INSTALLING PEDALS

LEVEL 1

Note that the right pedal axle is right-hand threaded (usual thread direction) and the left is left-hand (reverse) threaded. Both unscrew from the crank in the pedaling direction.

a. Removal

1. **Engage the wrench.** Slide a 15mm pedal wrench onto the wrench flats of the pedal axle (Fig. 13.3). Or if the pedal axle is designed to accept it, you can use a 6mm or 8mm hex key from the backside of the crankarm (Fig. 13.4). This design is particularly

13.3 Removing or installing pedal with a 15mm wrench

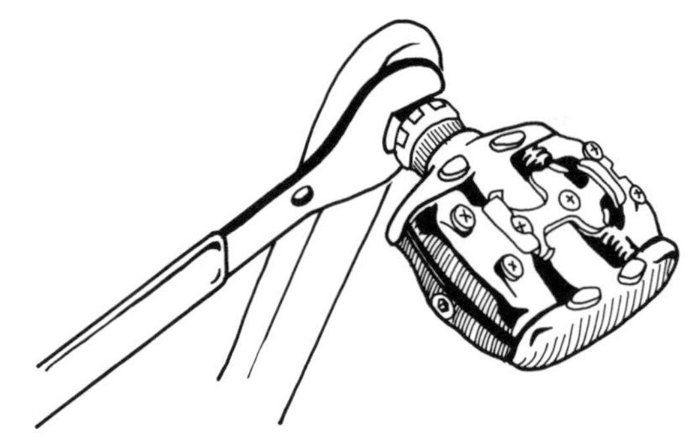

13.4 Removing or installing pedal with a hex key

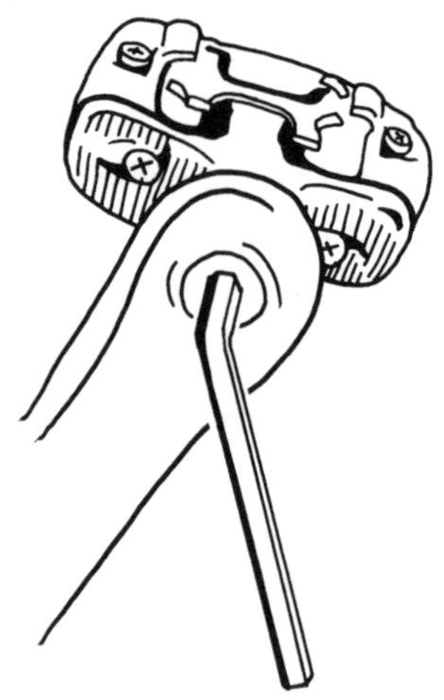

handy on the trail, because you probably won't be carrying a 15mm wrench. But if you are at home and the pedal is on really tight, it will probably be easier to use a standard pedal wrench. Some pedals have no wrench flats and can be removed only with a 6mm or 8mm hex key (Fig. 13.4).

2. **Unscrew the pedal in the appropriate direction.** The right (or drive-side) pedal unscrews counterclockwise when viewed from that side. The left-side pedal is reverse-threaded, so it unscrews in a clockwise direction when viewed from the left side of the bike. Once loosened, either pedal can be unscrewed quickly by turning the crank forward with the wrench engaged on the pedal spindle and the rear wheel positioned off the ground so that it can turn.

b. Installation

1. **Use a rag to wipe clean the threads on the pedal axle and inside the crankarm.**
2. **Grease the pedal threads.**
3. **Start screwing the pedal in with your fingers.** Turn clockwise for the right pedal, counterclockwise for the left one.
4. **Tighten the pedal.** Use a 15mm pedal wrench (Fig. 13.3) or a 6mm or 8mm hex key (Fig. 13.4). This can be done quickly by turning the cranks backward with the wrench engaged on the pedal spindle.

SETTING UP CLIP-IN PEDALS

Setting up clip-in pedals involves installation and adjustment of the cleats on the shoes and adjusting the pedal-release tension.

13-2

INSTALLING AND ADJUSTING PEDAL CLEATS ON THE SHOES

LEVEL 1

The cleat is important because its position determines the fore-and-aft, lateral (side-to-side), and rotational position of your foot. If your feet aren't properly oriented on the pedals, you could eventually develop hip, knee, or ankle problems.

1. **If the shoe has a precut piece of rubber covering the cleat-mounting area, remove it.** Cut around the cover's outline with a knife, pry an edge up with a screwdriver (Fig. 13.5), and yank it off with some pliers. If the rubber is hard to pull off, warm it up with a hair dryer to soften the glue.
2. **Put the shoe on, and mark the position of the ball of your foot.** It's the big bump behind your big toe; mark on the side of the shoe toward the crankarm. This will help you position the cleat fore and aft. Take the shoe off, and continue drawing the line straight across the bottom of the shoe (Fig. 13.6). If you have big feet (size 46 or larger), forgo this step, because you're generally better off putting the cleat as far back as it will go (in the back set of threaded holes). This far-back cleat position minimizes wasted energy by reducing the length of the lever (your foot from heel to cleat) that your calf has to control. It also reduces hot spots—pain under and between the metatarsals, which is more of an issue the heavier and stronger you are.
3. **If there are threaded holes in the shoe sole to accept cleat screws, skip to step 4.** If there are no threaded shoe holes, you must install the backing plate and threaded cleat plate that came with the pedals. Remove the shoe's sock liner, put the rectangular backing plate inside the shoe over the two slots, and put the threaded plate on top of it with the threaded protuberances sticking out through the slots, rather than up at your foot.
4. **Screw the cleat that came with the pedals onto the shoe.** Grease the cleat screw threads first, and use a 4mm hex key to install the screw. Make sure you orient the cleat in the appropriate direction. Some cleats have an arrow indicating forward (Fig. 13.6); if yours do not, the instructions accompanying the pedals probably specify which direction the cleat should point. Also note if the right and left cleats are different (see note under next step).

NOTE: *The cleat-engagement bars of Crank Brothers, Time, and Look mountain-bike pedals can press grooves into the shoe sole ahead of and behind the cleat; this can cause some shoe soles to crack, particularly carbon soles. To prevent this, put a steel shoe* *shield plate under the cleat; it, rather than the shoe sole, will rest on the cleat-engagement bars. Crank Brothers sells shoe shields, as do a few other shoe manufacturers.*

5. **Position the cleat.** Put it in the middle of its lateral- and rotational-adjustment range, and line up the center of the cleat over or (preferably) 1cm behind the mark you made in step 2 (Fig. 13.6). It is actually the position of the ball of the foot relative to the pedal spindle you are interested in, and with some systems the center of the cleat is not actually at the center of the pedal. Adjust these cleats accordingly to establish your desired relationship between the pedal spindle and the ball of the foot (usually you want the spindle directly under or 1–2cm behind the ball of the foot).

13.5 Removing rubber cover concealing the cleat holes

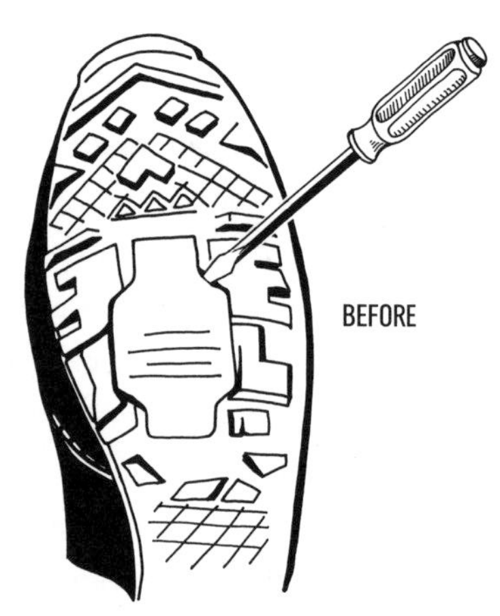

13.6 Cleat setup on an SPD-compatible shoe

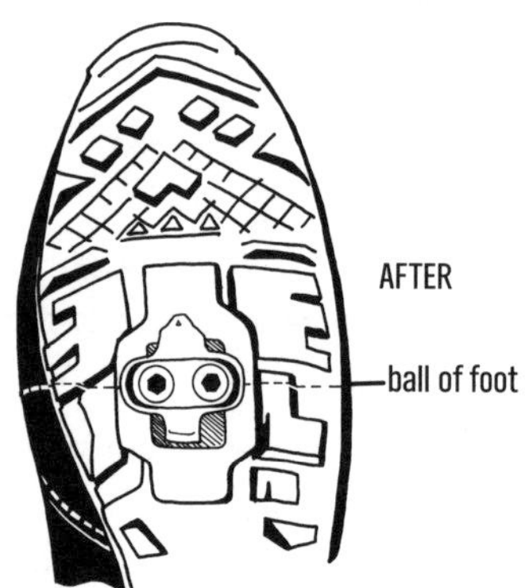

NOTE: *Cleats for Time ATAC pedals (Figs. 13.14 and 13.20), Crank Brothers pedals (Eggbeater [Fig. 13.19], Candy, Acid, Mallet, and Double Shot), and Look 4 × 4, Quartz and S-Track pedals have no lateral or rotational adjustment; just set the screws at your mark and tighten the cleat down, making sure the arrow on the Time and Look Quartz and S-Track cleat points forward. Crank Brothers and Look 4 × 4 cleats are symmetrical, so mount either end forward. Put the Time cleat with the imprinted stars (or the "G/L" imprint) onto the left shoe for less float range and earlier (13-degree) release angle; put it on the right shoe for more float and wider (17-degree) release angle. Conversely, put the Crank Brothers (or Look 4 × 4) cleat with the imprinted circles onto the right shoe for less float range and earlier (15-degree) release angle; put it on the left shoe for more float and wider (20-degree) release angle. Look Quartz cleats are available in either 15-degree or 20-degree release angle. You may now tighten the screws, skip the remaining steps, and go for a ride!*

As for the positioning of cleats for release angle, I recommend setting either system, especially Crank Brothers and Look, for the earlier release angle. That way, you are more likely to get out in a hurry when you need to. There is no spring-retention adjustment on Crank Brothers, Look, and pre-2004 Time pedals (Time ATAC XS pedals, introduced in 2004, do have a spring-tension adjustment), so the only way to get more retention is to increase the release angle. If you find yourself pulling out before you want to, for instance, when descending fast on rocky terrain, interchange the cleats to increase the release angle.

6. **Snug down the screws enough that the cleat won't move when clipped in or out of the pedals.** Don't tighten fully yet. Follow the same steps with the other shoe.
7. **Set the lateral cleat position.** Put the shoes on, sit on the bike, and clip into the pedals. Ride around a bit. Notice the position of your feet; you don't want them in so far that they bump the cranks. Take off the shoes and adjust the cleats laterally, if necessary, to move the feet side to side. Get back on the bike and clip in again. (Some cleats have no lateral adjustability.)

NOTE: *If you can't clip into the pedals, the shoe treads may be so tall that they are preventing the cleats from reaching the pedal's cleat-engagement bars. In that case, put as many plastic cleat shims (which are packaged with many pedals) under the cleat as necessary to get engagement. As the shoe tread wears, you may not need the shims when you next replace your cleats.*

8. **Set the rotational cleat position.** Ride around some more. Notice whether your feet feel twisted and uncomfortable. You may feel pressure on either side of your heel from the shoe. If necessary, remove your shoes and rotate the cleat slightly. Some pedals offer free-float, allowing the foot to rotate freely for a few degrees before releasing. Precise rotational cleat adjustment is less important if the pedal is free-floating.

NOTE: *For Speedplay Frogs (Fig. 13.16), angle the cleat slightly toward the outside of the shoe, and tighten the mounting screws just enough that the cleat can still turn. Clip into the pedal, and rotate the heel inward until it just touches the crankarm. Tighten the cleat in this position. Frogs have no inward release; this procedure sets the inward stop.*

9. **Once cleat position feels right, trace the cleats with a pen or scribe.** That way, you can tell if the cleat stays put.
10. **While holding the cleat in place, tighten the bolts firmly.** Hold the hex key close to the bend so that you do not exert too much leverage and strip the threads.
11. **If the cleat holes are open to the inside of the shoe, seal them.** Remove the insole, place a waterproof sticker over the opening inside, and replace the insole.
12. **When riding, take the 4mm wrench along.** You may want to fine-tune the cleat adjustment over the course of a few rides.
13. **Check cleat bolt tightness every few months.** You don't want to lose a screw (or a cleat) while riding. You may have to pick dirt out of the screw head and hammer the 4mm hex key into the hex hole so that the wrench won't slip. Ideally, use a torque wrench (4–5 N-m) with a hex bit.

NOTE: *If the shoe tread is thin or worn, the tread may not contact the pedal, and the shoe can rock inefficiently from side to side on the pedal. Some shoes have replaceable, screw-on tread; new tread is one way to eliminate rocking, as are new shoes. Otherwise, you can install tread-contact sleeves on Crank Brothers pedals. On Candys, the square, cup-shaped sleeves (Fig. 13.25) snap right on, but on Eggbeaters, you have to remove the axle (13-6) in order to get them on, because they won't go over the axle flange. The cylindrical contact sleeves are made for current Eggbeaters with cylindrical ends (Fig. 13.24); they won't fit on old Eggbeaters with square ends (Fig. 13.19; you would have to wrap tape around the ends of those to shim them up to contact your shoe tread). You also have to remove the colored plastic ring on either end of the Eggbeater pedal body; the raised lip on one end of each contact sleeve will engage the groove the ring was in. Warm the contact sleeves in hot water or with a hair dryer to ease pushing them on.*

13-3

ADJUSTING THE RELEASE TENSION ON CLIP-IN PEDALS

LEVEL 1

If you find the factory release setting to be too loose or too restrictive, you can adjust the release tension on most clip-in pedal brands; exceptions are Crank Brothers, Look 4 × 4, Speedplay, and pre-2004 Time. The adjusting screws are usually located at the front and rear of the pedal (Fig. 13.7). The screws affect the tension of the nearest set of clips.

13.7 Release tension adjustment

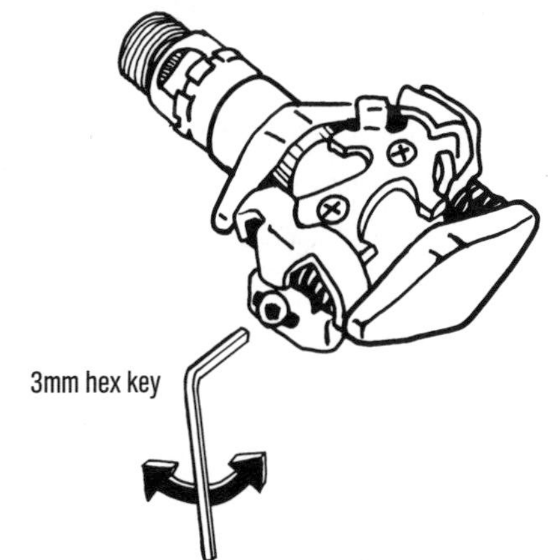

The adjusters are usually operated with a small (usually 3mm) hex key. Old Onza H.O. (Fig. 13.8) pedals are adjusted differently; see 13-4.

Before starting, clean the shoe cleats and the pedal clips. Lubricate the clip edges and cleat ends with dry lubricant (so that you don't track it onto your carpet) and the pedal springs with wet chain lube (Fig. 13.27). Whenever you have trouble getting in or out, start with this step.

1. **Locate the tension-adjustment screws.** They are usually on either end of the pedal, fore and aft; you can see the screw in Figures 13.7 and 13.9 to 13.13. Time ATAC XS (2004 and later) spring-adjustment screws are located on the outboard side at the front and back of the pedal.
2. **Loosen or tighten the release tension as you see fit.** To loosen the tension adjustment, turn the screw counterclockwise. To tighten it, turn it clockwise (Fig. 13.7). It's the standard "lefty loosey, righty tighty" approach. There usually are click stops in the rotation of the screw. Tighten or loosen one click at a time (one-quarter to one-half turn), and go for a ride to test the adjustment. Many types include an indicator that moves with the screw to show relative adjustment. Make certain that you do not back the screw out so far that it comes out of the spring plate.

NOTE: *With early Ritchey, Scott, Girvin, Topo, Wellgo, and other dual-rear-clip–dual-rear-spring pedals, you will decrease the amount of free-float in the pedal as you increase the release tension.*

13-4

ADJUSTING THE RELEASE TENSION OF PEDALS THAT HAVE NO ADJUSTMENT SCREW

(a) Time (pre-2004, Figs. 13.14 and 13.20), Crank Brothers (Fig. 13.19), and Look 4 × 4 pedals have no tension adjustment. They offer high retention and lots of float, yet they require low entry and release force and therefore require no adjustment. As mentioned in 13-2, you can change retention by interchanging the right and left cleats to obtain a wider or narrower release angle.

13.8 Onza H.O. clip-in pedal, pre-1997

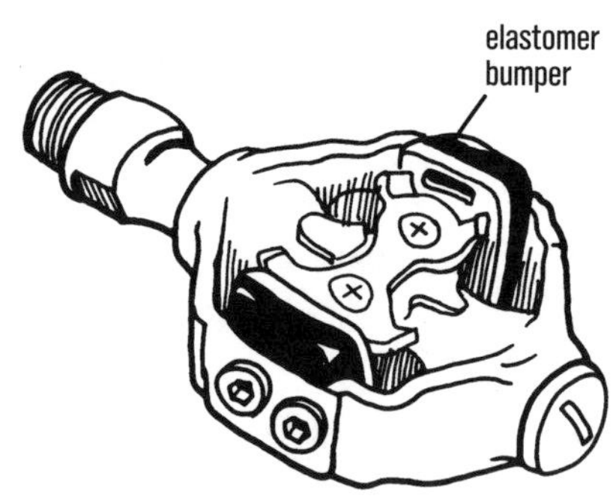

(b) Old Onza H.O. clip-in pedals (Fig. 13.8) rely on elastomer bumpers to provide release tension. Adjust them by changing the elastomer. Bumpers of varying hardness are included with the pedals. Onza's black bumpers are the hardest, and the clear ones are the softest. There are several grades in between. The harder the bumper, the greater the release tension. To replace bumpers, unscrew the two hex bolts holding each bumper on (Fig. 13.21). Pull out the old bumper and put in the new one. While you are at it, make sure that the Phillips screws that hold in the cleat guides are tight, because they have a tendency to loosen and fall out. In fact, it wouldn't hurt to put a small dab of Loctite on the threads while you're checking them.

(c) Speedplay Frogs (Fig. 13.16) have no tension adjustment; ease of release can be adjusted by rotating the cleat on the shoe sole to change the release angle.

OVERHAULING PEDALS

Just like a hub or bottom bracket, pedal bearings and bushings need to be cleaned and regreased regularly. Most pedals have a lip seal around the axle where it enters the pedal. Pedals without one get dirty inside very quickly, which can wear out the bushings and spindle.

First remove the pedal from the bike (13-1a, Figs. 13.3 and 13.4) and inspect it to figure out which section of directions to follow for overhauling.

There is wide variation in mountain bike pedal designs. This book is not big enough to go into great detail about the inner workings of every single model. In general, pedal guts fall into two broad categories: those that have cartridge bearings and/or bushings (Figs. 13.13 to 13.16 and 13.19 to 13.21), and those that have loose ball bearings (Figs. 13.11, 13.12, and 13.22).

Many pedals are closed on the outboard end and have a nut surrounding the axle on the inboard end (Figs. 13.9 to 13.13). The axle assembly installs into the pedal as a unit and is accessed by this inboard nut. The axle assemblies on other pedal designs are accessed from the outboard end by removing a dust cap (Figs. 13.18 to 13.22).

Before you start, figure out how the pedal is put together so that you will know how to take it apart; the following paragraphs and the illustrations on subsequent pages should help. In a few cases, the configuration of the pedal guts may not be clear until you have completed step 1 in the overhaul process.

Shimano pedals usually have two sets of loose bearings and a bushing that comes out as a complete axle assembly (Figs. 13.11 and 13.12). You will see the tiny ball bearings at the small end of the axle (Fig. 13.17).

13-5

OVERHAULING PEDALS CLOSED ON THE OUTBOARD SIDE

1. **Make sure the pedal does not have a dust cap or screw cover on the outboard end.** If it does, skip to 13-6.
2. **Remove the pedal body.** This requires either an open-end wrench, a splined tool designed for the pedal, or snapring pliers. On most varieties, removal is accomplished by unscrewing the nut surrounding the axle where it enters the inboard side of the pedal (Figs. 13.9 and 13.10). See note under (a) regarding thread direction. High-end pre-2004 Time ATACs with carbon-filled plastic bodies, later butterfly-shaped Ritcheys and current Ritchey V3 Pros, and old Time TMTs use a different approach. These pedals mostly rely on a snapring on the inboard end (Fig. 13.14) that must be removed with snapring pliers (Fig. 1.3). In place of the snapring, 2000–2003 Time ATACs with carbon-filled plastic bodies have an aluminum cup threaded onto the inboard end of the pedal.
   (a) Shimano (except M959), Look Quartz and S-Track, pre-2002 Look, and Exus take a plastic splined tool, but the Look tool is not compatible with the other brands. Use a large adjustable wrench to turn the tool (Fig. 13.9). Most other closed-end pedals (including Shimano M959) take a 17mm or 18mm open-end wrench (Fig. 13.10). The nut is often plastic and can crack if you turn it the wrong way, so be careful. Hold the pedal body with your hand or a vise while you unscrew the assembly. The fine threads take many turns to unscrew.

NOTE: *The threads inside the pedal body are reversed from the crankarm threads on the axle; the internal threads on the drive-side pedal are left-hand threaded,*

13.9 Removing the axle from a Shimano clip-in pedal

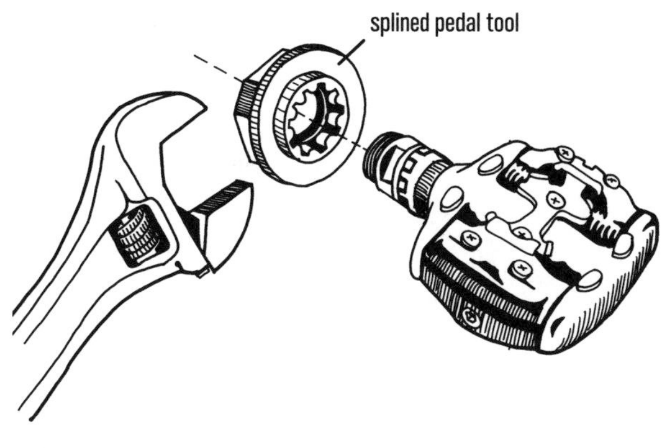

13.10 Removing the axle from a Scott pedal

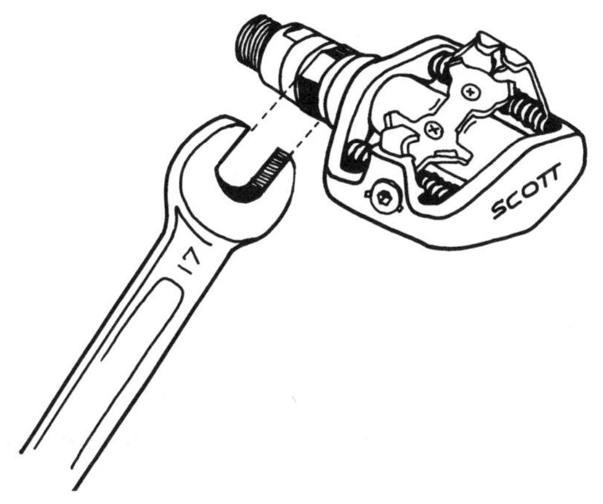

and vice versa. That means the right axle assembly unscrews clockwise and the left axle assembly unscrews counterclockwise. It's confusing, but like bottom-bracket threads, pedal bodies are threaded so that pedaling forward works to unscrew the assembly.

(b) Most Look, recent high-end Ritchey, and pre-2004 high-end Time mountain pedals have a large inboard cartridge bearing and an outboard needle-bearing cartridge.

(c) Axles in original Time plastic-body ATACs, recent high-end Ritchey, and old Time TMTs are removed via a snapring on the crank side (Fig. 13.14). Popping the snapring out requires inward-squeezing snapring pliers. Now skip to step 3 with any of these. On 2000–2003 Time ATACs with carbon-filled plastic bodies, unscrew the aluminum cup from the inboard end of the pedal body. You can try an adjustable pin tool or snapring pliers on the small pinholes on its face, or you can carefully clamp the cup in a vise or grab it with pliers. Skip to step 3.

(d) Speedplay Frogs have a cartridge bearing and a needle bearing, which can be regreased without opening the pedal. Remove the Phillips screw from the outboard end of the pedal body, and squirt grease in with a fine-tip bicycle grease gun (Fig. 1.2) until it comes out the axle end. If you decide to open a Frog (reminds you of junior-high biology, doesn't it?), the pedal comes apart like a clamshell when you unscrew the single bolt on each side with a 2.5mm hex key (Fig. 13.16). Before you put the pedal back together, put a thin bead of automotive gasket sealer all the way around the edge of one pedal half to seal water out.

3. **Examine the axle assembly and the bore of the pedal body.** Is there grease inside? Is it relatively clean? Are all internals free of rust? If there is a cylindrical sleeve surrounding the spindle (Figs. 13.11 and 13.12), does it spin smoothly on the ball bearings without back-and-forth play? If the answer to any of these questions is no, a thorough overhaul is recommended; skip to step 5 for that. Otherwise, simply perform step 4.

4. **Perform quick clean and lube.**

(a) **With a rag, wipe the outside of the spindle assembly.** Do not remove any nuts from the spindle and disassemble anything attached to it.

(b) **Clean inside the pedal bore.** Shove the thin end of a rag in there and twist it around.

(c) **Grease inside the pedal bore.** Put a glob of grease inside the pedal body about equal to ⅓ or ½ the volume of the pedal bore.

(d) **Push the spindle assembly back in.** If it won't go all of the way in, you may have too much grease or an air bubble behind the grease is stopping it, so remove some grease and try again. Pushing the axle assembly into the hole provides enough pressure on the grease to squeeze it through bearings like in Figures 13.11 and 13.12, as well as through open cartridge bearings.

(e) **Tighten the collar nut.** Avoid cross-threading, something easy to do with a plastic collar nut. You're done! Skip to step 10.

5. **Disassemble the axle assembly.** You will notice zero, one, or two nuts on the thin end of the axle. These nuts serve to hold the bearings and/or bushings in place.

(a) No nut? Skip to step 6.

(b) If the axle has just a single nut on the end (Figs. 13.13 and 13.16), simply hold the axle's large end with a 8mm hex key or 15mm pedal wrench and unscrew the little nut with a 9mm wrench (or whatever size fits it). The nut may be very tight, because it has no locknut.

(c) If the axle has two nuts on the end, they are tightened against each other. To remove them, hold the inner nut with a wrench while you unscrew the outer nut with another wrench (Fig. 13.17). Shimano and older Tioga pedals use two nuts in this fashion. On Shimanos, the inner nut acts as a bearing cone; be careful not to lose the tiny ball bearings as you unscrew the cone.

13.11–13.16 Clip-in pedals without outboard dust caps

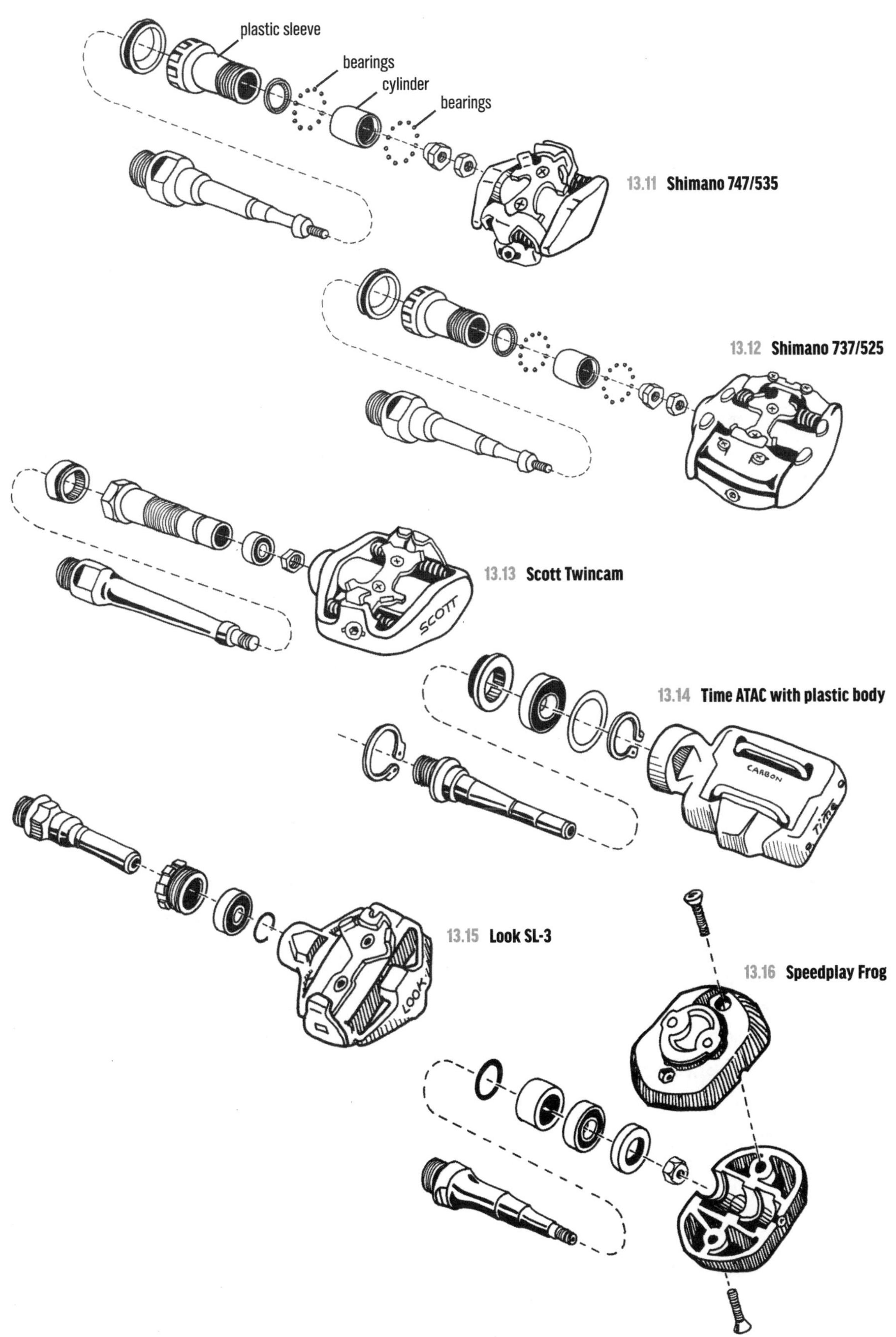

13.17 Tightening a Shimano locknut

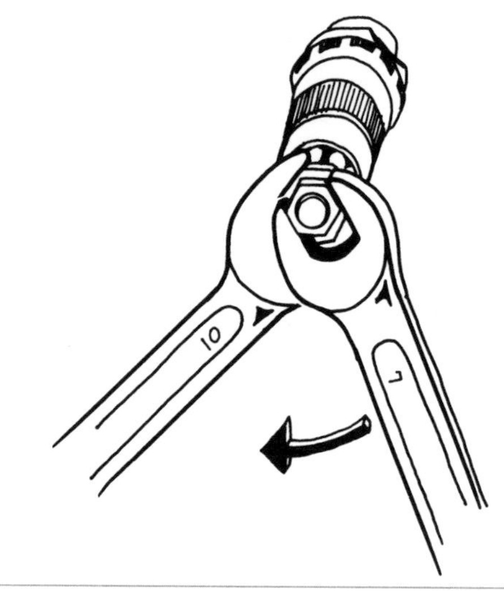

6. **Clean all the parts.**
   (a) If the bearings are loose, use a rag to clean the ball bearings, the cone, the inner ring that the bearings ride on at the end of the plastic sleeve (it looks like a washer), the bearing surfaces on either end of the little steel cylinder, the axle, and the inside of the plastic axle sleeve (Figs. 13.11 and 13.12). To get the bearings really clean, wash them in the sink in soap and water with the sink drain plugged; the motion is the same as washing your hands, and it results in both the bearings and your hands being clean for a sterile reassembly. Blot dry.
   (b) If, on a pedal with a cartridge bearing (Figs. 13.13 to 13.16), the bearing is dirty or worn out, clean it if you can; otherwise, replace it. These bearings often have steel bearing covers that cannot be pried off without damaging them, nor can the covers be replaced. If the pedal has plastic bearing covers, pry them off with a razor blade (Fig. 8.20), clean the bearing with solvent, let it dry, and repack it with grease.
   (c) Needle bearings (Time, Look, some Crank Brothers, Ritchey, and Coda) can be cleaned with a solvent and a thin bottlebrush slipped inside the pedal-body bore.
   (d) On a bushing-only pedal, such as older Tioga, just wipe down the axle and the inside of the bushings.

7. **Grease and reassemble.** This is a simple process with all pedals closed on the outboard end. Exceptions are pedals with loose bearings that are too rusty, dirty, or maladjusted to perform step 4 on.
   (a) With a loose-bearing pedal, it is exacting work to place the bearings on their races and screw the cone on while keeping them in place. On a Shimano pedal (Figs. 13.11 and 13.12), grease the bushing inside the plastic axle sleeve, and slide the axle into the sleeve. Slide the steel ring, on which the inner set of bearings rides, down onto the axle and against the end of the sleeve. Make sure that the concave bearing surface faces out, away from the sleeve. Coat the ring with grease, and stick half of the bearings onto the outer surface of the ring. Slip the steel cylinder onto the axle so that one end rides on the bearings. Make sure that all of the bearings are seated properly and that none are stuck inside of the sleeve.
   (b) To prevent the bearings from piling up on each other and ending up inside the sleeve instead of on the races, grease the cone and start it on the axle a few threads. Place the remaining half of the bearings on the flanks of the cone. Being careful not to dislodge the bearings, screw the cone in until the bearings come close to the end of the cylinder but do not touch it. While holding the plastic sleeve, push the axle inward until the bearings seat against the end of the cylinder. Make sure that the first set of bearings is still in place. Screw the cone in without turning the axle or cylinder. Tighten it with your fingers only, and loosely screw on the locknut.

8. **Adjust the axle assembly.** (For cartridge-bearing pedals with no end nut, skip this step.)
   (a) Pedals with a small cartridge bearing and a single nut on the end of the axle, such as Exus, Coda, VP, Wellgo, Topo, Girvin, Speedplay, and Scott, simply require that you tighten the nut against the cartridge bearing while holding the other end of the axle with a 15mm pedal wrench. This approach secures the inner ring

of the cartridge bearing against the shoulder on the axle, and proper adjustment is ensured.

(b) On pedals with two nuts on the end of the axle, hold the cone or inner nut with a wrench and tighten the outer locknut against it (Fig. 13.17). Check the adjustment for freedom of rotation, and be sure there is no play. Readjust as necessary by tightening or loosening the cone or inner nut and retightening the locknut.

9. **Replace the axle assembly in the pedal body.**

(a) Smear grease on the inside of the pedal hole; this will ease insertion and act as a barrier to dirt and water. Screw the sleeve back in place with the same wrench you used to remove it (Figs. 13.9 and 13.10).

NOTE: *Pay attention to proper thread direction (see note in 13-5, step 2). Tighten carefully; it is easy to cross-thread or overtighten and strip or crack the plastic nut.*

(b) On original Time plastic-body ATACs, recent Ritcheys, and old Time TMTs, after replacing the axle assembly, pop the snapring back into its groove just inside the inboard lip of the pedal body. For 2000–2003 Time plastic-body ATACs, screw the aluminum retaining cup back onto the pedal body to hold the axle assembly in place.

10. **Put the pedals back on your bike.**

Go for a ride.

13-6

OVERHAULING CARTRIDGE-BEARING PEDALS WITH AN OUTBOARD DUST CAP

LEVEL 2

Original Time ATACs (Fig. 13.14) were closed on the end, but since about 2004, all Time ATAC pedals (Fig. 13.20) have had an outboard dust cap. Crank Brothers (Fig. 13.19), Look 4 × 4, older Ritchey, Onza H.O. (Figs. 13.18 and 13.21), some Wellgo, some VP, Nashbar, and Norco, among others, also have an outboard dust cap.

1. **Take off the dust cap.** Some take a 5mm or 6mm hex key, others require a coin or a screwdriver, and many Time ATACs take snapring pliers or a pin spanner (both Fig. 1.2; you can file the pins down to make them fit the tiny holes in the dust cap). The

13.18 Removing or tightening the locknut on an Onza H.O. pedal

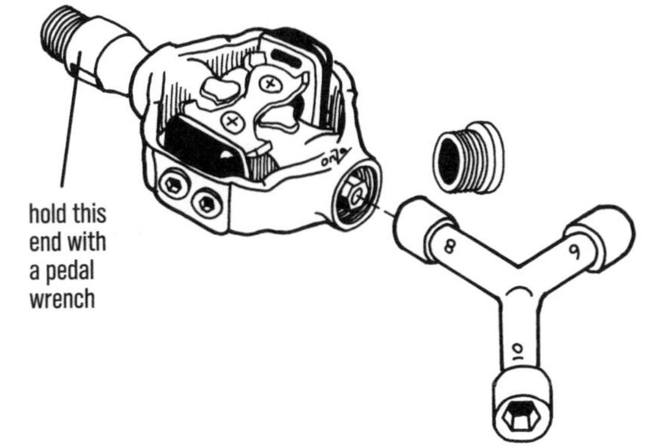

13.19 Original Crank Brothers Eggbeater, exploded

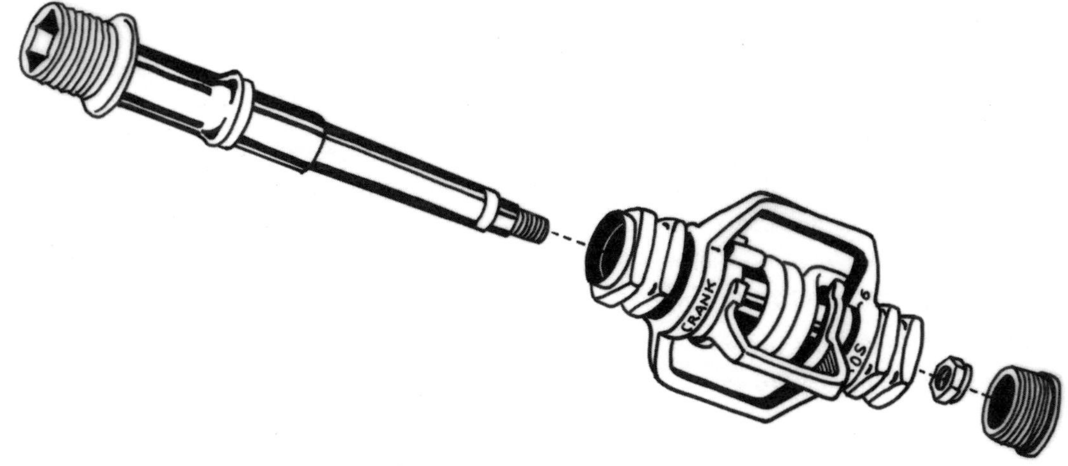

13.20 Time ATAC Alium, exploded

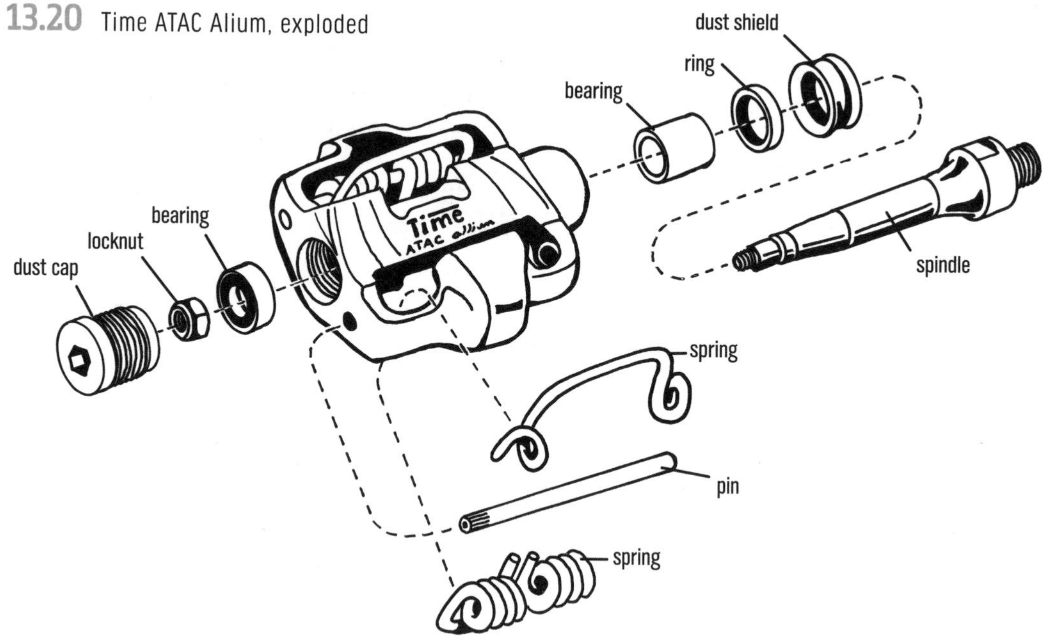

13.21 Onza H.O., exploded

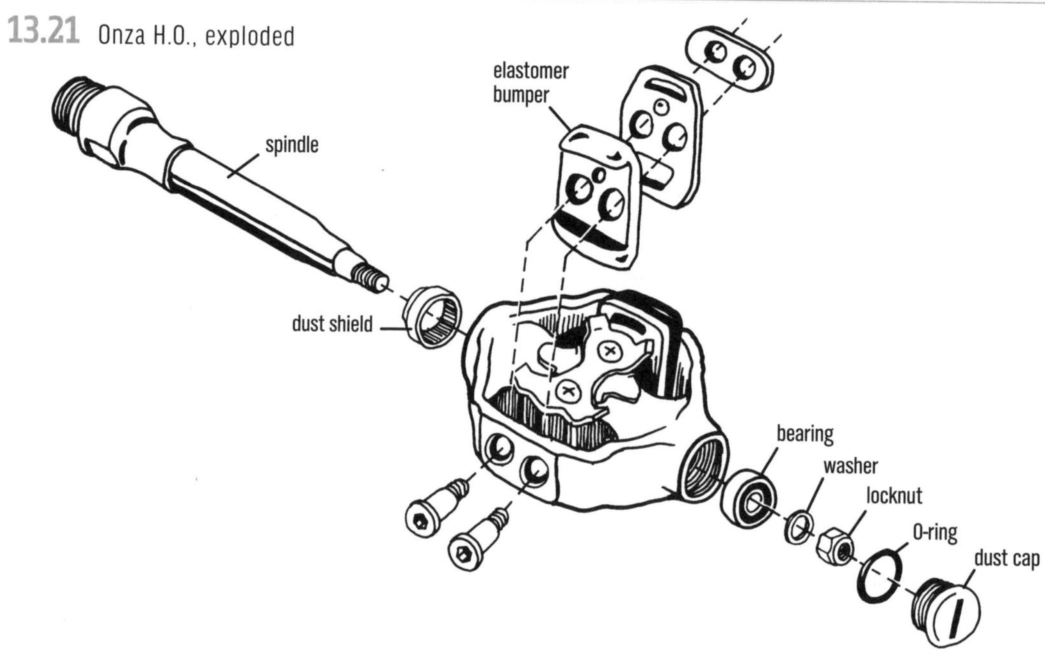

outer portion of Crank Brothers 50/50 pedals also fulfills the duties of a dust cap; with a Torx T25 key, unscrew the two long bolts from the inboard end to remove the outer part, and continue with step 2.

2. **Unscrew the outboard nut.** Use a socket wrench (Fig. 13.18), usually 8mm, 9mm, or 10mm. Hold the crank end of the axle with a 6mm or 8mm hex key or a 15mm pedal wrench.
3. **Push the axle out the inboard end.** It may take a tap or two on the end of the axle with a drift punch while holding the pedal body in a vise. This frees the outboard cartridge bearing. The guts should look similar to Figures 13.19 to 13.21.
4. **Clean and regrease the axle and the inside of the pedal body hole.** Replace the cartridge bearing if necessary. On many pedals, the brass or plastic bushings inside the pedal body are also replaceable; the idea is the same as in 13-8, but you need the correct size drift punch to push them out, and replacement bushings may not be available.

All Crank Brothers pedals since the time that the Eggbeater, Candy, and Mallet had model numbers (i.e., 1, 2, 3, 7, or 11) have bearings or bushings on the inboard end that you can replace yourself, and parts are readily available. These pedal models originally came with needle bearings inside, which tended to freeze solid with rust in wet conditions; around 2012, a running change to Igus bearings (plastic bushings) solved it. If you have a needle-bearing version (or worn-out Igus bearings), see 13-8 for instructions on pushing out the needle bearings (or worn-out Igus bearings) and installing new Igus bearings.

5. **Reassemble the pedal.** Push the axle back into the pedal body, slip the cartridge bearing onto the outboard end of the axle, and thread on the end nut.
6. **Tighten the little end nut down against the cartridge bearing.** Hold the crank end of the axle with a 15mm pedal wrench or a 6mm or 8mm hex key.

NOTE: *Some pedals will still have side play at this point, since their dust cap is an integral part of the assembly. Once it is tightened down, the play goes away.*

7. **Replace the dust cap.**

13-7

OVERHAULING LOOSE-BEARING PEDALS WITH AN OUTBOARD DUST CAP

NOTE: *Many non–clip-in pedals are not worth the effort to overhaul, and not all economical pedals are accessible to overhaul. Assess the value of the pedals and your time before continuing.*

LEVEL 2

1. **Remove the dust cap with the appropriate tool.** Use pliers, a screwdriver, a coin, a hex key, an adjustable pin tool, or a splined tool made especially for the pedals. Different dust cap styles are shown in Figures 13.18 to 13.22 (although not all on loose-bearing pedals), and it should be easy to figure out which tool is needed to remove the cap. If you see a cartridge bearing inside rather than loose balls, go back to to 13-6.
2. **Unscrew the locknut.** Hold the inboard end of the axle with a pedal wrench or hex key, and use the appropriate size socket wrench (as shown in Fig. 13.18) on the locknut.
3. **Unscrew the cone.** Hold the pedal over a rag to catch the bearings. Keep the bearings from the two ends separate in case they differ in size or in number. Count them so that you can put the right amount back in when you reassemble the pedal. The guts should look like Figure 13.22.
4. **With a rag, clean the parts.** Wipe off the bearings, cones, and bearing races. Clean the inside of the pedal body by pushing the rag through with a screwdriver. If there is a dust cover on the inboard end of the pedal body, either clean that in place or after popping it out with a screwdriver.
5. **To get the bearings really clean, wash them in a plugged sink with soap and water.** The motion is the same as washing your hands, and it results in both the bearings and your hands being clean for a sterile reassembly. Blot dry.
6. **If you removed it, press the inboard dust cover back into the pedal body.**
7. **Replace the inboard-side bearings.** Smear a thin layer of grease in the inboard bearing cup first. Once all the bearings are in place, there will be a gap equal to about half the size of one bearing.
8. **Replace the outboard-side bearings** (Fig. 13.23). Drop the axle in and turn the pedal over first so that the outboard end is up, and smear grease in that end.
9. **Bring the cones to the bearings.** Screw the outboard cone in until it almost contacts the bearings, and then push the axle straight in (from the outboard end) to bring the cone and bearings together; this prevents the bearings from piling up and getting spit out as the cone turns down against them. Without turning the axle (which would knock the inboard bearings about), screw the cone in until it is finger-tight.
10. **Slide on the washer and screw on the locknut.** While holding the cone with a cone wrench, tighten the locknut (similar to Fig. 13.17, but you will be holding the cone with a 13mm or so cone wrench, not a 10mm standard open-end wrench).
11. **Check that the pedal spins smoothly without play.** Readjust as necessary by loosening the locknut, tightening or loosening the cone, and retightening the locknut.
12. **Replace the dust cap.**

13.22 Loose-bearing pedal, exploded view

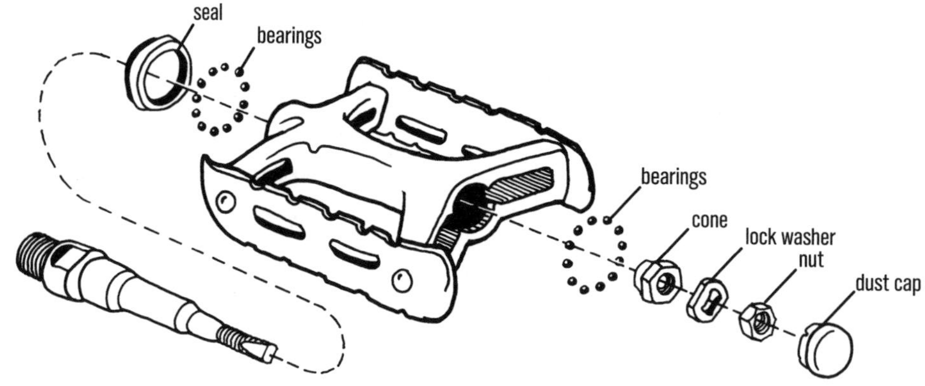

13.23 Replacing the ball bearings

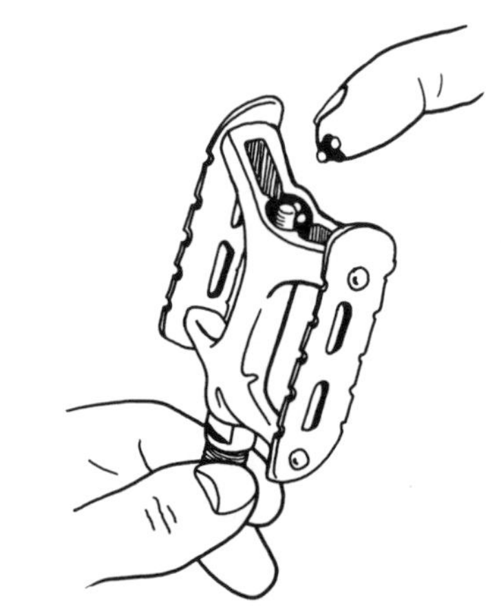

13-8

UPGRADING INBOARD BEARINGS IN CRANK BROTHERS PEDALS

LEVEL 2

When Crank Brothers Candy pedal bodies became aluminum, rather than plastic, and Eggbeater bodies became cylindrical on their ends rather than square (Fig. 13.19), the Eggbeater, Candy, and Mallet had model numbers (i.e., 1, 2, 3, 7, or 11), and the pedals started coming with needle bearings inside. The needle bearings tended to rust solid in wet conditions, and, around 2012, in a running change, all Crank Brothers pedals started to come instead with Igus bearings (plastic bushings), which do not rust.

If you have a needle-bearing version, you can push out the needle bearings and install the Igus bearings (Fig. 13.24). You simply buy a Crank Brothers Pedal Refresh Kit, which includes a chromed rod that you will use to drive out the bearing, and all new parts for the interior, other than the spindle (new spindles, including ones of different lengths and materials, are also available from Crank Brothers). The kit includes, for each pedal, a new outboard cartridge bearing, the Igus LL-Glide Bearing, the spindle locknut, an inner seal as on the original pedals, an outer seal that did not exist on the original pedals, and the dust cap. It also includes O-rings and inner and outer bushings, which do not get used on Candy 1s or Eggbeaters, but which are part of the Candy 2, 3, and 11, and the Mallet 1, 2, 3, and DH. This refresh kit is also useful as a maintenance item for replacing the Igus bearings, especially when replacing spindles.

1. **Remove the dust cap.** Some take a hex key, and others require a screwdriver. The outer portion of Crank Brothers 50/50 pedals also fulfills the duties of a dust cap; with a Torx T25 key, unscrew the two long bolts from the inboard end to remove the outer part, and continue with step 2.
2. **Unscrew the spindle locknut.** This requires an 8mm socket wrench on the locknut and an 8mm hex key in the end of the spindle.
3. **Pull out the spindle.** The cartridge bearing may just fall out now, or you can push it out with the spindle tip.
4. **Slide the seal off the spindle.** With an Eggbeater (Fig. 13.24), a Candy 1, or a 50/50 2 or 3, skip to step 9.
5. **Unscrew the two bolts holding the pedal body together.** Use a Torx T25 key.
6. **Pull the pedal body apart.** Pull the inner and outer halves of the pedal body away from the central spring clip section (Fig. 13.25).

13.24 **Top:** original needle bearing and single seal on worn spindle. **Bottom:** new spindle with Igus bearing and double seal; optional tread contact sleeves installed.

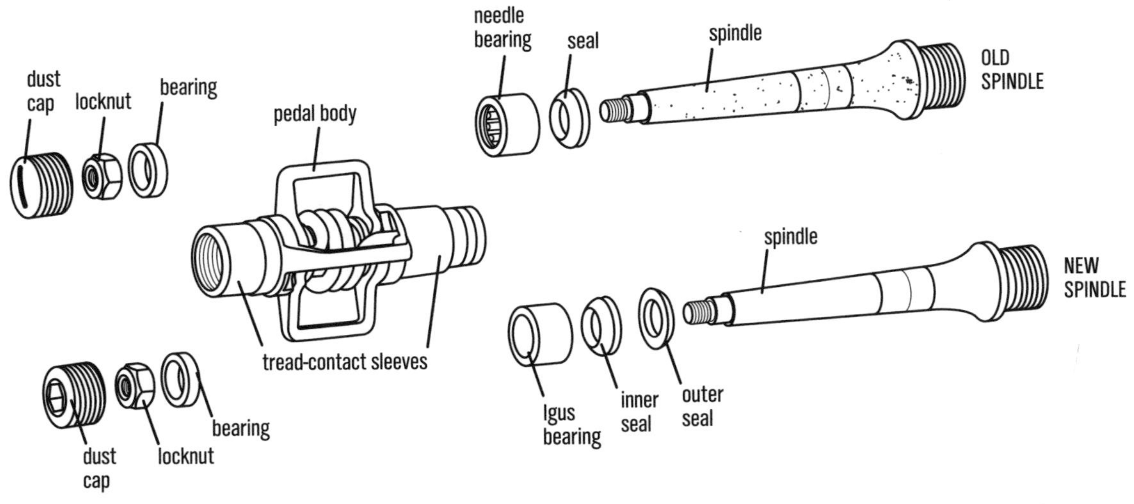

13.25 Exploded Candy 3 pedal, including tread-contact sleeves

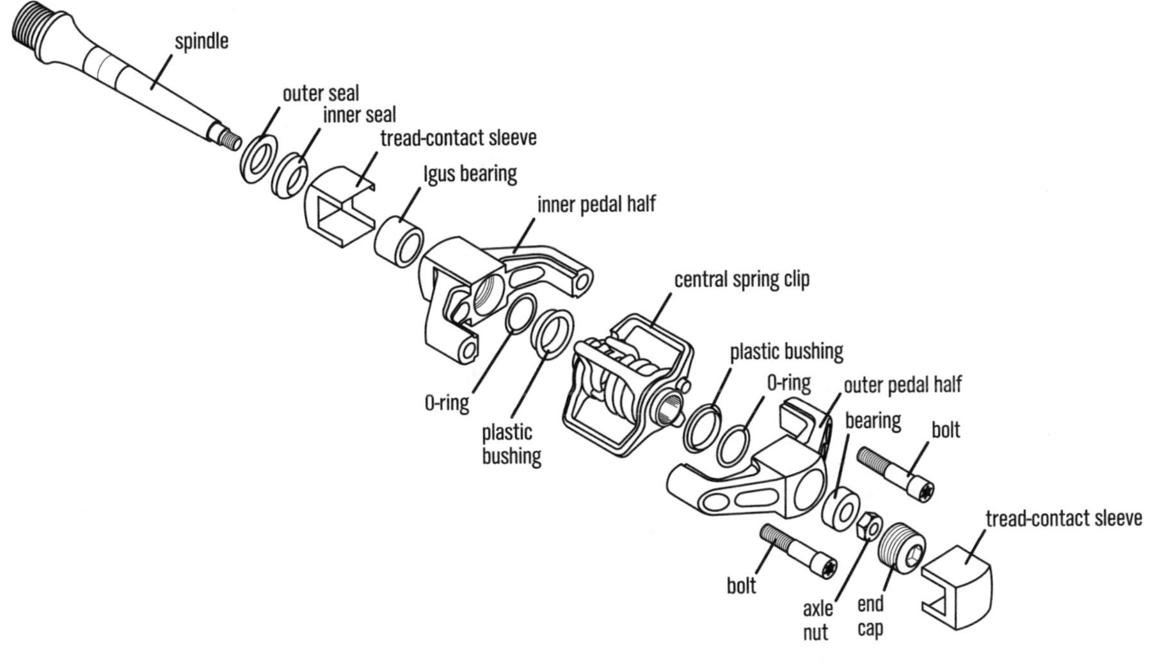

7. **Remove the inner and outer plastic bushings.** These are little top-hat-shaped plastic parts that fit in the medial ends of the spindle hole in the inner and outer pedal halves (Fig. 13.25). Use a knife to get under the brim of each one and work around it, prying it out.
8. **Remove the O-rings.** The O-rings are underneath the plastic bushings (Fig. 13.25).
9. **Push out the inboard bearing.** Slip the drift punch that came with the Refresh Kit into the pedal hole from the outboard end until it contacts the bearing. Set the inboard end of the pedal hole over the open end of a 14mm socket. The pedal hole should sit directly over the socket and the pedal body should sit on the socket walls. Tap the drift with a hammer (Fig. 13.26) until the bearing pops into the socket.

13.26 Driving the old needle bearing out of an Eggbeater pedal body

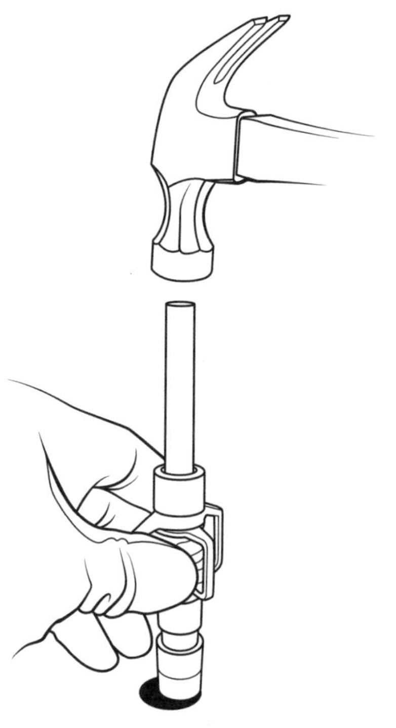

10. **Press the new Igus bearing into the pedal hole.** Set the plastic Igus bearing over the inboard end of the pedal hole. Set a 10mm socket against it, and tap it with a hammer to drive the plastic bushing into the pedal body until it hits the stop that the original bearing butted up against. With an Eggbeater, a Candy 1, or a 50/50 2 or 3, skip to step 13.
11. **Install the new O-rings and inner and outer plastic bushings.** The O-rings go in first. The longer bushing goes into the inboard pedal half (Fig. 13.25). Tap them in with a hammer if you can't push them in fully with your thumb.
12. **Bolt the pedal body back together.** Push the inner pedal half onto the longer end of the central spring sleeve. Push the outer pedal half onto the shorter end of the central spring sleeve. Slide the bolts in, and tighten them with a Torx T25 key; torque setting is 6 N-m.
13. **Slide the outer and inner seal onto the spindle.** The outer seal has its rounded end toward the crank, and the lips of the inner and outer seal face each other.
14. **Apply grease.** Grease the spindle, seals, and the ends of the pedal bore.
15. **Push in the outer bearing.** It will just drop in with finger pressure.
16. **Install the spindle.**
17. **Screw on the locknut.** Suggested torque is 4 N-m.
18. **Tighten the dust cap.** Suggested torque is 3 N-m, but there is a lot of threadlock compound on the new dust cap, so you might reach that torque before it is screwed all of the way in. Tighten it in until it is flush. Again, the outer portion of 50/50 2 and 3 pedals fulfills the duties of a dust cap; tighten the bolts holding the body together with a Torx T25 key; suggested torque is 2.5 N-m.
19. **Push the inner and outer seals snug in the inboard end of the pedal hole.** You're done! Enjoy your refreshed pedals!

13-9

UPGRADING THE PEDAL SPINDLE

LEVEL 2

You may want to replace worn spindles or lighten the pedals by substituting titanium spindles. Another reason to replace the axle is to use a shorter or longer one to decrease or increase the pedaling stance (distance between the feet when pedaling). With some pedals, you may be able to buy a separate axle that you can install into the same sleeve, bushings, and bearings as the axle it replaces. Other manufacturers sell axles only as a complete assembly, including the axle, sleeve, bushings, and bearings.

If you're going to install a lightweight or different-length aftermarket axle or axle assembly into the pedals, make sure that you purchase one intended for the pedal brand and model you have. If all you are doing is replacing the axle, follow the overhaul procedures outlined in 13-5 to 13-7. If you bought the entire assembly, just take out the old assembly. Again (I obviously feel the need to say this often), pay attention to the direction of the threads (see note in 13-5, step 1). Following the procedures in 13-5 to 13-7, install the new assembly.

Reinstall the pedals (Fig. 13.3 or 13.4). You'll be amazed how much lighter your bike feels . . . or is that your wallet?

TROUBLESHOOTING PEDAL PROBLEMS

13-10

CREAKING NOISE WHILE PEDALING

1. **The shoe cleats are loose, or they are worn and need to be replaced** (see 13-2).
2. **Pedal bearings need cleaning and lubrication** (see 13-5 to 13-7).
3. **The shoe sole is cracked.** See Note in 13-2, after step 4.
4. **The noise is originating from somewhere other than the pedals** (see Chapter 11, "Troubleshooting Crank and Bottom-Bracket Noise").

13-11

TOO EASY OR TOO HARD RELEASE OR ENTRY WITH CLIP-IN PEDALS

1. **Release tension needs to be adjusted** (see 13-3).
2. **Pedal-release mechanism needs to be cleaned and lubricated.** Clean off the mud and dirt, and then drip chain lubricant onto the springs (Fig. 13.27) and dry lubricant onto the cleat contacts on the clips.
3. **The cleats themselves need to be cleaned and lubricated.** Clean off dirt and mud, and put a dry chain lubricant or a dry grease (such as pure Teflon) on the contact ends of the cleats.
4. **The cleats are worn out.** Replace them (13-2).
5. **The knobs on the shoe sole that contact the pedal might be so tall that they prevent the cleat from engaging.** Install a shim or two under the cleat. Alternatively, locate where the pedal edges contact the sole, and trim some of the rubber with a knife.
6. **The clips on the pedal are bent down.** Straighten them if you can, or replace them. If you can't repair or replace the clips, you may have to replace the entire pedal.
7. **If it is hard to clip in, check the metal cleat guide plate at the center of the pedal.** It is held on with two Phillips screws, and they may be loose or may have fallen out.

13.27 Lubricating the release mechanism

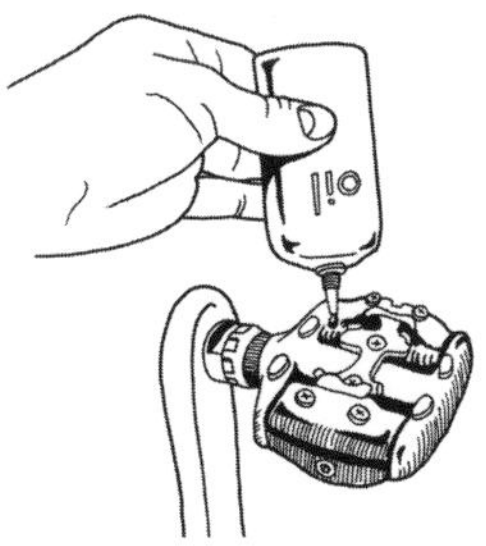

8. **Time ATAC spring clips are bent and not holding the cleat well.** They are replaceable by driving out the pin with a hammer and a punch (see Fig. 13.20).

13-12

KNEE AND JOINT PAIN WHILE PEDALING

1. **Cleat misalignment often causes pain on the sides of the knees** (see 13-2).
2. **You need more rotational float.** Pedals that offer the most are Crank Brothers, Time, and Speedplay.
3. **Your foot needs to tip one way or the other.** If your foot wants to roll inward (pronate), but your shoe and pedal force your foot to roll outward, then there is likely to be an increase in the tension on the iliotibial band, the tendon connecting the hip and calf. This will eventually cause pain on the outside of the knee. Alternatively, if your foot rolls inward so much that your knee falls in toward the top tube, you need to prop up the medial side of the foot. Either way, see a specialist who can make or prescribe custom footbeds (insoles), or wedges under the footbeds, to correct the problem.

NOTE: *Simply placing wedges under the cleats instead of the footbeds won't work with tiny mountain-bike cleats; they're too small to force the shoe to tip and correct the foot issue.*

4. **Fatigue and improper seat height can also contribute to joint pain.** Pain in the front of the knee right behind the kneecap can indicate that the saddle is too low. Pain in the back of the leg behind the knee suggests that the saddle is too high. (See Appendix C for seat-height guidelines.)

CAUTION: *If any of these problems result in chronic pain, consult a specialist.*

Even if you're on the right track, you'll get run over if you just sit there.

—WILL ROGERS

14

SADDLES AND SEATPOSTS

TOOLS

4mm, 5mm, and 6mm hex keys
Torx T10, T25 keys
Screwdriver
Grease
Assembly paste or spray
Snapring pliers

OPTIONAL

Soft hammer
Slide hammer
Vise
Penetrating oil
Hacksaw
Flex-hone
Electric drill
Cutting oil
RockShox Reverb bleed kit
Vise Grip pliers

After a few hours on the bike, I can pretty much guarantee that you will be most aware of one component on your bike: the saddle. It is the part of your bike with which you are most . . . uh . . . intimately connected. Nothing can ruin a good ride faster than a poorly positioned or badly fitting saddle.

The seatpost connects the saddle to the frame. It must hold the saddle firmly in the proper position without letting it tilt or slide down or back. It may also drop the saddle lower on command so that you can style drop-offs and rough or turny descents with aplomb, speed, and control. Some have shock-absorbing systems that cushion the ride.

14-1

CHOOSING A SADDLE

Most bike saddles are made up of a padded and covered flexible plastic shell suspended like a hammock between rails that attach at the tip and tail (Fig. 14.1). There are countless variations on (and a few notable exceptions to) this theme. Some have extra-thick foam or high-tech gel pads. Some are made with depressions, holes, or splits in the shell to reduce pressure in sensitive areas. Some have

14.1 Covered, padded, nylon-shell saddle

rails made of titanium, hollow cro-moly steel, carbon fiber, or even aluminum. Others have leather, synthetic leather, or Kevlar covers, or are made entirely of carbon fiber. You can expect to spend anywhere from $20 to more than $300 for a non-custom saddle, and more for a custom one. Price, however, may not be the best indicator of what makes a saddle really good—namely, comfort.

You have a lot of choices in saddles. My best advice is to ignore price, weight, fashion, and looks and instead choose a saddle that is comfortable. I could go on for pages about high-tech gel padding, scientifically designed shells that flex in just the right places at just the right moment, and all sorts of factors that engineers consider when designing a saddle. None of it would count for squat if, after reading it, you ran out and bought a saddle that turned out to be a giant pain in the rear. Saddles are different because people are different. Try as many as you can before buying one.

a. Width

A notion of saddle width that is currently widely held (pun intended) is that it should be around 2cm wider than the center-center distance between the rider's ischial tuberosities (sit bones) when in a moderate riding position. If the rider sits quite upright, then 3–4cm is added, and if he or she rides in a very aggressive position with the grips much lower than the saddle, then 0–1cm can be added. The theory is that the sit bones can handle pressure, and the soft tissues cannot, so getting a saddle on which the sit bones can perch will minimize numbness, blood-flow reduction, pain, and damage to the sensitive structures in the area. The saddle also must not restrict leg movement, a vote against having it too wide when riding in a low, aggressive position.

This obviously begs the question of how to measure sit bone width. It does not follow from body size; many very large people have narrow sit bone spacing, and vice versa. Two common shop tools for measuring this are shown in Fig. 14.2. By sitting on the Specialized memory-foam Ass-O-Meter, the rider temporarily leaves depressions in the memory foam, and the distance between the centers of these is measured. Similarly, by sitting on a piece of paper atop the sharp plastic points on SQLab's fit bench, the rider leaves a pattern of perforations through the paper where the sit bones were. Without these tools, you can also sit on a flat piece of cardboard on a low bench for a minute, get up, and then rub over the cardboard lightly with the side of a long piece of chalk to reveal the depressions the sit bones left.

14.2 Sit bone spacing measuring tools: SQLab saddle fit bench, Specialized memory-foam Ass-O-Meter

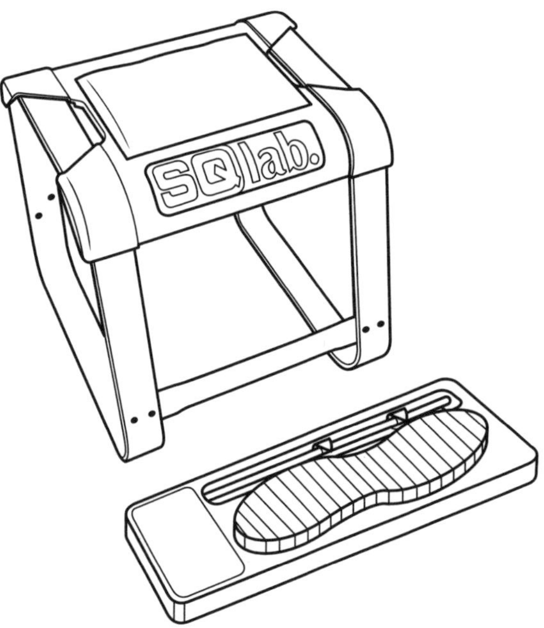

14.3 Saddle designed to not contact the perineum

b. Shape

There is no single saddle shape that works for everybody. Don't expect to hit it on the first try. Look for a saddle that does not compress the arteries and nerves running just along the medial sides of the sit bones, and that does not cause pain on other sensitive areas.

The marketing of saddles designed to prevent male impotency (Fig. 14.3) can limit a consumer's ability to select appropriately. If you buy a saddle out of fear and it is uncomfortable, you have done yourself a disservice. Don't take it on faith or scientific studies that your saddle is protecting you. If it hurts, or you get numb while riding on it, it isn't working for you. What works for one person won't necessarily work for another.

Determine which saddle shape and design are the most comfortable for your body, and then—and only then—start looking at titanium or carbon rails, fancy covers, and all the other things that improve a saddle and add to its cost. I know a lot of people who need 300g or 400g saddles with thick padding to feel comfortable on even a short ride. I know others who can ride for hours

14.4 Brooks leather saddle

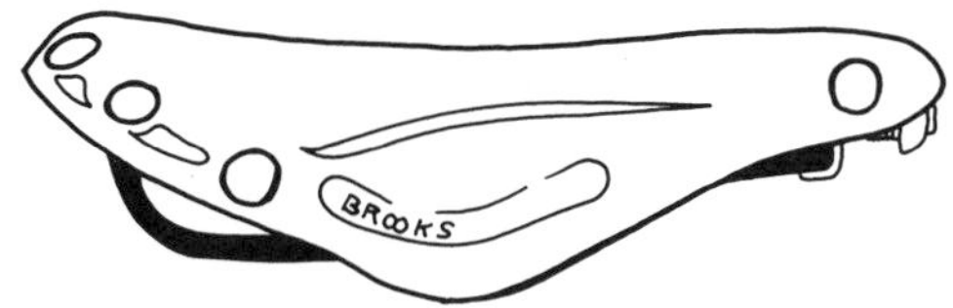

on a skinny little sub-100g saddle. It's a matter of preference. Any decent bike shop worth its weight in titanium should let you try a saddle for a while before locking you into a sale.

c. Shell and cover

Brooks, Selle Anatomica, and similar saddles have no plastic shell, foam padding, or cover. They are simply constructed from a single piece of thick leather attached to a steel frame with large brass rivets (Fig. 14.4). This was the main type of saddle up until the 1980s. Brooks still makes them and even offers them with titanium rails these days. This sort of saddle requires a long break-in period and frequent applications of a leather-softening compound that comes with the saddle or from a shoe store. Like a lot of old-style bike parts, you either love 'em or hate 'em. If you're not familiar with them by now, go out and buy a modern saddle (Figs. 14.1, 14.3).

A saddle with a plastic shell and foam padding requires little maintenance except keeping it clean; check periodically that the rails are not bent or cracked (a good sign that you need to replace the saddle).

14-2

POSITIONING THE SADDLE

Even if you have found the perfect saddle, it can still feel like some medieval torture device if it isn't properly positioned. Saddle placement is critical to finding a comfortable riding position. Not only does saddle position affect how you feel on the bike, but with the saddle in the right place, you also you become a much better rider. There are three basic elements to saddle position: tilt, fore-and-aft, and saddle height (Fig. 14.5).

14.5 Saddle adjustments

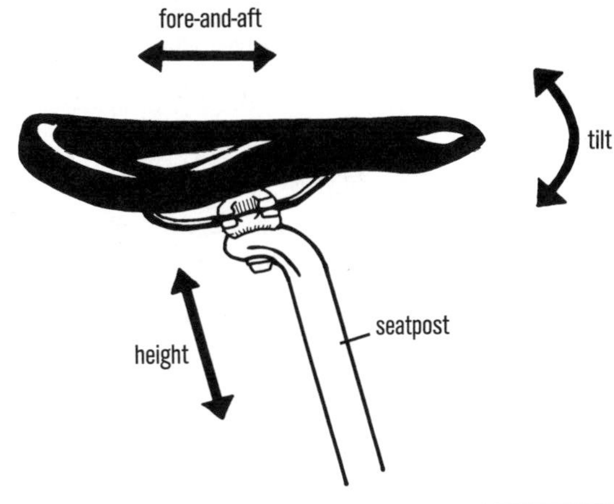

14.6 Knee bend at bottom dead center should be 25–30 degrees from straight for cross-country riding and bent much farther for gravity riding.

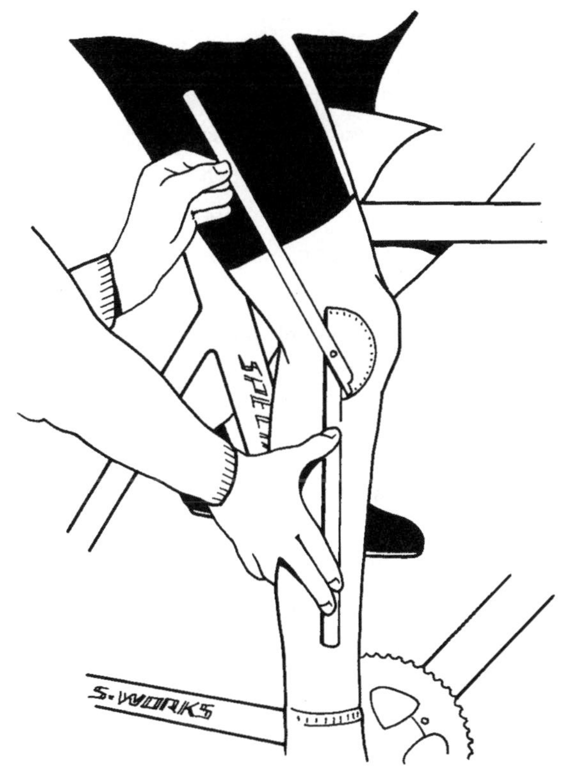

See Appendix C, C-3, for a detailed explanation of setting saddle and handlebar position. The following are some short guidelines.

Proper saddle height (Fig. 14.5) is key to transferring good power to the pedals. The ideal saddle height for cross-country mountain bike racing places your leg in a 90–95 percent extension (knee bend of 25–30 degrees) when your foot is at bottom dead center when you are riding (Fig. 14.6). However, you may find this position to be too high for riding technical singletrack, and it is way too high for gravity-driven mountain biking. That's the beauty of a height-adjust (dropper) seatpost (see section 14-9)—you can have it both ways! To improve your balance and center of gravity when riding a technical descent, bring the saddle height down—how far depends on you and the kind of riding you do. Pro downhill racers prefer very low saddle heights when compared with pro cross-country racers.

Make sure the seatpost is inserted deeper than the limit line inscribed on it.

A common cause of numb crotch and butt fatigue (and even sore arms and shoulders) is an improperly tilted saddle (Fig. 14.5). The general rule of thumb is that you should keep the saddle level when you first install it, although it may take a slight downward tilt (2 degrees) at the nose to accomplish this—the long seatpost extensions on many mountain bikes mean that there may be enough flex in the seatpost and saddle such that a saddle with a 2-degree downward slope becomes level when you sit on it. Other than perhaps for downhill riders, I strongly discourage making that tilt difference between nose and tail much more than ¼ inch (6mm) in height. Too much upward tilt places too much of your body weight on the nose of the saddle. Too much downward tilt will cause you to scoot down the saddle as you ride. That puts unnecessary pressure on your back, shoulders, and neck.

Fore-and-aft position (Fig. 14.5) determines where your butt sits on the saddle, the position of your knees relative to the pedals, and how much of your weight is transferred to your hands. Saddles are designed to have your butt centered over the widest part. If this is not where you sit, reposition the saddle or get a different one. You want to position the saddle so that you have a comfortable amount of bend in your arms without feeling too cramped or stretched out. If you find that your neck and shoulders feel tighter than usual and your hands are going numb, try redistributing your weight by moving the saddle back. Fore-and-aft saddle position also affects how your legs are positioned relative to the pedals. Ideally, your fore-and-aft position should be such that your knee pushes straight down on the forward pedal when the crankarms are horizontal. If the saddle will not go back as far as you wish and you have a short saddle or a leather saddle with a forward offset, try replacing it with a longer model or getting a seatpost with more offset. And vice versa.

Butt pain is intimately connected to handlebar position, as are other aches and pains. The shorter the upper-body reach and higher the handlebar, the more weight will go on the butt. The longer the reach and lower the handlebar, the more the top of the pelvis rotates forward and moves saddle pressure from the sit bones to the soft tissue of the perineum and genital area. As a general rule, a novice rider will want a shorter reach, a higher handlebar, and perhaps a correspondingly wider saddle than will an experienced rider. Once again, consult Appendix C.

14-3

MAINTAINING THE SEATPOST

A standard seatpost requires little maintenance other than removing it from the frame every few months. When you do that, wipe it down, regrease it, dry out and grease the inside of the frame's seat tube (turn the bike upside down to pour out any trapped water), and then reinstall the seatpost. Also do this after any ride in the rain. This maintenance keeps the seatpost clean and moving freely for the purpose of adjustability. It also should prevent the seatpost from getting stuck in the frame (a very nasty and potentially serious problem) and will prevent a steel seat tube from rusting out from the inside. The procedures for installing a new seatpost are in 14-8, and those for removing a stuck seatpost are in 14-13.

Dropper posts (sections 14-9 to 14-11) over time require readjustment of cable tension or bleeding of the hydraulic-activation system. They also require regular cleaning and lubrication of the inner shaft, and occasional replacement of the guide blocks that keep the post running smoothly up and down in its indexing grooves with little rotational waggle.

PRO TIP — CARBON SEATPOSTS

CARBON-FIBER SEATPOSTS ARE SUSCEPTIBLE to breakage if the seatpost-clamping mechanism digs in and cuts some fibers or makes a notch in the post. Point-loading on a thin carbon part is not a good idea. Unless you have a special seatpost binder clamp intended for carbon seatposts, you should reverse the binder clamp so that its slot is not lined up with the seat-tube slot. That way, it will pull more evenly around the circumference and not push the slot into the seatpost (Fig. 14.7). And if you have a slotted seat-tube shim tube that some bikes have to fit the seatpost, you should offset that slot from the other two slots as well. You can imagine how the top corners of the slot could dig into the back of the seatpost (where the load on it is also highest) if you crank down on the bolt with the slots of the binder clamp and seat tube and perhaps a seat-tube shim all stacked up there. This is where carbon seatposts break.

Carbon seatposts often slip down because the clear-coat layer on the outside is soft and gets compressed by the clamping pressure. Of course this also leads to the user tightening the bolt more, thereby increasing the probability of it snapping off. If the post slips, use some carbon assembly paste or spray on it (Fig. 14.7). This stuff works because the small plastic spheres or grit in the solution press back against the clamp to increase the pressure uniformly and avoid point-loading. In the absence of carbon assembly paste, don't grease the post. Unlike an aluminum or steel seatpost, a carbon post itself cannot corrode, so grease to prevent seizure is less of an issue. However, corrosion of a surrounding aluminum seat tube or seat-tube sleeve can nonetheless seize the post, so the best solution is carbon assembly compound.

14.7 Spraying Carbogrip carbon assembly compound on a carbon post

binder clamp reversed

seat tube slot

Suspension seatposts require periodic tune-ups (see 14-12). Regularly check any seatpost for cracks or bends so that you can replace it before the seatpost breaks with you on it.

If your bike has a carbon seatpost, be sure to read the Pro Tip on this subject.

14-4

INSTALLING THE SADDLE

Remember the heavy steel posts with the skinny section on top on kids' bikes (or on cheap adult bikes)? Those seatposts have a separate saddle clamp that slides down over the seatpost shaft, and a single horizontal bolt that pulls together a number of knurled washers with ears to hold the saddle rails. They are cheap to make, being simply a steel tube with some washers and a bolt, but they do not hold up well to adult use.

Much more secure (and generally lighter) is a seatpost with an integrated saddle clamp. Most posts have either one or two bolts for clamping the saddle.

A single-bolt seatpost can have either a vertical bolt or a horizontal bolt. The vertical bolt pulls two clamshell clamp halves together (Figs. 14.5 and 14.8). The horizontal bolt pulls two ears of the clamp toward each other to clamp the rails, and the ears may even slide down on angled ramps so that tightening the bolt pulls the rails

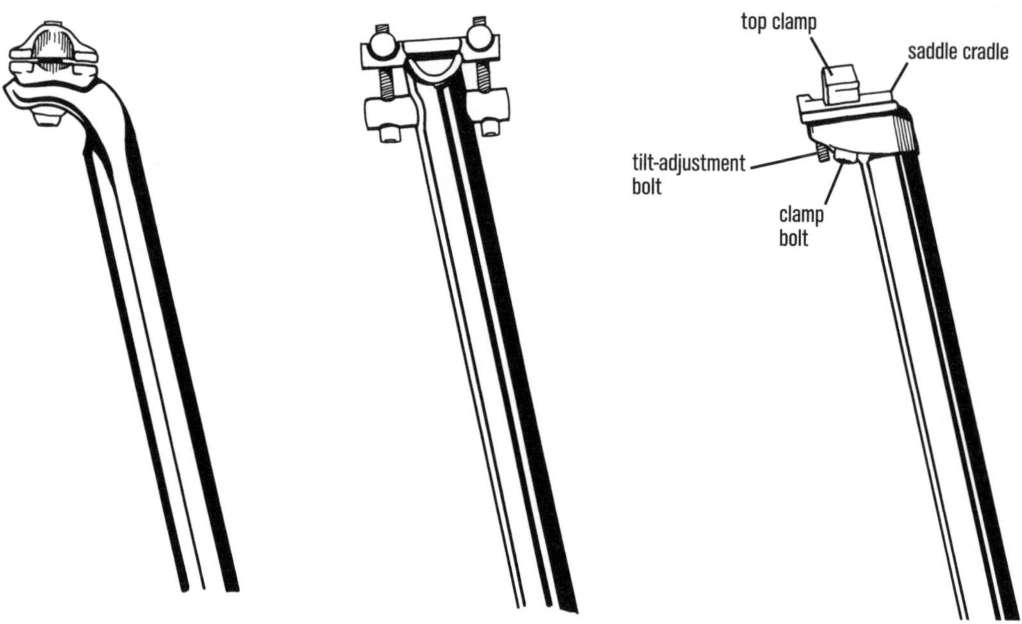

14.8 Single-bolt seatpost

14.9 Two-bolt seatpost

14.10 Single-bolt seatpost with small adjusting bolt

down onto the saddle cradle as well as against it from the sides.

The two-bolt posts can rely on one of four systems. In one system, the two bolts work together, pulling the saddle rails into the clamp (Figs. 14.9 and 14.12). In the second system, a smaller second bolt works to offset the force of the main bolt (Fig. 14.10). In the third system, a smaller second bolt allows the clamp to be swung up and down to adjust saddle tilt after the main bolt has been tightened (Fig. 14.20). In the fourth system, two side-by-side bolts hold the saddle clamp together.

14-5

INSTALLING THE SADDLE ON A SEATPOST WITH A SINGLE VERTICAL CLAMP BOLT

Systems with a single vertical bolt (Fig. 14.8) usually have a two-piece clamp that fastens onto the saddle rails. On most single-bolt models, moving the clamp and saddle along a serrated curved platform controls saddle tilt. Before you tighten the clamp bolt, make sure there is no second, much smaller bolt (or setscrew) that adjusts seat tilt. If you find a second bolt, skip to 14-6.

1. **Loosen the bolt.** It usually takes a 5mm or 6mm hex key. Loosen until there are only a couple of threads holding the upper clamp.

14.11 Saddle installation on a single-bolt seatpost

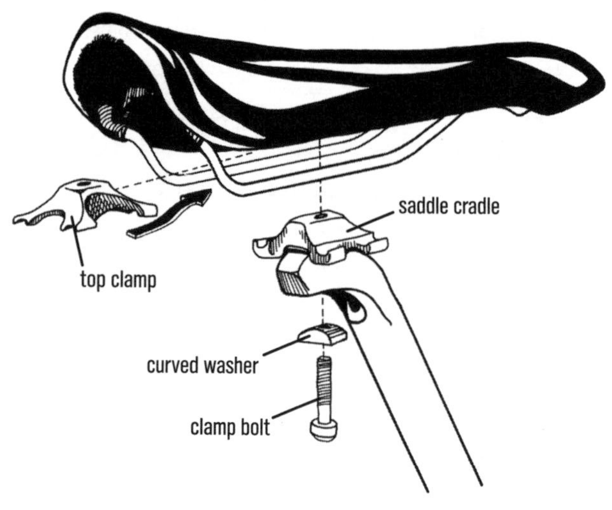

2. **Turn the top half of the clamp 90 degrees, and slide in the saddle rails.** Slide the saddle in from the back where the space between the rails is wider. You might need to remove the top of the clamp from the bolt completely if the clamp is too large. If you do disassemble the clamp, pay attention to the orientation of the parts so that you can put them back together the same way.
3. **Install the saddle.** Set the seat rails into the grooves in the saddle cradle piece, and set the top

clamp piece on top of the rails (Fig. 14.11). Slide the saddle to the desired fore-and-aft position.

4. **Tighten the bolt and check the seat tilt.** Readjust if necessary.

14-6

INSTALLING THE SADDLE ON A SEATPOST WITH TWO EQUAL-SIZED CLAMP BOLTS

This type of post is illustrated in Figure 14.9.

1. **Loosen or remove one or both of the bolts.** They usually take a 4mm or 5mm hex key. Loosen the bolts until you can open the clamp enough to perform step 2.
2. **Install the saddle.** Slide either the top clamp piece or saddle cradle piece out, set the saddle rails in their grooves in the piece remaining on the post, and then slide the piece you removed back in from the side.
3. **Slide the saddle to the desired fore-and-aft position.** Tighten down one or both of the clamp bolts completely.
4. **Set the saddle tilt.** Loosen one clamp bolt, and tighten the other to change the tilt of the saddle (Fig. 14.12). Repeat as necessary. Complete by tightening both bolts.

14.12 Saddle installation on a two-bolt seatpost

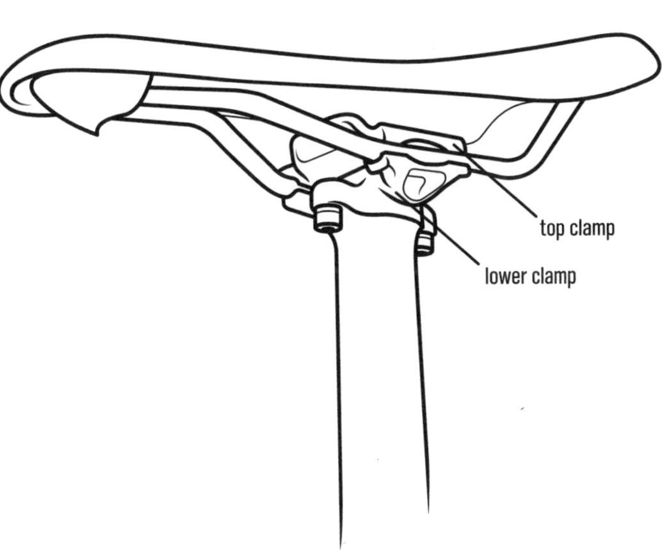

14-7

INSTALLING THE SADDLE ON A SEATPOST WITH A LARGE CLAMP BOLT AND A SMALL SETSCREW

This type of post is illustrated in Figure 14.10; it used to be much more common.

1. **Loosen the large clamp bolt.** It usually takes a 5mm or 6mm hex key. Loosen until the top part of the clamp can either be removed or rotated out of the way so that you can slide the saddle rails into place.
2. **Install the saddle.** Place the saddle rails into the grooves in the cradle, and set the grooves of the top clamp piece down over them. Slide the saddle to the desired fore-and-aft position. Tighten the large bolt.
3. **Adjust the saddle tilt.** Loosen the large clamp bolt, adjust the saddle angle as needed by turning the tilt-adjustment bolt, and retighten the clamp bolt. Repeat until the desired adjustment is reached.

NOTE: *The setscrew may be vertical or horizontal. With a vertical setscrew (Fig. 14.10), the screw is usually adjacent to the clamp bolt. A horizontal setscrew is usually positioned at the top front of the seatpost, pushing back on the clamp. With such a setscrew, push down on the back of the saddle with the clamp bolt loose to make sure the clamp and setscrew are in contact.*

14-8

INSTALLING THE SEATPOST INTO THE FRAME

1. **Check for irregularities, burrs, and other problems inside the seat tube.** Check visually and manually with your finger; if there are some, you may need to sand or otherwise clean out the inside of the seat tube. It may be necessary for a bike shop to ream the seat tube if a seatpost of the correct size will not fit.
2. **Grease the seatpost, the inside of the seat tube, and the binder bolt.** If you are using a sleeve or shim to adapt an undersized seatpost to fit the frame, grease it inside and out, and insert it.
3. **Insert the seatpost** (Fig. 14.13) **and tighten the seat binder bolt.** Some binder bolts are tightened

14.13 Seatpost installation into the frame

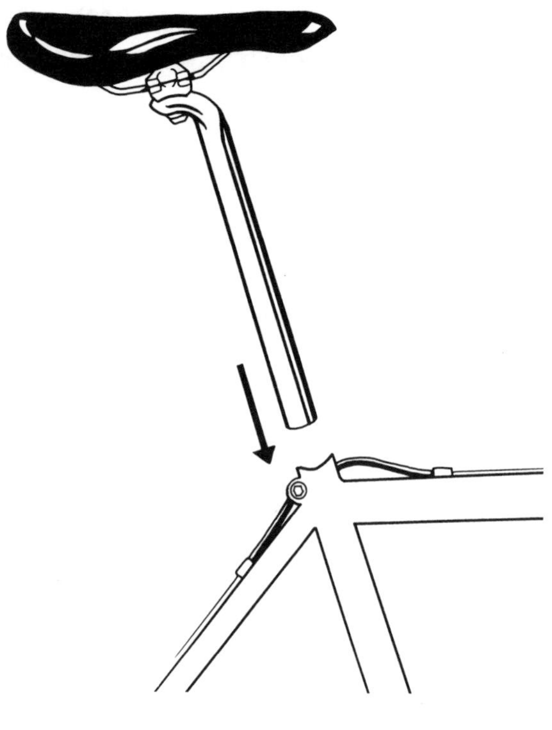

14.14 Closing a quick-release seatpost binder

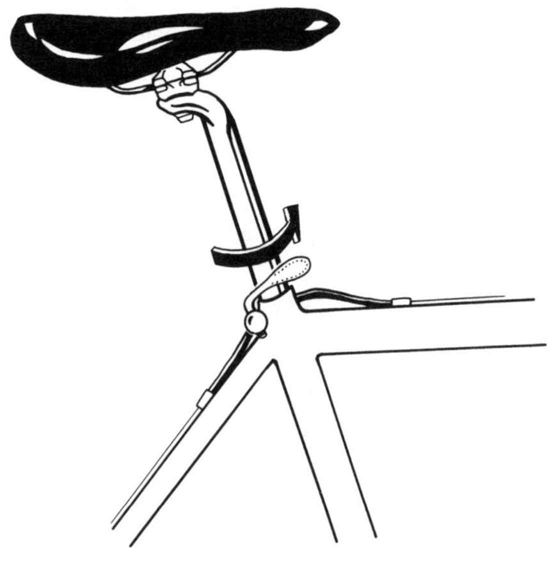

with a wrench (usually a 4mm or 5mm hex key), and some have a quick-release lever (Fig. 14.14). To tighten a quick-release, flip the lever open so that it is directly in line with the body of the bolt—in other words, about halfway open. Finger-tighten the nut on the other end, and then close the lever. It should be fairly snug, about tight enough to leave an impression in the heel of your hand for a few seconds. Open the lever, reposition the end nut, and close the lever again as necessary to get the right closing force.

4. **Adjust the seat height to your desired position.** It is a good idea to mark this height on the post with an indelible marker, paint pen, or piece of tape. This way, if you remove the seatpost, you can just slide it right back to the proper place.

IMPORTANT: *As mentioned in 14-3, periodically remove the seatpost, invert the bike to drain water out of the seat tube, and let it dry out. The frequency depends on the conditions you are riding in; certainly do it after a ride in pouring rain. For a steel frame, spray oil (or better yet, J. P. Weigle Frame Saver) into the seat tube to arrest the rusting process; this might help with oxidation inside aluminum and magnesium frames as well. Regrease the post and the inside of the seat tube, and reinstall the post.*

14-9

INSTALL AND ADJUST HEIGHT-ADJUSTABLE TELESCOPING ("DROPPER") SEATPOST

Dropper posts (Figs. 14.15 to 14.19) allow the rider to vary seat height with the flip of a lever, the push of a button, or the pull of a knob. Older dropper posts generally have only two positions: a high saddle for pedaling efficiency and a low saddle for safe descent of technical trails. Current dropper posts generally have multiple, or infinite, height positions into which they can lock.

To raise the saddle, push and hold the button or lever while standing up off the saddle; release the lever or button when the saddle reaches its high position. Some older posts require you to be sitting down when initially operating the lever or button to release the mechanism and then stand up. To lower the saddle, push and hold the button or lever while sitting on the saddle, and release it when the saddle reaches its low position.

Installation is fairly straightforward in the case of a dropper post with an external cable (or hydraulic hose). Unless you need a longer cable or hose or are shorten-

14.15 Dropper post with external cable routing

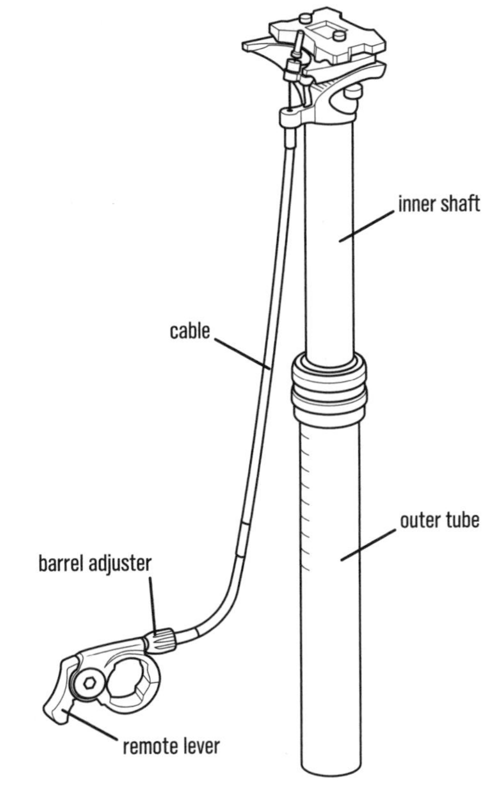

ing the one you have, it's not much more complicated than installing a standard seatpost.

Installation is more complicated with a stealth dropper post, in which the activating cable or hydraulic hose runs inside the frame and up the inside of the seat tube to the bottom end of the seatpost. In this case, you have to disconnect the cable or hose in order to feed it through the frame. This also means reconnecting the cable or hose, and in the latter case, bleeding the hydraulic system in order for the seatpost to release when you push the button on the handlebar.

1. **Clamp the post into the frame** (14-8). If the seatpost is a tight fit in the frame (i.e., if it takes some twisting to get it in), grease the seatpost; if it slides easily up and down, put friction paste on it. Set your seat height with the seatpost in its fully extended position. Don't tighten the seat binder too tightly, especially with a 30.9mm-diameter seatpost (its outer tube is thinner than the outer tube of the same model of seatpost in a 31.6mm diameter); the binder can squeeze the seatpost so tightly that the dropper mechanism will not slide up and down freely. Check it as you tighten to ensure that you are not binding the seatpost mechanism.

   Skip to step 3 with a hydraulically activated seatpost (RockShox Reverb) and to step 4 with an electronically activated one (Magura Vyron).

   With an internally routed (stealth) cable or hose, you'll have to run the cable or the hose through the frame to the seatpost first (with the remote detached from the handlebar so the cable or hose can slide in far enough to extend out of the top of the seat tube). Guide the cable or hose through the frame as in 5-15, connect it (step 2, below, for cable; go to 14-11 for a hydraulically activated—RockShox Reverb—seatpost), and then clamp the post into the frame.
2. **Tighten the cable.** The cable tension has to be spot-on for a dropper post to work properly. If the cable is too loose, you will run through the entire throw of the lever without releasing the post sufficiently that it snaps back up from the low position. If the cable is too tight, the saddle will go down as you pedal. There will generally be a barrel adjuster at the thumb lever (Fig. 14.15). After you've checked that the cable housing is the right length and have secured the cable at the cable-fixing bolt at the seatpost, back out the barrel adjuster (counterclockwise viewed from where the cable housing enters it) to tighten the cable, and screw it in to loosen the cable. If you run out of adjustment, loosen the cable-fixing bolt at the seatpost, pull the cable taut, and cinch down the cable-fixing bolt again. Be careful, because the screw is often tiny and can easily strip the threads it's screwed into. Skip to step 5 with a cable-activated seatpost.
3. **Bleed the Reverb.** With a RockShox Reverb hydraulically activated seatpost, the system has to be free of air bubbles or you will push the button all of the way in without releasing the post sufficiently to allow it to snap back up from the low position. Go to 14-11.
4. **Activate the Vyron.** With a Magura Vyron electronic seatpost, if you push the button and nothing happens, the battery probably needs charging.

There is a micro-USB charging port on the head of the seatpost; peel back a cover to get at it. If charging doesn't solve the problem, you may need to pair the post and the button again. Note that there is a small delay between button activation and release of the seatpost, so make sure you stay seated on the saddle long enough to push it down after pushing the button.

5. **Set the remote to your desired position**. Clamp it onto the handlebar where you can easily reach it with your thumb.
6. **Secure the cable.** See that the cable housing curves smoothly from the handlebar. With external routing, zip-tie it or hold it in guides on the frame. Ensure that the housing runs smoothly into the frame and isn't damaged on the edge of the hole; insert the rubber grommet in the frame hole, if supplied by the bike manufacturer.

Most dropper posts have an air spring inside with a Schrader valve at the bottom or the top. If the system loses pressure over time, reinflate it to manufacturer spec with a shock pump with a no-leak chuck, and plan on servicing it and replacing the air seals if it gets worse.

14-10

BLEED ROCKSHOX REVERB DROPPER POST

If the Reverb is not properly bled, the dropper mechanism will not release sufficiently when you push the button or lever. The bleed kit is identical to the SRAM/Avid brake-bleed kit (9-6b, Figs. 9.21–9.23), although the 2018 Reverb 1x remote lever requires the new Bleeding Edge tip on one syringe. The procedure for replacing and connecting the hose and guiding it through the frame is also identical to that of hydraulic brakes (9-4e). However, the Reverb takes 2.5-weight oil, not DOT fluid—don't contaminate the bleed kit with the wrong fluid because it will cause the seatpost seals to fail.

Note that when removing the remote for pushing the hose through the frame or shortening it, you can simply twist the remote unit counterclockwise to unscrew it from the hose. Cleanly cut the hose end prior to reinstallation. With the older pushbutton remote, you will probably need to tighten the collar nut with a wrench to get a good seal. With the 1x remote lever, you can screw the entire remote lever body onto the hose while pushing on it; it will seal and save you a lot of time. If you keep the

14.16 Zip-tying a RockShox Reverb Stealth hydraulic dropper post (internally routed hose) to the rear wheel

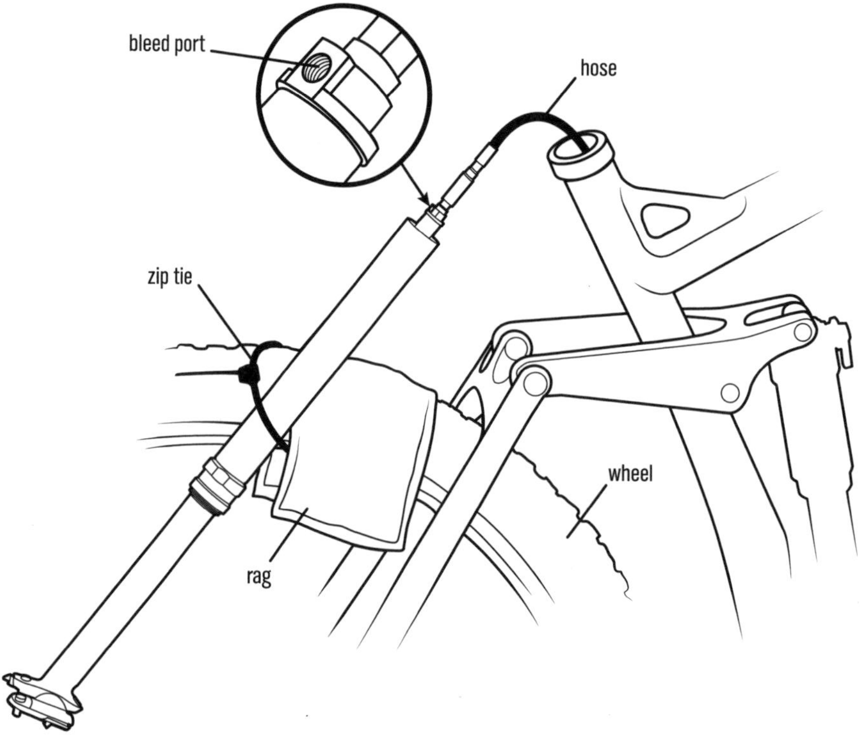

tip of the hose up while the remote is detached and only cut the end after you have guided it through the frame, bleeding will probably be unnecessary.

1. **Extend the seatpost to full height.** Push the button on the handlebar remote unit to get it to pop up.
2. **Secure the bike.** With a stealth (internally routed hose) seatpost, stand the bike up in a bike rack or display stand, because you will have to remove the seatpost and hence will not have a post to clamp with a bike stand. For an externally routed post you can also use a bike rack or wheel stand; if you are instead going to clamp it in a bike stand, loosen the seat binder and slide the seatpost up in the frame to the limit line. Gently tighten the seat binder just enough to keep the seatpost from sliding up. Clamp the lower, large portion of the seatpost in the bike stand below the collar nut (don't clamp it on the hose!). Skip to step 5 for an externally routed post.
3. **Pull out the stealth seatpost.** Remove the remote unit from the handlebar and push hose slack into the frame to avoid damaging it as you raise the seatpost. Loosen the seat binder and pull the seatpost out, pushing hose in through the frame as you pull it up.
4. **Secure the stealth seatpost to the rear wheel.** With a zip-tie, attach the seatpost to the top of the wheel (Fig. 14.16), bleed screw and hose up and saddle down.
5. **Rotate the handlebar remote unit so that its bleed screw is pointed up.** With a stealth seatpost, pull back some hose now that the seatpost is secured to the rear wheel, and clamp the remote to the handlebar, bleed screw up. With an externally routed post, just loosen the remote's clamp, rotate it on the bar, and tighten the bolt.
6. **Set the speed adjuster on the slowest position.** On the remote, turn the speed adjuster—a barrel adjuster on the pushbutton remote (Fig. 14.17) and a T25 screw under the rubber cover on the 1x remote lever—fully counterclockwise (opposite the etched arrow).
7. **Fill one syringe three-quarters full with RockShox Reverb hydraulic fluid.** Point it up and push out any air bubbles. Leave the other syringe empty; with the 1x remote lever, this will be the syringe with the Bleeding Edge tip.
8. **Remove the bleed screws on the seatpost and remote.** The bleed screw on the standard Reverb is under the saddle clamp; on the Stealth it is on the hose connection at the bottom. Use a T10 Torx key except on the 1x remote lever; on that unit, loosen and lightly retighten the bleed screw (yes, leave it in!) with a 3mm hex key.
9. **Plug the empty syringe into the remote bleed port.** With the older pushbutton remote, screw the syringe bleed tip into the bleed hole. With the 1x remote lever, engage the hex flats on the Bleeding Edge syringe tip into the 3mm hex hole in the bleed screw, snap it in so the O-ring seals in the bolt head, and then rotate the Bleeding Edge tip (by means of its wings) ½ to 1 turn counterclockwise to open the bleed port.
10. **Screw the three-quarters-full syringe into the seatpost bleed port.**
11. **Force fluid from the seatpost syringe to the remote syringe.** With both syringes pointed down so that any air stays up against the plungers, gently

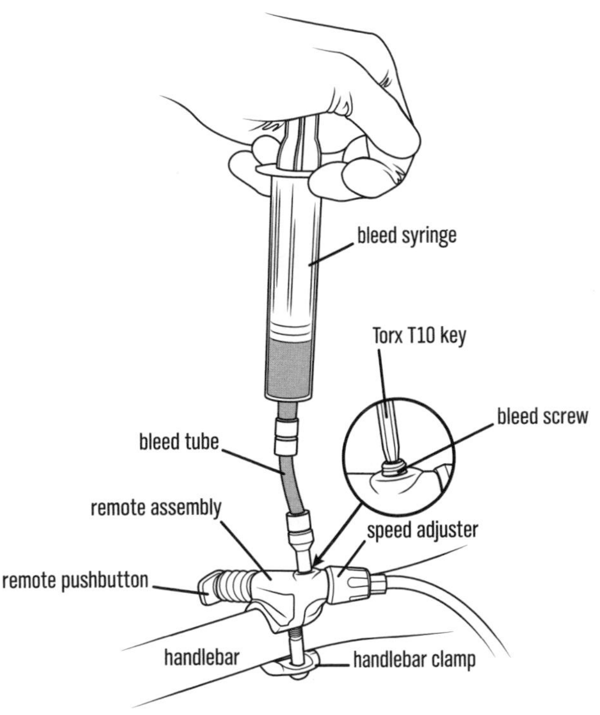

14.17 Bleeding RockShox Reverb hydraulic dropper-post system at the remote

14.18A Bleeding RockShox Reverb hydraulic system at the base connection

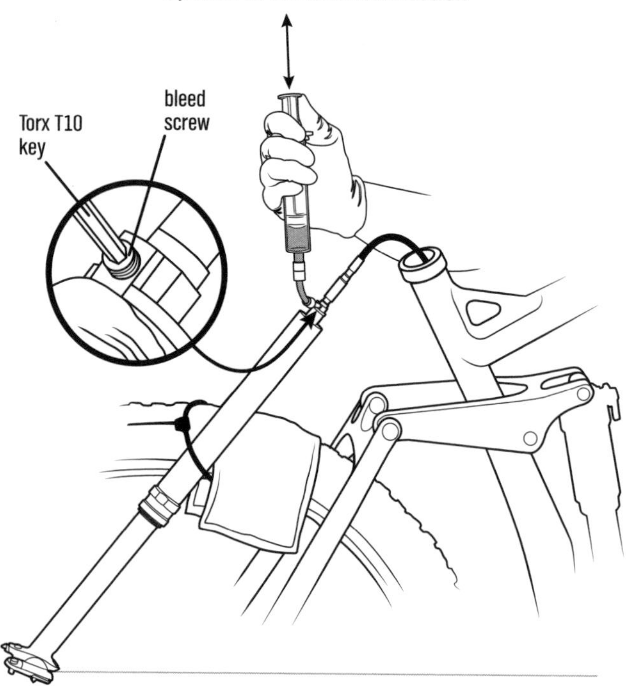

14.18B Bleeding RockShox Reverb hydraulic system at the seatpost head

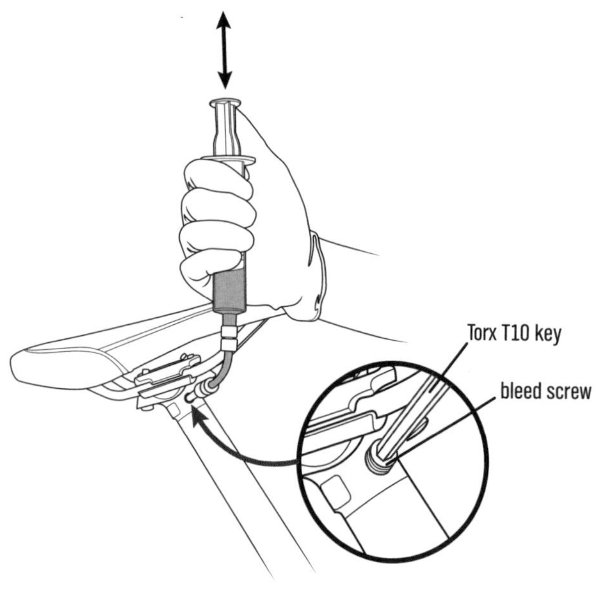

push the plunger on the seatpost syringe (Figs. 14.18A and 14.18B) while gently pulling up on the remote syringe plunger (Fig. 14.17) until both syringes are half full. Depress the remote lever, and release it slowly while pushing on the seatpost syringe plunger; keep pushing until almost no fluid remains in the seatpost syringe.

12. **Unscrew the seatpost syringe.**
13. **Replace the seatpost bleed screw.** Tighten with a Torx T10 (1.7 N-m torque).
14. **Install the seatpost to the correct height.** Cut the zip-tie on the Reverb Stealth. If the seatpost is a tight fit in the frame (that is, if it takes some twisting to get it in), grease the seatpost; if it slides easily up and down, put friction paste on it. Loosen the seat binder and slide the seatpost into the frame (with a Stealth post, gently tug the remote end of the hose out of the hole in the frame as you push the seatpost down). Tighten the seat binder, but not above 6.7 N-m (60 in-lbs) of torque, especially with the 30.9mm seatpost size (its outer tube is thinner than the 31.6mm model); the binder can squeeze the seatpost so tightly that the dropper mechanism will not slide up and down freely.
15. **Set the remote to your desired position.** With a Torx T25 key, loosen the clamp bolt on a separate clamp or on the Matchmaker adapter on the brake lever, rotate and slide the remote to where you want it, and snug up the clamp bolt.
16. **Bleed the remote.** With the button type, press the remote button eight times while pulling up on the syringe plunger and each time pushing the remote back out by pushing on the plunger (Fig. 14.17). Turn the speed adjuster fully in each direction four times. Pull up on the syringe plunger to pull air bubbles out of the remote assembly. Repeat until no more air bubbles come out into the syringe bleed tube. Leave the speed adjuster on the lowest setting. With the 1x lever remote, create a light vacuum by pulling up gently on the plunger, push the 1x lever, and release the plunger. Gently push on the plunger while slowly releasing the 1x lever. Repeat 6–10 times until bubbles largely no longer appear.
17. **Firmly push the syringe plunger.** This will push the remote button or lever out to its fully extended position. With the pushbutton remote, skip to step 19.
18. **While still pushing on the syringe plunger, close the bleed screw.** This only applies to the 1x lever; rotate the wings on the Bleeding Edge syringe tip clockwise fully to close the bleed valve. Create a light vacuum by pulling up gently on the syringe plunger.

19. **Remove the remote syringe.** With the pushbutton remote, unscrew the syringe tip. With the 1x lever, pull the Bleeding Edge syringe tip out of the bleed bolt and skip to step 21.
20. **Replace the bleed screw on the remote housing.** This only applies to the pushbutton remote; tighten the bleed screw with a Torx T10 (1.7 N-m).
21. **Test the system.** Check the thoroughness of the bleed by pushing the remote button or lever a few times. If the button or lever does not come all of the way back out, redo the bleed.
22. **Set the speed adjuster to your preference.** The seatpost will snap up faster as the adjuster is turned in the direction of the etched arrow.
23. **Clean up.** Wipe the hose, remote, and seatpost with a rag and rubbing alcohol. Close the rubber cover on the 1x remote lever body.
24. **Secure the hose.** The hose should curve smoothly from the handlebar. With external routing, zip-tie it to the frame or push it into the frame guides if present. With internal routing, check that it is running smoothly into the frame and not being damaged on the edge of the hole; insert a rubber grommet in the frame hole, if supplied by the bike manufacturer.

14-11

OVERHAUL DROPPER SEATPOST

I recommend following the same regular maintenance procedures you would use for a standard seatpost (14-3), with the addition of wiping and lightly lubricating the extending inner seatpost shaft after every ride or two.

The original dropper post, the GravityDropper, has a rubber boot over it and is easily given a quick lube. Here's how:

1. **Lift the lower portion of the rubber boot.** The post must be in the up position.
2. **Remove the top cap.** An open-end or adjustable wrench will do the trick.
3. **Push the handlebar switch or pull knob and hold it.**
4. **Gently pull out the inner post.** Be careful and wear safety glasses in case it pops out.
5. **Thoroughly clean and lubricate.** Use a light lithium grease.
6. **Push the handlebar switch or pull knob and hold it.**
7. **Reinstall the inner post and tighten the top cap.**

It's a good idea to overhaul dropper posts every 50–100 hours of riding during which you're using the dropper feature frequently. Ideally, have a replacement service kit at the ready with O-rings, bushings, guide blocks, and other replacement parts for your post. Most air-sprung posts are similar to the Maverick SpeedBall/Crank Brothers Joplin (Fig. 14.19): cylindrical with O-rings, bushings, and guide blocks (or keys) running in grooves

14.19 Maverick SpeedBall/Crank Brothers Joplin height-adjustable seatpost overhaul

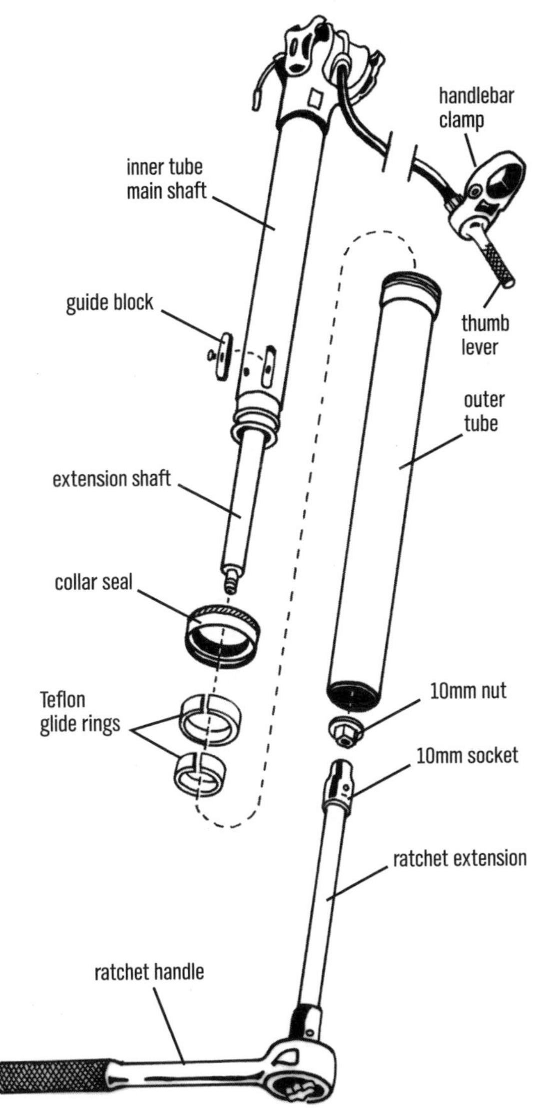

to keep it from rotating. Other posts may require different tools, like snapring pliers and a large wrench for a collar nut, and they may require you to deflate the seatpost. You don't want seatpost shafts flying around like missiles; wear safety glasses. On stealth (internal cable or hose routing) posts, the Schrader valve is on the top, so you need to remove the saddle to get at it.

Maverick SpeedBall/Crank Brothers Joplin service

1. **Remove the seatpost from the bike.** There's no need to remove the saddle.
2. **Remove the threaded collar nut.** Unscrew the nut by hand (you may need to carefully break it free with a pipe wrench or pliers), and slide it up the inner tube.
3. **Remove the nut from the bottom of the inner tube** (Fig. 14.19). Use a 10mm socket on an extension inside the end of the outer tube to reach the nut.
4. **Slide the inner tube out of the outer tube.**
5. **Clean the components thoroughly.** Use a light solvent/degreaser and/or a clean rag. Push the clean rag down into the outer tube with a dowel. Remove the Teflon glide rings (plastic bushings) and replace or clean them.
6. **Replace any worn parts.** If the seatpost wiggles back and forth, it needs a new guide block; unscrew the old one with a Torx T10 key. After removing the guide block, slide off the collar seal, and clean or replace the thick rubber seal inside, lubricate it, and slide it back up on the inner tube. Slide on the new or cleaned bushings. Screw in a new or cleaned guide block; put some blue Loctite on the threads.
7. **Apply a slippery, petroleum-based grease.** (Slick Honey works well.) Smear it all around inside the lip of the outer tube and on the inner tube, guide block, and Teflon glide rings (Fig. 14.19).
8. **Insert the inner tube back into the outer tube.** Line up the guide block with the groove in the outer tube.
9. **Thread the 10mm nut onto the compression shaft.** Tighten it to 4 N-m (35 in-lbs) of torque.
10. **Put the seatpost in the lowered position.**
11. **Screw on the collar seal.**

Worn guide blocks (keys) and bushings need to be replaced if the seatpost develops excessive rotational or fore-and-aft play.

Watch out for the post's main spring. The GravityDropper has a coil spring, while most dropper posts have an air spring; either type of spring can be dangerous if you are not careful, so get out of the way when working on it. You can fully deflate an air spring through its Schrader valve, but to get all of the air out, you must cycle the spring up and down using the release lever while the post is upside down. Once the block that guides the post has been removed, don't touch the release lever in case there is still air inside. On the GravityDropper, unscrew the bottom cap slowly, and keep your face away from the end of the tube where the spring will shoot out.

IMPORTANT: *See the "Important" note under 14-8.*

14-12

ADJUSTING SUSPENSION SEATPOSTS

Shock-absorbing seatposts come equipped with some sort of spring—either a steel coil, an elastic polymer ("elastomer"), or compressed air. The telescoping, elastomer-spring type (Fig. 14.20) is probably the most common. Some seatposts have linkages that swing the saddle on an arc, rather than up and down.

To adjust the "boing" in most suspension seatposts, you'll have to change the amount of preload on the spring, or replace the spring(s), or change the air pressure inside. With telescoping seatposts (Fig. 14.20), and even with some parallelogram-linkage posts, you must first pull the seatpost out of the frame.

If you look up inside a telescoping seatpost from the bottom, you will usually see a large slotted screw threaded into the walls of the post. If you tighten the screw clockwise, you will increase the preload on the spring and thereby stiffen the seatpost. If you loosen the screw (counterclockwise), you will reduce the preload and soften the ride. If it's an air shock, you'll instead find a Schrader valve.

To change springs, remove the screw and then the spring completely, make the switch (remember to grease the coils or elastomers!), and replace the screw.

14.20 Suspension seatpost

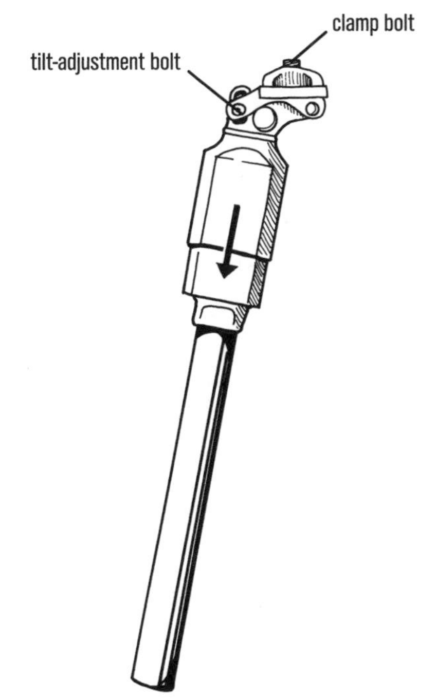

You will find that as you change preload or elastomer combinations, the height of your saddle changes, so expect to slide the seatpost up and down in the frame to adjust for this.

Some parallelogram-linkage posts can be adjusted by turning a preload screw behind the saddle clamp and/or pushing the elastomer out from the side and replacing it (or interchanging small elastomer plugs into a larger elastomer).

There are other suspension-seatpost designs available as well, and most shock-absorbing seatposts are tuned as described here.

I recommend following the same regular maintenance procedures you would use for a standard seatpost (14-3), in addition to maintenance of the suspension components.

IMPORTANT: *See the "Important" note under 14-8.*

14-13

REMOVING A STUCK SEATPOST

LEVEL 3

The seatpost can get stuck when you do not follow the advice in the "Important" note in 14-8. This is a level 3 task because of the risk involved. This may be a job best done by a shop, because if you make a mistake, you run the risk of destroying the frame. If you're not 100 percent confident in your abilities, go to someone who is—or at least to someone who will be responsible if he or she screws it up.

1. **Remove the seat binder bolt.** Slide the binder collar up the seatpost. Easy enough.
2. **Squirt ammonia around the seatpost, and let it sit overnight.** Ammonia dissolves aluminum oxide. Automotive antifreeze or Coca-Cola (yes, the soft drink) will often do so too; penetrating oil will work only with a steel seatpost in a steel frame. To get the most penetration, remove the bottom bracket (Chapter 11) and turn the bike upside down; pour ammonia, antifreeze, or Coke in from the bottom of the seat tube, and let the bike sit upside down overnight. Drain it the next morning and continue to the next step.
3. **Stand over the bike and twist the saddle.**
4. **If step 3 does not free the seatpost, use thermal expansion/contraction.** The idea is to get either the seatpost to shrink and pull away from the seat tube, or get the seat tube to expand and pull away from the seatpost, or do both. But you must be aware of the relative Coefficient of Thermal Expansion (CTE) of the materials the seatpost and seat tube are made of. The CTE of aluminum (and magnesium) is about double that of steel, and the CTE of carbon is almost zero.
   - (a) If you have a carbon-fiber seatpost stuck in a frame made out of any of these materials, warm up the seat-lug area with a hair dryer (or even a heat gun, if you're careful) to expand it.
   - (b) If you have an aluminum seatpost stuck inside a carbon or steel frame, cool the seatpost down. If there is a big enough hole from the bottom-bracket shell into the seat tube after you've removed the crankset, drop in small hunks of dry ice, and let the post get really cold. From the outside, discharge the entire cartridge of a CO_2 tire inflator at the joint of the seatpost and the seat collar to freeze the seatpost and shrink it. (Alternatively, cool the exposed seatpost with a plastic bag filled with dry ice.)

(c) If both parts are made of the same metal, you can still try to cool the post rapidly while heat ing the seat lug in hopes they'll shrink and expand in opposite directions.

(d) Now try twisting as in step 3.

5. **If step 4 does not free the seatpost, you will need to move into the difficult and risky part of this procedure:**

(a) Clamp the top of the seatpost into a large bench vise that is bolted to a very secure work-bench. Remove the saddle and all the clamps from the top of the seatpost first, and turn the bike upside down above the vise. You have just ruined your seatpost. Don't ever ride it again.

(b) Perform the thermal expansion/contraction trick from step 4.

(c) Grab the frame at both ends, and begin to carefully apply a twisting pressure. Be aware that you can easily apply enough force to bend or crack the frame, so be careful. If the seatpost finally releases, it often makes such a large pop that you will think you have broken many things!

6. **If that did not work, take the bike to a car repair shop.** Ask a mechanic to smack the underside of the seatpost clamp with an air impact hammer. If this maneuver works, it will take seconds and won't damage the frame. Be forewarned, however, that the action is loud and violent.

7. **If step 6 fails, cut off the seatpost a few inches above the seat lug.** There are a number of things you can try now:

(a) Warm up the seat-lug area with a hair dryer to expand it. Discharge the entire cartridge of a CO_2 tire inflator down inside the seatpost to freeze and shrink it. Now clamp what's left of the seatpost in a vise, and try twisting as in step 5.

(b) Get your hands on a slide hammer. Borrow one or or rent one from an auto body–related business. This is a tool for pulling dents out of car bodies and consists of a long rod with a heavy cylindrical weight (5 pounds is a good size for this) that slides along it. The end of the rod can be attached to what you are pulling by a number of means, and when you rapidly slide the weight toward the handle, it pulls the whole rod forcibly in the direction of the handle when it hits the handle. To attach the slide hammer rod end to the seatpost, either clamp Vise Grip pliers to the post or drill a transverse hole through it. Assemble the slide onto the slide hammer rod, and attach its end to either the pliers with clamping jaws or into the holes via hooks that point inward toward each other. Have someone hold the frame (the inertia of the frame makes it easy to hold even when you are in the act of freeing the post), and slam the weight toward the end away from the frame. It may pop right out.

(c) Drill a ½-inch (13mm) hole transversely through the seatpost, insert a long steel rod through it, have somebody hold the frame, and twist on it as hard as you can. The post will make a huge noise if it comes free.

8. **If the seatpost is still not free, try one of two (completely undesirable) alternatives:**

(a) Go to a machine shop and get the remaining seatpost reamed out of the seat tube. The chances of the shop managing to line everything up perfectly so that the cutter does not make a hole in the seat tube are not good.

(b) Cut the remaining seatpost out by hand. Sit down and think for a while before you take on this job, and then proceed carefully, because there is a risk of completely destroying the frame. Here's how to do it:

(i) Cut off the seatpost a little more than an inch above the seat lug on the frame. Use a hacksaw.

(ii) Remove the blade from the saw, and wrap a piece of tape around one end of the blade. You're making a handle with which to grab the blade.

(iii) Hold on to the taped end, and slip the other end into the center of the post.

(iv) Carefully (very carefully) make two outward radial cuts about 60 degrees apart.

Your goal is to remove a wedge from the hunk of seatpost stuck in the frame. Be careful—this is where many people cut too far and go right through the seatpost and into the frame. Of course you wouldn't do that, would you?

NOTE: *A much faster but also more dangerous method is to use a reciprocating handheld jigsaw with a long blade for wood. It will go through an aluminum seatpost very quickly. The question is, what will it do to your frame? On a steel or titanium frame, the wood blade may only polish the inside of the seat tube where it hits the tube. The blade will probably go through a carbon, aluminum, or magnesium frame just as easily as it goes through an aluminum seatpost, though.*

(v) Work the remaining piece out. Once you've made the cut, pry or pull this piece out with a large screwdriver or a pair of pliers. Be careful here, too. A lot of overenthusiastic home mechanics have damaged their frames by prying too hard. Curl in the edges with the pliers to free more and more of the seatpost from the seatpost walls. The seatpost should eventually work its way out.

9. **If all else fails, dissolve the seatpost.** If you happen to be someone with access to such things, a metal such as gallium that is liquid at near room temperature will dissolve aluminum (check that the gallium owner also has a disposal permit and materials for the residue). But if the frame is aluminum, then you're stuck. And whatever the frame is made out of, you'd better get a chunk of that material and put some of the liquid metal on it first to see what happens before you go near the frame with it.

With the post out of the frame, clean the inside of the seat tube thoroughly. A Flex-Hone, sold in auto parts stores (or loaned at rental stores) for reconditioning brake cylinders, is an excellent tool for the purpose. Put the hone in an electric drill, and be sure to use plenty of honing fluid or cutting oil as you work it up and down inside the seat tube as it spins. An alternative is to use sandpaper wrapped around your fingers, although you will not be able to reach very far into the frame.

Inasmuch as removing a stuck post is so miserable that no one wants to do it twice, I'm certain that I do not need to remind you to grease any metal seatpost (14-3) thoroughly before inserting it in the frame, and check it regularly thereafter as previously outlined. Carbon seatposts have a soft, clear-coated exterior and can mechanically lock into the frame or be held in by corrosion of the seat tube or seat-tube sleeve, so I recommend carbon assembly spray or paste to reduce these dangers (see the Pro Tip earlier in this chapter).

TROUBLESHOOTING PROBLEMS IN THE SEAT AND SEATPOST

14-14

LOOSE SADDLE

Check the seatpost clamp bolts. They are probably loose. Reestablish your fore-and-aft saddle position (14-2) and your saddle tilt, and tighten the bolts. Check for any damage to the clamping mechanism, and replace the post if necessary. If you need help, look up the instructions that apply to your seatpost.

The saddle could also be clamped properly but be wiggling side to side on a dropper post with worn keys or guide blocks that are no longer keeping it from rotating. Overhaul the seatpost and replace the keys or guide blocks (14-11).

14-15

STUCK SEATPOST

A stuck seatpost can be a serious problem. Follow the instructions in 14-13 carefully. Otherwise, you might damage your frame.

14-16

SADDLE SQUEAK WITH EACH PEDAL STROKE

The problem comes from the smooth leather or plastic moving against metal parts or from grit in the rail attachments.

1. **Lubricate the contact points of the saddle with the clamp and rails.** On saddles that extend low

on the sides, contact of the leather overlapping the saddle shell with the seatpost clamp or rails is likely the culprit. Greasing or powdering the contact area with talcum will eliminate the noise. Also, roughing up the leather at the point where it contacts metal will quiet it down, because smooth leather sliding on metal can squeak.

2. **Squirt chain lube into the three points where the rails are inserted into the plastic shell of the saddle.** Grit working at the rails could be making the noise.
3. **Grease the rails where they sit in the channels in the seatpost clamp**. Loosen the clamp, clean the rails and channels, and apply a thin film of grease.

14-17

CREAKING NOISES FROM THE SEATPOST

A seatpost can creak from movement of the clamp that holds the saddle, or movement of the shaft against the sides of the seat tube while you ride. A dry seatpost can also cause creaks, so first try greasing it.

1. **On frames with an internal collar, shorten the seatpost.** Some frames have an internal collar to adapt the seat tube to a certain seatpost diameter. Remember that the internal diameter of the seat tube is larger below the collar. I have seen bikes that creaked because the bottom of the seatpost rubbed against the sides of the seat tube below the extension of the collar. You can solve that problem by shortening the seatpost a bit with a hacksaw. If you do saw off the seatpost, make sure that you still have at least 3 inches of seatpost inserted in the frame for security.
2. **Grease the inside of the seat tube, the seatpost sizing shims if present, and the seatpost.** Movement between these parts can cause creaking. You also might try friction paste on the post.
3. **If the creaking originates from the seatpost head where the saddle is clamped, check the clamp bolts.** Lubricate the bolt threads, and you will be able to tighten them a bit more.
4. **If shock-absorbing seatposts squeak as they move up and down, try greasing the sides of the inner shaft.** Grease the elastomers inside too.

14-18

SEATPOST SLIPPING DOWN

1. **Tighten the frame binder bolt.**
2. **If the seatpost still slips, try using some friction paste.** This is available specifically for preventing seatpost slippage and is often called carbon assembly paste or spray; carbon seatposts are particularly prone to slippage (see Pro Tip earlier in the chapter). Grease and beach sand, or valve-grinding compound, on the seatpost is a poor second alternative.
3. **Check the seatpost diameter.** If the seat-binder lug is pinched closed, and you still can't get the post to not slip down, even with assembly (friction) paste, you may be using a seatpost with an incorrect diameter, or the seat tube on the bike is oversized or has been stretched. Try putting a larger seatpost in the frame, and replace if you find one that fits better. If the next size up is too big, you may need to shim the existing post. Cut a 1 × 3–inch piece of aluminum from a pop can. Pull the seatpost out, grease it and the pop-can shim, and insert both back into your frame. Bend the top lip of the shim over to prevent it from disappearing inside the frame. You may need to experiment with various shim dimensions until you find a piece that will go in with the seatpost and also prevent slippage. Go ahead—pop cans are cheap.

A child of five could understand this.
Fetch me a child of five.
—GROUCHO MARX

15

WHEELBUILDING

TOOLS

Screwdriver
Spoke wrench
Truing stand
Wheel dishing tool
13mm, 14mm, 15mm cone wrenches
17mm open-end wrench (or an adjustable wrench)
Spoke-prep compound

OPTIONAL

Slotted, bladed-spoke twist-prevention tool
Three-way spoke socket wrench for nipples internal to the rim
Bent-shaft nipple screw-driver
Truing-stand through-axle adapter
Linseed oil
Deep-wall socket

Congratulations. You have arrived at a task often used to gauge the talents of a bike mechanic. Next to building a frame or a fork, building a good set of wheels is the most critical, and perhaps most creative, of a bike mechanic's tasks. Despite the air of mystery surrounding the art of wheelbuilding, the construction of a good set of bicycle wheels is really a pretty straightforward task. I have not labeled any of the procedures in this chapter as level 1, 2, or 3 tasks, because apart from a few special tools, the art of wheelbuilding isn't terribly difficult. It only requires concentration, patience, and a love of mechanical objects.

Clearly, wheels are the central component of a bike. For any bike to perform well, its wheels must be well made and properly tensioned. Once you learn how, it is quite rewarding to turn a pile of small parts into a set of strong and light wheels that you can bash around with confidence. You will be amazed at what they can withstand, and you will no longer go through life thinking that building wheels is something that just the experts do. With practice, you can build wheels at your house that are just as good as any custom-made set and superior to those built by machine.

And if you have special needs, like wheels for a heavy rider, yours will be better than factory wheels. All you have to do is follow these instructions.

This chapter is not meant to be an exhaustive description of how to build all types of wheel spoking patterns, but you will learn here how to build the five spoking patterns that are used in virtually all mountain bike wheels. (If you are interested in a more comprehensive treatment of the subject of wheelbuilding, I recommend reading *Barnett's Manual* by John Barnett, *The Art of Wheelbuilding* by Gerd Schraner, or *The Bicycle Wheel* by Jobst Brandt and watching *Master Wheel Building I* and *Master Wheel Building II* DVDs by Bill Mould.) In this chapter, you will learn how to build wheels for disc brakes as well as for rim brakes in the classic three-cross spoking pattern, in which each spoke crosses over three other spokes (Fig. 15.1). But three-cross isn't the only spoking pattern to be learned here; section 15-7 details how to build radially spoked wheels (for rim brakes only); 15-8 points out a few details about rim-brake wheels; and 15-9 sets out the lacing of two-cross and one-cross wheels (for very small wheels or for wheels with huge hub flanges). Finally, section 15-10 discusses heavier-duty wheels for big riders.

15.1 The complete wheel with three-cross spoke pattern

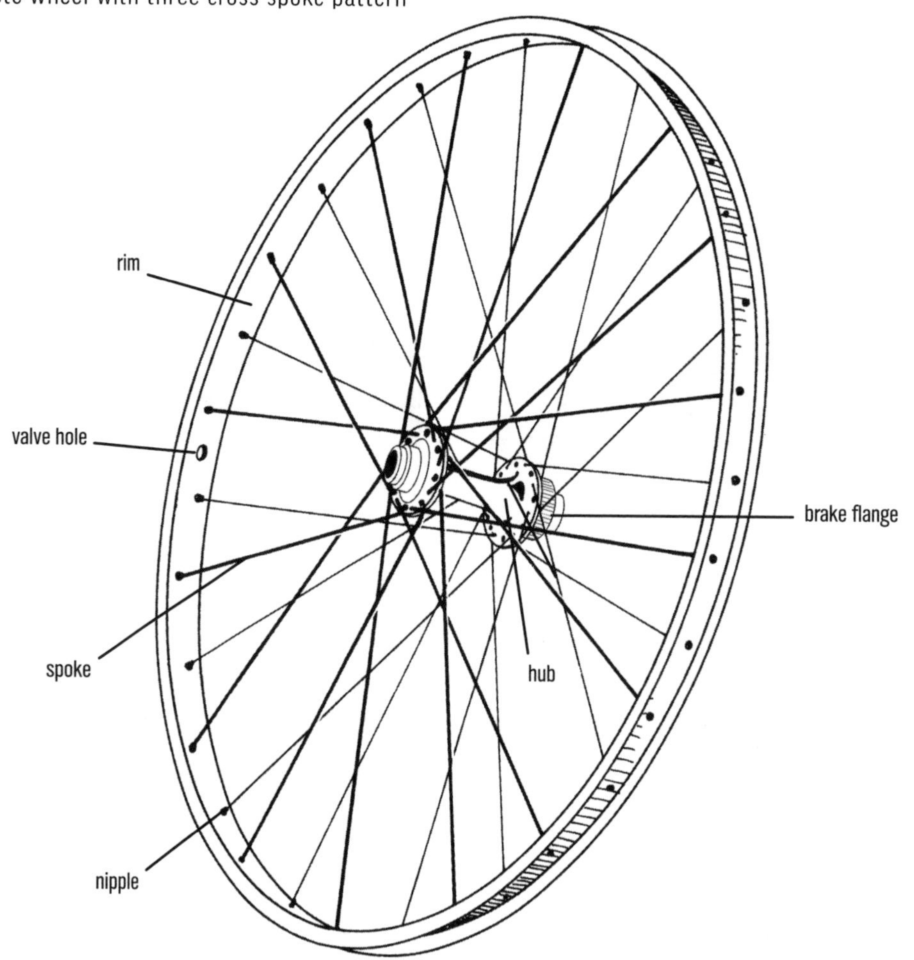

15.2 Spoke and nipple

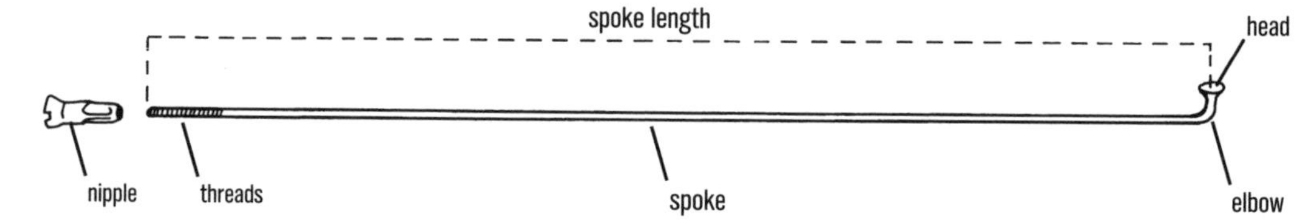

15-1

GETTING TOGETHER WHEELBUILDING PARTS

Here are the parts you need: a rim, a hub (make sure that the hub you are using has the same number of holes as the rim), properly sized spokes, and nipples to match. The illustrations and instructions assume you're using a disc-brake hub; if you're building a rim-brake wheel, all of the instructions are the same, provided you start with the orientation of the rear hub as explained in step 3 of the next section.

a. Spokes and nipples

I suggest getting the spokes from your local bike shop. That way, a mechanic can help make sure you are getting the right spoke lengths (Fig. 15.2) and can counsel you on what gauge (thickness) of spoke to buy, as well as what rim makes sense for your weight, budget, and the kind of riding you plan on doing. You can also use an online spoke calculator to figure out spoke length. Remember: When purchasing spokes or when using a spoke calculator, you must specify that you will be using a three-cross spoking pattern (unless you are building a radial wheel [15-7] or a one- or two-cross wheel [15-9]). Check that all of the spokes you buy are the specified lengths and thicknesses; spokes sometimes get mixed up in the boxes in a shop's inventory.

NOTE: *If you are building a wheel with a Rohloff or other internal-gear hub, or with a motorized hub for an e-bike, the hub flange may be too large to build a three-cross wheel, as the spokes would wrap across the other spokes on the hub flange, causing them undue stress. Also, to properly oppose the braking and driving forces, the spokes need to be tangential to the hub shell, and with a Rohloff or other hub with a huge flange, the spokes would be beyond tangential if laced three-cross on 26-inch, 27.5-inch, and 700C (29-inch) wheels. Rohloff says that 26-inch, 27.5-inch, and 700C (29-inch) wheels must always be laced in a two-cross pattern with its hubs, and 24-inch and smaller wheels must be laced only in a one-cross pattern. See 15-9 for instructions.*

For building disc-brake wheels, I recommend avoiding alloy nipples and superthin spokes unless the rider is small. Brass nipples and 14/15-gauge double-butted spokes (see 15-10a for more on this) will best take the extra braking loads.

NOTE ON SPOKE LENGTH: *Make sure that the spokes are long enough. This is particularly important with aluminum spoke nipples. If the spoke is not long enough to come up at least to a couple of turns into the bulbous part at the top of the nipple, the top of the nipple can snap off. This has the same effect as breaking a spoke, except you also end up with a piece rattling around in the rim. Using spokes that are too short does not pay. Of course, you don't want spokes that are too long, either; otherwise, the nipples could bottom out on the threads before the spoke is fully tensioned, or the spokes could poke up into or through the rim strip.*

NOTE ON USING OLD SPOKES: *If you are just replacing a rim on an old wheel, do not use the old spokes. You won't save much money reusing the old spokes, and the rounded-out nipples and weakened spokes will soon make you wish you had bought a new set.*

NOTE ON THIN SPOKES: *If you are using spokes that are 1.8mm (15-gauge) or thinner at the elbow (not recommended with disc brakes), there may be some play between the hub holes and the spokes. This play will work the spokes over time and bring on premature spoke breakage. DT Swiss sells spoke washers to go between each spoke head and the hub flange to take up this slack.*

15.3A Bent-handle nipple screwdriver

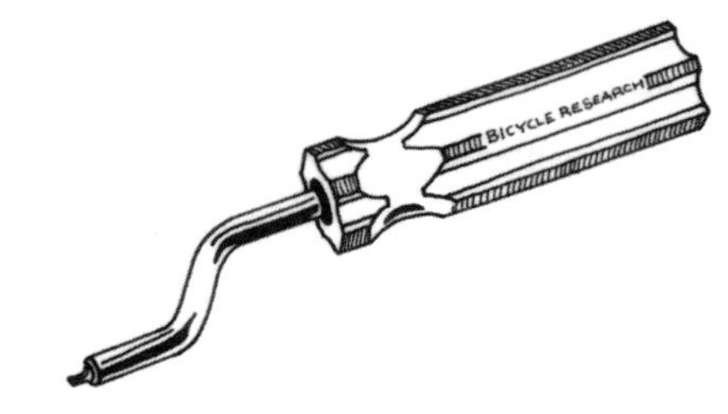

15.3B DT Swiss spoke wrench and antitwist tool for aero spokes

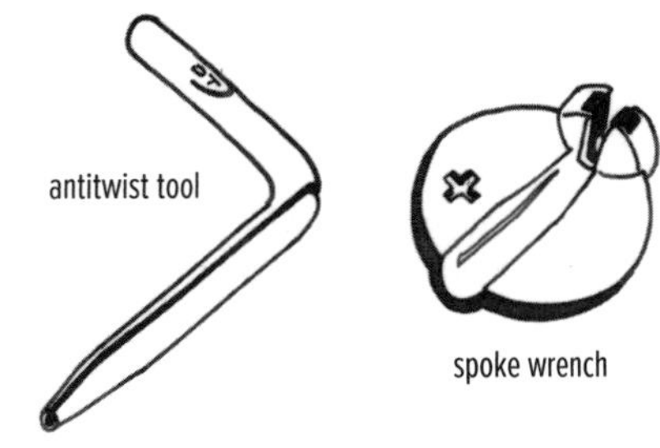

b. Tensioning tools

You only need a spoke wrench. If you are building lots of wheels, though, a bent-handle nipple screwdriver (Fig. 15.3A) speeds up tensioning.

c. Tools for aero (bladed) spokes

If you will be using bladed spokes, you will need a slotted tool to grab the flat of the spoke to keep it from twisting. DT Swiss makes a nice nesting pair of tools for this: a spoke wrench with a cone-shaped longitudinal groove that mates with the cone-shaped end of a slotted, L-shaped tool (Fig. 15.3B). With these tools, you can hold the spoke way down at the end of its flat section, close to the nipple, as you turn the spoke; with other slotted tools you will be grabbing the spoke above the spoke wrench, far from the nipple, and, with a thin spoke, you can easily leave a permanent kink—a twist in the middle of the flat section of the spoke.

d. Tools for deep rims

If you will be building onto a deep-section rim, you will need another special tool. Which tool that is will depend on whether the rim has standard-size spoke holes so the nipples extend out of the rim in the normal way, or the rim has tiny spoke holes so the nipples will be internal to the rim. In the former case, you need a way to hold the nipple as you extend it through the large hole in the rim bed to the hole in the inner wall without losing it in the open cavity of the rim. Special tools for this exist, but for building a few deep-section wheels, an extra spoke with another nipple threaded on it will work. Spin the nipple on far enough to leave about 6mm of thread exposed, and then fix it in place by crimping it onto the spoke. You'll thread this "tool" into the top of a nipple until it stops against the crimped nipple; reach through the rim with the tool and thread the nipple a few turns onto a spoke coming from the hub. This tool will also speed tensioning later, because the crimped-on nipple will only allow each nipple to thread onto its spoke a set number of turns before it stops. Having each nipple threaded on the same amount will eliminate tension disparities as you begin the tensioning process (you'll understand the importance of this when we reach that section).

If the deep rim has only small holes that will not let a nipple through, you will need a deep-wall socket with which to turn the internal nipples. Park makes a three-way square/5mm/5.5mm specialty spoke wrench for this purpose (Fig. 15.3C); the square drive is for using standard nipples upside down in the rim, and the 5mm and 5.5mm deep-wall sockets are for turning long hex nuts, including 3⁄16-inch nuts (equaling 4.7mm, which is close enough to 5mm to work).

15.3C Park SW-15 three-way spoke wrench for internal nipples

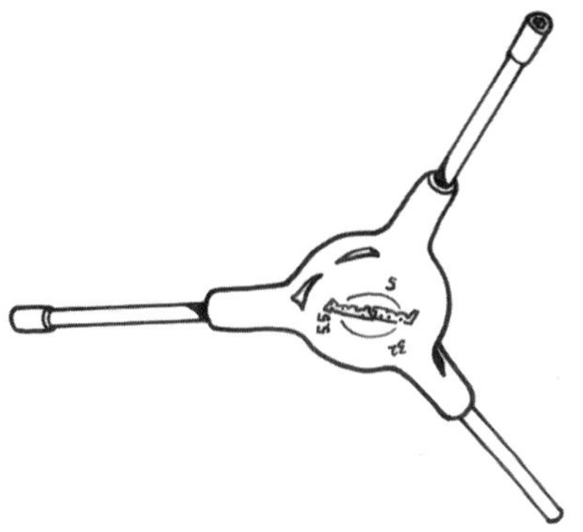

e. Spoke prep

For the sake of brevity and clarity, I do not mention using spoke prep compound with every instruction to thread a nipple onto a spoke. Although the use of thread compound is not mandatory, I think using it improves the wheel: Spoke prep encourages the nipples to thread on smoothly, it takes up some of the slop between the spoke and nipple threads, and its thread-locking ability discourages the nipples from vibrating loose. In place of a separate compound, you can use DT Pro Lock nipples, which contain a two-component adhesive in the nipple thread to prevent the spoke-nipple connection from loosening under the effect of operating loads (loading and unloading of the wheel during riding), thus ensuring constant spoke tension.

If you use spoke prep, apply it to the spoke threads before putting on the nipples. You do not want too much, as it will be hard to adjust the nipples months and years down the road; you just want the spoke prep in the valleys of the threads. You can get the right amount if you dip the threads of a pair of spokes into the prep compound, and then take two more dry spokes and roll the threads of all four spokes together with your fingers.

With DT Pro Lock nipples, you don't have to do any of this; just thread them onto the spoke. However, you will want to complete the wheel in one sitting. As you thread the DT nipples, you will be bursting little beads of the two glue components inside the nipple, like epoxy glue. If you finish the wheel while the glue is viscous, the nipples will hold better than if you let them harden and then turn them again in ensuing days. The nipples still offer benefits upon later wheel truing, since not all of the spheres will burst during initial wheel building, but the efficacy of the thread-locking will be reduced each time.

In the absence of spoke prep, at least dip the threads of each spoke in grease. Grease accomplishes everything spoke prep does, save for locking the threads.

A little linseed oil around the head of a nipple where it contacts the rim makes turning it easier as tension increases.

15-2

LACING THE WHEEL

This section shows how to build a three-cross wheel. See section 15-7 for building a radially spoked wheel, section 15-9 for two-cross and one-cross wheels, and section 15-10 for notes on wheels for big riders.

1. **Divide the spokes into four separate groups, two sets for each side of each hub flange.** Rubber-band each set together.

IMPORTANT: *If you are building a rear wheel or a front disc-brake wheel, you should be working with two different spoke lengths, because spokes on the drive side (or front brake-rotor side) are almost always shorter, due to different axle spacing and sometimes different hub flange diameters on the two sides. For a radial wheel, skip to 15-7.*

2. **Hold the rim on your lap with the valve hole away from you.** Some rims will have all of the spoke holes lined up along the centerline of the rim; with other rims, notice that the holes alternate being offset upward and downward from the rim centerline. I have written the lacing instructions for rims that have the holes offset; if your rim has all of its holes right along the centerline of the rim, ignore references to this up/down hole offset. With an off-center rim (Fig. 15.4), have the spoke holes offset toward the drive side with a front disc-brake wheel, and away from the drive side with a rear wheel.

NOTE: *If you are building a rear wheel with spoke holes drilled asymmetrically off center, make sure as you are lacing that you orient the rim so that the spoke holes are offset to the left (non-drive) side (see Fig. 15.4). If you are building a front disc-brake wheel with an off-center rim, make sure as you are lacing that you orient the rim so that the spoke holes are offset to the right (non-rotor) side. This type of rim is meant to reduce wheel dish by offsetting the nipples to reduce the otherwise very steep angle at which rear drive-side or front disc-side spokes normally hit the rim. The balanced left-to-right spoke tension is intended to increase the lifetime of the wheel, and the lower spoke angle moves the rear drive-side spokes away from the rear derailleur.*

15.4 Rear rim-brake wheel with asymmetrical rim laced correctly

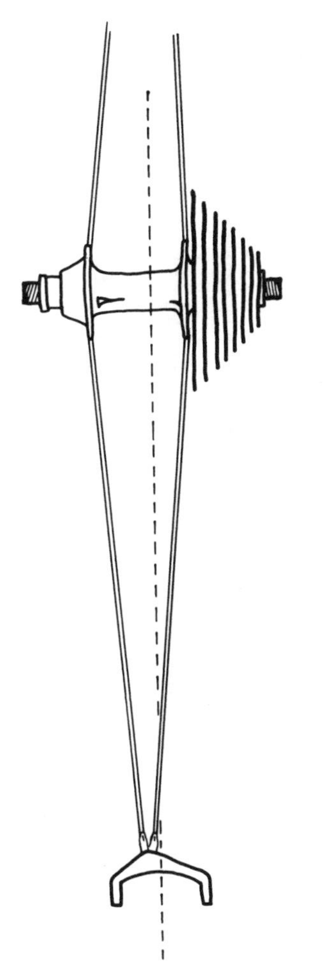

3. **Hold the hub in the center of the rim, with the non-drive (left) side of the hub pointing up.**

IMPORTANT: *Non-disc-brake front hubs are symmetrical; pick a side to be the left side. With this lacing pattern on a disc brake, the spokes on both sides that oppose the braking force on the rotor will come out of the outside of the hub flanges. The wheel will hence be stronger, because these pulling spokes come into the rim from a wider angle.*

a. First set of spokes

4. **Drop a spoke down into every other hole in the top (left-side) hub flange so that the spoke heads are facing up** (Fig. 15.5). If you're building a rear wheel, make sure to put the longer spokes on the left (non-drive) side. On a disc-brake front wheel, the shorter spokes will be on the left side. On some

15.5 First half of left-side spokes placed in hub

(older) hubs, half of the holes you are looking at will be countersunk deeper into the hub flange to provide a radius less stressful on the spoke elbow, so don't use those holes—use their neighbors. That said, most hubs today have the same countersinking on all holes so that they can also be used to build a completely symmetrical, radially spoked wheel.

5. **Put a spoke into the first hole counterclockwise from the valve hole, and screw the nipple on three turns** (Fig. 15.6). Notice that this hole is offset upward. On an asymmetrically drilled rim (Fig. 15.4), this means that the hole is offset upward from the centerline of the spoke holes, not the centerline of the rim. If the first hole that is counterclockwise from the valve hole isn't offset upward, you have a misdrilled rim, and you must return it or offset all instructions one hole.

NOTE: *With Mavic UST tubeless rims, first slide the threaded eyelet onto the spoke, then thread the nipple*

15.6 First spoke, left side up

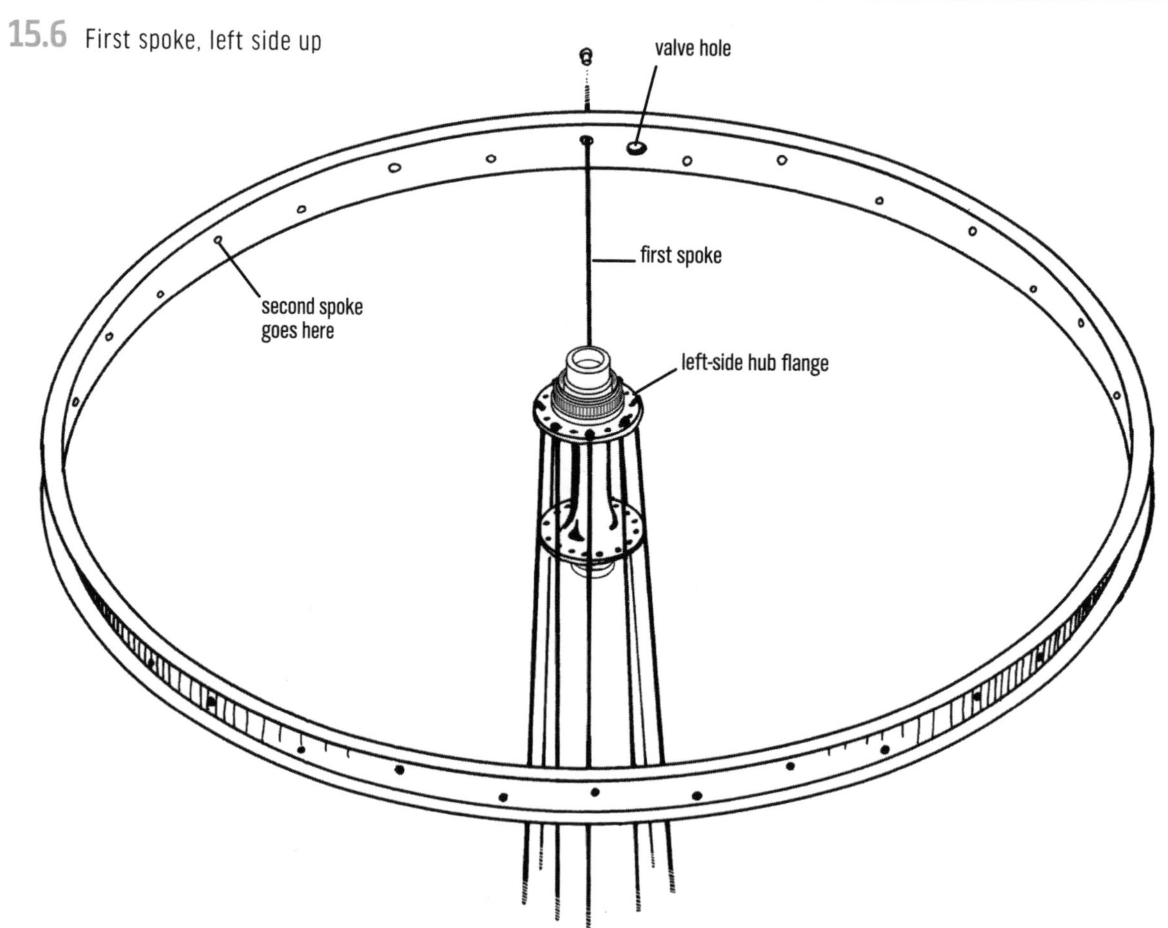

15.7 First set of spokes laced

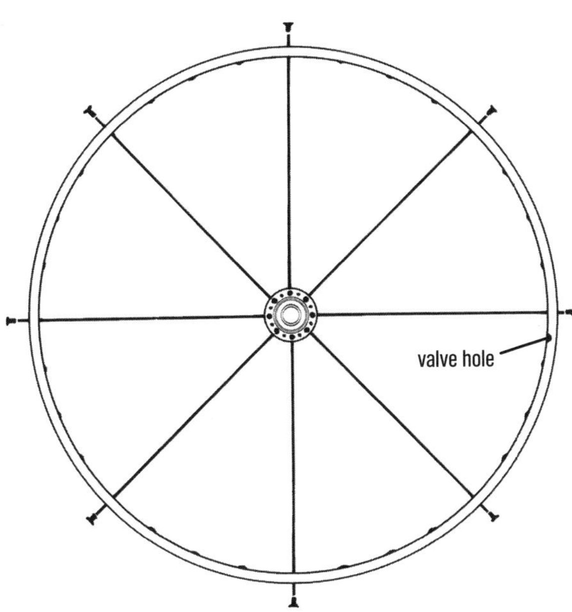

15.8 Spoke-hole offset

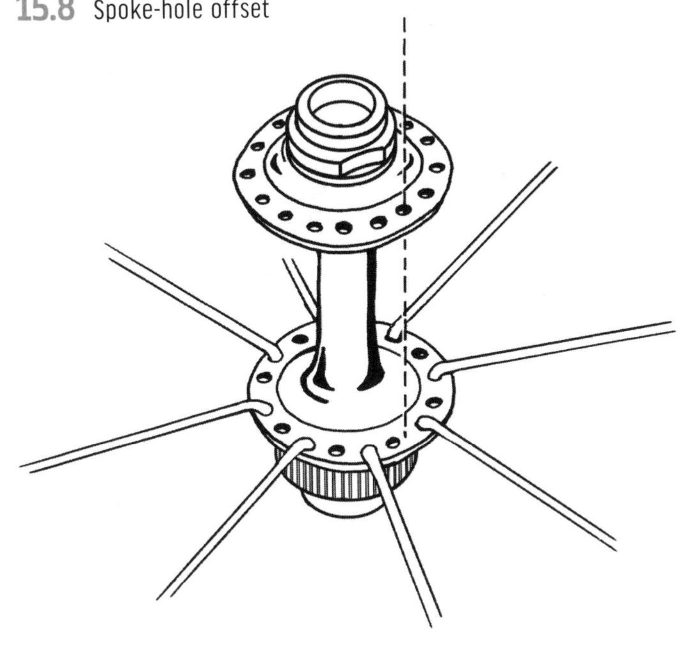

onto the spoke clockwise three turns, and then tighten the threaded eyelet into the rim with the special splined tool. Putting some spoke threadlock on the eyelet threads, as well as on the spoke threads, is a good idea.

6. **Working counterclockwise, put the next spoke on the hub into the hole in the rim four holes away from the first spoke, and thread the nipple onto the spoke three turns.** There should be three open rim holes between these spokes, and the hole you put the second spoke into should also be offset upward.
7. **Continue counterclockwise around the wheel in the same manner.** You should now have used half of the rim holes that are offset upward, and there should be three open holes between all spokes (Fig. 15.7). Check for symmetry. Your wheel should look like a pizza with all of the slices of the same size.
8. **Flip the wheel over.**

b. Second set of spokes

9. **Sight across the hub from one flange to the other flange.** Notice that the holes in one flange do not line up with the holes in the other flange; each hole lines up in between two holes on the opposite flange (Fig. 15.8).
10. **Drop a spoke down through the hole in the top flange that is immediately clockwise from the first spoke you installed (the spoke that is adjacent to the valve hole).** If this is a rear wheel, you are now using the shorter spokes. With a front disc-brake wheel, use the longer spokes. With a front rim-brake wheel, all of the spokes are the same length.
11. **Put this new spoke into the second hole clockwise from the valve hole, next to the first spoke you installed** (Figs. 15.9, 15.10). This hole will be offset upward from the centerline of the spoke holes.
12. **Thread the nipple clockwise onto the spoke three turns.**
13. **Double-check your work.** Make sure that the spoke you just installed starts at a hole in the hub's top (right-side) flange, which is half a hole space clockwise from the hole in the lower flange where the first spoke you installed started. When you look at the wheel from the side, these two spokes (your first spoke and the one you just installed) should not cross each other and should look like they are trending slightly away from each other (Fig. 15.10). In wheel-building parlance, these two spokes are called "diverging parallel" spokes.
14. **Drop a spoke down through the hole in the top (right-side) hub flange two holes away in either**

15.9 Lacing second set

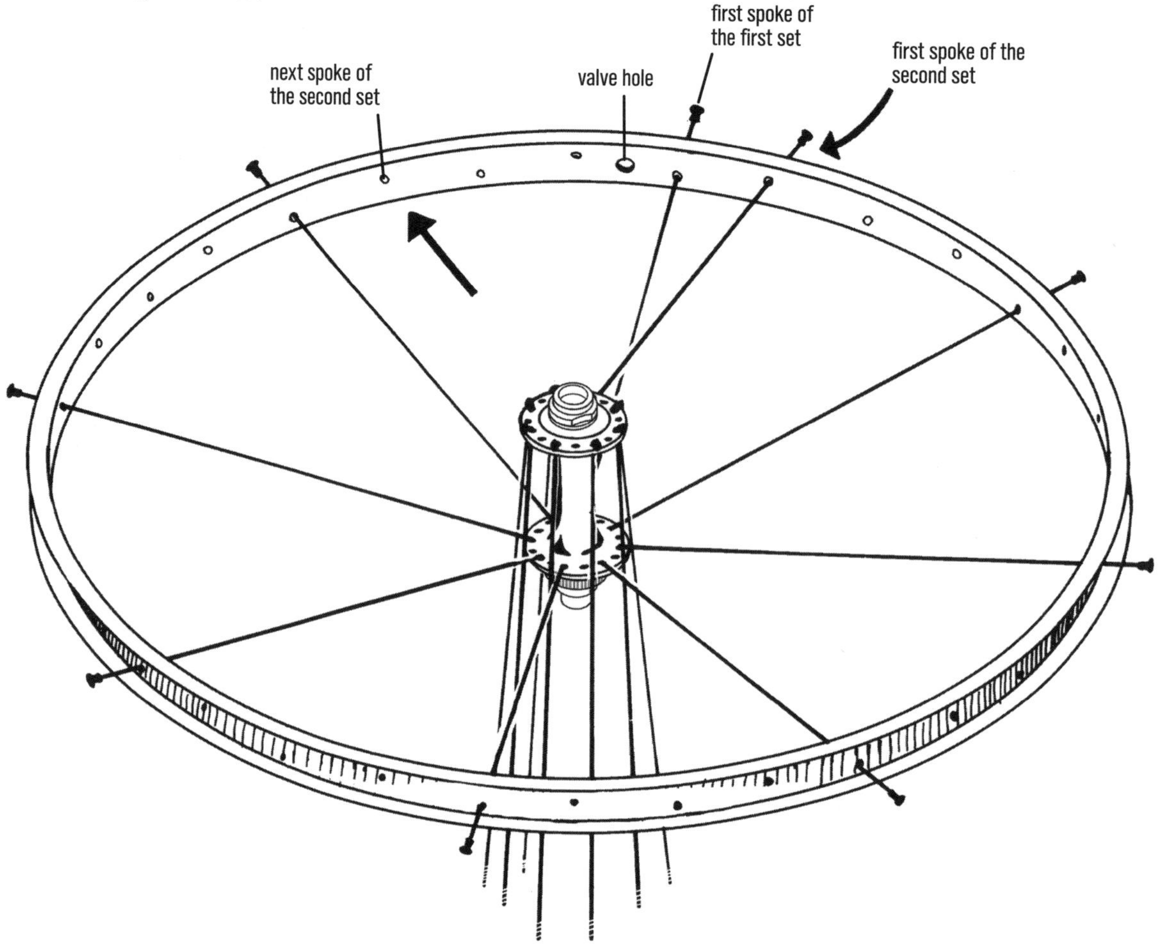

15.10 "Diverging parallel" spokes

15.11 Second set of spokes laced

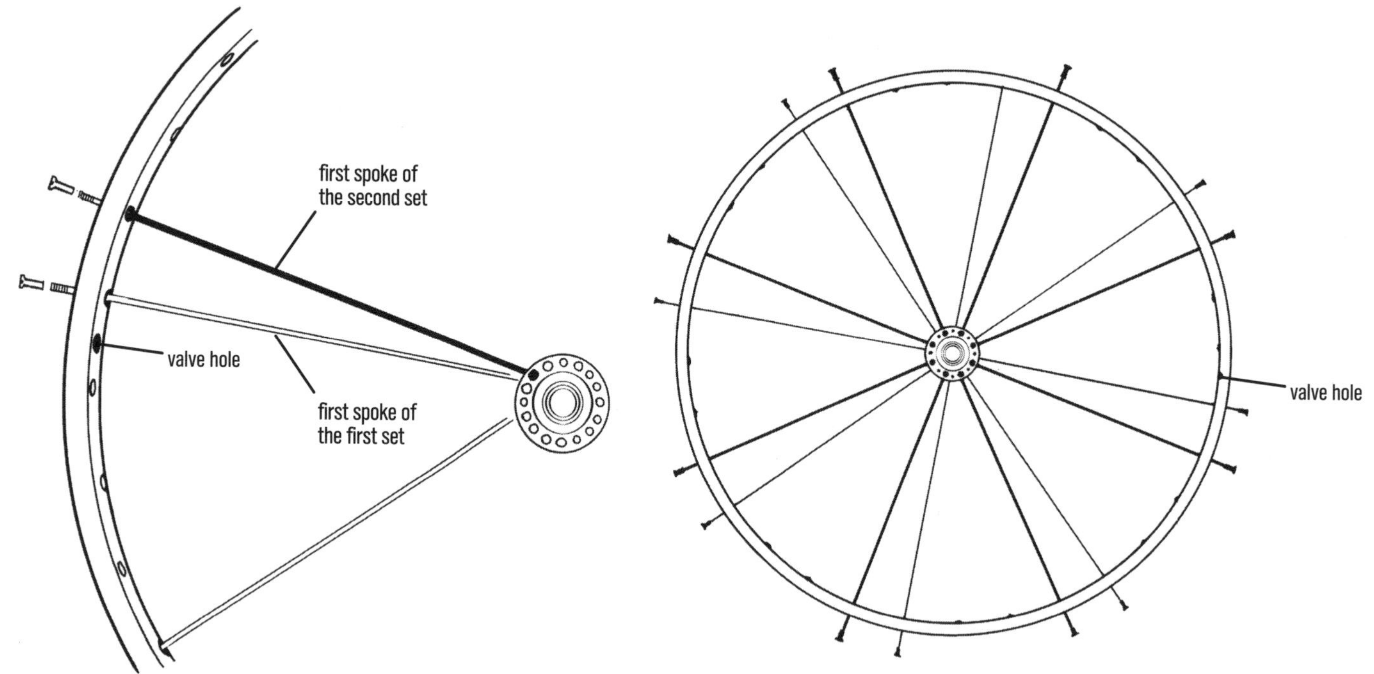

direction. Continue around until every other hole has a spoke hanging down through it (Fig. 15.9).

15. **Working counterclockwise, take the next spoke from the hub and put it in the rim hole that is three holes counterclockwise from the valve hole.** This hole should be offset upward and four holes counterclockwise from the spoke you just installed. Thread the nipple onto the spoke three turns.
16. **Follow this pattern counterclockwise around the wheel** (Fig. 15.11). You should have now used half of the rim holes that are offset upward, as well as half of the total rim holes. The second set of spokes should all be in upwardly offset holes, one hole clockwise from each spoke of the first set.

c. Third set of spokes

17. **Drop spokes through the remaining holes on the left side of the hub, from the inside of the hub out** (Fig. 15.12). Remember: If you are building a front disc-brake wheel, these should be the shorter spokes; if you are building a rear wheel, these will be the longer spokes.
18. **Flip the wheel over.** Grab the spokes you've just dropped through the hub to keep them from falling out.
19. **Fan the spokes out.** Now they cannot fall back down through the hub holes.
20. **Grab the hub shell and rotate it counterclockwise as far as you can** (Fig. 15.13).
21. **Pick any spoke on the top (non-drive-side) hub flange that is already laced to the rim.** Now find the spoke five hub holes away in a clockwise direction.
22. **Take this new spoke, cross it under the spoke you counted from (the one five holes away), and stick it into the rim hole two holes counterclockwise from that spoke** (Fig. 15.14). Thread the nipple onto the spoke three turns. Expect to flex the spokes some.
23. **Continue around the wheel, doing the same thing** (Fig. 15.15). You may find that some of the spokes don't quite reach far enough. If that's the case, at a point about an inch from the spoke elbow, push down on the spoke to help it reach.

15.12 Placing third set of spokes in hub

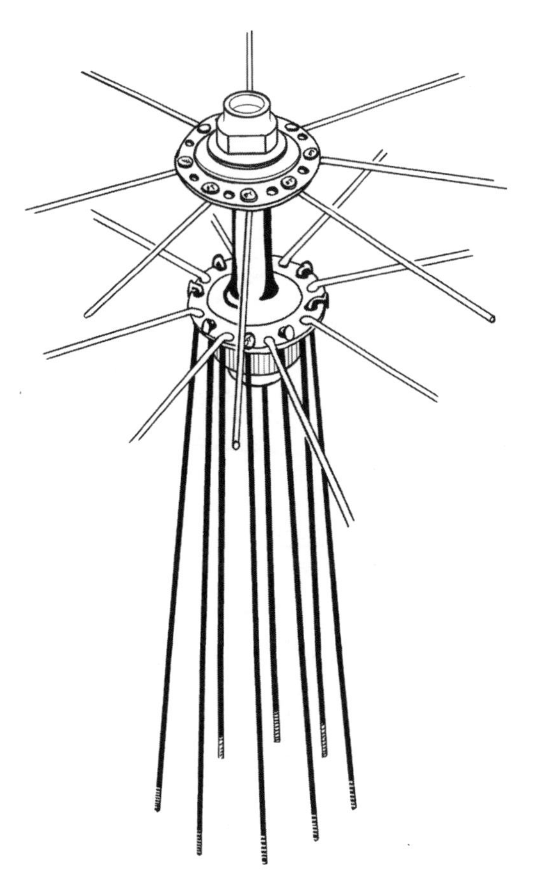

15.13 Rotating hub counterclockwise

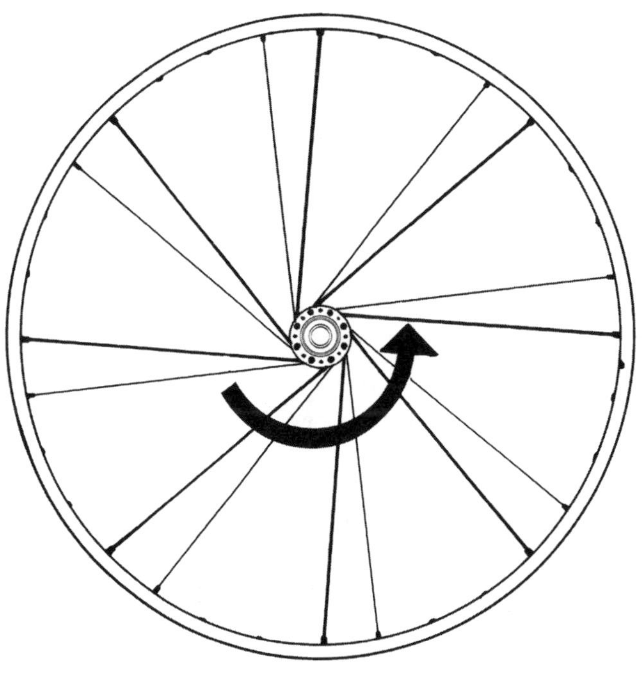

15.14 Lacing third set of spokes

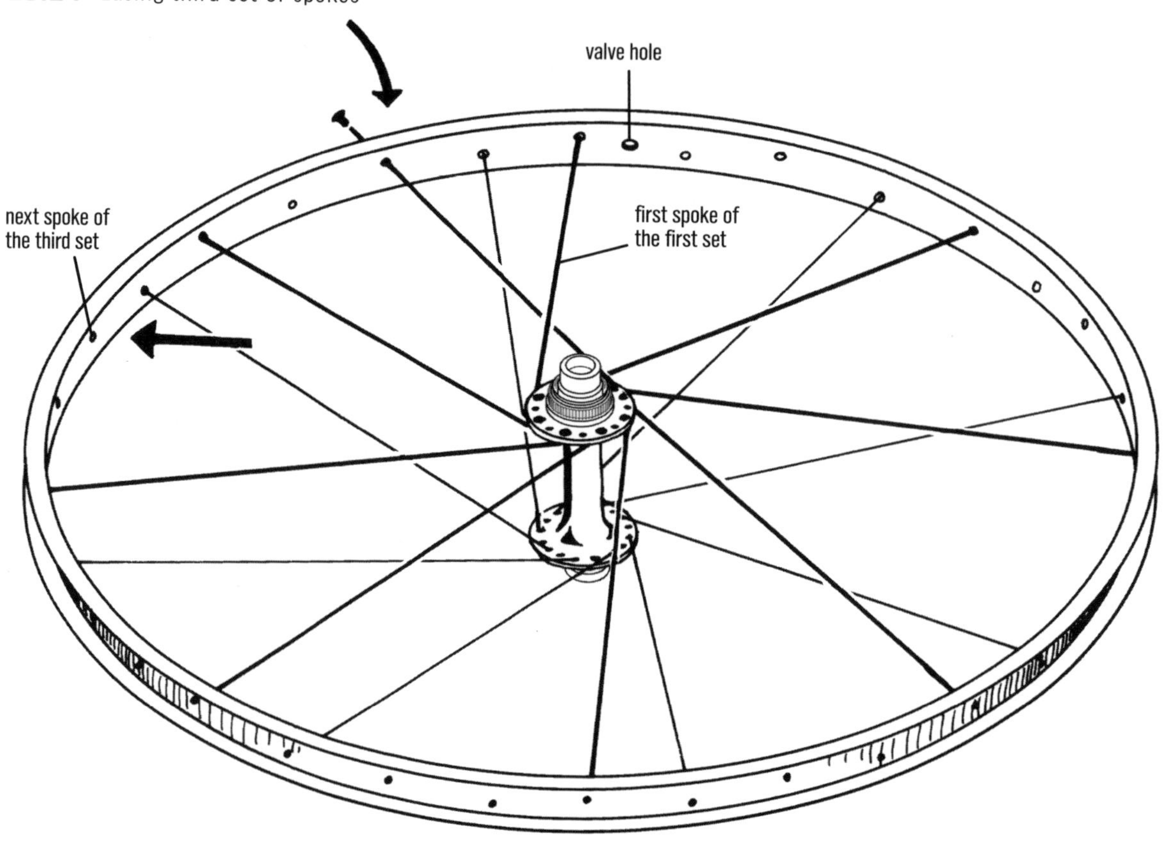

15.15 Third set laced

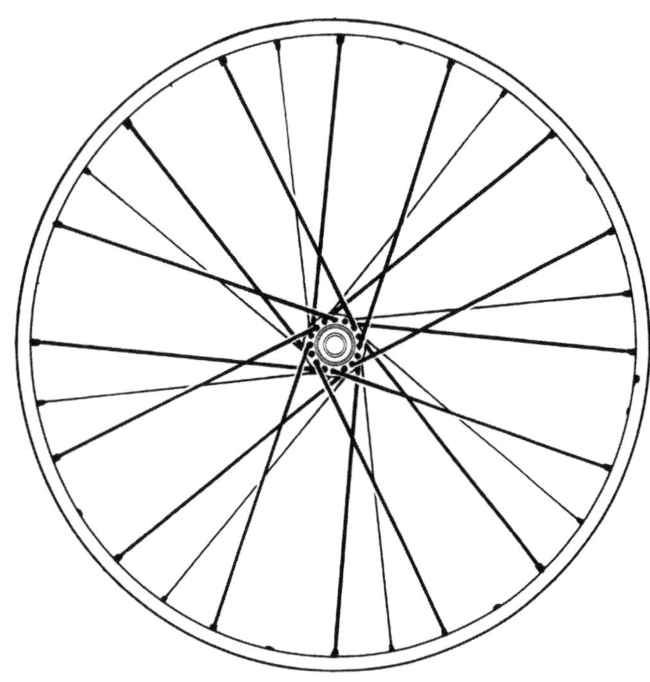

24. **Check your work.** Make sure that every spoke coming out of the upper side of the top flange (the spokes that come out toward you with their spoke heads hidden from view) crosses over two spokes and under a third. All three of these crossing spokes come from the underside of the same flange and have their spoke heads facing toward you. These crossing spokes begin one, three, and five hub holes counterclockwise from the spoke that you just inserted into the rim (Fig. 15.15). This is called a "three-cross" pattern because every spoke crosses three others on its way to the rim (over, over, under). Every upwardly offset hole should now be occupied on the rim.

d. Fourth (and final) set of spokes

25. **Drop spokes down through the remaining hub holes in the bottom flange from the inside out.** Do it as shown in Figure 15.12, but with the other

15.16 Lacing fourth set of spokes

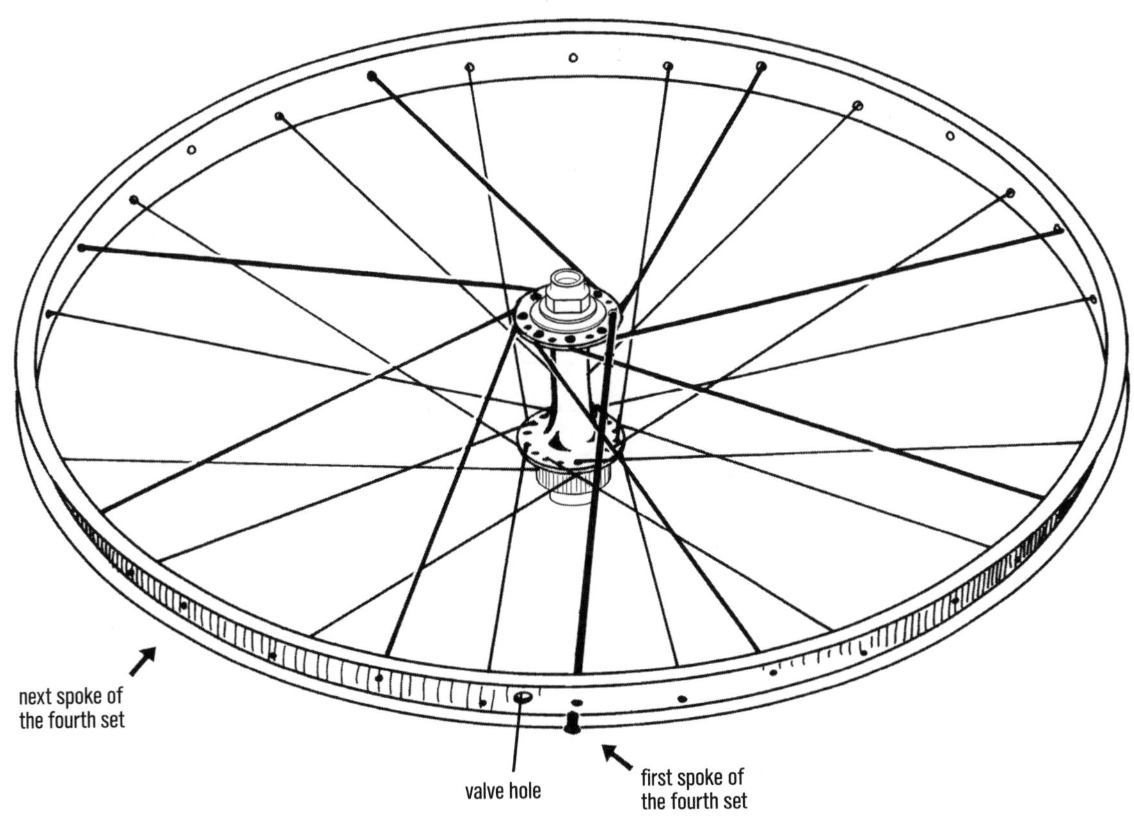

side of the hub up. On a front disc-brake hub, these are again the longer spokes, and vice versa for a rear hub.

26. **Flip the wheel over.** Grab the spokes to keep them from falling out.
27. **Fan the spokes out.**
28. **Pick any spoke on the top (drive-side) hub flange that is already laced to the rim.** Now find the spoke five hub holes away in a counterclockwise direction.
29. **Take that spoke, cross it over two spokes and under the spoke you counted from.** Stick the spoke into the rim hole two holes clockwise from the spoke it crosses under (Fig. 15.16). Thread the nipple onto the spoke three turns.
30. **Continue around the wheel, doing the same thing until the wheel is laced as shown in Figure 15.17.** You may find that some of the spokes don't quite reach far enough. If that's the case, push down on each one about 1 inch from the spoke elbow to help them reach.
31. **Check your work.** Make sure that every spoke coming out from the upper side of the top flange (the spokes that come out toward you with their spoke heads hidden from view) crosses over two spokes and under a third (Fig. 15.17). All three of these crossing spokes come from the underside of the same flange, with spoke heads facing you. The crossing spokes begin one, three, and five hub holes clockwise from each spoke emerging from the top of the upper (left) hub flange (Fig. 15.17). Every hole should now be occupied on the rim.

    When you look at the wheel from the side, the valve hole should be between two spokes (your first spoke and the first spoke of the fourth set) that do not cross each other but whose trajectories look like they are trending slightly toward each other. In other words, if these spokes were to

15.17 Correctly laced three-cross disc-brake wheel

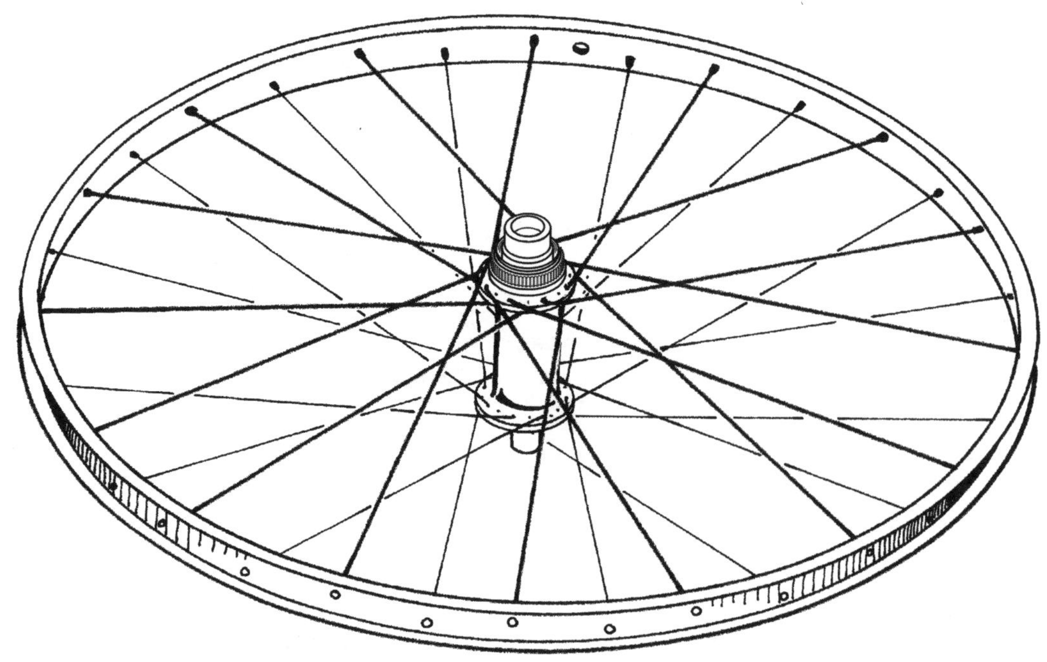

15.18 "Converging parallel" spokes

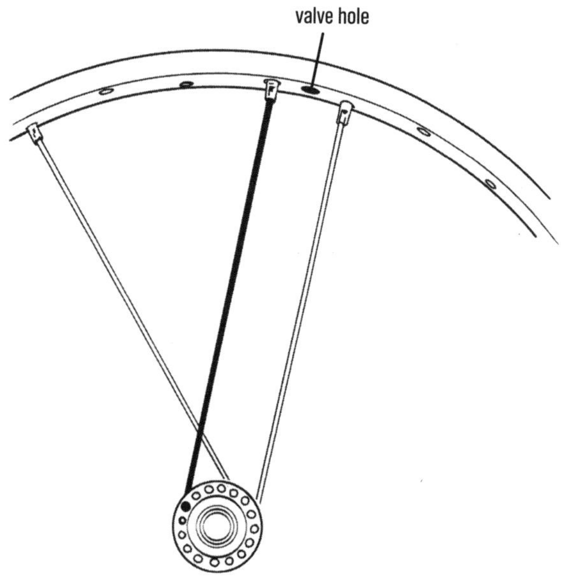

continue infinitely outward, their trajectories would eventually cross far beyond the rim (Fig. 15.18). In wheel-builder speak, these two spokes are called "converging parallel" spokes. Lacing the spokes this way around the valve hole will provide the maximum possible space between the spokes for the pump head when inflating the tire.

IMPORTANT: *If you are building a disc-brake wheel, note that the spokes coming out of the outside of the hub flange on both sides oppose the counterclockwise twist the brake applies on the hub (Fig. 15.17).*

15-3

TENSIONING THE WHEEL

1. **Put the wheel in the truing stand.** If you are using a through-axle hub, you can get adapters for many truing stands that allow you to secure the hub in the stand, or you can run a long screwdriver or an actual through-axle through the hub and set it in the stand's V-notches. If the wheel is so tall that it doesn't clear the truing stand bed (usually only an issue with 36-inch rims or when a 29er wheel is trued with a big tire on it), you can get extension pieces for the truing stand uprights.
2. **Tighten each nipple first with a screwdriver and then with a spoke wrench until only three threads are visible beyond the bottom of the nipple.** See Figure 15.19 for tighten rotation direction. The bent-shaft nipple screwdriver shown in Figure 15.3A speeds this process up immeasurably. The shaft

spins in the handle, which you turn like a crank; it's much faster than twisting a screwdriver.

Once a spoke is fairly tight, it will tend to twist whenever you turn its nipple, so every time you tighten or loosen a spoke nipple, turn it back the opposite direction one-eighth turn afterward. This unwinds the twist in the spoke that your tightening or loosening just caused. If you are using bladed spokes, prevent them from twisting with the tool shown in Figure 15.3B.

3. **Using your thumb, press each spoke coming outward from the outer side of each hub flange down at its elbow.** This straightens out its line to the rim. Spokes coming out of the inner side of the flange do not need this adjustment.
4. **Tighten each nipple a half-turn all the way around the wheel.** Do this uniformly, so that the wheel is not thrown out of true.
5. **Check to see whether the spokes are tight enough to give a tone when plucked.** Squeeze pairs of spokes toward each other, and compare their tension with that of a good wheel with spokes of the same gauge; your wheel should have considerably less tension at this point.
6. **Repeat steps 4 and 5 until the spokes all make a tone but are under less tension than an existing, good wheel.** Final tensioning will come with the remainder of the truing process.

NOTE ON BLADED SPOKES: *If you are using flat spokes, you will need to hold the spoke flats (Fig. 15.3B) to keep them from twisting.*

15-4

TRUING THE WHEEL

a. Lateral true

Side-to-side trueness is the most obvious wheel parameter when you spin the wheel.

1. **Make sure the hub axle has no end play.** If it does, adjust the hub (see 8-6d) to eliminate the end play.
2. **Optionally, put a drop of linseed oil around the top of each nipple where it seats in the rim.** The oil will lubricate the contact area between the nipple and the inside of the rim hole.
3. **Set the truing stand feelers so that one of them scrapes the side of the rim at the worst lateral wobble** (Figs. 15.19, 15.20).
4. **Working in the range of a few spokes on either side of where the rim scrapes, tighten the spokes coming from the opposite flange of the hub.** See Figs. 15.19, 15.20. Start with one-quarter turn on nipples at the center of the scraping area, and decrease the amount you turn each nipple as you move away in either direction. This process pulls the rim away from the feeler. If your adjustments do the opposite, you are turning the nipples in the wrong direction.

15.19 Lateral truing: Pull rim to the right

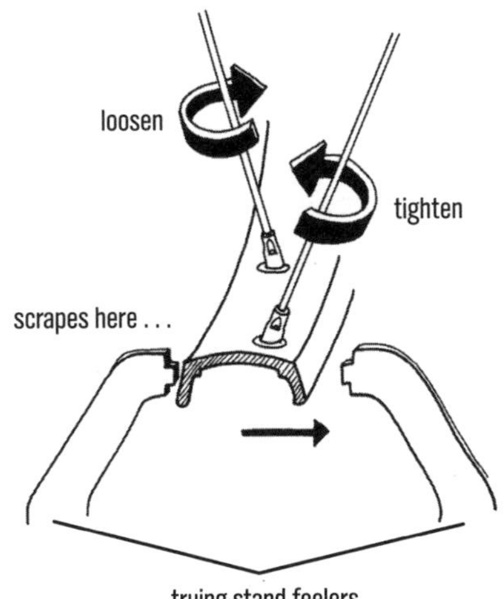

15.20 Lateral truing: Pull rim to the left

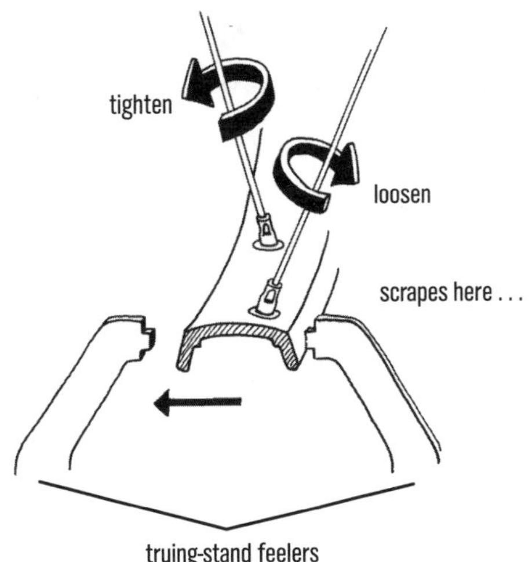

IMPORTANT: *You normally turn something to the right to tighten it and to the left to loosen it, but tightening and loosening spoke nipples at the bottom of the wheel are the opposite of what you would normally do (Figs. 15.19 to 15.22). This is because the nipple head is underneath the spoke wrench. This does not apply if you rotate the wheel so that you are looking down on the nipple from the top. Try opening a jar that is upside down, and you will immediately understand the principle involved. Think of the spoke as a bolt and the nipple as a nut.*

5. **Work around the wheel in this way.** As the wheel gets truer, bring the feelers in closer.

NOTE: *On disc-brake rims, you may need to scrape decals off the rim sides so that they won't hang up on the truing stand feelers and make it hard to tell where the real wobbles are.*

b. Radial true

While not as obvious visually as side-to-side trueness, out-of-roundness is more noticeable when riding, and it is more important to the longevity of the wheel, because a wheel with uniform tension lasts longer. Radial truing, however, can be somewhat slow and frustrating work. If you find yourself running out of patience for this job, step away for a while and then start again when you feel fresh and ready.

6. **Set the truing stand feelers so that they now contact the circumference of the rim, rather than the sides.**
7. **Bring the feelers in until they scrape against the highest spot on the rim** (Fig. 15.21).
8. **Tighten the spokes a one-quarter turn where the rim scrapes.** This will pull the rim inward. Decrease the amount of each turn (to a one-eighth turn and less) as you move away from the center of the scraping area.
9. **Work around the wheel this way, bringing the feelers in as the wheel becomes rounder.**
10. **Wherever there is a dip in the rim, loosen the spokes** (Fig. 15.22). If the spokes are too tight at this point, they will be hard to turn and will creak and groan as you do. When the spokes become hard to turn (i.e., the nipples feel on the verge of rounding off), loosen all of the spokes in the wheel a one-quarter turn before continuing. Compare tension with a good wheel with the spokes of the same gauge; tension at this point should still be lower in the wheel you are building.

15.21 Radial truing: Pull the rim in.

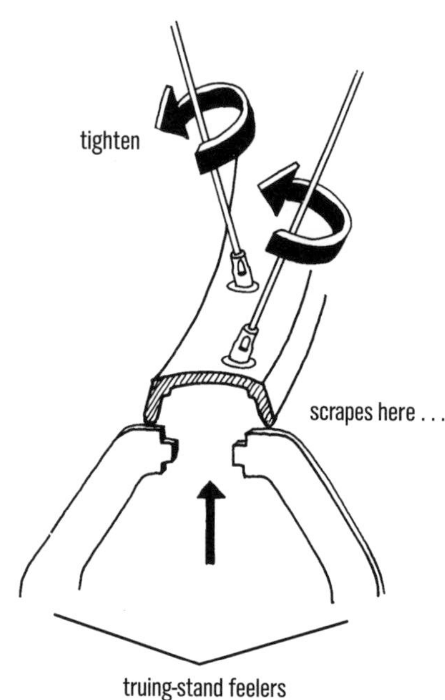

15.22 Radial truing: Let the rim out.

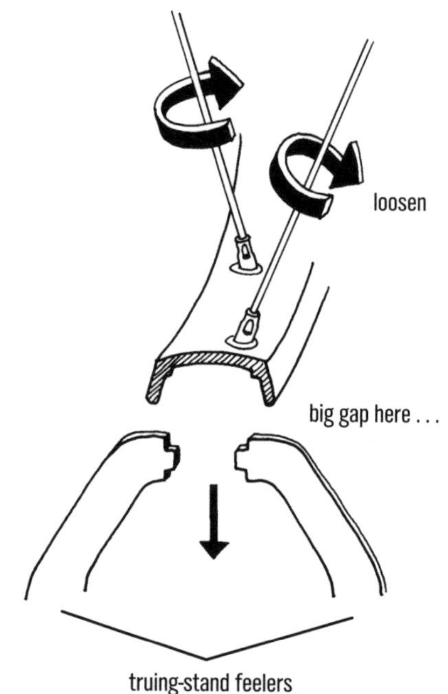

15-5

CENTERING (OR DISHING) THE WHEEL

The rim on a good wheel must be centered in the frame or fork (between the brake pads if you have rim brakes). On a rear hub or a disc-brake front hub, one hub flange is set inboard farther from the axle end (and hence from the frame or fork dropout) on that side than is the other flange. Although the wheel in Figure 15.23 (a front wheel for rim brakes) is symmetrical, an end view like this of either a rear wheel or a disc-brake front wheel would show the tighter spokes on the side with either the cogs or disc rotor to be much flatter (i.e., less angle to the rim) than the (looser) spokes on the other side. Thus, the wheel will be dish-shaped when the rim is centered. This is what is meant by wheel dish.

1. **Place the dishing tool across the right side of the wheel, bisecting the center** (Fig. 15.23).

15.23 Using the dishing tool to check the centering of the rim relative to the axle ends

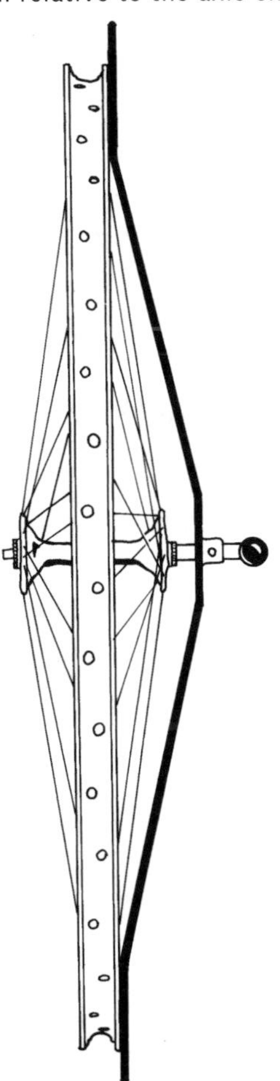

15.24 Checking wheel dish on the other side of the hub

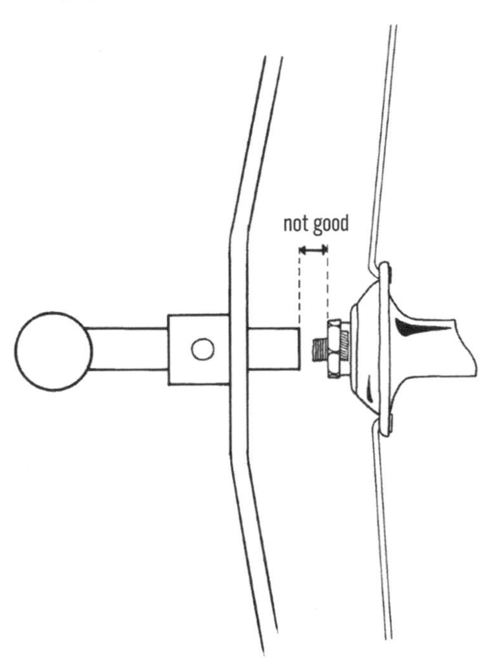

2. **Tighten or loosen the dishing gauge screw until the gauge contacts the outer face of the axle end nut** (Fig. 15.23).
3. **Flip the wheel over.**
4. **Place the dishing tool across the other side of the wheel.**
5. **Check the gap of the dishing gauge with this axle end nut face** (Fig. 15.24). Any gap between the dishing gauge and the axle end nut face indicates the amount that the rim is offset from the centerline of the wheel. If there is no gap, but an overlap instead, reset the dishing gauge on this side (the previously overlapped side).
6. **Flip the wheel over and check the other side.** Repeat steps 4 and 5 on the opposite side.
7. **Put the wheel back in the truing stand.**
8. **Pull the rim toward the center.** Tighten the spokes on the opposite side of the wheel from the axle end that had the gap between the axle end nut face and the dishing gauge to reduce the gap between the dishing tool and the nut. Tighten a half turn each. If the spokes start getting really tight (they will creak a lot when tightening, the nipples will start rounding off, and the spokes will feel much tighter than

the spokes in a comparable wheel), then loosen the spokes on the opposite side of the wheel a half turn each.

9. **Recheck the wheel with the dishing gauge by repeating steps 1–5.**
10. **If the wheel is still off-dish (there is still a gap between the dishing gauge and the end nut when you flip the wheel over), repeat steps 6–8.** Continue until the gap is zero (i.e., the dish is correct).
11. **Stress the spokes by squeezing each pair of spokes together with your hands** (Fig. 15.25). They will make a pinging noise as they unwind. If you followed my recommendation above of turning the nipple back the opposite direction one-eighth turn after each time you tighten or loosen it, the spokes should not be wound up much, and prestressing the wheel will be unnecessary (you'll know because the spokes won't ping when you stress them).
    (a) Leaning on the wheel is a quicker way to prestress the wheel, but this method has the potential to wreck the wheel if you are not careful. Do not press down with all your might; too much pressure can destroy your work. To proceed, set the axle end on the workbench and carefully press down on the rim with your hands at the 9 o'clock and 3 o'clock positions. This procedure will affect an area of about three spokes on each side, so rotate the wheel three spokes, press down again, rotate three more spokes in the same direction, press down again, and so on. After you finish one side, flip the wheel over and do the other side.
    (b) If prestressing throws the wheel way out of true, the spokes are probably too tight. Loosen them all a one-eighth turn. Note, though, that some loss of wheel trueness is normal. If the loss is minor, you can overlook it and continue with step 12.
12. **Repeat the process.** Repeat truing the wheel (15-4), followed by dishing the wheel (15-5), prestressing the spokes (and turning the nipples back a one-eighth turn from the rotation direction on each adjustment of one) frequently as you go. Keep improving the accuracy of the build this way.
13. **Bring up the tension to that of a comparable wheel by making small tightening adjustments to every nipple.** Adjust dish and true after each time around, until the wheel is as you want it.
14. **If the rim is oily, wipe it down with a citrus-based biodegradable solvent.**
15. **Congratulate yourself on building your wheel.** Show it off to your friends.

15.25 Relieving spoke tension

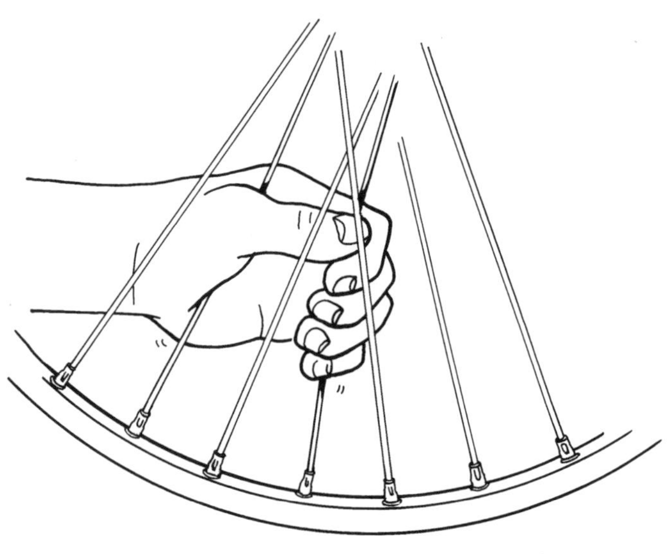

15-6

REVIEWING THE WHEEL BUILD

Your wheel (Fig. 15.17) has the braking pulling spokes, which oppose the twist on the hub when the disc brake is applied, to the outside of the hub flange. In plain speak, this means that you have a spoking pattern that best resists the twisting force on the hub produced by braking forces, due to the wider spoke stance angle.

When driven by the chain, the pulling (or dynamic) spokes relative to braking in the rear wheel are not pulling relative to the chain; rather, they are static spokes relative to pedaling forces. The pulling spokes relative to driving forces are instead the ones directed in such a way that the twist on the hub created by the chain force increases the tension in them. If you look at the wheel from the drive side, you will see what I am talking about. You will also see that the static spokes relative to the chain (which are the pulling spokes relative to braking) do not oppose a driving twist on the hub. In fact, their tension decreases when you stomp on the pedals. But their tension increases under braking.

By placing all of the pulling spokes relative to braking so that they come from the inside of the hub flanges out (i.e., the spoke heads are on the inward side of the flanges), you have made the spokes undergoing the largest tension changes lie across the hub flange and thus better support the spoke elbow. You have also increased their angle to the rim and hence their ability to oppose forces acting on the rim.

On a rear wheel, when the pulling spokes come out of the inside of the flange and cross around the outside of the static spokes, they pull all of the spokes slightly inward away from the rear derailleur under the driving force of the chain. This may someday prevent the rear derailleur from catching on the spokes when you are starting up in low gear out of the saddle, applying high torque to the hub.

If you have chosen the appropriate parts for your weight and riding style and have the proper spoke tension, then you should have a strong wheel that will last you a long time. Congratulations!

15-7

LACING RADIALLY SPOKED WHEELS

With the advent of stronger rim materials and stiffer rim cross-sections, radially spoked wheels—in which the spokes emanate radially from the hub out to the rim, rather than crossing each other (Fig. 15.26)—were very popular in the waning days of rim brakes on mountain bikes. They are very simple to lace up, and radial spoking offers a number of advantages, but a completely radial wheel can be used only on the front. On the rear, you must still use a crossing pattern on one side to oppose the twist on the hub caused by the chain (Fig. 15.27).

Similarly, you cannot use a radially spoked (or partially radially spoked) wheel with disc brakes, as the spokes cannot oppose the torque the brake applies

15.26 Radially spoked front wheel

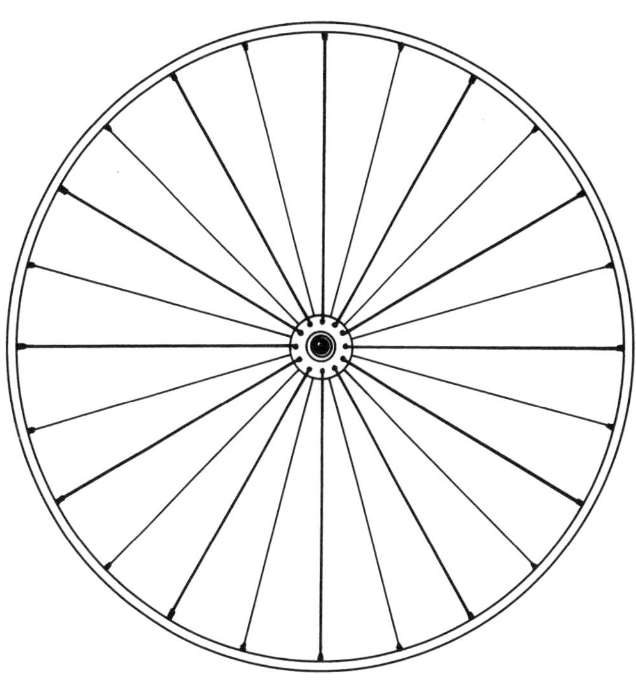

15.27 Radial/three-cross rear wheel with a radially spoked drive side

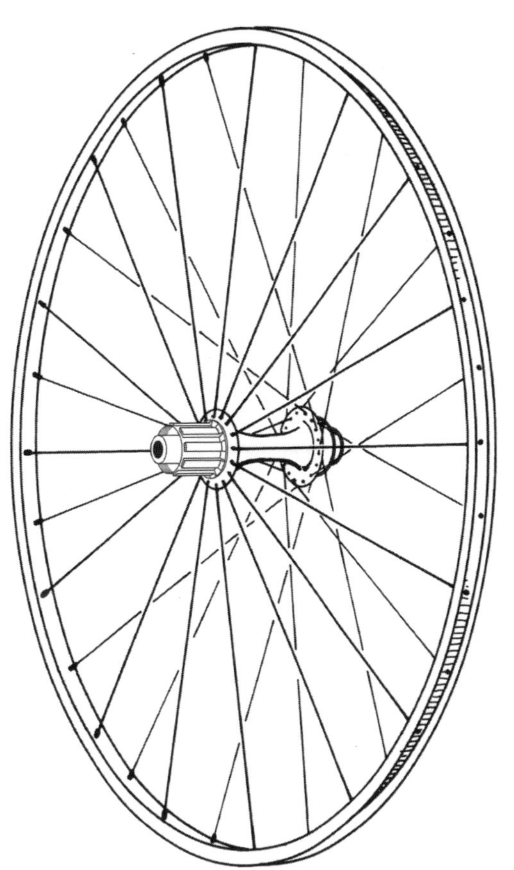

15.28 Radial/three-cross rear wheel with radial spokes on non-drive side

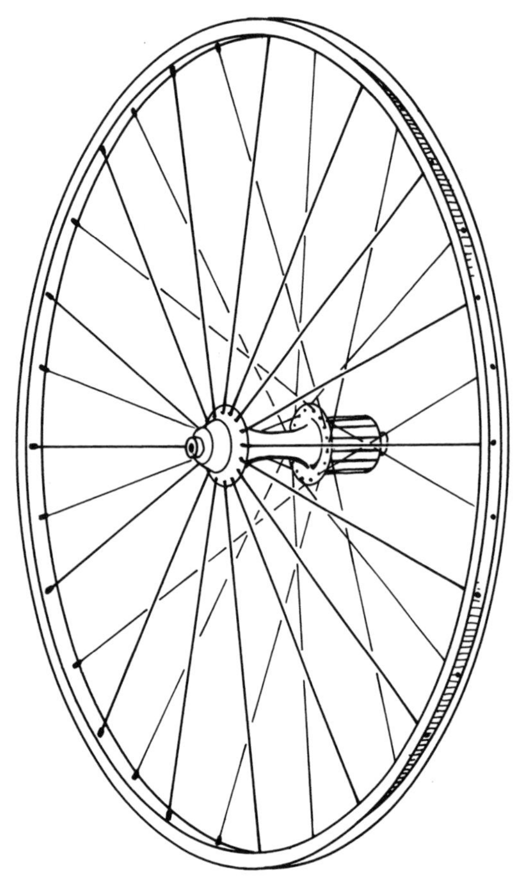

to the hub. In order to oppose the braking force, the spokes must be tangential to the hub flange, or the brake will keep twisting the hub until they are. This will eventually break radial spokes as well as spokes laced with too few crosses (i.e., one-cross or two-cross lacing patterns) for the size of the rim and hub flange.

A radially spoked wheel is vertically stiffer than a crossed one because radial spokes allow little opportunity for spokes to absorb energy in the spoking pattern. The radial wheel can be laterally stiffer too because all of the spokes can come to the outside of the hub flange and increase the pulling angle to the rim.

A radial wheel is lighter because the spokes are shorter. Further weight can be removed with fewer spokes, and radial spoking allows any even spoke count to be used (with nonradial patterns, the spoke count must be a multiple of four). Radial spoking allows the use of flangeless direct-pull hubs and nail-head or double-threaded spokes (straight spokes without elbows), eliminating a potential weak spot in each spoke.

Note that the warranty of some hubs is voided when spoked radially. Shimano has this stipulation, for instance. The stress is greater on hub holes with radial spoking because there is less material to resist the tearing of the hub holes when the spoke is pulling straight outward than if it is pulling tangentially along the hub flange.

a. How to lace a radial front wheel

Simply drop all the spokes from the inside of each flange outward, and lace the spokes straight to the rim (Fig. 15.26).

b. How to lace a rear wheel with a radial drive side and a three-cross non-drive side

A rear wheel will generally last longer with crossed spokes on both sides, but if you want to lace one side radially, you can choose which side to make radial. If the hub flange is stiff enough to carry the resistance to torque on the hub by using crossed spokes on the left side, there is an advantage to having the radial spokes on the drive side (Fig. 15.27). This advantage is due to the spoke angle to the rim on the drive side, which is reduced due to the drive-side hub flange being further inboard; the angle is maximized with radial spokes, since the spokes are shorter. This makes the tension of the spokes on the two sides of the wheel more even, which generally results in greater wheel longevity. If the hub shell is too skinny and flimsy to oppose the chain's driving force, you will need to reverse the below instructions and put the radial spokes on the non-drive side (Fig. 15.28), or lace it three-cross on both sides.

First lace the non-drive side following the instructions in 15-2a, steps 4–7, and 15-2c, steps 17–24. Now lace the drive-side spokes radially outward through the hub flange and straight to the rim (Fig. 15.27).

The tensioning and truing steps are the same for radial wheels and radial/three-cross wheels as for standard three-cross wheels, but radial spoke tension should be higher to help prevent the spokes from vibrating loose.

NOTE: *These instructions place the radial spokes to the outside of the hub flange so that their angle to the rim* *(and hence their ability to oppose lateral forces on it) is highest.*

15-8

LACING THREE-CROSS WHEELS FOR RIM BRAKES

There is no difference the way you lace a three-cross wheel whether it's made for rim brakes or for disc brakes. Follow all of the instructions in 15-2 for lacing the wheel, and in subsequent sections for tensioning, truing, and dishing the wheel. For a rear wheel, orient the hub as instructed with regard to the drive side and non-drive side. For a front wheel, both sides are the same, so orient the hub either way. The spokes for the front wheel will all be the same length on both sides.

15-9

LACING TWO-CROSS AND ONE-CROSS WHEELS

Two-cross lacing is a must for small wheels (24-inch) or for standard-size (26-inch, 27.5-inch, and 29-inch) rear wheels with a Rohloff SpeedHub or other internal-gear or internal-motor rear hubs with enormous hub flanges. One-cross lacing may be necessary for 20-inch wheels and for 24-inch wheels with Rohloff or other huge-flange hubs. This ensures that the spokes are approximately tangent to the hub flanges to best oppose driving and braking forces, and it prevents spokes from overlapping the heads of adjacent spokes.

Reducing the number of spoke crossings on any wheel shortens the spokes. This removes some weight and reduces the vertical compliance of the wheel. Obviously, you will need shorter spokes; this will be accounted for automatically in online spoke calculators when you input the number of crosses.

To lace a two-cross wheel, begin by following the steps in 15-2a and 15-2b exactly (or follow the steps in 15-8a and 15-8b exactly for a Rohloff or other rear disc-brake hub). In 15-2c, steps 21–22, count three hub holes over, not five. And step 24 in the case of a two-cross wheel now becomes the following:

24. **Check your lacing.** Make sure that every spoke coming out of the upper side of the top flange (the spokes that come out toward you with their spoke heads hidden from view) crosses over one spoke and under a second. Both of these crossing spokes come from the underside of the same flange and have spoke heads facing toward you. These crossing spokes begin one and three hub holes counterclockwise from the spoke that you just inserted into the rim. This is called a two-cross pattern because every spoke crosses two others on its way to the rim (over, under). Every upwardly offset hole should now be occupied on the rim.

In 15-2d, steps 25–26, count three hub holes over, not five, and cross over one spoke and under the spoke you counted from. Step 28 in the case of a two-cross wheel now becomes the following:

28. **Check your lacing.** Make sure that every spoke coming out from the upper side of the top flange (the spokes that come out toward you with their spoke heads hidden from view) crosses over one spoke and under a second. Both of these crossing spokes come from the underside of the same flange and have their spoke heads facing toward you. The crossing spokes begin one and three hub holes clockwise from each spoke emerging from the top of the upper (left) hub flange (Fig. 15.17). Every hole should now be occupied on the rim. The valve hole should be between converging parallel spokes (Fig. 15.18) to make room for the pump head when inflating the tire.

Lacing a one-cross wheel reduces by one the number of hub holes you count over to and the number of crosses the spoke makes (now each spoke crosses only under one other spoke and crosses over none).

15-10

BUILDING WHEELS FOR BIG RIDERS

Wheels for heavy and tall riders require greater lateral and vertical stiffness. The weight of the rider can more easily bend and laterally flex the rim, and it creates another problem as well. A heavy rider increases the detensioning of the spokes at the bottom of the wheel by making the rim more D-shaped at the bottom as it rolls. If the spokes are under less tension, or if the nipple flanges periodically lose contact with the bases of the rim holes, the nipples may unscrew and the wheel will fall apart. To achieve the higher strength necessary to resist this loosening, you can modify several characteristics.

a. Spoke count and thickness

The spoke count needs to be high: Thirty-six or more spokes are highly preferable for riders weighing more than 190 pounds. The spokes need to be thick at the ends, as thicker spokes are less prone to breakage. Although 14/15-gauge (2.0mm, or 14-gauge, on each end, and 1.8 mm, or 15-gauge, throughout the center section), double-butted spokes are thinner than straight 14-gauge spokes, DT Swiss testing has shown that the wheel will probably last longer with them. Because most breakage occurs at the nipple or the elbow and butted spokes are the same thickness there, spoke breakage will not increase. But butted spokes will stretch more, allowing the spoke nipples to better stay in contact with the rim as the rim changes shape while rolling.

b. Nipple type

Brass nipples are preferable to aluminum ones, due to the extra stress a big rider puts on the wheel. The added weight is slight, and the increase in durability can be significant.

c. Rim section

The deeper the rim, the higher its hoop strength (vertical stiffness and strength). Very deep V-section rims work with low spoke counts because of this high hoop strength. The strongest wheel can be made from a deep-section rim drilled for more spokes. Unfortunately for heavy riders, many deep V-section rims also have thinner walls to reduce weight and thus their strength is not as high as it could be.

d. Spoking pattern

With 8-, 9-, 10-, 11-, and 12-speed rear wheels, dish is high (one side of the wheel is flatter than the other), meaning that there is a great tension difference between spokes on the two sides. The loose spokes on the left may unscrew, especially under high pedaling forces, and the tight spokes on the right may break. As the chain twists the cogs clockwise, the spokes opposing the twist (the pulling spokes) get tighter, while the static spokes are under reduced tension and may unscrew.

An asymmetrical rim (one that is drilled off-center) can reduce spoke tension imbalance by reducing the wheel dish. The rim holes are offset to the left side (Fig. 15.4) so that the drive-side spokes come to the rim at a lower angle and can work with lower tension and more even tension between the two sides. The left-side spokes come to the rim at a higher angle and can be under higher tension without forcing the use of dangerously high tensions on the drive side. Before lacing an off-center rim, make sure you read the note in step 2 of section 15-2.

Using radial spokes on the drive side (Fig. 15.27) of a rim-brake rear wheel (with an extremely stiff hub shell) can counteract the problem of grossly uneven tension. This is because the spokes on the flatter (drive) side are shorter than the spokes coming in from the wider angle on the non-drive side. Thus, their relative angles to the rim are more similar, which makes the tension of the spokes on the two sides of the wheel more even, which generally results in greater wheel longevity. If the hub shell is too skinny and flimsy to oppose the chain's driving force, you will need to lace the wheel three-cross.

If you come to a fork in the road, take it.

—YOGI BERRA

16

FORKS

The fork connects the front wheel to the handlebar, allows the bike to be steered, and supports the front brake. It also offsets the front hub some distance forward of the steering axis (the centerline through the headset), and this offset distance (called the fork rake), combined with the angle of the steering axis (the head angle) and the wheel size, determines how your bike is going to handle and steer.

All forks, suspended (Figs. 16.1 and 16.2) or rigid (Fig. 16.3), provide at least a minimum amount of suspension by allowing the front wheel to move up and down. The steering axis angles the fork forward from vertical, and the offset sets the front hub forward; those two characteristics make it possible for a fork—even a so-called rigid fork—to flex along its length and absorb vertical shocks. Suspension forks add a much greater range of vertical wheel travel.

All mountain bike forks are made up of a steering tube, a fork crown, two fork legs (except the Cannondale Lefty, which has one), fork ends (also called dropouts or fork tips), and brake bosses (usually disc-brake mounts and/or cantilever/V-brake posts, but on really old steel forks you may find roller-cam/U-brake posts). Figures 16.1 to 16.3 illustrate these parts.

Most mountain bikes come equipped with suspension forks (Figs. 16.1 and 16.2). Their most distinguishing feature is the spring inside. That spring can be made of compressed air, steel or titanium coils, elastic polymer bumpers (elastomers), or a combination. Most suspension forks also include a damping system to control how fast the spring compresses and rebounds. The damper acts much in the same way as the device on your screen door that prevents it from slamming.

Hydraulic damping systems are the most common. They rely on the controlled movement of oil from one chamber to another through holes that slow the rate of flow. Some pneumatic dampers operate on a similar principle with compressed air instead of oil.

The most commonly used suspension-fork design has telescoping fork legs that consist of two sections: inner legs (stanchions) attached to the fork crown and steering tube, and outer legs (lower legs or casting) attached to the front hub that slide up and down over the inner legs (Fig. 16.4). Although this description applies to the vast majority of

TOOLS

- Shock pump
- Blunt tool for releasing air pressure
- Metric hex keys
- Metric open-end wrenches
- Adjustable wrench
- Needle-nose pliers
- Safety glasses
- Soft hammer
- Small and long screwdrivers
- Nonlithium suspension grease (Slick Honey)
- Citrus degreaser
- Isopropyl alcohol

OPTIONAL

- Torque wrench
- Ratchet handle
- Ruler or caliper
- Dropout-alignment tools
- Solid bench vise
- Snapring pliers
- Metric socket wrenches
- Bike stand
- 10mm, 11mm sockets
- Pick
- Awl
- 22mm, 24mm, 26mm, 28mm sockets with ends ground flat
- Wiper seal driver

CONTINUED ON P. 350

TOOLS, CONT.

Fox damper-removal tool or aluminum drift with bore hole
19mm open end wrench
Long wooden or plastic dowel
Soft bottle brush
Syringe or graduated cylinder
Medium-strength threadlock compound (Loctite 242 or similar)
Antiseize compound for titanium bolts
Suspension oil of correct weight for your fork
Air-cylinder oil (Fox Float Fluid)

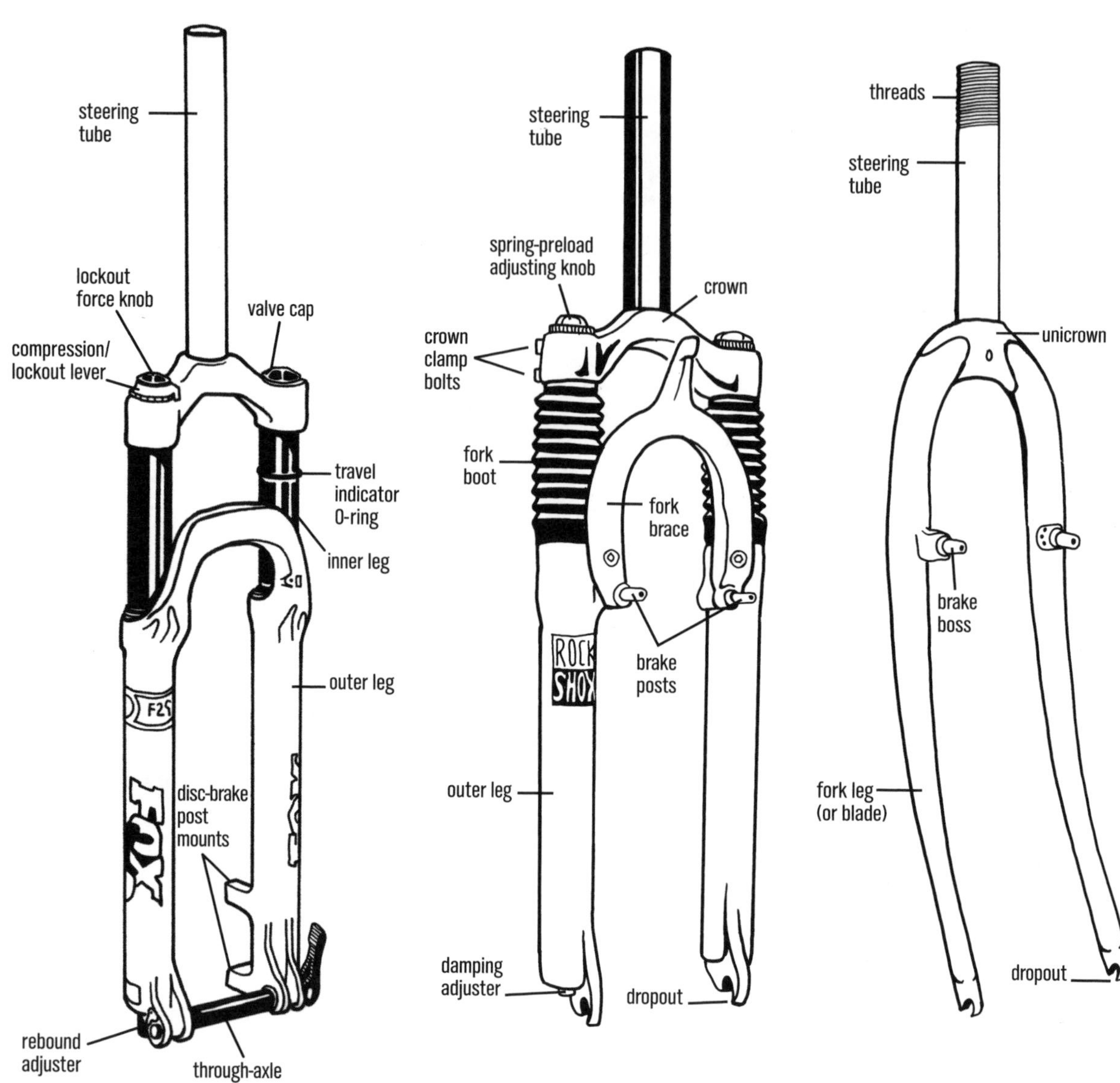

16.1 Disc-brake-specific air-sprung suspension fork with 15mm lever-operated through-axle

16.2 Cantilever/V-brake-specific elastomer-sprung suspension fork for 9mm quick-release axle

16.3 Rigid fork

suspension forks, there are variations that vie for a small piece of the fork market. Cannondale's Headshok design incorporates rigid fork legs attached to a single shock unit inside the head tube, and its Lefty fork has only a single telescoping, left leg. Upside-down telescoping forks have thinner lower legs sliding up and down inside fatter upper legs. There are also linkage suspension forks that use a system of pivots and movable arms attached to the fork legs and controlled via a spring.

16-1

INSPECTING THE FORK

LEVEL 1

For the most part, forks are pretty durable, but they do break sometimes. A fork failure can ruin your day because the means of controlling the bike is eliminated. Such loss of control usually involves the rapid transfer of your body directly onto the ground, resulting in a substantial amount of pain.

16.4 How a suspension fork works

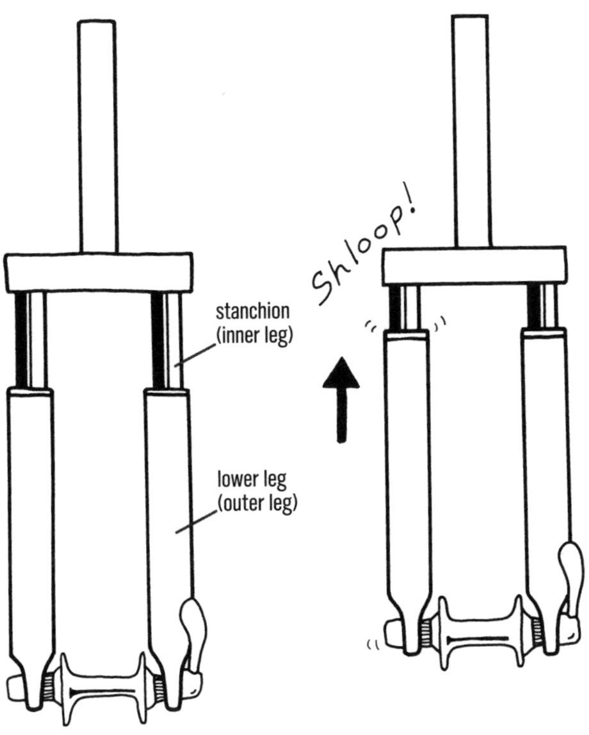

16.5 One messed-up fork

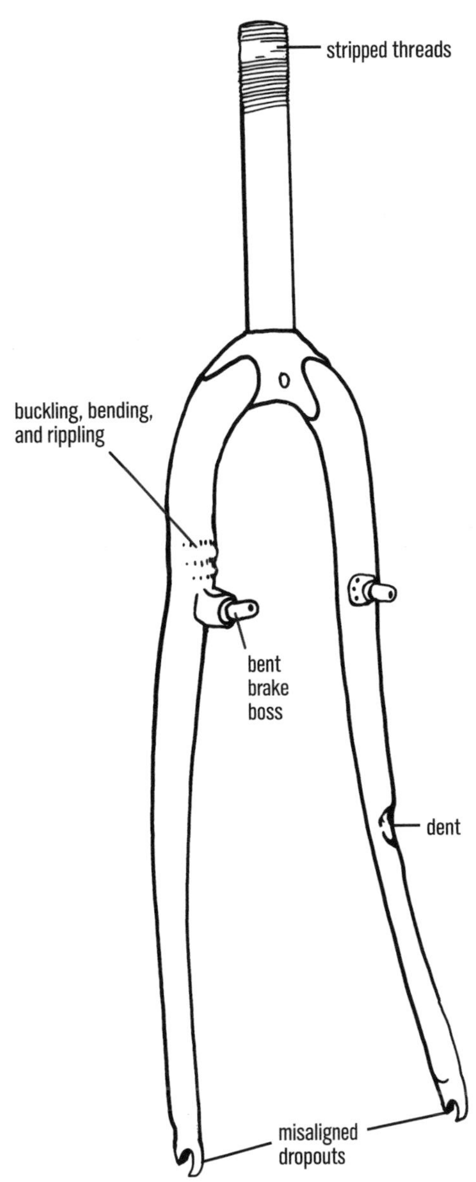

Ever since I first opened my frame-building shop, people have regularly brought me an amazing collection of forks that have broken, sometimes with catastrophic consequences. Some had steering tubes broken either at the fork crown or in the threads. Others had fork crowns that broke or separated, fork legs that folded, cantilever posts that snapped, fork braces supporting the brake cable that detached, and front dropouts that broke off. Top caps can fly off coil- or elastomer-sprung forks (and shoot up at your face!), and seals can blow on air-sprung forks; in either case, the fork immediately bottoms out. Pivots on linkage forks can break or fall apart. You can go a long way toward preventing problems such as these by regularly inspecting your bike's fork and addressing issues immediately.

With that in mind, get into the habit of checking the fork regularly for any warning signs of impending failure—bends, cracks, and stressed paint. If you have crashed your bike, give the fork a very thorough inspection. If you find any indication that the fork has been damaged, replace it. A new fork is cheaper than emergency room charges, brain surgery, or an electric wheelchair.

When you inspect a fork, remove the front wheel, clean the mud off, and look under the crown and between the fork legs. Carefully examine all the outside areas. Look for any spots where the paint or finish looks cracked or stretched. Look for bent parts, from little ripples in fork legs to skewed cantilever posts and bent dropouts (Fig. 16.5).

Put the wheel back in, and watch to see if the fork legs twist when you secure the wheel quick-release or axle nuts; twisting indicates bent dropouts. Check to make sure that a true wheel centers under the fork crown. If it doesn't, turn the wheel around and put it back in the fork. That way you can confirm whether the misalignment is in the fork or the wheel. If the wheel lines up off to one side when it is put in one way and off the same amount to the other side when it is put in the other way, then the wheel is off-center (out of dish; see 15-5) and the fork is straight. If the wheel is skewed off to the same side in the fork no matter which direction you place the wheel, then the fork is misaligned (16-17).

I recommend overhauling your bike's headset annually (see 12-18 and 12-19), and when you do, carefully examine the steering tube for any signs of stress or damage. Check for bent, cracked, or stretched areas; stripped threads on the steering tube (Fig. 16.5) or inside the cantilever brake bosses or disc-brake bosses (Fig. 16.1); bulges where the stem expands inside (threaded steering tube); or crimping where the stem clamps around the top (threadless steering tube).

With a threaded fork, hold the stem up next to the steering tube to make sure that when the stem quill (Fig. 12.9) is inserted to the depth at which you have been using it, the bottom of the quill is always more than 1 inch below the bottom of the steering-tube threads. If you secure the stem quill by tightening the expander bolt (Fig. 12.10) when it is in the threaded region, you are asking for trouble; the threads cut the steering-tube wall thickness down by about 50 percent, and each thread offers a sharp breakage plane along which the tube can cleave.

On telescoping suspension forks, if the fork has crown clamp bolts (older forks were made this way), check that they are tight (ideally, with a torque wrench to verify that they are tightened to the manufacturer's recommended torque). If the fork crown has titanium clamp bolts, consider replacing them (or get a new fork without clamp bolts); the heads of titanium fork-crown bolts have been known to snap off. Forks with crown bolts (Figs. 16.12–16.13 and 16.25–16.26) are no longer made, which is a good thing.

Check for oil leaks from the top or bottom of the outer leg. Check for torn, cracked, or missing seals around the top of the outer leg. Check that the air valve and the valve core inside it are tight. If the fork has stripped threads inside the disc-brake bosses (Fig. 16.1), it's more likely that the brake caliper has been attached with bolts that are too short than that the bolts were overtorqued, but be careful to avoid both.

On linkage forks, regularly check that all bolts are tight and all pins have their circlips or other retaining devices in place so that they do not fall out. Check for cracks and bends around the pivots.

If you have any doubts about anything on the fork, take it to the expert at your bike shop. Err on the side of caution—replace it before it fails.

16-2

REPAIRING FORK DAMAGE

If your inspection has uncovered damage that does not automatically require fork replacement, here are some guidelines to go by and some means of repair.

a. Dents

Not all fork dents threaten the integrity of the fork. On a rigid fork, particularly a steel one, a small dent usually poses little risk. A large dent (Fig. 16.5), of course, does. On a suspension fork, almost any dent can hurt the fork's operation, even if it does not pose a breakage threat. Fortunately, many suspension-fork parts are replaceable.

b. Fork misalignment

Within limits, a rigid fork made of steel can be realigned if it is slightly off-center (see 16-18). Rigid forks made out of any other material and suspension forks cannot be realigned. Don't try!

c. Stripped steering-tube threads

If the threads on the steering tube are damaged (Fig. 16.5) so that the headset slips when you try to tighten it, replace the fork.

d. Obvious bend, ripple, or crease in fork legs

Replace the fork if you feel, or see, ripples and bends in the legs (Fig. 16.5). The poor handling and potential breakage pose too great a threat to your safety to be worth saving a few bucks.

e. Bent or stripped disc-brake bosses or cantilever bosses

Post-mount disc-brake bosses (Fig. 16.1) are threaded inside and can become stripped. If the cause was the use of overly short bolts to mount the brake caliper, you may be able to tap the threads with an M6 tap and install the recommended-length bolts for your brake and fork. Otherwise, you probably need to replace the lower legs or the entire fork. Bent disc-brake mounts, whether they are post mounts (Figs. 16.1 and 9.9) or IS mounts (Fig. 9.11), mandate replacement of the lower legs or the entire fork.

On most suspension forks that have cantilever studs, as well as on some rigid forks, the studs can be unscrewed with an 8mm open-end wrench and replaced. It is a good idea to use a threadlock compound such as Loctite 242 on the threads when installing.

A bent or stripped cantilever boss on an old rigid fork (Fig. 16.5) usually means that you must buy a new fork. If you have a frame builder in your area, he or she may be able to weld or braze a new boss onto a steel fork and repaint it.

16-3

MAINTAINING RIGID FORKS

Beyond touching up the paint and performing the regular inspections just described, the only maintenance procedure for a rigid fork (Fig. 16.3) is to check the fork's alignment (16-17) and possibly improve on it (16-18), but only if it is a steel fork.

16-4

TUNING AND MAINTAINING SUSPENSION FORKS

Suspension forks (Figs. 16.1 and 16.2) offer a significant performance advantage over rigid forks and increase the bike's versatility. On all but the cheapest models, telescoping forks with air springs, coil springs, elastomer springs, or a combination of these inside generally are also equipped with hydraulic damping systems.

I address fork tuning in 16-7, and you should read that section carefully, as it also relates to tuning rear suspension. Specific tuning for specific fork types is in 16-8 and 16-9. Customization of tuning—changing the stacks of shims over orifices in the fork damper in order to change the fork's overall performance—is a task for the fork factory or for a custom suspension-tuning shop.

With such a vast and expanding array of fork models out on the trails, detailing the overhaul of every one is impossible here without turning this book into a multivolume encyclopedia. Most telescoping forks share enough similarities that they lend themselves to common basic service steps, but, at a minimum, you need to obtain recommended oil weights and volumes from the manufacturer (these are often available online). Often a different oil volume will be required in each side of the fork, and getting it wrong can lock up the fork, so guessing is a poor idea.

You may be on your own with a linkage-style fork; if you can't find online information, I hope you kept your owners' manual! Be especially vigilant about inspecting these forks regularly.

16-5

FREQUENT MINOR TELESCOPING SUSPENSION FORK MAINTENANCE

LEVEL 1

On old fork models, if you do minor maintenance frequently and keep the inner legs covered with fork boots (Fig. 16.2), you can greatly increase the life of the seals as well as the time between fork overhauls. A dry or dirty dust seal rubbing on a dry or dirty inner leg usually causes stickiness in suspension forks.

Newer fork models don't have fork boots and rely on wiper seals to keep dirt out of the lower legs. If a bit of dirt does get past the wiper seal and lodges in one of the bushings the inner leg slides on, it can rapidly wear through the hard-anodized surface of the inner leg, leading to permanent performance loss. You don't want to have to buy a new fork any sooner than necessary, and helping the wider seals do their job of keeping dirt and water out of the inner legs by keeping them wiped clean will increase the fork's life span.

1. **Wipe clean the exposed upper tubes and the seal atop each lower leg.** If the fork has protective boots

(Fig. 16.2), remove them for these steps if they are the flat type that wraps around and attaches with Velcro; slide them up if they are the rubber bellows type (Fig. 16.2).

2. **Put a thin coat of Teflon-fortified lubricant on the outside of the seals and inner legs.**
3. **If the fork has boots, put them back into position.** You may need to stretch the bottom of each rubber bellows–type fork boot with needle-nose pliers to get it in the groove around the top of the outer leg and behind the fork brace.

16-6

MEASURING FORK TRAVEL

You can't make any useful ride adjustments to a fork (a.k.a. fork tuning) without knowing the fork's total travel, its sag (a.k.a. ride height), and the amount of its travel used on a given trail.

a. Measure maximum possible travel

The owners' manual should state the total fork travel, but if the fork has internally adjustable travel, you might not know how it's set, depending on whether or not an internal travel spacer (16-14, Fig. 16.34) is installed.

To measure total available travel, do the following:

1. **Measure the distance down the upper fork leg tube from the lower edge of the fork crown to the top of the lower leg.** On forks without adjustable travel, this measurement will usually be close to the total available travel. Pull up on the handlebar to make sure that a negative spring is not preventing the crown from coming up any higher.
2. **Remove the main spring(s).** With an air fork, fully deflate the top spring through the valve atop the fork crown (Fig. 16.6). Push down on the fork while doing so to push out maximum air. With a coil or elastomer fork, unscrew each spring top cap and pull the spring out (Figs. 16.7 and 16.26), or just let the spring stick out the top of the leg. Some top caps are knurled and can be unscrewed by hand (Figs. 16.7 and 16.26), whereas others require removal with a 22mm or larger socket, ground flat on the face (Fig. 16.8). On older forks that have

16.6 Deflating the positive chamber in an air-sprung suspension fork

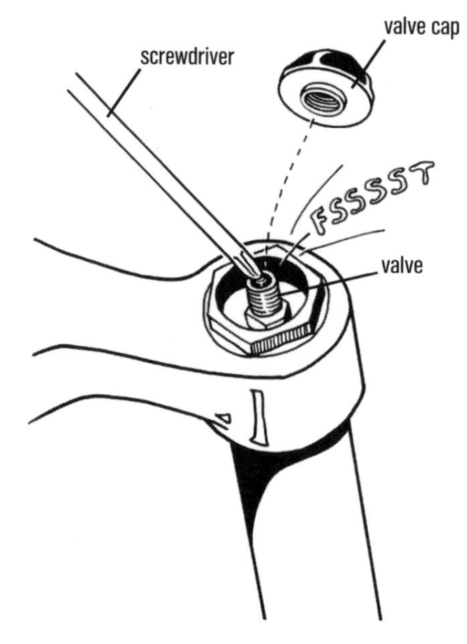

16.7 Removing the spring stack from a Manitou SX suspension fork

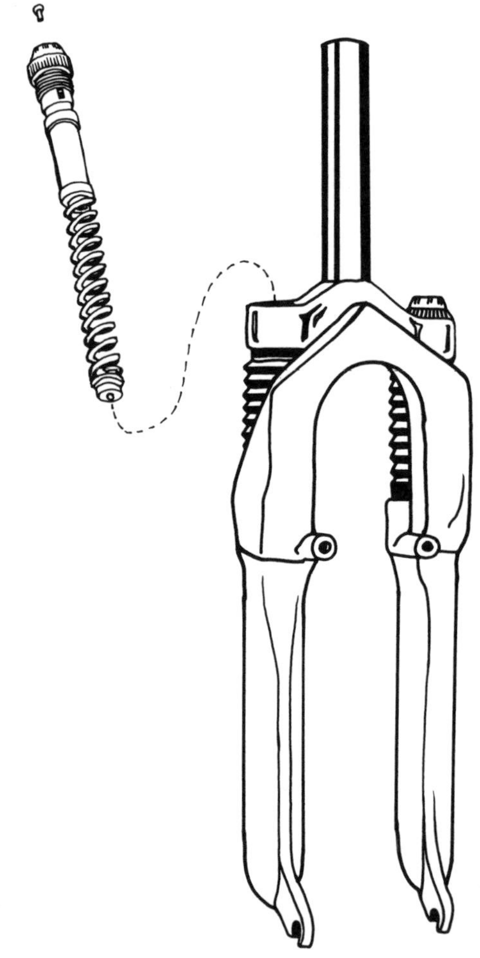

16.8 Removing an air top cap with a large socket ground flat on the end

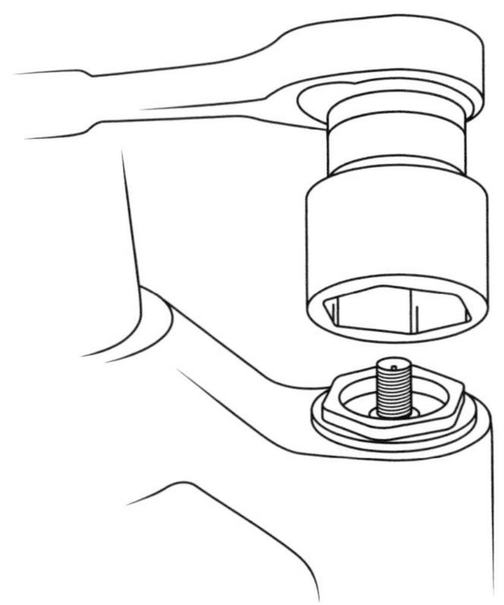

16.9 Lockout and compression damping adjustments

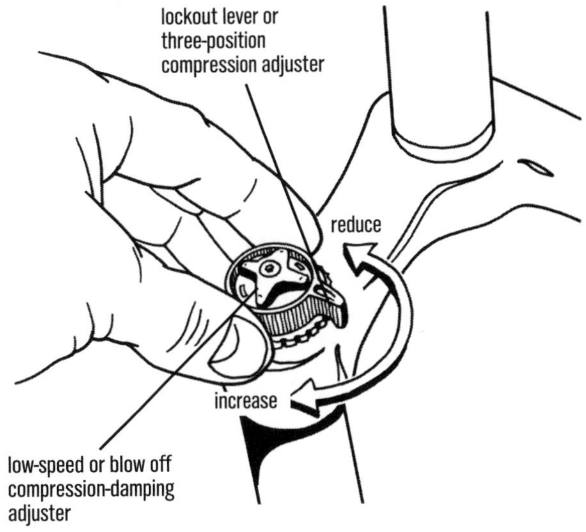

crown bolts and a split crown, loosening the crown bolts makes it easier to unscrew the cap. If there are springs in both legs, remove or deflate both of them.

3. **Measure the distance along the upper fork leg tube from the lower edge of the fork crown to the top of the lower leg.** Push down on the handlebar to make sure the fork won't compress further.
4. **Subtract the second measurement from the first measurement.** This is the fork's total available travel.

IMPORTANT: *To ensure that the tire cannot hit the fork crown during a ride (which is highly undesirable), measure the distance from the top of the tire to the bottom of the fork crown. Compare this measurement to the total fork travel, or remove the springs and push the fork down as hard as you can toward the tire. If the crown can hit the tire on full compression, it can stop you dead (literally!). Use a tire size that ensures that the tire cannot hit the crown. Many fork manuals specify the largest tire you can safely use with a specific fork.*

b. Measure sag

Sag, also called ride height, is the amount of fork compression that occurs when the rider sits on the bike without moving. It is a critical measurement you will need in order to properly tune a fork. Although you can measure the sag of a rear shock by yourself, you can't properly measure fork sag without someone else holding you up, so enlist a friend to assist you. If the fork has boots, they must be removed for this procedure. (Velcro ones are easy to remove; remove rubber bellows as in 16-11.)

1. **Adjust the compression damping to its lightest setting.** If the fork or rear shock has a lockout lever, make sure it is switched off (turned fully counterclockwise if it's atop the crown, flipped to its open position if it's a remote lever or button on the handlebar). If the fork has a compression-damping adjuster or blow-off knob (also atop the drive-side fork leg), turn it fully counterclockwise until it stops (Fig. 16.9).
2. **Establish a way to mark the travel used.** If the fork or rear shock already has an indicator O-ring around at least one upper tube (Fig. 16.1), you're good to go. If not, smear a ring of dark-colored grease around one upper tube where it comes out of the seat atop the lower leg, or tighten a plastic zip-tie around it (Fig. 16.10), and slide it down against the seal (same idea for rear shock shaft).
3. **Have your friend hold your bike upright while you sit on it.** Wear the full riding gear you normally use, including full hydration pack.
4. **Standing on the pedals, bounce up and down to activate the suspension.** Minimize using the brakes while avoiding running over or head-butting your friend.

16.10 Using a zip-tie to measure travel or sag

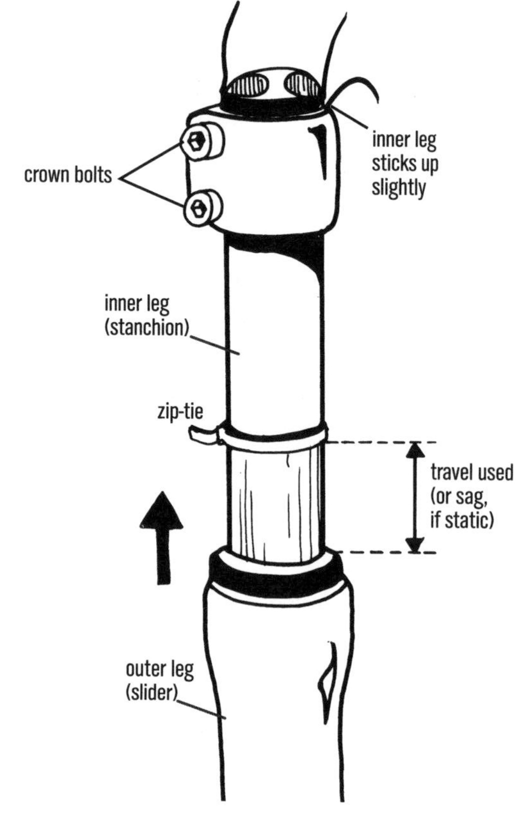

5. **Settle into your desired position.** The rear shock sags much farther when seated than when standing and vice versa for the fork, so base your position on the type of riding you do. For cross-country riding on a shorter-travel bike, check sag while seated; for trail/enduro riding, set sag seated for moderate riding and standing for bike-park riding; for downhill, set sag in a standing, aggressive attack position. For the fork, measure sag while standing. If you have an intelligent, or pedal platform, fork or rear shock that automatically distinguishes between bumps and pedaling forces, it may take some time to fully sag. You might have the impression that it does not sag, but there is a small bleed port in the damper that allows oil to pass by the piston. So wait 30 seconds or so for the fork or rear shock to settle to ride height.
6. **Have your friend push the indicator O-ring, zip-tie, or grease glob down against the upper seal.** Stay motionless in your chosen position while he or she does this.
7. **Gently climb off the back of the bike.** Avoid compressing the suspension further; doing so would negate your careful placement of the marker for sag height.
8. **Measure the sag.** Measure the distance between the marker and the seal. You want to have the bike sag to a certain percentage of your fork's or rear shock's total travel, so compare your sag measurement to total travel measured in section a. This is how you determine the spring setting (see 16-7).

c. Measure travel used

Start your ride with an indicator O-ring (Fig. 16.1), zip-tie (Fig. 16.10), or grease glob on an upper tube pushed down against the top of the outer leg. At the end of the ride, measure the distance from this mark to the top of the outer leg (Fig. 16.10). Compare this to your total travel. This helps you determine the proper spring rate, preload, and damping to use. If the fork is adjusted properly for the course you are riding, it will bottom out (i.e., use the full travel) at least once on the course; otherwise, you are not using the fork's full potential. The tuning sections (16-7 to 16-9) go into more detail on this subject.

16-7

TUNING SUSPENSION FORKS

LEVEL 1

This section sets out what to adjust on a fork and why. Subsequent sections explain how to actually perform those adjustments. Most of this advice applies equally to rear suspension, so use this guide for tuning rear shocks as well.

Pick out a short test course with at least 100 meters of rough trail, a hill, and a sharp turn. Go back and forth, climbing and descending, keeping your riding style consistent, to see how each change you make affects performance. A short course is good, because it encourages frequent retesting and you can remember everything that happened.

Make one change at a time, and ride the course in between so that you can isolate the effect of each change. Record your observations immediately after every ride, rather than waiting until you have time to work on your bike and have forgotten what it was that you wanted to change.

Familiarize yourself with your fork adjustments by riding the test course with each variable at its maximum setting and then at its minimum setting (again, changing one thing at a time), including tire pressure. This allows you to really appreciate what, say, rebound damping does and what it feels like when the damping is too high or too low. The downside is that you'll become so obsessed with how your fork works that your friends will harass you about fiddling with it during rides!

Before adjusting the fork, keep in mind these three important caveats (you'll discover the truth of them if you ride your test course with, one at a time, tire pressure, positive spring rate, high-speed compression damping, and rebound damping at maximum or minimum settings):

1. **Tires are your first line of suspension.** Softer tires (up to a point) grant improved traction and lower rolling resistance on rough terrain. The softer the tires, the lower the small-bump compliance you will need from the fork and rear suspension. And once you find a tire pressure you like, ride it that way until you retune your shocks, because your suspension will feel different with different tire pressure.
2. **A bottoming sensation might not be caused by the fork actually bottoming out.** Check for an overly stiff spring, or for excessive damping for the bike, the rider, and the terrain.
3. **A harsh sensation may or may not be caused by soft springs.** Check for a spring rate that's too soft or rebound damping that's too high for the bike, rider, and terrain. These cause the suspension to ride with much of the travel compressed (i.e., the fork is packed up).

After setting tire pressure, adjust the fork in the following sequence:

a. Spring rate

Rate is how much length the spring shortens in response to a given force. You change the rate on an air fork by changing the air pressure (Figs. 16.11 and 16.12), and you change the rate on a coil or elastomer fork by interchanging springs (Fig. 16.7) and/or varying the spring preload (Fig. 16.13).

16.11 Adjusting pressure (spring rate) in an air-spring fork

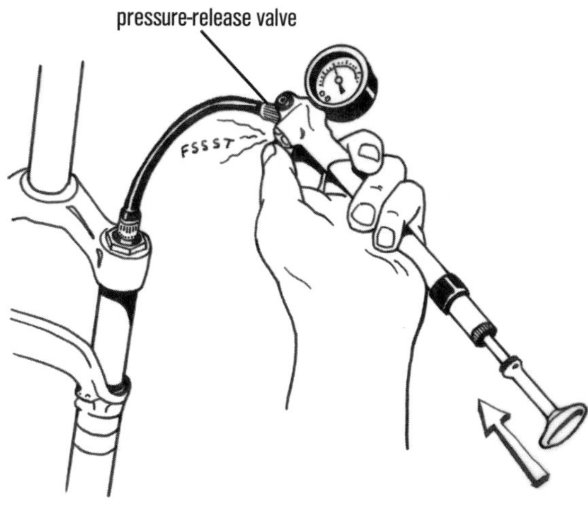

16.12 Adjusting air pressure in a RockShox Mag or early SID fork through its ball-needle valve

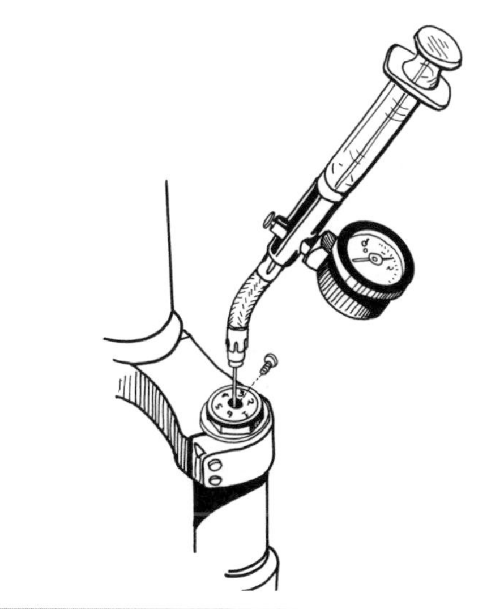

16.13 Adjusting spring preload on coil spring or elastomer spring fork

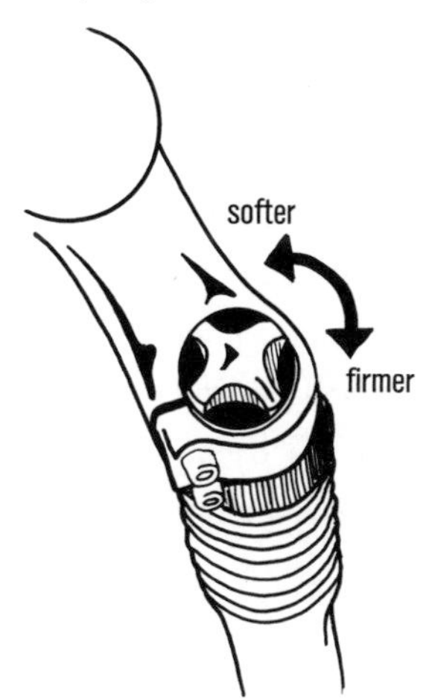

Some air forks have suggested pressure imprinted or etched on a fork leg. Lacking that, consult the manufacturer's online manual. Otherwise, a good starting pressure for many forks is to put the numerical value of your weight in kilograms into the fork in psi (e.g., if you weigh 80kg, try 80psi in the fork). Remember the sag measurement you made in 16-6b? This is where to use it; you want the fork to settle into a certain percentage of its travel when it is just rolling along on smooth ground. The rear shock sags further when seated than when standing, and vice versa for the fork.

- For cross-country racing, total travel is usually 4–5 inches, and you want it stiff. Measure rear sag while seated and front while standing, and set the spring rate so that sag is 20–25 percent of travel.
- Trail bikes with 4–6+ inches of travel work well with sag set at 25–30 percent of travel. Set rear sag while seated for moderate riding and standing for bike-park riding.
- Set enduro or all-mountain bikes with 5–7 inches of travel to 25–35 percent sag, while standing.
- Downhill and other gravity-driven riding with 7–10 inches of travel benefit from 30–40 percent sag, measured while standing in an aggressive, low-shoulder, attack position.

1. **If the spring rate is too soft, you will easily bottom the fork (i.e., go through its full travel), need high preload (coil and elastomer forks), and have too low of a front end on rough downhills.**
2. **If the spring rate is too hard, the fork will rarely or never bottom, meaning that you're not absorbing the bumps as you could, costing you energy and comfort.**

b. Spring preload (coil-spring and elastomer forks only)

Adjust spring rate before you adjust preload. Ideally, you want to hit the spring rate right on and have minimal preload, because preload makes the spring rate ramp up faster and respond more harshly.

Without springs to interchange, preload is the only spring rate adjustment; use it alone to set sag depth. This setting obviously overlaps with the spring rate.

You can often adjust preload while riding, because the knob(s) is (are) easily accessible (Fig. 16.13). This is how you know how to change it:

1. **If the spring preload (or spring rate) is too low, the bike will sag too much, the front end will be too low entering turns, and oversteering will occur.**
2. **If the spring preload (or spring rate) is too high, the bike will sag insufficiently, the fork will feel stiff and/or harsh, and it will understeer, especially when attempting a tight turn at low speeds.**

c. Air volume (air forks only)

Varying air volume in an air fork is similar to varying preload in coil and elastomer forks. Reducing air volume makes the fork stiffen up faster as it moves (that is, the spring rate ramps up faster), and vice versa. For instance, if you find an air pressure that works well through its initial and intermediate travel, but you bottom the fork too often, you can decrease the air volume to stiffen the fork sooner with the same air pressure.

Ideally, your fork will have volume spacers (Figs. 16.14 and 16.15) with which you can change air volume. In the infancy of air-sprung forks, riders decreased the volume in an air cylinder by unscrewing the air valve and pouring in some oil. However, due to foaming of the oil, adjusting the air volume with spacers is highly preferable.

16.14 Fox tongue-and-groove snap-together air-volume spacers

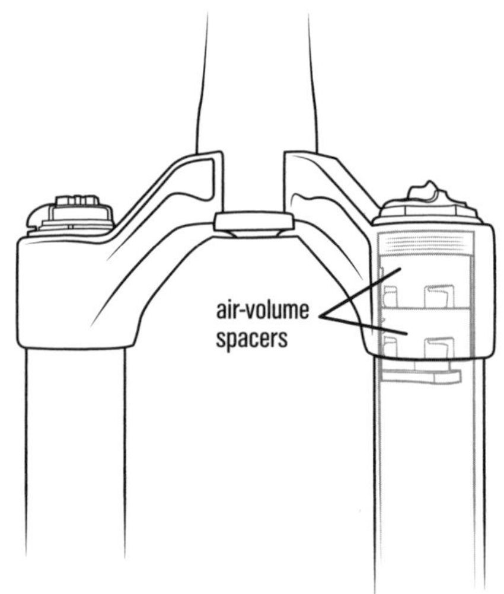

16.15 RockShox screw-together Bottomless Token air-volume spacers

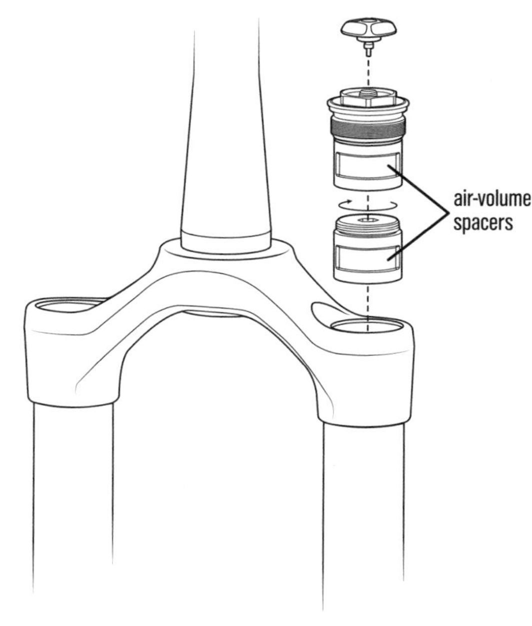

d. Negative spring rate

The negative spring in an air fork works against the main air (positive) spring to compress the fork. This makes the fork more compliant over small bumps and makes it behave more like a coil spring throughout its travel. Early air forks did not have negative springs, and riders suffered on stutter bumps.

Air springs have a progressive spring rate, which means that as the spring is compressed, the force it takes to move the next increment of travel goes up exponentially, rather than linearly. A coil spring, in contrast, has a linear spring rate through much of its stroke; it takes the same increase in force to move 1mm farther, whether you are at the beginning or middle of the spring's compression. Thus, it takes more force to get an air fork to move initially on a little bump, and the force required to keep moving farther into the stroke on bigger bumps ramps up. By pulling the fork down, a negative spring decreases the fork's net spring rate when the fork is high in its travel, helping the fork react quickly to small bumps. A negative air spring—a separate air chamber below the positive-spring chamber—will also start in its fully compressed state when the fork is fully extended. As the fork compresses, pressure in the positive spring builds while pressure in the negative air spring drops; its force opposing the main spring ramps down rapidly while the main spring's force is ramping up rapidly, so the net spring rate of the fork is fairly linear.

NOTE: *Because elastomers also are progressive springs, they are often paired with coil springs to ramp up at the end to avoid bottoming (Fig. 16.7).*

The negative-spring adjustment is generally via a Schrader valve. Some forks have the positive and negative spring chambers connected via a bypass port around the piston seal when the fork is fully extended, so there is only a single valve through which you pump both chambers. Whenever the fork is at full length, the bypass around the seal equalizes the pressure in both chambers. As soon as the fork is compressed even a small amount, the piston seal is above the bypass port, so the positive and negative chambers are separate. The negative-spring adjustment can instead be a second valve at the bottom of the leg or at the top, adjacent to the positive-spring valve.

Because the negative air chamber pulls the fork down when filled with compressed air, make sure if you have two valves that you pump up the positive spring first. If you mistakenly pump up the negative spring first while the positive spring pressure is zero, then as you pump the positive spring, the pressure in the negative spring will go up and up as the fork lengthens. In that situation, you can try pumping the positive spring to a zillion psi, and you still will not get the fork to full length!

Instead of an air negative spring, some air forks have a negative coil spring at the bottom of the fork, which may or may not be adjustable.

Many forks with positive and negative spring chambers are designed so that if the pressure in the positive and negative springs is equal, the bump force required to start the fork moving is close to zero, meaning that small-bump compliance is optimized. If you have a separate valve for the negative spring, start with equal pressure, and once you find the pressure in your positive spring that you like for big hits, tweak the negative spring pressure to get the small-bump compliance you seek.

e. Rebound damping

Rebound damping controls the speed of return of the spring. Have you ever checked a car strut (shock absorber) by pushing down on a fender and releasing it to see if the car bounces a single time or keeps boinging up and down? That's a check of rebound damping, and you want a single bounce from your bike.

High-end forks generally have a rebound-damping adjuster knob at the bottom of the damper (non-disc-brake) side, often red in color (Fig. 16.16).

Back the rebound adjuster out fully (counterclockwise; often it's indicated on the adjuster as turning it toward "faster" versus "slower"). Ride off a curb and check how many bounces you get. Keep tightening the adjuster until you get only a single bounce off the curb. Another test is to see what happens when you load the suspension and quickly release it, allowing the handlebar and saddle to rebound freely. If you're just checking the fork, stand on the ground straddling the bike, throw your full weight onto the handlebar grips, and then let go of them right when the suspension compresses as far as it's going to go. The rebound setting you want is so the fork rebounds quickly but is slowed down just enough that the wheel doesn't leave the ground.

There is an ideal rebound setting for every type of terrain at a given speed, so play around with this adjustment. It's important for control and comfort.

16.16 Adjusting rebound damping

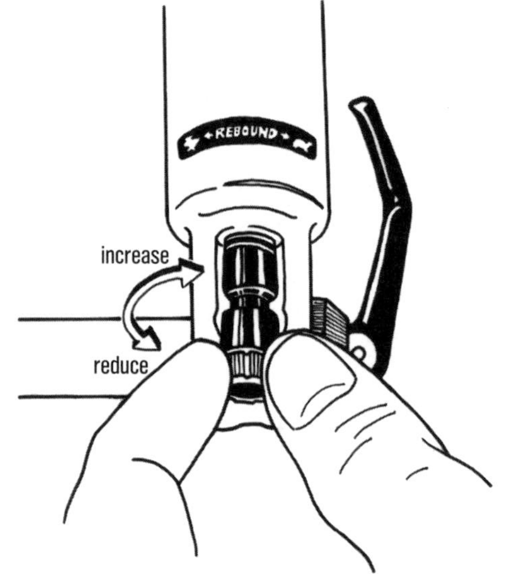

1. **If rebound damping is set too low, the fork will spring back too fast.** It's set too low if the wheel springs up from the ground after landing a jump and feels as if you're returning from a bungee jump. You will also have trouble maintaining a straight path through rocks, you will understeer, the front wheel will try to climb the berm while cornering, and the ride height will be too high in rough terrain.
2. **If rebound damping is set too high, the fork will not spring back fully after each hit.** When bumps come rapidly, the fork will pack up—it will get progressively shorter so that it will have less and less travel available for successive rapid hits. This will feel quite harsh, and the fork will bottom out after several successive large hits, even if compression damping and spring rate are set correctly. The fork will oversteer and won't rebound after landing a jump.

f. Low-speed compression damping

Compression damping in general controls the speed at which the spring compresses, and low-speed compression damping in particular controls the speed at which the spring compresses when the fork shaft is moving slowly (i.e., not hitting a big bump). Low-speed compression damping is added to the fork to help it resist bobbing due to pedaling, and it keeps the suspension up during hard braking and fast, bermed turns, when the G-force would otherwise push the rider and the bike down. If the fork has a separate low-speed compression-damping adjuster, it will generally be a knob atop the damper (drive-side) leg in the center of the three-position compression adjuster or lockout lever (Fig. 16.9).

As with all damping, don't overdo it; use just enough low-speed damping to keep the suspension up when you are braking, pedaling, and riding berms and deep dips. However, if you're going to overdo one of them, it's better to overdo the low-speed compression-damping setting than it is the high-speed setting, especially if you ride big bumps and drops at speed. That's because as your suspension goes deeper into its travel upon hitting a big bump or landing from a drop, the inward movement of the fork stanchions or the rear shock shaft slows down. Consequently, the low-speed compression damping takes over near the end of the travel and controls the shock movement, rather than the high-speed compression damping. So if you have the high-speed compression damping set more open, the shock will absorb the big impacts quickly and feel nice and plush at high speed over sharp hits, and the low-speed compression damping will still prevent you from bottoming out harshly. Combined with a low rebound-damping setting, the shock will quickly absorb and return from big impacts and be back up high, ready for the next bump.

1. **If low-speed compression damping is set too low, the symptoms include bobbing when pedaling, fork dive while braking, and oversteering.**
2. **If low-speed compression damping is set too high, the ride height is high despite a soft spring and/or little preload, the fork does not absorb stair steps well, and understeering is common.**

g. High-speed compression damping

High-speed compression damping is for the big stuff; it controls the speed at which the spring compresses when the fork shaft is moving fast (i.e., hitting a big bump). Without it, the fork would bottom out hard. This adjuster will often be a blue three-position dial atop the damper (drive-side) leg (Fig. 16.9).

Set the adjuster all the way out unless the fork bottoms out; you want the adjuster set low enough that it absorbs bumps quickly through the system and lets the spring work. Even with the perfect spring rate, if high-speed compression damping is set too high, the piston will move so slowly through the oil that you'll already be past the obstacle before the fork has moved. Your goal in setting compressing damping is to have the fork travel used up when you are at the highest point of the rock you're rolling over.

1. **If high-speed compression damping is set too low, the symptoms include bottoming out and instability.**
2. **If high-speed compression damping is set too high, the fork will not react to small bumps, will feel harsh, and will rarely or never bottom out.**

h. Bottoming control damping

Some forks have this adjustment; it's an additional bit of high-speed compression damping. Use as little bottoming control as possible. But if you ride very aggressively and take a lot of big hits yet still want a super-plush ride, this adjustment could give you everything you want.

i. Inertia valve

A fork with a pedal platform may have an inertia valve that opens when the fork shaft moves fast from hitting a bump hard. If the fork has an adjuster for this valve, it will generally be atop the damper (drive-side) leg, like a lockout blow-off adjuster (Fig. 16.17). You can set the inertia valve adjustment to determine how big a bump it will take to blow open the compression-damping circuit.

Set the adjustment high at first so that it gives you the pedaling performance you want on a smooth climb. Then ride your test course and see how much bump compliance you've sacrificed by setting the inertia valve for high blow-off, and back it off until you find your happy medium.

1. **If the inertia-valve adjustment is set too high, the fork will not respond to small bumps.**
2. **If the inertia-valve adjustment is set too low, the fork will bob when pedaling on smooth surfaces.**

16.17 RockShox Gate lockout blow-off sensitivity adjuster

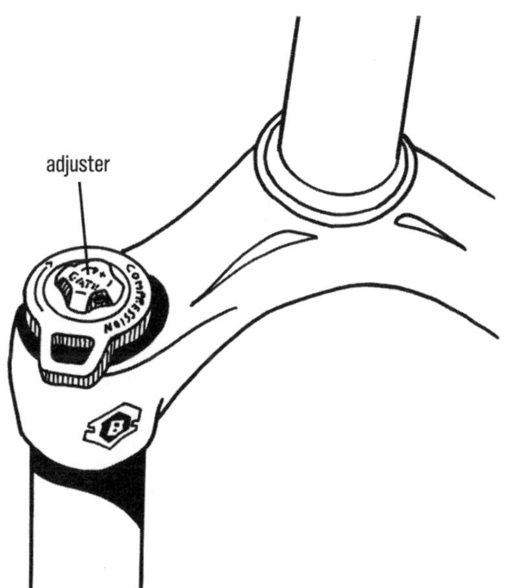

16.18 Adjusting RockShox travel with U-Turn knob

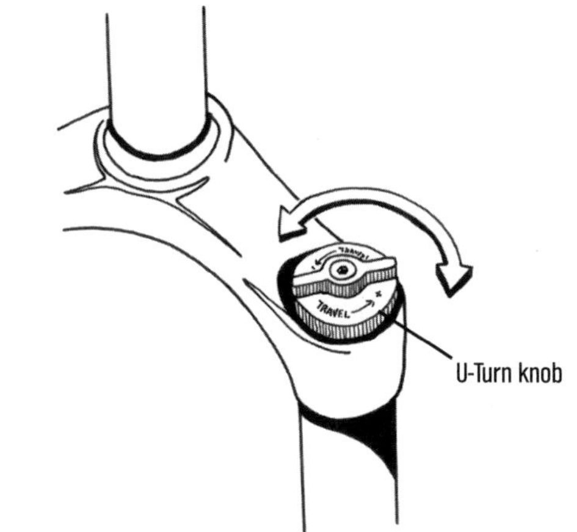

j. Travel

If fork travel is adjustable, increase it for rough courses and decrease it for climbing and riding smooth trails or roads. Some forks, both air and coil-spring types, have a knob atop the spring leg to adjust travel on the fly (Fig. 16.18), with marks etched on the outside of the upper tube indicating travel. Other forks require removing the outer legs and replacing the air-spring shaft with one of a different length (16-14d) or removing or adding a spacer on the fork shaft below the piston or plunger inside the inner leg (16-14c, Figs. 16.34, 16.40, and 16.41). Spacers below the plunger restrict travel by preventing the shaft from extending as far out of the bottom of the inner leg.

A shorter fork is an advantage aerodynamically when you are riding long stretches of road; it helps get you down out of the wind. And when climbing, a shorter fork keeps the front end down for improved ergonomics and maintaining more weight on the front wheel for better steering.

A longer fork is an advantage on big bumps at high speed, as it allows the bump force to be absorbed more completely.

k. General fork-tuning guidelines

- **Speed up:** Because rebound and compression damping are speed-sensitive, don't worry about settings that feel good at low speeds being too light for high speeds; the fork will get stiffer as you hit things faster. At all speeds, you want the fork to pop back as quickly as possible without kicking back, and you want it to compress into the travel as quickly as possible without bottoming.
- **Heat up:** Damping is temperature-sensitive. Oil is thick and sluggish in the cold, but when the weather gets hot, the fork gets really lively. You will need to adjust accordingly in summer. Choose a stiffer spring rate and firmer damping adjustments.
- **Cool off:** In the cold of winter, overall speeds are slower, the grease and oil in the fork are thicker, and the elastomers and coil springs are stiffer. Consequently, lighten up the spring and damping adjustments.

l. Common ride symptoms and suspension-tuning fixes for them

Fork too hard	Fork too soft
Decrease compression damping	Increase spring rate
Decrease rebound damping	Increase compression damping
Decrease spring rate	Increase oil viscosity
Decrease oil viscosity	Replace old damper oil
Increase spring rate*	Put oil in (empty) damper
Decrease inertia valve adjustment	

*If you are running a spring rate that is too soft for your weight and ability, you could be misled into thinking that the spring rate is too stiff. This is because you are using up the fork travel before you begin to ride. Furthermore, the fork is working in a stiffer spring-rate range on smaller hits, giving the impression that the fork is harsh and stiff. This is where the ride-height (sag) adjustment is important.

Front end understeers/ nervous descending	Front end knifes/ oversteers
Increase rebound damping	Decrease rebound damping
Decrease spring preload	Increase spring preload
Decrease spring rate	Increase spring rate
Decrease compression damping	Increase compression damping

Front end pushes or washes out in turns	No response to small bumps
Increase rebound damping	Decrease compression damping
Decrease spring preload	Decrease spring preload
Decrease spring rate	Decrease spring rate
Decrease compression damping	Increase negative spring rate
	Decrease rebound damping
	Overhaul dirty fork
	Decrease inertia valve adjustment

16-8

TUNING AIR-SPRING FORKS

Lightweight forks use compressed air as a spring. It would, after all, be hard to come up with a spring lighter than one made of air!

After you understand how to adjust an air-spring fork from reading this section, you can fully utilize the tuning tips in 16-7.

a. Adjusting positive air pressure

Greater air pressure in the positive spring chamber means a stiffer fork, and vice versa. It is a good idea to check the air pressure every couple of weeks; all forks lose pressure over time.

Do not use a tire pump on the fork; the large stroke volume is poor for adjusting low volumes of air at high pressure, the gauge won't tell you how much air is left in the fork, and unless the fork takes a ball-inflation needle, you will lose most of the pressure when removing the pump head. Use a shock pump with a no-leak fitting, because a Schrader valve can lose air when the pump is removed. Good shock pumps have a no-leak fitting built into the head (Fig. 16.11) that prevents leakage by maintaining its seal around a Schrader valve as the head is unscrewed until the valve's center pin is no longer depressed.

Before shock pumps had no-leak heads, an alternative no-leak-upon-pump-removal was sometimes built into the fork; RockShox Mag-series forks, early SIDs, and the air assist on post-2002 Judys use valves that require a ball-inflation needle (Fig. 16.12). SID forks from 1999 to 2000, and early Marzocchi air forks, require a special adapter that fits down in the recessed Schrader valve and prevents air from escaping as the adapter is removed.

When you screw the pump head onto the valve, air from the fork fills the shock pump's hose, thus dropping the pressure in the fork. Therefore, the pump gauge is not an accurate indicator of what pressure was in the fork, since an indeterminate amount of pressure was lost in the process. However, the no-leak pump head does guarantee that its gauge reading is the pressure that will remain in the fork after removing the pump. Pump to the pressure needed to give you the desired sag—see 16-7a, or consult the fork manual.

IMPORTANT: *If the fork has an air negative spring with a separate valve, always pump the positive spring chamber first. Pumping the negative spring first will shorten the fork, increasing the volume of the negative spring chamber. Subsequently pumping the positive spring will increase pressure in both chambers without bringing the fork up to full length.*

The pump for old RockShox Mag forks is just a plastic syringe with a dial gauge on it (Fig. 16.12). The air valve on each leg is located beneath either a Phillips screw or a plastic pry-off cap on top of the compression-damping adjustment knob. Tighten the compression adjustment before inserting the pump needle to avoid pinching the rubber valve on the top of the adjuster rod inside, causing it to leak. Moisten the needle, and insert it into the valve hole (Fig. 16.12).

b. Adjusting the negative spring

The negative spring in an air fork works against the main air spring to actually compress the fork.

A second Schrader valve for an air negative spring will be either at the bottom of the left leg or at the top, adjacent the positive-spring valve. Again, pump the positive spring before pumping the negative spring. Your fork may have only one Schrader valve, yet it may still have an air negative spring (see 16-7d). Some air forks have a coil-type negative spring, and many of these are not adjustable. Early (1998) SID forks have a coil-type negative spring on top of the cartridge shaft under the right-hand piston that is adjustable. There is a circlip constraining the top end of the spring, and you can clip it into any of six grooves to vary the compression of the negative spring. You must take the fork apart and remove the cartridge to do it, though (16-13 and 16-16a).

c. Adjusting damping

On both front and rear shocks, blue-anodized knobs and levers are likely to be compression-damping adjustments, and red ones are rebound-damping adjusters. Compression-damping adjusters are commonly found on the top of the right (drive-side) fork leg, and rebound adjusters are most commonly found at the bottom of the same fork leg. Turning the knobs or levers counterclockwise opens the damper orifices to decrease damping, and turning them clockwise increases damping (slows down the shock movement). Knobs generally have click stops under them, so that you can count the number of clicks to get to a proscribed adjustment.

A common setup is to have a red, rebound-damping adjuster knob at the bottom of the right leg (Fig. 16.16) and a blue, rotating three-position compression-adjustment or lockout lever atop the right leg (Figs. 16.9 and 16.17), with or without a separate low-speed compression-damping adjustment knob atop it. The three-position lever controls both low-speed and high-speed compression damping. Generally, it leaves the compression damping fully open in the fully counterclockwise position as well as in the middle position, but in the fully clockwise (firm or locked out) position, it has closed off both the low-speed and high-speed compression-damping circuits. In order that you don't break your fork (and yourself) if you hit a hard bump at high speed with the fork in the locked-out position, there is a blow-off feature, where the force of the oil trying to get through the hole(s) in the damper flexes (or pushes a spring that is behind) thin, flexible steel shim washers covering the hole(s), thus allowing the fork to compress and save itself and you. The magnitude of the impact required to allow oil to flow through the blow-off hole(s) is sometimes adjustable (Fig. 16.17). Some high-end forks also have a separate low-speed compression-damping adjustment knob atop the three-position lever (Fig. 16.9).

Turn each adjuster knob all of the way until it stops in one direction. Then turn it all of the way in the other direction, but this time, count how many clicks to go from one end of the adjustment range to the other. Go back half that number of clicks to set the adjustment in the middle of its range. This is a good place to start. Adjust from there based on what you feel when you ride.

High-speed compression damping

- For a softer, more plush ride, turn the knob or lever counterclockwise.
- For slower fork movement and a firmer ride, turn the knob or lever clockwise.

Low-speed compression damping

- For less damping and a softer, more plush ride, turn the knob counterclockwise.
- For less bobbing when pedaling and a higher ride height on berms and dips, turn the knob clockwise.

Rebound damping

- For faster rebound (higher riding speeds with generally smaller drops and hits) and more traction (keeping the tires on the ground), turn the knob counterclockwise.
- For slower rebound (slower riding speeds with bigger drops and hits), turn the knob clockwise.

Some forks have a 2- or 3-position remote lever (or electronic button) on the handlebar to control compression damping; it is sometimes paired with a second lever that controls the rear shock.

d. Making other adjustments

Reducing air volume in an air-sprung fork makes the fork stiffen up faster as it moves (i.e., the spring rate ramps up faster). You can decrease the volume in an air cylinder of a modern fork by putting in volume spacers to take up space in the positive air chamber. Deflate the fork, unscrew the air cap and snap on (Fig. 16.14) or screw in (Fig. 16.15) another volume spacer or two under it. Alternatively, by removing volume spacers, you will give the fork a more linear spring rate that goes through its travel more easily. On older air forks without air-volume spacers, you can decrease positive-spring air volume by removing the air top cap and pouring some oil into the chamber. Obviously, do not try this with a modern fork with a negative air spring and only one inflation valve; the oil will flow down through the bypass around the air piston and instead decrease the volume of the negative spring chamber and leave the positive-spring volume unchanged!

On the 1998 RockShox SID, you can change air volume by changing piston height: Increase the volume by tightening the piston deeper into the fork and decrease it by unscrewing the piston. You get at the piston by releasing the air with a ball needle (Fig. 16.12) and unscrewing the top nut with a big socket ground flat (Fig. 16.8). Screw the piston in or out with an 8mm hex key.

If you have the fork apart (16-13), you can change oil viscosity (16-16a and 16-16b) to change the speed of the fork shaft on compression or rebound.

Some forks have an inertia valve overriding the compression-damping adjustment that distinguishes between pedaling forces and bump forces. The fork will be highly damped until a bump of a certain impact magnitude is encountered, at which point the inertia valve opens (a heavy mass that was covering the oil hole in the damper bounces up) and the fork moves freely.

You can also change the blow-off threshold (Fig. 16.17) when the fork is the firm setting for climbing. You may be able to vary travel (Fig. 16.18) on the fly on some models.

16-9

TUNING COIL-SPRING AND ELASTOMER FORKS

Telescoping coil/elastomer-spring suspension forks (Figs. 16.7 and 16.25) are simple in principle and are generally straightforward to adjust. After you understand how to adjust your coil/elastomer-spring fork from reading this section, you can fully utilize the tuning tips in 16-7.

a. Setting spring preload

Spring preload, the amount of compression of the spring at rest, can be adjusted on many coil-spring or elastomer forks, but it should be done only after you have installed the correct spring for you. On many forks, you can adjust the preload simply by turning the adjuster knobs on the top of the fork crown (Fig. 16.13)—even while riding, as you encounter terrain variations. On forks that have springs in only one leg (usually the left leg), there is only one preload adjuster. Note that the U-Turn knob (Fig. 16.18) on some RockShox coil-spring models adjusts travel, not preload.

Rotating the adjuster knob(s) clockwise gives a firmer ride by tightening (and thus shortening) the spring stack. Rotating the adjuster knob(s) counterclockwise softens the ride. Make sure the top cap surrounding the knob does not unscrew from the fork crown; you may need to hold it tight with one hand (or a wrench) when you loosen the adjuster knob. Check the top cap occasionally to make sure it is not unscrewed or being forced out because of stripped threads. If its threads seem to be stripped, get a new top cap right away; if the top cap pops off, the spring can shoot up into your face at high velocity.

Preloading uses up some of the spring's length and therefore makes the spring stiffen up faster as the fork moves. Varying the preload changes the fork's sag, and it shortens the life of the spring, as it is being compressed even while the bike is hanging in the garage. It is better to change the spring stack (see 16-19b) to get the ride you want and minimize preload; use the preload adjustment only as a way to change the fork quickly during a particular ride.

b. Replacing elastomers and coil springs

To make major changes in the fork's spring rate, you must change the springs inside of the fork (Fig. 16.7). Manufacturers usually color-code the elastomers—and often coil springs as well—for stiffness, although you can tell the difference between stiff and soft elastomer bumpers by squeezing them between your fingers. (Some manufacturers refer to the elastomers as MCUs, for micro-cellular urethane, referring to small air voids trapped inside the urethane spring.) Extra springs usually come with the fork, or you can buy them at your bike shop.

1. **Unscrew the top cap (or caps, if there are springs in both legs) counterclockwise.** On its old, pre-1998 forks, RockShox recommends that you first loosen the crown bolts to relieve inward pressure on the fork legs before unscrewing the caps. On some forks, the top caps can be unscrewed with your fingers, whereas others require a flat-ground socket (Fig. 16.8) to unscrew the top cap; sometimes you must remove a knob covering it first. Manitou TPC (Twin Piston Cartridge—Fig. 16.46) and RockShox Pure damping systems have springs in only one leg.
2. **Pull the spring(s) out of the fork** (Fig. 16.7). Oftentimes, springs will come out attached to the top cap(s), and several springs may be snapped together with plastic connectors. On many older (pre-1996 or so) forks, the top cap is connected to a rod (the skewer in Fig. 16.25) that runs through each of the elastomer bumpers. If the springs do not come out with the top caps, turn the bike upside down or compress the fork to get them out.
3. **Clean any old grease off the parts.**

4. **Choose the coil springs and/or elastomers that you intend to use.** If any of the elastomers are misshapen or look squished or worn in any way, replace them. Some manufacturers provide a nominal and replacement length for coil springs and elastomers. Measure the springs and check them!
5. **Apply a new coating of grease to the new parts and everything you just cleaned.** Make sure to grease the outside of the coil springs to reduce the noise of the springs rubbing inside the legs. However, in forks with open-bath dampers, hydraulic oil sloshes all around in the spring chamber, so there is no need to grease the springs, but you should replace the oil bath (Fig. 16.32). Pour out the old oil, and replace it with the same volume and weight of new oil (check specs in the fork manual). Don't let any dirt fall down into the leg.
6. **Put the spring stack in the fork leg** (Fig. 16.7), **and screw the cap down.** Be sure to retighten the crown bolts to the required torque if you loosened them.

c. Adjusting damping

Damping controls the speed at which the spring compresses or extends; the damping system moves oil through or around a piston that is being forced through the oil chamber (in rare cases, compressed air is used instead of oil). Varying the size of the hole or the thickness of the oil varies how easily the piston can move through the oil.

16.19 Damping adjuster on a Manitou SX fork

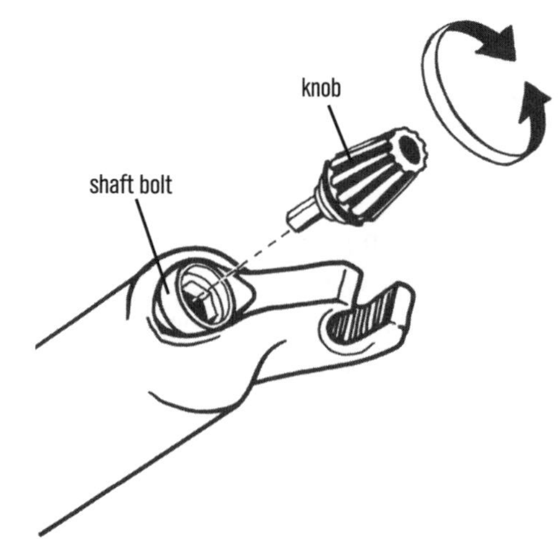

16.20 Damping adjustment on early RockShox Judy fork

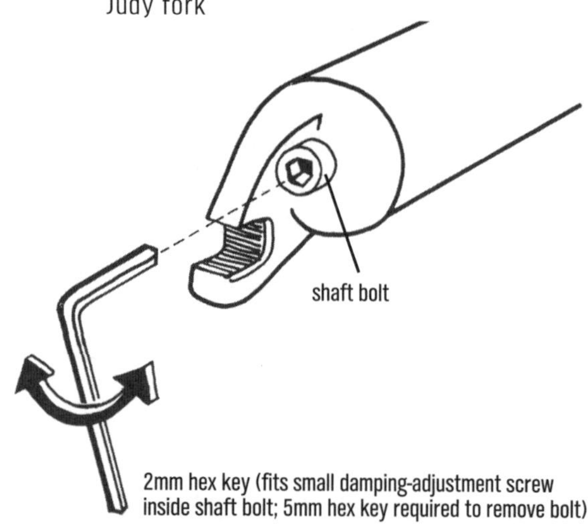

The locations of compression- and rebound-damping adjusters are described in 16-7, but on older forks, they were a bit different. The rebound-damping adjuster, if it existed, was through the shaft bolt at the bottom of the fork leg with a tiny hex key on a knob (Fig. 16.19) or separate (Fig. 16.20).

Additionally, some forks have a remote lockout or three-position compression-adjustment lever on the handlebar.

16-10

REMOVING FORK LEGS FROM THE FORK CROWN

LEVEL 1

This discussion applies to double-crown (a.k.a. triple-clamp) forks—forks with a crown above and below the head tube (Fig. 16.21)—as well as to single-crown forks with a slotted crown. In newer single-crown forks, the stanchions are pressed into the crown and cannot be removed. If the fork has a slotted crown, pulling the inner legs out of the crown is a more convenient way to add or replace dust boots (which are important on old forks) than pulling the outer legs off the inner legs (16-13).

When pulling the inner legs out of a single slotted crown or out of both slotted crowns of a double-crown fork, you can leave the fork steering tube and crown in the bike. It is not necessary to remove the fork legs from the crown to replace or adjust springs or dampers.

16.21 Upper crown height and orientation of triple-clamp forks for head tubes of different lengths (example shown is for RockShox)

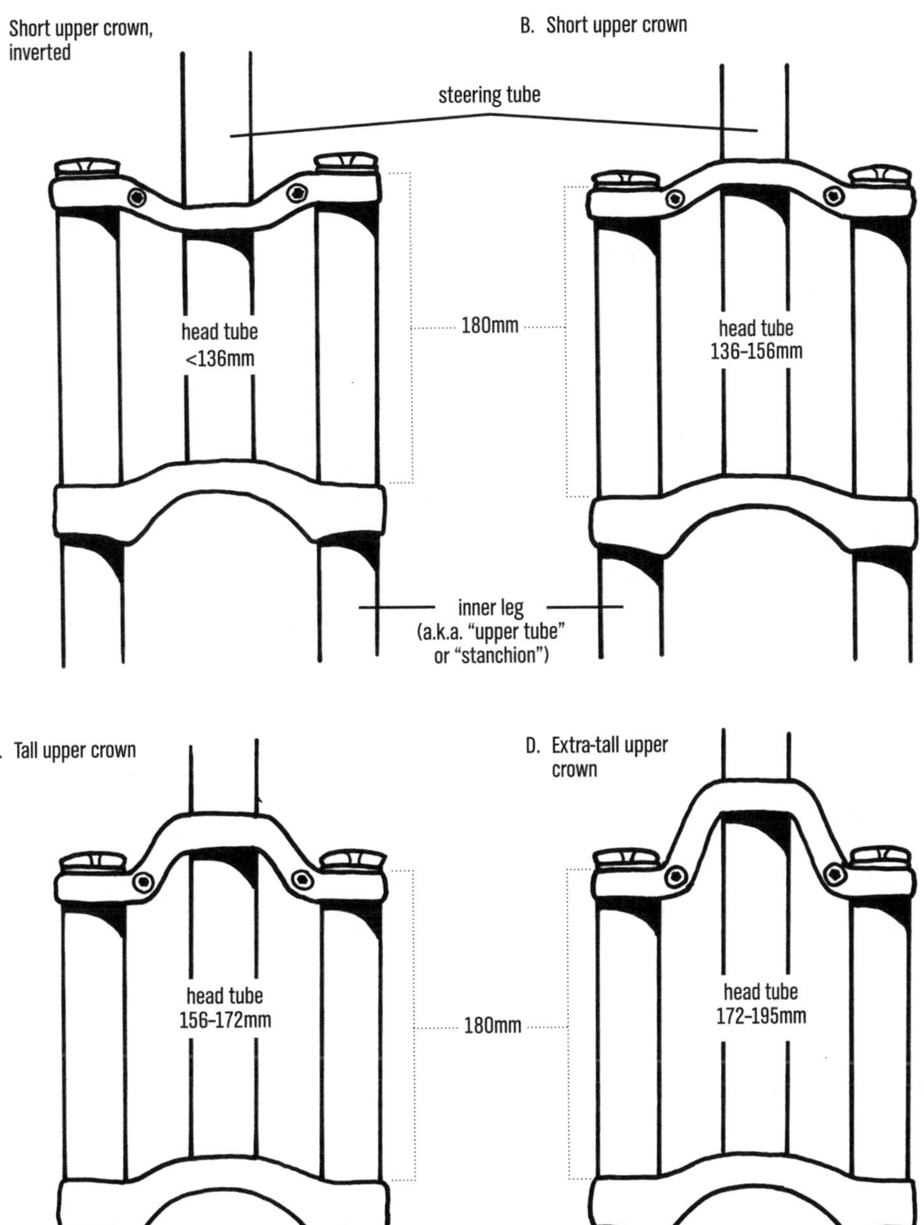

1. **Disconnect the brake.** Unbolt a hydraulic brake. Disconnect a cable-actuated brake by removing the cable end from the brake lever (the reverse of 10-6, steps 8 to 10).
2. **Loosen the crown bolts.** They are shown removed in Figures 16.25 and 16.26. If there are two bolts on each side of the fork crown, do not completely loosen one bolt while leaving the other fully tightened; doing so places a great deal of clamping force on the tight bolt. Instead, go back and forth, unscrewing each bolt a bit at a time.
3. **Pull both fork legs out of the crown(s) using a gentle rocking motion.** Be careful; you can dent the stanchions if you rock them too hard.

16-11

REMOVING AND INSTALLING FORK BOOTS

Fork boots (Figs. 16.2 and 16.7) keep dirt off the inner legs and the seals at the top of the outer legs. To check the fork's travel (16-7), you will need to remove the boots if installed. Fork boots are generally unnecessary on modern forks, thanks to effective multiple sealing systems.

1. **Remove the inner or outer legs.** With a slotted-crown fork or a double-crown model, remove the fork legs from the crown (see 16-10). With pressed-in legs, pull off the outer legs—see 16-13, steps 1–11.
2. **Pull the fork boots off.**
3. **Slide the boots onto the inner legs, large end down toward the outer legs.** Make sure that you are using boots designed for the fork.
4. **Pull the lip of each fork boot into the groove in the top of the outer leg.** You may need to stretch the boot with needle-nose pliers to get it to slide over the outer leg behind the fork brace.
5. **Replace the inner or outer legs.** For a slotted crown, see 16-12; for a pressed-on crown, start with step 12 in 16-13.

16-12

INSTALLING INNER LEGS IN A FORK CROWN

This procedure applies only to single-crown forks prior to 1998 as well as to double-crown (or triple-clamp) forks (Fig. 16.21).

1. **Wipe the upper tubes clean, and make sure that the fork-crown bolts are loose.** If applicable, install fork boots (16-11).
2. **Insert the upper tubes into the fork crown.**
   (a) Single-crown forks: Some forks have a lip against which the inner leg is supposed to rest. Slide the inner leg up into the crown until it hits the lip. Otherwise, on single-crown forks, push the inner leg through the crown until the top of the leg sticks up no more than 2mm above the top of the crown (Fig. 16.10).
   (b) Triple-clamp forks: Slide the inner legs up through both crowns. Make sure that the upper crown is the right shape and orientation to work with the bike's head-tube length (Fig. 16.21). For instance, the manual for the RockShox triple-clamp forks in Figure 16.21 specifies that the lower crown must be positioned to allow 180mm of exposed upper tubes above the lower crown. Upper crowns come in different heights (i.e., short, tall, extra-tall) for different frame sizes. They can be installed either right side up or inverted so that the crowns clamp the upper tubes just below the top while ensuring that the specified length of upper tube extends above the lower crown (Fig. 16.21).

IMPORTANT FOR DOUBLE-CROWN FORKS: *If more than the manufacturer's specified length of upper tubes is above the lower crown, the lower crown will be too close to the tire and could hit it during large impacts, which can stop you and the bike dead and leave you that way. Furthermore, some triple-clamp forks have shims to be inserted into the lower crown's clamping slots; other triple-clamp forks have reinforcements around the upper tubes under the lower crown. To avoid fork failure, make sure to include whatever the manufacturer intended; consult the fork manual.*

3. **Tighten the crown bolts to the manufacturer's specified torque**. If the crown has paired bolts (Fig. 16.10), alternately tighten the two bolts on each side. If the fork requires shims at the clamps, make sure they're in place. Use an antiseize compound on titanium bolts and a medium-strength thread-lock compound on steel bolts.

IMPORTANT: *Fork-crown bolts are critical bolts—make sure to tighten them to the required torque specified in Appendix D.*

16-13

OVERHAULING LOWER LEGS OF FORKS WITH BOLTS OR NUTS AT THE BOTTOM

Even on top-quality suspension forks, the oil bath inside can rapidly become contaminated, leading to ruined stanchions, bushings, and seals. This lower-leg service is an important maintenance procedure to restore small-bump sensitivity, reduce friction, and increase the life span of lower-leg bushings. I recommend doing it every 50 hours of bumpy trail riding.

The fork needs immediate overhaul if, as you gradually lean harder on the handlebar, it is hard to get it to compress, and when the fork finally does compress, it goes down chunk, chunk, chunk; in fact, it may be too late by that point. The fork needs an overhaul and new seals if the air chamber does not hold consistent pressure, if oil leaks out of the fork, and especially if it's reached the point where you have no damping—the fork just acts like a pogo stick. It needs new bushings if the lower legs clunk back and forth on the upper tubes; due to the special tools involved, this service is not included in this book and is a factory or suspension-shop service.

On Manitou forks with Microlube grease fittings on the backs of the legs, you can delay a full disassembly and cleaning with a few squirts of the proper (thin) Manitou Microlube grease with a fine-tipped grease gun into the grease fitting. It only takes a bit; if you fill the fork with grease, you will lock it up!

These instructions apply to forks with a bolt head or nut at the bottom of each outer leg (Figs. 16.19, 16.20, and 16.23). Forks without bottom bolts or nuts are generally very old but can be disassembled. You need either snapring pliers to remove the snapring at the top of each outer leg or a long hex key to get at the head of the compression bolt way down inside after removing the springs.

This is an oily job; wearing rubber gloves is a very good idea. Keep track of everything, and put all the parts back together the same way after cleaning and lubricating. Collect the old oil and recycle it at an auto parts store or through your county's hazardous waste facility. Do not pour it down a drain or sewer, or into the ground in your back yard.

1. **Remove the front brake** (Chapter 9 or 10) **and the front wheel** (2-2). I recommend removing the fork from the bike as well (12-18 and 12-19).

NOTE: *On a fork with a removable fork brace (Fig. 16.2), do not remove the brace.*

2. **Clean the outside of the fork.** The last thing you want is to get dirt inside. Clean well around the upper legs, the dust seals atop the outer legs, and the bolts and adjusters at the bottoms of the fork legs.
3. **Record settings.** Record your air pressure (or spring preload with a coil/elastomer fork) and rebound-damping setting so you can go back to them after you're done.
4. **Turn the fork upside down.** This keeps oil from pouring out.
5. **Remove adjuster knobs from the bottom of the fork legs.** On some older forks, these will pull out of the center of the shaft bolt (Fig. 16.19); on newer forks, they will be attached with a setscrew to a thin shaft sticking out through the bottom nut. With the latter, loosen the setscrew with a tiny hex key (2mm or 2.5mm), and slide the knob off (Fig. 16.22).

NOTE: *Some air forks have a negative-spring air valve at the bottom of one leg; deflate it (careful: oil will spurt out!).*

6. **Unscrew the bolts or nuts on the bottoms of the fork legs.** With bolts, leave a few threads engaged; with nuts, completely remove them and the crush washers under them. If the bolts or nuts are not backing out, you are turning the shafts inside the fork along with them. To prevent the shaft from spinning, pump more air into an air fork (Figs. 16.11 and 16.12), or tighten the preload adjuster on a

16.22 Removing/installing a rebound adjuster knob from the bottom of the lower leg

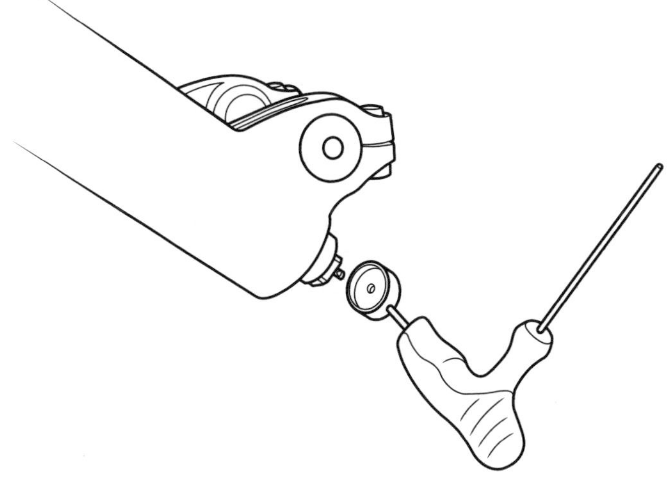

coil-spring or elastomer fork (Fig. 16.13). On some forks, this bolt unscrews clockwise, rather than counterclockwise, so check for that if it isn't loosening up.

7. **Deflate or remove the springs from the fork.** Deflating (Figs. 16.6, 16.11, and 16.12) or removing springs (Fig. 16.7) is a good precaution in case you go further once you've pulled off the outer legs. You don't want a projectile flying at you or around the room.
8. **Place a bucket or oil pan under the fork ends.**
9. **Free each shaft with a tap of a mallet.** On forks with a nut on the bottom, do not hit either the bare shaft or a socket placed over the nut; in the former case, you will mushroom the aluminum threads or break off the rebound adjuster shaft, and in the latter case, you'll deform the aluminum nut so that the rebound knob will no longer fit over it. Instead, tap on (ideally) the fork maker's tool for this (Fig. 16.24), a socket on a used bottom nut, or an aluminum drift tool you can make. Fox's damper removal tool is an internally threaded rod with a head on it; screw it onto the shaft a few turns and bop it on the head with a soft hammer (Fig. 16.24). Lacking the special tool, a good drift for this is an aluminum tube (or an aluminum rod with a hole drilled down the center) that is small enough to fit inside of the bottom nut you removed and whose bore is large enough to fit over the thin, rebound-adjuster shaft sticking out of the end of the shaft. If the bolts are steel and accessible, you can strike them with a soft hammer directly (Fig. 16.23); otherwise, hit the end of the hex key inserted into the bolt, and remove the bolts. You'll know when the inner leg shafts pop free from the lower legs.
10. **Rotate the fork down with a bucket positioned underneath.** Let the oil drain.
11. **Pull the lower leg assembly off the inner legs** (Figs. 16.25 and 16.26). Let any oil pour out into the bucket. If the lower legs will not easily slide off completely, the press fit of the shaft in the lower leg may still be engaged; put the bolt back in a few turns (remove the washer and O-ring, if present), and whack the end of the hex key in the bolt again until it disengages; with a bottom nut, whack the end of the threaded shaft tool (Fig. 16.24), a socket on a used bottom nut, or a homemade drift again.
12. **Clean the inner legs and the shafts sticking out of them.** Use a clean, lint-free rag and citrus degreaser. Check for scratches and surface

16.23 Freeing an inner shaft from the lower leg

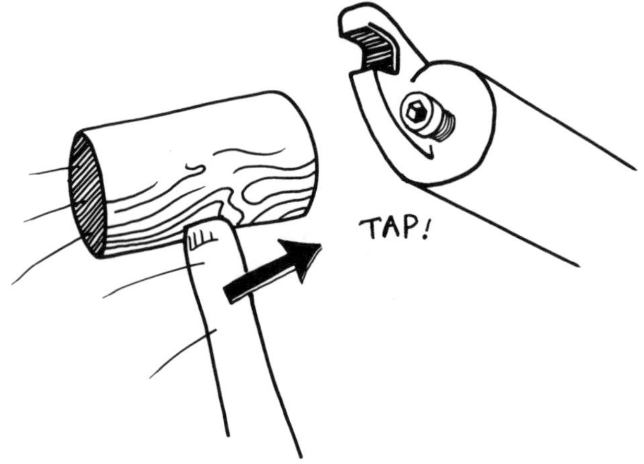

16.24 Tapping a Fox damper removal tool to free the inner shaft

16.25 1995 RockShox Judy fork (inner leg shortened for clarity)

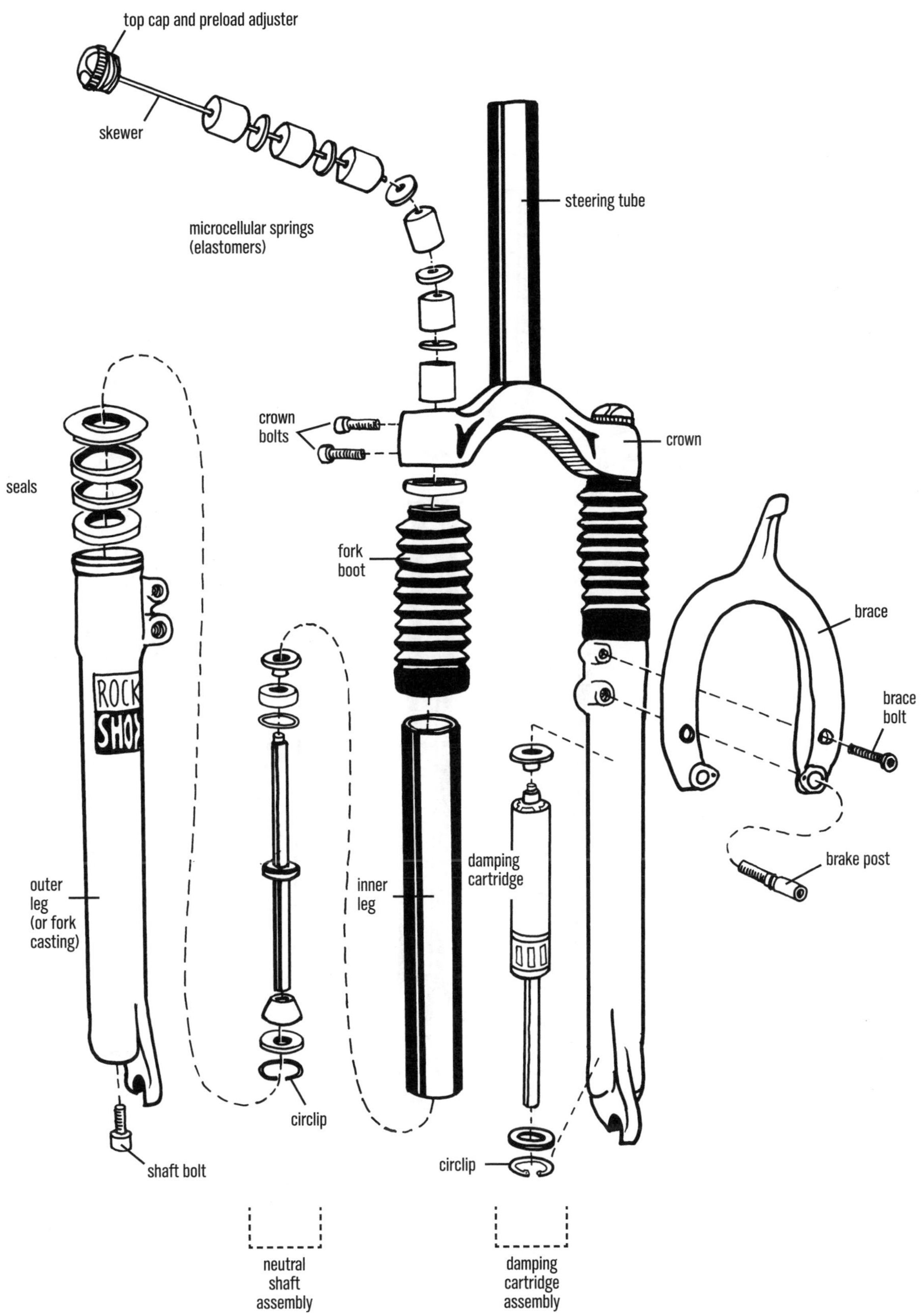

16.26 1997 Manitou SX

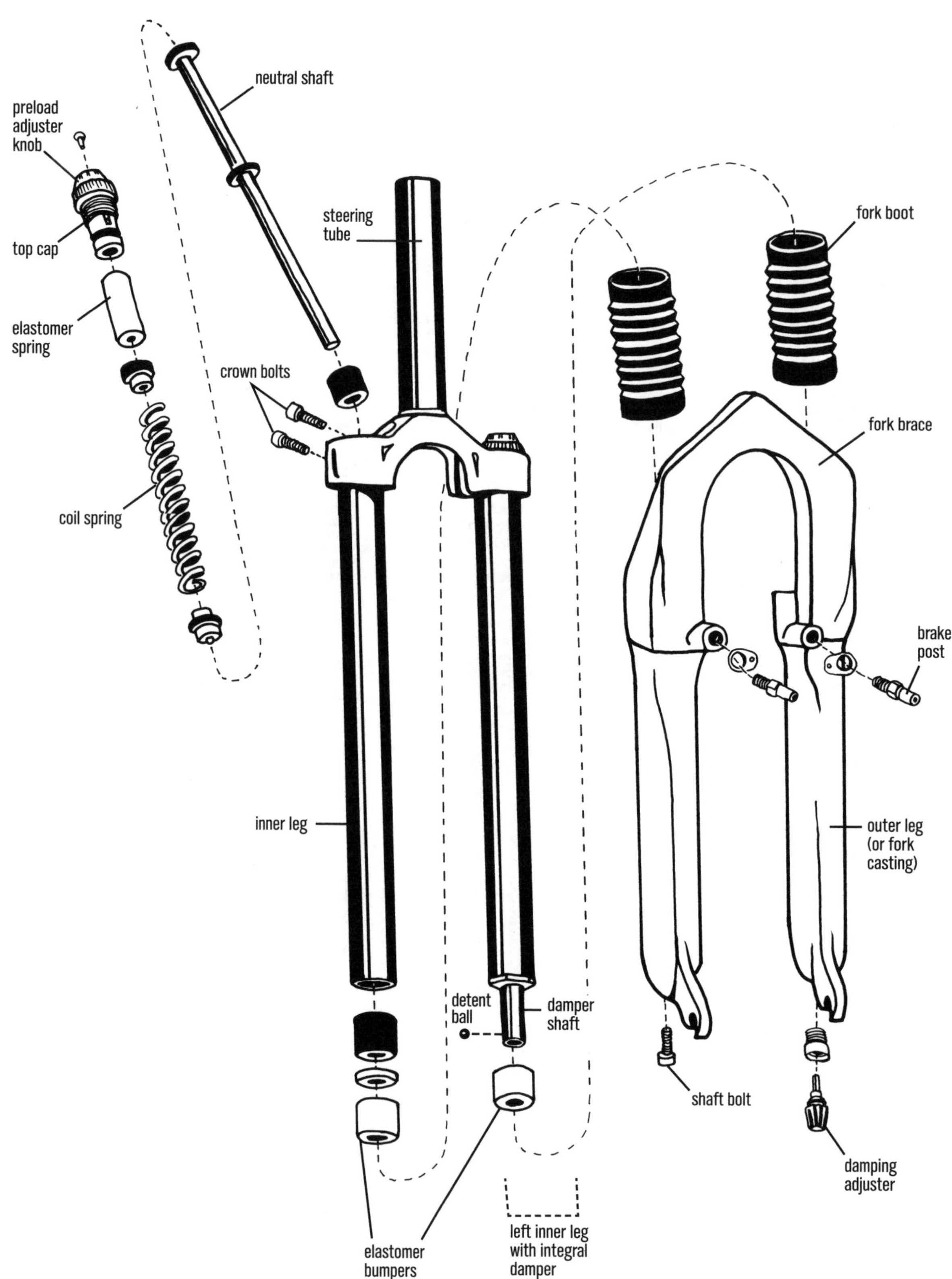

abrasion. Check the thin shafts (Figs. 16.25 and 16.26) for bending and for oil leaks. If you want to change travel or service the air spring or damper, skip now to 16-14, 16-15, or 16-16.

NOTE: *On many older Manitou forks, there is an elastomer around the damper shaft that extends from the inner leg (Fig. 16.26). Remove it to clean and grease the shaft, but be ready to catch the steel ball that will fall out. This is the detent ball for the damper; it helps put the clicks in the damper adjustment.*

13. **Clean or replace the wiper seals atop the outer legs.** If replacing, pry out the wiper seals with a plastic dowel (Fig. 16.27), with the open end of a 19mm combination wrench, or with a downhill (i.e., heavy-duty) tire lever. To clean, remove the elastic, coil-spring-wrapped O-ring and wipe the seal and all of its nooks and crannies well with alcohol and a clean rag; replace the elastic ring.
14. **Remove and clean or replace the foam rings.** If there is a foam ring below the wiper seal, pull it out. If you're not replacing the ring, blot it dry with a rag or paper towel, clean it well with isopropyl alcohol, wring it out, and let it dry.

16.27 Prying out a wiper seal with a plastic tube

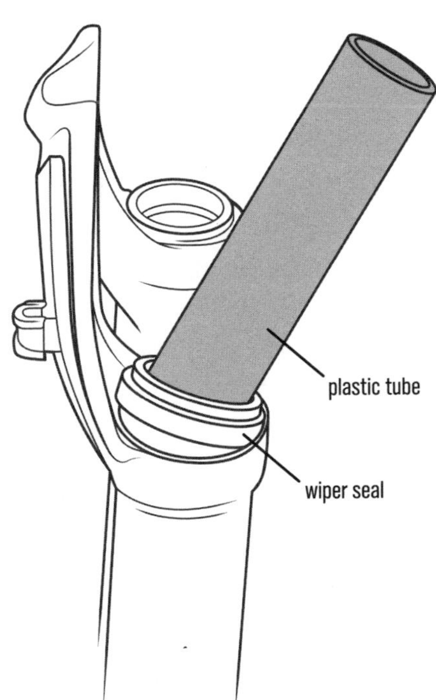

15. **Clean the bushings inside the outer legs.** There are two bushings in each outer leg: one at the top and one halfway down. Reach the bottom one with a rag wrapped around a long rod (Fig. 16.29). If the bushings are shot (indicated by movement of the inner leg wobbling inside the bushings), they may be replaceable by a shop with the tools, or you can buy new lower legs.
16. **Replace the foam rings.** Soak the foam ring in shock oil prior to reinstallation.
17. **Replace the wipers.** If you removed them for cleaning or replacement, push them back in fully. A wiper seal driver is a plastic tool for this job; it looks like a cylindrical drinking cup, and you can push down on the tool by hand or tap it with a mallet. Without the tool, you can push down with the side of a screwdriver shaft laid flat on the edge of the seal, working around until its lip is flush with the top of the leg.
18. **Lubricate the wipers and bushings in the lower legs.** On an oil-bath fork, wipe a bit of shock oil on the wipers and upper bushing only. On a fork without an oil bath, use nonlithium grease (such as Slick Honey). To grease the lower bushings, slather grease on the end of a long rod, and reach down to the lower bushings with it (Fig. 16.30). Don't grease between the upper and lower bushings.
19. **Lubricate the inner legs.** On an oil-bath fork, wipe a bit of shock oil on the inner legs. On a fork without an oil bath, use the same nonlithium grease (Fig. 16.31).
20. **Slide on fork boots or a travel O-ring.** Older forks with questionable seals do well with boots. And without them, it's nice to measure sag or the amount of travel you're using by the position of the O-ring (Fig. 16.1).
21. **Slide the outer legs gently over the inner legs.** Rotate the fork in the bike stand so that the shafts angle upward (Fig. 16.28), and take care not to dislodge the spring-wrapped O-rings or damage the wiper seals or the lower bushings. To get the lower legs on completely, it may help to spread the lower leg assembly slightly while you rock it side to side to engage the bushings on the inner

16.28 Sliding the lower legs onto the stanchions

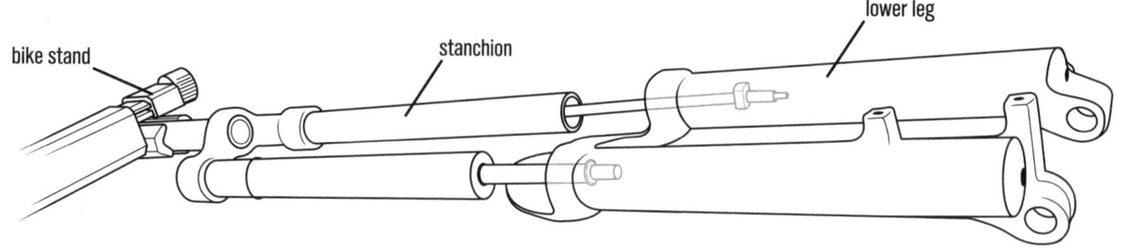

legs. Don't push it on far enough for the damper shaft and neutral shaft to reach the holes in the bottoms of the outer legs.

22. **Replace the oil bath.** Skip this step if your fork has no oil bath. Look on the manufacturer's website for oil bath specs—oil volume and weight—for your fork. There will generally be a different volume of oil in each leg. Draw up the correct amount of the correct-weight oil into a syringe, and squirt it into the hole in the bottom of the outer leg. Do not exceed the recommended oil volume, as this could damage the fork. If the shaft is in the way of the oil going in, pull the outer legs out a bit further before pushing in the oil; on a leg with a bottom bolt, make sure you're not pouring oil into the end of the shaft.
23. **Push the outer legs in fully.** The inner shafts should come to the holes, and, in the case of bottom nuts, should stick out of the holes. You may need to reach in with a thin implement to line them up.
24. **Replace the shaft bolts or nuts.** Install new crush washers under the nuts, and replace the washer and O-ring on the bolt, if you removed them. Engage the threads on or in the shaft, and screw in the nut or the bolt (Fig. 16.20). If the bolt threads do not engage or the shaft threads do not emerge, push the outer legs on farther.

16.29 Cleaning bushings

16.30 Greasing bushings

16.31 Greasing inner leg

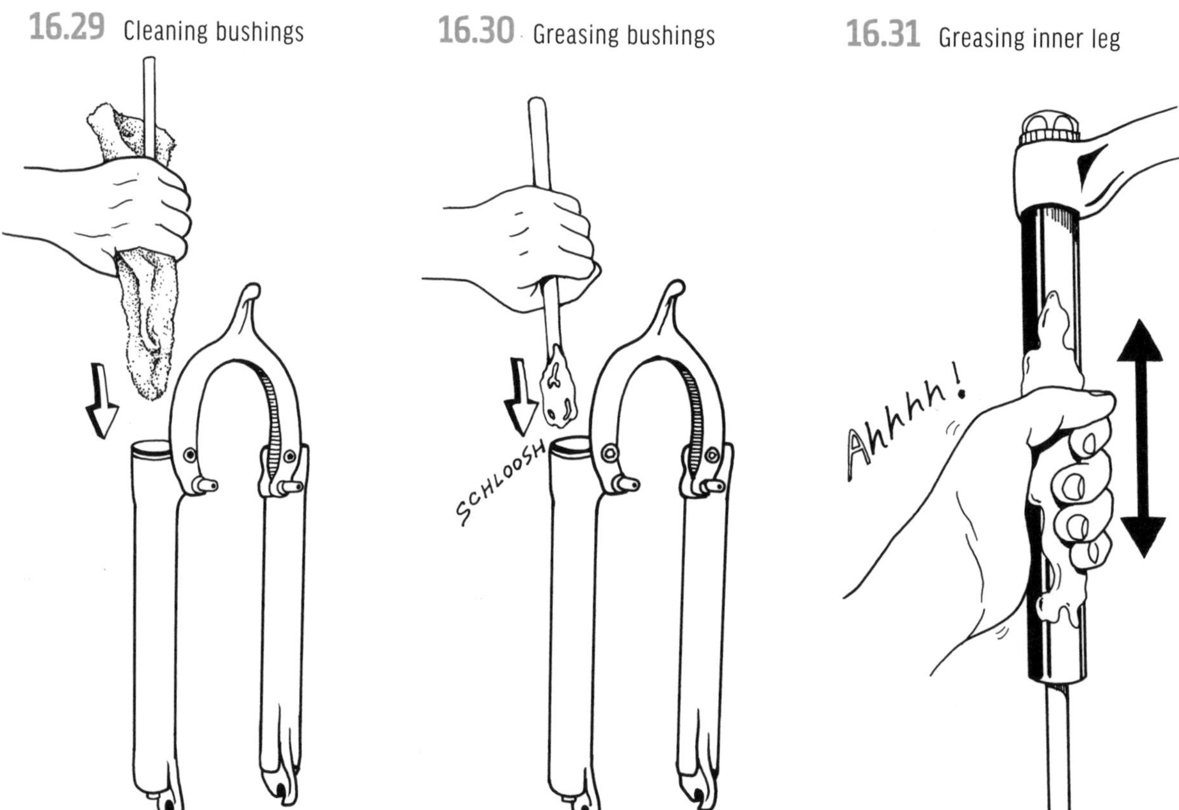

16.32 Filling an oil bath

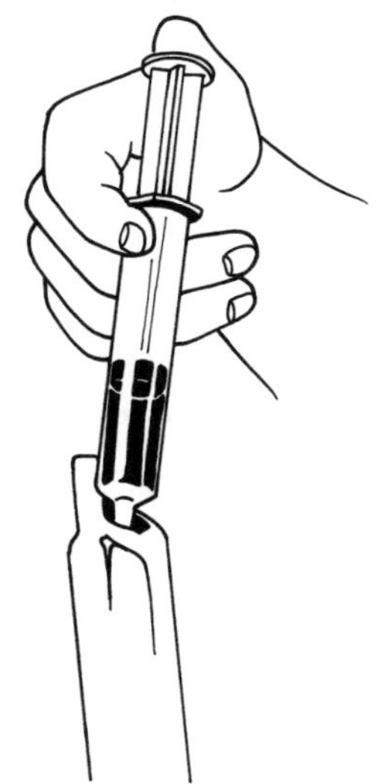

25. **Tighten the bolts or nuts.** See the fork manufacturer's website, the owners' manual, or Appendix D for torque specs. If the shaft spins without tightening, pump up the fork to 60 psi or replace the coil/elastomer springs and tighten the preload adjuster knob; now tighten the bolt or nut (Fox torque is 50 in-lbs).
26. **Replace the adjuster knobs.** Put them on the way you took them off (Fig. 16.19 or 16.22). With a setscrew, make sure it is lined up with the notch in the adjuster shaft (Fig. 16.22).
27. **Replace the springs.** Pump up the fork (Figs. 16.11 and 16.12), or drop in the coil/elastomer springs and screw on the top cap (Fig. 16.7).
28. **Reinstall the brake and the wheel.**

    You're done!

16-14

CHANGING FORK TRAVEL

Changing travel ranges from being very easy, with an external knob or lever, to quite complicated, with adding or interchanging internal travel spacers, or interchanging or reversing the air piston and neutral shaft, and voilà, different travel. On-the-fly travel adjustments are for terrain changes; travel adjustments involving fork disassembly are to change bike geometry.

Check the total travel as in 16-6a and ensure that there is no way the fork crown can hit the tire when the fork is compressed fully. Although changing travel does not make the crown come down farther, increasing travel doesn't mean you can use a bigger tire.

a. Travel-adjust knob

To increase or decrease travel, turn the large knob or lever on top of the left leg. The idea is to lower the front end for uphills and raise it for descending. You can do it while riding carefully, but it's recommended to only do it when stopped.

On RockShox Coil U-Turn, the big coil spring itself screws down or up along large threads around the outside of a big plastic plunger to shorten or lengthen the fork. You can shorten the fork while riding by turning the U-Turn knob clockwise, but you must take your weight off the spring to turn it counterclockwise and lengthen it. RockShox 2-Step Air (only two possible travel settings) and Air U-Turn compensate for the travel change and keep the same spring rate.

PRO TIP — OLD FORKS WITHOUT AN OIL BATH

IF THE FORK HAS EXTERNAL SHAFT BOLTS and does not have an oil bath or the Manitou Microlube system, you can keep it lubricated longer by putting an oil bath in. Squirt in 15cc or so of automatic transmission fluid through the bottom bolt holes before replacing the bolts (Fig. 16.32). This keeps lubricant sloshing around up to the upper bushings as you ride. Use an aluminum, brass, or plastic crush washer under the bottom bolt to prevent leakage.

WARNING: Don't put an oil bath in forks with internal shaft bolts you reach from the top (there is no bolt on the bottom of the outer leg), as you could crack the legs by tightening the bolts down onto trapped oil.

On Fox TALAS, rotating the adjuster lever surrounding the air valve changes total travel available over a range of up to 30mm (i.e., from 160mm to 130mm). To lengthen the fork, turn it clockwise, toward the "+" symbol; it will not come up until you unweight the handlebar. Turning it counterclockwise, toward the "–" symbol, shortens the fork, if weighted. You can reduce the available amount of travel change in 5mm increments by stacking 5mm spacers under the air cap as described below.

Reducing or increasing available change of Fox TALAS travel adjustment

1. **Wipe the left side of the fork crown clean.**
2. **Remove the valve cap.** Just unscrew the cap and do not deflate it; you need pressure in the air cylinder to perform the travel-adjustment reduction or increase.
3. **Remove the TALAS lever retaining nut.** With an 11mm socket, unscrew the little nut surrounding the valve.
4. **Remove the TALAS lever.** Pull it up.
5. **Pull the TALAS unit partially out.** Using a large socket ground flat on the end, unscrew the top cap (Fig. 16.8). If the socket is not ground flat, it will not get enough grip on the low-profile aluminum cap, and it will slide off, stripping the nut. Pull up on the top cap to reveal the shaft under the cap (Fig. 16.33).

16.33 Installing travel-adjustment reduction spacer on a Fox TALAS fork

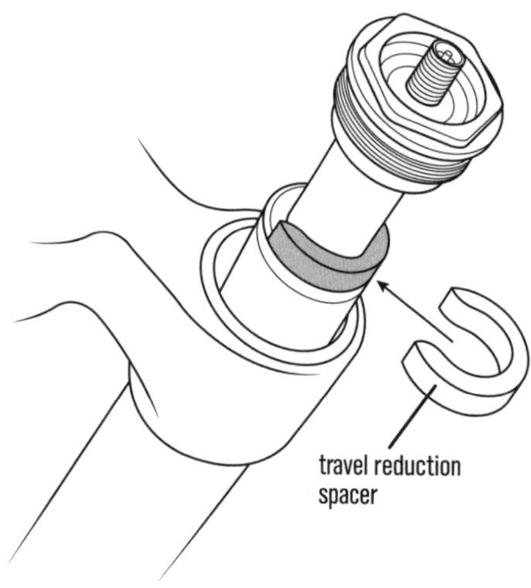

6. **Install or remove spacers.** Clip one or more spacers onto the shaft under the top cap (Fig. 16.33). Each spacer reduces the travel adjustment by 5mm. Obviously, to increase the travel adjustment, remove spacers, if present.
7. **Replace the TALAS unit.** Drop it back down into the fork leg, and torque the top cap to 25 N-m (Fig. 16.8).
8. **Replace the TALAS lever.** Set the lever back in place, and tighten the 11mm nut surrounding the valve to hold it down.
9. **Replace the valve cap.** No need to adjust air pressure; just screw the valve cap back on.

b. Plunger-height screw under the spring

The Vari-Travel system on some RockShox Psylo forks simply requires removing the left top cap (Fig. 16.8) and removing the coil spring. Reach down inside with a long screwdriver, and turn the screw on the top of the plunger; each turn changes the travel by 1mm.

c. Internal spacers

A common way to change travel is by disassembling the fork and inserting or removing plastic travel spacers. By putting spacers below the plunger on a fork shaft, you can reduce travel, because the shaft cannot extend as far; the spacer hits the cap in the bottom of the inner leg and prevents the plunger from coming as far down—look at Fig. 16.34 to understand the concept. To increase travel, remove spacers from below the plunger.

Organize all of the parts you remove in the same orientation and order they came out to ensure proper reinstallation.

Change travel by adding or removing internal spacers

1. **Remove the lower legs as in 16-13, steps 1–11.**

NOTE: *On some older forks, you may be able to avoid removing the lower legs and just take the top cap off the spring side of the fork crown with a big, flat-end socket (Fig. 16.8). Pull out (or turn the fork over and drop out) the spring shaft, install or remove travel spacers, and replace.*

16.34 Changing travel on a RockShox Judy with All Travel spacers

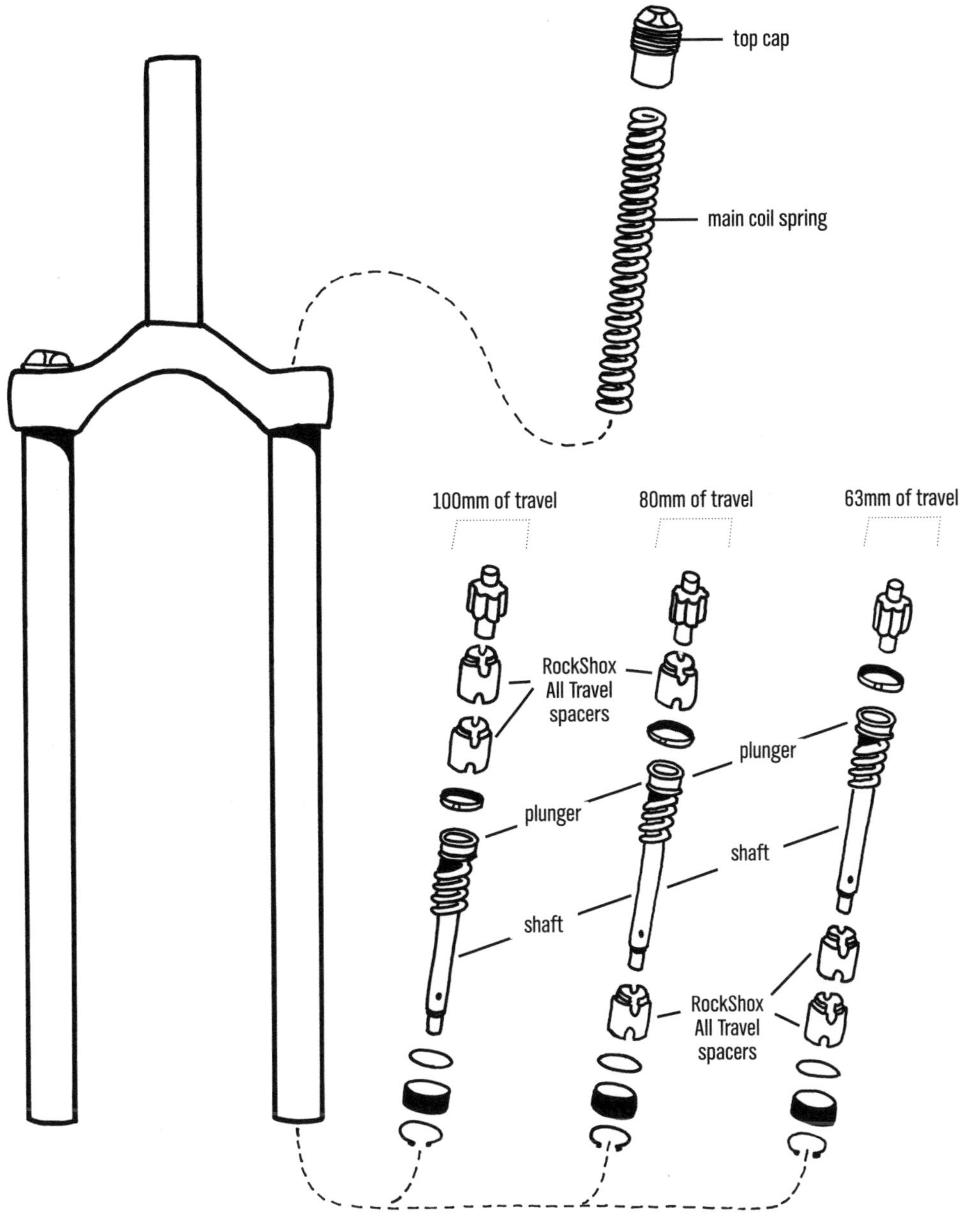

2. **Remove the spring.** In an air fork, if you haven't already done so, release pressure from the valve(s)—(Fig. 16.6 or 16.12); oil may spurt out. Remove the elastomer and/or coil spring (Fig. 16.7 or 16.34) in that fork type.
3. **Remove the top cap.** If you have a coil/elastomer fork, you would have already removed the top cap in step 2. Unscrew an air top cap with a 6-point socket of the appropriate size ground flat on the end (Fig. 16.8). Volume spacers (Figs. 16.14 and 16.15) may be attached to the cap, and you may want to change the number of them based on the new travel; for instance, with less travel, you would want more volume spacers to get the fork to ramp up faster and avoid bottoming out at lower air pressure settings, and vice versa. If there is a volume spacer down inside the stanchion that didn't come up with the air cap, push up on the air shaft while plugging the hole in the volume spacer with your finger so that air will push it up (Fig. 16.35A) until you can grab it and pull it out. Snap the volume spacer back onto the air cap.

16.35A Plugging the hole in a Fox volume spacer

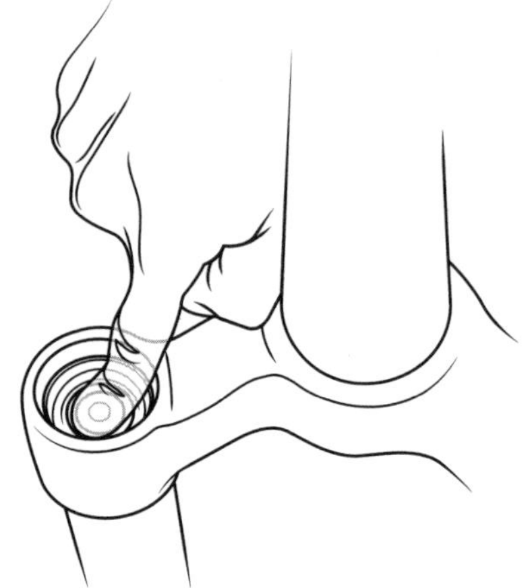

4. **Drain oil from the air cylinder.**
5. **Remove the circlip from inside the lower end of the stanchion.** For RockShox and other forks with eyes at the ends of the snapring, use inward-squeezing snapring pliers (Fig. 16.36). See 16-14d, step 4 for other tips on circlip removal from some RockShox forks. On a Fox fork, the circlip is a thin split ring. Remove it by prying the outer, notched tip inward with a thin screwdriver while sliding an awl or thin screwdriver tip under the outer ring and lifting its end up above the groove (Fig. 16.37); continue working around with the first screwdriver, prying up on the ring as you slide it clockwise around the edge of the stanchion. Remove the washer and second, single-loop circlip, if present, by prying inward and up on the notch at one tip with a thin screwdriver.

NOTE: *Instead of using a circlip, some RockShox SID forks from 1999 and the early 2000s require a special SID cartridge-removal tool: a hollow 15mm hex key that fits over the shaft and engages the threaded cap at the bottom of the inner leg. Put a 15mm socket or box wrench on the other end of the tool, and unscrew the cap clockwise. Yes, this SID cap is left-hand-threaded!*

6. **Pull out any loose neg plates, if present.** They may look like cylindrical cups or flat washers.
7. **Pull out the shaft assembly.** On an air fork, there will be an air piston on the top end with a sliding seal head below it (Fig. 16.38); a coil/elastomer fork's shaft will look similar (Fig. 16.34).

16.35B Pushing the spacer up in order to pull it out

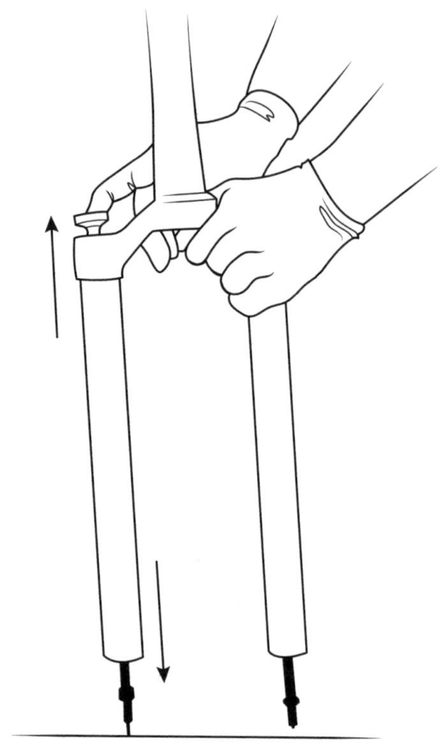

16.36 Removing cartridge-retaining circlip from RockShox inner leg

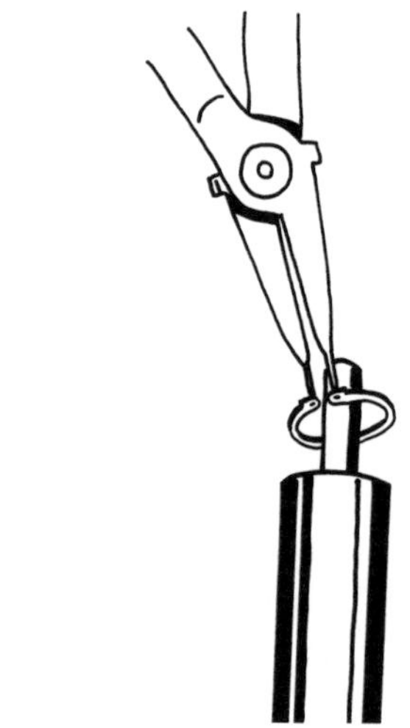

16.37 Removing a Fox split-ring circlip

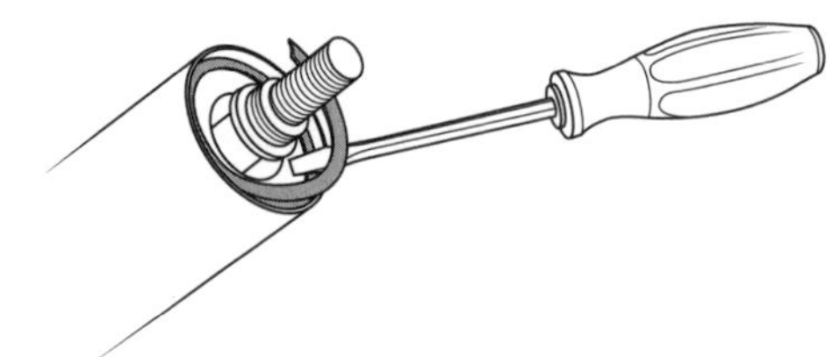

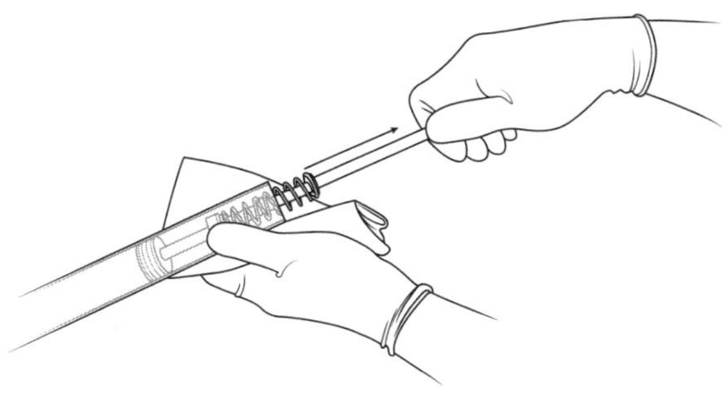

16.38 Removing an air spring shaft

8. **Install or remove a travel spacer (or two).** Either the (C-shaped, or slotted, spool-shape) spacers will snap onto the spring shaft from the side (Figs. 16.39 and 16.40), or cylindrical spacers will slide up onto the shaft (Figs. 16.41 and 16.34); the lips of the spacers may snap into each other or into the bottom of the small spring attached to the plunger. If you want to reduce travel by the length of the spacer, install a travel spacer onto the shaft under the plunger. To increase travel, remove a spacer from below the plunger. The thickness of the travel spacer will be the amount you add or subtract from the fork's travel. If need be, pull down on the coil springs to make room for the spacer under the piston. On some Fox Float forks, the springs are attached to the piston, and cylindrical travel spacers snap into the plastic, cup-shaped neg plate below it (Fig. 16.41A). Alternatively, some Float forks have a blue, cylindrical shuttle bumper slip up over a cylindrical plastic sleeve surrounding the shaft below the coil springs (Fig. 16.41B); slide off the shuttle bumper and replace it with a longer or shorter one to decrease or increase fork travel.
9. **Meticulously clean the inside of the stanchion.** Push a clean, lint-free rag or paper towel through it with a wood or plastic dowel. Any scratches inside the stanchion will cause air leakage.
10. **Grease inside the lower end of the stanchion.** Wipe Slick Honey inside the stanchion end with your finger.
11. **Install the piston shaft assembly.** Grease the piston O-ring with Slick Honey. Wiggle the piston into the end of the stanchion, and push the plastic seal head and any washers and spacers between the

16.39 Various Fox Float travel-reduction spacers and air spring shafts

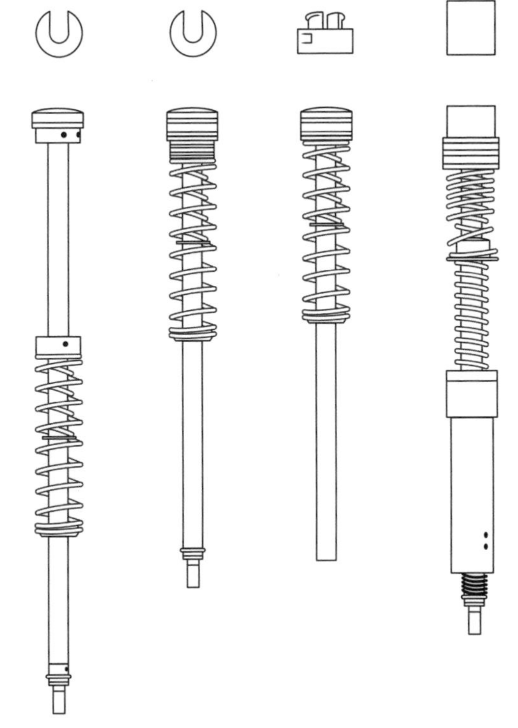

16.40 Installing a Fox C-shaped travel-reduction spacer on the air spring shaft

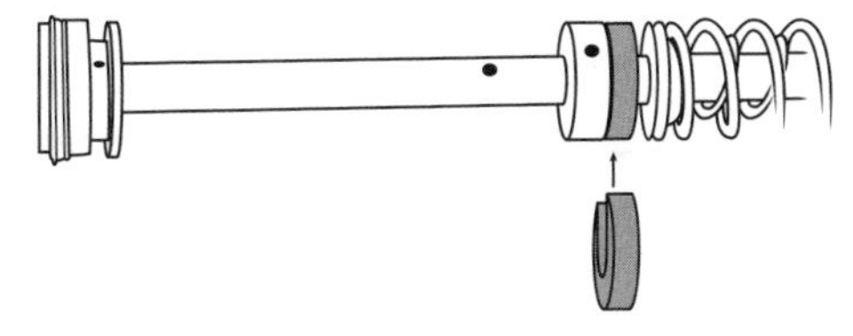

16.41A Installing Fox cylindrical travel-reduction spacers

16.41B Installing Fox shuttle bumper

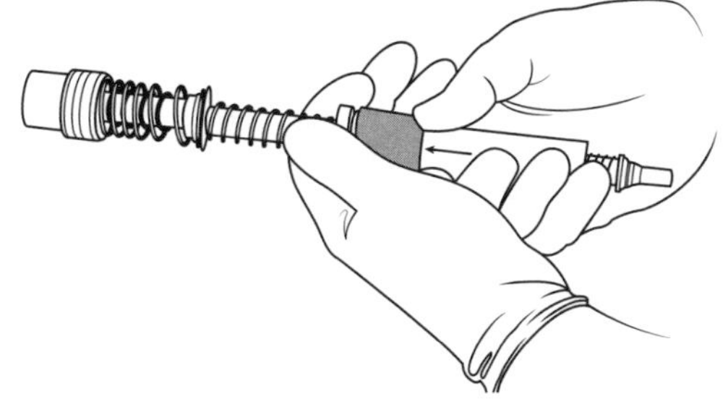

seal head and piston into the stanchion (Fig. 16.38). Slide the shaft all of the way in to protect it from scratching or bending while installing the snapring.

12. **Install the snapring.** Line up the snapring eyes on either side of the bump on the face of the seal head, if present. Orient the snapring so that its sharp edge faces outward. Squeeze the eyes together with the snapring pliers (Fig. 16.36), and push the snapring down with a screwdriver tip until it drops into its groove all of the way around. Or, with a Fox Float fork, install the split-ring snapring, but if yours is a fork with a second, deeper-in single-loop snapring, install that first in the groove further down inside the stanchion (push it in place with the tip of a screwdriver), followed by the washer (same orientation as on disassembly). To install a Fox split-ring snapring, push one end of it into the groove, then push the ring from the side toward the opposite stanchion wall as you work it around and into the groove with your thumb (Fig. 16.42).
13. **Check that the shaft assembly is secure.** Yank on the shaft to ensure that the snapring is installed securely.
14. **Add air-volume spacers.** If shortening travel in an air fork, you'll probably want to add volume spacers under the top cap, and vice versa if lengthening travel. RockShox Bottomless Tokens (Fig. 16.15) screw into each other and the air cap; other brands have volume spacers that interlock with a tongue-and-groove system (Fig. 16.14).

16.42 Installing the split-ring circlip in a Fox Float fork stanchion

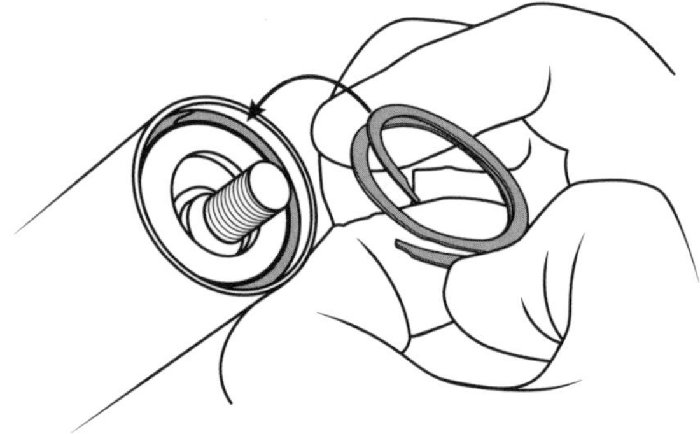

15. **Add oil to the air cylinder.** Add just a tiny bit to keep the seal lubricated; Fox recommends 5 mL of its Float Fluid.
16. **Install the top cap.** Grease the top-cap O-ring of an air fork (and a volume-spacer O-ring, if present), and tighten the top cap in with the ground-flat socket and a torque wrench (Fig. 16.8); RockShox torque is 28 N-m; Fox is 25 N-m. On coil/elastomer forks, install the entire spring/top cap assembly and tighten the top cap.
17. **Install the lower legs.** Follow instructions in 16-13, starting with step 12.

RockShox SID XC All Travel

This variation of the spacer system from the early 2000s requires reversing the neutral shaft. The SID XC packaging includes a single All Travel spacer identical to those on the Judy (Fig. 16.34).

1. **Remove the lower legs as in 16-13, steps 1–11.**
2. **Remove both shafts from the inner legs.** Using snapring pliers (Fig. 16.36) and a small-blade screwdriver, pry out the snapring. The negative spring shaft is reversible; if you want maximum travel, leave the All Travel spacer out of the fork and set up the shaft so that the end with a circumscribed line around it points down; screw the piston into the opposite end. The scribed end is longer from the glide ring to the end, so having it down allows more shaft to extend out of the inner leg and—*voilà!*—more travel. The negative spring goes on the longer end, below the glide ring. Once you get it apart, you'll see what I mean.
3. **If you want less travel, slide the All Travel spacer up onto the damper shaft (the shaft from the right leg) and snap it into the spring below the piston.** Remove and save the thin spring guide that had been snapped into the spring.
4. **Unscrew the piston from the short end (without the circumscribed mark) of the left-side shaft.** Pull the negative spring off the long end.
5. **Screw the piston into the longer, scribed end, and put the negative spring onto the shorter, unscribed end.** The negative spring has a tight-fitting plastic spring guide snapped into the piston

end, so it does not slide easily. I find it simpler to snap the spring off the guide and move them separately. With the short end of the neutral shaft down and the All Travel spacer limiting the downward extension of the damper shaft, travel is shortened.

6. **Reassemble the fork as in 16-13, starting with step 12.**

d. Changing the air-spring shaft

interchanging the air-spring shaft assembly with another one of a different length changes the travel on a RockShox Solo Air fork. Before starting, have the new, shorter or longer air piston shaft assembly in hand. The travel length of the shaft will differ depending on if it is a 29-, 27.5-, or 26-inch fork, and these lengths may be etched into the end of the shaft.

1. **Remove the lower legs as in 16-13, steps 1–11.** Line up all parts in order and in proper orientation for reassembly.
2. **Remove the air top cap.** Unscrew it with a 6-point socket of the appropriate size (often 24mm) ground flat on the end. Volume spacers (Fig. 16.14) may be attached to the cap, and you may want to change the number of them based on the new travel; for instance, with less travel, you would want more volume spacers to get the fork to ramp up faster and avoid bottoming out at lower air pressure settings, and vice versa.
3. **Inspect and protect the air shaft.** Push the shaft all of the way into the stanchion so you don't scratch it or bend it while removing the snapring.
4. **Remove the circlip securing the seal head within the stanchion.** On RockShox, there is often a bump on the face of the plastic seal head in between the eyes of the circlip; the bump prevents the circlip from being removed, so push in on it with a screwdriver and rotate it slightly under one of the eyes. With inward-squeezing snapring pliers, remove the snapring (Fig. 16.36).
5. **Pull out the air piston shaft assembly.**
6. **Meticulously clean the inside of the stanchion.** Push a clean, lint-free rag or paper towel through it with a wood or plastic dowel. Any scratches inside the stanchion will cause air leakage.
7. **Install the new air piston shaft assembly.** Grease the air seal on the new piston with Slick Honey. Wiggle the piston into the end of the stanchion (Fig. 16.38), and push the plastic seal head and any washers and spacers between the seal head and piston into the stanchion. Slide the shaft all of the way in to protect it from scratching or bending while installing the snapring.
8. **Install the snapring.** Line up the snapring eyes above the bump on the face of the seal head, if present. Orient the snapring so that its sharp edge faces outward. Squeeze the eyes together with the snapring pliers (Fig. 16.36), and push the snapring down with a screwdriver tip until it drops into its groove all of the way around. Ensure that the bump on the face of the seal head ends up between the snapring eyes.
9. **Check that the shaft assembly is secure.** Yank on the shaft to ensure that the snapring is installed properly and is retaining it securely.
10. **Add volume spacers.** If shortening travel, you'll probably want to add volume spacers under the top cap, and vice versa if lengthening travel. RockShox Bottomless Tokens screw into each other and the air cap (Fig. 16.15); other brands have volume spacers that interlock with a tongue-and-groove system (Fig. 16.14).
11. **Install the air top cap.** Tighten it with the ground-flat socket and a torque wrench (Fig. 16.8); RockShox torque is 28 N-m.
12. **Install the lower legs.** Follow instructions in 16-13, starting with step 12.

16-15

OVERHAULING THE AIR SPRING

This is exactly the same as the service in 16-14c and 16-14d, except you won't be changing the travel; you're just cleaning the air cylinder and replacing or cleaning the O-rings on the air piston, the air top cap, and the sealed air volume spacer, if present. With a Fox Float fork, follow the steps in 16-14c, and with a RockShox or other fork brand, 16-14d is more applicable. Replace piston seals in the same configuration as the old ones came off, as they are often not symmetrical—there may be a wiper edge that protrudes more than the other edge.

16-16

DAMPER OIL CHANGE

Even on a fully sealed damping cartridge, operating heat and wear of internal parts on each other will turn beautifully colored, clear damping oil into murky brown or gray stuff. Changing the oil will not only make the fluid look beautiful again, it will also improve performance and extend the life of the damper.

Dampers vary so much from model to model and brand to brand that they don't lend themselves to writing out detailed service steps that apply to them all. That said, the steps to changing oil in a damper are no more complicated than many of the other steps in this chapter; it's simply unrealistic to include the specific ones here for all dampers. This first section provides general oil-change instructions for modern damping cartridges. The three sections that follow it describe changing oil in open-bath dampers and replacing old damping cartridges.

NOTE: *Old Marzocchi and RockShox HydraCoil open-bath dampers get an oil change automatically when you overhaul the lower legs as described in 16-13.*

a. General instructions for modern damping cartridge

Since I can't include step-by-step instructions for all dampers here, these are general instructions that apply fairly universally. As long as you are careful whenever you take anything apart and keep all of the parts in order and in proper orientation so you can reassemble them the same way, you should be okay. Otherwise, seek out specific instructions online for your damper, or take your fork in for factory service or send it to a specialty suspension shop.

LEVEL 3

1. **Remove the lower legs (16-13 steps 1–11).** Or at least free the damper shaft by following steps 1–10 in 16-13 on the damper (drive) side.

16.43 Using the lockout lever as a wrench to unscrew the lockout assembly from a Fox 36 FIT RLC damping cartridge

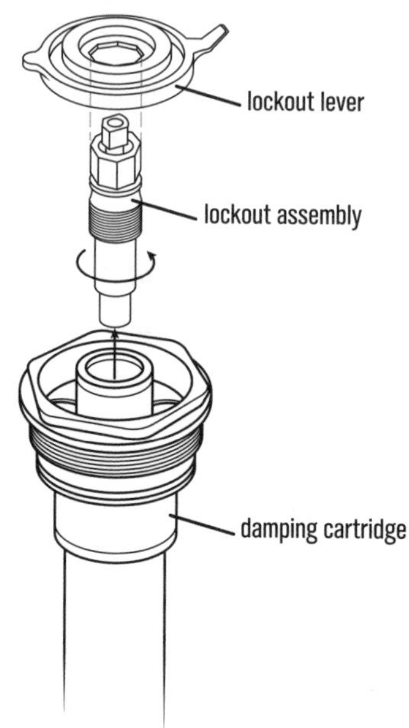

2. **Remove the damping-adjuster knobs atop the right side of the fork crown.** Hold the knob while unscrewing its mounting bolt to avoid damaging the internal adjuster shaft by twisting it against its stop. Account for all detent balls that give the knobs their click stops.
3. **Pull out the damping cartridge from the top.** Unscrew it with a large, flat-ended socket (Fig. 16.8). If required to free it, remove the circlip retaining the bottom of the cartridge (16-14).
4. **Open the top of the cartridge.** You're on your own here; figure out if you remove a circlip or unscrew (Fig. 16.43) the top shaft of the cartridge (lockout assembly) that attaches to the compression-damping adjuster atop the fork leg.
5. **Pour out the old oil.** Turn the cartridge upside down over a bucket so oil pours out of the hole in the top. Stroke the shaft repeatedly, and, if the damper has a flexible (rubber) bladder, squeeze it repeatedly as well.
6. **Pour in new oil.** Ensure from online or printed specs that it's the correct weight for your fork. Press the neck of a funnel into the hole in the top of the damper and cut it or put a hose on it so that it

16.44 Filling a Fox damper through a MixMizer syringe, plunger removed, force-threaded into the hole for the lockout assembly

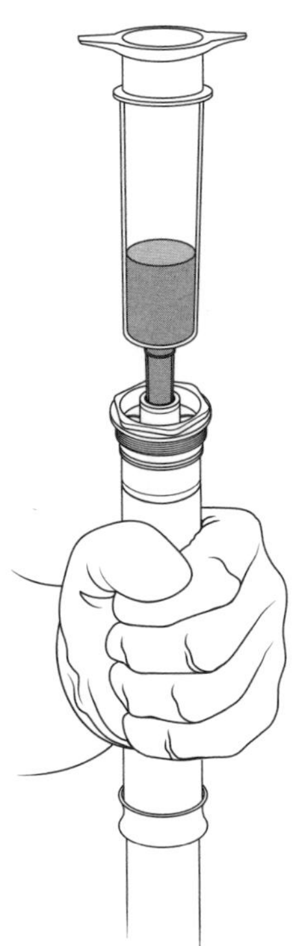

16.45 Hand-tighten old-style (pre-TPC) Manitou damper nut with O-ring slipped down around inner leg

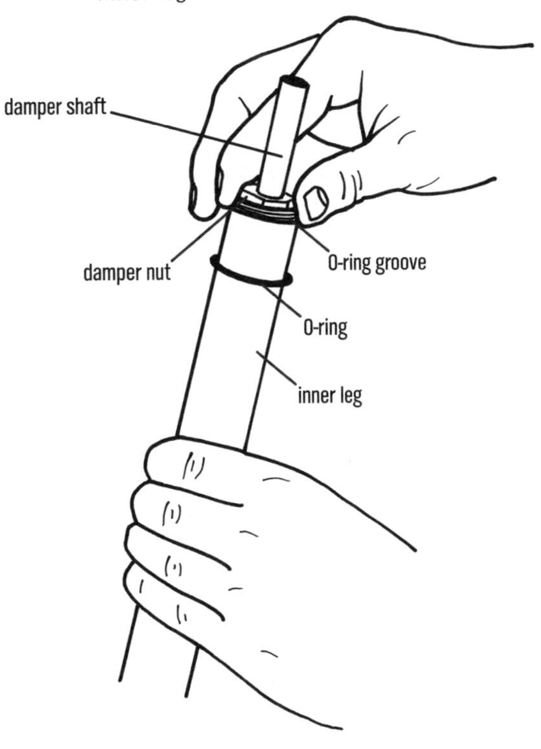

fits tightly and seals in the hole; a MixMizer syringe (available at automotive parts stores) without the plunger in it works well (Fig. 16.44). If the damper has a flexible bladder, squeeze it so that it doesn't fill and bulge out. Stroke the shaft and squeeze the bladder repeatedly until no more air bubbles come up into the funnel.

7. **Close the top of the cartridge.** Replace parts the reverse of how you removed them.
8. **Install the cartridge in the upper leg.** Tighten the top cap to torque spec (Fig. 16.8).
9. **Replace the lower legs as in 16-13, starting with step 12.**

b. Changing the damper oil on pre-1998 Manitou damper

This section applies to Manitou pre-TPC dampers (Fig. 16.26). Skip to 16-16c, below, for Manitou TPC and RockShox Pure damping systems.

1. **Remove the lower legs** (16-13 steps 1-11). It will look like Figure 16.26.
2. **Remove the springs.** Remove coil/elastomer springs (16-9b) to prevent a shaft from shooting out at you when you remove its retainer.
3. **Unscrew the damper seal nut at the bottom of the leg** (Fig. 16.45). Do this with the fork upside down; this will prevent oil from pouring out.
4. **Pour out the old oil.** Hold the leg right side up over a bucket.
5. **Rinse with new oil.** Automatic transmission fluid (ATF) is 15-weight and works fine for pre-TPC Manitou forks if you don't have fork hydraulic oil of the correct weight. Slosh some ATF around inside to rinse the damper clean, and pour it back out.
6. **Fill with oil.** Fill the damper with new fork oil or ATF to the top of the inverted inner leg. Stroke the shaft a few times to get the air bubbles out, and top off with oil again.
7. **Slide the nut back down onto the shaft and start it in the threads.**
8. **Roll the O-ring down off the nut so that it surrounds the fork inner leg** (Fig. 16.45).

9. **Tighten the end nut by hand until the O-ring groove is just showing above the tube** (Fig. 16.45). Air and excess oil vent out through a hole under the nut's lip when the O-ring is not covering it.
10. **Roll the O-ring into its groove in the nut.**
11. **Tighten the nut with a wrench.**
12. **Continue with reassembly of the fork** (16-13, step 12).

c. Changing oil in a Manitou TPC or RockShox Pure damper

LEVEL 2

Manitou turn-of-the-millennium TPC dampers are easy to service. The RockShox Pure damping system, as found in post-2001 SID SL and high-end Psylo forks, is very similar to Manitou TPC, and almost the same instructions apply. If you don't find decals with these damper names on the outer leg, look for springs (air, coil, or elastomer) in only one leg of the fork; that's a hint that the Manitou is TPC or the RockShox is Pure, although some 1999–2001 low-end RockShox forks have a single-sided spring with a single-piston cartridge in the other leg that's not Pure.

The springs should be installed (or inflated) in the other leg so that the fork is at its full length.

1. **Unscrew the adjuster-knob cap on the top of the damper leg.** It's usually on the right side (Fig. 16.46), but early TPC dampers were in the left leg. Keep the fork right side up. You may need to remove the lockout or adjustment lever first. The upper (compression-damping) piston is attached to the cap—jiggle it as you pull the piston up and through the threads (Fig. 16.46).
2. **Pour out the old oil.** If you don't have the fork owners' manual, note the height of the oil below the top of the crown before pouring it out. You will need to fashion a dipstick out of a dowel rod or wire.
3. **Add a little new 5-weight fork oil.** Both TPC and Pure take 5-weight. Slosh it around inside to rinse the damper clean, and pour it back out.
4. **Pour in new 5-weight oil to the proper level.** If you have the fork owners' manual, fill to the level below the top of the crown as specified. If not, fill to the height it was before. Keep the fork straight up and down.

16.46 Getting the Manitou TPC's top piston in and out

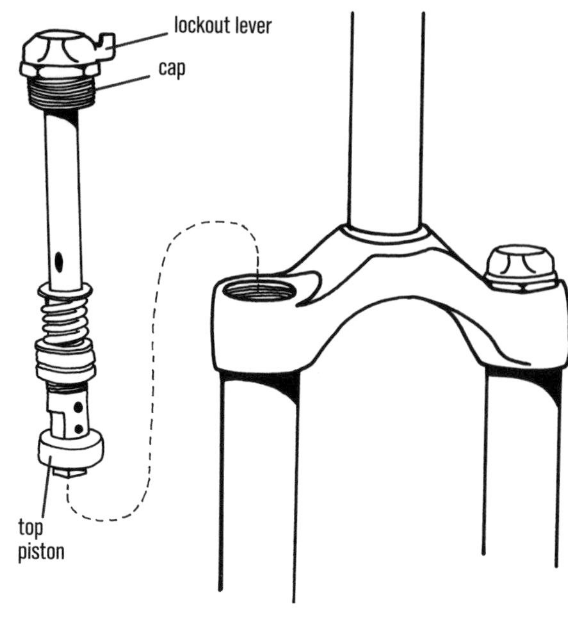

5. **Slip the piston back in** (Fig. 16.46), **and tighten the cap back on.** That's it for Manitou.
6. **For RockShox Pure, bleed the fork as detailed in steps 7–12.** Bleeding is required to squeeze the air out, because the Pure damper, unlike the TPC, has no air in it.
7. **Half-fill a small plastic syringe (without a plunger in it) with 5-weight fork oil, and stick it into the hole in the top cap.**
8. **Remove the springs from the other leg or deflate it.**
9. **Stroke the fork up and down slowly.** This pushes air bubbles up into the syringe and pulls oil down out of it. Continue until no more air comes up.
10. **Replace the springs or inflate the other leg to make the fork full length.**
11. **Remove the syringe.**
12. **Screw the adjuster lever back on.** Remember to put the little detent ball back onto its spring in the cap—it makes the lever click.

d. Replacing RockShox Judy and early SID damping cartridge

LEVEL 3

This section applies to early RockShox cartridge dampers (Fig. 16.25), which were intended to be replaced when they failed rather than be serviced. If you can locate a replacement cartridge, here's how you change it. The cartridge has no air in it, so if you open it, drain it, replace O-rings, and intend to fill it yourself, simply reassemble it while it's submerged in a bucket of clean, 5-weight ATF or fork oil; stroke the shaft back and forth until no more air bubbles emerge.

1. **Start by removing the lower legs as described in 16-13 through step 11.**
2. **Remove the springs.** Remove coil/elastomer springs (16-9b) or deflate the air springs (Fig. 16.11) to prevent the cartridge from shooting out at you when you remove its retainer. Starting in 1999, some air-spring (SID) models require a hollow 15mm hex key for removing the cartridge. Be careful when you unscrew the cartridge retainer—it is left-hand-threaded.
3. **Remove the circlip at the bottom of the inner leg.** Use inward-squeezing snapring pliers (Fig. 16.36).
4. **Pull out the cartridge.** Inspect it for leakage, scratches on the shaft, and other damage. Replace it, unless you're refilling it as described above.
5. **Slide in the new or refilled cartridge.** Slide the washer onto its shaft.
6. **Replace the circlip.** Orient the snapring so that its sharp edge faces outward. If it instead has the 15mm hex retainer, tighten it counterclockwise, as it is left-hand threaded.
7. **Install the lower legs as in 16-13, starting with step 12.**

16-17

CHECKING FORK ALIGNMENT

LEVEL 3

Checking the alignment may help explain bike-handling problems. You will need a ruler, a true front wheel, and, for a rigid steel fork, dropout-alignment tools (Fig. 1.4). With any type of fork other than a rigid steel one, this procedure is diagnostic only, because you should not try to realign any other type of fork.

If you find the alignment to be off more than a couple of millimeters in any direction with any fork other than a steel, unsuspended one, you need a new fork. If the fork is new, misalignment should be covered as a warranty item.

If a steel fork is more than 8mm off in any direction, you ought to get a new fork. If the dropouts of a steel fork are slightly bent, you can realign them. You can also take a moderately bent (between 2mm and 8mm) steel fork to a frame builder or a bike shop for realignment. Make sure that whomever you take it to is properly equipped with a fork jig or alignment table and is well versed in the art of cold setting (a fancy term for bending) steel forks.

1. **Remove the wheel and the fork** (12-18 and 12-19).
2. **Measure the spacing between the faces of the dropouts** (Fig. 16.47). Adult and high-quality children's bikes should have a spacing of 100mm between the inner surfaces of the dropouts. (Some low-end kids' bikes have narrower spacing—about 90mm or so, and fat-bike and Boost forks have wider spacing.) Measure the distance between the flat surfaces, not between wheel-retaining bumps. Dropout spacing up to 102mm and down to 99mm is acceptable. Beyond that in either direction means you need a new fork. If you have a rigid fork made of steel, you can go to a bike shop or frame builder for realignment.
3. **Clamp the steering tube of the fork in a bike stand or a padded vise.**
4. **Install the dropout-alignment tools** (Fig. 16.48). These, of course, do not work in through-axle forks. The tools are made to be used on quick-release dropouts on either the fork or the rear triangle of the bike, so they have two axle diameters and spacers for use in the (wider-spaced) rear dropouts. Move all the spacers to the outside of the fork ends so that only the cups of the tools are placed inside the dropouts. Install the tool so that the shaft is seated up against the top of the dropout slot. Tighten the handles.
5. **Check how the tool's cups line up.** Ideally, the ends of the cups should be parallel and lined up with each other (Fig. 16.49). The cups of Campagnolo dropout-alignment tools are nonadjustable and are nominally 50mm in length; the ideal space between their ends is 0.1–0.5mm. The cups on Park dropout-aligning tools (illustrated in Figs. 16.48 to 16.50) can be threaded in and out so that you can bring the

16.47 Measuring dropout spacing

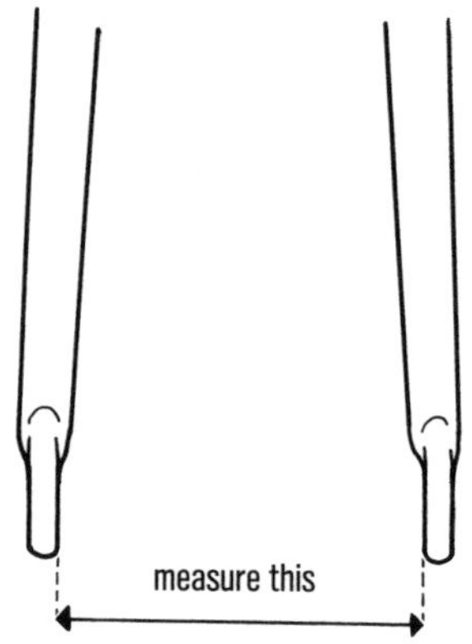

16.48 Aligning dropout with dropout-alignment tool

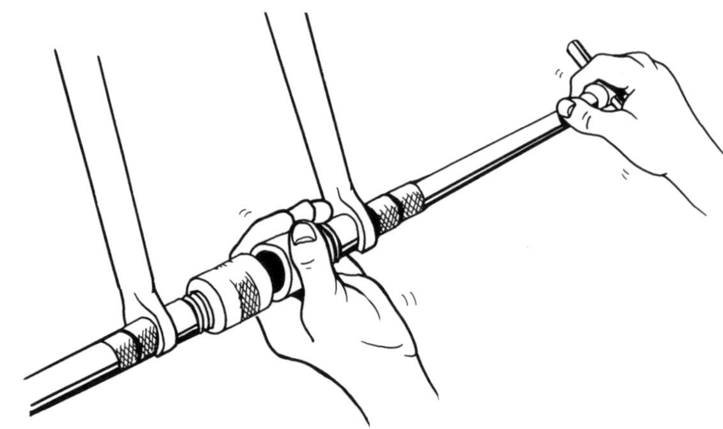

faces up close to each other no matter the dropout spacing. If they are lined up (Fig. 16.49), and the dropouts are spaced between 99mm and 102mm apart, continue on to step 6. If the dropouts on your rigid fork made of steel are not lined up straight across with each other (Fig. 16.50) and are within the 99–102mm spacing range, skip to 16-18 to align them; then come back to here after they're aligned.

NOTE: *It is crucial that the fork dropout faces be parallel before you continue with step 6, or the rest of the alignment procedures will be a waste of time. Clamping the hub into misaligned dropouts will force the fork legs to twist, and any measurement of the side-to-side and fore-and-aft alignment of the fork legs with misaligned dropouts will not be accurate.*

6. **Remove the tire from a front wheel.** Make sure the wheel is true and properly dished (15-4 and 15-5).
7. **Install the wheel in the fork.** Make sure the axle is seated against the top of the dropout slot on either side and that the quick-release skewer is tight. Lightly push the rim from side to side to be certain that there is no play in the front hub. If there is play, you must first adjust the hub (8-6d).
8. **Check the alignment visually.** Look down the steering tube and through the rim's valve hole to the bottom side of the rim (Fig. 16.51). The steering tube should be lined up with this line of sight through the wheel (Fig. 16.52). When you are sighting through the steering tube and the valve hole, you should see the same amount of space

16.49 Correct dropout alignment

16.50 Incorrect dropout alignment (dropout is twisted or right fork leg is bent back)

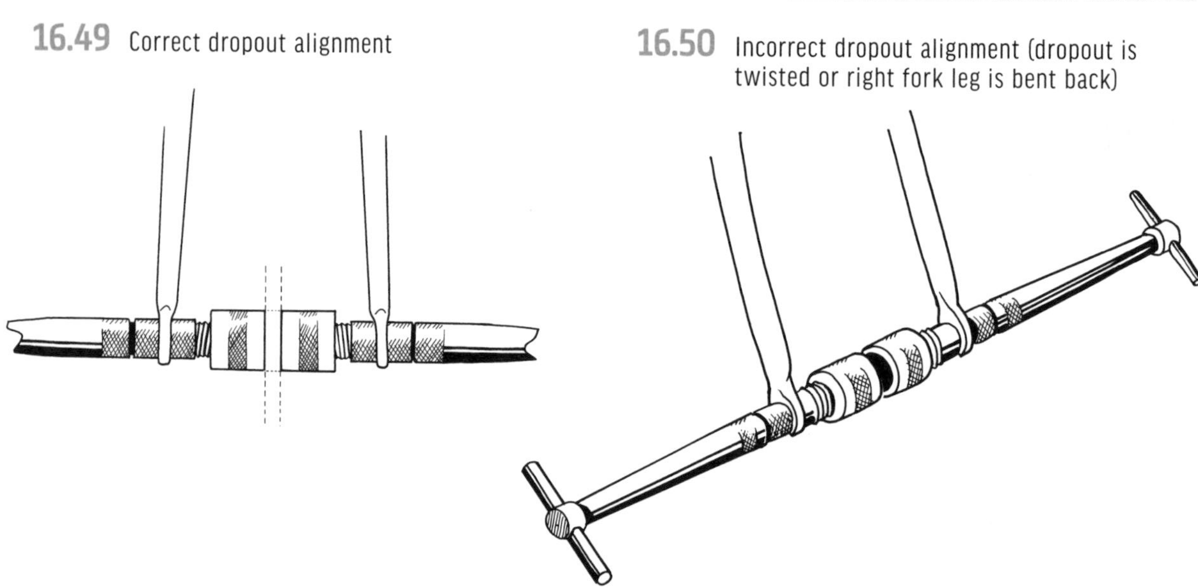

between each side of the rim and the sides of the steering tube while you see the seam side of the rim centered through the valve hole.

(a) Turn the wheel around, and install it again so that what was the right end of the hub is now the left. Sight through the steering tube and the wheel valve hole again. Placing the wheel in the fork both ways corrects for deformation in the axle or any wobble in the wheel. If the wheel is true and properly dished and the axle is in good shape, the wheel should line up exactly as it did before. If it does not line up, but the wheel is off by the same amount to one side as it is to the opposite side when the wheel is turned around, the wheel is off and the fork is fine side to side.

(b) If this test indicates that the fork is up to 2–3mm off to the side, that is close enough; continue on. If it is off by more than 3mm, get a new fork or have it aligned by a frame builder (as long as it is steel; by now you know not to try to align suspension, titanium, carbon-fiber, or aluminum forks).

NOTE: *If you are sighting through the wheel in this way and you cannot see the bottom side of the rim through the valve hole because the hub is in the way, the fork has big problems. For the bike to handle properly, the front hub must have some forward offset from the steering axis (Fig. 16.51). This offset, or rake, is usually around 4 cm on a mountain bike. If you sight through the steering tube and see the front hub, the fork is bent backward so much that it has little or no offset. If this is the case, you need a new fork.*

9. **Place a ruler on edge across the fork legs just below the fork crown** (Fig. 16.53). Make sure the ruler is perpendicular to the steering tube.
10. **Lift the fork so that you are sighting across the ruler and the front hub toward a light source.** The

16.51 Sighting through the steering tube to check fork alignment

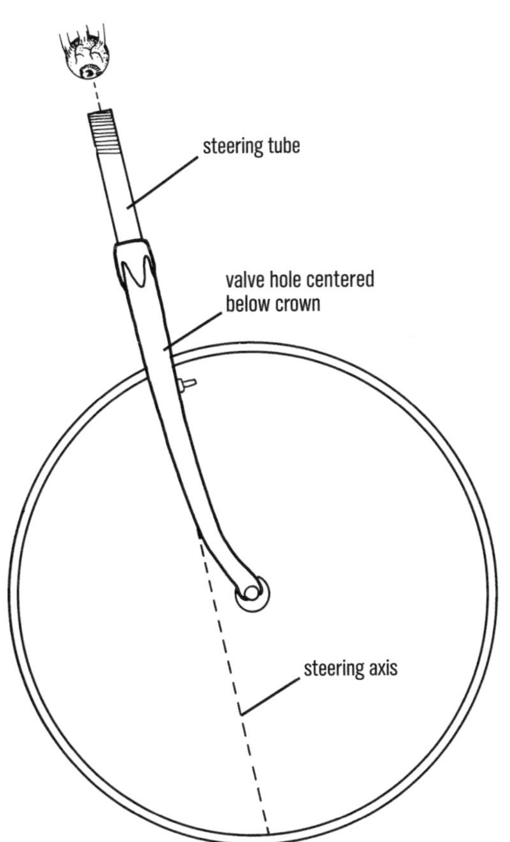

16.52 Correct alignment of valve hole in a straight fork

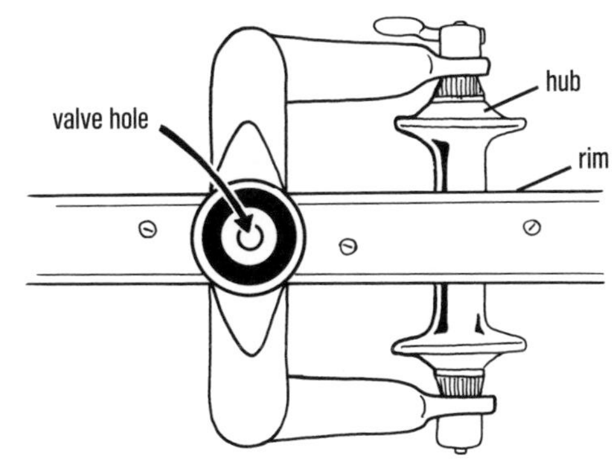

16.53 Checking fork alignment with a ruler

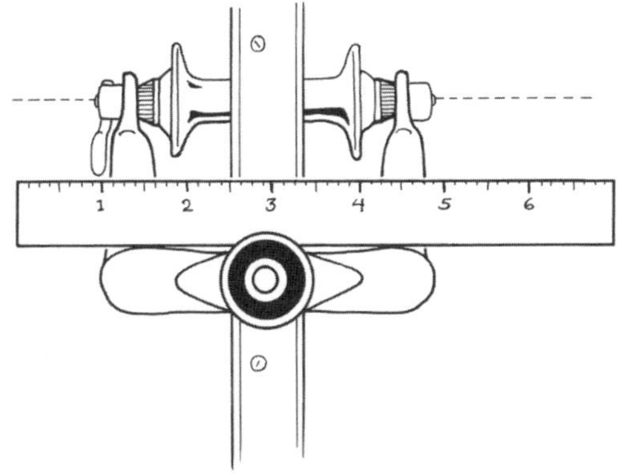

ruler's edge should line up parallel with the axle ends sticking out of either end of the hub (Fig. 16.53). This test will tell you if one fork leg is bent back relative to the other. If the axle lines up parallel to the ruler, or very close to that, fork alignment has checked out completely, and you can put it back in the bike. If one fork leg is considerably behind the other, get a new fork or have this one aligned (if it is steel).

16-18

ALIGNING DROPOUTS ON A STEEL FORK

You should do this only with a steel non-suspension fork.

Dropouts are easy to tweak out of alignment; simply pulling the bike off a roof rack and failing to lift it high enough to clear the rack skewer will do it. A fork may also have come with misaligned dropouts when new.

NOTE: *If the dropout is bent more than 7 degrees or so, or if the paint is cracked at the dropout where it is bent, bending the dropout back is too dangerous. Replace the fork.*

1. **Install dropout-alignment tools, and check the alignment as in 16-17, steps 5–7.** If the tool cups are not lined up with each other (Fig. 16.50) and the fork spacing is between 99 and 102mm, continue with steps 2–4 to align the dropouts. If the fork spacing is wider than 102mm or less than 99mm, there is no point in aligning the dropout faces, because you must bend the fork legs as well to correct the spacing. Without an alignment table or fork jig, you cannot do this accurately. Get a new fork or have a qualified mechanic or frame builder align the fork.
2. **Clamp the crown** (or unicrown, Fig. 16.3) **of the fork very tightly.** Do this between two wood blocks in a well-anchored vise (Fig. 11.40).
3. **Bend each dropout until the open faces of the dropout-alignment tools are parallel.** Grab the end of the dropout-alignment tool handle with one hand and the cup of the tool with the other, and turn the tool to bend the dropout (Fig. 16.48). Recheck. Repeat until the edges line up straight with each other (Fig. 16.49).
4. **Remove the tools and continue with 16-17, step 6.**

Come to kindly terms with your Ass for it bears you.

—JOHN MUIR, *HOW TO KEEP YOUR VOLKSWAGEN ALIVE*

17

FRAMES

TOOLS

String
Bike stand
Metric wrenches and hex keys
Torque wrench

OPTIONAL

Dropout-alignment tools
Derailleur-hanger alignment tool
Vise
Pliers
Easy out (screw extractor)
Tap handle
Metric taps
English-thread bottom-bracket tap set
Sockets of various sizes
Hammer
Piece of old inner tube
Drill and drill bits
16mm cone wrench
Suspension grease
Grease gun
Thread-cutting oil
Dropout saver

Your Volkswagen is not a donkey . . . and your mountain bike is not a Volkswagen. Still, you'd be well served to follow Muir's sage advice and stay on good terms with your bike. In doing so, pay close attention to the frame, because it is the most important part of your bike. It is the one part of your bike that is nearly impossible to fix on the trail if it breaks, and if it does fail, the consequences can be serious. Therefore, get to know your frame. Come to kindly terms with it . . . for it bears your ass . . . or something like that.

17-1 FRAME DESIGN

The traditional diamond, or double-diamond, mountain bike frame design evolved from a combination of postwar cruiser bikes and road-racing bikes. The rigid design of a road bike relies on a front triangle and a rear triangle (Fig. 17.1); and although the basic concept is similar, there are some notable differences between road and mountain bike geometries. Mountain bike frames feature a higher bottom bracket for more ground clearance; a longer and wider rear triangle for more tire clearance and a wider rear axle; a shorter seat tube (and correspondingly lower top tube) for more stand-over clearance; brake bosses (first for rim brakes, now for disc brakes); larger-diameter tubing; and, often, suspension.

In the pursuit of low frame weight and price coupled with high frame strength, durability, comfort, control, and performance, mountain bike frame design has changed radically in the relatively short time since its inception. Testament to that is comparing modern mountain bikes, even fully rigid ones (Figs. i.3 and 17.1), with Marin County Repack-style bikes of the late 1970s or Crested Butte off-road cruisers of the same era patterned after circa-1940 Schwinn-style cruisers. And full suspension (Figs. i.4 and 17.2), now ubiquitous even on cheap department store bikes, is a whole new breed of animal.

17-2 SUSPENSION FRAME DESIGN

Rear suspension (also called full suspension because it is usually combined with a suspension fork) involves a design totally different from the

17.1 Rigid frame

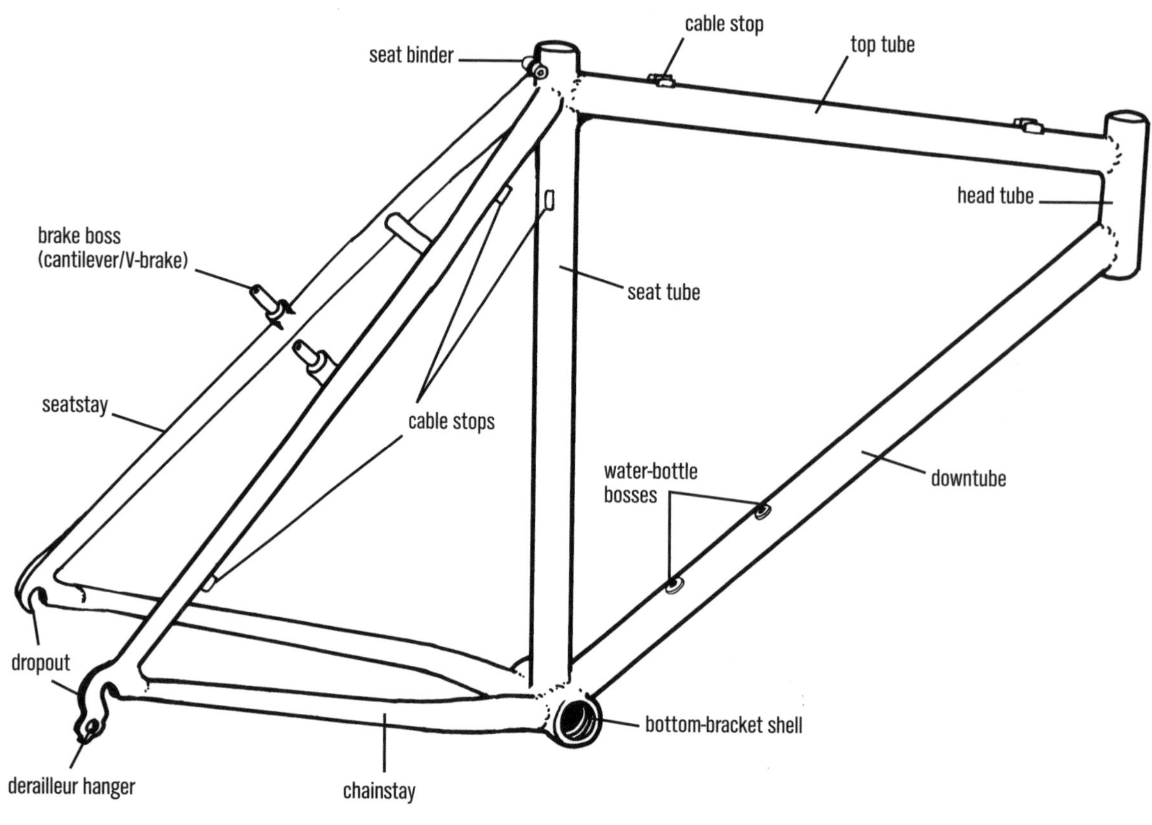

17.2 Frame with rear suspension; a newer frame would include a tapered head tube, disc-brake mounts, and a replaceable derailleur hanger

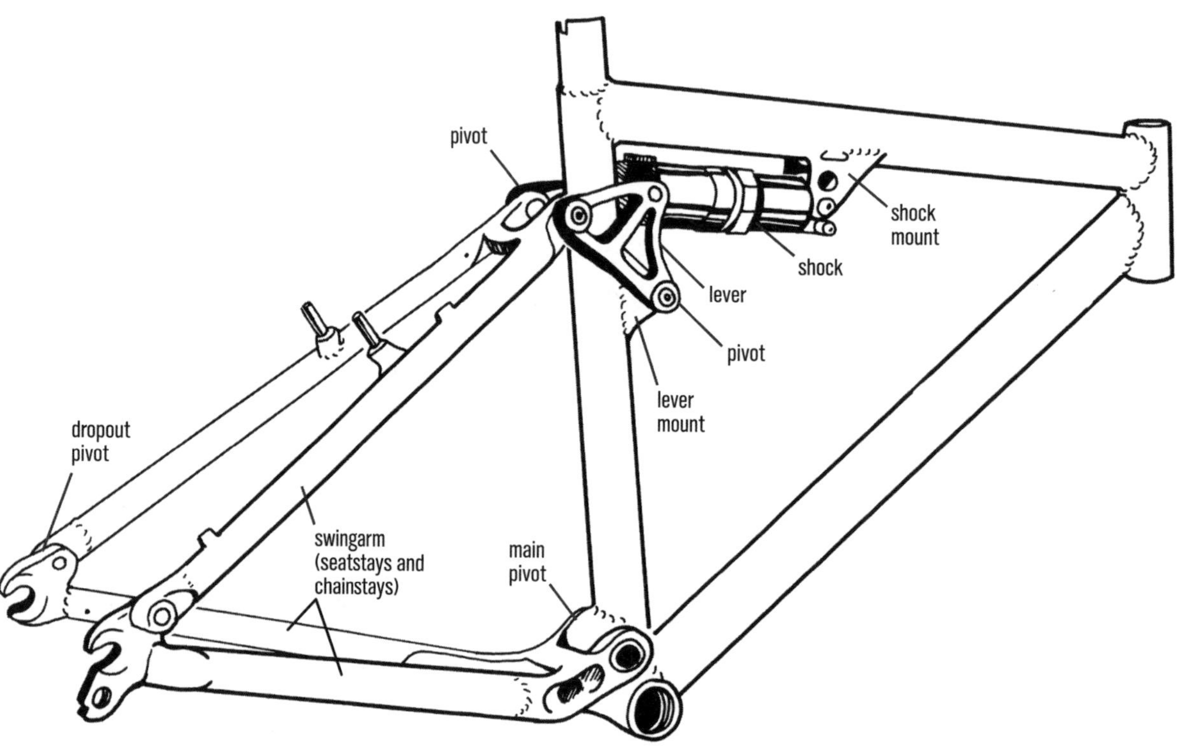

traditional double diamond. Most suspension frames have a front triangle and a rear swingarm (Fig. 17.2).

There are almost as many rear-suspension designs (and names for them) as there are suspension-frame designers. Over the past few years, the changes have been fast and furious, and it would be pointless to go on at length about specifics of current designs, as they will change. Instead, this chapter divides suspension frames into broad categories and gives general maintenance guidelines for them.

17-3

EVOLUTION OF FRAME MATERIALS

The evolution of bicycle-frame materials has been going on since the birth of the bicycle. Wood was the material of choice for the first bicycles, but that was soon replaced by steel, aluminum, and even bamboo. Aluminum, carbon fiber, steel, and titanium are the materials now most commonly used to build mountain bike frames. Because these materials come in a variety of grades with varying costs and physical properties, assume that I am talking about the highest grades for the materials used in bicycles. For example, the aluminum used in window frames is a lot different from the heat-treated 6000- and 7000-series aluminum used in high-end bicycle frames.

Steel has the highest modulus of elasticity (a principal determiner of stiffness) as well as the highest density and tensile strength of all of the metals commonly used in frames. The modulus of elasticity, the density, and the tensile strength of aluminum are much lower than those values for steel, and titanium has values between the two. Metal tubing characteristics are maximized for bicycles by (1) butting (i.e., placing thickness where it is needed and not where it adds useless weight), (2) increasing diameter to add stiffness, and (3) heat-treating and alloying to boost certain physical properties of the metal. With intelligent use of materials, long-lasting frames of comparable stiffness-to-weight and/or strength-to-weight ratios can be built out of any of these metals.

Carbon fiber and similar composite frame materials consist of fibers embedded in a resin (plastic) matrix. These materials can be extremely light, strong, and stiff. Bike frames can be built by molding them in a single piece (monocoque construction); by gluing large molded carbon subassemblies together (also sometimes misleadingly called monocoque; each piece is, of course, a single piece, but the whole is a number of pieces); by bonding "tube to tube" where carbon tubes are mitered to fishmouth around each other and then wrapped with carbon fibers to hold them together; or by gluing carbon-fiber tubes into lugs (usually made of carbon fiber or aluminum). A big advantage of molded carbon composites is that they can be made thicker or thinner in specific areas where extra strength is or is not needed. And contrary to widely held notions, they can be repaired when they break; a number of small carbon frame-building shops offer such a service.

17-4

INSPECTING THE FRAME

LEVEL 1

You can avoid potentially dangerous or at least ride-shortening frame failures by inspecting the frame frequently. If you find damage and you are not sure how dangerous the bike is to ride, take it to a bike shop for advice.

1. **Clean the frame every few rides.** That way, you can spot problems early.
2. **Inspect all tubes for cracks, bends, buckles, dents, and paint stretching or cracking.** Look especially near the joints where stress is highest. If in doubt, take the frame to an expert for advice.
3. **Perform the coin test on questionable areas of carbon frames.** Using the edge of a coin, tap on the frame in any area that looks cracked, damaged, or otherwise suspicious, as well as over the rest of the frame for comparison. If, instead of a satisfying "clack" sound consistent over the tubes, you hear a dead "thwap" sound, there may be cracks or delaminated areas hidden within the structure. Take the frame to an expert for analysis.
4. **Inspect the frame joints and all pivot and mounting points for cracks.** Check that the dropouts, the derailleur hanger on the right dropout, brake mounts, cable stops, and suspension pivot points

are not bent or cracked. (See Figs. 17.1 and 17.2 for names and locations of frame parts.) Some derailleur hangers, dropouts, cantilever/V-brake bosses (Fig. 17.1), and even disc-brake mounts (Fig. 17.3) bolt to the frame and are replaceable. Otherwise, badly bent or broken dropouts, rim- or disc-brake mounts, and cable stops require having a new one welded, riveted, brazed, or carbon-wrapped on; a frame builder in your area may be able to do it.

5. **Look for corrosion on steel and aluminum frames.** Every six months, and also after every wet ride, remove the seatpost and invert the bike to let water pour out of the seat tube, and allow everything to dry fully before replacing the post. Look and feel for deep rusted areas inside or for rust falling out. I recommend squirting a rust protective spray designed for bicycle frames (Frame Saver), WD-40, or oil inside the tubes periodically (after letting the frame dry out upside down with the seatpost removed). Remember to grease both the seatpost and the inside of the seat tube when you reinsert the post. After sanding off the rust on any external areas where the paint has come off, touch them up with touch-up paint or nail polish (hey, it's available in lots of cool colors, but be advised that it is not as durable as good paint and requires periodic retouching).

6. **On suspension frames, check the swingarm movement and the shock condition.** Disconnect the shock and verify that it pivots freely at each end; grease the mounting points and shock eyelets and do not over-torque mounting hardware (torque maximum is often 60 in-lbs). Move the swingarm up and down, and flex it laterally, feeling for play or binding in the pivots. Verify that the frame's body-eye mount and shaft-eye mount are aligned with each other so that the shock shaft will compress straight into the shock body; side-loading can cause a shock to fail. Check the shock for leaking oil, cracks, a bent shaft, or other damage. Shock maintenance is in 17-8.

7. **Check that a true and properly dished wheel sits straight in the frame.** It should be centered between the chainstays and seatstays, and lined up in the same plane as the front triangle. Tightening the hub skewer should not result in bowing or twisting of the chainstays or seatstays.

17.3 International Standard (IS) disc-brake mount (shown on a sliding dropout for a Rohloff SpeedHub)

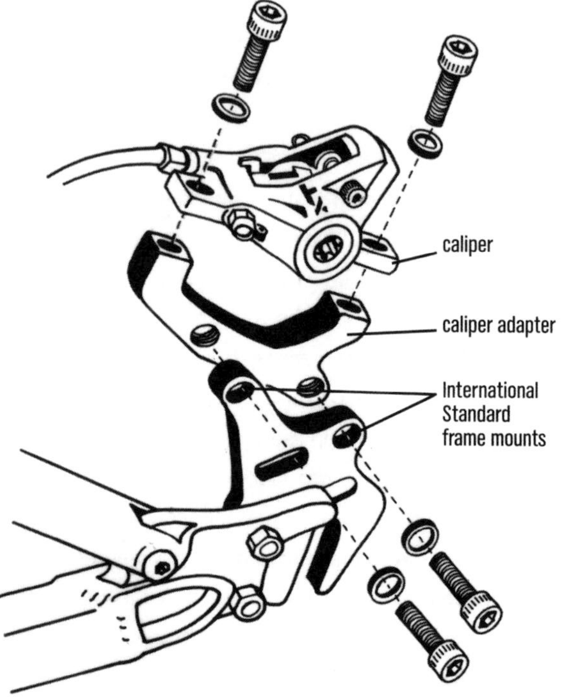

17-5

CHECKING AND STRAIGHTENING THE REAR-DERAILLEUR HANGER

1. **Thread a derailleur-hanger alignment tool into the derailleur hanger on the right dropout** (Fig. 17.4).
2. **Install a true rear wheel without a tire on it.**
3. **Swing the tool around, measuring the spacing between its arm and the rim all the way around.** The arm of the tool should be the same distance from the rim at all points. The Park tool in Figure 17.4 has an adjustable extension rod extending at right angles from the arm held by a hand-turned setscrew to check the spacing to the rim.
4. **If the tool has play in it, keep it pushed inward lightly as you perform all of the measurements.** Otherwise, you will get inconsistent data.

17.4 Checking derailleur-hanger alignment

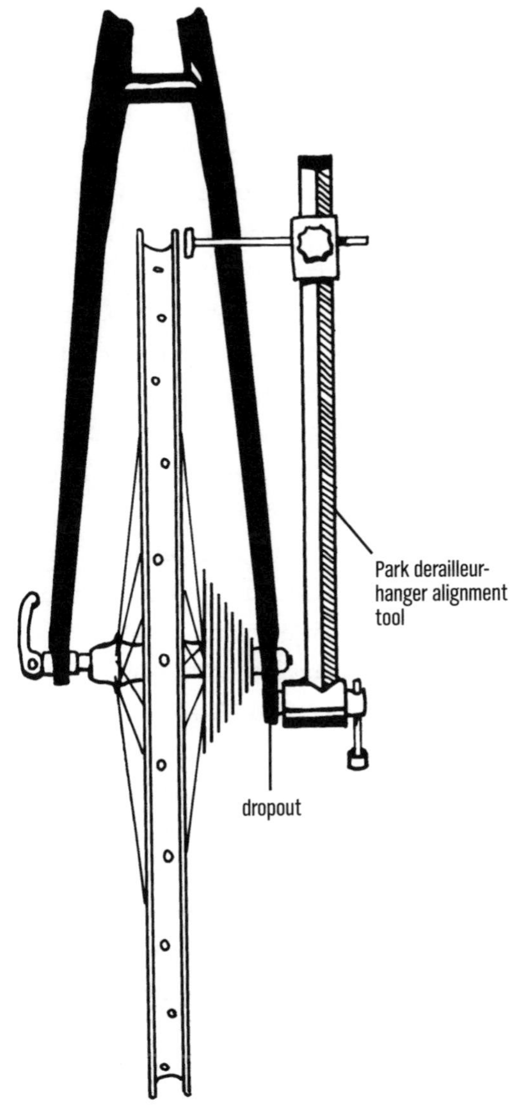

17.5 Installing replacement rear-derailleur hanger

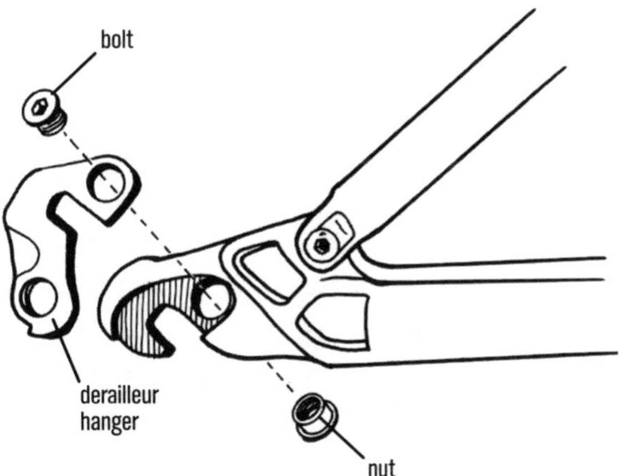

17.6 Inserting dropout saver

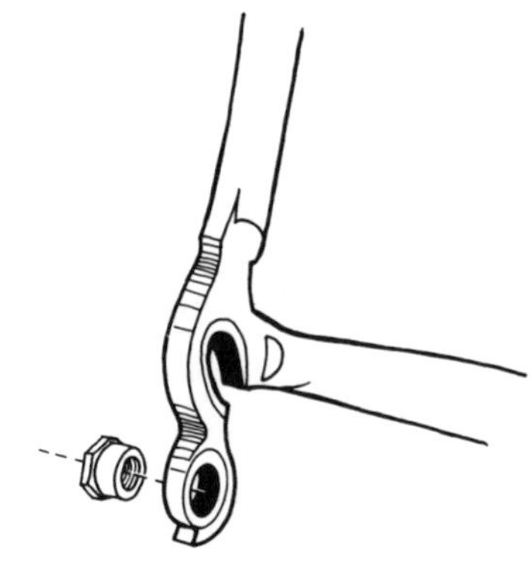

5. **Correct any variation in spacing between the tool arm and the rim.** If the tool varies in its lateral proximity to the rim by more than 1mm or so all the way around, carefully bend the hanger by pulling outward on the arm of the tool lightly where it is closest to the rim.
6. **If the derailleur hanger is really bent, be careful.** You may not be able to align it without breaking it (you may even have trouble threading the tool in because the threaded hole has become oval). If the derailleur hanger bolts onto the frame, replace it with a new one (Fig. 17.5), available from your bike dealer or wheelsmfg.com.
7. **If the threads are too damaged to use and the derailleur hanger is not replaceable, see 17-6 and Figure 17.6 for other derailleur hanger options.**

17-6

FIXING DAMAGED THREADS

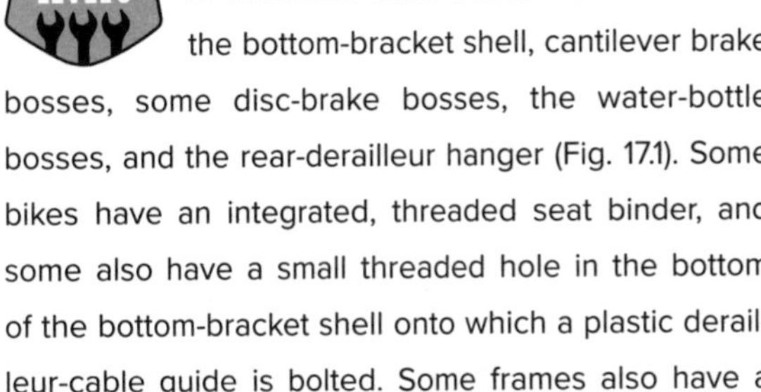

A mountain bike frame has threads in the bottom-bracket shell, cantilever brake bosses, some disc-brake bosses, the water-bottle bosses, and the rear-derailleur hanger (Fig. 17.1). Some bikes have an integrated, threaded seat binder, and some also have a small threaded hole in the bottom of the bottom-bracket shell onto which a plastic derailleur-cable guide is bolted. Some frames also have a threaded front-derailleur mount or two.

1. **Whenever you have to retap any threads, first brush them clean and then use oil on the tap.** With titanium threads, use canola oil.

2. **Chase (retap) any threads on the frame that are stripped or cross-threaded.** Use a thread tap of the appropriate size.
3. **Turn the tap forward (clockwise) a bit, then turn it back, then forward (two steps forward and one back), and so on.** This will prevent the tap from binding and possibly breaking. The exception is left-hand-threaded, drive-side bottom-bracket threads, with which you do the same thing, but in the opposite direction. Be aware that taps are made of very hard and brittle steel. If you put any side forces on small taps, they can break off easily. Be careful. If the tap breaks, you'll have a real mess, because the broken tap in the hole is harder than the frame, so it's impossible to drill the broken tap out. If you break off a tap in the frame, do not try to get it out yourself. Take the frame to a bike shop, a machine shop, or a frame builder before you break off what little is left sticking out. Unless you put the tap in crooked, breaking one should not be a problem when you retap damaged frame threads; these threads will be so worn that getting the tap to find any metal to bite into will probably be your biggest problem.

IMPORTANT: *Tapping a bottom-bracket shell takes a good amount of expertise. You really need expert supervision if you have never done it before and still want to do it yourself. In addition to making sure that you place the correct tap in the correct end of the shell, be certain that the taps go in straight. Most bottom-bracket taps have a shaft between the two taps to keep them parallel to each other (Fig. 1.4). They must both be started at the same time from both ends. If you mess it up, you may ruin the frame. If in doubt, ask an expert.*

The following tap sizes are commonly found on mountain bikes:

Water-bottle bosses	M5 (5mm × 0.8)
Bottom-bracket shell hole for shift-cable guide	M5 (5mm × 0.8)
Seat binders and brake bosses	M6 (6mm × 1)
Disc-brake post-mounts	M6 (6mm × 1)
Derailleur hanger	M10 (10mm × 1)
Bottom-bracket shells	1.37 inches × 24 tpi*

*tpi = threads per inch

NOTE: *Remember, chain-side bottom-bracket threads are left-hand-threaded; the other side is right-hand-threaded.*

4. **Replace the old bolt with a new one.** Exceptions are bottom-bracket cups and rear derailleurs, as long as the threads look perfect.
5. **If tapping the threads and using a new bolt do not work, try these remedies:**
   (a) For cantilever/V-brake bosses, some brake posts are replaceable. Replaceable posts have wrench flats (usually 8mm) at the base, and they thread into a boss welded to the frame. If they are not configured like this, you must take your bike to a frame builder to get a new boss welded on.
   (b) Among disc-brake bosses, international standard (IS) mounts (Fig. 17.3) are standard on frames, while post-mounts (Fig. 9.9) are standard on forks. About the only thing that can happen to robust IS tabs is that they can get bent. But threaded disc-brake mounts (i.e., post-mounts) are found on some frames, and if the threads are stripped, it is often because overly short bolts were employed to mount the disc brake (generally, 10–12mm of thread engagement is required; this much bolt length should stick through the brake's mounting tab; see Fig. 9.9). If so, you may be able to tap out the threads and use new, longer bolts without further problems. Otherwise, the entire dropout, swingarm, or frame may need to be replaced.
   (c) For water-bottle bosses, some bike shops have a tool that rivets bottle bosses into the frame. Check for this possibility first, because you can avoid a new paint job that way. Otherwise, if the frame is metal, take it to a frame builder to get a new boss welded or brazed on.
   (d) For derailleur-hanger threads, if your bike has a replaceable rear-derailleur hanger, bolt a new one onto the frame (Fig. 17.5). This is a good thing to carry with you on the trail.

   Without that, and if the threads are so badly damaged that an M10 tap won't repair them, another option is to use a Dropout Saver derailleur-hanger backing nut (Fig. 17.6) made by Wheels Manufacturing and available at bike shops. A Dropout Saver is simply a sleeve threaded the same as your dropout was, with 16mm wrench flats on one end. Drill out the hole in the damaged derailleur hanger with a 15⁄32-inch drill bit, push the Dropout Saver in from the backside, and screw in the derailleur while holding the Saver with a 16mm box wrench. Dropout Savers come in two lengths, according to the thickness of the dropout.

Yet another option is to saw off the derailleur hanger with a hacksaw and use a separate derailleur hanger from a really cheap bike that fits flat against the outside of the dropout and is held in by the hub axle bolts or quick-release. You could also get a new frame or swingarm, or you may be able to have the dropout replaced by a frame builder.

(e) For seat binders with a separate binder clamp, just replace the whole clamp rather than mess with its threads. With a welded-on binder, you can drill it out and use a quick-release or a bolt and nut, or you can saw and file it off and use a binder clamp. Threads rarely get stripped on welded-on seat binders, however; it is usually the bolt that is the problem.

(f) For bottom-bracket cable-guide bolt-hole threads, a new hole in the bottom of the bottom bracket can be drilled and tapped, or the stripped hole can be tapped out with larger threads for a larger screw. Make sure the screw you use is short enough that it does not protrude into the inside of the bottom-bracket shell.

17-7

REPAIRING CHIPPED PAINT AND SMALL DENTS

Fixing chips is simply a matter of cleaning the area and touching it up. Sand any chipped paint or rust completely away before touching up the spot. Use touch-up paint, model paint, or fingernail polish.

Small dents can be filled with automotive body putty, but there is little point to filling them if you are doing only a touch-up, because the area probably won't look that great anyway.

There are frame painters around the country who can fill dents, repaint frames, and even match original decals; search online.

17-8

MAINTAINING THE SHOCK

a. Daily maintenance

Keep the shock shaft, shaft seals, and bottom-out bumper clean. Clean them after every ride (you know—when you wipe down and lube the chain), but do not use a pressure washer on them; the pressure can blow the seals inward and contaminate the shock. Lightly lubricate the shaft or shock body (the part that slides in and out).

b. Every-40-hours maintenance: Air-can service

Remove and lubricate the air can (or air sleeve—see Fig. 17.7) from an air shock after every 40 hours of riding unless you are riding in very clean conditions.

This may seem like a ridiculously frequent schedule of maintenance, but think about it for a second. You don't think twice (or I hope you don't) about changing the oil in your car's engine every 3,000 miles, do you? Well, 3,000 miles is 50 hours of driving at 60 mph, and the engine has a lot of oil volume and an oil filter to keep pulling contaminants out of the oil as it circulates.

Now consider your bicycle shock. Its piston is constantly going up and down as you ride, it has less than a teaspoon of lubricant in it, and it has no room for an oil filter. Forty hours does not seem so extreme in that context, does it? If you don't do the air-can service yourself every 40 hours, have a qualified mechanic do it for you.

Regrease the eyelet bushings after every 40 hours of riding along with doing the air-can service.

Air-can overhaul

Because the air can should be serviced frequently, these instructions will show you how to overhaul it in

17.7 Rear-shock parts and adjustments

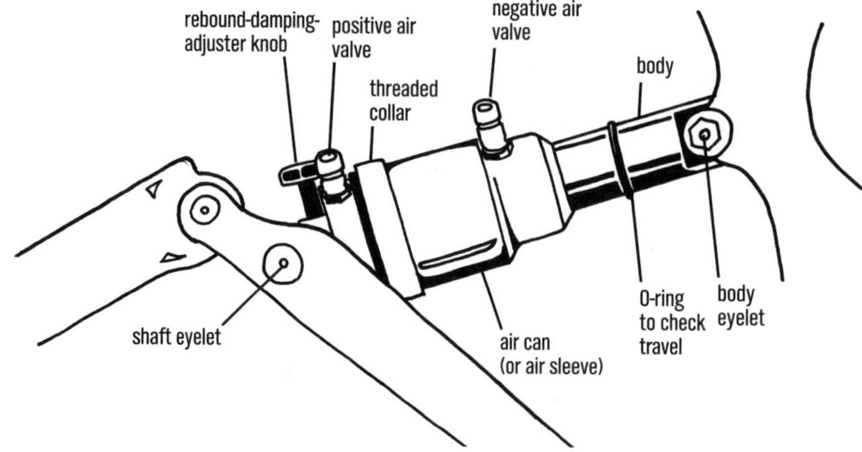

a hotel room or parking lot, without a vise. And even a vise may not be enough to push the can on far enough to screw it back on anyway, so reassembling the air can this way, on the bike, may be a must in any case.

1. **Deflate the shock.** Sit on the bike while depressing the Schrader valve pin with a blunt implement (hold a rag over it to keep oil from spurting on you) to release as much air as possible; do it slowly in case it has a negative air spring (see 17-11a).
2. **While sitting on the bike, grasp the air can with your hands and unscrew it** (Fig. 17.8). Turn it counterclockwise, as if it were a lid on a jar. If you cannot get enough of a grip to twist the air can, wrap an inner tube around it first.

NOTE: *Alternatively, after removing the shock as in step 3, you can clamp the faces of the shaft eyelet (the eyelet at the big end of the shock—see Fig. 17.7) between soft jaws or wood or aluminum blocks in a vise (Fig. 17.9) and unscrew the air can there.*

3. **Remove the shock from the bike.** This can be as simple as removing the two mounting bolts and pulling the shock off. But if the shock doesn't pull out of its mounting tabs when you remove the bolts, there may be a hollow, threaded shaft through the eyelet that overlaps into the holes in the frame tabs. In this case, install one bolt a few threads and tap it inward with a hammer to knock the sleeve out of the near tab. Remove the bolt, and tap the sleeve the rest of the way out with a hex key or the like.
4. **Slide the air can off** (Fig. 17.9). If the mounting hardware in the body eyelet bushing is too wide to

17.8 Unscrewing a Fox air can from the shaft eyelet while it's on the bike

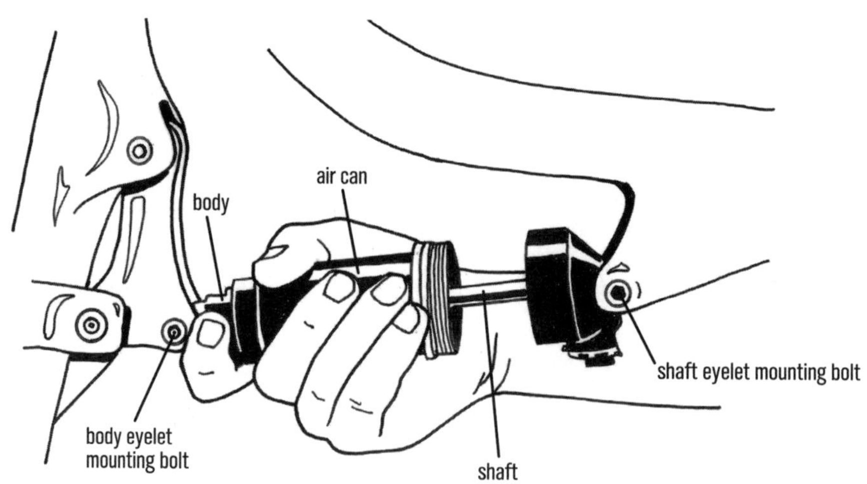

allow the can to slide off, you'll need to remove it. Pull top-hat-shaped aluminum spacer/reducers out with pliers or a vise, or by twisting them out with an easy out (spiral-flute screw extractor) on a tap handle. Alternatively, if you have a long sleeve through the eyelet bushing with an aluminum spacer on either end, pull off the spacers and push out the sleeve with your fingers or a vise.

5. **Clean all seals and contact areas.** With a clean, lint-free rag and isopropyl alcohol if need be, thoroughly clean the seals at the narrow end of the can, the seals around the piston, the inside of the can, the threads, and the O-ring that the can screws up against (Fig. 17.9). Wipe down the piston shaft and the shock body as well.

6. **Inspect the seals and replace if needed.** If any are damaged, or if you have had significant air or oil leakage, buy a seal kit for the shock, and install those seals. Remove O-rings and rubber square-cross-section seals (square seals or quad seals) from the shock body (piston) by squeezing and pushing them with your thumbs until a loop pops up that you can grab and roll off with your thumb. The backing rings surrounding the seal are split and will pull off easily. Dig the seals out of their grooves in the narrow end of the can with a sharpened pick of some sort. A couple of long nails you sharpened by spinning them in a drill against a belt sander will do fine; leave one straight, and bend the other one near the tip so that you have some options for digging the seals out of their grooves. Rather than going in under the edge of each seal with your tool, which could scratch the shock's anodized coating and create an air leak, stab each rubber seal and Teflon backing ring with your sharp pick, and stretch the seal up over the lip of its groove and pull it out. Press the new O-rings, rubber square seals, glide rings, seal-backing rings, and so on back in place where the old ones were.

7. **Lubricate the seals.** Wipe some Slick Honey grease on the piston seals and the seals in the end of the air can. Better yet, squeeze Fox Float Fluid from a 5mL blister pack on the seals, lay a bead of it on the shock body, and squirt the rest of it inside the can just before it meets the threads in step 8, below.

17.9 DT Swiss air shock disassembled in vise for air-can service

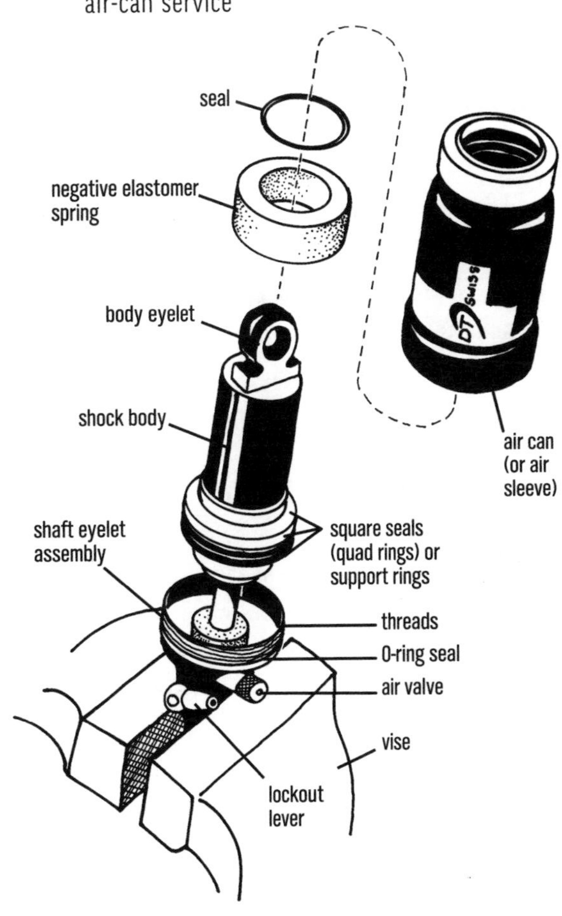

8. **Slide the air can back onto the shock body and, if you can, up against the threads.** If you have the shock in a vise or can do so in your hands, screw it on (clockwise) hand tight. If you can't get the air can screwed on, you will in step 10. Not being able to press the sleeve far enough to engage the threads is actually a good sign, indicating the seals are effective at building pressure behind the piston. Slip the travel-indicator O-ring onto the shock body.

9. **Install the shock in the bike.** First grease the eyelets and reinstall the aluminum spacer/reducers or mounting sleeves and spacers (press them in with a vise if they won't go in by hand alone). Replace any bolts, washers, and sleeves you removed.

10. **Sit on the saddle to compress the shock and screw the air can back on.** Using the leverage of the bike

to compress the shock is the most effective way to overcome the air pressure building behind the piston as the can slides on and get the threads on the can and shaft eyelet cup to meet up. While sitting on the saddle, grab the air can and screw it on clockwise (Fig. 17.8), as if it were a jar screwing into its lid. Turn it by hand until it stops with the seam of the wraparound label lined up behind the shock.

11. **Inflate the shock to your desired air pressure.**
12. **Slide the sag-indicator O-ring back against the can seal.**

    You're ready to ride another 40 hours!

c. 6–12-month service: Replace shock eyelet bushings

LEVEL 2

To maximize longevity, grease the bushings every 40 hours, after removing the mounting hardware (step 2), when you service the air can.

1. **Remove the shock from the frame.**
2. **Remove the shock-mounting hardware.** If you have a two-piece system of top-hat-shaped aluminum spacer/reducers pressed into the eyelet bushing (Fig. 17.10), pull each one out with pliers or a vise, or push an appropriately sized easy out on a tap handle into the hole and twist it out counterclockwise. If you have a three-piece system (or five-piece, counting O-rings) consisting of a long sleeve through the eyelet bushing, an aluminum spacer on either end, and O-rings between them, pull off the spacers and O-rings and push out the sleeve with your fingers or a vise (Fig. 17.11). If the sleeve is still stuck once it's flush on one end, push it through with the vise using a socket smaller than the diameter of the sleeve.
3. **Determine what kind of eyelet bushing you have.** The standard DU bushing is a one-piece steel or bronze sleeve, often lined with slippery plastic. Two-piece plastic Igus bushings press in from either side and have a lip against the face of the eyelet.
4. **Push out the eyelet bushing.**

    (a) **Push out a standard, one-piece metal (DU) bushing.** These push out from one side, ideally with a vise and a bushing-removal tool, but if you don't have this, sockets will substitute; the methodology is the same. Find a socket whose outside diameter (OD) is just slightly less than the OD of the DU bushing (or the ID of the eyelet hole) and a second socket whose inside diameter (ID) is slightly larger than the OD of the bushing. Between vise jaws, place the larger socket against the top of the eyelet so its open end surrounds the bushing and the smaller socket with its closed end against the opposite end of the bushing; tighten the vise handle until the smaller socket pushes the bushing through until it pops out of the eyelet and into the larger socket. It's a bit tricky to hold all three pieces in the vise by yourself; another person or a couple of strong magnets to hold the sockets in place will help. Another alternative that is easier to hold together and requires no vise is to put a long

17.10 Pushing top-hat-shaped aluminum spacer/reducer into a shock eyelet bushing

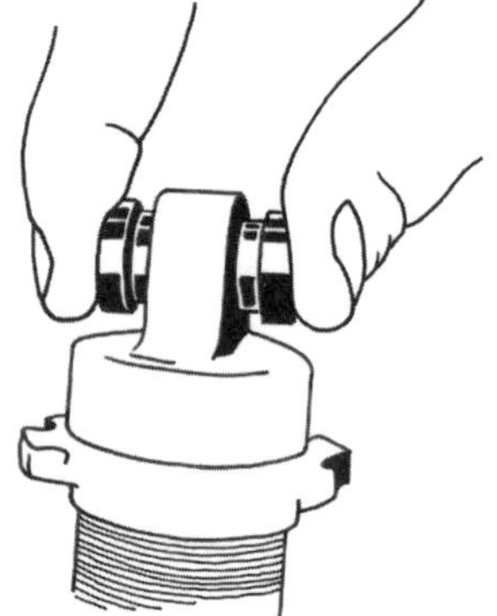

17.11 Pushing out (or in) a shock-mounting sleeve with a vise

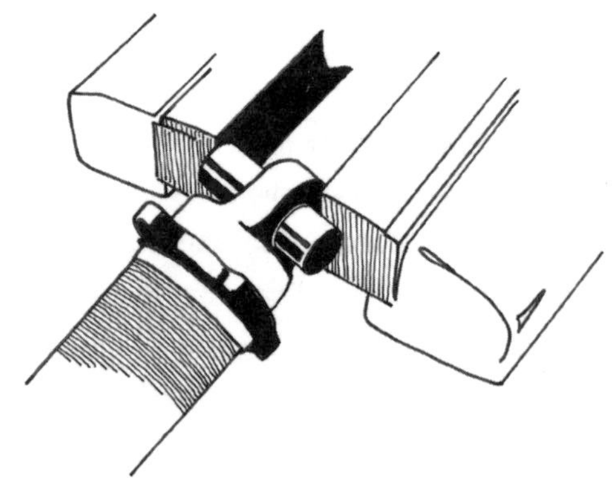

bolt through the two sockets and the eyelet and put a washer and nut on the other end. Line up the sockets properly, and tighten the bolt until the bushing pops out into the larger socket (Fig. 17.12).

(b) **Remove two-piece plastic (Igus) bushings.** Get the first one out by reaching inside the eyelet and setting the edge of a small punch or screwdriver against the edge of the opposite bushing where the bushings meet in the middle of the eyelet (Fig. 17.13) and tapping it out, working your way around. Don't do this unless you have a replacement on hand, because it gouges up the bushing. Get the second one out the same way as in the previous step (4a, above) by pushing it out in a vise (or with a bolt, washer, and nut through the whole shebang) with a socket just smaller than the ID of the eyelet into a socket whose opening is just bigger than the OD of the lip of the bushing (Fig. 17.12).

5. **Push in the new bushing.** Press the DU bushing or the pair of Igus bushings into the eyelet with the soft jaws of a vise (Fig. 17.14).
6. **Replace the shock-mounting hardware.** Press in the pair of top-hat-shaped aluminum spacer/reducers (Fig. 17.10) or the bolt sleeve (Fig. 17.11) into the eyelet bushing with a vise. With a sleeve type, replace the O-rings, if present, and the spacers against them on either side.
7. **Remount the shock in the frame.**

d. Annual maintenance: Damper oil change

Just as in a high-quality suspension fork, a rear shock has a spring to compress and rebound over bumps, and it has a damper to slow the movement of the spring both when it compresses and when it rebounds. You've seen in section b how to take the air can (the spring) off an air shock; the damper is inside the shock body (Fig. 17.9).

The oil inside the shock body degrades over time, reducing damping. If you notice a damping loss, you need an oil change. This is generally a factory operation; send it in during the winter when you're not riding,

17.12 Using a bolt, washer, and nut to push a bushing out of a shock eyelet with a small socket so it pops out into a larger socket

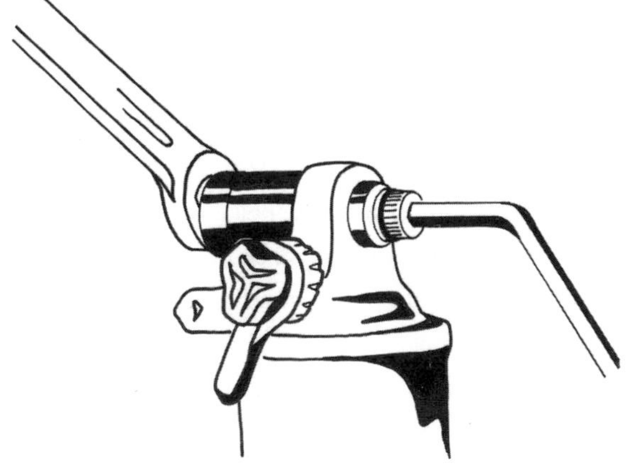

17.13 Tapping out the first Igus bushing from a shock eyelet with a punch

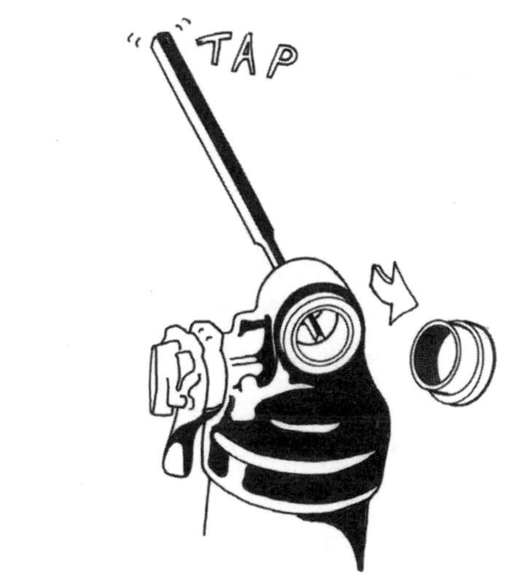

17.14 Pushing in a new DU bushing with a vise and wood blocks

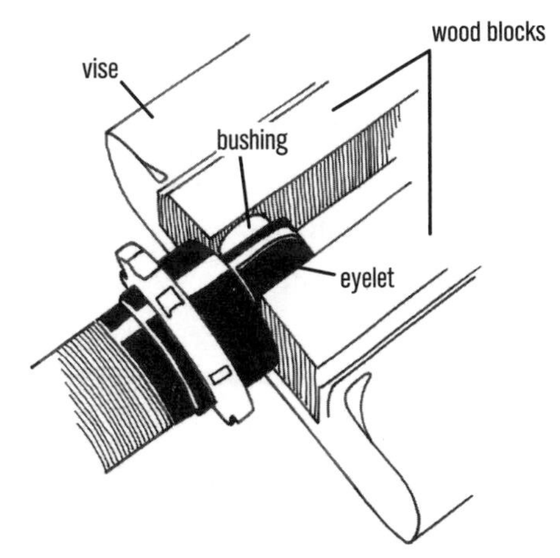

unless you have a spare shock to substitute. Because special tools are recommended for many shock services, such as replacing air valves, glide rings, shaft bushings, shaft seals, and eyelet bushings, it may not make sense for you to own these tools, in which case a good relationship with a shock-literate shop is in order.

That said, some shocks are friendlier to home mechanic service than others. Following is an example of a simple-to-service rear shock damper. Unless you have a shock like this, find the technical manual on your shock manufacturer's website.

To alter damping, shim stacks (on shock pistons) and oil weight can be changed. Shock shims are thin discs with springs behind them that cover the holes through the damper piston that allow oil to pass. The smaller the piston orifice, the slower the shock movement; thus, if a shim keeps a hole closed longer, damping is higher. Shims vary in flexibility and diameter; the smaller or more flexible the shim, the more easily oil can force its way past it.

Tune kits containing shims, seals, glide rings, springs, valve cores, a piston, and so on, as well as the shock oil you need, can be obtained from the manufacturer. The specific shim stack and oil weight in the shock will have been selected by the manufacturer to meet the requirements of your specific bike frame. In other words, all shocks of a certain length of a given model are not the same; the damper internals will have been chosen to match the leverage ratio of the bike for which it's intended. So unless you really know what you're doing, don't interchange parts inside your shock's damper. Yes, high-end tuning shops can get more performance out of a shock for your specific application by changing the shim stack, but don't try it without expert advice.

Damper service on a RockShox Monarch rear shock

LEVEL 3

Since 2010, both the RockShox Monarch and Ario shock dampers have been specifically designed for service by a shop mechanic or a motivated home mechanic. Have your parts kit in hand with replacement seals and valve core and RockShox Monarch air-fill adapter before you start.

1. **Open the rebound and the Gate.** Turn the red rebound adjuster knob all the way counterclockwise (toward the icon of the rabbit on the knob). Flip open the Gate lever (akin to a lockout lever) if present.
2. **Remove the air can.** Follow the instructions in 17-8 section b, above, through step 4.
3. **Release the air from the damper.** With a standard Schrader valve-core remover (Fig. 1.3), unscrew the little plastic air cap covering the damper air-fill port on the side of the body eyelet. Depress the valve with a pick; wear safety glasses because it is at 250 psi. Remove the valve core with the valve-core remover.
4. **Unscrew the air piston and pour out the oil.** Use a pair of adjustable wrenches, one on the 17mm flats atop the air piston (a.k.a. seal head) and one on the body eyelet (Fig. 17.15). Alternatively, you can clamp the body eyelet in a vise between aluminum or wood soft jaws. Unscrew the seal head completely from the top of the shock body over a bucket or pan; dump out the oil.
5. **Pump out the internal floating piston (IFP).** Screw a Monarch air-fill adapter into a shock pump (Fig. 17.16); the adapter that was required to pump up older Marzocchi Bomber and RockShox SID air forks looks similar but is slightly different. Screw the adapter into the air-fill port (where the valve core was in the side of the body eyelet). Drape a rag over the top of the shock body, and pump the IFP (a brass piston) out into the rag (Fig. 17.16).
6. **Remove the plastic compression ball from the seal-head bleed hole.** With a 2.5mm hex key, unscrew the setscrew from the bleed hole atop the air piston (Fig. 17.15), and with a 1.5mm hex key, push the small plastic ball out of the hole from the bottom.
7. **Remove the seals.** To avoid scratching the parts, remove the rubber O-rings, square seals and Teflon glide rings from around the air piston, the fixed oil piston at the end of the shaft, and the IFP by squeezing and pushing them with your thumbs until a loop pops up that you can grab. As described above in section b, step 6, and without scratching the metal, stab and remove the air piston's inner seal that seals the top of the shock body.

8. **Clean the parts.** Use isopropyl alcohol and a clean rag.
9. **Install new seals and glide rings.** Grease them lightly with Slick Honey first.
10. **Install the IFP to the proper depth inside the shock body.** Orient the IFP so that its recessed side is facing you. Measure the outside diameter of the body. Then, using the depth gauge of a measuring caliper (Fig. 1.4) in the bottom of the piston's recess, measure to the top of the shock body. If the outside diameter of the body measures 31mm, the IFP should be inserted to a depth of 48mm. Similarly, for progressively larger and longer shocks, the depth for a given body diameter will be 54mm for 38mm OD, 59mm for 44mm OD, 65mm for 51mm OD, 70mm for 57mm OD, 76mm for 63mm OD, and 78mm for 66mm OD.
11. **Install a new Schrader valve core into the damper air-fill port.** Use the Schrader valve-core remover to do so.
12. **Fill the damper with oil.** Clamp the body eyelet in a vise, and stand the shock body vertically. Fill it to the brim with 7-weight suspension oil. Scrape air bubbles off the oil surface.
13. **Install the shaft.** Push the damper piston (at the end of the shaft) down into the shock body slowly until oil comes out of the hole in the shaft above the piston, and then pull the piston back out, keeping it vertical. Again, the rebound and Gate adjusters must be fully open. Refill the body to the brim with oil, and scrape off the air bubbles. Slide the seal head (air piston) down to the bottom of the shaft against the fixed damper piston. Screw the seal head into the shock body, and tighten it with a 17mm or adjustable wrench; oil will come out of the bleed hole atop the seal head. Make sure that the shaft does not slide down, because if the fixed damper piston moves away from the seal head now or at any point during the rest of the assembly process, it will displace oil from the shock body. If you have a torque wrench and a 17mm crowfoot socket (Fig. 1.4), with the crowfoot turned at 90 degrees from the torque wrench handle in order not to multiply the torque, tighten the seal head to 250 in-lbs.

17.15 Unscrew seal head from shock body

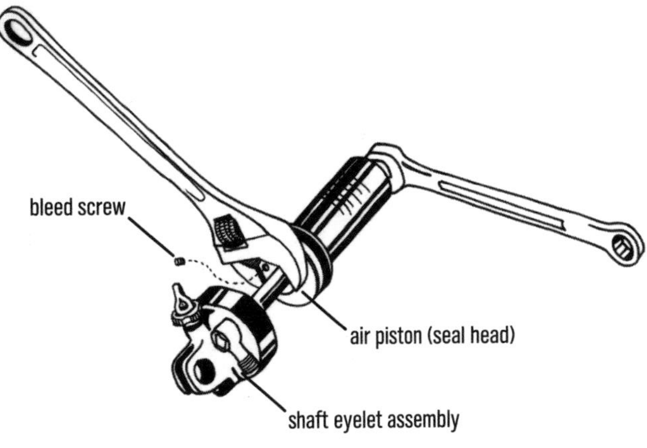

17.16 Pressurizing damper

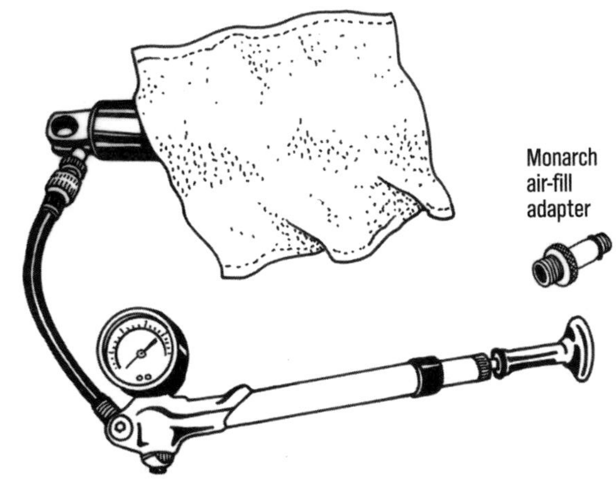

14. **Install a new compression ball.** Insert the new ball into the bleed hole atop the seal head. Follow it with the setscrew, and tighten with a 2.5mm hex key until you feel the setscrew just touch the ball. Tighten the setscrew an additional one-half turn.
15. **Pressurize the damper.** Using the shock pump with the Monarch air-fill adapter on it, pump air into the damper air-fill valve (Fig. 17.16). The pressure will be 250 psi for all Monarch and Ario shocks except the Ario 3.2, the pressure for which will be 500 psi. Ideally, you'd fill the damper with pure nitrogen (air is 78 percent nitrogen), but you'd need specialty shock fill equipment for that.
16. **Replace the air-fill port cap.** Use the Schrader valve-core remover to do so.
17. **Complete the air-can (air-sleeve) service.** Return to 17-8 section b, above, step 5.

17-9

MAINTAINING THE SUSPENSION FRAME

Because suspension frames vary in complexity, so does the level of necessary maintenance. In addition to the regular inspections described early in this chapter, maintenance is needed on the shock and the pivots.

a. Evaluation of the condition of the suspension

While the bike is standing still, you can tell if the suspension needs some lubrication. Stand next to the bike, apply the rear brake, and push down lightly on the saddle. Gradually increase the pressure. Notice how much pressure it takes before the bike finally compresses. If the suspension does not compress as you push harder and harder, and then it finally goes down in steps, chunk, chunk, chunk, you have a dry system. Clean and lubricate the pivots and the shock. There is no point in tuning the suspension until you have it moving smoothly.

If the swingarm begins compressing smoothly under a relatively gentle push, then you have a pretty clean, well-lubricated system.

Also remove the rear wheel and the shock, and lift the swingarm and let it drop. If the swingarm falls under its own weight, the pivots are not sticky. If you must forcefully move the swingarm through its travel, you had better get to work on those pivots. New frames usually take a little force to move until the pivot bushings and bearings work in; if your frame is like this and there is no noticeable lateral wiggle to the swingarm, the pivots are in decent shape.

b. Pivot maintenance

The pivots on any suspension frame require periodic attention. Pivots usually rely on cartridge bearings or bushings (usually steel or brass, but some are made of ceramic or plastic). They are held together by through-bolts, by clamps with pinch bolts surrounding the pivot shaft, or by pins secured with cotters or snaprings. High-quality bushings often let you know they are wearing by getting sticky before they get loose. Crummy bushing material just wears thinner and gets loose. Bearings fail if the seal fails, allowing the bearings to get dirty or lose their lubricant, or they get overloaded. Side loads can wreck a bearing, in particular a bearing not designed to take forces from the side. This is more likely to happen on installation or removal than on usage on the bike, however, unless some suspension members are way out of alignment.

The shock also pivots at its eyelets at either end (Fig. 17.7) on brass or plastic-lined steel bushings. Check that the shock pivots smoothly and without play. If the eyelet bushings are worn, replace them (17-8c).

Plan on cleaning and greasing pivot bushings on the order of every 40 hours of riding. Inspect them for wear frequently. Check for wear by feeling for lateral play and binding with the shock deflated or disconnected. If the rear end is loose from side to side, or if it is so sticky that you can hardly move it, go through the system by taking the linkage apart one piece at a time (this usually just requires removing screws and bolts). Test each joint for wiggle and bind once you have isolated it from the other joints. Check the bushings for scoring and deformation into an oval shape, and lubricate any squeaky ones. Avoid using solvents on plastic bushings, as some materials can swell. Bearings usually wear longer than bushings, as long as they are kept greased. If not yet ruined, they can be opened and repacked with grease as is done for other cartridge bearings (Fig. 17.17). When they are worn out, they can be replaced with stock sizes that you can find at an automotive bearing store or from wheelsmfg.com or Enduro.

If there are grease fittings on any of the pivots, by all means put a grease gun on them frequently and pump some grease in (you can spoon bicycle suspension grease, like Slick Honey, into the grease gun).

17.17 Removing bearing seal

To replace worn bushings and bearings, you will need to push them out of the holes in the ends of the linkage arms in which they are seated. Ideally, you would use bearing extractors, a bearing press, and drifts from wheelsmfg.com. For a home mechanic not willing to invest in seldom-used tools, some socket wrenches, box-end wrenches, a vise, thick washers, a quick-release skewer, and even a hammer and a punch can often do the trick.

1. **Push out the old bushing or bearing.**
   (a) If the linkage member has a through-hole of the same size on both sides (i.e., the bushing or bearing can go in from either side), then you can push out the bearing or bushing with a vise and a socket. Select a socket just slightly smaller than the outer diameter of the bushing or bearing, and set it against the bushing or bearing between the jaws of a vise (Fig. 17.18). If the bushing is a tiny one at the bottom of the seatstay smaller than any socket, you can use a bolt slightly smaller than its outer diameter to push it out; use a large nut or wood or aluminum block with a hole in it bigger than the bushing on the other side to back it up when you smack the bolt with a hammer or squeeze it in a vise. Never apply pressure to the inner bore or against the seals of a cartridge bearing (at least one you want to keep—the old trashed one you may treat as you like); the socket should be against the outer ring of the bearing. Against the opposite jaw of the vise, place the box end of a wrench or a socket just bigger than the bushing or bearing so that it surrounds the bushing or bearing up against the face of the linkage member (Fig. 17.18). Tighten the vise until it pushes the bushing or bearing out (into the box wrench or socket).
   (b) If the bushing or bearing is up against a seat in the linkage arm—in other words, the hole is not the same diameter on both sides as the bushing or bearing but instead is bored in from one side such that the bushing or bearing has a pocket (it seats against the flat bottom of the hole)—you must employ a different technique.

17.18 Pushing out a swingarm bearing or bushing with a vise, using a socket as a drift and a box wrench for clearance

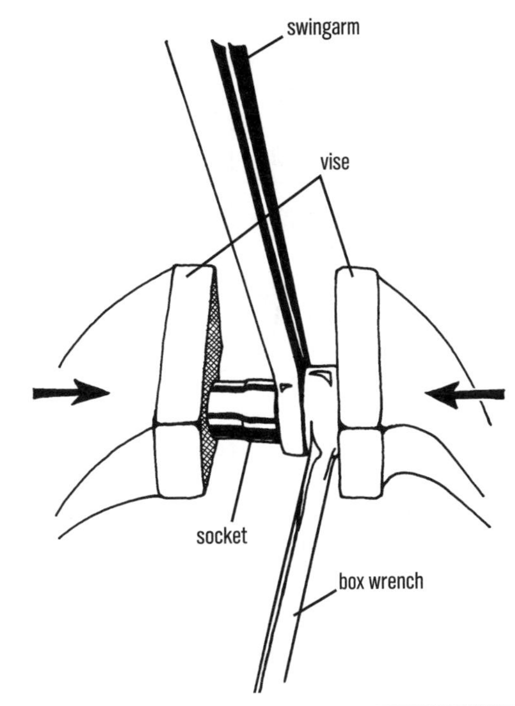

The ideal tool for this is a bearing-extraction kit; wheelsmfg.com offers one. In the absence of that tool, line up, from the other side, a punch, screwdriver, or bolt through the small hole (for the pivot bolt to pass through) against the inside diameter of the bearing. Support the linkarm somehow (perhaps on a socket just bigger than the bearing), and tap out the bearing into it with a hammer or by squeezing with a vise as in Fig. 17.18. A short quick-release skewer with some spacers on it could also do the trick.

2. **Grease the hole the new bushing or bearing will go into, and press it in.**
   (a) On linkarms that have parallel faces at the bushing or bearing hole, just place the bushing or bearing against the hole in the linkarm, and press it in with the flat jaws of the vise; put aluminum or brass plates against the vise jaws if they do not have smooth faces. Place the old bushing or bearing (or a socket just smaller than its OD—Fig. 17.19) against the new one if it needs to be pressed in farther than flush with the arm's face.

17.19 Pushing a cartridge bearing into a swingarm eyelet with a vise, using a socket as a drift

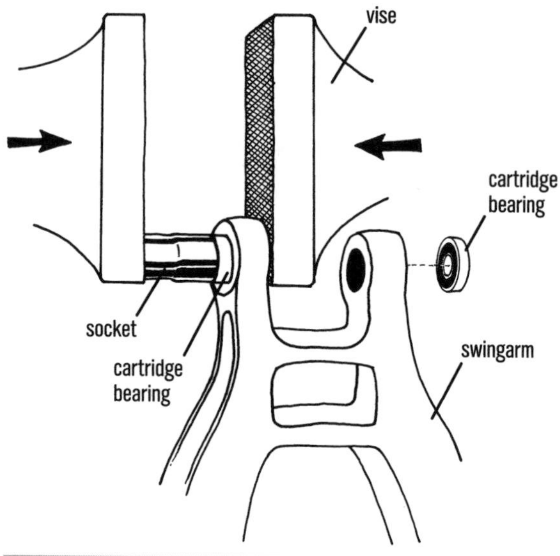

(b) You must press in the bushing or bearing with an appropriately sized socket on linkarms that have deep bearing pockets or that do not have parallel faces at the bushing or bearing hole. Obviously, a flat vise cannot push a bearing into a configuration like this; you need to place something behind the bearing, such as a socket, that can enter the pocket and still be pulled back out. (You could perhaps use the old bearing to push the new one in, but not if the pocket is deeper than the thickness of one of the bearings, because the old bearing will get stuck in the pocket.) Select a socket slightly smaller than the outer diameter of the bearing. Squeeze the bushing or bearing with one (smooth) jaw of the vise against the back of the linkarm eyelet pocket and the other jaw against the socket (Fig. 17.19).

17-10

MAINTAINING A PIVOTLESS SUSPENSION FRAME

LEVEL 1 Suspension frames without moving pivots fall into two categories: beam bikes, and bikes with a shock that depends on the flex of the chainstay rather than on pivots.

The principal beam suspension found on mountain bikes is the Softride beam. Inspect the beam-mounting points on the frame periodically for indications of fatigue (stretched, bulged, or cracking metal or paint). Softride beams themselves require no maintenance.

One of the simplest and lightest rear-suspension designs relies on a small shock behind the seat tube and flexing chainstays. Beyond checking the chainstays for indications of fatigue (stretched, bulged, or cracking metal or paint), you really need only to keep the shock serviced as described previously and tuned to your weight and riding style as described next.

17-11

TUNING THE REAR SUSPENSION

LEVEL 2 There are three main variables in the setup of the rear-suspension system: sag, compression damping, and rebound damping. A more complex adjustment is to the volume of the air chamber. Shock-mounting positions or shock length can also be changed to alter bike geometry.

The four main types of shocks are air-oil (Figs. 17.7 and 17.20), air-air, coil spring (or coil over), and elastomer (or elastomer over). In both air-oil and air-air shocks, compressed air acts as the spring. Coil-over and elastomer-over shocks use either a coil spring or an elastomer spring surrounding a damper cylinder (the coil or elastomer fits over the damper—thus the name).

Most shocks rely on the flow of oil through a small opening separating two chambers to slow the suspension movement; stacks of thin washers (shims) cover the hole and flex to control oil flow. The oil provides the damping, and there is often pressurized gas behind an additional, movable piston pushing against the oil chamber. Nitrogen is commonly used as the gas in shock dampers, as it is less likely to emulsify if it mixes with the oil. Air-air shocks, however, operate on the same basic principle as a hydraulic damper but damp the suspension through the movement of compressed air, rather than oil, through metering holes in a piston controlled by shim stacks.

Air-oil and air-air shocks are tuned for the spring rate by varying the air pressure and the volume of the

17.20 Shock adjustments on Fox RP23

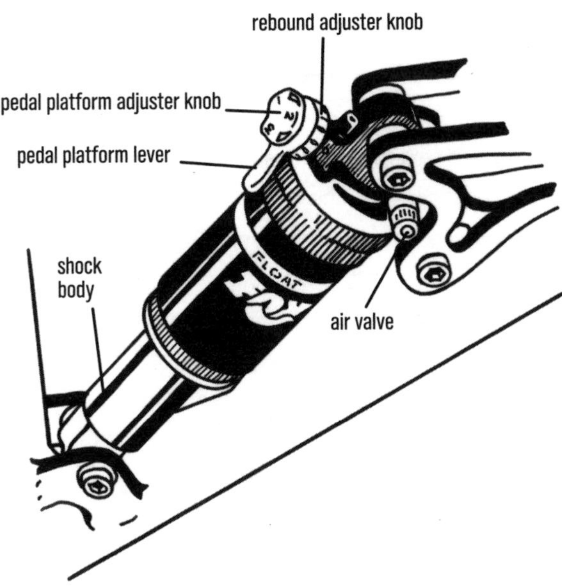

air chamber. You must have a shock pump with a no-leak head (Fig. 17.16). When a no-leak pump head is unscrewed, it maintains its seal around the valve until it backs out far enough to release the valve pin so that the valve closes, preventing any air from escaping the shock. A pump like this is a must for pumping air into shocks, because the air volume is so small and the air pressure is so high that the slightest air release dramatically reduces internal air pressure. You can trust the reading on the gauge to equal the shock pressure when you remove the pump, but no pump is ideal for checking the air pressure already in a shock, because you lose pressure filling the pump hose when you screw the pump head on.

You can change the spring rate of most coil-over and elastomer-over shocks by turning a threaded preload collar around the shock body or by replacing the spring.

Start with the air pressure, coil, or elastomer (or spring preload) recommended by the bike manufacturer for your weight, and experiment from there. Because the shock location, leverage ratio, and pressure requirements vary from bike to bike, the recommendations will come from the bike manufacturer and not from the shock manufacturer. The leverage ratio is the amount the rear axle moves vertically with a given amount of shock movement. For instance, a 2:1 leverage ratio indicates that the rear wheel will move up 2 inches if the shock compresses 1 inch (or 6 inches of rear-wheel travel with 3 inches of shock compression). Obviously, the lower the leverage ratio, the lower the spring rate you can use and get similar performance.

On rear shocks with hydraulic (or compressed-air) damping systems, damping is adjusted by varying the size of the orifices through which the oil (or compressed air) flows; this is accomplished by changing the shims (thin steel discs of varying diameter and flexibility) that cover the holes in the piston and are held in place by a spring. Alternatively, changing the viscosity of the oil (or the air pressure in the damper) changes the damping. On many models, oil flow through the damping orifices is adjusted with knobs (Fig. 17.20). Many shocks have a pedal-platform lever (Fig. 17.20) or a lockout lever (Fig. 17.9). Tuning techniques vary from shock to shock, so be sure to read the owners' manual that came with yours.

Please review 16-7 through 16-9 for an explanation of suspension spring rates, preload, compression damping, rebound damping, and other considerations—the same recommendations for setting up forks apply to rear suspension.

The recommendations that follow apply to cross-country, trail, and enduro bikes as well as to downhill versions, although more specific downhill considerations are found in 17-12.

a. Setting sag

Sag, or ride height, is the amount the bike compresses when you just sit or stand on it. You want the bike to sag so that the shock is preloaded and forces the wheel into the ground when the bike is unweighted over bumps, thereby increasing tire contact and traction in rough terrain. Ride height is not dependent on damping because there is no movement involved; it is dependent only on the spring rate (and preload, with a coil shock). On an air shock, a good starting pressure in psi is equal to your weight in pounds (e.g., if you weigh 160 pounds, try 160 psi in the shock). Ideal shock pressure depends on the leverage ratio of the frame as well as on rider weight, so properly setting sag is critical.

NOTE: *Most shocks have a maximum allowable pressure of 300 psi.*

Changing the spring and/or the spring preload adjustment changes the sag. A good rule of thumb is to set the spring so that sag uses up one-quarter of the bike's travel, or 30 percent sag for a more plush ride. Consult 16-7 through 16-9 for more specific recommendations based on your riding style.

For an air shock, measure the travel and sag with an O-ring or zip-tie around the shock body (Fig. 17.7), just as for a fork (16-6). Find the shock's full travel length by deflating the shock and sitting on the saddle (and deflating it more while sitting on it). Note the position of the end of the air can, and then inflate the shock and measure from there to the end of the air-can seal to find the total travel. To find the sag, sit on the bike (if it's a downhill bike, stand in the attack position), push the O-ring against the air can, and then carefully get off toward the front of the bike. Measure from the O-ring to the end of the air-can seal, and divide by the full travel length to get the percentage of sag. For instance, if your shock has 38mm (1.5 inches) of total travel, 25 percent sag will be approximately 10mm, and 30 percent sag will be approximately 11mm. Some rear shocks have sag values etched on the shock body.

Adjust the sag in an air shock by adjusting the air pressure with a pump with a no-leak head (Fig. 17.15). If (or in case) you have a negative air spring (see 16-7d), leave the pump attached, and, by gently sitting on and off the saddle, cycle the shock slowly 10–20 times through the top 25 percent of its travel with every increase of 50 psi. This will allow air to go through a bypass port around the piston to equalize the pressure in the positive and negative springs; you will hear a little pssst and the reading on the pump gauge will drop when the chambers equalize. If you find the shock to be very harsh on small bumps, you may have neglected to do this.

Similarly, if (or in case) you have a negative air spring (see 16-7d), release air slowly in order to let air out of both the positive and negative springs. If you let it out too fast, the negative spring will have much more pressure than the positive spring and will pull the shock down. If the shock is shorter than its full length and won't extend fully, this is what happened. To alleviate it, pump the shock until it gets to full length, then cycle it through its top 25 percent of travel as described above to equalize pressure in the positive and negative springs.

On a coil or elastomer shock, measure the shock shaft length or the eye-to-eye length of the entire shock when you are off the bike. Have someone else measure it again when you are sitting on the bike. The difference between the measurements is the sag. If you're going for a sag that is, say, 25 percent of travel on a coil or elastomer shock, if less than 75 percent of the shaft length is still showing, increase the preload or the spring rate (by interchanging springs). If more than 75 percent of the shaft length is showing, decrease the preload or spring rate. In both of these systems, the preload is usually set by turning a threaded collar surrounding the shock body that compresses the coil spring or the elastomers. Depending on the shock and the spring used, if you have used more than two preload turns of the spring collar to reach 25 percent travel usage in ride height, you need a stiffer spring.

IMPORTANT: *Excessive preload on a coil shock can cause the shock to fail due to coil bind, where there is no space between coils—each loop of the coil is stacked up against the next one. If you must preload the shock more than two full turns to set the sag, you are in danger of coil bind and need a stiffer spring. Consider, for instance, if the bottom-out of the rear suspension occurs at 2 inches of shock travel and the shock has 2⅜ (2.375) inches of total travel. If you tighten the preload collar down more than 0.375 inch, the coil will bind and stop the shock before the swingarm bottoms out. Coil bind puts tremendous stress on the shock and breaks important parts that you would like to keep. It is more of an issue with stiffer springs and occurs at fewer preload-collar turns, because the thicker wire of the spring leaves less space between coils.*

b. Checking the front-rear balance

You want to have the front and rear suspension balanced so that everything works together like a beautiful symphony in motion. Check the front-rear balance by standing next to the bike on level ground, lightly applying the front brake, and stepping straight down on the pedal closest to you while the crankarm is at bottom dead center. If the top tube doesn't tip forward or backward as the suspension is compressed, the spring rates are well balanced. Next, stand on the bike in riding position. If one end drops noticeably more than the other, you need to increase the spring preload and/or the spring rate on the end that dropped farther (or soften the spring on the end that dropped less).

On steep downhill courses, more of your weight will be shifted to the front of the bike, so more sag in the rear is a good idea.

c. Adjusting rebound damping

If adjustable, set the rebound damping as low as you can without causing the bike to pogo (bounce repeatedly after a bump). Do the curb test, starting with the rebound fully open (the rebound knob turned counterclockwise until it stops). Ride off the curb, and note how many times the shock bounces. You want only one bounce. Turn the rebound knob (Fig. 17.20; rebound knobs are usually red) clockwise one-quarter turn, and ride off the curb again, repeating until you get only one bounce. Record the number of turns (i.e., clockwise) of the knob from fully open.

If you can adjust rebound on the fly and have no lockout lever, turn the rebound damping up when you climb. It does not need to throw you up as much when you hit things, as you are going much slower anyway. You will climb faster this way. (If you have a lockout- or three-position compression-damping lever, tighten that down [clockwise] instead on smooth climbs.) Remember to reduce the rebound damping or open the lockout when you head back down.

See the rebound tests in 16-7e. See what happens when you load the suspension and quickly release it, allowing the handlebar and saddle to rebound freely. One way is to stand next to the bike, throw your full weight onto the near pedal and handlebar grip, and then jump off the pedal and let go of the handlebar right when the suspension compresses as far as it's going to go. Another is to stand on the pedals, quickly load your hands and feet with as much force as you can (have the saddle in the lowest position of your dropper post), and then simultaneously let go of the handlebars and jump off of the pedals; stand up immediately and move forward over the bike, getting out of the way of the saddle as it comes up. Seek a rebound setting at which the shock rebounds quickly but is slowed down just enough that the wheel doesn't leave the ground.

d. Adjusting compression damping

If the shock has a compression-damping adjustment, set it as light as possible without bottoming out the shock more than once or twice on the course. You want to absorb the bumps with the springs as much as you can. The low-speed compression-damping adjustment, which is generally the only adjustment the shock will have, is to stiffen the bike against pedaling loads and riding into berms and smooth dips; use the lockout lever to lock out both low- and high-speed compression when climbing, and the middle position of a three-position lever to stiffen the low-speed compression while leaving high-speed compression fairly open. On downhills, open the lever fully.

e. Setting the inertia valve or pedal platform

Some modern shocks have an inertia valve on the compression-damping system designed to distinguish between bump forces and pedaling forces; a Specialized Brain shock is an example. It has a mass sitting in oil closing off the compression damping, and when a hard bump is encountered the mass bounces up off the orifice it is covering, allowing oil to flow through until the mass settles back down and seals it off again. The threshold bump force magnitude required to open the inertia valve and make the shock fully active may be adjustable with a knob.

More common is to heavily damp low-speed compression while pedaling by means of a lever (Fig. 17.20; compression adjusters are usually blue). Such a system is often said to provide a pedal platform for the rider to push against while pedaling without bobbing up and down.

On some shocks, you can flip a lever like a lockout lever to go in and out of the pedal platform on the fly (Fig. 17.20). If adjustable, set the threshold as low as you can (so it blows open most easily when hitting a bump), yet high enough that when you pedal aggressively on a smooth surface the rear suspension does not bob up and down.

The following steps detail how to adjust the ProPedal knob on a Fox RP23 (Fig. 17.20). Later Fox Factory shock models have a three-position lever (firm, medium, and open) with a three-position open-mode-tuning knob on the front of it you can pull out and turn to vary the stiffness of the pedal platform; the effect is similar to below:

1. **Turn the ProPedal lever to the PROPEDAL position.** If you are looking at the lever from the front, turn it clockwise, the same as you would any lock-out lever.
2. **Pull the ProPedal knob out toward you.**
3. **Turn the ProPedal knob to the setting you want.** There are three numbers on the knob: 1 = light (less heavily damped), 2 = medium, and 3 = firm. Turn the knob clockwise until the number you want lines up with the ProPedal lever. As the ProPedal knob turns, it clicks twice per setting: It clicks as you leave the current setting, and it clicks into the new setting.
4. **Push the ProPedal knob in.** This locks in the setting.

f. Adjusting air volume

Like an air fork, changing the volume (16-8d) of the air chamber changes how rapidly the air spring ramps up as the shock goes deeper into its travel. Some shocks offer air cans of different diameters (Fig. 17.21) or interchangeable volume spacers; either one requires shock disassembly to change. Some shocks also allow you to change the volume in the negative air spring.

Less air volume in the positive air spring makes the shock ramp up faster on big hits. More air volume in the positive air spring allows the shock to move more easily through its travel at the same pressure setting. If your sag is set correctly and you're bottoming out too easily, install a smaller air can or a larger volume spacer. If your sag is set correctly and you're not using full travel, install a larger air can or a thinner (or no) volume spacer.

The negative air spring is at its highest pressure and pulls down hardest on the shock when the shock is fully extended, so that small bumps compress it more easily while at full ride height. Increasing the volume of the negative air spring makes the net spring rate more linear on a graph of spring force versus travel used. This results in increased compliance on small bumps (the net spring force is lower for the first quarter of the travel, giving it a softer initial stroke and better traction), increased mid-stroke support, and bottom-out resistance (the net spring force is higher in the middle and later parts of the stroke).

NOTE: *Check that the bike has room for the larger air can or an extra-volume negative-air sleeve before installing one.*

17.21 Air cans of varying diameter for a Fox shock

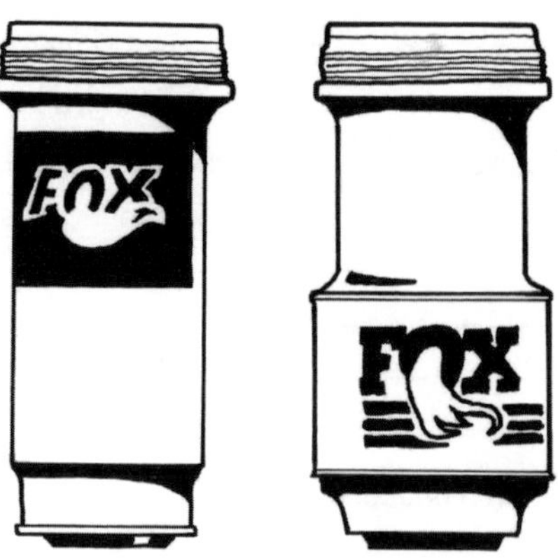

I. Changing air volume in the positive air spring with a different-size air can

1. **Remove the air can.** Follow steps 1-6 in 17-8b (Figs. 17.8 and 17.9).
2. **Install the new air can.** Obviously, get one of a different size made specifically for your shock (Fig. 17.21). Continue with step 7 in 17-8b.

II. Changing air volume in the positive air spring with volume spacers

1. **Unscrew the air can** (Fig. 17.8). Leave the shock mounted on the bike for safety and ease of unscrewing the can. Follow the first two steps in 17-8b and pull the can down away from the shaft eyelet assembly.
2. **Move parts away from the volume spacer.** With your fingers, slide the bottom-out bumper O-ring and plate (washer) down the shaft (Fig. 17.22).
3. **Remove the volume spacer.** Use a blunt instrument to pry it out.
4. **Grease inside the eyelet.** Put a thin layer of grease in the shaft eyelet assembly cup and its threads.

17.22 Sliding the plate away from the volume spacer

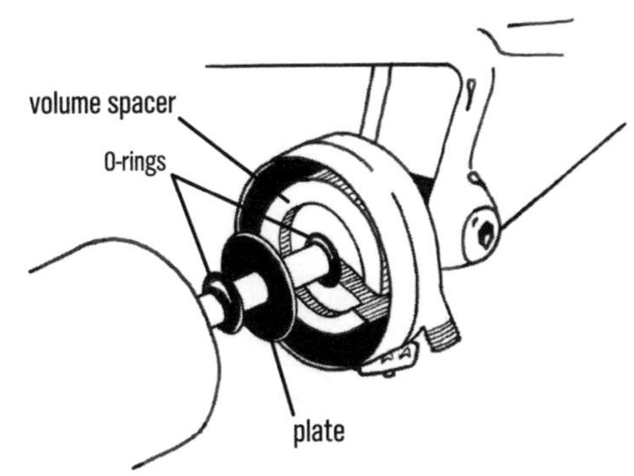

17.23 Installing the volume spacer

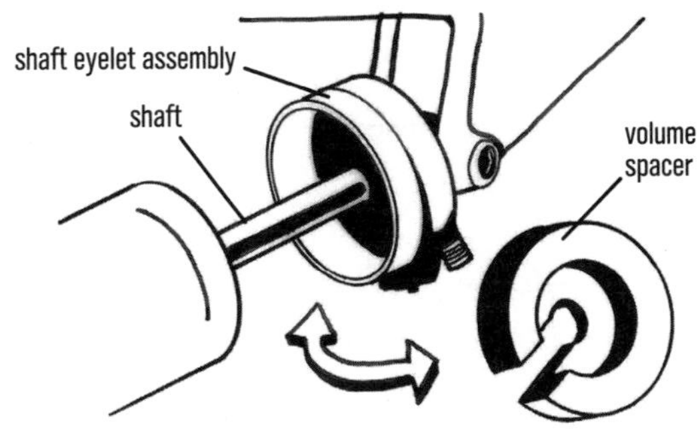

5. **Pop in a new volume spacer.** The recessed face of the volume spacer should face away from the shaft eyelet cup, toward the shock body (Fig. 17.23). Press it firmly up into the eyelet assembly cup until it snaps in place around the top of the shaft.
6. **Slide the bottom-out bumper O-ring and plate in place.** They will nest inside the recess in the volume spacer (Fig. 17.22).
7. **Screw the air can back on** (Fig. 17.8). Continue with 17-8b, starting with step 10.

III. Changing air volume in the negative air spring with a different-size extra-volume air sleeve

You can do this by replacing a standard air can with an air can that has an extra-volume (Evol) sleeve on it or by changing the extra-volume sleeve with one of different diameter.

To interchange a standard air can for an Evol can, follow the instructions in 17-8b for air can service, substituting the Evol can upon reassembly.

Interchanging Evol sleeves

1. **Deflate the shock.** With the shock still on the bike and while depressing the Schrader valve pin with a blunt implement (use a rag over it to keep oil from spurting on you), push up and down on the saddle around the 25 percent travel area. This should deflate the negative spring, since the bypass hole that connects the positive and negative air chambers is near the typical sag position.
2. **Remove the shock from the bike.**
3. **Remove the circlip around the base of the extra-volume sleeve.** With a tiny screwdriver, get under the tip of the wire circlip and lift it out of its groove (Fig. 17.24).
4. **Pull the extra-volume sleeve off the air can.** Two O-rings (Fig. 17.25) create the resistance you have to overcome. If you didn't deflate the negative spring, however, it will be harder to pull off, and when it does come off, you might get a blast of air and Float Fluid in your eyes. Wear safety glasses.

17.24 Removing the circlip retaining a Fox Evol negative-spring extra-volume sleeve

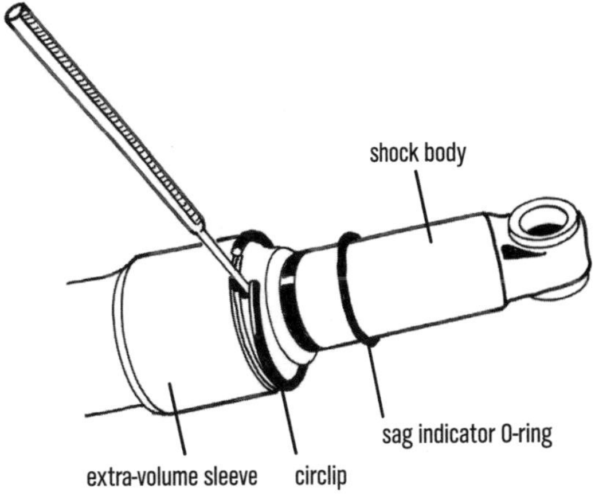

17.25 Removing a Fox Evol negative-spring extra-volume sleeve

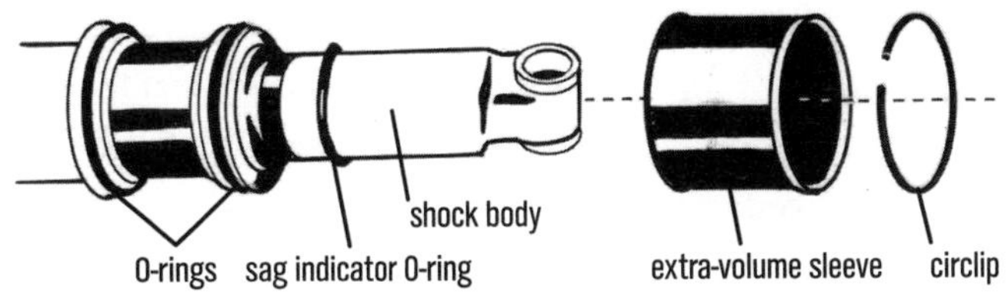

5. **Apply grease.** Smear Slick Honey grease on the O-rings and just inside either end of the new extra-volume sleeve. You can also decrease the volume of your existing Evol sleeve by slathering grease between the O-rings around the can before reinstalling the sleeve.
6. **Push on the new extra-volume sleeve** (Fig. 17.25).
7. **Replace the circlip** (Fig. 17.24).
8. **Reinstall the shock on the bike.**
9. **Inflate the shock.** Leave the pump attached, and, by gently sitting on the saddle then getting off it, cycle the shock slowly 10–20 times through the top 25 percent of its travel with every increase of 50 psi. This will allow air to go through a bypass port around the piston to equalize the pressure in the positive and negative springs; you will hear a little pssst, and the reading on the pump gauge will drop each time the chambers equalize.

g. Adjusting shock mounts

Some shocks have adjustable attachment positions. Usually, these have a number of different mounting holes for one eye of the shock, but some frames mount the body end of the shock via a threaded collar that can be turned to vary the position of the shock. Varying the shock position varies the head angle, bottom-bracket height, and ride height of the bike, and it may change the rear travel length as well. Adjusting to produce a shallower head angle usually also produces a lower bottom bracket, and makes the bike more stable at high speed, plus it gets the front wheel farther out ahead for steep drops. Making the head angle steeper (and the bottom bracket higher) offers improved maneuverability at low speeds.

Some frames also have adjustable head angles to accomplish similar things.

h. Offset bushings

Offset bushings can be used to create a similar effect to changing the shock-mounting points as above in section g. They, along with aluminum spacers that fit over their ends, replace the top-hat-shaped spacer/reducers or the bushing sleeve that adapt the eyelet bushings of the shock to fit the mounting bolt (usually M6 or M8) and the width of the mounts. Offset bushings are bored off-center for the mounting bolt and change the effective length of the shock.

17.26 Offset bushings with the holes offset toward the shock to shorten its effective length

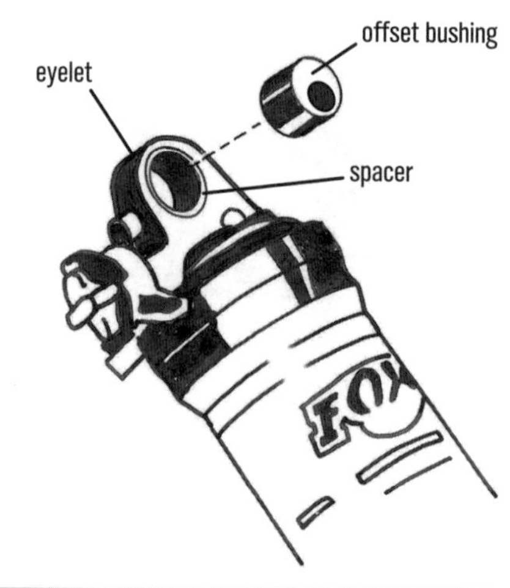

If the bushing's holes are offset toward the shock (Fig. 17.26), its length is effectively shorter, which lowers the bottom bracket and decreases the head angle for greater high-speed stability and performance over drops. Alternatively, offsetting the holes away from the shock raises the bottom bracket and increases the head angle for improved low-speed maneuverability.

To figure out the correct offset bushings and aluminum ring spacers to get, measure the width of your shock eyelet and spacers when installed and the diameter of the mounting bolt.

1. **Remove shock-mounting hardware.** See 17-8c.
2. **Slide in the offset bushing.** It should slide in through the eyelet bushing with your fingers. If it doesn't, push it in with a vise with one soft jaw on the edge of the eyelet and the other on the end of the offset bushing (Fig. 17.11).
3. **Rotate the bushing the way you want it.** To shorten the effective shock length, have the hole toward the shock, and to lengthen the effective shock length, have the hole away from the shock.
4. **Install spacers.** Put the aluminum ring spacers that came with the offset bushing on either side of the eyelet. If the bushing sticks out of one end, push it in flush with the spacers on both ends using soft vise jaws.

5. **Bolt the shock back onto the bike.** Check full travel range with the shock deflated to ensure that this change in effective shock length won't lead to tire interference or other clearance issues.

i. Riding and tweaking

Again, see 16-7 through 16-9, and follow the guidelines about picking a test course and taking notes. You want to bottom out a couple of times on the front and rear on a course. If the suspension is never bottoming out, the spring is too stiff or the compression damping is too high.

Change settings in small increments. It is easy to overadjust. Make only one adjustment at a time. Also, once you have balanced the front and rear ends, any adjustment you make to the front you should also make to the rear, and vice versa. Read the bike manual as well as the shock manual for adjustment methods and recommendations.

Suspension tuning is affected by (1) rider weight, (2) rider ability, (3) riding speeds, (4) course conditions, (5) rider style, (6) rider position on the bike, and (7) temperature. If any change, adjust the tuning.

Use the softest springs you can with little preload; you want to bottom out occasionally, but not frequently. If you are bottoming out too much, change the compression damping, the spring rate, or the volume of the air negative spring. If the compression is slow, yet you are still bottoming out, the spring is too soft. You will feel beaten up on the intermediate hits, or when you are bottoming out on a big hit it will be harsh through the entire stroke. Stiffen the spring rate, and lighten the compression damping. The ride height (sag) you've chosen dictates some of the spring rate.

Preload makes the spring rate ramp up faster. If you can use a stiffer spring and back off on preload, big impacts will be less harsh.

Set the compression damping to blow off quickly; your plush spring won't bottom out harshly anyway! The compression damping should be set low enough that on big hits you use up all the travel, but the saddle shouldn't smack you in the butt when you hit bottom. Tighten up compression damping if you blow through the stroke and get bounced too hard.

Decrease the rebound damping to return quickly. You want a lively rebound, because a sluggish return will allow the suspension to pack up (as you go over stutter bumps, water bars, or closely spaced rocks, the bike will ride lower and lower).

Increase the rebound damping if the bike springs back too fast. The rebound should not be so quick that you are getting bounced (remember the curb test—17-11c). Tighten the rebound for climbs if you have an adjuster and no lockout lever.

Again, if you have no damping adjuster, change the oil viscosity (17-8d).

Damping is speed sensitive. Don't worry about settings that feel good at low speeds being too light for high speeds; the shock will get stiffer as you hit things faster. At all speeds, you want the shock to pop back as quickly as possible without kicking back.

"Pedal platform" is a distinct type of low-speed compression damping designed to resist low-frequency (not just low-speed) shock-shaft movement; it keeps the bike riding higher, and on the fork, it resists dive when braking. If you have an adjustable pedal platform threshold, set it as low as possible yet with the firmness you want when pedaling aggressively on smooth surfaces (a high setting gives the firmest locked-out feel). The low threshold setting will allow the shock's compression-damping system to open up and move freely on smaller bumps.

Damping is also temperature sensitive. Oil is thick and sluggish when cold, but when it gets hot, the shock becomes livelier. Adjust accordingly in summer, with a stiffer spring and firmer damping adjustments. And in the cold of winter, when overall speeds are slower, the grease and oil in the shock are thicker, and coil and elastomer springs are stiffer, lighten up the spring and damping adjustments.

If there is no damping adjuster, you can vary the oil viscosity. Look up the stock oil weight in the shock tech manual. Heavier oil slows the shock; lighter oil speeds it up. And replacing the oil in the shock periodically is a good idea, even if you like its performance, because oil breaks down with use and has little worn bits of the shock floating in it. For most people, this is a shop service; see 17-8d.

17-12

ADJUSTING WHEN RIDING DOWNHILL COURSES

Everything in 17-11 applies, with a few additions here. Again, you are looking for a setup in which the front and rear shocks bottom out on the biggest bump on the course, but make sure you are riding at full speed before making adjustments.

On rougher courses, increasing the spring rate (or preload) will keep you from bottoming out so much. Compensate for small bumps by reducing rebound damping to keep the shock from packing in on successive hits. If the bike is bucking, increase rebound damping a bit.

NOTE: *Too much rebound damping, not just too little, will sometimes cause bucking. Heavily damped shocks will respond so slowly that they will pack in over repeated bumps, giving you a rigid bike and low ride height (i.e., the suspension will be fully compressed and won't return).*

On smooth courses, try decreasing the spring rate (or preload) and increasing damping. Negotiating turns will usually be the major challenge, and the lower ride height (sag) provided by the softer spring will keep you closer to the ground. The greater sag will also increase the available amount of negative shock travel (the amount the wheel can go down), which will help maintain tire traction when the bike is turning and when it is braking. Higher compression damping, although making the shock absorption slower, will still be fast enough to deal with isolated bumps and will eliminate the harshest bottoming-out. Higher rebound damping will reduce the bouncing of the bike after the isolated bumps.

Where there is no general rough or smooth characterization of the course, set up the suspension to perform best on the sections in which you have the most trouble for the most elapsed time. In other words, don't set the suspension for a tricky section you get through in a couple of seconds; set it up for a challenging section on which you will spend half a minute.

17-13

CHECKING FRAME ALIGNMENT AND ADJUSTING DROPOUT ALIGNMENT

LEVEL 3

These are inexact methods for determining frame alignment. If the alignment is way off, these methods will tell you. If you find alignment problems, other than perhaps moderately bent derailleur hangers (17-5) or dropouts, do not attempt to correct them. Adjusting frame alignment is a difficult and delicate task; if it can be done at all, only someone who is practiced at aligning frames should perform this task with an accurate frame-alignment fixture. Remove both wheels from the frame before beginning this inspection.

1. **Stretch a string around the head tube from dropout to dropout.** With the frame clamped in a bike stand, tie the end of a string to one rear dropout. Stretch it tightly around the head tube, and tie it symmetrically to the other dropout (Fig. 17.27).

17.27 Checking frame alignment with a string

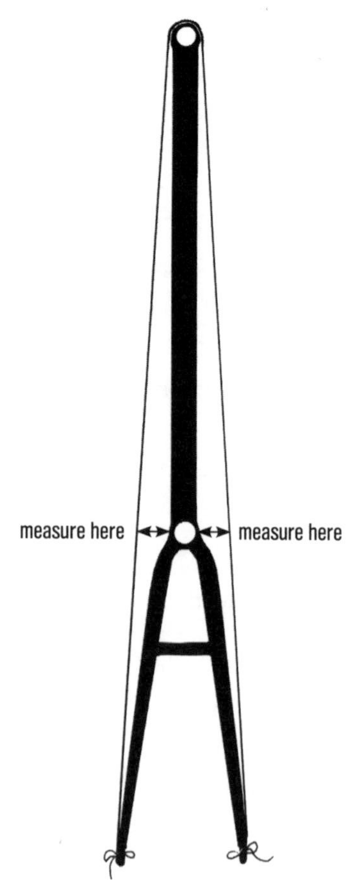

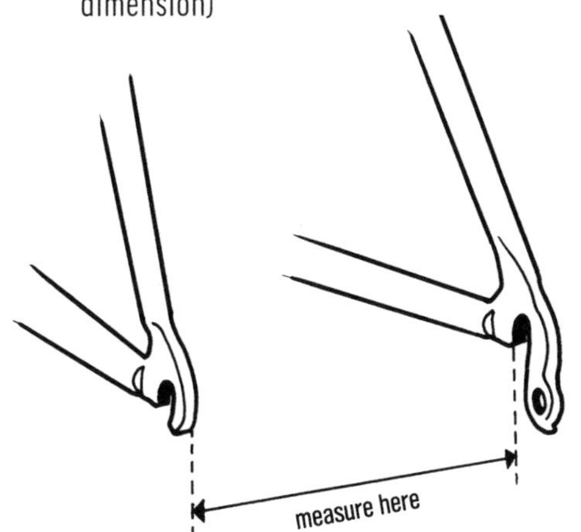

17.28 Measuring dropout width (or axle overlock dimension)

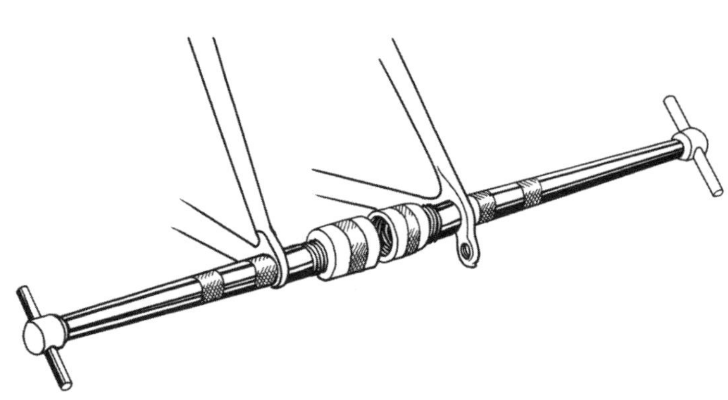

17.29 Using dropout-alignment tools on rear dropouts

2. **Measure from the string to the seat tube on each side** (Fig. 17.27). The measurement should be at least within 1mm of being the same on both sides.
3. **Put a true and properly dished rear wheel in the frame.** Check that it lines up behind the seat tube in the same plane as the front triangle. Make certain that the wheel is centered between the seatstays and chainstays (or swingarms). The hub should slide in easily without requiring you to pull outward or push inward on the dropouts. Tightening the hub quick-release or through axle should not result in bowing or twisting of frame members.
4. **Remove the wheel, and measure the spacing between the dropouts** (Fig. 17.28). On most mountain bikes with quick-release hubs made since 1990, this spacing should be 135mm, but wider spacings are common on downhill and freeride bikes. Through-axle mountain bikes start at 142mm spacing, although this is the axle length and is measured into the axle pockets in the dropouts; the faces of the dropouts are still 135mm apart. Boost spacing increases the axle length to 148mm. Dropout spacing on fat bikes continues to grow and as of this printing is generally either 170mm or 190mm. Mountain bikes made between 1984 and 1990 or so should have a rear spacing of 130mm. Mountain bikes made prior to 1984 are likely to have a rear spacing of 125mm. Measure the width of the rear hub with a caliper to see what the rear-end spacing of the frame should be. No matter what the nominal measurement for the frame should be, if it is between 1mm less and 1.5mm more than the nominal, it is acceptable. For instance, for a frame whose rear spacing should be 135mm, acceptable spacing is 134–136.5mm.
5. **With quick-release dropouts, put dropout-alignment tools in them.** Ensure that their shafts are fully seated into the dropout slots (Fig. 17.29). Arrange the tool spacers (and the cups if they are adjustable) so that the faces of the cups are within 1mm of each other. Tighten the handles on the tools. The tool cups should line up straight across from each other, with their faces parallel. If the tools do not line up with each other, one or both dropouts are bent. If the frame has replaceable bolt-on dropouts, replace them. If the frame has a composite or bonded rear triangle and the dropouts are misaligned, there is nothing you can do about it if the bike is not equipped with replaceable dropouts. If the frame has a steel rear triangle, you can align the dropouts by bending them carefully with the dropout-alignment tools. Hold the cup of the tool with one hand, and push or pull on the handle with the other. Aluminum or titanium rear dropouts can sometimes be aligned, but it is something you should have a shop do. Titanium is hard to bend because it keeps springing back, and you run a great risk of breaking aluminum by bending it.
6. **On suspension frames, check for smooth swing-arm movement without play.** See 17-10.

APPENDIX A
TROUBLESHOOTING INDEX

This index is intended to assist you in finding the right sections of this book to help you fix your bike. If you already know where the problem lies, consult the Table of Contents for the chapter covering that part of the bike. If, however, you are not sure which part of the bike is affected, this troubleshooting index can be of assistance. It is organized alphabetically, but because people's descriptions of the same problem vary, you may need to look through the entire list to find your symptom. If more than one symptom applies, you will need to examine all possible causes.

This index can assist you with a diagnosis and can recommend a course of action. After each recommended action is a list of chapter numbers to which you can refer for the repair procedure to fix the problem.

TABLE A.1 — TROUBLESHOOTING BIKE PROBLEMS

SYMPTOM	LIKELY CAUSES	ACTION	CHAPTER
bent wheel	misadjusted spokes	true wheel	8
	broken spoke	replace spoke	8
	bent rim	replace rim	15
bike pulls to one side	wheels not true	true wheels	8
	tight headset	adjust headset	12
	pitted headset	replace headset	12
	bent frame	replace or straighten	17
	bent fork	replace or straighten	16
	loose hub bearings	adjust hubs	8
	tire pressure very low	inflate tires	2, 7
bike shimmies at high speed	frame cracked	replace frame	17
	frame bent	replace or straighten	17
	wheels overly flexible	replace wheels or tighten spokes	15
	wheels are out of true or dish	true and dish wheels	8
	loose hub bearings	adjust hubs	8
	headset too loose	tighten headset	12
	flexible frame/heavy rider	replace frame	17
	poor frame design	replace frame	17
bike vibrates when braking	see "chattering and vibration when braking" under "Strange Noises"		
brake doesn't stop bike	misadjusted brake	adjust brake	9, 10
	worn brake pads	replace pads	9, 10
	wet rims	keep braking	10

Continues >>

SYMPTOM	LIKELY CAUSES	ACTION	CHAPTER
brake doesn't stop bike, cont.	greasy rims	clean rims	10
	sticky brake cable	lube or replace cable	10
	steel rims in wet weather	use aluminum rims	15
	brake damaged	replace brake	9, 10
	sticky or bent brake lever	lube or replace lever	9, 10
	air in hydraulic brake	bleed brake	9
	worn disc-brake pads	replace pads	9
	grease on brake pads	clean pads	9, 10
	grease on rotor	clean rotor and pads	9
	air in hydraulic brake	bleed brake	9
	brake pads missing	install pads	9, 10
brake rotor rubs	bent rotor	align rotor	9
	caliper mounted askew	center caliper over rotor	9
	sticking piston	clean and lube piston	9
	pads pushed out too far	push pistons back	9
brake rubs on rim (see also "bent wheel")	brake misaligned	adjust brake	10
chain falls off in front	misadjusted front derailleur	adjust front derailleur	5
	chainline off	adjust chainline	11
	chainring bent or loose	replace or tighten	11
chain jams in front between chainring and chainstay ("chain suck")	dirty chain	clean chain	4
	bent chainring teeth	replace chainring	11
	chain too narrow	replace chain	4
	chainline off	adjust chainline	11
	stiff links in chain	free links, lube chain	4
	thick inner chainring teeth	use thinner chainring	11
chain jams in rear	misadjusted rear derailleur	adjust derailleur	5
	chain too wide	replace chain	4
	small cog not on spline	reseat cogs	8
	poor frame clearance	return to dealer	17
chain skips	tight chain link	loosen tight link	4
	worn out chain	replace chain	4
	misadjusted derailleur	adjust derailleur	5
	worn rear cogs	replace cogs and chain	8, 4
	dirty or rusted chain	clean or replace chain	4
	bent rear derailleur	replace derailleur	5
	bent derailleur hanger	straighten hanger	17
	loose derailleur jockey wheels	tighten jockey wheels	5
	bent chain link	replace chain	4
	sticky rear shift cable	replace shift cable	5

SYMPTOM	LIKELY CAUSES	ACTION	CHAPTER
chain slaps chainstay	chain too long	shorten chain	4
	weak rear derailleur spring	replace spring or derailleur	5
	clutch setting too low	tighten RD-cage clutch	5
	terrain very bumpy	ignore noise or slow down	n/a
derailleur hits spokes	misadjusted rear derailleur	adjust derailleur	5
	broken spoke	replace spoke	8
	bent rear derailleur	replace derailleur	5
	bent derailleur hanger	straighten or replace	17
knee pain	poor shoe-cleat position	reposition cleat	13
	saddle too low or high	adjust saddle	14
	clip-in pedal has no float	get floating pedal	13
	foot rolled in or out	replace shoes or get orthotics	n/a
pain or fatigue when riding, particularly in the back, neck, and arms	incorrect seat position	adjust seat position	14
	too much riding	build up miles gradually	n/a
	incorrect stem length	replace stem	12
	poor frame fit	replace frame	17
	incorrect handlebar height	adjust stem height or get stem with different angle	12
	not enough suspension	increase travel or get new fork or frame	16, 17
	suspension too harsh	tune suspension	16, 17
pedal entry difficult (with clip-in pedals)	mud in cleat or pedal	clean cleat and pedal	13
	spring tension set high	reduce spring tension	13
	shoe sole knobs too tall	trim knobs	13
	loose cleat	tighten cleat	13
	dry cleat and pedal	lubricate cleat and pedal clips	13
	cleat guide on pedal loose or gone	tighten or replace	13
pedal moves laterally, clunks, or twists while pedaling	loose crankarm	tighten crank bolt	11
	pedal loose in crankarm	tighten pedal to crank	13
	bent pedal axle	replace pedal or axle	13
	loose bottom bracket	adjust bottom bracket	11
	bent bottom-bracket axle	replace bottom bracket or axle	11
	bent crankarm	replace crankarm	11
	loose pedal bearings	adjust pedal bearings	13
pedal release difficult (with clip-in pedals)	spring tension set high	reduce spring tension	13
	loose cleat on shoe	tighten cleat	13
	dry pedal spring pivots	oil spring pivots	13
	dirty pedals	clean and lube pedals	13
	bent pedal clips	replace pedals or clips	13
	dirty cleats	clean, lube cleats	13
	worn cleat	replace cleat	13

Continues >>

SYMPTOM	LIKELY CAUSES	ACTION	CHAPTER
pedal release too easy (with clip-in pedals)	release tension set too low	increase release tension	13
	cleats worn out	replace cleats	13
rear shifting poor (see also "chain jams in rear" and "chain skips")	misadjusted derailleur	adjust derailleur	5
	sticky or damaged cable	replace cable	5
	loose rear cogs	seat and tighten cogs	8
	worn rear cogs	replace cogs	8
	worn/damaged chain	replace chain	4
resistance while coasting or pedaling	tire rubs frame or fork	adjust axle and/or true wheel	2, 8
	brake drags on rim	adjust brake	10
	tire pressure very low	inflate tire	2, 7
	hub bearings too tight	adjust hubs	8
	hub bearings dirty/worn	overhaul hubs	8
	mud packed around tires	clean bike	2
resistance while pedaling only	bottom bracket too tight	adjust bottom bracket	11
	bottom bracket dirty/worn	overhaul bottom bracket	11
	chain dry/dirty/rusted	clean/lube or replace	4
	pedal bearings too tight	adjust pedal bearings	13
	pedal bearings dirty/worn	overhaul pedals	13
	bent chainring rubs frame	straighten or replace	11
	chainring rubs frame	adjust chainline	11
stiff steering	tight headset	adjust headset	12
suspension problems, front or rear	fork needs tuning	tune fork	16
	fork needs overhaul	overhaul fork	16
	rear suspension misadjusted	tune rear shock	17
	rear shock dirty	overhaul rear shock	17
	suspension pivots worn/dirty	overhaul pivots	17
tire loses air or is flat	deflated tire	pump tire	7
	hole(s) in tube	patch or replace tube	7
	leaky valve	replace tube or valve core	7
	hole in tubeless tire	patch or replace tire	7
	leaky seal around tubeless tire	seal, replace, or add sealant	7
STRANGE NOISES: WEIRD NOISES CAN BE HARD TO LOCATE; USE THIS LIST TO ASSIST IN LOCATING THEM.			
chattering and vibration when braking	bent or dented rim	replace rim	15
	loose headset	adjust headset	12
	brake pads toed-out	adjust brake pads	10
	brake pads worn out	replace brake pads	9, 10
	bent or scratched rotor	straighten or replace rotor	9, 10
	loose rotor	tighten rotor mounting bolts	9

SYMPTOM	LIKELY CAUSES	ACTION	CHAPTER
STRANGE NOISES, cont.			
chattering and vibration when braking, cont.	wheel out of round	true wheel	8
	loose wheel spokes	tighten spokes	15
	greasy sections of rim	clean rim	8
	oily disc-brake rotor or pads	clean rotor and pads/replace pads	9
	loose caliper mounting bolts	tighten brake mounting bolts	9, 10
	rim worn out, ready to collapse	replace rim ASAP!	15
	cracked seatstay	replace frame or seatstay	17
	flexible seatstays at disc-brake mount	replace frame or reinforce brake mount	17
	overly flexible fork	replace fork	12, 16
	cantilever brake paired with head-set cable hanger	mount cable hanger on fork crown, or replace cantilever with V-brake	10
clicking noise	cracked shoe cleats	replace cleats	13
	cracked shoe sole	replace shoes	13
	loose bottom bracket	tighten BB	11
	loose crankarm	tighten crankarm	11
	loose pedal	tighten pedal	13
	loose chainrings	tighten chainring bolts	11
clunking from fork	headset loose	adjust headset	12
	suspension-fork bushings worn	replace bushings	16
creaking noise (see also "squeaking noise")	dry handlebar/stem joint	grease handlebar and inside stem clamp	12
	hard-anodizing of stem and handlebar	roughen stem clamp interior with sandpaper	12
	dry stem/steering-tube joint	grease steering tube and inside stem clamp	12
	cartridge BB moves inside cup	grease inside cup	11
	loose seatpost	tighten seatpost	14
	loose shoe cleats	tighten cleats	13
	loose crankarm	tighten crankarm bolt	11
	loose front derailleur clamp	tighten clamp	5
	cracked frame	replace frame	17
	dry, rusty seatpost	grease seatpost	14
	loose front quick-release or front axle	tighten QR or axle	8
rubbing or scraping noise when pedaling	crossed chain	avoid extreme gears	5
	front derailleur rubbing	adjust front derailleur	5
	chainring rubs frame	longer bottom bracket or move bottom bracket over	11

Continues >>

SYMPTOM	LIKELY CAUSES	ACTION	CHAPTER
STRANGE NOISES, cont.			
rubbing, squealing, or scraping noise when coasting or pedaling	tire dragging on frame	straighten wheel	2, 8
	tire dragging on fork	straighten wheel	2, 8
	brake dragging on rim	adjust brake	10
	mud packed around tires	clean bike	2
	dry, dirty hub dust seals	clean dust seals	8
	one pad hitting rotor	center caliper over rotor	9
	both pads dragging on rotor	push pistons back or loosen cable	9
squeaking noise	dry hub or BB bearings	overhaul hubs or BB	8, 11
	dry pedal bushings	overhaul pedals	13
	squeaky saddle	grease leather-rail contact and oil rail attachments	14
	dry suspension pivots	overhaul suspension	16, 17
	rusted or dry chain	lube or replace chain	4
	dry suspension fork	overhaul fork	16
	dry suspension seatpost	overhaul seatpost	14
	loose cassette lockring	tighten to torque	8
	broken freehub pawl spring	replace spring or freehub	8
squealing noise when braking	brake pads toed-out	adjust brake pads	10
	greasy rims	clean rims and pads	10
	loose brake arms	tighten brake arms	10
	flexible seatstays	use brake booster plate	10
	oily disc-brake rotor	clean rotor and pads	9
	loose cassette lockring	tighten to torque	8
	broken freehub pawl spring	replace spring or freehub	8

APPENDIX B
GEAR DEVELOPMENT

The gear tables on the following pages are based on 26-inch (66cm), 27.5-inch (69cm), and 29-inch (72cm) tire diameters. Your gear-development numbers may be slightly different if the diameter of your bike's rear tire—at inflation, with your weight on it—is not 26, 27.5, or 29 inches. However, these numbers will be very close—unless, of course, your bike has 20-inch, 24-inch, 32-inch, or 36-inch wheels or some other nonstandard size. For these other sizes, use the measurement system described below.

If you want to have totally accurate gear development numbers for the tire on your bike at a specific inflation pressure, you can measure the tire diameter very precisely following the procedure in Figure B.1 and the steps on the next page. You can then create your own gear chart by plugging your tire's diameter into the gear development formula above the first chart, or by multiplying each number in the first (26-inch) chart by the ratio of your tire's diameter (in inches) divided by 26 inches (the tire diameter we used in the first gear chart).

Here are the steps to measure the diameter of the tire (have someone assist you in making the marks):

1. Sit on the bike with the tire pumped to your desired pressure.
2. Mark the spot on the rear rim that is at the bottom, and mark the floor adjacent to that spot.
3. Roll bike forward one wheel revolution, and mark the floor again where the mark on the rim is at the bottom (Fig. B.1).
4. Measure the distance between the marks on the floor; this is the tire circumference at pressure with your weight on it.
5. Divide this number by π (π = 3.14159) to get the diameter.

NOTE: *This roll-out procedure is also the method used to measure the wheel size with which to calibrate your bike's computer, except that most computers use the front wheel for data.*

GEAR FORMULA

Gear development = (number of teeth on chainring) × (wheel diameter) ÷ (number of teeth on rear cog).

To find out how far you travel with each pedal stroke in a given gear, multiply the gear development by π (3.14159).

B.1 Rolling out the wheel to measure its circumference. For gear development, measure the rear wheel. For cyclocomputer setup, measure the front wheel.

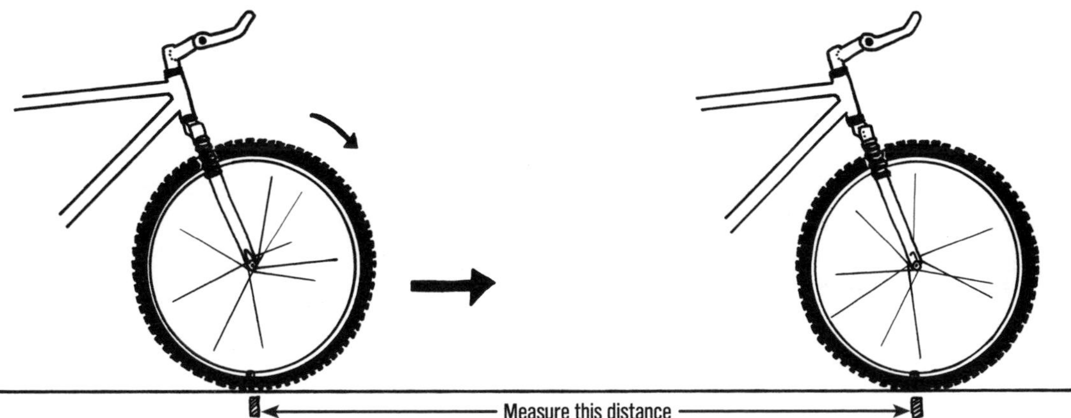

TABLE B.1 — GEAR DEVELOPMENT FOR 26-INCH WHEELS

REAR HUB COGS	CHAINRING GEAR TEETH													
	20	21	22	23	24	25	26	27	28	29	30	31	32	33
9	58	61	64	66	69	72	75	78	81	84	87	90	92	95
10	52	55	57	60	62	65	68	70	73	75	78	81	83	86
11	47	50	52	54	57	59	61	64	66	69	71	73	76	78
12	43	45	48	50	52	54	56	58	61	63	65	67	69	71
13	40	42	44	46	48	50	52	54	56	58	60	62	64	66
14	37	39	41	43	45	46	48	50	52	54	56	58	59	61
15	35	36	38	40	42	43	45	47	49	50	52	54	55	57
16	32	34	36	37	39	41	42	19	45	47	49	50	52	54
17	31	32	34	35	37	38	40	41	43	44	46	47	49	50
18	29	30	32	33	35	36	38	39	40	42	43	45	46	48
19	27	29	30	31	33	34	36	37	38	40	41	42	44	45
20	26	27	29	30	31	32	34	35	36	38	39	40	42	43
21	25	26	27	28	30	31	32	33	35	36	37	38	40	41
22	24	25	26	27	28	30	31	32	33	34	35	37	38	39
23	23	24	25	26	27	28	29	31	32	33	34	35	36	37
24	22	23	24	25	26	27	28	29	30	31	32	34	35	36
25	21	22	23	24	25	26	27	28	29	30	31	32	33	34
26	20	21	22	23	24	25	26	27	28	29	30	31	32	33
27	20	20	21	22	23	24	25	26	27	28	29	30	31	32
28	19	19	20	21	22	23	24	25	26	27	28	29	30	31
29	19	19	20	21	22	22	23	24	25	26	27	28	29	30
30	18	18	19	20	21	22	23	23	24	25	26	27	28	29
31	17	18	18	19	20	21	22	23	23	24	25	26	27	28
32	16	17	18	19	19	20	21	22	23	24	24	25	26	27
33	16	17	17	18	19	20	20	21	22	23	24	24	25	26
34	15	16	17	18	18	19	20	21	21	22	23	24	24	25
35	15	16	16	17	18	19	19	20	21	22	22	23	24	24
36	14	15	16	17	17	18	19	19	20	21	22	22	23	24
37	14	15	15	16	17	18	18	19	20	20	21	22	22	23
38	14	14	15	16	16	17	18	18	19	20	21	21	22	23
39	13	14	15	15	16	17	17	18	19	19	20	21	21	22
40	13	14	14	15	16	16	17	18	18	19	19	20	21	21
42	12	13	14	14	15	15	16	17	17	18	19	19	20	20
50	10	11	11	12	12	13	14	14	15	15	16	16	17	17

CHAINRING GEAR TEETH																	
34	35	36	37	38	39	40	41	42	43	44	45	46	47	48	49	50	51
98	101	104	107	110	113	115	118	121	124	127	130	133	136	139	141	144	147
88	91	94	96	99	101	104	107	109	112	114	117	120	122	125	127	130	133
80	83	85	87	90	92	94	97	99	102	104	106	109	111	113	116	118	120
74	76	78	80	82	84	87	89	91	93	95	97	100	102	104	106	108	110
68	70	72	74	76	78	80	82	84	86	88	90	92	94	96	98	100	102
63	65	67	69	71	72	74	76	78	80	82	84	85	87	89	91	93	95
59	61	62	64	66	68	69	71	73	74	76	78	80	81	83	85	87	88
55	57	58	60	62	63	65	67	68	70	71	73	75	174	78	80	81	83
52	53	55	57	58	60	61	63	64	66	67	69	70	72	73	75	76	78
49	51	52	53	55	56	58	59	61	62	64	65	66	68	69	71	72	74
46	48	49	51	52	53	55	56	57	59	60	62	63	64	66	67	68	70
44	45	47	48	49	51	52	53	55	56	57	58	60	61	62	64	65	66
42	43	45	46	47	48	49	51	52	53	54	56	57	58	59	61	62	63
40	41	43	44	45	46	47	48	50	51	52	53	54	56	57	58	59	60
38	40	41	42	43	44	45	46	47	49	50	51	52	53	54	55	56	58
37	38	39	40	41	42	43	44	45	47	48	49	50	51	52	53	54	55
35	36	37	38	39	41	42	43	44	45	46	47	48	49	50	51	52	53
34	35	36	37	38	39	40	41	42	43	44	45	46	47	48	49	50	51
33	34	35	36	37	38	38	39	40	41	42	43	44	45	46	47	48	49
32	32	33	34	35	36	37	38	39	40	41	42	43	44	45	45	46	47
30	31	32	33	34	35	36	37	38	39	39	40	41	42	0	44	45	46
29	30	31	32	33	34	35	36	36	37	38	39	40	41	42	42	43	44
28	29	30	31	32	33	34	34	35	36	37	38	39	39	40	41	42	43
28	28	29	30	31	32	32	33	34	35	36	37	37	38	39	40	41	41
27	28	28	29	30	31	31	32	33	34	35	35	36	37	38	39	39	40
26	27	28	28	29	30	31	31	32	33	34	34	35	36	37	37	38	39
25	26	27	27	28	29	30	30	31	32	33	33	34	35	36	36	37	38
25	25	26	27	27	28	29	30	30	31	32	32	33	34	35	35	36	37
24	25	25	26	27	27	28	29	29	30	31	32	32	33	34	34	35	36
23	24	25	25	26	27	27	28	29	29	30	31	31	32	33	34	34	35
23	23	24	25	25	26	27	27	28	29	29	30	31	31	32	33	33	34
22	23	23	24	25	25	26	27	27	28	29	29	30	31	31	32	32	33
21	22	22	23	24	24	25	25	26	27	27	28	28	29	30	30	31	32
18	18	19	19	20	20	21	21	22	22	23	23	24	24	25	25	26	27

TABLE B.2 — GEAR DEVELOPMENT FOR 27.5-INCH WHEELS

REAR HUB COGS	CHAINRING GEAR TEETH													
	20	21	22	23	24	25	26	27	28	29	30	31	32	33
9	60	63	66	69	72	75	78	81	85	88	91	94	97	100
10	54	57	60	62	65	68	71	73	76	79	81	84	87	90
11	49	52	54	57	59	62	64	67	69	72	74	77	79	81
12	45	48	50	52	54	57	59	61	63	66	68	70	72	75
13	42	44	46	48	50	52	54	56	59	61	63	65	67	69
14	39	41	43	45	47	49	50	52	54	56	58	60	62	64
15	36	38	40	42	43	45	47	49	51	53	54	56	58	60
16	34	36	37	39	41	42	44	20	48	49	51	53	54	56
17	32	34	35	37	38	40	42	43	45	46	48	50	51	53
18	30	32	33	35	36	38	39	41	42	44	45	47	48	50
19	29	30	31	33	34	36	37	39	40	41	43	44	46	47
20	27	29	30	31	33	34	35	37	38	39	41	42	43	45
21	26	27	28	30	31	32	34	35	36	38	39	40	41	43
22	25	26	27	28	30	31	32	33	35	36	37	38	40	41
23	24	25	26	27	28	30	31	32	33	34	35	37	38	39
24	23	24	25	26	27	28	29	31	32	33	34	35	36	37
25	22	23	24	25	26	27	28	29	30	32	33	34	35	36
26	21	22	23	24	25	26	27	28	29	30	31	32	33	34
27	21	21	22	23	24	25	26	27	28	29	30	31	32	33
28	20	20	21	22	23	24	25	26	27	28	29	30	31	32
29	19	20	21	22	22	23	24	25	26	27	28	29	30	31
30	19	19	20	21	22	23	24	24	25	26	27	28	29	30
31	18	18	19	20	21	22	23	24	25	25	26	27	28	29
32	17	18	19	20	20	21	22	23	24	25	25	26	27	28
33	16	17	18	19	20	21	21	22	23	24	25	26	26	27
34	16	17	18	18	19	20	21	22	22	23	24	25	26	26
35	16	16	17	18	19	19	20	21	22	23	23	24	25	26
36	15	16	17	17	18	19	20	20	21	22	23	23	24	25
37	15	15	16	17	18	18	19	20	21	21	22	23	23	24
38	14	15	16	16	17	18	19	19	20	21	21	22	23	24
39	14	15	15	16	17	17	18	19	20	20	21	22	22	23
40	14	14	15	16	16	17	18	18	19	20	20	21	22	22
42	13	14	14	15	16	16	17	17	18	19	19	20	21	21
50	11	11	12	12	13	14	14	15	15	16	16	17	17	18

CHAINRING GEAR TEETH																	
34	**35**	**36**	**37**	**38**	**39**	**40**	**41**	**42**	**43**	**44**	**45**	**46**	**47**	**48**	**49**	**50**	**51**
103	106	109	112	115	118	121	124	127	130	133	136	139	142	145	148	151	154
92	95	98	101	103	106	109	111	114	117	120	122	125	128	130	133	136	139
84	86	89	91	94	96	99	101	104	106	109	111	114	116	119	121	123	126
77	79	81	84	86	88	91	93	95	97	100	102	104	106	109	111	113	115
71	73	75	77	79	81	84	86	88	90	92	94	96	98	100	102	104	107
66	68	70	72	74	76	78	80	81	83	85	87	89	91	93	95	97	99
62	63	65	67	69	71	72	74	76	78	80	81	83	85	87	89	91	92
58	59	61	63	65	66	68	70	71	73	75	76	78	182	81	83	85	87
54	56	58	59	61	62	64	66	67	69	70	72	74	75	77	78	80	81
51	53	54	56	57	59	60	62	63	65	66	68	69	71	72	74	75	77
49	50	51	53	54	56	57	59	60	61	63	64	66	67	69	70	71	73
46	48	49	50	52	53	54	56	57	58	60	61	62	64	65	67	68	69
44	45	47	48	49	50	52	53	54	56	57	58	60	61	62	63	65	66
42	43	44	46	47	48	49	51	52	53	54	56	57	58	59	61	62	63
40	41	43	44	45	46	47	48	50	51	52	53	54	56	57	58	59	60
38	40	41	42	43	44	45	46	48	49	50	51	52	53	54	55	57	58
37	38	39	40	41	42	43	45	46	47	48	49	50	51	52	53	54	55
36	37	38	39	40	41	42	43	44	45	46	47	48	49	50	51	52	53
34	35	36	37	38	39	40	41	42	43	44	45	46	47	48	49	50	51
33	34	35	36	37	38	39	40	41	42	43	44	45	46	47	48	49	49
32	33	34	35	36	37	37	38	39	40	41	42	43	44	0	46	47	48
31	32	33	34	34	35	36	37	38	39	40	41	42	43	43	44	45	46
30	31	32	32	33	34	35	36	37	38	39	39	40	41	42	43	44	45
29	30	31	31	32	33	34	35	36	37	37	38	39	40	41	42	42	43
28	29	30	30	31	32	33	34	35	35	36	37	38	39	40	40	41	42
27	28	29	30	30	31	32	33	34	34	35	36	37	38	38	39	40	41
26	27	28	29	29	30	31	32	33	33	34	35	36	36	37	38	39	40
26	26	27	28	29	29	30	31	32	32	33	34	35	35	36	37	38	38
25	26	26	27	28	29	29	30	31	32	32	33	34	35	35	36	37	37
24	25	26	26	27	28	29	29	30	31	31	32	33	34	34	35	36	36
24	24	25	26	26	27	28	29	29	30	31	31	32	33	33	34	35	36
23	24	24	25	26	26	27	28	29	29	30	31	31	32	33	33	34	35
22	23	23	24	25	25	26	27	27	28	28	29	30	30	31	32	32	33
18	19	20	20	21	21	22	22	23	23	24	24	25	26	26	27	27	28

TABLE B.3 — GEAR DEVELOPMENT FOR 29-INCH WHEELS

REAR HUB COGS	CHAINRING GEAR TEETH													
	20	21	22	23	24	25	26	27	28	29	30	31	32	33
9	63	66	69	72	75	78	81	85	88	91	94	97	100	103
10	56	59	62	65	68	70	73	76	79	82	85	87	90	93
11	51	54	56	59	62	64	67	69	72	74	77	79	82	85
12	47	49	52	54	56	59	61	63	66	68	70	73	75	78
13	43	46	48	50	52	54	56	59	61	63	65	67	69	72
14	40	42	44	46	48	50	52	54	56	58	60	62	64	66
15	38	39	41	43	45	47	49	51	53	54	56	58	60	62
16	35	37	39	41	42	44	46	21	49	51	53	55	56	58
17	33	35	36	38	40	41	43	45	46	48	50	51	53	55
18	31	33	34	36	38	39	41	42	44	45	47	49	50	52
19	30	31	33	34	36	37	39	40	42	43	45	46	47	49
20	28	30	31	32	34	35	37	38	39	41	42	44	45	47
21	27	28	30	31	32	34	35	36	38	39	40	42	43	44
22	26	27	28	29	31	32	33	35	36	37	38	40	41	42
23	25	26	27	28	29	31	32	33	34	36	37	38	39	40
24	23	25	26	27	28	29	31	32	33	34	35	36	38	39
25	23	24	25	26	27	28	29	30	32	33	34	35	36	37
26	22	23	24	25	26	27	28	29	30	31	33	34	35	36
27	22	22	23	24	25	26	27	28	29	30	31	32	33	34
28	21	21	22	23	24	25	26	27	28	29	30	31	32	33
29	20	20	21	22	23	24	25	26	27	28	29	30	31	32
30	19	20	21	22	23	23	24	25	26	27	28	29	30	31
31	18	19	20	21	22	23	24	25	25	26	27	28	29	30
32	18	18	19	20	21	22	23	24	25	26	26	27	28	29
33	17	18	19	20	21	21	22	23	24	25	26	26	27	28
34	17	17	18	19	20	21	22	22	23	24	25	26	27	27
35	16	17	18	19	19	20	21	22	23	23	24	25	26	27
36	16	16	17	18	19	20	20	21	22	23	23	24	25	26
37	15	16	17	18	18	19	20	21	21	22	23	24	24	25
38	15	16	16	17	18	19	19	20	21	22	22	23	24	24
39	14	15	16	17	17	18	19	20	20	21	22	22	23	24
40	14	15	16	16	17	18	18	19	20	20	21	22	23	23
42	13	14	15	15	16	17	17	18	19	19	20	21	21	22
50	11	12	12	13	14	14	15	15	16	16	17	17	18	19

CHAINRING GEAR TEETH																	
34	**35**	**36**	**37**	**38**	**39**	**40**	**41**	**42**	**43**	**44**	**45**	**46**	**47**	**48**	**49**	**50**	**51**
106	110	113	116	119	122	125	128	132	135	138	141	144	147	150	153	157	160
96	99	101	104	107	110	113	116	118	121	124	127	130	132	135	138	141	144
87	90	92	95	97	100	103	105	108	110	113	115	118	120	123	126	128	131
80	82	85	87	89	92	94	96	99	101	103	106	108	110	113	115	117	120
74	76	78	80	82	85	87	89	91	93	95	98	100	102	104	106	108	111
68	70	72	74	77	79	81	83	85	87	89	91	93	95	97	99	101	103
64	66	68	70	71	73	75	77	79	81	83	85	86	88	90	92	94	96
60	62	63	65	67	69	70	72	74	76	78	79	81	189	85	86	88	90
56	58	60	61	63	65	66	68	70	71	73	75	76	78	80	81	83	85
53	55	56	58	60	61	63	64	66	67	69	70	72	74	75	77	78	80
50	52	53	55	56	58	59	61	62	64	65	67	68	70	71	73	74	76
48	49	51	52	54	55	56	58	59	61	62	63	65	66	68	69	70	72
46	47	48	50	51	52	54	55	56	58	59	60	62	63	64	66	67	68
44	45	46	47	49	50	51	53	54	55	56	58	59	60	62	63	64	65
42	43	44	45	47	48	49	50	51	53	54	55	56	58	59	60	61	63
40	41	42	43	45	46	47	48	49	51	52	53	54	55	56	58	59	60
38	39	41	42	43	44	45	46	47	48	50	51	52	53	54	55	56	58
37	38	39	40	41	42	43	44	46	47	48	49	50	51	52	53	54	55
35	37	38	39	40	41	42	43	44	45	46	47	48	49	50	51	52	53
34	35	36	37	38	39	40	41	42	43	44	45	46	47	48	49	50	51
33	34	35	36	37	38	39	40	41	42	43	44	45	46	0	48	49	50
32	33	34	35	36	37	38	39	39	40	41	42	43	44	45	46	47	48
31	32	33	34	35	35	36	37	38	39	40	41	42	43	44	45	45	46
30	31	32	33	33	34	35	36	37	38	39	40	41	41	42	43	44	45
29	30	31	32	32	33	34	35	36	37	38	38	39	40	41	42	43	44
28	29	30	31	32	32	33	34	35	36	36	37	38	39	40	41	41	42
27	28	29	30	31	31	32	33	34	35	35	36	37	38	39	39	40	41
27	27	28	29	30	31	31	32	33	34	34	35	36	37	38	38	39	40
26	27	27	28	29	30	30	31	32	33	34	34	35	36	37	37	38	39
25	26	27	27	28	29	30	30	31	32	33	33	34	35	36	36	37	38
25	25	26	27	27	28	29	30	30	31	32	33	33	34	35	35	36	37
24	25	25	26	27	27	28	29	30	30	31	32	32	33	34	35	35	36
23	23	24	25	26	26	27	28	28	29	30	30	31	32	32	33	34	34
19	20	20	21	21	22	23	23	24	24	25	25	26	26	27	28	28	29

APPENDIX C
MOUNTAIN BIKE FITTING

Fit should be the primary consideration when selecting a bike. You can adapt to heavier bikes and bikes painted a color that's not your favorite, but your body will eventually protest if you're riding one that doesn't fit. The simple need to protect your more sensitive parts should keep you away from a bike without sufficient stand-over clearance (Fig. C.1), but there are a lot of other factors to consider. You need to make certain that your bike has enough reach to ensure that you don't bang your knees on the handlebar. You also need to check that your weight is properly distributed over the wheels so that you don't end up going over the handlebar on downhill stretches or find yourself unweighting the front end on steep climbs. An improperly sized bike is both inefficient and terribly uncomfortable. Therefore, take some time to learn how you can pick a properly sized bike.

I've outlined two methods for finding your frame size. The first is a simple method of checking your fit on bikes at your local bike shop. The second is a bit more elaborate, as it involves taking body measurements. This more detailed approach will allow you to calculate the proper frame dimensions whether the bike is assembled or not.

C.1 Bike height

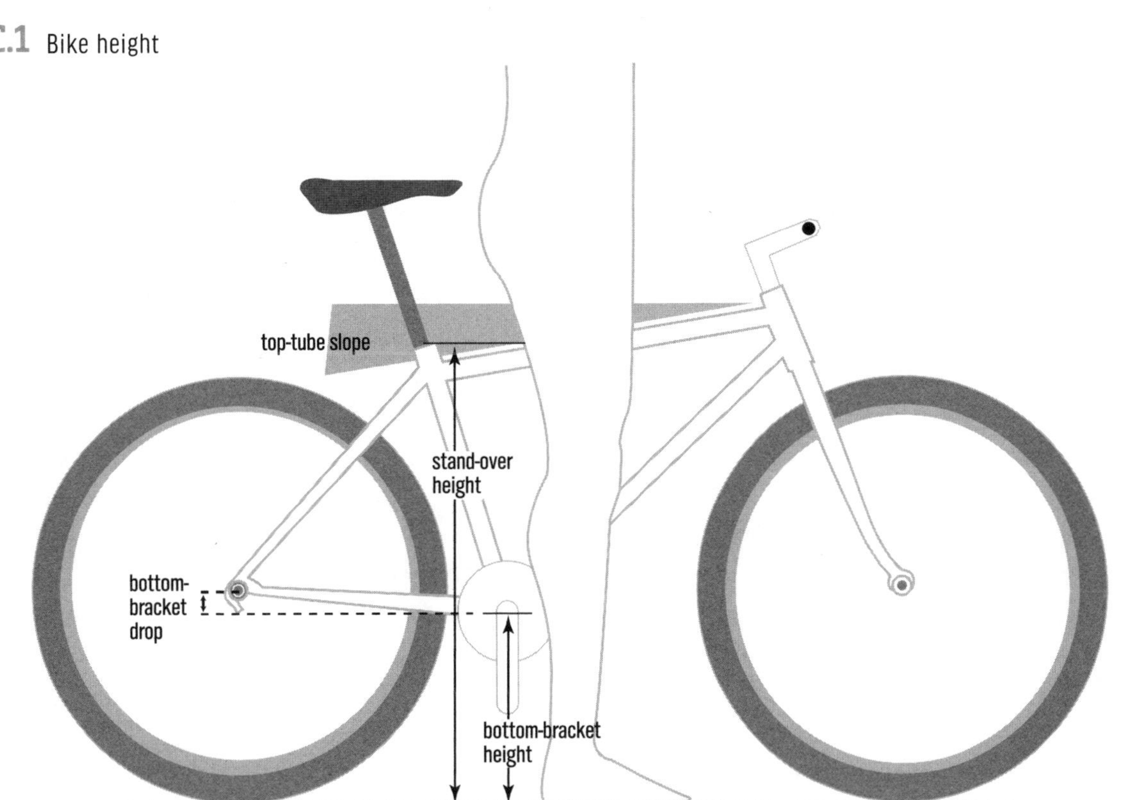

C-1

SELECTING THE CORRECT SIZE OF AN ASSEMBLED BIKE

a. Stand-over height

Stand over the bike's top tube and lift the bike straight up until the top tube hits your crotch. The wheels should be at least 2 inches off the ground to ensure that you can jump off the bike safely without hitting your crotch. There is no maximum dimension here. Though it may seem like a lot, 5 inches or more of stand-over height is fine, as long as the top tube is long enough for you and the handlebar height can be set properly for you.

NOTE: *If you have 2 inches of stand-over clearance on one bike, you should not assume that a bike from a different manufacturer with the same listed frame size will also offer you the same stand-over clearance. Manufacturers often measure frame size in different ways. They also slope and/or curve their top tubes differently and use different bottom-bracket heights (Fig. C.1), all of which affect the final stand-over height.*

All manufacturers measure the frame size up the seat tube from the center of the bottom bracket, but the location of the top end of the measurement varies. Some measure to the center of the top tube (center-to-center measurement), some measure to the top of the top tube (center-to-top measurement), and others measure to the top of the seat tube (also called center-to-top measurement), even though there is wide variation in the length of the seatpost collar above the top tube. Obviously, each of these methods will give you a different frame size for the same frame.

No matter how the frame size is measured, the stand-over height of a bike depends on the slope of the top tube and the bottom-bracket height (Fig. C.1). Top tubes that slant up to the front are common, so stand-over clearance is obviously a function of where you are standing. If the bike has an up-angled top tube, straddle it 1 or 2 inches forward of the nose of the saddle, and then lift the bike up into your crotch to measure stand-over clearance.

A bike with a 125mm-travel suspension fork will have a higher front end than a bike with an 80mm- or 100mm-travel suspension fork, or a bike with a rigid fork, because the frame has to accommodate the up-and-down movement of the front wheel. And a full-suspension bike is automatically even taller, because the pedals need to be higher off the ground (higher bottom bracket) to ensure that they still clear obstacles when the suspension compresses. Complicating it even more are different wheel sizes. For instance, a 29er bike (bike with 29-inch wheels) will have a higher front end than a 26er with the same fork travel. All of this makes it difficult to compare listed frame sizes from even the same manufacturer to determine stand-over height. Unless the manufacturer lists the stand-over height in its catalog or on its website and you know your inseam length, you need to actually stand over the bike.

NOTE: *If you are short and cannot find a frame size small enough for you to get at least 2 inches of stand-over clearance, consider a bike with 24-inch wheels.*

b. Knee-to-handlebar clearance

Make sure your knees cannot hit the handlebar (Fig. C.2). Check this while standing out of the saddle as well as when seated and with the front wheel turned slightly. Be certain that your knees will not hit when you are in the most awkward pedaling position you might use.

c. Handlebar reach and drop

Ride the bike. See whether the reach feels comfortable when you are holding the handlebar grips or the bar ends. Make sure that you can grab the brake levers easily and that your knees do not hit your elbows as you pedal. Check to see that the stem can be raised or lowered enough to achieve a comfortable handlebar height.

NOTE: *Threadless headsets (the standard on all bikes today) allow very limited adjustment of stem height). Large changes in height require a change of stems.*

ANOTHER NOTE: *If you are tall and cannot find a bike with the handlebar as high as you need, consider a bike with 29-inch wheels instead of 27.5-inch or 26-inch wheels. Bigger wheels mean a taller fork, which will get your handlebar up higher.*

d. Pedal overlap

Pedal overlap is a common bike-shop term, but it is a misnomer because you are actually interested in

C.2 Knee and toe clearance

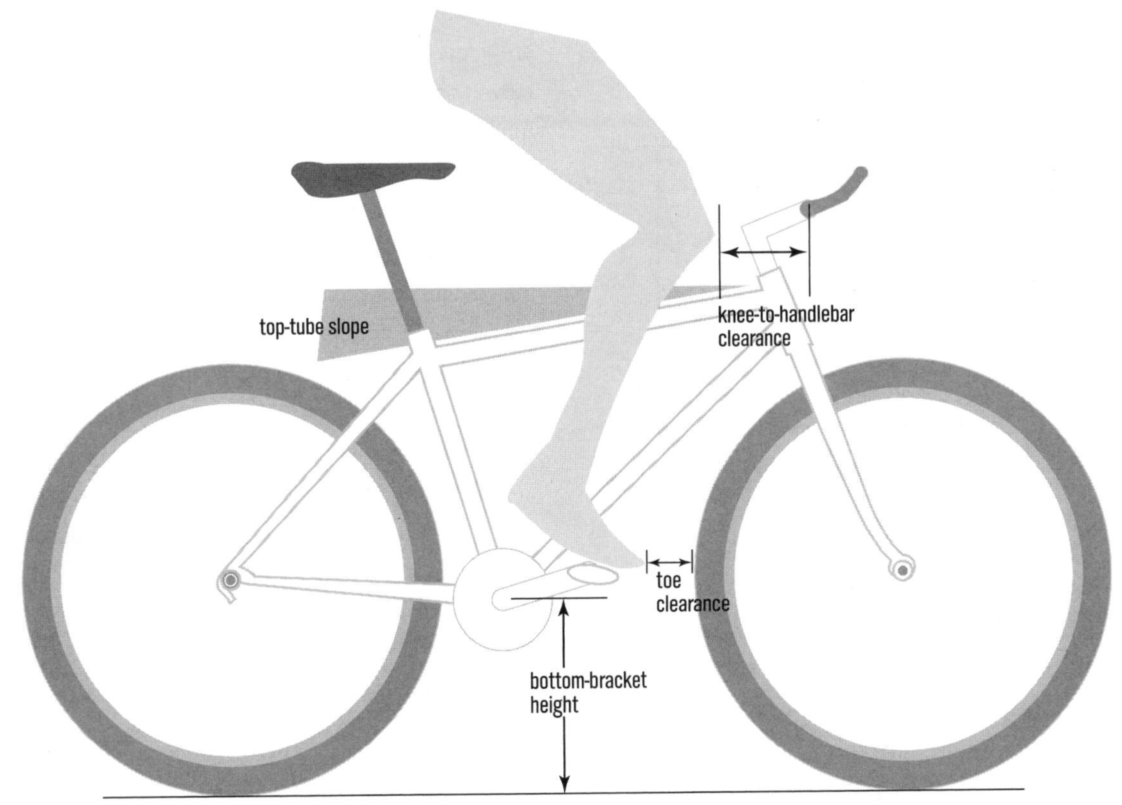

whether your toe, not the pedal, can hit the front tire when you are turning sharply at low speeds. Sitting on the bike with the crankarms horizontal and your foot on the pedal, turn the handlebar and check that your toe does not hit the front tire (Fig. C.2). Toe overlap is to be avoided for any kind of slow-speed, technical riding, as pedaling up rocky terrain slowly can often result in the front wheel turning sharply back and forth as the feet pass by. Toe overlap is not an issue for riding at high speeds, because you cannot turn the front wheel at enough of an angle to hit the foot when the bike is rolling fast, even in a sharp turn.

C-2

CHOOSING FRAME SIZE FROM YOUR BODY MEASUREMENTS

You will need a second person to assist you. This method is for cross-country and trail bikes; bikes for gravity riding (downhill, freeride, dual slalom, four-cross, jumping, etc.) need to be considerably smaller.

By taking the three easy measurements shown in Figure C.3, most people can get a very good frame fit. The procedure I use when I design a custom bike frame for a client is more complex than this, involving more measurements, but the following method works well for picking an off-the-shelf bike. To skip the trouble of making these calculations yourself, go to the free "Bicycle Fit Calculator" page at zinncycles.com; it will automatically calculate your frame size from these measurements.

a. Measure your inseam

Stand with your stocking feet about 2 inches apart, and measure up from the floor to a three-foot carpenter's level or broomstick held level and pulled firmly up into your crotch with one hand in front and one in back. You can also use a large book and slide it up a wall to keep the top edge horizontal—as you pull it up as hard as you can—into your crotch. You can mark the top of the level, broomstick, or book on the wall and measure up from the floor to the mark. With the book, it is harder to pull up enough to compress the soft tissue up against

C.3 Body measurements

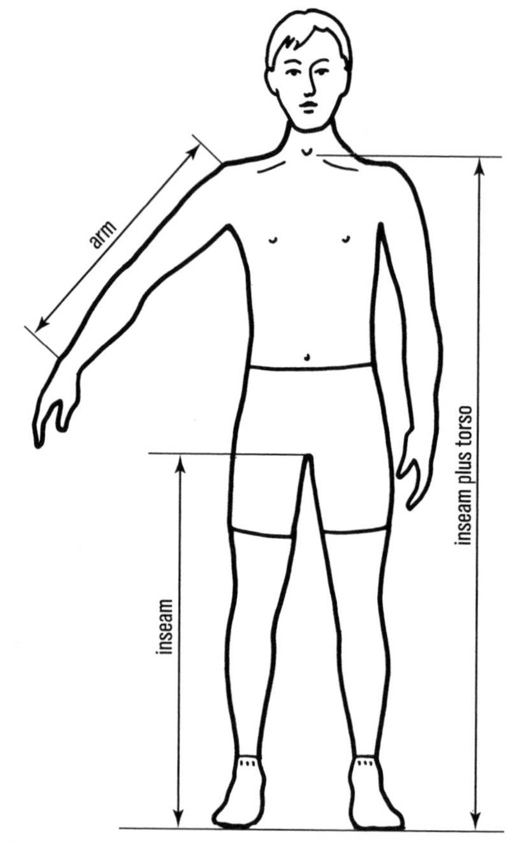

the bottom edge of the pelvis than with a broomstick, so pull up hard.

b. Measure your inseam-plus-torso length

Hold a pencil horizontally against your jugular notch, the U-shaped bone depression just below your Adam's apple. Standing up straight while facing a wall, mark the wall with the horizontal pencil. Measure up from the floor to the mark.

c. Measure your arm length

Hold your arm out from your side at a 45-degree angle with your elbow straight, as shown in Figure C.3. Measure from the sharp bone point directly behind and above your shoulder joint (the lateral tip of the acromion) to the wrist bone on your little finger side.

d. Find your frame size

Subtract 36–42cm (13.5–16.5 inches) from your inseam length. This length is your frame size measured from the center of the bottom bracket to the top of a horizontal top tube; subtract another ¾ inch to get an approximate center-to-center size. If the frame you are interested in has a sloping top tube, rather than a curved one, you need a bike with an even shorter seat-tube length to get sufficient stand-over clearance. With a curved- or sloping-top-tube bike, project a horizontal line back to the seat tube (or seatpost) from the top of the top tube at the center of its length (Fig. C.4). Mark the seat tube or seatpost at this line. Measure from the center of the bottom bracket to this mark; this length should be 36–42cm less than your inseam measurement. Also, if the bike has a bottom bracket higher than 29cm (11½ inches), subtract the additional bottom-bracket height from the seat-tube length as well.

Generally, smaller riders will want to subtract closer to 36cm from their inseam, while taller riders will subtract closer to 42cm. However, full-suspension bikes generally have higher bottom brackets, which reduces stand-over height, so the more suspension travel your bike has, the shorter the seat-tube length you will want. There is considerable range here. The top-tube length (next step) is more important than a specific frame size, and if you have a short torso and short arms, you can use a small frame to get the right top-tube length, as long as you can raise your bar as high as you need it. Be aware that interrupted-seat-tube configurations of some full-suspension frames make measuring frame size challenging.

You really want to be sure that you have plenty of stand-over clearance, so do not subtract less than 36cm from your inseam to obtain your seat-tube length. If you are short and cannot find a bike small enough to get at least 2 inches of stand-over clearance, look for one with a curved top tube, or consider one with 26-inch wheels or 24-inch wheels. And if you are tall and cannot find a bike big enough for you without installing a super-long seatpost that will cantilever you out over the rear wheel, causing you to wheelie on climbs, consider one with 29-inch wheels instead of 27.5- or 26-inch wheels.

NOTE: *A step-through frame (i.e., "women's frame," "mixte frame," or "girl's bike") having a steeply up-angled top tube meeting near the bottom-bracket shell makes seat-tube length for stand-over clearance nearly irrelevant. With a step-through bike, the only considerations will be horizontal and vertical reach to the bar.*

C.4 Bike dimensions

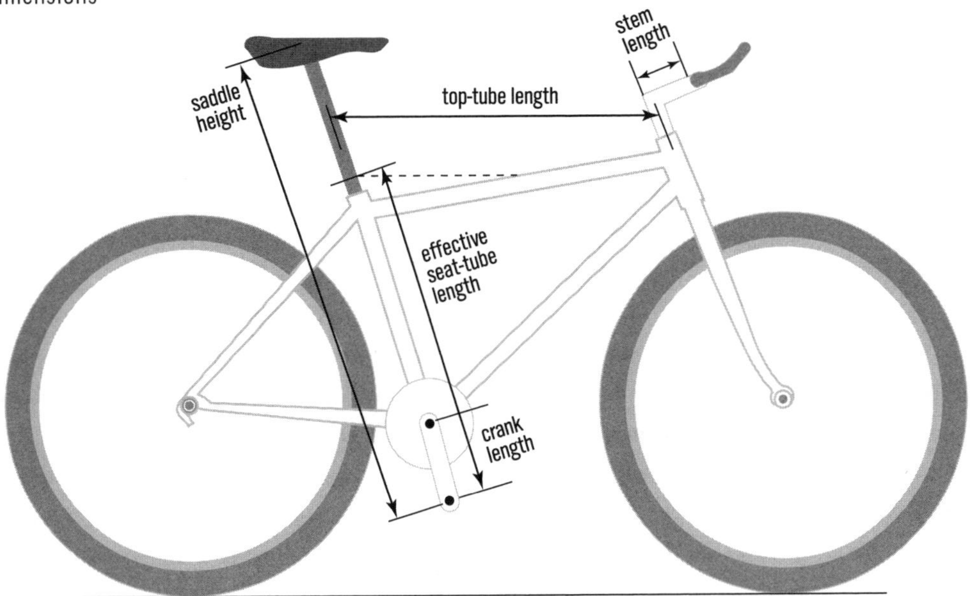

e. Find your top-tube length

To find your torso length, subtract your inseam measurement (found in step a) from your inseam-plus-torso measurement (found in step b). Add this torso length to your arm-length measurement (found in step c). To find the top-tube length, multiply this arm-plus-torso measurement by a factor in the range between 0.47 and 0.5. If you are a casual rider, use 0.47; if you are a very aggressive rider, use 0.5; if you are in between, use a factor in between.

The top-tube length is measured horizontally from the center of the seat tube (or seatpost) to the center of the head tube (Fig. C.4). Obviously, the horizontal top-tube length is greater than the length found by measuring along the top tube on a sloping-top-tube bike, so don't just measure along a sloping top tube. Your body position is dictated by the horizontal distance of your hands reaching forward from your butt. Measuring along a sloping line does not give you useful information.

NOTE: *Full-suspension bikes with an interrupted seat tube often have a seatpost clamp that angles the seatpost back sharply along a line that would not intersect the bottom bracket. If you are tall and prefer to have your seat high, your seat would end up far back from where it normally would be on a bike of that size, and vice versa. You need to account for this shallow seatpost angle by estimating where the center of your virtual seat tube would be by extrapolating a line from the center of the saddle to the center of the bottom bracket. Measure from this imaginary line horizontally forward to the center of the head-tube to find your top-tube length.*

f. Find your stem length

Multiply the arm-plus-torso length you found in step e by a factor in the range between 0.085 and 0.115 to find the stem length. A casual rider should multiply by 0.085 or so, while an aggressive rider should multiply by closer to 0.115. This is a starting stem length. Finalize the stem length once you are sitting on the bike and see what feels best.

If you will have to accept a top-tube length different from one that is ideal for you, you will need to make a corresponding adjustment in stem length.

g. Determine crankarm length

Most mountain bikes come with 175mm cranks (measured hole to hole—see Fig. C.4), and it is rare to find another crank length on one. But tall riders will often be better off with 180mm crankarms or even longer custom cranks, while short riders will want 170mm or even shorter custom cranks; see more about crank length at zinncycles.com.

C-3

POSITIONING OF YOUR SADDLE AND HANDLEBAR

The frame fit is only part of the equation. Except for the stand-over clearance, a good frame fit is relatively meaningless if the seat setback, seat height, handlebar height, and handlebar reach are not set correctly for you.

a. Saddle height

When your foot is at the bottom of the pedal stroke, lock your knee without rocking your hips. Do this sitting on your bike with someone holding you up, or do it with your bike mounted on a stationary trainer stand; have someone else observing. Your foot should be approximately level.

A second way to determine seat height uses your inseam measurement (Fig. C.3), found in step a under section C-2 above. Multiply your inseam length by 1.09; this is the length from the center of the pedal spindle (when the pedal is down) to one of the points on the top of the saddle where it contacts your sit bones (ischial tuberosities) (Fig. C.4). Adjust the seat height (Chapter 14) until you get the proper height.

NOTE: *These two methods yield similar results, although the measurement-multiplying method is dependent on shoe sole and pedal thicknesses. Both methods yield a biomechanically efficient pedaling position, but if you do a lot of technical riding and descending, you will want a dropper post (14-9) or a lower saddle for better bike-handling control.*

b. Saddle setback

Sit on your bike on a stationary trainer with the crankarms horizontal and your foot at its normal angle when pedaling at that point. Have a friend drop a plumb line from a point just below your kneecap on the front of your knee. You may use a heavy ring or washer tied to a string for the plumb line. The bottom end of the plumb line should touch the end of the crankarm (thus approximating the center of rotation of the knee over the center of rotation of the pedal) or pass up to 2cm behind it (Fig. C.5). A saddle positioned fore-and-aft in this manner encourages smooth pedaling at high revolutions per minute, while 2cm farther back encourages powerful seated climbing. You may also wish to experiment with a saddle position farther forward; this can keep the front wheel on the ground on steep climbs.

C.5 Saddle and stem position

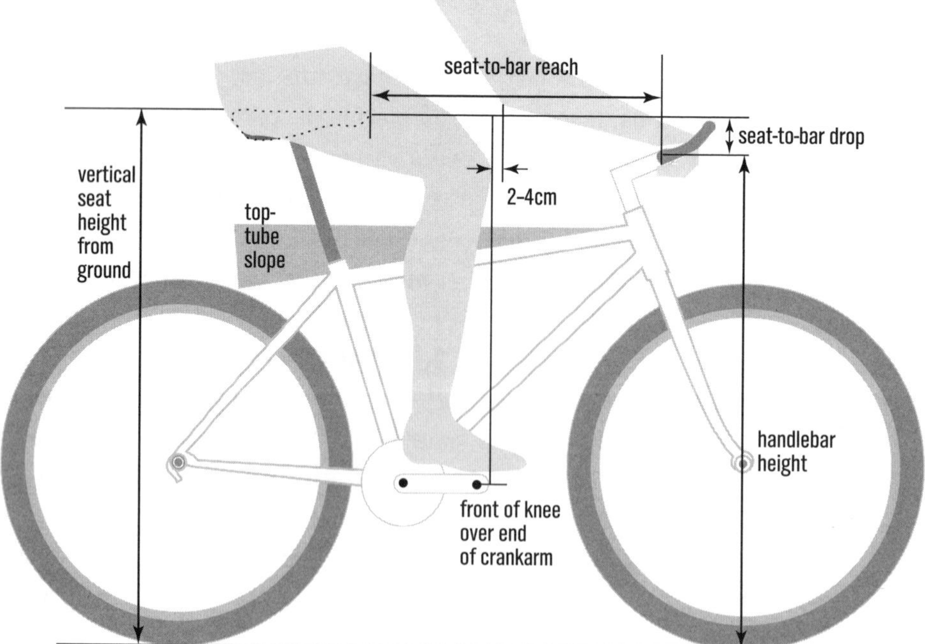

You also will want to make sure that your cleat position (Chapter 13) is set properly. Generally, you will want your foot deep enough into the pedal that the ball of your foot is right over the pedal spindle or up to 2cm ahead of it. Riders with big feet will want their cleats farther back, often as far back as possible, while riders with short feet will often want to move the cleats far forward.

Slide the saddle back and forth on the seatpost (Chapter 14) until you achieve the desired fore-and-aft saddle position. Set the saddle level or very slightly tipped down at the nose so that it comes to level with seatpost and saddle flex when you're sitting on it (downhill racers and big-air free-riders sometimes tip their saddles up steeply at the nose, but their saddles are set very low, and you cannot pedal in an efficient, high seat position this way). Recheck the seat height in step a above, as fore-and-aft saddle movements affect seat-to-pedal distance, too.

c. Handlebar height

Measure the handlebar height relative to the saddle height by measuring the vertical distances of the saddle and handlebar up from the floor (Fig. C.5) and subtracting one from the other. How much higher you set the saddle than the handlebar (or vice versa) depends on your flexibility, riding style, overall size, and the type of riding you prefer.

Aggressive and/or tall cross-country riders will prefer to have their saddle at least 10cm higher than their bar. Shorter riders will want proportionately less drop, as will less aggressive riders. Riders doing lots of gravity-driven riding will want their bars higher; handlebars on downhill and dual-slalom bikes are commonly considerably higher than the saddle. Generally, people beginning mountain bike riding will like the bars high and can lower them as they become more comfortable with the bike, with faster speeds, and with riding more technical terrain.

If in doubt, start with 4cm of drop for general cross-country riding and vary it from there. The higher the bar, the greater the tendency of the front wheel to pull up off the ground when climbing, and the more wind resistance you can expect, while, up to a point, the more comfort and control you will have going down technical terrain with drop-offs. Change the handlebar height by raising or lowering the stem (Chapter 12), or by switching stems of different angles and/or bars of different bends.

Again, threadless headsets allow only limited handlebar-height adjustment without substitution of a different stem or handlebar.

d. Setting handlebar reach

The ideal reach from the saddle to the handlebar is also dependent on personal preference. Aggressive riders will want a more stretched-out position than will casual riders. This length is subjective, and I find that I need to look at the rider on the bike and get a feel for how he or she would be comfortable and efficient before suggesting a length for handlebar reach.

A useful starting place is to drop a plumb line from the back of your elbow with your arms bent in a comfortable riding position. This plane determined by your elbows and the plumb line should be 2–4cm horizontally ahead of each knee at the point in the pedal stroke when the crankarm is horizontally forward (Fig. C.5). Select a position you find comfortable and efficient; pay attention to what your body wants.

Vary the saddle-to-handlebar distance by changing stem length (Chapter 12), not by changing the seat fore-and-aft position, which is based on pedaling efficiency (step b above) and not on reach.

NOTE: *There is no single formula for determining handlebar reach and height. I can tell you that using the all-too-common method of placing your elbow against the saddle and seeing if your fingertips reach the handlebar is close to useless. Similarly, the oft-suggested method of seeing whether the handlebar obscures your vision of the front hub is not worth the brief time it takes to look, as it largely depends on elbow bend, the inclination of your neck, and the front-end geometry of the bike. Another method, involving dropping a plumb bob from the rider's nose, is dependent on both the handlebar height and elbow bend and thus does not lend itself to a prescribed relationship for all riders.*

e. Bar-end position

Bar ends are an optional item, have largely fallen out of favor, and generally only appear on cross-country

C.6 Bar-end angle for performance riders

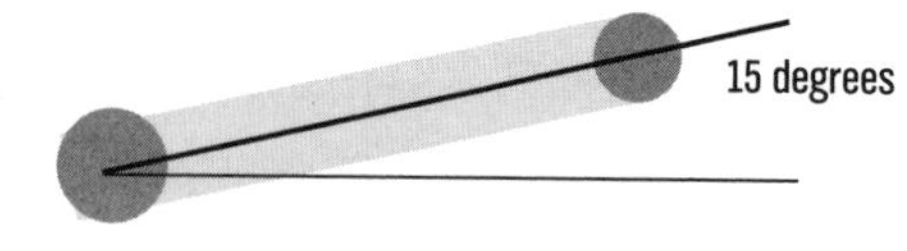

C.7 Bar-end angle for casual riders

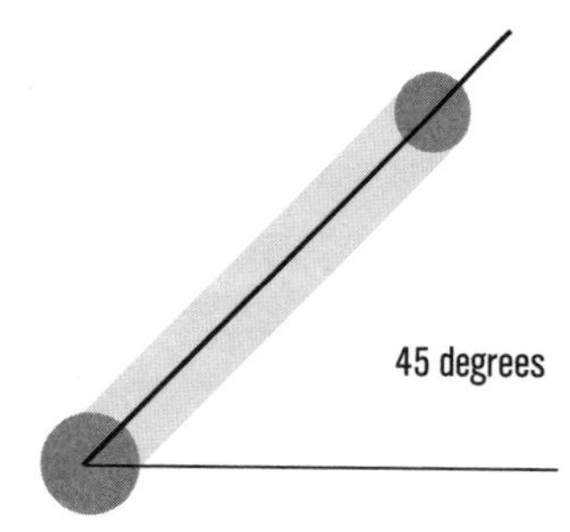

bikes, if at all. Performance riders using them should position the bar ends in the range between horizontal and 15 degrees up from horizontal (Fig. C.6). Bar-end angles in this range allow powerful pulling on the bar ends when you're climbing out of the saddle, because the bar ends are perpendicular to the forearms when you're standing. This bar-end angle also makes for a lower, more extended position when you're seated and grabbing the hooks of the bar ends. Use the bar-end position you find comfortable for pulling against when climbing standing or seated and for pedaling on extended paved stretches while seated.

Casual riders often prefer a higher angle (Fig. C.7) in order to pull with a straight wrist and closed fist while seated. Leave the bar-end mounting bolts a bit loose, sit on the bike, and grab the bar ends comfortably. Tighten them down in that position.

NOTE: *Bar ends are not to be used to adapt a poor-fitting bike. Do not use the bar ends to raise your hand position by pointing them straight up. If you want a higher hand position, get a taller or more up-angled stem and/or a higher-rise handlebar. Bar ends are not meant to be positioned straight up and used for cruising along while you're sitting up high, because steering is compromised and you will not be able to reach the brakes when you need them.*

f. Inner-bar-end position

Inner bar ends placed inboard of the grips (12-6) make a lot more sense on a wide, riser bar than standard ("outer") bar ends on the ends of the handlebar. Outer bar ends on a bar like that are too wide to properly support the shoulders. Inner bar ends support the shoulders with the arms in a relaxed position (Fig. C.8).

Place the inner bar ends inboard of the grips and tilt them so they are comfortable. You can rest the crotch of your thumb and forefinger on them while resting the other fingers on the brake lever (Fig. C.9). That way, you can use the brakes from a comfortable, and more aerodynamic, position akin to riding on the hoods of the brake levers on a road bike with a drop bar.

C.8 Inner bar ends provide a narrow, comfortable, efficient, and aerodynamic position for riding nontechnical terrain

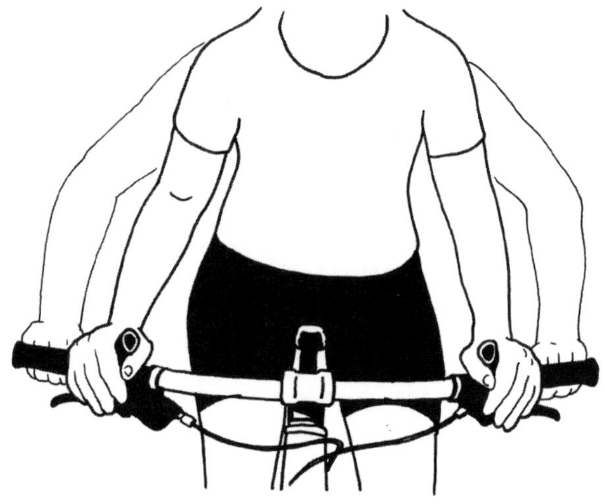

C.9 The hands can be on the brakes while riding on the inner bar ends

APPENDIX D
TORQUE TABLE

One of the single biggest sources of mechanical problems (and breakage) is the overtightening or undertightening of fasteners, particularly on lightweight equipment. It is great to have a "feel" for what is tight enough, but many people either do not have this sense or overestimate their sensitivity to it; "feel" should only supplement torque measurement.

With some parts, particularly today's superlight stems and handlebars, it is important to tighten them to their exact torque specification to prevent them from breaking while riding, which would result in an immediate and terrifying loss of control. Even experienced mechanics, with their sense of feel well developed from years of practice, sometimes overtighten the small bolts on lightweight stems.

That said, I do recommend that you try to develop that feel for bolt tightness. For small bolts, choke up on the wrench or hex key so you can tell more easily how hard you are twisting it. (Torque = Force × Radius; when you choke up on the wrench you reduce the radius at which you apply force, so that you have to apply more force to get the same torque on the bolt. That in turn makes you aware of the effort it takes.) When you think the bolt is tight enough, check the tightness with a torque wrench to calibrate your sense of feel.

There is also a danger in undertightening fasteners. The handlebar in an undertightened stem clamp can come loose and twist, or an undertightened brake cable can pull free when you yank hard on the brakes. Also, an undertightened bolt suffers more fatigue during use than one that is preloaded.

The standard method for calculating a torque specification is to load the fastener to 80 percent of its yield strength. This method works on rigid joints. High bolt preload ensures that the fastener is always in tension to prevent metal fatigue in the fastener. However, many bike parts are not rigid, and high torques can overcompress or crush components. This is especially important when you are using parts of different brands, eras, or materials together, since a stem manufacturer's torque specification for a handlebar clamp may not have anticipated that a carbon handlebar would be used; what works for an aluminum bar can crush the carbon one. There is no springback in a rigid joint, but if parts flex under tightening (a handlebar is a good example), that flex may provide the preload that the bolt needs at a considerably lower torque setting than if it were bolted through solid steel parts.

Torque wrenches usually have a knob at the base of the handle to pull tension on an internal spring. Set the desired torque by twisting the knob and reading the torque setting on a vernier scale or in an indicator window. When the set torque is reached, the head of the wrench snaps over to the side or ratchets freely. An older style of torque wrench, called a beam wrench, has a needle arm parallel to the wrench shaft that moves across a scale. When using either type of torque wrench, hold it at the handle and pull smoothly.

You actually need two torque wrenches for working on bikes. Big ones cannot measure torques accurately for small bolts. Small ones have a limited capacity and cannot tighten a bottom bracket or crank bolt sufficiently.

Using a torque wrench is not a guarantee against a screwup; it simply reduces the chances of one. First, you must make sure that the torque setting you use is the one recommended for the bolt you are tightening. The torque table in this appendix includes a lot of bolts, but it obviously cannot include all bolts from all manufacturers, so if you can consult an owner's manual or find the correct torque on the manufacturer's website, do so. Also, manufacturers often adjust torque specifications following changes in design or materials, so always check the instruction manual for torque settings when possible, even if the bolt is listed in this chart.

Second, lubrication of the bolt, temperature, and a variety of other variables will affect torque readings as well. Bicycle specifications generally assume that the bolt threads have received lubrication or threadlock compound (which provides lubrication before it dries), but that the underside of the bolt head is dry. Lubricating under the bolt head allows the bolt to turn farther at the same torque setting than the same bolt without lubrication under the head, and it thus increases the tension on the bolt.

Third, the torque reading will depend on whether the bolt is turning or you are starting a stationary bolt into motion, since its coefficient of static friction will be higher than its coefficient of dynamic (sliding) friction. If you try to determine the torque of a bolt by checking the torque required to unscrew the bolt, you will have estimated a higher torque than the actual one, particularly if the bolt has been in place for some time and has corrosion or dirt around it. This may be the best you can do in some circumstances, but proceed with caution.

Fourth, the reading on the torque wrench assumes that the wrench head is centered over the bolt; the torque reading will be low if you have a radius multiplying the torque. For example, measuring tightening torque on a pedal axle (if not using a hex key in the hex hole in the axle end) requires a "crow's foot" 15mm open-end wrench attachment on a torque wrench. If extending straight out, the crow's foot creates an offset between the axle centerline and the tool head centerline, which multiplies the torque setting displayed on the wrench handle (i.e., it will make the wrench—the radius—effectively longer). The decimal by which you must multiply the torque reading on the wrench to determine the actual torque applied to the bolt will usually be imprinted on the crow's foot. You must use this torque multiplication factor if you have the crow's foot extending straight out from the torque wrench.

However, if you keep the crow's foot at 90 degrees from the torque wrench, provided the crow's foot is short relative to the length of the torque wrench, you can use the torque settings as is on the wrench (since the hypotenuse and the long side of the right triangle will be close to the same length).

Finally, torque wrenches are not 100 percent accurate, and their accuracy changes over time. Most torque wrenches can be calibrated; automotive parts stores and some hardware stores can do this for you. Ultimately, your feel and common sense are also necessary to ensure safety.

It will be worth your while to review section 2-19 (in Chapter 2) to help you develop a feel for bolt tightness. Whether or not you have "the touch," a torque wrench is a wonderful thing, as long as you know how tight the bolt is supposed to be.

Listed in the table that follows are tightening torque recommendations from many component manufacturers. Where there is only a maximum torque listed, you can assume the minimum torque should be about 80 to 90 percent of that number.

Most torques are for steel bolts; where possible, aluminum and titanium bolts are described as such in the table. Note that it is particularly important to use a copper-filled lubricant like Finish Line Ti-Prep on titanium bolts to prevent them from binding and galling; the same goes for installing any bolt into threads in a titanium component or bike frame.

CONVERSION BETWEEN UNITS

Table D.1 is in inch-pounds (in-lbs), foot-pounds (ft-lbs), and Newton-meters (N-m) (the latter being the one I find easiest to use, since the numbers tend to be nice, round one- or two-digit numbers).

Unit Conversion Factors for Table D.1

- Divide in-lbs settings by 12 to convert to foot-pounds (ft-lbs).
- Multiply in-lbs settings by 0.113 to convert to Newton-meters (N-m).
- Multiply kgf-cm settings by 0.098 to convert to Newton-meters (N-m).

BOLT SIZES

- **M5 bolts** are 5mm in diameter and take a 3mm or 4mm hex key or T25 Torx key (derailleurs often take a 5mm hex key or an 8mm box wrench).

- **M6 bolts** are 6mm in diameter and generally take a 5mm hex key or T25 Torx key.
- **M7 bolts** are 7mm in diameter and generally take a 6mm hex key.
- **M8 bolts** are 8mm in diameter and generally take a 6mm hex key.
- **M10 bolts** are 10mm in diameter and on bikes will likely take a 5mm or 6mm hex key (rear derailleur mounting bolt).

The designation M in front of the bolt size number means millimeters and refers to the bolt shaft size, not to the hex key that turns it; an M5 bolt is 5mm in diameter, an M6 is 6mm, and so on, but there may be no relation to the wrench size. For example, an M5 bolt usually takes a 4mm hex key (or in the case of a hex-head style, an 8mm box-end or socket wrench), but M5 bolts on bicycles often accept nonstandard wrench sizes. M5 bolts attach bottle cages to the frame, and while some accept a 4mm hex key, many have a rounded "cap" head and take a 3mm hex key or sometimes a 5mm hex key. The M5 bolts that clamp a front derailleur around the seat tube or that anchor the cable on a front or rear derailleur also take a nonstandard hex key size, namely a 5mm. And M5 disc-brake rotor bolts often take a Torx T25 key. Conversely, the big single pinch bolts found on old stems usually take only a 6mm hex key, but they may be M6, M7, or even M8 bolts.

Generally, tightness can be classified in four levels:

1. **Snug (10–30 in-lbs, or 1–3 N-m):** Small setscrews, bearing preload bolts (as on threadless headset top caps) and screws going into plastic parts need to be snug.
2. **Firmly tightened (30–80 in-lbs, or 3–9 N-m):** This refers to small M5 bolts, like shoe cleat bolts, brake and derailleur-cable anchor bolts, derailleur band clamp bolts, small stem faceplate, or stem steerer clamp bolts. Some M5 and M6 seatpost clamp bolts need to be firmly tightened.
3. **Tight (80–240 in-lbs, or 9–27 N-m):** Wheel axle nuts, old-style single-bolt stem bolts (M6, M7, M8), and some seatpost binder bolts and seatpost saddle clamp bolts need to be tight.
4. **Really tight (280–600 in-lbs, or 31–68 N-m):** Crankarm bolts, pedal axles, cassette lockring bolts, and bottom-bracket cups are large parts that need to be really tight. The load on them is so high that they will creak or loosen if they are not tight enough.

TABLE D.1 — MOUNTAIN BIKE FASTENER TORQUE TABLE

(See "Unit Conversion Factors for Table D.1" on page 438)

GENERAL TORQUE SPEC FOR STEEL BOLT THREADED INTO AN ALUMINUM PART	in-lbs	N-m	ft-lbs
M5 bolt	60	7	5
M6 bolt	120	14	10
M6 bolt clamping a carbon part	100	11	8
M7 bolt	180	20	15
M8 bolt	220	25	18

	in-lbs		N-m		ft-lbs	
BOTTOM BRACKETS AND CRANKS	**min**	**max**	**min**	**max**	**min**	**max**
Bontrager Big Earl ISIS crankarm fixing bolts, M12		420		55		35
Bontrager chainring fixing bolt, aluminum	50	70	6	8	4	6
Bontrager chainring fixing bolt, steel	70	95	8	11	6	8
Bontrager Select, Race, Race Lite Earl ISIS crank bolts, M15		420		55		35
Bontrager square-taper (Sport) crankarm fixing bolts, M8	320	372	36	42	27	31
FSA aluminum Allen chainring fixing bolt		87		10		7
FSA aluminum cartridge bottom-bracket cups	347	434	39	49	29	36
FSA aluminum Torx chainring fixing bolt		104		11		9
FSA BB30 crankarm fixing bolt	345	434	39	49	29	36
FSA M5 pinch bolt, split aluminum crankarm, BB90, BB86	97	133	11	15	8	11
FSA M5 pinch bolt, split aluminum crankarm, MegaExo	106	115	12	13	9	10
FSA M8 steel crankarm fixing bolt	304	347	34	39	25	29
FSA M12 steel crankarm fixing bolt	434	521	49	59	36	43
FSA M14 aluminum crankarm fixing bolt	391	434	44	49	33	36
FSA M14 steel crankarm fixing bolt	434	521	49	59	36	43
FSA M15 aluminum crankarm fixing bolt	434	521	49	59	36	51
FSA M15 steel crankarm fixing bolt	434	521	49	59	36	43
FSA M17 crankarm fixing bolt, carbon crank, BB90, BB86	398	487	45	55	33	41
FSA M18 bearing preload bolt, BB90, BB86	6	13	0.7	1.5	0.5	1.1
FSA M18 bearing preload bolt, MegaExo	4	6	0.4	0.7	0.3	0.5
FSA M18 crankarm fixing bolt, carbon crank, MegaExo	398	487	45	55	33	41
FSA MegaExo bottom-bracket cups	345	434	39	49	29	36
FSA steel Allen chainring fixing bolt	80	106	9	12	7	9
Race Face X-Type crankarm fixing bolt	363	602	41	68	30	50
Shimano chainring fixing bolt, steel	70	95	8	11	6	8
Shimano crankarm fixing bolt (Octalink/Hollowtech)	305	435	35	50	25	36
Shimano Hollowtech 2 left crankarm fixing pinch bolts (M5)	88	132	12	14	7	11
Shimano integrated-spindle (Hollowtech 2) bearing cups	305	435	35	50	25	36
Shimano left crankarm bearing preload cap (Hollowtech 2)	4	6	0.5	0.7	0.3	0.5
Shimano loose-ball-bearing bottom bracket fixed cup	609	695	69	79	51	58

BOTTOM BRACKETS AND CRANKS, CONT.	in-lbs		N-m		ft-lbs	
	min	max	min	max	min	max
Shimano loose-ball-bearing bottom bracket lockring	609	695	69	79	51	58
Shimano square/Octalink cartridge bottom bracket cups	435	608	50	70	36	51
Shimano square-taper crankarm fixing bolt (M8 steel)	305	391	34	44	25	33
Shimano XTR FC-M960 left crankarm bearing preload cap	6	13	0.7	1.5	0.5	1.1
Shimano XTR FC-M960 left crankarm fixing pinch bolts (M5)	106	132	12	15	9	11
Shimano XTR FC-M970 adjustment nut	9	13	1.0	1.5	0.7	1.1
Shimano XTR FC-M970 adjustment nut fixing bolt	9	10	1.0	1.2	0.8	0.9
Shimano XTR FC-M970 crankarm fixing bolt (8mm hex key)	392	479	44	54	33	40
SRAM/Truvativ aluminum chainring bolts	71	80	8	9	6	7
SRAM/Truvativ GXP external bearing cups	301	363	34	41	25	30
SRAM/Truvativ GXP left crank bolt	416	478	47	54	35	40
SRAM/Truvativ GXP self extractor cup (16mm hex key)	106	133	12	15	9	11
SRAM/Truvativ Howitzer ISIS external bearing cups	301	363	34	41	25	30
SRAM/Truvativ ISIS cartridge bearing cups	301	363	34	41	25	30
SRAM/Truvativ M8 crank bolts, square taper	336	372	38	42	28	31
SRAM/Truvativ M12 crank bolts, ISIS	381	425	43	48	32	35
SRAM/Truvativ M15 crank bolts, ISIS	381	425	43	48	32	35
SRAM/Truvativ self-extractor cup, ISIS or square taper (10mm hex key)	106	133	12	15	9	11
SRAM/Truvativ steel chainring bolts	106	124	12	14	9	10
Surly bottom-bracket cups		354		40		30
Surly chainring bolts		106		12		9
Surly left crankarm pinch bolts		106		12		9
Trek tandem eccentric	75	100	8	11	6	8
Zinn Zinn-tegrated external bearing cups	301	363	34	41	25	30
Zinn Zinn-tegrated left crank bolt	400	450	45	51	33	38
Zinn Zinn-tegrated self extractor cup (10mm hex key)	106	133	12	15	9	11

BRAKES: RIM BRAKES	in-lbs		N-m		ft-lbs	
	min	max	min	max	min	max
CANTILEVER BRAKES, V-BRAKES, AND OTHER RIM BRAKES						
Avid Arch Supreme arch-mounting bolt	35	40	4	5	2.9	3.3
Avid split-clamp lever-mounting bolts	28	36	3	4	2	3
brake arm–mounting bolt, M6	40	60	5	7	3	5
brake-cable anchor bolt, M5	50	70	6	8	4	6
brake-lever clamp bolt, M6	50	70	6	8	4.2	5.8
brake-lever clamp—slotted screw	22	26	3	3	2	2
cantilever brake pad bolt	70	78	8	9	6	7
Shimano V-brake leverage-adjuster bolt	9	13	1.0	1.5	0.8	1.1
straddle-cable yoke nut	35	43	4	5	3	4
V-brake pad nut	50	70	6	8	4	6

Continues >>

BRAKES: RIM BRAKES, CONT.	in-lbs		N-m		ft-lbs	
	min	max	min	max	min	max
MAGURA HYDRAULIC RIM BRAKE						
bleed screws		35		4		3
brake-line sleeve nuts		35		4		3
M5 housing clamp bolt		35		4		3
M6 center bolt		52		6		4

BRAKES: DISC BRAKES	in-lbs		N-m		ft-lbs	
	min	max	min	max	min	max
AVID/SRAM DISC BRAKE						
banjo bolt	50	55	5	6	4	5
cable-fixing bolt	40	60	5	7	3.3	5
caliper adapter–mounting bolts, M6	80	90	9	10	7	8
caliper–mounting bolts, M6	80	90	9	10	7	8
CPS caliper-mounting bolts to adapter or post mount, M6	70	90	8	10	6	8
MatchMaker lever clamp bolt	43	52	4.9	5.9	3.6	4.3
rotor-mounting bolts, M5, Torx		55		6		5
single-lever clamp bolt	30	40	4	5	2.5	3.3
two-bolt lever clamp bolt	2.5	3.0	0.3	0.3	0.2	0.3
CODA DISC BRAKE						
caliper-mounting bolts, M6	69	78	8	9	6	7
hose sleeve	69	78	8	9	6	7
lever clamp bolt	72	108	8	12	6	9
rotor-mounting bolts, M5	40	50	5	6	3.3	4.2
DIATECH DISC BRAKE						
mounting pins	62	80	7		5	7
FORMULA DISC BRAKE						
caliper-mounting bolts, M6	76	84	9	9	6	7
rotor-mounting bolts, M5	42	47	5	5	4	4
valve couplers	101	111	11	13	8	9
HAYES DISC BRAKE						
caliper bleeder	25	35	3	4	2	3
caliper-bridge bolts	100	120	11	14	8	10
caliper-mounting bolts, 74mm caliper with mount bracket M6	100	120	11	14	8	10
caliper-mounting bolts, 74mm caliper with post-mount forks M6	75	85	8	10	6	7
lever pins	14.0	20.0	1.6	2.3	1.2	1.7
master cylinder bleeder	18	22	5	6	2	2
master cylinder jam nut	45	55	5	6	4	5
master cylinder pivot set screw	12.0	16.0	1.4	1.8	1.0	1.3
pad pin (Ryde)	10	14	1.1	1.6	0.8	1.2
reservoir cap screws	4.3	5.3	0.5	0.6	0.4	0.4
rotor-mounting bolts, M5	45	55	5	6	4	5

BRAKES: DISC BRAKES, CONT.	in-lbs		N-m		ft-lbs	
	min	max	min	max	min	max
HAYES HOSE CONNECTIONS						
banjo bolt	55	65	6	7	5	5
caliper—G1	40 + 1 turn		4.5 + 1 turn		3 + 1 turn	
caliper—G2	55	65	6	7	5	5
master cylinder (HFX-9, Sole, El Camino, Stroker, and later)	55	65	6	7	4.6	5.4
master cylinder (HFX-Mag, Mag Plus)	40 + 1 turn		4.5 + 1 turn		3 + 1 turn	
HAYES MASTER CYLINDER (BRAKE LEVER) CLAMP BOLTS						
98/99, DH Purple (1-pc clamp)	15	20	1.7	2.3	1.3	1.7
handlebar master cylinder clamp screw	25	35	3	4	2	3
HFX-9, Sole (1-pc clamp)	30	35	3.4	4	2.5	2.9
Mag, Mag plus, EC, HFX-9 (2-pc clamp)	15	20	1.7	2.3	1.3	1.7
HOPE DISC BRAKE						
rotor-mounting lockring		310		35		26
MAGURA DISC BRAKE						
master cylinder (brake lever) clamp bolts		34		4		3
master cylinder hose fitting		34		4		3
master cylinder reservoir cover screws		5		1		0
rotor-mounting bolts, M5		34		4		3
ROCKSHOX DISC BRAKE						
cable-guide hardware		50		6		4
caliper-mounting bolts, M6		51		6		4
caliper hose fitting, 0-degree		51		6		4
caliper hose fitting, 90-degree (banjo bolt)		51		6		4
rotor-mounting bolts, M5		50		6		4
SHIMANO HYDRAULIC DISC BRAKE, 2-PISTON LX/XT/XTR TYPE						
banjo bolt	44	60	5	7	4	5
caliper bleed nipple	35	53	4	6	3	4
caliper-mounting bolts, M6	53	69	6	8	4	6
lever clamp bolt	53	69	6	8	4	6
lever hose-sleeve nut	44	60	5	7	4	5
rotor-mounting bolts, M5	18	35	2	4	2	3
rotor-mounting splined lockring		350		40		29
SHIMANO HYDRAULIC DISC BRAKE, OLD 4-PISTON XT TYPE						
banjo bolt	44	60	5	7	4	5
caliper bleed nipple	27	44	3	5	2	4
caliper mounting bolts, M6		55		6		5
lever clamp bolt		70		8		6
pad axle bolt	20	35	2	4	2	3

Continues >>

BRAKES: DISC BRAKES, CONT.	in-lbs		N-m		ft-lbs	
	min	max	min	max	min	max
SHIMANO HYDRAULIC DISC BRAKE, OLD 4-PISTON XT TYPE, CONT.						
reservoir screws	3	5	0.3	1	0.2	0.4
rotor-mounting bolts, M5	18	35	2	4	2	3

CHAIN GUIDES	in-lbs		N-m		ft-lbs	
	min	max	min	max	min	max
SRAM/Truvativ M4, Box Guide		40		5		3
SRAM/Truvativ M5, Box Guide		40		5		3
SRAM/Truvativ M6, Box Guide Box Kit		71		8		6
SRAM/Truvativ M6, Box Guide Team Hardware		97		11		8

DERAILLEURS AND SHIFTERS	in-lbs		N-m		ft-lbs	
	min	max	min	max	min	max
Shimano front-derailleur cable anchor bolt, M5	44	60	5	7	4	5
Shimano front-derailleur clamp bolt, M5	44	60	5	7	4	5
Shimano rear-derailleur cable anchor bolt, M5	44	60	5	7	4	5
Shimano rear-derailleur mounting bolt, M10	70	90	8	10	6	8
Shimano rear-derailleur pulley center bolts, M5	27	34	3	4	2	3
Shimano Rapidfire shifter clamp bolt, M6	53	69	6	8	4	6
Shimano thumb-shifter clamp bolt, Allen, M6	53	69	6	8	4	6
Shimano thumb-shifter clamp bolt, slotted screw	22	26	3	3	2	2
Shimano thumb-shifter parts anchor screw	22	24	3	3	2	2
Shimano XT/XTR lever cable-access screw cover	3	4	0	1	0	0
SRAM 3.0 front-derailleur clamp bolt, M5		70		8		4
SRAM front-derailleur cable anchor bolt, M5		44		5		4
SRAM Grip Shift lever-mounting screw		17		2		1
SRAM rear-derailleur cable anchor bolt, M5	35	45	4	5	4	5
SRAM rear-derailleur cage-stop screw		13		2		2
SRAM rear-derailleur mounting bolt, M10	70	85	8	10	3	4
SRAM rear-derailleur pulley center bolts, M5		22		3		7
SRAM trigger lever-mounting bolt		44		5.0		3.7
SRAM X-Gen front-derailleur clamp bolt, M5	44	62	5	7	0	6

HUBS, CASSETTES, QUICK-RELEASE SKEWERS	in-lbs		N-m		ft-lbs	
	min	max	min	max	min	max
bolt-on steel skewer (Control Tech)		65		7	0	5
bolt-on titanium skewer (Control Tech)		85		10	0	7
Bontrager freehub-body mounting bolt		400		45		33
Bontrager quick-release axle locknut		150		17		13
Cannondale Lefty front-axle bolt		133		15.0		11
Chris King quick-release axle locknut		100		11		8
Mavic cassette cog lockring		354		40		30
nutted front hub		180		20	0	15
nutted rear hub	260	390	29	44	22	33

HUBS, CASSETTES, QUICK-RELEASE SKEWERS	in-lbs		N-m		ft-lbs	
	min	max	min	max	min	max
Shimano cassette cog lockring	261	434	30	50	22	36
Shimano freehub-body mounting bolt	305	434	35	50	25	36
Shimano hub quick-release lever closing	43	65	5	7	4	5
Shimano quick-release axle locknut	87	217	10	25	7	18
SRAM cassette cog lockring	310	350	35	40	26	29

MISCELLANEOUS	in-lbs		N-m		ft-lbs	
	min	max	min	max	min	max
AheadSet bearing preload, M6		22		3	0	2
fender to frame bolts, M5	50	60	6	7	4	5
water-bottle cage bolts, M5	25	35	3	4	2	3

PEDALS AND SHOES	in-lbs		N-m		ft-lbs	
	min	max	min	max	min	max
Crank Brothers pedal axle to crankarm	301	363	34	41	25	30
Crank Brothers shoe-fixing cleat bolt, M5	35	44	4	5	3	4
Shimano pedal axle to crankarm	304	355	34	40	25	30
Shimano shoe-fixing cleat bolt, M5	41	52	5	6	3	4
shoe spike, M5		34		4	0	3
Speedplay Frog spindle nut	35	40	4	5	3	3
SRAM/Truvativ pedal spindle into crankarm	186	301	21	34	16	25
Time pedal axle to crankarm		310		35	0	26
toeclips to pedals, M5	25	45	3	5	2	4

SEATPOSTS AND SEAT BINDERS	in-lbs		N-m		ft-lbs	
	min	max	min	max	min	max
Easton seatpost saddle-rail clamp bolts	95	105	11	12	8	9
Oval Concepts M6 saddle-rail clamp bolts		133		15		11
Ritchey saddle-rail clamp bolt: Comp, Old Pro, M8		400		45		33
Ritchey saddle-rail clamp bolt: WCS, New Pro, M6		165		19		14
RockShox Reverb bleed screw	13	19	1.5	2.2	1.1	1.6
RockShox Reverb remote clamp	44	53	5	6	3.7	4.4
RockShox Reverb seatpost binder clamp		59		6.7		4.9
Salsa LipLock M6 seatpost collar clamp bolt		70		8		4
seatpost saddle rail–clamp bolt, M8	175	345	20	39	15	29
seat-tube clamp binder bolt, M6	105	140	12	16	9	12
SRAM/Truvativ M6 two-bolt		62		7		5
SRAM/Truvativ M8 single bolt		80		9		7
steel seatpost band–clamp bolt	175	345	20	39	15	29
Thomson Elite saddle-rail clamp bolt, M5		60		7		5
Thomson M5 seatpost collar clamp bolt		25		3		2
Thomson Masterpiece saddle-rail clamp bolt, M5		45		5		4
two-piece seat binder bolt, M6	35	60	4	7	3	5

Continues >>

STEMS AND BAR ENDS	in-lbs		N-m		ft-lbs	
	min	max	min	max	min	max
3T M5 bolts (front clamp, steering-tube clamp)	80		9		7	0
3T M6 bolts (front clamp, steering-tube clamp)		130		15	0	11
3T M8 bolts (single handlebar clamp)		220		25	0	18
3T M8 bolts (single steering-tube clamp; expander)		175		20	0	15
bar end M6 bolt	120	140	14	16	10	12
Bontrager M8 steering-tube clamp bolts		200		23	0	17
Deda M5 steel bolts (bar clamp, steering-tube clamp)		90		10	0	8
Deda M5 titanium bolts (bar clamp, steering-tube clamp)		70		8	0	6
Deda M6 bolts (bar clamp, steering-tube clamp)		160		18	0	13
Deda M8 bolts (quill expander)		160		18	0	13
Dimension one-bolt handlebar clamp, M8 bolt	205	240	23	27	17	20
Dimension two-bolt face-plate bar clamp, M6	80	90	9	10	7	8
Dimension two-bolt steering-tube clamp, M6	80	90	9	10	7	8
Easton EA50, 70 bar and steering-tube clamp bolts	60	70	7	8	5	6
Easton EM90 (single M8) steering-tube clamp bolt	70	80	8	9	6	7
Easton MG60, EM90 bar clamp bolts	50	60	6	7	4	5
Easton MG60 (two M6) steering-tube clamp bolts	50	60	6	7	4	5
FSA M5 chromoly bolts		78		9	0	7
FSA M5 titanium bolts—use Ti prep		68		8	0	6
FSA M6 chromoly bolts		104		12	0	9
FSA M8 chromoly bolts		156		18	0	13
ITM aluminum M6 bolts in magnesium stem	44	53	5	6	4	4
ITM M5 bolts (bar clamp, steering-tube clamp) 2 front bolts	62	70	7	8	5	6
ITM M5 bolts (bar clamp) 4 front bolts	35	44	4	5	3	4
ITM M6 bolts (fork collar)	88	105	10	12	7	9
ITM M7 bolts (single-bolt front clamp)	106	120	12	14	9	10
ITM M8 bolts (single-bolt clamp or expander)	150	160	17	18	13	13
Oval Concepts M6 clamp bolts for alloy steering tubes		93		11	0	8
Oval Concepts M6 clamp bolts for carbon steering tubes		58		7	0	5
Oval Concepts M6 faceplate bolts for alloy bars		93		11	0	8
Oval Concepts M6 faceplate bolts for carbon bars		58		7	0	5
Oval Concepts titanium M5 faceplate bolts for alloy bars		84		10	0	7
Oval Concepts titanium M5 faceplate bolts for carbon bars		49		6	0	4
Oval Concepts titanium M6 clamp bolts for alloy steering tubes	84		10		7	0
Oval Concepts titanium M6 clamp bolts for carbon steering tubes	53		6		4	0
Ritchey WCS M5 faceplate bolts for alloy bars	26	52	3	6	2	4
Ritchey WCS M5 faceplate bolts for carbon bars		35		4	0	3
Ritchey WCS M6 clamp bolts for alloy steering tubes	52	86	6	10	4	7
Ritchey WCS M6 clamp bolts for carbon steering tubes		78		9	0	7
Salsa one-bolt handlebar clamp, M6 bolt		140		16	0	12
Salsa one-bolt steering-tube clamp, M6 bolt	100	110	11	12	8	9

STEMS AND BAR ENDS, CONT.	in-lbs		N-m		ft-lbs	
	min	max	min	max	min	max
Salsa SUL two-bolt face-plate bar clamp, M6	120	130	14	15	10	11
single stem handlebar clamping bolt, M8	145	220	16	25	12	18
SRAM/Truvativ M6 bolts—bar		60		7	0	5
SRAM/Truvativ M6 bolts—steering-tube		80		9	0	7
SRAM/Truvativ M7 bolts		120		14	0	10
Thomson Elite, X2, X4 steering-tube clamp bolts, M5		48		5	0	4
Thomson Elite handlebar clamp bolts, M5		48		5	0	4
Thomson X4 handlebar clamp bolts, M5		35		4	0	3
Wedge expander bolt for quill stems, M8	140	175	16	20	12	15

SUSPENSION: FORKS	in-lbs		N-m		ft-lbs	
	min	max	min	max	min	max
FOX						
air tank valve	40	50	5	6	3	4
all 32, 36, and 40 bottom bolts	45	55	5	6	4	5
all 32, 36, and 40 bottom nuts	45	55	5	6	4	5
all 32, 36, and 40 rebound knob screws	10	12	1	1	1	1
all rebound piston bolts	50	60	6	7	4	5
axle pinch-bolts on 36 and 40 lower leg	14	24	2	3	1	2
base valve assembly to cartridge tube	50	60	6	7	4	5
base valve bolt R & RL-RLC	50	60	6	7	4	5
brake post	75	85	8	10	6	7
cartridge tube to seal head	50	60	6	7	4	5
crown pinch bolts on 40 upper and lower crowns		30		3		3
DH axle to lower leg on 36 and 40 Forx	14	24	2	3	1	2
F80X compression cylinder	105	115	12	13	9	10
F80X IV compression piston bolt 8-32 × 0.250-in.	50	60	6	7	4	5
F80X IV shaft	115	125	13	14	10	10
F80X IV shaft extension	115	125	13	14	10	10
IV body to slim cartridge tube	40	50	5	6	3	4
lockout threshold knob set screw (RLT & RLC)	10	12	1	1	1	1
low- & high-speed compression knob setscrew (36, 40)	10	12	1	1	1	1
LW aluminum damper shaft to rebound piston insert & upper insert (all F80-F100X & 05 R-RL-RLC dampers)	50	60	6	7	4	5
LW Al float air shaft to lower inserts (Loctite 262)	65	75	7	8	5	6
LW Al vanilla plunger shaft to upper & lower inserts (Loctite 262)	65	75	7	8	5	6
rebound adjuster screw to rebound rod	4	6	0	1	0	0
Schrader valve core	3	5	0	1	0	0
Slim cartridge tube to slim sealhead	40	50	5	6	3	4
TALAS ball screw fitting to top cap (Loctite 242)	50	60	6	7	4	5
TALAS base stud (Loctite 242, 1 drop)	50	60	6	7	4	5

Continues >>

SUSPENSION: FORKS, CONT.	in-lbs		N-m		ft-lbs	
	min	max	min	max	min	max
TALAS hex fitting (Loctite 242)	17	19	2	2	1	2
TALAS IFP shaft bolt to IFP shaft (Loctite 242)	50	60	6	7	4	5
TALAS lower piston bolt (Loctite 242, 1 drop)	50	60	6	7	4	5
TALAS M2 screw tank valve to hex adapter (Loctite 242)	4	6	0	1	0	0
TALAS tank valve (Loctite 262, 360 degrees)	35	45	4	5	3	4
TALAS top cap		220		25		18
TALAS top cap to IFP shaft (Loctite 242, 1 drop)	50	60	6	7	4	5
TALAS T-port end cap (Loctite 242, 1 drop)	4	6	1	1	0	1
top cap to old chrome damper shaft upper insert	70	80	8	9	6	7
top cap to LW aluminum damper shaft assembly (all F80-F100X & 05 R-RL-RLC dampers)	70	80	8	9	6	7
top cap, Vanilla preload knob screw (inside top cap)	10	12	1	1	1	1
top caps, 2011 and later 32, 36, 40 (Damper, Preload, Air, TALAS)		220		25		18
Top caps, pre-2010 32, 36, 40 (Damper, Preload, Air, TALAS)	160	170	18	19	13	14
MANITOU						
adjuster caps and top caps	35	50	4	6	3	4
brake post	90	110	10	12	8	9
comp rod screw	13	53	2	6	1	4
damper screw	13	20	2	2	1	2
EFC/Mach 5/SX cartridge bolt	10	30	1	3	1	3
EFC/Mach 5/SX cartridge cap	30	50	3	6	3	4
fork-brace bolt	90	110	10	12	8	9
leg caps	25	35	3	4	2	3
M6 (5mm key) crown clamp bolt		60		7		5
M8 (6mm key) crown clamp bolt	110	130	12	15	9	11
neutral shaft bolt	10	30	1	3	1	3
MARZOCCHI						
26mm top caps	80	97	9	11	7	8
brake post	71	88	8	10	6	7
cartridge foot nut and pump rod	80	97	9	11	7	8
monster cartridge foot nut	204	230	23	26	17	19
upper and lower crown bolts	44	62	5	7	4	5
upper and lower crown bolts–Monster, Shiver	80	97	9	11	7	8
ROCKSHOX						
air valve assembly, Schrader type	20	40	2.3	5	2	3
air valve core, Schrader type	8	12	0.9	1.4	1	1
axle bolt, Boxxer	40	60	4.5	7	3	5
axle pinch bolt, Boxxer	20	30	2.3	3.4	2	3
blackBox Lever clamp bolt	6	10	0.7	1.1	1	1
bottom bolt, 8mm, Boxxer	45	75	5	8	4	6
bottom bolt, 8mm, dual air	35	55	4	6	3	5

SUSPENSION: FORKS, CONT.	in-lbs		N-m		ft-lbs	
	min	max	min	max	min	max
ROCKSHOX, CONT.						
bottom bolt, 8mm, hollow	45	75	5	8	4	6
bottom bolt, 8mm, solid	45	70	5	8	4	6
brake post	65	95	7	11	5	8
crown bolt, Boxxer	50	80	6	9	4	7
knob screw, 4mm, U-turn/Pure Climb-It knob	10	14	1.1	1.6	1	1
PopLoc clamp bolt	18	22	2	2.5	2	2
pure compression piston bolt	30	50	3.4	6	3	4
pure rebound piston bolt	30	50	3.4	6	3	4
pure remote knob cable set-screw	6	10	0.7	1.1	1	1
pure remote cable-set clamp screw	6	10	0.7	1.1	1	1
SID upper-tube threaded retainer	45	75	5	8	4	6
top cap, aluminum, all	55	75	6	8	5	6
top cap, plastic, all	55	75	6	8	5	6
top cap, U-turn air	115	145	13	16	10	12
RST						
Mozo brake arch bolt	70	80	8	9	6	7
Mozo fork crown clamp bolt	70	80	8	9	6	7

SUSPENSION: REAR SHOCKS	in-lbs		N-m		ft-lbs	
	min	max	min	max	min	max
MANITOU						
air canister	13	21	2	2	1	2
Schrader valve core	4	9	1	1	0	1
ROCKSHOX						
air-can lockring	55	75	6	8	5	6
damper seal head		250		28	0	21
Schrader valve core	8	12	1	1	1	1
Schrader valve housing	25	35	3	4	2	3
shaft-eyelet assembly	100	110	11	12	8	9
U-Turn air-can assembly	60	70	7	8	5	6

GLOSSARY

adjustable cup the non-drive-side cup in the bottom bracket (Fig. 11.23). This cup is removed for maintenance of the bottom-bracket spindle and bearings, and it adjusts the bearings. The term is sometimes also applied to the top cup of the headset (Figs. 12.20–12.21).

AheadSet (a trademark of Dia-Compe and Cane Creek; *or* "threadless headset"): a style of headset that allows the use of a fork with a threadless steering tube (Fig. 12.6).

Allen key (*or* "Allen wrench") (*see* "hex key").

all-terrain bike (ATB) another term for mountain bike.

anchor bolt (*or* "cable anchor" *or* "cable anchor bolt" *or* "cable-fixing bolt"): a bolt securing a cable to a component (Fig. 5.3).

Answer Products an American bicycle- and motorcycle-component company and the parent company of Manitou.

Avid a brake manufacturer, subsidiary of SRAM.

axle a shaft around which a part turns, usually on bearings or bushings.

axle overlock dimension a length of a hub axle from dropout to dropout, referring to the distance from locknut face to locknut face (Fig. 17.28).

ball bearing a set of balls, generally made out of steel or ceramic, rolling in a track to allow a shaft to spin inside a cylindrical part; may also refer to one of the individual balls.

bar end a short handlebar extension clamped onto the end of the handlebar and extending approximately perpendicular to it (Fig. 12.7).

barrel adjuster a threaded cable stop that allows for fine adjustment of cable tension. Barrel adjusters are commonly found on rear derailleurs, shifters, and brake levers (Figs. 5.3, 5.20–5.21, 10.1) and dropper-post remote levers.

BB (*see* "bottom bracket").

bearing (*see* "ball bearing").

bearing cone a conical part with a bearing race around its circumference. The cone presses the ball bearings against the bearing race inside the bearing cup (Fig. 8.6).

bearing cup (or "headset cup") a polished, dish-shaped surface inside which ball bearings roll. The bearings roll on the outside of a bearing cone that presses them into their track inside the bearing cup (Figs. 8.6, 11.5, 12.19).

bearing race a track or surface on which the bearings roll. The race can be inside a cup, on the outside of a cone, or inside a cartridge bearing.

binder bolt a bolt clamping a seatpost in a frame (Fig. 14.7), a bar end to a handlebar (Fig. 12.7), a stem to a handlebar (Fig. 12.5), or securing a threadless steering tube (Fig. 12.6).

bonk (1) v. to run out of fuel for the human body so that the ability to continue strenuous activity is impaired. (2) n. the state of having such low blood sugar from insufficient intake of calories that the ability to perform vigorous activity is impaired.

bottom bracket (BB) an assembly that allows the crank to rotate (Fig. 11.13). Generally the bottom-bracket assembly includes bearings and an axle and on older bikes may include a fixed cup, an adjustable cup, and a lockring.

bottom-bracket drop the vertical distance between the center of the bottom bracket and a horizontal line passing through the wheel-hub centers. Drop is equal to the wheel radius minus the bottom-bracket height (Appendix C, Fig. C.1).

bottom-bracket height the height of the center of the bottom-bracket spindle above the ground (see Appendix C, Fig. C.1).

bottom-bracket shell a cylindrical housing at the bottom of a bicycle frame through which the bottom-bracket axle passes (Fig. 11.13).

brake a mechanical device that decelerates or stops the motion of the wheel (and hence of the bicycle and rider) through friction.

brake block (*see* "brake pad").

brake booster an arch-shaped part bolted to the ends of the brake bosses to reduce the flex of the bosses and seatstays when the cantilever or V-brakes are applied (Fig. 10.33).

brake boss (*or* "brake pivot," *or* "brake post," *or* "cantilever boss," *or* "cantilever pivot," *or* "cantilever post"): a fork- or frame-mounted pivot for a brake arm (Figs. 16.2–16.3, 17.1).

brake caliper a brake part fixed to the frame or fork containing moving parts attached to brake pads that stop or decelerate a wheel (Figs. 9.11, 9.13, 9.14, 9.17, 9.19, 9.20, 9.22, 9.24, 9.26, 9.27–9.29).

brake pad (*or* "brake block"): a block of rubber or similar material used to slow the bike by creating friction on the rim, hub-mounted disc, or other braking surface (Figs. 9.1–9.3, 9.5, 9.24, 9.27–9.29).

brake pivot (*see* "brake boss").

brake post (*see* "brake boss").

brake shoe a metal pad holder that secures the brake pad to the brake arm (Fig. 10.14).

braze-on boss a generic term for most metal frame attachments, even those not brazed but rather welded or glued to the frame.

brazing a method commonly used to construct steel bicycle frames. Brazing involves the use of brass or silver solder to connect frame tubes and attach various "braze-on" items, such as brake bosses, cable guides, and rack mounts, to the frame. Although it is rarely done, it is also possible to braze aluminum and titanium.

bushing a metal or plastic sleeve that acts as a simple bearing in pedals, suspension forks, rear shocks and shock-mounting points, suspension swingarms, derailleur pivots, and jockey wheels.

butted tubing a common type of frame tubing with varying wall thicknesses. Butted tubing is designed to accommodate high-stress points; the ends of the tubes are thicker and other sections are thinner to reduce weight.

cable (*or* "inner wire"): wound or braided wire strands used to operate brakes and derailleurs.

cable anchor (*see* "anchor bolt").

cable anchor bolt (*see* "anchor bolt").

cable boss (*see* "cable stop").

cable end cap a cap on the end of a cable that keeps it from fraying (Fig. 5.26).

cable hanger cable stop on a stem, headset washer, fork, or seatstay arch used to stop the brake cable housing for a cantilever or U-brake (Figs. 10.4–10.6).

cable housing (*or* "outer wire"): a metal-reinforced exterior sheath through which a cable passes (Fig. 5.26).

cable stop (*or* "cable boss," *or* "cable-housing stop," *or* "outer wire stop"): a fitting on the frame, fork, or stem at which a cable-housing segment terminates (Fig. 17.1).

cable-housing stop (*see* "cable stop").

cage two guiding plates through which the chain travels. Both the front and rear derailleurs have cages. The cage on the rear also holds the jockey pulleys. Also, a water-bottle holder.

caliper (*see* "brake caliper" and "measuring caliper").

Campagnolo an Italian bicycle-component company.

Cane Creek (originally Dia-Compe USA): American bicycle-component company and originator of the threadless headset.

cantilever boss (*see* "brake boss").

cantilever brake a cable-operated rim brake consisting of two opposing arms pivoting on frame- or fork-mounted posts. Pads mounted to each brake arm are pressed against the braking surface of the rim via cable tension from the lever (Figs. 10.16–10.32).

cantilever pivot (*see* "brake boss").

cantilever post (*see* "brake boss").

cartridge bearing ball bearings encased in a cartridge consisting of steel inner and outer rings, ball retainers, and sometimes bearing covers (Figs. 8.5, 8.23, 11.31).

cassette a group of cogs that mounts on a freehub (Fig. 8.23); also, a group of chainrings that mounts on a spiderless crankarm (Fig. 11.12).

cassette hub (*see* "freehub").

casting (*see* "outer leg").

chain a series of metal links held together by pins and used to transmit energy from the crank to the rear wheel (Fig. 4.1).

chain link a single unit of bicycle chain consisting of four plates with a roller on each end and in the center (Fig. 4.7).

chain suck a dragging of the chain by the chainring past the release point at the bottom of the chainring. The chain can be dragged upward until it is jammed between the chainring and the chainstay (Fig. 4.27).

chain whip (*or* "chain wrench"): a flat piece of steel, usually attached to two lengths of chain (Fig. 1.2). This tool is used to remove the rear cogs on a free-hub or freewheel. (See also "Vise Whip," a more robust and secure substitute for this tool.)

chainline an imaginary line connecting the center of the middle chainring with the middle of the cogset. This line should, in theory, be straight and parallel to the vertical plane passing through the center of the bicycle. The chainline is measured as the distance from the center of the seat tube to the center of the middle chainring (5-50, Fig. 5.61).

chainring a multiple-tooth sprocket attached to the right crankarm (Fig. 11.1).

chainring-nut tool a tool used to secure the chainring nuts while tightening the chainring bolts (Fig. 1.2).

chainstay a frame tube on a bicycle connecting the bottom-bracket shell to the rear dropout and hence to the rear hub axle (Figs. 17.1–17.2).

chase (*see* "goose chase").

circlip (*or* "Jesus clip" or "snapring"): a C-shaped or spiral ring that fits in a groove to hold two cylindrical parts together.

clip-in pedal (*or* "clipless pedal"): a pedal that relies on spring-loaded clips to grip a cleat attached to the bottom of the rider's shoe without the use of toeclips and straps (Fig. 13.2).

clipless pedal (*see* "clip-in pedal").

cog a sprocket located on the drive side of the rear hub (Fig. 8.23).

compression damping a diminishment of the speed of compression of a spring on impact by hydraulic or mechanical means.

cone a threaded conical nut that serves to hold a set of bearings in place and also provides a smooth surface upon which those bearings can roll (Fig. 8.6); can also refer to the conical (or male) member of any cup-and-cone ball-bearing system (*see also* "bearing cone").

crank bolt (*see* "crankarm anchor bolt").

crank length the distance between the centerline of the bottom-bracket spindle and the centerline of the pedal axle (Appendix C, Fig. C.4).

crankarm a lever attached at the bottom-bracket spindle and to the pedal used to transmit a rider's energy to the chain (Fig. 11.1).

crankarm anchor bolt (*or* "crank bolt"): a bolt attaching the crank to the bottom-bracket spindle on a cotter-less drivetrain (Fig. 11.1).

crankset an assembly that includes a bottom bracket, two crankarms, a chainring set, and accompanying nuts and bolts (Fig. 11.1).

cross-three (*see* "three-cross").

crowfoot socket (*see* "crowfoot wrench").

crowfoot wrench (*or* "crowfoot socket" or "crow's foot"): a flat, open-end wrench head with a square hole at its base to accept the drive stub of a socket wrench or torque wrench (Fig. 1.3).

crow's foot (*see* "crowfoot wrench").

cup a cup-shaped bearing surface that surrounds the bearings in a bottom bracket (Fig. 11.13), headset (Fig. 12.19), or hub (Fig. 8.6) (*see also* "bearing cup"). Also, the upper part of the shaft-eyelet assembly of a rear shock (the big end of the shock).

damper (*or* "damping cartridge"): a mechanism in a suspension fork or shock that reduces the speed of the spring's oscillation movement (Fig. 16.25).

damping a reduction in speed of the oscillation of a spring, as in a suspension fork or shock.

damping cartridge (*see* "damper").

derailleur a gear-changing device that allows a rider to move the chain from one cog or chainring to another while the bicycle is in motion (Figs. 5.3, 5.17–5.21).

derailleur hanger a metal extension of the right rear dropout through which the rear derailleur is mounted to the frame (Fig. 17.1).

Di2 model name of Shimano electronic-shifting components.

diamond frame a traditional bicycle frame shape (Fig. 17.1).

disc brake a brake that stops the bike by squeezing brake pads attached to a caliper mounted to the frame or fork (Figs. 9.9–9.11) against a circular disc attached to the wheel (Figs. 9.6–9.8).

dish (*or* "wheel dish"): a difference in spoke tension on the two sides of the rear wheel (Fig. 15.23).

dishing (*or* "wheel dishing"): a centering of the rim in the frame or fork by adjustment of spoke tension in a wheel.

dishing tool a tool to check the centering of a wheel rim relative to the axle ends.

double a two-chainring drivetrain setup (as opposed to a three-chainring, or "triple," setup).

down tube a frame tube that connects the head tube and bottom-bracket shell together (Fig. 17.1).

drift a flat-ended rod used for driving out bearings and bushings.

drivetrain the crankarms, chainrings, bottom bracket, front derailleur, chain, rear derailleur, and freewheel (or cassette).

drop (1) the difference in height between two parts (*see also* "bottom-bracket drop"). (2) a terrain discontinuity you may or may not want to ride off. (3) something not to do with your tools.

dropouts (*or* "fork ends" *or* "fork tips"): slots in the fork and rear triangle where the wheel axles attach (Figs. 16.2, 17.1).

dropper post a telescoping seatpost whose length can be adjusted on the fly, while riding.

DT (a.k.a. DT Swiss): a manufacturer of spokes, other bicycle components, and tools.

dust cap a protective cap keeping dirt out of a part.

easy-out a cone-shaped, hardened-steel tool with coarse, reverse threads to remove broken bolts. To remove a broken bolt with one, a hole is drilled into the center of the bolt, the easy-out is inserted into the hole, and the easy-out is then turned with a tap handle in a counterclockwise direction.

elastomer a urethane spring sometimes used in suspension forks (Fig. 16.26), rear shocks, suspension seatposts, and saddles; also called an MCU for the material and construction (microcellular urethane).

electronic shifting (*see also* "Di2"): a system for shifting gears on a bicycle in which the power to shift the derailleurs comes not from the pull on a cable, but rather from an electric signal turning a servo motor on and off.

endo a (usually unintentional) rotation of the bike and rider forward over the front wheel.

expander bolt a bolt that when tightened pulls a wedge up inside or alongside the part into which the bolt is anchored to provide outward pressure and secure said part inside a hollow surface. Expander bolts are found inside quill stems (Figs. 12.9–12.10) and some handlebar-end plugs and handlebar-end shifters.

expander wedge (*or* "wedge"): a part threaded onto an expander bolt and usually used to secure a quill stem inside the fork steering tube or handlebar-end plugs or handlebar-end shifter inside a handlebar. An expander wedge is threaded down its center axis to accept the expander bolt and is either cylindrical in shape and truncated along an inclined plane (Figs. 12.9–12.11) or conical in shape and truncated parallel to its base.

ferrule a cap for the end of cable housing (Fig. 5.26).

fixed cup a nonadjustable cup of the bottom bracket located on the drive side of the bottom bracket (Fig. 11.13).

flange largest diameter of the hub, where the spoke heads are anchored (Fig. 15.4).

fork a part that attaches the front wheel to the frame (Figs. 16.1–16.3).

fork casting (*see* "outer leg").

fork crown a crosspiece connecting the fork legs to the steering tube (Figs. 16.1–16.2).

fork ends (*see* "dropouts").

fork rake (*or* "offset," "rake," or "wheel offset"): perpendicular offset distance of the front axle from an imaginary extension of the steering-tube centerline (*see also* "steering axis").

fork steerer (*see* "steering tube").

fork tips (*see* "dropouts").

fork trail (*or* "trail"): the distance measured on the ground between the vertical line passing through the center of the front-hub axle (i.e., the center of the wheel contact patch) and the extension of the centerline of the head tube.

Fox a bicycle-suspension manufacturer that makes forks, rear shocks, and dropper posts. Parent company of RaceFace and Easton.

frame a central structure of a bicycle to which all of the parts are attached (Figs. 17.1–17.2).

freehub (*or* "cassette hub"): a rear hub that has a built-in freewheel mechanism to which the rear cogs are attached (Fig. 8.23).

freewheel a mechanism through which the rear cogs are attached to the rear wheel on a derailleur bicycle (Figs. 8.23–8.25). The freewheel is locked to the hub when turned in the forward direction, but it is free to spin backward independently of the hub's movement, thus allowing the rider to stop pedaling and coast as the bicycle is moving forward (*see also* "freehub").

friction shifter a traditional (nonindexed) shifter attached to the frame or handlebar. Cable tension is maintained by a combination of friction washers and bolts.

front triangle (*or* "main triangle"): the head tube, top tube, down tube, and seat tube of a bike frame (Fig. 17.1).

FSA (Full Speed Ahead): a manufacturer of bicycle components.

girl's bike (*see* "step-through frame").

goose chase (*see* "wild goose chase").

granny gear the lowest gear, generally of a triple drivetrain. In the granny gear the chain is on the largest rear cog and the innermost (usually of three) front chainrings.

Grip Shift a twist shifter of the SRAM Corporation that is integrated with the handlebar grip of a mountain bike (Figs. 5.34–5.37). The rider shifts gears by twisting the grip (*see also* "twist shifter").

handlebar a curved tube, connected to the fork via the stem, that the rider holds in order to turn the fork and thus steer the bicycle. The brake levers and shift levers are attached to it (Fig. 12.1).

head angle an acute angle formed by the centerline of the head tube and the horizontal.

headset a bearing system, consisting of a number of separate cylindrical parts installed into the head tube and onto the steering tube, that secures the fork and allows it to spin and swivel in the frame (Figs. 12.19–12.22).

headset cup (*see* "bearing cup").

headset top cap (*see* "top cap").

head tube the front tube of the frame through which the steering tube of the fork passes (Fig. 17.1). The head tube is attached to the top tube and down tube and contains the headset.

hex key (*or* "Allen key" or "Allen wrench"): a hexagonal wrench that fits inside a hexagonal hole in the head of a bolt (Fig. 1.1A).

hub the central part of a wheel to which the spokes are anchored and through which the wheel axle passes (Figs. 8.1, 8.5–8.7).

hub brake a disc, drum, or coaster brake that stops the wheel with friction applied to a braking surface attached to the hub.

Hurricane Components a bicycle-component company.

Hutchinson a French tire company.

hydraulic brake a type of brake that uses fluid pressure to move the brake pads against the braking surface (Figs. 9.11, 10.33).

index shifter a shifter that clicks into fixed positions as it moves the derailleur from gear to gear.

inertia valve a valve on the compression-damping system on a front or rear shock that opens upon hard impacts and otherwise stays closed, in order to distinguish between bump forces and pedaling forces and prevent the shock from bobbing up and down during pedaling. The inertia valve is similar to a lockout lever, but unlike a lockout, it allows the shock to still be fully active for bump absorption while engaged.

inner (*see* "inner leg").

inner leg on a telescoping suspension fork, a tube, usually clamped into the fork crown (except in the case of an "upside-down fork"), that slides in and out of the larger-diameter outer leg as the fork compresses and rebounds (Fig. 16.26). On a standard (non-upside-down) fork, it is also called an "upper tube," "inner," or "stanchion."

inner wire (*see* "cable").

integrated headset a headset in which the bearing seats are integrated into the head tube (rather than requiring separate headset cups) and the bearings are completely concealed within the head tube (Fig. 12.20).

Jesus clip (*see* "circlip").

jockey pulley (*see* "jockey wheel").

jockey wheel (*or* "jockey pulley"): a circular, cog-shaped pulley attached to the rear derailleur that is used to guide, apply tension to, and laterally move the chain from rear cog to rear cog (Fig. 5.47).

knobby tire an all-terrain tire used on mountain bikes (Fig. 7.1).

lawyer tabs (*see* "wheel-retention devices").

leverage ratio amount the rear axle moves vertically on a full-suspension bike with a given amount of movement of the shock shaft.

link a pivoting steel hook on a V-brake arm that the cable-guide "noodle" hooks into (Fig. 10.12) (*see also* "chain link").

lock washer a notched or toothed washer that serves to hold surrounding nuts and washers in position.

locknut a nut that serves to hold the bearing adjustment in a headset, hub, or pedal.

lockout a valve on the compression-damping system on a front or rear shock that prevents the shock from compressing. Modern shocks usually have a "blow-off" system that will allow the compression-damping circuit to open with a large impact to prevent the shock from being damaged on big hits while the lockout is engaged.

lockring a large circular locknut. On a bottom bracket, it is the outer ring that tightens the adjustable cup against the face of the bottom-bracket shell (Fig. 11.13). On a rear shock, the lockring is the threaded ring that tightens the coil spring on a coil-over shock or is used to secure the fore-aft position of the shock body on some air shocks. On a freehub, the lockring holds the cogs on (Fig. 8.23). On a CenterLock disc brake–compatible hub, the lockring secures the rotor to the hub shell (Fig. 9.8).

Low Normal (originally "Rapid Rise"): a style of rear derailleur pioneered by Shimano in which the return spring is connected to the opposite vertices of the rear derailleur's parallelogram linkage elements compared to the setup for a standard rear derailleur. This arrangement results in the derailleur's moving to the low-gear position (the largest, most inboard rear cog) when the cable tension is removed, rather than to the high-gear position (the smallest, most outboard cog), as on a standard rear derailleur.

Magura a German brake company.

main triangle (*see* "front triangle").

Manitou an American suspension-fork and component company, subsidiary of Answer Products.

Marzocchi an Italian suspension-fork and component company.

master cylinder a piston chamber at the lever end of a hydraulic brake system (Figs. 9.15, 10.33).

master link a detachable link that holds the chain together. The master link can be opened by hand without a chain tool (Fig. 4.14).

Mavic a French wheel and bicycle-component company.

MCU (*see* "elastomer").

measuring caliper a tool for measuring the outside dimensions of an object or the inside dimensions of a hole by means of movable jaws (Fig. 1.4).

Michelin a French tire company.

mixte frame (*see* "step-through frame").

mounting bolt a bolt that mounts a part to a frame, fork, or component (*see also* "pivot bolt").

needle bearing a steel cylindrical cartridge with rod-shaped rollers arranged coaxially around the inside walls (Fig. 8.20).

nipple (1) a thin nut designed to receive the end of a spoke and seat it in the holes of a rim (Figs. 15.1–15.2). (2) a flared tip of a hydraulic caliper bleed fitting onto which a bleed hose can be attached (Fig. 9.26).

noodle a curved cable-guide pipe on a V-brake arm that stops the cable housing and directs the cable to the cable anchor bolt on the opposite arm (Fig. 10.12).

NoTubes (*or* "NoTubes.com") (*see* "Stan's NoTubes").

NoTubes.com (*see* "Stan's NoTubes").

offset (*see* "fork rake").

outer (*see* "outer leg").

outer leg in a telescoping suspension fork, a tube, often cast from magnesium and attached to the front-wheel axle (except in the case of an "upside-down fork"), that slides up and down over the smaller-diameter inner leg as the fork compresses and rebounds (Fig. 16.2). On a standard (non-upside-down) fork, it is also called the "casting," "fork casting," "outer," or "slider."

outer wire (*see* "cable housing").

outer wire stop (*see* "cable stop").

pedal a platform the foot pushes on to propel the bicycle (Figs. 13.1–13.2).

pedal overlap (*or* "toe overlap" or "toeclip overlap"): an overlapping of the toe with the front wheel while pedaling (Appendix C, Fig. C.2).

pedal platform a highly damped low-speed compression circuit on a rear shock or suspension fork designed to reduce pedal-induced bobbing as well as keep the suspension high during braking and while riding berms and dips.

pedaling stance the lateral distance between the feet while pedaling. It's the distance measured between the two vertical planes defined by the inboard side of each shoe at the first metatarsal as they move around the pedaling circle.

pin spanner a V-shaped wrench with two tip-end pins. The pin spanner is often used for tightening the adjustable cup of the bottom bracket or other lockrings (Fig. 1.2).

pivot a pin about which a part rotates through a bearing or bushing. The pivot is found on brakes, derailleurs, and rear-suspension systems.

pivot bolt a bolt on which another part pivots.

preload (*see* "spring preload").

Presta valve a thin, metal tire valve that uses a locking nut to prevent air from escaping from the inner tube or tire (Fig. 1.1B).

pulley (*see* "jockey wheel").

Q-factor the distance from the outer face of one crankarm at the pedal hole to the plane formed by the outer face of the other crankarm at the pedal hole as it spins. Q-factor is measured normal to this plane. In practice, the easiest way to measure Q-factor is to install the two crankarms on the spindle so that they are parallel to each other (at 0 degrees, rather than at 180 degrees from each other) and measure from the outer face of one crankarm at the pedal hole to the outer face of the other crankarm at the pedal hole.

quick-release (1) a tightening lever and shaft used to attach a wheel to the fork or rear dropouts without using axle nuts (Figs. 8.5–8.6). (2) a quick-opening lever and shaft pinching the seatpost inside the seat tube in lieu of a wrench-operated bolt. (3) a quick cable release on a brake. (4) a fixing mechanism that can be quickly opened and closed, as on a brake cable or wheel axle. (5) any anchor bolt that can be quickly opened and closed by a lever.

quill a vertical tube of a stem for a threaded headset system that inserts into the fork steering tube. It has an expander wedge and bolt inside to secure the stem to the steering tube (Fig. 12.9).

quill stem a stem with a quill to insert inside a threaded fork steering tube (Fig. 12.9).

race a circular track on which bearings roll freely (*see also* "bearing race").

Race Face a Canadian bicycle-component company.

rake (*see* "fork rake").

Rapid Rise (*see* "Low Normal").

Rapidfire shifter an indexing shifter manufactured by Shimano for use on mountain bikes with two separate levers operating each shift cable (Figs. 5.24, 5.37).

ratchet (*see* "socket wrench").

rear triangle a rear part of the bicycle frame that includes the seatstays, the chainstays, and the seat tube (Fig. 17.1).

rebound damping a diminishing of speed of return of a spring by hydraulic or mechanical means.

ride height (*see* "sag").

rim an outer hoop of a wheel to which the tire is attached (Fig. 15.1).

riser bar a handlebar with a double bend on each side of the stem clamp so that the grips are higher than the stem.

Ritchey an American bicycle and bicycle-component company.

RockShox an American suspension-fork and component company, subsidiary of SRAM.

roller-cam brakes a brake system using pulleys and a cam to force the brake pads against the rim surface (Fig. 10.40).

saddle (*or* "seat"): a platform made of leather and/or plastic upon which the rider sits (Fig. 14.1).

sag (*or* "ride height"): the amount the front or rear shock compresses with the rider's weight static on the bike. Its purpose is to preload the shock so that it forces the rear wheel down into the ground when the bike is unweighted after a bump, thus increasing tire contact and traction in rough terrain.

Schrader valve a high-pressure air valve with a spring-loaded air-release pin inside (Fig. 1.1B). Schrader valves are found on some bicycle inner tubes and tubeless tires, on air-sprung suspension forks and rear shocks, and on automobile tires and tubes.

sealant (see "tire sealant", Fig. 7.18).

sealed bearing a bearing enclosed in a protective seal in an attempt to keep contaminants out (Fig. 8.5) (*see also* "cartridge bearing").

seat (*see* "saddle").

seat angle an acute angle formed by the centerline of the seat tube and the horizontal.

seat cluster an intersection of the seat tube, top tube, and seatstays.

seat tube a frame tube into which the seatpost is inserted (Fig. 17.1).

seatpost a tubular member supporting, securing, and allowing height adjustment of the saddle (Fig. 14.5).

seatstay a frame tube on a bicycle connecting the seat tube or the rear shock to the rear dropout and hence to the rear hub axle (Figs. 17.1–17.2).

shim a thin element inserted between two parts to ensure that they are the proper distance apart. On bicycles, a shim can be a thin washer and can be used to space a disc-brake caliper away from the frame or fork or to space a bottom-bracket cup away from the frame's bottom-bracket shell. A shim can also be a thin piece of metal used to make a seatpost fit more tightly inside the seat tube. Shims can also be small, thin discs found inside suspension forks and rear shocks to control suspension movement by permitting or hindering passage of hydraulic fluid through an orifice.

Shimano a Japanese bicycle-component company and maker of XTR, XT, Saint, LX, and STX component lines as well as Rapidfire (shifters), SPD (pedals), and STI (shifting systems).

sidepull cantilever brake (*see* "V-brake," Figs. 10.11–10.14).

singletrack a trail with a single furrow made for feet or a two-wheeled vehicle, as opposed to a road or "doubletrack," which has a track for each pair of wheels on a four-wheeled vehicle.

skewer (1) a long rod. (2) a hub quick-release (Figs. 8.5–8.6). (3) a shaft passing through a stack of elastomer bumpers in a suspension fork (Fig. 16.25).

slave cylinder a piston chamber in the caliper of a hydraulic brake.

slider (*see* "outer leg").

Slime a brand of tire sealant consisting of chopped fibers in a liquid medium injected inside a tire or inner tube to flow to and fill small air leaks (Fig. 7.18).

snapring (*see* "circlip").

socket a cylindrical tool with a square hole in one end to mount onto a socket-wrench handle and with hexagonal walls inside the opposing end to grip a bolt head or nut (Fig. 1.2).

socket wrench (*or* "socket wrench handle" or "wrench handle"): a cylindrical wrench handle with a ratcheting square head extending at right angles to the handle onto which sockets or other wrench bits for turning bolts or nuts are installed (Fig. 1.2).

spacer on a bicycle, generally a thick washer, cylindrical in shape, intended to maintain a fixed distance between two parts. Spacers can be found between the headset and the stem and between the stem and the top cap on a threadless steering tube, and between the upper bearing cup and the top nut on a threaded steering tube. Spacers may also be used to space a bottom-bracket cup away from the frame's bottom-bracket shell.

spanner (British parlance): a wrench.

spider a star-shaped piece of metal that connects the right crankarm to the chainrings (Fig. 11.1).

spline one of a set of longitudinal grooves and ridges designed to interlock two mechanical parts (Figs. 8.23, 9.7–9.8).

spokes metal rods that connect the hub to the rim of a wheel (Figs. 15.1–15.2).

spring an elastic contrivance that when compressed returns to its original shape by virtue of its elasticity. In bicycle suspension applications, the spring used is normally either an elastic polymer cylinder, a coil of steel or titanium wire, or compressed air.

spring preload (*or* "preload"): an initial loading of a spring so that part of its compression range is taken up prior to impact.

sprocket a circular, multiple-toothed piece of metal that engages a chain (*see also* "cog" and "chainring").

SRAM an American bicycle-component company and maker of Grip Shift, Half Pipes, and ESP (derailleurs); owner of Sachs, RockShox, Avid, and Truvativ bicycle-component companies.

stanchion (*see* "inner leg").

stand-over clearance (*see* "stand-over height").

stand-over height (*or* "stand-over clearance"): the distance between the top tube of the bike and the rider's crotch when the rider is standing over the bicycle (Appendix C, Fig. C.1).

Stan's NoTubes (*or* "NoTubes" or "NoTubes.com"): a brand of tire upgrade system named after inventor Stan Koziatek that includes a latex-based tire sealant (Fig. 7.18) to convert a standard tire to a tubeless tire.

star bolt (*see* "Torx bolt").

star nut (*or* "star-fangled nut"): a pronged nut that is forced down into the steering tube and anchors the headset top cap bolt to adjust a threadless headset (Figs. 12.6, 12.19–12.21).

star wrench (*see* "Torx wrench").

star-fangled nut (*see* "star nut").

steerer (*see* "steering tube").

steering axis the imaginary line around which the fork rotates (Fig. 16.51).

steering tube (*or* "fork steerer" or "steerer"): a vertical tube on a fork that is attached to the fork crown, fits inside the head tube, and swivels within it by means of the headset bearings (Figs. 16.1–16.3). A steering tube can be threaded or threadless, meaning that the top headset cup can either screw onto the steering tube or slide onto it, and the stem can either insert inside the steering tube and clamp with an expander wedge (threaded) or clamp around the steering tube (threadless).

stem (*or* "gooseneck"): a connection element between the fork steering tube and the handlebar (Fig. 12.1).

stem length the distance between the center of the steering tube and the center of the handlebar measured along the top of the stem (Appendix C, Fig. C.4).

step-through frame (*or* "girl's bike," *or* "mixte frame," or "women's frame"): a bicycle frame with a steeply up-angled top tube connecting the bottom of the seat tube to the top of the head tube. The frame design is intended to provide ease of stepping over the frame and ample stand-over clearance.

straddle cable a short segment of cable connecting two brake arms together (Figs. 10.16–10.18).

straddle-cable holder (*see* "yoke").

swingarm a movable rear end of a rear-suspension frame (Fig. 14.2). (*see also* "chainstay").

tap (*or* "thread tap"): a threaded tool made of hardened steel to cut threads. It is shaped like a pointed bolt shaft, but it has lengthwise grooves cut across the threads to give the threads cutting edges. The tap has a square head that fits in a handle to provide leverage to turn the tap.

threaded headset a headset whose top bearing cup and top nut above it screw onto a threaded steering tube (Fig. 12.19).

threadless headset (*see* "AheadSet").

three-cross (*or* "cross-three"): a pattern used by wheel builders that calls for each spoke to cross three others in its path from the hub to the rim (Fig. 15.1).

through-axle a removable rod that forms not only the axle of a front or rear hub but also the system that secures the wheel into the fork or frame.

thumb shifter a thumb-operated shift lever attached on top of the handlebar (Fig. 5.31).

tire bead an edge of a tire that seats down inside the rim (Fig. 7.7). The bead's diameter is held fixed to established standards by means of a strong, stretch- and tear-resistant material—usually either steel or Kevlar. These strands alone are also referred to as the "bead."

tire lever a tool to pry a tire off the rim (Figs. 7.4–7.5).

tire sealant a liquid that, when put into a tire, plugs leaks when air under pressure forces it through the leaks. The sealant (Fig. 7.18) either hardens in air or blocks the hole with fibers and flakes suspended in the liquid, or both.

toe overlap (*or* "toeclip overlap") (*see* "pedal overlap").

toeclip overlap (*see* "pedal overlap").

top cap (*or* "headset top cap"): a round top part of a headset with a bolt passing through it that screws into the star nut to apply downward pressure on the stem to properly load and adjust the headset bearings on a threadless steering tube (Figs. 12.19–12.21).

top cup upper headset cup (see "bearing cup").

top tube a frame tube that connects the seat tube to the head tube (Fig. 17.1).

torque a rotational analogue of force. Torque is a vector quantity whose magnitude is the length of the radius from the center of rotation out to the point at which the force is applied, multiplied by the magnitude of the force directed perpendicular to the radius. On bicycles, we are primarily interested in the tightening torque applied to a fastener (this value can be measured with a torque wrench—see Appendix D) and the torque applied by the rider on the pedals to propel the rear wheel and hence the bicycle.

torque wrench a socket-wrench handle with a graduated scale and an indicator to show how much torque is being applied as a bolt is being tightened (Figs. 1.3–1.4, 2.18; see Appendix D).

Torx bolt (*or* "star bolt"): a bolt with a six-point star-shaped hole in its head.

Torx wrench (*or* "star wrench"): a tool with a star-shaped end that fits in the star-shaped hole in the head of a Torx bolt (Figs. 1.3, 1.5).

trail (1) where to ride your mountain bike. (2) (see "fork trail").

triple a three-chainring combination (Figs. 11.1–11.2, 11.4) attached to a right crankarm.

Truvativ a bicycle-component manufacturer, subsidiary of SRAM.

tubeless a system of rim and tire that stays inflated without an inner tube.

tubeless ready a tubeless system distinct from the UST tubeless system. Tubeless-ready (TR) indicates that the rim or tire or both require tire sealant in order to stay inflated. A TR rim requires an airtight rim strip to seal off the spoke holes in the rim bed. A TR tire is not guaranteed to be airtight without sealant inside to fill tiny pinholes in the tire casing; it is generally lighter than a UST tire, which is guaranteed to be airtight due to having a thin coating of rubber inside.

twist shifter a cable-pulling derailleur control handle surrounding the handlebar adjacent to the hand grip. It is twisted forward or back to cause the derailleur to shift (Fig. 5.28). (*See also* "Grip Shift.")

U-brake a mountain bike brake consisting of two arms shaped like inverted Ls affixed to posts on the frame or fork (Fig. 10.39).

unicrown a manufacturing method of nonsuspended (i.e., rigid) forks in which the fork legs curve toward each other and are welded directly to the steering tube (Fig. 16.3).

upper tube (*see* "inner leg").

upside-down fork a suspension fork whose lower legs (attached to the wheel axle) are the inner legs of the fork and move up and down within the upper, outer legs of the fork.

UST a tubeless-tire system originated by Mavic, Michelin, and Hutchinson in which an airtight tire with a sealing flap on its bead seals over a "hump" on the ledge inside a rim free of spoke holes in the rim bed (Fig. 7.7).

V-brake (*or* "sidepull cantilever brake"): a cable-operated cantilever rim brake consisting of two vertical brake arms pivoting on frame- or fork-mounted pivots pulled together by a horizontal cable. A brake pad is affixed to each arm, and there are a cable link and a cable-guide pipe on one arm and a cable anchor on the opposite arm (Figs. 10.11–10.14).

Vise Whip a Pedro's tool (Zinn designed) with a Vise-Grip handle used to remove the rear cogs on a freehub or freewheel (Fig. 1.2). (*See also* "chain whip.")

wedge (*see* "expander wedge").

wheel dish (*see* "dish").

wheel dishing (*see* "dishing").

wheel offset (*see* "fork rake").

wheelbase the horizontal distance between the two wheel axles.

wheel-dishing tool (*see* "dishing tool").

wheel-retention devices (*or* "lawyer tabs", *or* "wheel-retention tabs"): cast-in or separate fixtures at the fork ends designed to prevent the front wheel from falling out if the hub quick-release lever or axle and nuts are loose.

wheel-retention tabs (*see* "wheel-retention devices").

wild goose chase (*see* "chase").

women's frame (*see* "step-through frame").

wrench (*or* "spanner," in British parlance): a tool having jaws, a shaped insert, or a socket to grip the head of a bolt or a nut to turn it.

yoke (*or* "straddle-cable holder"): a part on a cantilever or U-brake attaching the brake cable to the straddle cable (Fig. 10.26); also, the part of a rear-suspension swingarm attached to the main pivot.

Zinn the author of this book, not to be confused with Zen.

BIBLIOGRAPHY

Barnett, John. *Barnett's Manual: Analysis and Procedures for Bicycle Mechanics*, 4th ed. Boulder, CO: VeloPress, 2000.

Brandt, Jobst. *The Bicycle Wheel.* Menlo Park, CA: Avocet, 1988.

Dushan, Allan. *Surviving the Trail.* Tumbleweed Films, 1993.

Langley, Jim. Bicycling Magazine*'s Complete Guide to Bicycle Maintenance and Repair*. Emmaus, PA: Rodale Press, 1994.

Leslie, David. *The Mountain Bike Book*. London: Ward Lock, 1996.

Lindorf, Werner. *Mountain Bike Repair and Maintenance*. London: Ward Lock, 1995.

Mould, Bill. *The Bicycle Wheel: Physics and Engineering*, video. Alexandria, VA: Bill Mould Wheels, 2017.

Muir, John, and Tosh Gregg. *How to Keep Your Volkswagen Alive: A Manual of Step-by-Step Procedures for the Compleat Idiot*. Santa Fe, NM: John Muir, 1969.

Pirsig, Robert. *Zen and the Art of Motorcycle Maintenance*. New York: William Morrow, 1974.

Schraner, Gerd. *The Art of Wheelbuilding: A Bench Reference for Neophytes, Pros, and Wheelaholics*. Denver, CO: Buonpane, 1999.

Stevenson, John, and Brant Richards. *Mountain Bikes: Maintenance and Repair.* Mill Valley, CA: Bicycle Books, 1994.

Taylor, Garrett. *Bicycle Wheelbuilding 101: A Video Lesson in the Art of Wheelbuilding*. Westwood, MA: Rexadog, 1994.

Van der Plas, Robert. *The Bicycle Repair Book*. Mill Valley, CA: Bicycle Books, 1993.

———. *Mountain Bike Maintenance*. San Francisco: Bicycle Books, 1994.

Zinn, Lennard. *Mountain Bike Performance Handbook*. Osceola, WI: MBI, 1998.

———. *Zinn and the Art of Road Bike Maintenance*, 5th ed. Boulder, CO: VeloPress, 2016.

INDEX

ILLUSTRATION INDEX

ABOUT THE AUTHOR

Lennard Zinn is a bike racer, frame builder, and technical writer. He grew up cycling, skiing, whitewater rafting, and kayaking—as well as tinkering with mechanical devices—in Los Alamos, New Mexico. After receiving his physics degree from Colorado College, he became a member of the US Olympic Development Cycling Team. He went on to work in Tom Ritchey's frame-building shop and has been producing custom mountain, road, and triathlon frames, as well as custom cranks and stems, at Zinn Cycles since 1982 (zinncycles.com).

Zinn has been writing for *VeloNews* since 1989 and is the senior technical writer for the magazine and its website, velonews.com. Other books by Zinn are *Zinn & the Art of Road Bike Maintenance* (VeloPress, 5th ed. 2016), *The Haywire Heart: How Too Much Exercise Can Kill You, and What You Can Do to Protect Your Heart* (VeloPress, 2017), *Zinn & the Art of Triathlon Bikes* (VeloPress, 2007), *Zinn's Cycling Primer* (VeloPress, 2004), *Mountain Bike Performance Handbook* (MBI, 1998), and *Mountain Bike Owner's Manual* (VeloPress, 1998).

ABOUT THE ILLUSTRATORS

Todd Telander is a former mechanic and bike racer who devotes most of his time these days to artistic endeavors. He attended the University of California at Santa Cruz, and, while earning degrees in Environmental Studies and Biology, he completed a graduate-level program in scientific illustration. He has since studied fine art in several western states and was awarded an artist's residency at Rocky Mountain National Park. In addition to drawing bike parts, he paints and draws wildlife and landscapes for publishers, museums, design companies, and individuals. You can see more examples of his work on his website, toddtelander.com.

Mike Reisel is a graphic designer who spends most of his time art directing magazines and websites, riding his bike, and ignoring the pleas to lubricate his drivetrain.